**MOMENTUM, ENERGY, AND MASS TRANSFER IN CONTINUA**

Second Edition

# Momentum, Energy, and Mass Transfer in Continua

John C. Slattery
*Professor of Chemical Engineering*
*Department of Chemical Engineering*
*Northwestern University, Evanston, Illinois*

ROBERT E. KRIEGER PUBLISHING COMPANY
HUNTINGTON, NEW YORK
1981

Original Edition 1972
Second Edition 1981

Printed and Published by
ROBERT E. KRIEGER PUBLISHING COMPANY, INC.
645 NEW YORK AVENUE
HUNTINGTON, NEW YORK 11743

Copyright © (original material) 1972 by
McGRAW-HILL BOOK COMPANY
Transferred to John C. Slattery 1980
Copyright © (new material) 1981 by
ROBERT E. KRIEGER PUBLISHING COMPANY, INC.

*All rights reserved. No reproduction in any form of this book, in whole or in part (except for brief quotation in critical articles or reviews), may be made without written authorization from the publisher.*

Printed in the United States of America

**Library of Congress Cataloging in Publication Data**

Slattery, John Charles, 1932-
   Momentum, Energy, and Mass Transfer in Continua.

   Includes bibliographical references and index.
   1. Energy transfer.   2. Mass transfer.   3. Heat—
Transmission.   I. Title.
[QC73.8.E53S58 1981]     531'.6     80-22746
ISBN 0-89874-212-9

# Contents

*Preface*   xix

**Chapter 1   KINEMATICS**   1

MOTION   2
    1.1.1   Body, Motion, and Material Coordinates   2
    1.1.2   Path Lines   7
    1.1.3   Streamlines   8
    1.1.4   Streak Lines   10

FRAME   12
    1.2.1   Changes of Frame   12
    1.2.2   Equivalent Motions   15

MASS   17
    1.3.1   Conservation of Mass   17
    1.3.2   Transport Theorem   18
    1.3.3   The Equation of Continuity   20
    1.3.4   Transport Theorem for a Region Containing a Singular Surface   22

|  |  |
|---|---|
| 1.3.5 Jump Mass Balance for a Phase Interface | 24 |
| *1.3.6 Stream Functions for Two-dimensional Motions | 25 |
| *1.3.7 Further Comments on Stream Functions for Two-dimensional Motions | 26 |

## DEFORMATION  28
    1.4.1 The Rate of Deformation Tensor Field  28
    1.4.2 Relative Spin and Vorticity  31

## Chapter 2  FOUNDATIONS FOR MOMENTUM TRANSFER  34

### FORCE  35
    2.1.1 Force  35

### EULER'S LAWS  37
    2.2.1 Euler's Laws  37
    2.2.2 The Stress Tensor  39
    2.2.3 Cauchy's First Law of Motion  41
    2.2.4 Cauchy's Second Law of Motion  43

### BEHAVIOR  45
    2.3.1 Behavior of Materials  45
    2.3.2 A Simple Constitutive Equation for Stress  47
    2.3.3 Two Useful Classes of Empirical Models  51
    2.3.4 The Noll Simple Fluid  54
    2.3.5 Surface Tension  57

### NEWTONIAN FLUID  58
    2.4.1 Cauchy's First Law for a Newtonian Fluid  58

### SUMMARY  59
    2.5.1 Summary of Useful Equations  59
    2.5.2 The Navier-Stokes Equation for a Two-dimensional Flow  65

## Chapter 3  APPLICATIONS OF THE DIFFERENTIAL BALANCES TO MOMENTUM TRANSFER  67

### PHILOSOPHY  68
    3.1.1 The Philosophy of Solving Problems in Fluid Mechanics  68

### COMPLETE SOLUTIONS  70
    3.2.1 Flow of an Incompressible Newtonian Fluid through a Tube  70
    3.2.2 Flow of an Incompressible Viscoelastic Fluid through a Tube  77

CONTENTS ix

    3.2.3  Tangential Annular Flow of an Incompressible Ellis Model Fluid   82
    3.2.4  A Wall Which Is Suddenly Set in Motion   87
    3.2.5  A Rotating Bucket   94

CREEPING FLOWS   99
    3.3.1  Creeping Flow   99
    3.3.2  Flow in a Cone-plate Viscometer of an Incompressible Power-model Fluid   101
    3.3.3  Flow past a Sphere   111

NONVISCOUS FLOWS   119
    3.4.1  Potential Flow   119
    3.4.2  The Bernoulli Equation   122
    3.4.3  The D'Alembert Paradox   126
    3.4.4  Potential Flow past a Sphere   130
    3.4.5  Plane Potential Flow in a Corner   134

BOUNDARY-LAYER THEORY   137
    3.5.1  Boundary-layer Theory for Plane Flow past a Flat Plate   137
    3.5.2  More on the Boundary Layer in Plane Flow past a Flat Plate   143
    3.5.3  Boundary-layer Theory for Plane Flow past a Curved Wall   145
    3.5.4  Solutions Found by Combination of Variables   151
    3.5.5  Plane Stagnation Flow   155
    3.5.6  Flow in a Convergent Channel   158
    3.5.7  Boundary-layer Theory for Flow past a Body of Revolution   161

**Chapter 4  APPLICATIONS OF INTEGRAL AVERAGING TECHNIQUES TO MOMENTUM TRANSFER**   168

TIME AVERAGING   169
    4.1.1  Turbulence   169
    4.1.2  Time Averages   170
    4.1.3  The Time Average of a Time-averaged Variable   173
    4.1.4  Empirical Correlations for the Reynolds Stress Tensor $\mathbf{S}^{(t)}$   174
    4.1.5  Turbulent Flow through a Tube   178

AREA AVERAGING   183
    4.2.1  Area Averaging   183
    4.2.2  Flow from Rest in a Circular Tube   183

|  |  |  |
|---|---|---|
| 4.2.3 | Approximate Boundary-layer Theory for Plane Flow past a Curved Wall | 186 |

## LOCAL VOLUME AVERAGING 193

|  |  |  |
|---|---|---|
| 4.3.1 | Flow through Porous Media | 193 |
| 4.3.2 | Local Volume Averaging | 194 |
| 4.3.3 | Theorem for the Local Volume Average of a Gradient | 196 |
| 4.3.4 | The Local Volume Average of the Equation of Continuity | 199 |
| 4.3.5 | The Local Volume Average of Cauchy's First Law | 200 |
| 4.3.6 | Empirical Correlations for **g** | 202 |
| 4.3.7 | Summary of Results for an Incompressible Newtonian Fluid | 206 |
| 4.3.8 | Averages of Volume-averaged Variables | 207 |
| 4.3.9 | Flow through a Packed Tube | 209 |
| 4.3.10 | Neglecting the Divergence of the Local Volume-averaged Extra Stress | 213 |
| 4.3.11 | Flow of a Viscoelastic Fluid through a Tube | 215 |

## INTEGRAL BALANCES 218

|  |  |  |
|---|---|---|
| 4.4.1 | The Integral Balances | 218 |
| 4.4.2 | The Integral Mass Balance | 219 |
| 4.4.3 | The Integral Mass Balance for Turbulent Flows | 222 |
| 4.4.4 | Integral Mass Balance: An Example | 224 |
| 4.4.5 | The Integral Momentum Balance | 225 |
| 4.4.6 | The Integral Momentum Balance for Turbulent Flows | 228 |
| 4.4.7 | Empirical Correlations for $\mathscr{F}$ and $\mathscr{G}$ | 231 |
| 4.4.8 | The Integral Momentum Balance: An Example | 235 |
| 4.4.9 | The Mechanical Energy Balance for Incompressible Materials | 238 |
| 4.4.10 | The Mechanical Energy Balance for Incompressible Materials in Turbulent Flow | 243 |
| 4.4.11 | Empirical Correlations for $\mathscr{E}$ | 246 |
| 4.4.12 | The Mechanical Energy Balance for Incompressible Materials: An Example | 247 |
| 4.4.13 | The Integral Moment of Momentum Balance | 254 |
| 4.4.14 | The Integral Moment of Momentum Balance for Turbulent Flows | 258 |
| 4.4.15 | The Integral Moment of Momentum Balance: An Example | 259 |
| 4.4.16 | Two Extremum Principles | 262 |
| 4.4.17 | The Velocity Extremum Principle | 263 |
| 4.4.18 | The Stress Extremum Principle | 267 |
| 4.4.19 | Physical Interpretation of the Velocity and Stress Extremum Principles | 269 |
| 4.4.20 | Upper and Lower Bounds to the Drag Coefficient for a Sphere | 271 |

CONTENTS                                                              xi

**Chapter 5  FOUNDATIONS FOR ENERGY TRANSFER              278**

THERMODYNAMICS                                                        279
    5.1.1  Classical Thermostatics                 279
    5.1.2  Thermodynamics                          283

ENERGY TRANSMISSION                                                   290
    5.2.1  Energy Transmission                     290

ENERGY BALANCE                                                        292
    5.3.1  The Energy Balance                      292
    5.3.2  The Differential Energy Balance         293

BEHAVIOR                                                              296
    5.4.1  More on the Behavior of Materials       296
    5.4.2  Fourier's Law                           297

ENTROPY INEQUALITY                                                    299
    5.5.1  The Entropy Inequality                  299
    5.5.2  The Differential Entropy Inequality     302

SUMMARY                                                               304
    5.6.1  Summary of Useful Equations             304

**Chapter 6  APPLICATIONS OF THE DIFFERENTIAL
BALANCES TO ENERGY TRANSFER                                  308**

PHILOSOPHY                                                            309
    6.1.1  The Philosophy of Solving Problems Involving the Energy
Balance                                                               309

CONDUCTION                                                            310
    6.2.1  Conduction in Solids                    310
    6.2.2  Cooling of a Semi-infinite Slab: Constant Surface Temperature  311
    6.2.3  Cooling a Semi-infinite Slab: Newton's Law of Cooling          313
    6.2.4  Cooling a Flat Sheet: Constant Surface Temperature             317

MORE COMPLETE SOLUTIONS                                               323
    6.3.1  Conduction, Convection, and Viscous Dissipation in Fluids      323
    6.3.2  Couette Flow of a Compressible Newtonian Fluid                 323
    6.3.3  Couette Flow of a Compressible Newtonian Fluid with Variable
Viscosity and Thermal Conductivity                                    330

## NO DISSIPATION — 336
- 6.4.1 When Viscous Dissipation Is Neglected — 336
- 6.4.2 Natural Convection between Vertical Heated Plates — 338

## NO CONVECTION — 345
- 6.5.1 When Convection Is Neglected (Creeping Flow) — 345

## NO CONDUCTION — 347
- 6.6.1 When Conduction Is Neglected — 347
- 6.6.2 Speed of Propagation of Sound Waves — 348

## BOUNDARY-LAYER THEORY — 352
- 6.7.1 Thermal Boundary-layer Theory for Plane Flow past a Flat Plate — 352
- 6.7.2 More on the Thermal Boundary Layer in Plane Flow past a Flat Plate — 356
- 6.7.3 Thermal Boundary-layer Theory for Plane Flow past a Curved Wall — 361
- 6.7.4 The Temperature Distribution in Flow past a Wedge — 363
- 6.7.5 Thermal Boundary-layer Theory for Flow past a Body of Revolution — 367
- 6.7.6 Energy Transfer in the Entrance of a Heated Section of Tube — 370

## DOWNSTREAM IN A TUBE — 379
- 6.8.1 More about Energy Transfer in a Heated Section of a Tube — 379
- 6.8.2 Still More about Energy Transfer in a Heated Section of a Tube — 381

## Chapter 7 APPLICATIONS OF INTEGRAL AVERAGING TECHNIQUES TO ENERGY TRANSFER — 386

### TIME AVERAGING — 386
- 7.1.1 Turbulent Energy Transfer — 386
- 7.1.2 The Time-averaged Differential Energy Balance — 386
- 7.1.3 Empirical Correlations for the Turbulent Energy Flux Vector $\mathbf{q}^{(t)}$ — 387
- 7.1.4 Turbulent Energy Transfer in a Heated Section of a Tube — 391

### AREA AVERAGING — 396
- 7.2.1 Area Averaging in Energy Transfer — 396
- 7.2.2 A Straight Cooling Fin of Rectangular Profile — 396
- 7.2.3 Approximate Thermal Boundary-layer Theory for Plane Flow past a Curved Wall — 400

CONTENTS    xiii

LOCAL VOLUME AVERAGING    405
    7.3.1  Energy Transfer in Porous Media    405
    7.3.2  The Local Volume Average of the Differential Energy Balance    406
    7.3.3  Empirical Correlations for **h**    408
    7.3.4  Summary of Results for a Nonoriented, Uniform-porosity Structure    412
    7.3.5  Transpiration Cooling    414

INTEGRAL BALANCES    419
    7.4.1  More on Integral Balances    419
    7.4.2  The Integral Energy Balance    420
    7.4.3  The Integral Energy Balance for Turbulent Flows    427
    7.4.4  Empirical Correlations for $\mathscr{Q}$    429
    7.4.5  The Integral Energy Balance: An Example    432
    7.4.6  More about the Mechanical Energy Balance    436
    7.4.7  The Integral Entropy Inequality    441
    7.4.8  The Integral Entropy Inequality for Turbulent Flows    443
    7.4.9  The Integral Entropy Inequality: An Example    444

**Chapter 8  FOUNDATIONS FOR MASS TRANSFER**    446

VIEWPOINT    447
    8.1.1  Viewpoint in Considering Multicomponent Materials    447
    8.1.2  Body, Motion, and Material Coordinates of a Species    447

MASS BALANCE    450
    8.2.1  The Species Mass Balance    450
    8.2.2  Concentrations, Velocities, and Mass Fluxes    452

REVISED POSTULATES    457
    8.3.1  Previous Postulates    457
    8.3.2  Conservation of Mass    458
    8.3.3  Euler's First Law    460
    8.3.4  Euler's Second Law    461
    8.3.5  The Energy Balance    461
    8.3.6  The Entropy Inequality    466

BEHAVIOR    471
    8.4.1  Behavior of Multicomponent Materials    471
    8.4.2  More on the Behavior of Multicomponent Materials    471
    8.4.3  Constitutive Equations for the Stress Tensor    473
    8.4.4  Constitutive Equations for the Energy Flux Vector    474
    8.4.5  Constitutive Equations for the Mass Flux Vector    475

|  |  |  |
|---|---|---|
| 8.4.6 | Constitutive Equations for the Mass Flux Vector in Binary Solutions | 480 |
| 8.4.7 | Constitutive Equations for the Mass Flux Vector: Limiting Cases in Ideal Solutions | 483 |

## INTRINSICALLY STABLE EQUILIBRIUM 485
- 8.5.1 Stable Equilibrium 485
- 8.5.2 An Isolated System Approaching Equilibrium 486
- 8.5.3 Implications of Eq. (2-21) in Sec. 8.5.2 490
- 8.5.4 Implications of Inequality (2-22) in Sec. 8.5.2 493
- 8.5.5 Specific Results for a Single-component System 496
- 8.5.6 Relation to Classic Questions of Thermostatics 498

## ALTERNATIVE VIEWPOINT 499
- 8.6.1 An Alternative Approach to the Mechanics of Multicomponent Systems 499

## Chapter 9 APPLICATIONS OF THE DIFFERENTIAL BALANCES TO MASS TRANSFER 501

### PHILOSOPHY 501
- 9.1.1 The Philosophy of Solving Mass-transfer Problems 501

### COMPLETE SOLUTIONS 503
- 9.2.1 Unsteady-state Evaporation 503
- 9.2.2 Unsteady Diffusion with a First-order Homogeneous Reaction 511
- 9.2.3 Unsteady Diffusion with a Slow Catalytic Reaction 517
- 9.2.4 Thermal Diffusion in the Two-bulb Experiment 520
- 9.2.5 Pressure Diffusion in a Natural Gas Well 522
- 9.2.6 Forced Diffusion in Electrochemical Systems 525
- 9.2.7 Steady-state Evaporation through a Multicomponent Stagnant Film 529
- 9.2.8 Condensation of Mixed Vapors 532

### NO CONVECTION 538
- 9.3.1 When Diffusion Induced Convection Is Neglected 538
- 9.3.2 More on Unsteady-state Evaporation 540
- 9.3.3 More on Unsteady Diffusion with a First-order Homogeneous Reaction 541
- 9.3.4 More on Unsteady Diffusion with a Slow Catalytic Reaction 542

CONTENTS xv

FORCED CONVECTION 545
    9.4.1  When Natural Convection Can Be Neglected with Respect to Forced Convection 545
    9.4.2  Similarities between Energy and Mass Transfer 546
    9.4.3  Gas Absorption in a Falling Film with Chemical Reaction 547

**Chapter 10 APPLICATIONS OF INTEGRAL-AVERAGING TECHNIQUES TO MASS TRANSFER** 552

TIME AVERAGING 552
    10.1.1  Turbulent Mass Transfer 552
    10.1.2  The Time-averaged Differential Equation of Continuity for Species A 553
    10.1.3  Empirical Correlations for the Turbulent Mass Flux Vector $\mathbf{j}_{(A)}^{(t)}$ 554
    10.1.4  Turbulent Diffusion from a Point Source in a Moving Stream 556

AREA AVERAGING 559
    10.2.1  Area Averaging in Mass Transfer 559
    10.2.2  Longitudinal Dispersion 560

LOCAL VOLUME AVERAGING 565
    10.3.1  Mass Transfer in Porous Media 565
    10.3.2  The Local Volume Average of the Equation of Continuity for Species A 565
    10.3.3  When Fick's First Law Applies 566
    10.3.4  Empirical Correlations for $\boldsymbol{\delta}_{(A)}$ and $\boldsymbol{\Delta}_{(A)}$ 568
    10.3.5  Summary of Results for a Liquid or Dense Gas in a Non-oriented, Uniform-porosity Structure 572
    10.3.6  When Fick's First Law Does Not Apply 574
    10.3.7  Knudsen Diffusion 576
    10.3.8  The Local Volume Average of the Equation of Continuity 577
    10.3.9  The Effectiveness Factor for Spherical Catalyst Particles 578

INTEGRAL BALANCES 584
    10.4.1  Still More on Integral Balances 584
    10.4.2  The Integral Mass Balance for Species A 584
    10.4.3  The Integral Mass Balance for Species A Appropriate to Turbulent Flows 586
    10.4.4  Empirical Correlations for $\mathscr{J}_{(A)}$ 588
    10.4.5  The Integral Mass Balance for Species $A$: An Example 591
    10.4.6  The Integral Overall Mass Balance 594
    10.4.7  The Integral Overall Momentum, Mechanical Energy, and Moment of Momentum Balances 595

|  |  |
|---|---|
| 10.4.8 The Integral Overall Energy Balance | 595 |
| 10.4.9 The Integral Overall Energy Balance: An Example | 596 |
| 10.4.10 The Integral Overall Entropy Inequality | 600 |

## Appendix A  TENSOR ANALYSIS — 602

### SPATIAL VECTORS — 603
- **A.1.1 Spatial Vectors — 603
- A.1.2 Position Vectors — 605
- A.1.3 Spatial Vector Fields — 605
- A.1.4 Bases — 606
- A.1.5 Bases for the Spatial Vector Fields — 607
- A.1.6 Bases for the Spatial Vectors — 609
- A.1.7 The Summation Convention — 610

### DETERMINANT — 611
- A.2.1 Determinant — 611

### GRADIENT OF A SCALAR — 613
- **A.3.1 The Gradient of a Scalar Field — 613

### CURVILINEAR COORDINATES — 615
- **A.4.1 Curvilinear Coordinates — 615
- *A.4.2 The Dual Basis — 621
- *A.4.3 Covariant and Contravariant Components of Spatial Vector Fields — 623

### SECOND-ORDER TENSORS — 626
- **A.5.1 Second-order Tensor Fields — 626
- **A.5.2 Components of Second-order Tensor Fields — 627
- **A.5.3 The Transpose of a Second-order Tensor Field; Symmetric, Skew-Symmetric, and Orthogonal Tensor Fields — 632
- A.5.4 The Inverse of a Second-order Tensor Field — 636
- **A.5.5 The Trace of a Second-order Tensor Field — 637

### GRADIENT OF VECTOR — 638
- **A.6.1 The Gradient of a Vector Field — 638
- *A.6.2 Covariant Differentiation — 640

### THIRD-ORDER TENSORS — 645
- A.7.1 Third-order Tensor Fields — 645
- A.7.2 Components of Third-order Tensor Fields — 646
- A.7.3 Another View of Third-order Tensor Fields — 649

CONTENTS

| | |
|---|---|
| GRADIENT OF TENSOR | 650 |
|     A.8.1   The Gradient of a Second-order Tensor Field | 650 |
|    *A.8.2  More on Covariant Differentiation | 651 |
| VECTOR PRODUCT AND CURL | 653 |
|   **A.9.1  The Vector Product and Curl | 653 |
|     A.9.2  More on the Vector Product and Curl | 655 |
| DETERMINANT OF TENSOR | 658 |
|     A.10.1  The Determinant of a Second-order Tensor Field | 658 |
| INTEGRATION | 660 |
|   **A.11.1  Integration of Spatial Vector Fields | 660 |
|     A.11.2  Green's Transformation | 661 |
|     A.11.3  Change of Variable in Volume Integrations | 663 |
| **Appendix B  MORE ON THE TRANSPORT THEOREM** | **665** |
| ALTERNATE DERIVATIONS | 665 |
|     B.1.1  First Alternate Proof of the Transport Theorem | 665 |
|     B.2.1  Second Alternate Proof of the Transport Theorem | 666 |
| **Appendix C  DERIVATION OF INEQUALITY (4-1) of Sec. 8.5.4** | **669** |
| *Name Index* | 671 |
| *Subject Index* | 675 |

# Preface

This book is intended to serve as an integrated introduction to four subjects that are often taught separately and with their own viewpoint: fluid mechanics, thermodynamics, heat transfer, and mass transfer. Since the subject matter covers such a wide range of topics, many of which have been worthy of important monographs, the result must be seen as a survey.

Yet I do not wish to write a text that says a little bit about nearly everything with the result that there is no time to talk about some ideas in depth. I believe there are two important things a student can obtain from a study of these areas: an understanding of the common philosophy that is the basis for these four subjects and a certain skill in representing the important aspects of a real physical problem in mathematical terms. For this reason, I have devoted the major portion of the text to emphasizing the common foundations for these areas and to developing a consistent approach to limiting cases and approximate solutions.

Some compromises in subject matter had to be made. Radiation is not included because I felt that a *brief* summary consistent with the rest of the text was not possible. Interfacial effects such as surface tension have been given only passing mention. Although this is an area of particular interest to me and I feel that a strong argument can be given in support of its importance, I don't believe that it is a topic to which every first-year graduate student *must* be introduced.

Little is said about kinetic theory, for the following reasons: there are two models for real materials—the particulate model used in statistical mechanics and the continuum model of continuum mechanics. In the particulate model, we visualize that matter is associated with discrete particles and that mass is a discontinuous function of position. In the continuum model, we assume that matter is continuously distributed through space and that mass is a continuous function of position, at least within a single phase. There is no question that we all believe the particulate model involving molecules, atoms, electrons, etc., is a much more realistic picture of the details of real materials. Furthermore, from the point of view of statistical mechanics, averages can be defined that exactly satisfy the laws of balance introduced in the continuum model [1]. But the continuum model has the distinct advantage of simplicity when one wishes to describe gross phenomena. Using the continuum model, nonequilibrium statistical mechanics has told us a great deal about the behavior of real materials, particularly dilute gases. I was very tempted to summarize some of these results, but decided not to when I saw that it would be difficult to improve upon the brief surveys given by Bird, Stewart, and Lightfoot [2, pp. 19, 253, and 508].

The reader may have somewhat mixed emotions about continuum mechanics. He is eager to discover more about fluid mechanics, thermodynamics, etc., but is discouraged when he sees that he must first learn a little tensor analysis. Tensor analysis is the mathematical language through which this subject is most easily grasped. Yet, at this point, most beginning first-year graduate students in engineering have had an introduction to only certain aspects of the subject, generally referred to as *vector analysis*. (This situation seems to be rapidly changing. Here at Northwestern University, all engineering undergraduates are required to take a one-quarter introduction to linear algebra, of which tensor analysis is a special application.) For this reason I have included an appendix that treats those aspects of the subject of interest to both the beginner and the advanced student.

Appendix A has been written with three different groups of students in mind. The beginner is often anxious to skip over to the "practical applications" presented in Chapters 3, 4, 6, 7, 9, and 10. He has to see that there really is something worth doing before he becomes motivated to devote much attention to the theory or foundations. I suggest that he not attempt to read the entire appendix. It will be sufficient if he reads the portions of the appendix marked with a double asterisk(**). As a matter of fact it is not even necessary to read all of these portions before starting Chapter 1. It may be preferable to keep referring back to Appendix A as new concepts are required in the course of reading the first few chapters. However, there are two major points that are lost in reading only those sections marked with a double asterisk in Appendix A. First, the derivation of the transport theorem given in Sec. 1.3.2 cannot be understood: the alternate derivations given in Appendix B are suggested instead. Second, one cannot derive the required differential balances in curvilinear coordinates; this is not a major drawback, since all of the differential equations required are tabulated in the text for both cylindrical and spherical coordinates.

A somewhat more advanced or better motivated student will be curious to understand the rigorous derivation of the transport theorem and may feel somewhat

# PREFACE

hampered in reading the literature. I would encourage him to read the unmarked sections of Appendix A as well.

Finally, the sections marked with a single asterisk (*) are intended for those who expect to do quite a lot of research in those areas. These sections allow one to express scalar, vector, and tensor equations in terms of any desired curvilinear coordinate system. They also allow the student to understand those portions of the literature that are written in terms of vector and tensor components with respect to arbitrary curvilinear coordinate systems.

Beginning students may wish to read just the highlights of Chapters 1 and 2 before plunging into Chapter 3. In that case, I would strongly suggest Sec. 1.1.1, 1.3.1 to 1.3.3, 2.1.1, 2.2.1 to 2.2.3, 2.3.3, 2.4.1, and 2.5.1. As questions arise or as curiosity develops, the student can fill in with the other sections at his leisure. Although I have organized this book with the theory before the applications, I don't feel that one should be rigid about studying it in this manner.

I would like to warn the reader that not everything I feel is important has been placed in the body of the text. Many significant points are developed in the exercises and referred to later. I feel that this is a subject where one learns best by first reading how someone else thinks about a problem and then trying to do something possibly not too dissimilar himself. For this reason, I have worded most of the exercises in such a manner that the answer is revealed. Perhaps the most important thing to master here is the ability to think about real physical phenomena in abstract mathematical terms. It is important to have the answer so that you can check your reasoning. Answers also make the text a little more useful as a reference.

The exercises suggested were chosen to illustrate or to amplify the topics in the text. I have nearly always limited myself to problems requiring no more mathematical sophistication than that for which I feel every engineering graduate student should strive. As I stated at the beginning of this preface, I hope to convey here an understanding of the common philosophy that is the basis for fluid mechanics, thermodynamics, heat transfer, and mass transfer. I would also like the reader to develop a degree of skill in representing the important aspects of a real physical problem in mathematical terms. My intention has *not* been to write a "how-to-solve-them" book on partial differential equations. Rather, I will be pleased if the reader is encouraged to sample further the offerings of his mathematics department.

Although I do hope that this book will be useful as a reference, I have written it primarily as a textbook to be used either by a class or for self-study. For this reason, I have not been slavish in referring to original sources but have preferred to suggest references that give the most readable or most careful explanation of the point under discussion. I have attempted to provide a reference whenever the discussion does not involve either original work or a straightforward explanation found in every text.

I'm always curious about how one gets around to writing a book, so perhaps you are as well. I have had the privilege of teaching for our department the first-year graduate introductory course in this area since 1965. In 1966, I began writing up my lectures, together with some other material that I felt should be available for reference or for the reader seeking further information. Since 1967, updated Xerox copies of these notes have been sold for the cost of duplication to the classes and used as supplements to my lectures. The critical comments of my students as well as a thoughtful review by Professor R. B. Bird have been invaluable in prompting me to revise, rewrite, reorganize, and clarify various aspects of the text. I owe them a great deal for their help.

This text is not merely the product of my relationship with the graduate students here at Northwestern University. In a very real sense, it is the direct product of my schooling, the freedom I have been given here at Northwestern to pursue my research and teaching interests, and several excellent books to which I often refer in this text. I hasten to add that only the best aspects of the result are attributable to these relationships; any errors are mine.

Finally, I would like to express my gratitude to our several department secretaries who have had to cope with my midwestern accent on the tape recorder. But my fondest thanks go to my wife, Bea; she has cheerfully typed the manuscript and its many revisions.

*John C. Slattery*

## PREFACE TO SECOND EDITION

This represents primarily a correction and clarification of the first edition resulting from many questions, comments, and suggestions by students and colleagues. I would like to take this opportunity to thank them all for their encouragement and concern.

*John C. Slattery*
*August 10, 1980*

## REFERENCES

1. Noll, W.: *J. Rational Mechanics and Analysis*, **4**: 627 (1955).
2. Bird, R. B., W. E. Stewart, and E. N. Lightfoot: "Transport Phenomena," 7th printing, Wiley, New York, 1960.

# 1
# Kinematics

This entire chapter is introductory in much the same way as Appendix A is. In Appendix A, I introduce the mathematical language that we shall be using in describing physical problems. In this chapter, I indicate some of the details involved in representing from the continuum point of view the motions and deformations of real materials. This chapter is important not only for the definitions introduced, but also for the viewpoint taken in some of the developments. For example, the various forms of the transport theorem will be used repeatedly throughout the text in developing differential equations and integral balances from our basic postulates.

Perhaps the most difficult point for a beginner is to properly distinguish between the continuum model for real materials and the particulate or molecular model. I think we can all agree that the most factually detailed picture of real materials requires that they be represented in terms of atoms and molecules. In this picture, mass is distributed discontinuously throughout space; mass is associated with the protons, neutrons, electrons, etc., which are separated by relatively large voids. In the continuum model for materials, mass is distributed continuously through space, with the exception of surfaces of discontinuity, which represent phase interfaces or shock waves.

The continuum model is less realistic than the particulate model, but far simpler. For many purposes, the detailed accuracy of the particulate model is unnecessary. To our sight and touch, mass appears to be continuously distributed throughout the water which we drink and the air which we breathe. The problem is analogous to a study of traffic patterns on an expressway. The speed and spacings of the automobiles are important, but we probably should not worry about whether the automobiles had four, six, or eight cylinders.

This is not to say that the particulate theories are of no importance. Information is lost in a continuum picture. It is only through the use of statistical mechanics that a complete a priori prediction about the behavior of the material can be made. I will say more about this in the next chapter.

## 1.1 MOTION

**1.1.1 Body, motion, and material coordinates** My goal in this book is to lay the foundation for understanding a wide variety of operations employed in the chamical and petroleum industries. To be specific, consider the extrusion of a molten polymer to produce a fiber, catalytic cracking in a fluidized reactor, the production of oil and gas from a sandstone reservoir, the flow of a coal slurry through a pipeline. One important feature that these operations have in common is that at least some of the materials concerned are undergoing deformation and flow.

Let us begin by asking how we might describe a body of material it deforms. Figure 1.1.1-1 shows a rubber ball in three configurations as it strikes a wall and rebounds. How should we describe the deformation of this rubber body from its original configuration as a sphere? How should velocity be defined in order to take into account that it must surely vary as a function of position within the ball as well as time as the ball reaches the wall and begins to deform? We need a mathematical description for a body that allows us to describe where its various components go as function of time.

Let us begin by rather formally defining a body to be a set, any element $\zeta$ of which is called a particle or a material particle. A one-to-one continuous mapping of this set onto a region of the space $E$ studied in elementary geometry exists and is called a *configuration* of the body:

$$z = X(\zeta) \tag{1-1}$$

$$\zeta = X^{-1}(z) \tag{1-2}$$

The point $z = X(\zeta)$ of $E$ is called the place occupied by the particle $\zeta$, and $\zeta = X^{-1}(z)$, the particle whose place in $E$ is $z$.

It is completely equivalent to describe the configuration of a body in terms of the position vector **z** of the point $z$ with respect to the origin $O$ (Sec. A.1.2):

$$\mathbf{z} = \boldsymbol{\chi}(\zeta) \tag{1-3}$$

$$\zeta = \chi^{-1}(\mathbf{z}) \tag{1-4}$$

[1.1] MOTION 3

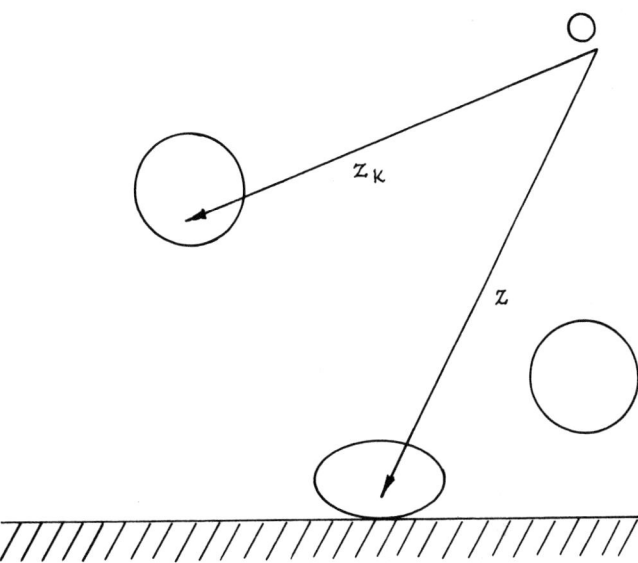

**Fig. 1.1.1-1** A rubber ball in three configurations as it strikes a wall and rebounds. The particle that was in the position z in the reference configuration is at time $t$ in the position z.

Here $\chi^{-1}$ indicates the inverse mapping of $\chi$. With an origin $O$ having been defined, it is unambiguous to refer to $\mathbf{z} = \chi(\zeta)$ as the place occupied by the particle $\zeta$ and $\zeta = \chi^{-1}(\mathbf{z})$ as the particle whose place is $\mathbf{z}$.

In what follows, we choose to refer to points in $E$ by their position vectors relative to a previously defined origin $O$.

A *motion* of a body is a one-parameter family of configurations; the real parameter $t$ is time. We write

$$\mathbf{z} = \chi(\zeta, t) \tag{1-5}$$

and

$$\zeta = \chi^{-1}(\mathbf{z}, t) \tag{1-6}$$

I recognize that I have introduced the material particle as a primitive concept, without definition but with a description of its attributes. A set of material particles is defined to be a body; there is a one-to-one continuous mapping of these particles onto a region of the space $E$ in which we visualize the world about us. But clearly we need a link with what we can directly observe.

While the body $B$ should not be confused with any of its spatial configurations, nevertheless, it is available to us for observation and study only in those configurations. For many purposes it is convenient to reflect this fact by using positions in some particular configuration $\varkappa$ as a means of specifying the particles of the body. This reference configuration may be, but need not be, one actually occupied by the body in the course of its motion. The place of a particle in $\varkappa$ will be denoted by

$$\mathbf{z}_\varkappa = \varkappa(\zeta) \tag{1-7}$$

The particle at the place $\mathbf{z}_\varkappa$ in the configuration $\varkappa$ may be expressed as

$$\zeta = \varkappa^{-1}(\mathbf{z}_\varkappa) \tag{1-8}$$

If $\chi$ is a motion of the body, then

$$\mathbf{z} = \chi(\zeta,t) = \chi_\varkappa(\mathbf{z}_\varkappa,t) \equiv \chi(\varkappa^{-1}(\mathbf{z}_\varkappa),t) \tag{1-9}$$

Referring to Fig. 1.1.1-1, we find that the particle which was in the position $\mathbf{z}_\varkappa$ in the reference configuration is at time $t$ in the position $\mathbf{z}$. This expression defines a family of *deformations* from the reference configuration. The subscript $\varkappa$ is to remind you that the form of $\chi_\varkappa$ depends upon the choice of reference configuration $\varkappa$.

The position vector $\mathbf{z}_\varkappa$ with respect to the origin $O$ may be written in terms of its rectangular cartesian coordinates:

$$\mathbf{z}_\varkappa = z_{\varkappa i}\mathbf{e}_i \tag{1-10}$$

The $z_{\varkappa i}$ ($i = 1, 2, 3$) are referred to as the *material coordinates* of the material particle $\zeta$. They locate the position of this material particle $\zeta$ relative to the origin $O$ when the body is in the reference configuration $\varkappa$. In terms of these material coordinates, we may express Eq. (1-9) as

$$\mathbf{z} = \chi_\varkappa(\mathbf{z}_\varkappa,t) = \chi_\varkappa(z_{\varkappa 1},z_{\varkappa 2},z_{\varkappa 3},t) \tag{1-11}$$

Let $A$ be any quantity, scalar or tensor. We shall have occasion to talk about the time derivative of $A$ following the motion of a particle. We define

$$\dot{A} \equiv \frac{d_{(m)}A}{dt} \equiv \left(\frac{\partial A}{\partial t}\right)_{\mathbf{z}_\varkappa} \equiv \left(\frac{\partial A}{\partial t}\right)_{z_{\varkappa 1},z_{\varkappa 2},z_{\varkappa 3}} \tag{1-12}$$

For example, the velocity vector $\mathbf{v}$ represents the time rate of change of position of a material particle:

$$\mathbf{v} \equiv \dot{\mathbf{z}} \equiv \frac{d_{(m)}\mathbf{z}}{dt} \equiv \left[\frac{\partial\chi_\varkappa(\mathbf{z}_\varkappa,t)}{\partial t}\right]_{\mathbf{z}_\varkappa} \equiv \left[\frac{\partial\chi_\varkappa(z_{\varkappa 1},z_{\varkappa 2},z_{\varkappa 3},t)}{\partial t}\right]_{z_{\varkappa 1},z_{\varkappa 2},z_{\varkappa 3}} \tag{1-13}$$

We may refer to the operation $d_{(m)}/dt$ as the *material* derivative (or substantial derivative [1, p. 73]).

We are involved with several derivative operations in the chapters that follow. Bird, Stewart, and Lightfoot [1, p. 73] have suggested some homely examples that serve to illustrate the differences.

*The partial time derivative $\partial c/\partial t$* Suppose we are in a boat which is anchored securely in a river, some distance from the shore. If we look over the side of our boat and note the concentration of fish as a function of time, we observe how the fish concentration changes with time at a fixed position in space:

[I.I] MOTION

$$\frac{\partial c}{\partial t} \equiv \left(\frac{\partial c}{\partial t}\right)_z \equiv \left(\frac{\partial c}{\partial t}\right)_{z_1, z_2, z_3}$$

*The material derivative* $d_{(m)}c/dt$  Suppose we now pull up our anchor and let our boat drift along with the river current. As we look over the side of our boat, we report how the concentration of fish changes as a function of time while following the water (the material):

$$\frac{d_{(m)}c}{dt} = \frac{\partial c}{\partial t} + \nabla c \cdot \mathbf{v} \qquad (1\text{-}14)$$

*The total derivative* $dc/dt$  We now switch on our outboard motor and race about the river, sometimes upstream, sometimes downstream, or across the current. As we (recklessly) peer over the side of our boat, we measure fish concentration as a function of time while following an arbitrary path across the water:

$$\frac{dc}{dt} = \frac{\partial c}{\partial t} + \nabla c \cdot \mathbf{v}_{(b)} \qquad (1\text{-}15)$$

Here $\mathbf{v}_{(b)}$ denotes the velocity of the boat.

## REFERENCE

1. Bird, R. B., W. E. Stewart, and E. N. Lightfoot: "Transport Phenomena," 7th printing, Wiley, New York, 1960.

## EXERCISES

**1.1.1-1**  Let $A$ be any real scalar field, spatial vector field, or second-order tensor field. Show that[1]

$$\dot{A} = \frac{\partial A}{\partial t} + (\nabla A) \cdot \mathbf{v}$$

**\*1.1.1-2**  Let $\mathbf{a} = \mathbf{a}(\mathbf{z},t)$ be some vector field which is a function of position and time.
   (a) Show that

$$\dot{\mathbf{a}} \equiv \frac{d_{(m)}\mathbf{a}}{dt} = \frac{\delta_{(m)}a^n}{\delta t}\mathbf{g}_n$$

where

$$\frac{\delta_{(m)}a^n}{\delta t} \equiv \frac{\partial a^n}{\partial t} + a^n{}_{,i}v^i$$

   (b) Show that

$$\dot{\mathbf{a}} \equiv \frac{d_{(m)}\mathbf{a}}{dt} = \frac{\delta_{(m)}a_n}{\delta t}\mathbf{g}^n$$

[1] Where we write $(\nabla A) \cdot \mathbf{v}$, some say instead $\mathbf{v} \cdot (\nabla A)$. When $A$ is a scalar, there is no difference. When $A$ is either a vector or second-order tensor, the change in notation is the result of a different definition for the gradient operation. See Secs. A.6.1 and A.8.1.

where

$$\frac{\delta_{(m)} a_n}{\delta t} \equiv \frac{\partial a_n}{\partial t} + a_{n,i} v^i$$

The quantity $\delta_{(m)} a^n/\delta t$ is referred to as the *material intrinsic derivative* of the contravariant vector component $a^n$; it should be viewed as the contravariant component of the vector field $d_{(m)}\mathbf{a}/dt$. In the same manner the quantity $\delta_{(m)} a_n/\delta t$ may be designated as the material intrinsic derivative of the covariant vector component $a_n$; it is the covariant component of the vector field $d_{(m)}\mathbf{a}/dt$.

**\*1.1.1-3** The concept of the material intrinsic derivative introduced in Exercise 1.1.1-2 may be extended readily to higher-order tensors. Consider the second-order tensor field $\mathbf{T} = \mathbf{T}(\mathbf{z},t)$.

(a) Show that

$$\frac{d_{(m)}\mathbf{T}}{dt} = \frac{\delta_{(m)} T^{ij}}{\delta t} \mathbf{g}_i \mathbf{g}_j$$

where

$$\frac{\delta_{(m)} T^{ij}}{\delta t} \equiv \frac{\partial T^{ij}}{\partial t} + T^{ij}{}_{,k} v^k$$

(b) Show that

$$\frac{d_{(m)}\mathbf{T}}{dt} = \frac{\delta_{(m)} T^i{}_j}{\delta t} \mathbf{g}_i \mathbf{g}^j$$

where

$$\frac{\delta_{(m)} T^i{}_j}{\delta t} \equiv \frac{\partial T^i{}_j}{\partial t} + T^i{}_{j,k} v^k$$

**\*1.1.1-4** Show that

$$\frac{d_{(m)}(\mathbf{a} \cdot \mathbf{b})}{dt} = \frac{d_{(m)}}{dt}(a^i b_i)$$

$$= \frac{\delta_{(m)} a^i}{\delta t} b_i + a^i \frac{\delta_{(m)} b_i}{\delta t}$$

**\*1.1.1-5** Let $\mathbf{a} = \mathbf{a}(\mathbf{z},t)$ be some vector which is a function both of position and time. Show that

$$\frac{d\mathbf{a}}{dt} = \frac{\delta a^m}{\delta t} \mathbf{g}_m$$

where $d\mathbf{a}/dt$ denotes a total derivative of the vector field $\mathbf{a}$ with respect to time not necessarily following a material particle. The *intrinsic derivative* of $a^m$ with respect to time $\delta a^m/\delta t$ is defined as follows:

$$\frac{\delta a^m}{\delta t} \equiv \frac{\partial a^m}{\partial t} + a^m{}_{,i} \frac{dx^i}{dt}$$

**1.1.1-6** (a) Starting with the definition for the velocity vector, prove that

$$\mathbf{v} = \frac{d_{(m)} x^i}{dt} \mathbf{g}_i$$

[1.1] MOTION

(b) Determine that, with respect to the cylindrical coordinate system defined in Exercise A.4.1-4,

$$\mathbf{v} = \frac{d_{(m)}r}{dt}\mathbf{g}_{\langle 1 \rangle} + r\frac{d_{(m)}\theta}{dt}\mathbf{g}_{\langle 2 \rangle} + \frac{d_{(m)}z}{dt}\mathbf{g}_{\langle 3 \rangle}$$

(c) Determine that, with respect to the spherical coordinate system defined in Exercise A.4.1-5,

$$\mathbf{v} = \frac{d_{(m)}r}{dt}\mathbf{g}_{\langle 1 \rangle} + r\frac{d_{(m)}\theta}{dt}\mathbf{g}_{\langle 2 \rangle} + r\sin\theta\frac{d_{(m)}\varphi}{dt}\mathbf{g}_{\langle 3 \rangle}$$

**1.1.2 Path lines** The curve in space along which the material particle $\zeta$ travels is referred to as the *path line* for the material particle $\zeta$. The path line may be determined from the motion of the material as described in Sec. 1.1.1:

$$\mathbf{z} = \boldsymbol{\chi}(\mathbf{z}_\varkappa, t) \tag{2-1}$$

Here $\mathbf{z}_\varkappa$ represents the position of the material particle $\zeta$ in the reference configuration $\varkappa$; time $t$ is a parameter along the path line which corresponds to any given position $\mathbf{z}_\varkappa$.

The path lines may be determined conveniently from the velocity distribution, since velocity is the derivative of position with respect to time following a material particle. The parametric equations of a particle path are the solutions of the differential system

$$\frac{d\mathbf{z}}{dt} = \mathbf{v} \tag{2-2}$$

or

$$\frac{dz_i}{dt} = v_i \quad \text{for } i = 1, 2, 3 \tag{2-3}$$

The required boundary conditions may be obtained by choosing the reference configuration to be a configuration that the material assumed at some time $t_0$.

For example, let the rectangular cartesian components of $\mathbf{v}$ be

$$v_1 = \frac{z_1}{1+t} \qquad v_2 = \frac{z_2}{1+2t} \qquad v_3 = 0 \tag{2-4}$$

and let the reference configuration be that which the material assumed at time $t = 0$. For Eqs. (2-4), Eqs. (2-3) are separable:

$$\int_{z_{\varkappa_1}}^{z_1} \frac{dz_1}{z_1} = \int_0^t \frac{dt}{1+t} \tag{2-5}$$

$$\int_{z_{\varkappa_2}}^{z_2} \frac{dz_2}{z_2} = \int_0^t \frac{dt}{1+2t} \tag{2-6}$$

and

$$z_3 = z_{\varkappa 3} \tag{2-7}$$

From Eqs. (2-5) and (2-6), we have

$$z_1 = z_{\varkappa 1}(1 + t) \tag{2-8}$$

and

$$z_2 = z_{\varkappa 2}(1 + 2t)^{\frac{1}{2}} \tag{2-9}$$

Equations (2-7), (2-8), and (2-9) are the parametric equations for the path of that particle which at time $t = 0$ occupies the position whose rectangular cartesian coordinates are $(z_{\varkappa 1}, z_{\varkappa 2}, z_{\varkappa 3})$.

If we eliminate the parameter time, then in the plane $z_3 = z_{\varkappa 3}$ the particle paths have the form

$$\frac{z_2}{z_{\varkappa 2}} = \left(2\frac{z_1}{z_{\varkappa 1}} - 1\right)^{\frac{1}{2}} \tag{2-10}$$

Equation (2-10) is shown in Fig. 1.1.4-2 for several initial positions $(z_{\varkappa 1}, z_{\varkappa 2})$.

### 1.1.3 Streamlines
The streamlines for time $t$ form that family of curves to which the velocity field is everywhere tangent at time $t$. The parametric equations for the streamlines are solutions of the differential equations

$$\frac{d\mathbf{z}}{d\alpha} = \mathbf{v} \tag{3-1}$$

or

$$\frac{dz_i}{d\alpha} = v_i \quad \text{for } i = 1, 2, 3 \tag{3-2}$$

Alternatively, we may think of the streamlines as solutions of the differential system

$$\frac{d\mathbf{z}}{d\alpha} \wedge \mathbf{v} = 0 \tag{3-3}$$

or

$$e_{ijk}\frac{dz_j}{d\alpha}v_k = 0 \quad \text{for } i = 1, 2, 3 \tag{3-4}$$

In these equations, $\alpha$ is an arbitrary parameter measured along the curves and time $t$ is a constant.

As an example, let us determine the streamlines corresponding to the velocity distribution studied in Sec. 1.1.2:

$$v_1 = \frac{z_1}{1+t} \quad v_2 = \frac{z_2}{1+2t} \quad v_3 = 0 \tag{2-4}$$

From Eqs. (3-2) and (2-4), we have

$$\int_{z_{1(0)}}^{z_1} \frac{dz_1}{z_1} = \frac{1}{1+t}\int_0^\alpha d\alpha \tag{3-5}$$

$$\int_{z_{2(0)}}^{z_2} \frac{dz_2}{z_2} = \frac{1}{1+2t} \int_0^\alpha d\alpha \tag{3-6}$$

and

$$z_3 = z_{3(0)} \tag{3-7}$$

Equations (3-5) and (3-6) give

$$z_1 = z_{1(0)} \exp \frac{\alpha}{1+t} \tag{3-8}$$

and

$$z_2 = z_{2(0)} \exp \frac{\alpha}{1+2t} \tag{3-9}$$

We refer to Eqs. (3-7), (3-8), and (3-9) as being the parametric equations of the streamline corresponding to time $t$ which passes through the point whose rectangular cartesian coordinates are $z_{1(0)}, z_{2(0)}, z_{3(0)}$.

If we eliminate the parameter $\alpha$ between Eqs. (3-8) and (3-9), we find

$$z_2 = z_{2(0)} \left(\frac{z_1}{z_{1(0)}}\right)^{(1+t)/(1+2t)} \tag{3-10}$$

Streamlines corresponding to time $t = 0$ and as $t \to \infty$ are shown in Fig. 1.1.4-3a and b, respectively, for several different points $[z_{1(0)}, z_{2(0)}]$.

Experimentalists sometimes sprinkle particles over a gas-liquid phase interface and take a photograph in which the motion of the particles is not quite stopped (see Figs. 3.5.1-1 and 3.5.1-3). The traces left by the particles are proportional to the velocity of the fluid at the surface (so long as we assume that very small particles move with the fluid). For a steady-state flow, such a photograph may be used to construct the particle paths. For an unsteady-state flow, it depicts the streamlines, the family of curves to which the velocity vector field is everywhere tangent.

In two-dimensional flows, the streamlines have a special significance. They are curves along which the stream function (Secs. 1.3.6 and 1.3.7) is a constant. See in particular Exercise 1.3.7-1.

## REFERENCE

1. Aris, Rutherford: "Vectors, Tensors and the Basic Equations of Fluid Mechanics," Prentice-Hall, Englewood Cliffs, N.J., 1962.

## EXERCISES

**1.1.3-1** [1, p. 81] Let $v$ be the magnitude of the velocity field. If $\mathbf{v}/v$ is independent of time, show that the streamlines and path lines coincide.

**1.1.3-2** For potential flow past a cylinder of radius $a$, the physical components of velocity in cylindrical coordinates are

$$v_r = V\left(1 - \frac{a^2}{r^2}\right)\cos\theta$$

$$v_\theta = -V\left(1 + \frac{a^2}{r^2}\right)\sin\theta$$

and

$$v_z = 0$$

Show that the family of streamlines is described by

$$\left(1 - \frac{a^2}{r^2}\right)r\sin\theta = \text{constant}$$

**1.1.3-3** Repeat the problem of the text employing Eq. (3-4) rather than Eq. (3-2).

## 1.1.4 Streak lines

The streak line through the point $z_{(0)}$ at time $t$ represents the positions at time $t$ of the material particles which at any time $\tau \leq t$ have occupied the place $z_{(0)}$.

Experimentally we might visualize that smoke, dust, or dye are continuously injected into a fluid at a position $z_{(0)}$ and that the resulting trails are photographed as functions of time. Each photograph shows a streak line corresponding to the position $z_{(0)}$ and the time at which the photograph was taken.

We saw in Sec. 1.1.1 that the motion $\chi$ describes the position $z$ at time $t$ of the material particle that occupied the position $z_\chi$ in the reference configuration:

$$z = \chi(z_\chi, t) \tag{4-1}$$

In constructing a streak line, we focus our attention on those material particles that were in the place $z_{(0)}$ at any time $\tau \leq t$:

$$z_\chi = \chi^{-1}(z_{(0)}, \tau) \tag{4-2}$$

The parametric equations of the streak line through the point $z_{(0)}$ at time $t$ are obtained by eliminating $z_\chi$ between Eqs. (4-1) and (4-2):

$$z = \chi(\chi^{-1}(z_{(0)}, \tau), t) \tag{4-3}$$

Time $\tau \leq t$ is the parameter along the streak line.

As an example, let us consider a streak line that arises in the velocity-distribution studies in Secs. 1.1.2 and 1.1.3:

$$v_1 = \frac{z_1}{1+t} \quad v_2 = \frac{z_2}{1+2t} \quad v_3 = 0 \tag{4-4}$$

From the path lines of Sec. 1.1.2, the particles that occupied the position $z_{(0)}$ at any time $\tau \leq t$ are described by

$$z_{\chi 1} = \frac{z_{1(0)}}{1+\tau} \quad z_{\chi 2} = \frac{z_{2(0)}}{(1+2\tau)^{\frac{1}{2}}} \quad z_{\chi 3} = z_{3(0)} \tag{4-5}$$

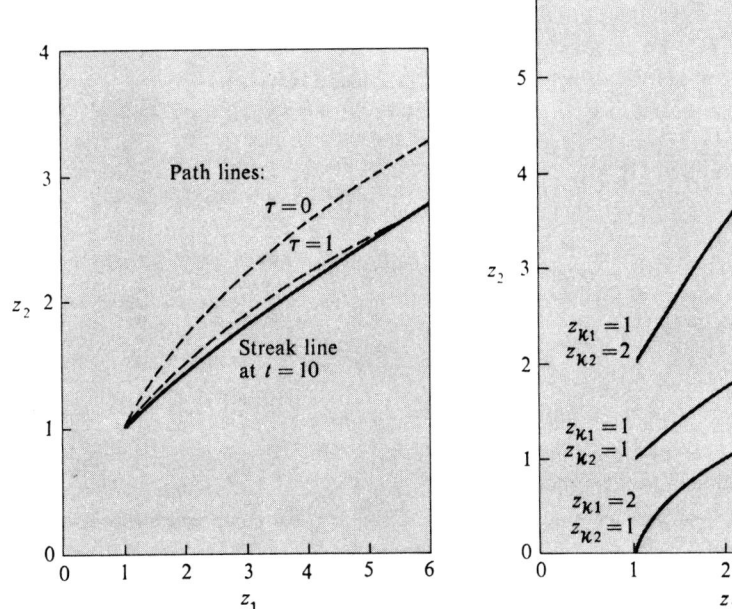

**Fig. 1.1.4-1** Streak line [Eq. (4-6)] at $t = 10$ with two corresponding path lines [Eq. (2-10)].

**Fig. 1.1.4-2** Path lines [Eq. (2-10)].

Again from the path lines of Sec. 1.1.2, the parametric equations for the streak line through $\mathbf{z}_{(0)}$ at time $t$ are

$$z_1 = z_{1(0)}\left(\frac{1+t}{1+\tau}\right) \quad z_2 = z_{2(0)}\left(\frac{1+2t}{1+2\tau}\right)^{\frac{1}{2}} \quad z_3 = z_{3(0)} \tag{4-6}$$

This streak line, together with some of the particle paths that contribute to it, is shown in Fig. 1.1.4-1.

For comparison, we display in Fig. 1.1.4-2 several other path lines for this velocity distribution as described by Eq. (2-10). In Fig. 1.1.4-3a and b, we illustrate several streamlines corresponding to time $t = 0$ and as $t \to \infty$, respectively, as described by Eq. (3-10).

## EXERCISE

**1.1.4-1** Show that, for a velocity distribution that is independent of time, the path lines, streamlines, and streak lines coincide.

*Hint:* In considering the path line, take as the boundary condition:

At $t = t_1$:    $\mathbf{z} = \mathbf{z}_{(0)}$

This suggests the introduction of a new variable $\alpha = t - t_1$, which denotes time measured since the particle passed through the position $\mathbf{z}_{(0)}$.

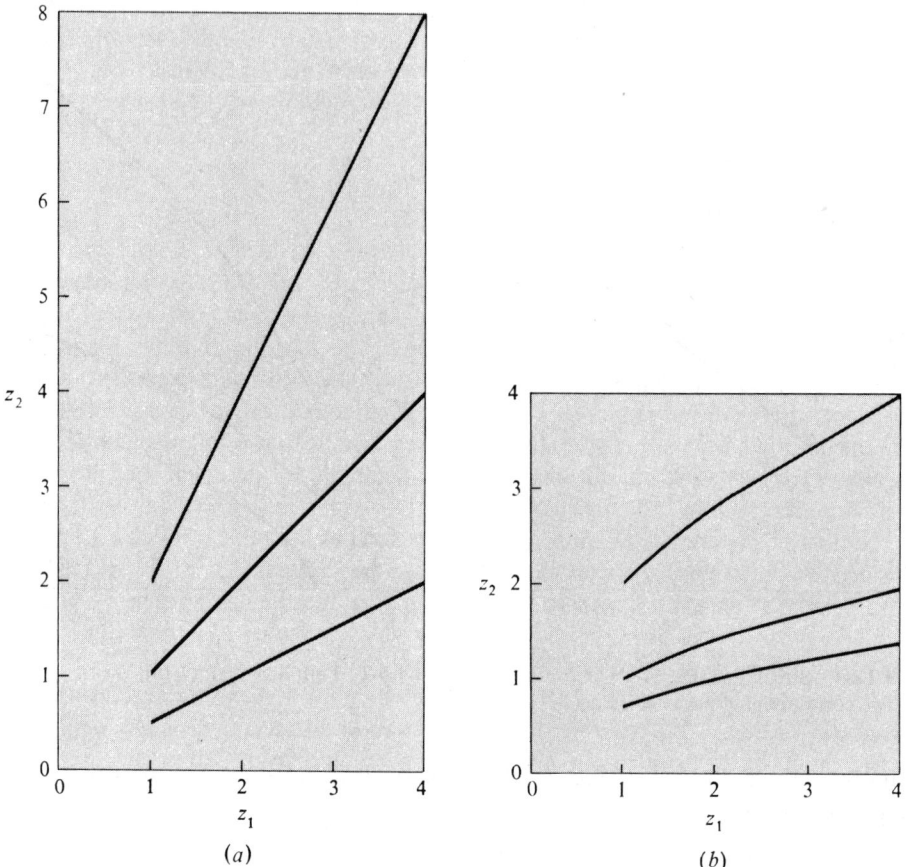

**Fig. 1.1.4-3** (*a*) Streamlines [Eq. (3-10)] at $t = 0$; (*b*) streamlines [Eq. (3-10)] as $t \to \infty$.

## 1.2 FRAME

**1.2.1 Changes of frame** [1, p. 22; 2, p. 41; 3, p. 437] The two fundamental quantities that we measure in kinematics are distances and time intervals. The position in space of some occurrence (for example, the position at which a particular velocity is achieved by a material) may be determined only with respect to a given frame of reference or observer. A frame of reference might be the walls of a laboratory, the fixed stars, or the shell of a space capsule that is following an arbitrary trajectory; it is a set of objects whose mutual distances remain unchanged during the period of observation. The time of some occurrence may be specified only with respect to the time of occurrence of some other event, such as the time at which a stopwatch was started or an electric circuit closed. Such a reference is part of the specification

[1.2] FRAME

of a frame of reference. Truesdell and Noll [2, p. 41] state that "a frame of reference may be described as a possible way of relating physical reality to a three-dimensional euclidean point space and a real time axis."

Consider the collection of all pairs $\{z,t\}$, where $z$ is a position in euclidean point space and $t$ a time. We will refer to the collection of all such pairs as space-time. A change of frame is a one-to-one mapping of space-time onto itself in such a manner that distances, time intervals, and temporal order remain unchanged.

Let $\mathbf{z}$ and $t$ denote a position and time in the old frame, $\mathbf{z}^*$ and $t^*$ the corresponding position and time in the new frame. The most general change of frame is of the form

$$\mathbf{z}^* = \mathbf{c}(t) + \mathbf{Q}(t) \cdot \{\mathbf{z} - \mathbf{z}_0\} \tag{1-1}$$

$$t^* = t - a \tag{1-2}$$

The fixed point $\mathbf{z}_0$ is mapped into $\mathbf{c}(t)$. If we choose $\mathbf{z}_0 = 0$, then $\mathbf{c}(t)$ represents the position vector of the old origin in the new frame. The time-dependent orthogonal second-order tensor $\mathbf{Q}(t)$ represents a rotation and, possibly, a reflection. A reflection allows for the possibility that an observer in the new frame looks at the old frame through a mirror. Alternatively, a reflection allows for the possibility that two observers orient themselves oppositely, one choosing to work in terms of a right-handed frame of reference and the other in terms of a left-handed frame of reference (see Exercise A.10.1-7). The quantity $a$ is a real number.

We speak of a quantity as being *frame indifferent* if it remains unchanged or invariant under all changes of frame. A frame-indifferent scalar $b$ does not change its value,

$$b^* = b \tag{1-3}$$

A frame-indifferent spatial vector remains the same directed line element under a change of frame in the sense that, if

$$\mathbf{v} = \mathbf{z}_1 - \mathbf{z}_2$$

then

$$\mathbf{v}^* = \mathbf{z}_1^* - \mathbf{z}_2^*$$

From Eq. (1-1),

$$\mathbf{v}^* = \mathbf{Q} \cdot \{\mathbf{z}_1 - \mathbf{z}_2\} = \mathbf{Q} \cdot \mathbf{v} \tag{1-4}$$

A frame-indifferent second-order tensor is one that transforms frame-indifferent spatial vectors into frame-indifferent spatial vectors. If

$$\mathbf{v} = \mathbf{T} \cdot \mathbf{w}$$

$$\mathbf{v}^* = \mathbf{Q} \cdot \mathbf{v}$$

and

$$\mathbf{w}^* = \mathbf{Q} \cdot \mathbf{w}$$

then

$$\mathbf{v}^* = \mathbf{T}^* \cdot \mathbf{w}^*$$

This means that

$$\mathbf{Q} \cdot \mathbf{v} = \mathbf{T}^* \cdot \mathbf{Q} \cdot \mathbf{w} = \mathbf{Q} \cdot \mathbf{T} \cdot \mathbf{w}$$

or

$$\mathbf{T}^* = \mathbf{Q} \cdot \mathbf{T} \cdot \mathbf{Q}^{\mathrm{T}} \tag{1-5}$$

It is important to carefully distinguish between a frame of reference and a coordinate system. When we defined a frame of reference as a set of objects whose mutual distances remain unchanged during the period of observation, it was understood that this set of objects did not all lie in a plane. Consequently, we may visualize replacing the set of objects by three mutually orthogonal unit vectors $(\mathbf{a}_{(1)}, \mathbf{a}_{(2)}, \mathbf{a}_{(3)})$. Any coordinate system whatsoever can be used to locate points in space with respect to these three vectors and their intersection. It is convenient to restrict admissible coordinate systems to those whose axes have a time invariant orientation with respect to $(\mathbf{a}_{(1)}, \mathbf{a}_{(2)}, \mathbf{a}_{(3)})$. Let $(z_1, z_2, z_3)$ be a coordinate system associated with the frame of reference $(\mathbf{a}_{(1)}, \mathbf{a}_{(2)}, \mathbf{a}_{(3)})$; similarly, let $(z_1', z_2', z_3')$ be a coordinate system associated with another frame of reference $(\mathbf{a}_{(1)}', \mathbf{a}_{(2)}', \mathbf{a}_{(3)}')$. We will say that these two coordinate systems are the *same*, if the orientation of the $z_i$ coordinate axes with respect to the vectors $\mathbf{a}_{(j)}$ is identical with the orientation of the $z_i'$ coordinate axes with respect to the vectors $\mathbf{a}_{(j)}'$. We generally find it convenient to use the same coordinate system in discussing two different frames of reference as we do in the next section.

The importance of changes of frame will become apparent in Sec. 2.3.1, where the principle of material frame indifference is introduced. This principle will be used repeatedly in discussing representations for material behavior and in preparing empirical data correlations.

## REFERENCES

1. Truesdell, C.: "The Elements of Continuum Mechanics," Springer-Verlag, New York, 1966.
2. Truesdell, C., and W. Noll: In S. Flügge (ed.), "Handbuch der Physik," vol. 3/3, Springer-Verlag, Berlin, 1965.
3. Truesdell, C., and R. A. Toupin: In S. Flügge (ed.), "Handbuch der Physik," vol. 3/1, Springer-Verlag, Berlin, 1960.

## EXERCISES

**1.2.1-1** Let $T$ be a frame-indifferent scalar field. Starting with the definition of the gradient of a scalar field in Sec. A.3.1, show that the gradient of $T$ is frame indifferent:

$$\nabla T^* \equiv (\nabla T)^* = \mathbf{Q} \cdot \nabla T$$

**1.2.1-2** In order that $\boldsymbol{\epsilon}$ (see Exercise A.7.2-11) be a frame-indifferent third-order tensor field, prove that

$$\boldsymbol{\epsilon}^* = (\det \mathbf{Q}) e_{ijk} \mathbf{e}_i \mathbf{e}_j \mathbf{e}_k$$

[1.2] FRAME

### 1.2.2 Equivalent motions [1, p. 42; 2, p. 22]
In Sec. 1.1.1, we described the motion of a material with respect to some frame of reference by

$$\mathbf{z} = \boldsymbol{\chi}(\mathbf{z}_\varkappa, t) \tag{2-1}$$

where we now understand that the form of this relation depends upon the choice of reference configuration $\varkappa$. According to our discussion in Sec. 1.2.1, the same motion with respect to some new frame of reference is represented by

$$\mathbf{z}^* = \boldsymbol{\chi}^*(\mathbf{z}_\varkappa^*, t^*) = \mathbf{c}(t) + \mathbf{Q}(t) \cdot [\boldsymbol{\chi}(\mathbf{z}_\varkappa, t) - \mathbf{z}_0] \tag{2-2}$$

We will say that any two motions $\boldsymbol{\chi}$ and $\boldsymbol{\chi}^*$ related by an equation of the form of Eq. (2-2) are *equivalent motions*.

Let us write Eq. (2-2) in an abbreviated form:

$$\mathbf{z}^* = \mathbf{c} + \mathbf{Q} \cdot (\mathbf{z} - \mathbf{z}_0) \tag{2-3}$$

The material derivative of this equation gives

$$\dot{\mathbf{z}^*} = \dot{\mathbf{c}} + \dot{\mathbf{Q}} \cdot (\mathbf{z} - \mathbf{z}_0) + \mathbf{Q} \cdot \dot{\mathbf{z}} \tag{2-4}$$

or

$$\dot{\mathbf{z}^*} - \mathbf{Q} \cdot \dot{\mathbf{z}} = \dot{\mathbf{c}} + \dot{\mathbf{Q}} \cdot (\mathbf{z} - \mathbf{z}_0) \tag{2-5}$$

In view of Eq. (2-3), we may write

$$\mathbf{z} - \mathbf{z}_0 = \mathbf{Q}^T \cdot \mathbf{Q} \cdot (\mathbf{z} - \mathbf{z}_0) = \mathbf{Q}^T \cdot (\mathbf{z}^* - \mathbf{c}) \tag{2-6}$$

This allows us to express Eq. (2-5) as

$$\begin{aligned}\dot{\mathbf{z}^*} - \mathbf{Q} \cdot \dot{\mathbf{z}} &= \dot{\mathbf{c}} + (\dot{\mathbf{Q}} \cdot \mathbf{Q}^T) \cdot (\mathbf{z}^* - \mathbf{c}) \\ &= \dot{\mathbf{c}} + \mathbf{A} \cdot (\mathbf{z}^* - \mathbf{c}) \end{aligned} \tag{2-7}$$

where

$$\mathbf{A} \equiv \dot{\mathbf{Q}} \cdot \mathbf{Q}^T \tag{2-8}$$

We refer to the second-order tensor $\mathbf{A}$ as the angular velocity tensor of the starred frame with respect to the unstarred frame [2, p. 24].

Since $\mathbf{Q}$ is an orthogonal tensor,

$$\mathbf{Q} \cdot \mathbf{Q}^T = \mathbf{I}^* \tag{2-9}$$

Taking the material derivative of this equation, we have

$$\begin{aligned}\mathbf{A} = \dot{\mathbf{Q}} \cdot \mathbf{Q}^T &= -\mathbf{Q} \cdot \dot{\mathbf{Q}^T} \\ &= -\mathbf{Q} \cdot (\dot{\mathbf{Q}})^T \\ &= -\mathbf{A}^T \end{aligned} \tag{2-10}$$

In this way we see that the angular velocity tensor is skew symmetric.

The angular velocity vector of the unstarred frame with respect to the starred frame $\boldsymbol{\omega}$ is defined as

$$\boldsymbol{\omega} \equiv \tfrac{1}{2}\boldsymbol{\epsilon}^* : \mathbf{A} \tag{2-11}$$

The third-order tensor $\boldsymbol{\epsilon}$ is introduced in Exercises A.7.2-11 and A.7.2-12 (see also Exercise 1.2.1-2); the double-dot notation is defined in Sec. A.7.3. Let us consider the following spatial vector in rectangular cartesian coordinates:

$$\begin{aligned}
\boldsymbol{\omega} \wedge (\mathbf{z}^* - \mathbf{c}) &= \boldsymbol{\epsilon}^* : [(\mathbf{z}^* - \mathbf{c})\boldsymbol{\omega}] \\
&= \boldsymbol{\epsilon}^* : [(\mathbf{z}^* - \mathbf{c})(\tfrac{1}{2}\boldsymbol{\epsilon}^* : \mathbf{A})] \\
&= e_{ijk}(z_k^* - c_k)(\tfrac{1}{2} e_{jmn} A_{nm}) \mathbf{e}_i^* \\
&= \tfrac{1}{2}(z_k^* - c_k)(A_{ik} - A_{ki})\mathbf{e}_i^* \\
&= (z_k^* - c_k)A_{ik}\mathbf{e}_i^* \\
&= \mathbf{A} \cdot (\mathbf{z}^* - \mathbf{c}) \tag{2-12}
\end{aligned}$$

We may consequently write Eq. (2-7) in terms of the angular velocity of the unstarred frame with respect to the starred frame [3, p. 437]:

$$\dot{\mathbf{z}}^* = \dot{\mathbf{c}} + \boldsymbol{\omega} \wedge [\mathbf{Q} \cdot (\mathbf{z} - \mathbf{z}_0)] + \mathbf{Q} \cdot \dot{\mathbf{z}} \tag{2-13}$$

## REFERENCES

1. Truesdell, C., and W. Noll: In S. Flügge (ed.), "Handbuch der Physik," vol. 3/3, Springer-Verlag, Berlin, 1965.
2. Truesdell, C.: "The Elements of Continuum Mechanics," Springer-Verlag, New York, 1966.
3. Truesdell, C., and R. A. Toupin: In S. Flügge (ed.), "Handbuch der Physik," vol. 3/1, Springer-Verlag, Berlin, 1960.

## EXERCISES

**1.2.2-1** (a) Show that velocity is not frame indifferent.

(b) Show that at any position in euclidean point space a difference in velocities with respect to the same frame is frame indifferent.

**1.2.2-2** *Acceleration* (a) Determine that [2, p. 24]

$$\ddot{\mathbf{z}}^* = \ddot{\mathbf{c}} + 2\mathbf{A} \cdot (\dot{\mathbf{z}}^* - \dot{\mathbf{c}}) + (\dot{\mathbf{A}} - \mathbf{A}^2) \cdot (\mathbf{z}^* - \mathbf{c}) + \mathbf{Q} \cdot \ddot{\mathbf{z}}$$

(b) Prove that [3, p. 440]

$$\ddot{\mathbf{z}}^* = \ddot{\mathbf{c}} + (\dot{\mathbf{A}} + \mathbf{A}^2) \cdot \mathbf{Q} \cdot (\mathbf{z} - \mathbf{z}_0) + 2\mathbf{A} \cdot \mathbf{Q} \cdot \dot{\mathbf{z}} + \mathbf{Q} \cdot \ddot{\mathbf{z}}$$

(c) Prove that

$$\boldsymbol{\omega} \wedge [\boldsymbol{\omega} \wedge (\mathbf{z}^* - \mathbf{c})] = \mathbf{A}^2 \cdot (\mathbf{z}^* - \mathbf{c})$$
$$\dot{\boldsymbol{\omega}} \wedge (\mathbf{z}^* - \mathbf{c}) = \dot{\mathbf{A}} \cdot (\mathbf{z}^* - \mathbf{c})$$

## [1.3] MASS

and

$$\omega \wedge (Q \cdot \dot{z}) = A \cdot Q \cdot \dot{z}$$

(d) Conclude that [3, p. 438]

$$\ddot{z}^* = \ddot{c} + \dot{\omega} \wedge [Q \cdot (z - z_0)]$$
$$+ \omega \wedge \{\omega \wedge [Q \cdot (z - z_0)]\}$$
$$+ 2\omega \wedge (Q \cdot \dot{z}) + Q \cdot \ddot{z}$$

**1.2.2-3** Give an example of a scalar that is *not* frame indifferent. *Hint:* What vector is not frame indifferent?

**1.2.2-4** *Motion of a rigid body* Determine that the velocity distribution in a rigid body may be expressed as

$$\dot{z}^* = \dot{c} + \omega \wedge (z^* - c)$$

What is the relation of the unstarred frame to the body in this case?

## 1.3 MASS

### 1.3.1 Conservation of mass

This discussion of mechanics is based upon several postulates. The first is *conservation of mass*.

The mass of a body is independent of time.

Physically, this means that, if we follow a portion of a material body through any number of translations, rotations, and deformations, the mass associated with it will not vary as a function of time. If $\rho$ is the mass density of the body, the mass $M$ of the body may be represented as

$$M = \int_{V_{(m)}} \rho \, dV \tag{1-1}$$

Here $V_{(m)}$ denotes that the integration is to be performed over the region of space occupied by the body in its current configuration; in general $V_{(m)}$, or the limits on this integration, is a function of time. The postulate of conservation of mass says that

$$\dot{M} \equiv \frac{d}{dt} \int_{V_{(m)}} \rho \, dV = 0 \tag{1-2}$$

Our next object will be to determine a relationship that expresses the idea of conservation of mass at each point in a material. In order to do this, we will find it necessary to interchange the operations of differentiation and integration in Eq. (1-2). Yet the limits on this integral describe the boundaries of the body in its current configuration and generally are functions of time. The next section explores this problem in more detail.

### 1.3.2 Transport theorem (see also Appendix B)   Let us consider the operation

$$\frac{d}{dt}\int_{V_{(m)}} \Psi\, dV$$

Here $\Psi$ is any scalar-, vector-, or tensor-valued function of time and position. Again, $V_{(m)}$ denotes that the integration is to be performed over the region of space occupied by the body in its current configuration. In general, we should expect that $V_{(m)}$, or the limits on this integration, is a function of time.

If we look at this volume integration in the reference configuration $\varkappa$, the limits on the volume integral are no longer functions of time; the limits are now expressed in terms of the material coordinates of the bounding surface of the body. This means that we may interchange differentiation and integration in the above operation. In terms of a rectangular cartesian coordinate system, let $(z_1, z_2, z_3)$ denote the current coordinates of a material point and let $(z_{\varkappa 1}, z_{\varkappa 2}, z_{\varkappa 3})$ be the corresponding material coordinates. Using the results of Sec. A.11.3, we may say that

$$\frac{d}{dt}\int_{V_{(m)}} \Psi\, dV = \frac{d}{dt}\int_{V_{(m)\varkappa}} \Psi J\, dV$$

$$= \int_{V_{(m)\varkappa}} \left( \frac{d_{(m)}\Psi}{dt} + \frac{\Psi}{J}\frac{d_{(m)}J}{dt} \right) J\, dV$$

$$= \int_{V_{(m)}} \left( \frac{d_{(m)}\Psi}{dt} + \frac{\Psi}{J}\frac{d_{(m)}J}{dt} \right) dV \tag{2-1}$$

where

$$J \equiv \sqrt{\left[ \det \frac{\partial z_i}{\partial z_{\varkappa j}} \right]^2} \tag{2-2}$$

The quantity $J$ may be thought of as the volume in the current configuration per unit volume in the reference configuration. It will generally be a function of both time and position. Here $V_{(m)\varkappa}$ indicates that the integration is to be performed over the region of space occupied by the body in its reference configuration $\varkappa$.

We call

$$\mathbf{F} = \frac{\partial z_i}{\partial z_{\varkappa j}} \mathbf{e}_i \mathbf{e}_j \tag{2-3}$$

the deformation gradient. From Exercise A.10.1-5,

$$\dot{J} = \frac{d_{(m)}J}{dt} = J\,\text{tr}\,(\mathbf{F}^{-1}\cdot\dot{\mathbf{F}}) \tag{2-4}$$

It is easy to show that

## [1.3] MASS

$$\mathbf{F}^{-1} = \frac{\partial z_{\varkappa m}}{\partial z_n} \mathbf{e}_m \mathbf{e}_n \tag{2-5}$$

Using the definition of velocity from Sec. 1.1.1, we have

$$\dot{\mathbf{F}} = \frac{\partial \dot{z}_i}{\partial z_{\varkappa j}} \mathbf{e}_i \mathbf{e}_j = \frac{\partial v_i}{\partial z_r} \frac{\partial z_r}{\partial z_{\varkappa j}} \mathbf{e}_i \mathbf{e}_j \tag{2-6}$$

Consequently,

$$\mathrm{tr}\,(\mathbf{F}^{-1} \cdot \dot{\mathbf{F}}) = \frac{\partial z_{\varkappa j}}{\partial z_i} \frac{\partial v_i}{\partial z_r} \frac{\partial z_r}{\partial z_{\varkappa j}}$$

$$= \frac{\partial v_i}{\partial z_i} = \mathrm{div}\,\mathbf{v} \tag{2-7}$$

and

$$\frac{1}{J} \frac{d_{(m)} J}{dt} = \mathrm{div}\,\mathbf{v} \tag{2-8}$$

Equation (2-8) allows us to write Eq. (2-1) as

$$\frac{d}{dt} \int_{V_{(m)}} \Psi \, dV = \int_{V_{(m)}} \left( \frac{d_{(m)} \Psi}{dt} + \Psi \, \mathrm{div}\,\mathbf{v} \right) dV \tag{2-9}$$

This may also be expressed as

$$\frac{d}{dt} \int_{V_{(m)}} \Psi \, dV = \int_{V_{(m)}} \left[ \frac{\partial \Psi}{\partial t} + \mathrm{div}\,(\Psi \mathbf{v}) \right] dV \tag{2-10}$$

or, by Green's transformation (Sec. A.11.2), we may say

$$\frac{d}{dt} \int_{V_{(m)}} \Psi \, dV = \int_{V_{(m)}} \frac{\partial \Psi}{\partial t} \, dV + \int_{S_{(m)}} \Psi \mathbf{v} \cdot \mathbf{n} \, dS \tag{2-11}$$

By $S_{(m)}$ we mean the closed bounding surface of $V_{(m)}$; like $V_{(m)}$, it will in general be a function of time. Equations (2-9) to (2-11) are three forms of the *transport theorem* [1, p. 347].

We will have occasion to ask about the derivative with respect to time of a quantity while following a system which is not necessarily a material body. For example, let us take as our system the air in a child's balloon and ask for the derivative with respect to time of the volume associated with the air as the balloon is inflated. Since material (air) is being continuously added to the balloon, we are not following a set of material particles as a function of time. On the other hand, there is nothing to prevent us from defining a particular set of fictitious system particles to be associated with our system. The only restriction we shall make upon this set of imaginary system particles is that the normal component of velocity of any system particle at the boundary of the system be equal to the normal component of velocity of the boundary of the system. Equations (2-8) to (2-11) remain valid if we replace (1) derivatives with respect to time while following material particles, $d_{(m)}/dt$, by derivatives with respect to time while following fictitious system particles, $d_{(s)}/dt$, and (2) the velocity vector for a material particle, $\mathbf{v}$, by the velocity vector for a fictitious system particle, $\mathbf{v}_{(s)}$. This means that

$$\frac{d}{dt}\int_{V_{(s)}} \Psi \, dV = \int_{V_{(s)}} \frac{\partial \Psi}{\partial t} \, dV + \int_{S_{(s)}} \Psi \mathbf{v}_{(s)} \cdot \mathbf{n} \, dS \tag{2-12}$$

Here $V_{(s)}$ signifies that region of space currently occupied by the system; $S_{(s)}$ is the closed bounding surface of the system. We will refer to Eq. (2-12) as the *generalized transport theorem* [1, p. 347].

## REFERENCE

1. Truesdell, C., and R. A. Toupin: In S. Flügge (ed.), "Handbuch der Physik," vol. 3/1, Springer-Verlag, Berlin, 1960.

## EXERCISES

**1.3.2-1** If both the current coordinates $(x^1,x^2,x^3)$ and the material coordinates $(x_\varkappa^1,x_\varkappa^2,x_\varkappa^3)$ of a material point are expressed with respect to a curvilinear coordinate system, show that

$$J = \frac{\sqrt{g_{(x)}}}{\sqrt{g_{(\varkappa)}}} \sqrt{\left[\det\left(\frac{\partial x^i}{\partial x_\varkappa{}^j}\right)\right]^2}$$

Here $g_{(x)}$ and $g_{(\varkappa)}$ denote the determinants of the $g_{ij}$ evaluated at the positions of the material particle in the current and reference configurations, respectively.

**1.3.2-2** Show that

$$\frac{d}{dt}\int_{V_{(m)}} dV = \int_{V_{(m)}} \operatorname{div} \mathbf{v} \, dV = \int_{S_{(m)}} \mathbf{v} \cdot \mathbf{n} \, dS$$

Here $S_{(m)}$ is the (time-dependent) closed bounding surface of $V_{(m)}$.

**1.3.3 The equation of continuity** Going back to the postulate of conservation of mass in Sec. 1.3.1,

$$\frac{d}{dt}\int_{V_{(m)}} \rho \, dV = 0 \tag{3-1}$$

and employing the transport theorem in the form of Eq. (2-9), we have that

$$\int_{V_{(m)}} \left(\frac{d_{(m)}\rho}{dt} + \rho \operatorname{div} \mathbf{v}\right) dV = 0 \tag{3-2}$$

But this statement is true for any body or for any portion of a body (since a portion of a body is a body). We conclude that the integrand itself must be identically zero:

$$\frac{d_{(m)}\rho}{dt} + \rho \operatorname{div} \mathbf{v} = 0 \tag{3-3}$$

## [1.3] MASS

By Exercise 1.1.1-1, this may also be written as

$$\frac{\partial \rho}{\partial t} + \text{div}\,(\rho \mathbf{v}) = 0 \tag{3-4}$$

Equations (3-3) and (3-4) are two forms of the *equation of continuity*. These equations express the requirement that mass be conserved at every point in the continuous material.

The equation of continuity is presented in Table 2.5.1-1 for rectangular cartesian, cylindrical, and spherical coordinates.

If the density following a fluid particle does not change as a function of time, Eq. (3-3) reduces to

$$\text{div}\,\mathbf{v} = 0 \tag{3-5}$$

Such a motion is said to be *isochoric*. If, for the flow under consideration, density is independent of both time and position, we will say that the fluid is *incompressible*. A sufficient, though not necessary, condition for an isochoric motion is that the fluid is incompressible.

### REFERENCE

1. Kaplan, Wilfred: "Advanced Calculus," Addison-Wesley, Cambridge, Mass., 1952.

### EXERCISES

**1.3.3-1** *Another form of the transport theorem (see also Appendix B.1.2)* Show that, if we assume that mass is conserved, for any $\hat{\Psi}$

$$\frac{d}{dt}\int_{V_{(m)}} \rho\hat{\Psi}\,dV = \int_{V_{(m)}} \rho\,\frac{d_{(m)}\hat{\Psi}}{dt}\,dV$$

**\*1.3.3-2** Derive Eqs. (B) and (C) of Table 2.5.1-1.

**1.3.3-3** (a) From Eq. (2-1) and the postulate of conservation of mass, determine that

$$\frac{d_{(m)}}{dt}\,[\ln(\rho J)] = 0$$

(b) Integrate this equation to conclude

$$\rho J = \rho_0$$

where $\rho_0$ denotes the density distribution in the reference configuration.

**1.3.3-4** Let us examine the argument that must be supplied in going from Eq. (3-2) to (3-3). We can begin by considering the analogous problem in one dimension. It is clear that

$$\int_0^{2\pi} \sin\theta\,d\theta = 0$$

does not imply that $\sin\theta$ is identically zero. But

$$\int_0^x f(y)\, dy = 0 \tag{1}$$

does imply that

$$f(y) = 0 \tag{2}$$

*Proof:* The Leibnitz rule for the derivative of an integral states that [1, p. 220]

$$\frac{d}{dx}\int_{a(x)}^{b(x)} g(x,y)\, dy = g(x,b(x))\frac{db}{dx} - g(x,a(x))\frac{da}{dx} + \int_{a(x)}^{b(x)} \frac{\partial g}{\partial x}\, dy$$

If we apply the Leibnitz rule to

$$\frac{d}{dx}\int_0^x f(y)\, dy$$

Eq. (2) follows immediately.

Let us now consider the analogous problem for

$$\int_{\zeta_1}^{\zeta_2}\int_{\eta_1(z)}^{\eta_2(z)}\int_{\xi_1(y,z)}^{\xi_2(y,z)} g(x,y,z)\, dx\, dy\, dz = 0 \tag{3}$$

where $\xi_1(y,z)$, $\xi_2(y,z)$, $\eta_1(z)$, $\eta_2(z)$, $\zeta_1$, and $\zeta_2$ are completely arbitrary. Prove that this implies

$$g(x,y,z) = 0 \tag{4}$$

**1.3.3-5** Consider flow of an incompressible fluid through a duct of arbitrary cross section. The velocity field may be time dependent, but at all times the duct is filled with fluid. Prove that, at any instant in time, the volume flow rate through any given cross section of the duct is the same as that passing through any other cross section of the duct.

**1.3.3-6** Consider the steady-state flow of a fluid such that streak lines are orthogonal everywhere to the density gradient. Prove that the motion of the fluid is isochoric.

## 1.3.4 Transport theorem for a region containing a singular surface [1, p. 525]

In this text, we view the phase interface as a singular surface, a surface that is discontinuous with respect to one or more quantities such as density and velocity. The phase interface in general is not material; it is common for mass to be transferred across a phase interface. As an ice cube melts, as water evaporates, as solid carbon dioxide sublimes, the phase interface moves through the material.

We wish to generalize the transport theorem developed in Sec. 1.3.2, in order that it might apply to a material region through which a phase interface is moving. Instead of a phase interface, we say that the material region is divided by a surface which is discontinuous (singular) with respect to a quantity $\Phi$ and possibly also with respect to velocity **v**. We assume that this surface of discontinuity may be in motion through the material with an arbitrary speed of displacement. If **u** denotes the velocity of a point on the surface,

$$u_{(\xi^+)} \equiv \mathbf{u}\cdot\boldsymbol{\xi}^+ \tag{4-1}$$

[1.3] MASS

is the speed of displacement of the surface measured in the direction $\xi^+$ in Fig. 1.3.4-1 and

$$u_{(\xi^-)} \equiv \mathbf{u} \cdot \xi^- \tag{4-2}$$

is the speed of displacement of the surface measured in the direction $\xi^-$ [1, p. 499].

A typical material region exhibiting a singular surface $S_{(sing)}$ is illustrated in Fig. 1.3.4-1. The quantities $\Phi$ and $\mathbf{v}$ are assumed to be continuously differentiable in the regions $V^+$ and $V^-$. Since in general the singular surface $S_{(sing)}$ is not material, the regions and surfaces $V^+, V^-, S^+, S^-$ are not material. We may write

$$\frac{d}{dt}\int_{V_{(m)}} \Phi \, dV = \frac{d}{dt}\int_{V^+} \Phi \, dV + \frac{d}{dt}\int_{V^-} \Phi \, dV \tag{4-3}$$

To each term on the right of Eq. (4-3), we may apply the generalized transport theorem of Sec. 1.3.2 to obtain

$$\frac{d}{dt}\int_{V^+} \Phi \, dV = \int_{V^+} \frac{\partial \Phi}{\partial t} dV + \int_{S^+} \Phi \mathbf{v} \cdot \mathbf{n} \, dS - \int_{S_{(sing)}} \Phi^+ u_{(\xi^+)} dS \tag{4-4}$$

and

$$\frac{d}{dt}\int_{V^-} \Phi \, dV = \int_{V^-} \frac{\partial \Phi}{\partial t} dV + \int_{S^-} \Phi \mathbf{v} \cdot \mathbf{n} \, dS - \int_{S_{(sing)}} \Phi^- u_{(\xi^-)} dS \tag{4-5}$$

By $\Phi^+$ and $\Phi^-$ we mean the limits of the function $\Phi$ obtained as any point $\mathbf{z}$

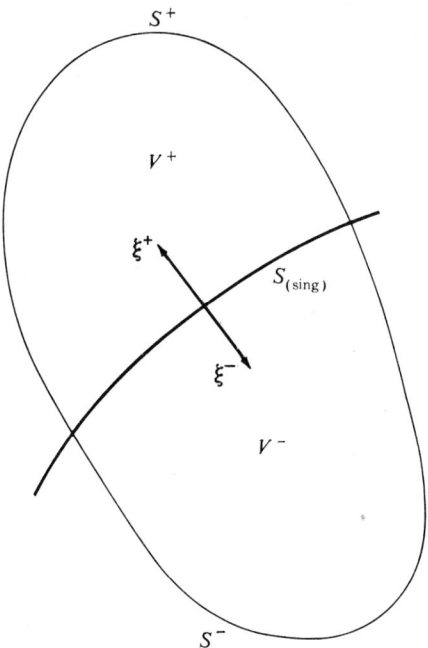

Fig. 1.3.4-1 Region containing a singular surface $S_{(sing)}$.

approaches a point $z_0$ on $S_{(sing)}$ while remaining within $V^+$ and $V^-$, respectively. Substituting these expressions into Eq. (4-3), we conclude that

$$\frac{d}{dt}\int_{V_{(m)}} \Phi \, dV = \int_{V_{(m)}} \frac{\partial \Phi}{\partial t} \, dV + \int_{S_{(m)}} \Phi \mathbf{v} \cdot \mathbf{n} \, dS - \int_{S_{(sing)}} [\Phi u_{(\xi)}] \, dS \qquad (4\text{-}6)$$

where the boldface brackets denote the jump of the quantity enclosed in crossing the interface:

$$[A\xi] \equiv A^+\xi^+ + A^-\xi^- \qquad (4\text{-}7)$$

We will refer to Eq. (4-6) as the transport theorem for regions containing a singular surface.

## REFERENCE

1. Truesdell, C., and R. A. Toupin: In S. Flügge (ed.), "Handbuch der Physik," vol. 3/1, Springer-Verlag, Berlin, 1960.

### 1.3.5 Jump mass balance for a phase interface

We will represent the phase interface as a singular surface, rather than as a three-dimensional region of some thickness. We neglect all interfacial effects such as surface tension that are attributable to the behavior of the material within this interfacial region. A discussion of various models for the interfacial region has been given elsewhere as well as a treatment of interfacial behavior [1].

The phase interface in general is not material. We observe mass moving across a phase interface when an ice cube melts. Here the speed of displacement of the phase interface is controlled by the rate of heat transfer to the system. Sometimes the speed of displacement of the phase interface might be specified by the rate of a chemical reaction. In general, the speed of displacement is given in the problem statement, or it is one of the unknowns which must be determined.

In Sec. 1.3.3, we found that the equation of continuity expresses the requirement that mass is conserved at every point within a continuous material. We wish to examine the implications of the first postulate of mechanics at a phase interface represented by a singular surface.

The material body to be considered contains a phase interface and may be represented by Fig. 1.3.4-1. The first postulate of mechanics, Sec. 1.3.1, says that

$$\frac{d}{dt}\int_{V_{(m)}} \rho \, dV = 0 \qquad (5\text{-}1)$$

By the transport theorem for regions containing a singular surface, Sec. 1.3.4, Eq. (5-1) may be written as

$$\int_{V^+ + V^-} \frac{\partial \rho}{\partial t} \, dV + \int_{S^+ + S^-} \rho \mathbf{v} \cdot \mathbf{n} \, dS - \int_{S_{(sing)}} [\rho u_{(\xi)}] \, dS = 0 \qquad (5\text{-}2)$$

The equation of continuity, Sec. 1.3.3, applies everywhere within each phase. Let us integrate it over the region $V^+$ and apply Green's transformation from Sec.

## [1.3] MASS

A.11.2:

$$\int_{V^+} \left\{ \frac{\partial \rho}{\partial t} + \text{div}(\rho \mathbf{v}) \right\} dV = 0$$

$$\int_{V^+} \frac{\partial \rho}{\partial t} dV + \int_{S^+} \rho \mathbf{v} \cdot \mathbf{n}\, dS - \int_{S_{(\text{sing})}} \rho^+ \mathbf{v}^+ \cdot \boldsymbol{\xi}^+ dS = 0 \tag{5-3}$$

Similarly, for the region $V^-$,

$$\int_{V^-} \frac{\partial \rho}{\partial t} dV + \int_{S^-} \rho \mathbf{v} \cdot \mathbf{n}\, dS - \int_{S_{(\text{sing})}} \rho^- \mathbf{v}^- \cdot \boldsymbol{\xi}^- dS = 0 \tag{5-4}$$

Subtracting Eqs. (5-3) and (5-4) from (5-2), we find that

$$\int_{(\text{sing})} [\rho \mathbf{v} \cdot \boldsymbol{\xi} - \rho u_{(\xi)}]\, dS = 0 \tag{5-5}$$

This must be true for any portion of a body containing a phase interface, no matter how large or small the body is. We conclude that the integrand itself must be zero:

$$[\rho(\mathbf{v} \cdot \boldsymbol{\xi} - u_{(\xi)})] = 0 \tag{5-6}$$

This is known as the *jump mass balance* for a phase interface which is represented by a singular surface when all interfacial effects are neglected.

### REFERENCES

1. Slattery, J. C.: *Ind. Eng. Chem., Fundamentals*, **6**:108 (1967); *ibid.*, **7**:672 (1968).
2. Truesdell, C., and R. A. Toupin: In S. Flügge (ed.), "Handbuch der Physik," vol. 3/1, Springer-Verlag, Berlin, 1960.

### EXERCISES

**1.3.5-1** Discuss how one concludes that, when there is no mass transfer across the phase interface, Eq. (5-6) reduces to

$$u_{(\xi^+)} = \mathbf{v}^+ \cdot \boldsymbol{\xi} = \mathbf{v}^- \cdot \boldsymbol{\xi}$$

**1.3.5-2** [2, p. 526] Write the postulate of conservation of mass for a material region which instantaneously contains a phase interface. Employ the transport theorem for a region containing a singular surface (Sec. 1.3.4). Deduce the jump mass balance, Eq. (5-6), by allowing the material region to shrink around the phase interface (surface of discontinuity).

**\*1.3.6 Stream functions for two-dimensional motions** By a two-dimensional motion, we mean here one such that in some coordinate system the velocity field has only two nonzero components.

Let us further restrict our discussion to incompressible fluids, so that the equation of continuity of Sec. 1.3.3 reduces to

$$\text{div } \mathbf{v} = 0 \tag{6-1}$$

Consider a motion such that in spherical coordinates

$$v_r = v_r(r,\theta) \qquad v_\theta = v_\theta(r,\theta) \qquad v_\varphi = 0 \tag{6-2}$$

From Table 2.5.1-1, Eq. (6-1) takes the form

$$\frac{1}{r^2}\frac{\partial}{\partial r}(r^2 v_r) + \frac{1}{r \sin \theta}\frac{\partial}{\partial \theta}(v_\theta \sin \theta) = 0 \tag{6-3}$$

Multiplying by $r^2 \sin \theta$, we may also write this as

$$\frac{\partial}{\partial r}(v_r r^2 \sin \theta) = \frac{\partial}{\partial \theta}(-v_\theta r \sin \theta) \tag{6-4}$$

Upon comparing Eq. (6-4) with

$$\frac{\partial^2 \psi}{\partial r\, \partial\theta} = \frac{\partial^2 \psi}{\partial \theta\, \partial r} \tag{6-5}$$

we see that we may define a *stream function* $\psi$ such that

$$v_r = \frac{1}{r^2 \sin \theta}\frac{\partial \psi}{\partial \theta} \tag{6-6}$$

and

$$v_\theta = -\frac{1}{r \sin \theta}\frac{\partial \psi}{\partial r} \tag{6-7}$$

The advantage of such a stream function $\psi$ is that in this way the equation of continuity is satisfied identically for the assumed flow described by Eq. (6-2).

Expressions for velocity components in terms of a stream function are presented for several situations in Table 2.5.2-1.

### EXERCISE

**1.3.6-1** Using arguments similar to those employed in this section, for each assumed form of velocity distribution in Table 2.5.2-1, express the nonzero components of velocity in terms of a stream function.

**\*1.3.7 Further comments on stream functions for two-dimensional motions** We continue to restrict our discussion to incompressible fluids, so that the equation of continuity of Sec. 1.3.3 reduces to

$$\text{div } \mathbf{v} = v^i_{,i} = 0 \tag{7-1}$$

Another form for the divergence of a vector is (Exercise A.8.2-4)

$$v^i_{,i} = \frac{1}{\sqrt{g}}\frac{\partial}{\partial x^i}(\sqrt{g}\, v^i) \tag{7-2}$$

If our two-dimensional motion is such that

$$v^3 = 0 \tag{7-3}$$

Eqs. (7-2) and (7-3) allow the equation of continuity to be written as

$$\frac{\partial}{\partial x^1}(-\sqrt{g}\, v^1) = \frac{\partial}{\partial x^2}(\sqrt{g}\, v^2) \tag{7-4}$$

But this is a necessary and sufficient condition for the existence of a *stream function* $\psi$ such that

$$\begin{aligned} \sqrt{g}\, v^1 &= -\frac{\partial \psi}{\partial x^2} \\ \sqrt{g}\, v^2 &= \frac{\partial \psi}{\partial x^1} \end{aligned} \tag{7-5}$$

The effect of the stream function in a two-dimensional problem is to allow the equation of continuity to be satisfied identically. If we think in terms of solving a boundary-value problem, we are able to reduce the number of variables by one.

For further discussion of these ideas as well as an extension of the concept of a stream function to steady compressible flows, see Truesdell and Toupin [1, p. 477].

## REFERENCES

1. Truesdell, C., and R. A. Toupin: In S. Flügge (ed.), "Handbuch der Physik," vol. 3/1, Springer-Verlag, Berlin, 1960.
2. Ericksen, J. L.: In S. Flügge (ed.), "Handbuch der Physik," vol. 3/1, Springer-Verlag, Berlin, 1960.
3. Brand, Louis: "Vector and Tensor Analysis," Wiley, New York, 1947.
4. Yih, Chia-Shun: *La Houille Blanche*, **12**:445 (1957).

## EXERCISES

**1.3.7-1** Let $\alpha$ be a parameter along any curve in a surface $\psi = $ constant such that $dx^3/d\alpha = 0$. Then

$$0 = \frac{d\psi}{d\alpha} = \frac{\partial \psi}{\partial x^i}\frac{dx^i}{d\alpha}$$

By Eq. (7-5), we find that

$$0 = \sqrt{g}\, v^2 \frac{dx^1}{d\alpha} - \sqrt{g}\, v^1 \frac{dx^2}{d\alpha}$$

Show that such a curve satisfies Eq. (3-3) and, consequently, is a streamline.

**1.3.7-2** *Stream functions for three-dimensional motions* Any continuously differentiable vector field **v** such that

$$\text{div } \mathbf{v} = 0$$

may be represented by two scalar functions $F$ and $G$ [2, p. 823; 3, sec. 104]:

$$\mathbf{v} = \nabla F \wedge \nabla G \tag{1}$$

In considering a three-dimensional motion of an incompressible fluid, the equation of continuity is automatically satisfied by such a representation for the velocity vector. From the viewpoint of the boundary-value problem to be solved, this has the effect of reducing both the number of unknowns and the number of equations by one. Equation (1) has found limited use in the solution of problems since it is nonlinear.

Thinking in terms of Eq. (1), we may treat a two-dimensional motion such as that discussed in the text by taking

$$F = \psi \quad \text{and} \quad G = -x^3$$

Determine that

$$v^1 = -\frac{1}{\sqrt{g}} \frac{\partial \psi}{\partial x^2}$$

$$v^2 = \frac{1}{\sqrt{g}} \frac{\partial \psi}{\partial x^1}$$

which is in agreement with the conclusions in the text.
For more details, see [1, p. 479; 4].

## 1.4 DEFORMATION

### 1.4.1 The rate of deformation tensor field [1, p. 347]
Our object here is to arrive at some characterization of the *rate of deformation* of a material.

Let us visualize the two intersecting material curves shown in Fig. 1.4.1-1. They are material curves in the sense that they should be regarded as being attached to the material particles. For an experimentalist, they might be two intersecting lines of dye that he has drawn through the body.

At the point of intersection of these curves, the rate of deformation of the material is described by the instantaneous rates at which the lengths of these curves change and by the rate of change of the angle between them. With this thought in mind, we shall follow the curves in Fig. 1.4.1-1 as the material is deformed in some arbitrary fashion.

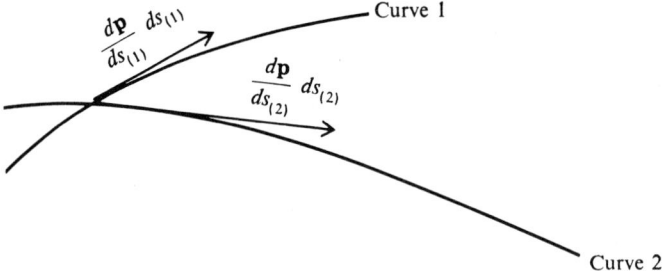

**Fig. 1.4.1-1** Intersecting material curves.

## [1.4] DEFORMATION

Let $s_{(1)}$ be the arc length measured along curve 1 in the present configuration; similarly, for curve 2. Let $S_{(1)}$ and $S_{(2)}$ be arc lengths measured along curves 1 and 2, respectively, in the reference configuration. We should be able to learn something about the local rate of deformation of the body by examining

$$\frac{d_{(m)}}{dt}\left(\frac{d\mathbf{p}}{ds_{(1)}} \cdot \frac{d\mathbf{p}}{ds_{(2)}} ds_{(1)} ds_{(2)}\right)$$

$$= \frac{d_{(m)}}{dt}\left(\frac{d\mathbf{p}}{dS_{(1)}} \cdot \frac{d\mathbf{p}}{dS_{(2)}} dS_{(1)} dS_{(2)}\right)$$

$$= \frac{d\mathbf{v}}{dS_{(1)}} \cdot \frac{d\mathbf{p}}{dS_{(2)}} dS_{(1)} dS_{(2)} + \frac{d\mathbf{p}}{dS_{(1)}} \cdot \frac{d\mathbf{v}}{dS_{(2)}} dS_{(1)} dS_{(2)}$$

$$= \frac{d\mathbf{v}}{ds_{(1)}} \cdot \frac{d\mathbf{p}}{ds_{(2)}} ds_{(1)} ds_{(2)} + \frac{d\mathbf{p}}{ds_{(1)}} \cdot \frac{d\mathbf{v}}{ds_{(2)}} ds_{(1)} ds_{(2)} \quad (1\text{-}1)$$

We can see how to simplify this to some extent by writing everything in terms of components. We have

$$\frac{d\mathbf{v}}{ds_{(1)}} = \frac{\partial \mathbf{v}}{\partial z_i} \frac{dz_i}{ds_{(1)}} = \frac{\partial v_j}{\partial z_i} \frac{dz_i}{ds_{(1)}} \mathbf{e}_j \quad (1\text{-}2)$$

by means of which we can express

$$\frac{d\mathbf{v}}{ds_{(1)}} \cdot \frac{d\mathbf{p}}{ds_{(2)}} = \frac{\partial v_j}{\partial z_i} \frac{dz_i}{ds_{(1)}} \frac{dz_j}{ds_{(2)}} \quad (1\text{-}3)$$

This allows us to express Eq. (1-1) as

$$\frac{d_{(m)}}{dt}\left(\frac{d\mathbf{p}}{ds_{(1)}} \cdot \frac{d\mathbf{p}}{ds_{(2)}} ds_{(1)} ds_{(2)}\right) = \left(\frac{\partial v_j}{\partial z_i} + \frac{\partial v_i}{\partial z_j}\right)\frac{dz_i}{ds_{(1)}}\frac{dz_j}{ds_{(2)}} ds_{(1)} ds_{(2)}$$

$$= 2 D_{ji} \frac{dz_i}{ds_{(1)}}\frac{dz_j}{ds_{(2)}} ds_{(1)} ds_{(2)} \quad (1\text{-}4)$$

where

$$D_{ij} = \frac{1}{2}\left(\frac{\partial v_i}{\partial z_j} + \frac{\partial v_j}{\partial z_i}\right) \quad (1\text{-}5)$$

Another way of expressing Eq. (1-4) is to write

$$\frac{d_{(m)}}{dt}\left(\frac{d\mathbf{p}}{ds_{(1)}} \cdot \frac{d\mathbf{p}}{ds_{(2)}} ds_{(1)} ds_{(2)}\right) = 2 \frac{d\mathbf{p}}{ds_{(2)}} \cdot \left(\mathbf{D} \cdot \frac{d\mathbf{p}}{ds_{(1)}}\right) ds_{(1)} ds_{(2)} \quad (1\text{-}6)$$

in which

$$\mathbf{D} = \tfrac{1}{2}[\nabla \mathbf{v} + (\nabla \mathbf{v})^T] = D_{ij}\mathbf{e}_i\mathbf{e}_j \tag{1-7}$$

We call $\mathbf{D}$ the rate of deformation tensor field. It is clear that $\mathbf{D}$ serves to measure the instantaneous rates of change of length and angle of material elements in a deforming body.

Arc length is that parameter such that [2, p. 6]

$$\frac{d\mathbf{p}}{ds}\cdot\frac{d\mathbf{p}}{ds} = \frac{dz_i}{ds}\frac{dz_i}{ds} = 1 \tag{1-8}$$

When curve 1 instantaneously coincides with curve 2 in Fig. 1.4.1-1, Eqs. (1-4) and (1-8) give

$$\frac{d_{(m)}}{dt}\left[\frac{d\mathbf{p}}{ds}\cdot\frac{d\mathbf{p}}{ds}(ds)^2\right] = \frac{d_{(m)}}{dt}(ds)^2$$

$$= 2D_{ji}\frac{dz_i}{ds}\frac{dz_j}{ds}(ds)^2 \tag{1-9}$$

This may be rewritten as

$$\frac{d_{(m)}}{dt}(\log ds) = D_{ji}\frac{dz_i}{ds}\frac{dz_j}{ds} \tag{1-10}$$

where the derivative on the left is known as the *rate of stretching* in the direction $d\mathbf{p}/ds$. If we assume, for example, that $d\mathbf{p}/ds$ instantaneously is tangent to the $z_1$ coordinate curve, we have for the stretching in the $z_1$ direction

$$\frac{d_{(m)}}{dt}(\log ds_{(1)}) = D_{11} \tag{1-11}$$

In this way we obtain a physical interpretation for the diagonal rectangular cartesian components of the rate of deformation tensor field.

Let us return to Eq. (1-4) and denote by $\beta_{12}$ the angle between the tangents to the two material curves at their intersection. In view of Eq. (1-8), we have

$$\frac{d\mathbf{p}}{ds_{(1)}}\cdot\frac{d\mathbf{p}}{ds_{(2)}} = \cos\beta_{12} \tag{1-12}$$

We may write Eq. (1-4), consequently, as

$$-\sin\beta_{12}\dot{\beta}_{12}\,ds_{(1)}\,ds_{(2)} + \cos\beta_{12}\,\overline{ds_{(1)}}\,ds_{(2)} + \cos\beta_{12}\,\overline{ds_{(1)}}\,ds_{(2)}$$

$$= 2D_{ji}\frac{dz_i}{ds_{(1)}}\frac{dz_j}{ds_{(2)}}ds_{(1)}\,ds_{(2)} \tag{1-13}$$

This may be rearranged to read

## [I.4] DEFORMATION

$$-\sin \beta_{12} \dot{\beta}_{12} = 2D_{ji} \frac{dz_i}{ds_{(1)}} \frac{dz_j}{ds_{(2)}}$$

$$-\cos \beta_{12} \left[ \frac{d_{(m)}(\log ds_{(1)})}{dt} + \frac{d_{(m)}(\log ds_{(2)})}{dt} \right] \quad (1\text{-}14)$$

The rate of decrease in the angle $\beta_{12}$ is called the *rate of shear* of the directions $d\mathbf{p}/ds_{(1)}$, $d\mathbf{p}/ds_{(2)}$. If curves 1 and 2 are instantaneously orthogonal in Fig. 1.4.1-1, we find that

$$-\dot{\beta}_{12} = \frac{-d_{(m)}\beta_{12}}{dt} = 2D_{ji} \frac{dz_i}{ds_{(1)}} \frac{dz_j}{ds_{(2)}} \quad (1\text{-}15)$$

If $d\mathbf{p}/ds_{(1)}$ and $d\mathbf{p}/ds_{(2)}$ are instantaneously tangent to the $z_1$ and $z_2$ coordinate curves, we have from Eq. (1-15)

$$-\dot{\beta}_{12} = 2D_{21} \quad (1\text{-}16)$$

To summarize, Eqs. (1-11) and (1-16) show us that the rectangular cartesian components of the rate of deformation tensor field may be interpreted as equal to the rates of stretching and the halves of the rates of shearing in the coordinate directions:

$$[D_{ij}] = \begin{bmatrix} \overline{\log ds_{(1)}}^{\cdot} & -\tfrac{1}{2}\dot{\beta}_{21} & -\tfrac{1}{2}\dot{\beta}_{31} \\ \cdot & \overline{\log ds_{(2)}}^{\cdot} & -\tfrac{1}{2}\dot{\beta}_{32} \\ \cdot & \cdot & \overline{\log ds_{(3)}}^{\cdot} \end{bmatrix} \quad (1\text{-}17)$$

### REFERENCES

1. Truesdell, C., and R. A. Toupin: In S. Flügge (ed.), "Handbuch der Physik," vol. 3/1, Springer-Verlag, Berlin, 1960.
2. Willmore, T. J.: "An Introduction to Differential Geometry," Oxford University Press, London, 1959.

### EXERCISE

*1.4.1-1 For an arbitrary orthogonal curvilinear coordinate system, give analyses which are similar to those concluding with Eqs. (1-11) and (1-16). Show that with respect to this coordinate system the *physical* components (see Sec. A.5.2) of the rate of deformation tensor field may be interpreted as equal to the rates of stretching and the halves of the rates of shearing in the coordinate directions.

### 1.4.2 Relative spin and vorticity
In contrast with what we did in Sec. 1.4.1, let us examine as a function of time a single material curve in a deforming body. Let $s$ be the arc length measured along this curve in the present configuration; similarly, let $S$ be the arc length measured in the reference configuration.

At some point along this curve, let us inquire how the angle between the tangent to this curve and some fixed direction **a** (we take **a** to have unit magnitude) changes as a function of time. Let $\varphi$ denote the angle between **a** and the tangent vector $d\mathbf{p}/ds$. We may begin by noting that

$$\frac{d_{(m)}}{dt}\left(\frac{d\mathbf{p}}{ds} \cdot \mathbf{a}\, ds\right) = \frac{d_{(m)}}{dt}(\cos \varphi\, ds) = -\dot{\varphi} \sin \varphi\, ds + \cos \varphi\, \dot{\overline{ds}} \tag{2-1}$$

Here we make use of the fact that $d\mathbf{p}/ds$ is a unit vector (see Sec. 1.4.1). We may also write

$$\frac{d_{(m)}}{dt}\left(\frac{d\mathbf{p}}{ds} \cdot \mathbf{a}\, ds\right) = \frac{d_{(m)}}{dt}\left(\frac{d\mathbf{p}}{dS} \cdot \mathbf{a}\, dS\right)$$

$$= \frac{d\mathbf{v}}{dS} \cdot \mathbf{a}\, dS$$

$$= \frac{d\mathbf{v}}{ds} \cdot \mathbf{a}\, ds \tag{2-2}$$

Looking at this in terms of components, we can rearrange it somewhat as

$$\frac{d\mathbf{v}}{ds} \cdot \mathbf{a}\, ds = \frac{\partial v_j}{\partial z_i} \frac{dz_i}{ds} a_j\, ds \tag{2-3}$$

Equations (2-1) to (2-3) imply that

$$-\dot{\varphi} \sin \varphi\, ds + \cos \varphi\, \dot{\overline{ds}} = \frac{\partial v_j}{\partial z_i} \frac{dz_i}{ds} a_j\, ds \tag{2-4}$$

Let us assume that $d\mathbf{p}/ds$ and **a** are instantaneously tangent to the $z_i$ and $z_j$ coordinate curves, respectively, at the point under consideration. Consequently, Eq. (2-4) reduces to

$$-\dot{\varphi}_{ij} = \frac{\partial v_j}{\partial z_i} \tag{2-5}$$

By $-\dot{\varphi}_{ij}$ we mean the rate at which a material element instantaneously pointing along the $z_i$ axis is turning toward the $z_j$ axis.

Let us define the second-order vorticity tensor **W** as

$$\mathbf{W} = \tfrac{1}{2}[\nabla \mathbf{v} - (\nabla \mathbf{v})^T] \tag{2-6}$$

In terms of rectangular cartesian components, we have

$$W_{ij} = \frac{1}{2}\left(\frac{\partial v_i}{\partial z_j} - \frac{\partial v_j}{\partial z_i}\right) \tag{2-7}$$

For a right-handed, rectangular cartesian coordinate system, the quantity

## [1.4] DEFORMATION

$$W_{21} = \frac{1}{2}\left(\frac{\partial v_2}{\partial z_1} - \frac{\partial v_1}{\partial z_2}\right) \tag{2-8}$$

is one-half the sum of the rate of the right-handed rotation of elements in the $z_1$ and $z_2$ coordinate directions about an axis in the $z_3$ direction.

The axial vector field (see Sec. A.7.3)

$$\begin{aligned}
\mathbf{w} \equiv \boldsymbol{\epsilon}:\mathbf{W} &= e_{ijk}W_{kj}\mathbf{e}_i \\
&= \frac{1}{2}\left(e_{ijk}\frac{\partial v_k}{\partial z_j} - e_{ijk}\frac{\partial v_j}{\partial z_k}\right)\mathbf{e}_i \\
&= e_{ijk}\frac{\partial v_k}{\partial z_j}\mathbf{e}_i \\
&= \text{curl } \mathbf{v}
\end{aligned} \tag{2-9}$$

is referred to as the *vorticity* spatial vector field. Referring to our physical picture of the components of the second-order vorticity tensor $\mathbf{W}$, we may think of the direction of the vorticity vector as being the local axis of spin. The *first theorem of Cauchy* follows immediately: The component of the vorticity spatial vector field in any direction is the sum of the rates of right-handed rotation about that direction of elements in any two directions perpendicular to it and to each other.

A motion in which the vorticity spatial vector field vanishes is referred to as being *irrotational*. A motion in which the vorticity spatial vector field does not vanish is said to be *rotational*.

A necessary and sufficient condition that a motion be irrotational is that the velocity be representable in terms of a scalar potential $P$ [1, Sec. 33]:

$$\mathbf{v} = -\nabla P \tag{2-10}$$

That it is a sufficient condition is shown in Exercise A.9.1-1.

### REFERENCE

1. Ericksen, J. L.: In S. Flügge (ed.), "Handbuch der Physik," vol. 3/1, Springer-Verlag, Berlin, 1960.

### EXERCISE

\* **1.4.2-1** For an arbitrary orthogonal curvilinear coordinate system, give an analysis that is similar to that concluding with Eq. (2-5). Show that with respect to this coordinate system $-W_{\langle 12\rangle}$ is the sum of the rate of the right-handed rotation of elements in the $x^1$ and $x^2$ coordinate directions about an axis in the $x^3$ direction.

# 2
# Foundations for Momentum Transfer

In what follows, the principal tools for studying fluid mechanics are developed. I begin by introducing the concept of force. Notice that force is not defined; it is a primitive concept in the same sense as is the material particle in Chap. 1. This forms the basis for introducing our second and third postulates: Euler's first and second laws. The stress tensor is introduced in order to derive the equations that describe at each point in a material the local balances for momentum and moment of momentum. These two local balances, Cauchy's first and second laws, together with the equation of continuity from Chap. 1, form the major foundations for fluid mechanics.

We conclude our discussion with an outline of what must be said about real material behavior, if we are to analyze any practical problems. It is especially at this point that statistical mechanics, based upon the molecular viewpoint of real materials, can be used to supplement the concepts developed in continuum mechanics. In continuum mechanics, we can indicate a number of rules that constitutive equations for the stress tensor must satisfy (the principle of determinism, the principle of local action, the principle of material frame indifference, etc.) but from first principles we cannot derive an explicit relationship between stress and deformation. If we work strictly within the bounds of continuum mechanics, we can derive such a relationship only by making some sort of assumption about its form. My personal feelings are that the most interesting advances in the way of describing material behavior will result in this way: considerations based upon simple molecular models will be generalized through the use of the statements about material behavior that have been postulated in continuum mechanics. For an excellent brief summary of what can be said about material behavior from a molecular point of view, see Bird, Stewart, and Lightfoot [1, chap. 1].

While our direct concerns here are for momentum transfer, practically all of the ideas developed will be applied again in examining energy and mass transfer. I firmly believe that the best foundation for energy and mass-transfer studies is a clear understanding of fluid mechanics.

## REFERENCE

1. Bird, R. B., W. E. Stewart, and E. N. Lightfoot: "Transport Phenomena," 7th printing, Wiley, New York, 1960.

## 2.1 FORCE

**2.1.1 Force** [1, p. 97; 2, pp. 531 and 536; 3, p. 39] Corresponding to each body $B$, there is a distinct set of bodies $\mathring{B}$ such that the mass of the union of these bodies is the mass of the universe. We refer to $\mathring{B}$ as the exterior or the surroundings of the body $B$.

A system of forces is a vector-valued function $\mathbf{f}(B,C)$ of pairs of bodies. The value of $\mathbf{f}(B,C)$ is called the force exerted on the body $B$ by the body $C$. A system of forces is restricted by the following two properties or axioms.

1. For a specified body $B$, $\mathbf{f}(C,\mathring{B})$ is an additive function defined over the subbodies $C$ of $B$.
2. Conversely, for a specified body $B$, $\mathbf{f}(B,C)$ is an additive function defined over the subbodies $C$ of $\mathring{B}$.

This is another way of saying that the parts of a body are subjected to forces from various parts of the environment that are independent of one another.

There are three types of forces with which we may be concerned:

1. *External forces.* These arise at least in part from outside the body and act upon the material particles of which the body is composed. One example is the uniform force of gravity. Another example would be the electrostatic force between two charged bodies. Let $P$ indicate a portion of a body $B$ as illustrated in Fig. 2.1.1-1. Taking $\mathbf{f}_e$ to be the external force per unit mass that the surroundings $\mathring{B}$ exert on the body $B$, we write the total external force acting on $P$ in terms of an integral over the volume of $P$: $\int_{V_p} \rho \mathbf{f}_e \, dV$. In general, the external force per

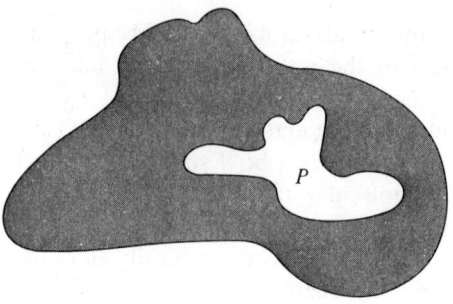

Fig. 2.1.1-1 The body $B$ of which $P$ is a portion.

unit mass is a function of position and $\mathbf{f}_e$ should be regarded as a spatial vector field.

2. *Mutual forces.* These arise within a body and act upon pairs of material particles. The small newtonian gravitational force between any two portions of a single body qualifies as a mutual force. We can imagine a body in which there is a distribution of electrostatic charge; we would speak of the electrostatic force between one portion of the body with a net positive charge and some other element of the body with a net negative charge as being a mutual force. Let $\mathbf{f}_m$ be the mutual force per unit mass that $B - P$ [we define $B - P$ to be such that $B = (B - P) \cup P$ and $(B - P) \cap P = 0$] exerts upon $P$; the total mutual force acting upon $P$ may be represented as an integral over the volume of $P$: $\int_{V_P} \rho \mathbf{f}_m \, dV$. We should expect the mutual force per unit mass generally to be a function of position within the material; $\mathbf{f}_m$ should be viewed as a spatial vector field. (We recognize here that the sum of the mutual forces exerted by any two parts of $P$ upon each other is zero [2, p. 533].)

3. *Contact forces.* These forces are not assignable as functions of position, but are to be imagined as acting upon the bounding surface of a portion of material in such a way as to be equivalent to the force exerted by one portion of the material upon another beyond that accounted for through mutual forces. In typing you exert a contact force upon the keys of the typewriter. If we deform some putty in our hands, during the deformation any one portion of the putty exerts a contact force upon the remainder at their common boundary. Let $\mathbf{t} = \mathbf{t}(\mathbf{z},P)$ represent the force per unit area that $B - P$ exerts upon the boundary of $P$ at the position $\mathbf{z}$. This force per unit area is usually referred to as *stress*. The total contact force that $B - P$ exerts upon $P$ may be written as an integral over the bounding surface of $P$: $\int_{S_P} \mathbf{t} \, dS$.

The stress principle specifies the nature of the contact load.

*Stress principle* There is a vector-valued function $\mathbf{t}(\mathbf{z},\mathbf{n})$ defined for all points $\mathbf{z}$ in a body $B$ and for all unit vectors $\mathbf{n}$ such that the stress that $B - P$ exerts upon any portion $P$ of $B$ is given by

$$\mathbf{t}(\mathbf{z},P) = \mathbf{t}(\mathbf{z},\mathbf{n}) \tag{1-1}$$

Here **n** is the unit normal that is outwardly directed with respect to the closed bounding surface of $P$. The spatial vector $\mathbf{t} = \mathbf{t}(\mathbf{z,n})$ is referred to as the stress vector at the position **z** acting upon the oriented surface element with normal **n**; **n** points into the material that exerts the stress **t** upon the surface element.

In any particular problem, we regard the forces exerted upon a body as being given a priori to all observers; all observers would assume a priori the same set of forces in a given problem. In prescribing these forces, we specify a particular dynamic problem. We, consequently, assume that all forces are independent of the observer or are frame indifferent [4, p. 27] (see Sec. 1.2.1):

$$\mathbf{f}_e^* = \mathbf{Q} \cdot \mathbf{f}_e \qquad \mathbf{f}_m^* = \mathbf{Q} \cdot \mathbf{f}_m \tag{1-2}$$

and

$$\mathbf{t}^* = \mathbf{Q} \cdot \mathbf{t} \tag{1-3}$$

## REFERENCES

1. Truesdell, C.: "Six Lectures on Modern Natural Philosophy," Springer-Verlag, New York, 1966.
2. Truesdell, C., and R. A. Toupin: In S. Flügge (ed.), "Handbuch der Physik," vol. 3/1, Springer-Verlag, Berlin, 1960.
3. Truesdell, C., and W. Noll: In S. Flügge (ed.), "Handbuch der Physik," vol. 3/3, Springer-Verlag, Berlin, 1965.
4. Truesdell, C.: "The Elements of Continuum Mechanics," Springer-Verlag, New York, 1966.

## 2.2 EULER'S LAWS

**2.2.1 Euler's laws** The second and third postulates we make in our development are Euler's two laws [1, p. 97; 2, pp. 531 and 537; 3, p. 39].[1]

*Euler's first law* The time rate of change of the momentum of a body relative to the fixed stars is equal to the sum of the forces acting on the body.

Let the volume and closed bounding surface of a body or any portion of a body be denoted, respectively, as $V_{(m)}$ and $S_{(m)}$. Referring to our discussion of forces in Sec. 2.1.1, in a frame of reference that is stationary with respect to the fixed stars, we may express Euler's first law as

$$\frac{d}{dt}\int_{V_{(m)}} \rho\mathbf{v}\,dV = \int_{S_{(m)}} \mathbf{t}\,dS + \int_{V_{(m)}} \rho\mathbf{f}\,dV \tag{1-1}$$

Here **f** is the field of external and mutual forces per unit mass:

$$\mathbf{f} \equiv \mathbf{f}_e + \mathbf{f}_m \tag{1-2}$$

In all but the unusual cases the effect of mutual forces can be neglected with respect to external forces. Hereafter we will assume that mutual forces have been dismissed and we will refer to **f** as the field of external forces per unit mass.

[1] Truesdell and Toupin [2, pp. 531 and 534] point out that "the laws of Newton ... are neither unequivocally stated nor sufficiently general to serve as a foundation for continuum mechanics."

In writing Eq. (1-1), it is our understanding that all spatial vector fields are expressed in a frame of reference (see Sec. 1.2.1) that is stationary with respect to the fixed stars. This requirement is satisfied if we express these spatial vector fields with respect to a coordinate system that is stationary with respect to the fixed stars. In meteorology and oceanography, it is important to distinguish between a frame of reference that is fixed with respect to the earth and one that is stationary with respect to the fixed stars. For most laboratory experiments and industrial operations, the differences are negligible.

*Euler's second law*  The time rate of change of the moment of momentum of a body relative to the fixed stars is equal to the sum of the moments of all the forces acting on the body.

In a frame of reference that is stationary with respect to the fixed stars, Euler's second law assumes the form

$$\frac{d}{dt} \int_{V_{(m)}} \rho(\mathbf{p} \wedge \mathbf{v}) \, dV = \int_{S_{(m)}} \mathbf{p} \wedge \mathbf{t} \, dS + \int_{V_{(m)}} \rho(\mathbf{p} \wedge \mathbf{f}) \, dV \tag{1-3}$$

In writing Euler's second law in this manner we confine our attention to the so-called nonpolar[1] case, i.e., we assume that all torques acting on the body are the result of forces acting on the body [2, pp. 538 and 546; 4 to 6]. For example, it is possible to induce a local source of moment of momentum by a suitable rotating electric field [7, 8]. In such a case it might also be necessary to account for the flux of moment of momentum at the bounding surface of the body; one should expect this, if there were a gradient in the moment of momentum associated with individual molecules at the boundary of the body. Effects of this type have not been investigated thoroughly, but are thought to be negligibly small for all but unusual situations. Consequently, they are neglected here.

**REFERENCES**

1. Truesdell, C.: "Six Lectures on Modern Natural Philosophy," Springer-Verlag, New York, 1966.
2. Truesdell, C., and R. A. Toupin: In S. Flügge (ed.), "Handbuch der Physik," vol. 3/1, Springer-Verlag, Berlin, 1960.
3. Truesdell, C., and W. Noll: In S. Flügge (ed.), "Handbuch der Physik," vol. 3/3, Springer-Verlag, Berlin, 1965.
4. Curtiss, C. F.: *J. Chem. Phys.*, **24**:225 (1956).
5. Livingston, P. M., and C. F. Curtiss: *J. Chem. Phys.*, **31**:1643 (1959).
6. Dahler, J. S., and L. E. Scriven: *Nature*, **192**:36 (1961).
7. Lertes, Peter: *Z. für Physik*, **4**:315 (1921); *ibid.*, **6**:56 (1921); *Physik. Z.*, **22**:621 (1921).
8. Grossetti, E.: *Nuovo Cimento*, **10**:193 (1958); *ibid.*, **13**:350 (1959).

[1] When molecules are referred to as *nonpolar*, it indicates that their dipole moment is zero. This is an entirely different use of the word than that intended here, where *nonpolar* means that all torques acting on the material are the result of forces.

[2.2] EULER'S LAWS                                                                                   39

**EXERCISE**

**2.2.1-1** *Cauchy's lemma*  Consider two neighboring portions of a continuous body.  Apply Euler's first law to each portion and to their union.  Deduce that on their common boundary

$$t(z,n) = -t(z, -n)$$

*Cauchy's lemma*  The stress vectors acting upon opposite sides of the same surface at a given point are equal in magnitude and opposite in direction.

**2.2.2 The stress tensor**  We ask here how $t(z,n)$ varies as the position $z$ is held fixed and $n$ changes.

At any point in a body consider the tetrahedron shown in Fig. 2.2.2-1. Three sides are mutually orthogonal and coincide with a set of rectangular cartesian coordinate planes intersecting at $z$; the fourth side has an outwardly directed normal $n$. Let the altitude of the tetrahedron be $h$; the area of the inclined face, $A$.  In terms of the rectangular cartesian basis fields, we may write $n = n_i e_i$. Since the cosines of the angles between the cartesian coordinate planes and the inclined plane are $n_1$, $n_2$, and $n_3$, respectively, the areas of the faces of the tetrahedron lying in the coordinate planes are $n_1 A$, $n_2 A$, and $n_3 A$.

Let us apply Euler's first law, Eq. (1-1), to the material in the tetrahedron at time $t$ to obtain

$$\frac{d}{dt}\int_{V_{(m)}} \rho v \, dV = \int_{S_{(m)}} t \, dS + \int_{V_{(m)}} \rho f \, dV \qquad (2\text{-}1)$$

Applying the form of the transport theorem introduced in Exercise 1.3.3-1 to the term on the left and using the theorem of mean value to evaluate the surface integral,

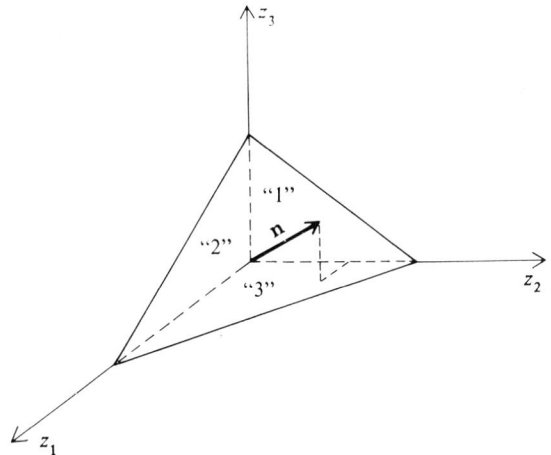

**Fig. 2.2.2-1**

we have

$$\int_{V_{(m)}} \left(\rho \frac{d_{(m)}\mathbf{v}}{dt} - \rho\mathbf{f}\right) dV = A\left(\mathbf{t}^* + \frac{A_1}{A}\mathbf{t}_1^* + \frac{A_2}{A}\mathbf{t}_2^* + \frac{A_3}{A}\mathbf{t}_3^*\right) \tag{2-2}$$

Here $\mathbf{t}^*$ represents the contact stress that the surroundings exert upon the inclined face; $\mathbf{t}_i^*$ is the contact stress that the surroundings exert upon the face whose inwardly directed unit normal is $\mathbf{e}_i$. Applying the theorem of mean value to the term on the left, we write

$$\left(\rho \frac{d_{(m)}\mathbf{v}}{dt} - \rho\mathbf{f}\right)^* \frac{hA}{3} = A(\mathbf{t}^* + n_1\mathbf{t}_1^* + n_2\mathbf{t}_2^* + n_3\mathbf{t}_3^*) \tag{2-3}$$

Divide through by $A$. Consider a sequence of geometrically similar tetrahedrons; in the limit as $h$ approaches zero, we obtain

$$\mathbf{t} = -(\mathbf{t}_1 n_1 + \mathbf{t}_2 n_2 + \mathbf{t}_3 n_3) \tag{2-4}$$

where all stress vectors are now evaluated at the point $\mathbf{z}$. From their definitions, the quantities $\mathbf{t}_1, \mathbf{t}_2, \mathbf{t}_3$ do not depend upon $\mathbf{n}$.

With the convention that $T_{km}$ is the $k$th component of the stress vector acting upon the *positive* side of the plane $z_m =$ constant, by *Cauchy's lemma* (Exercise 2.2.1-1) we write

$$\begin{aligned}-\mathbf{t}_1 &= T_{i1}\mathbf{e}_i \\ -\mathbf{t}_2 &= T_{i2}\mathbf{e}_i \\ -\mathbf{t}_3 &= T_{i3}\mathbf{e}_i\end{aligned} \tag{2-5}$$

Equation (2-4) may, therefore, be written as

$$\mathbf{t} = n_j T_{ij}\mathbf{e}_i \tag{2-6}$$

The matrix $[T_{ij}]$ defines a second-order tensor $\mathbf{T}$:

$$\mathbf{T} \equiv T_{ij}\mathbf{e}_i\mathbf{e}_j \tag{2-7}$$

which allows us to write Eq. (2-6) as

$$\mathbf{t} = \mathbf{T}\cdot\mathbf{n} \tag{2-8}$$

Remember in using Eq. (2-8) that $\mathbf{n}$ is that unit normal which is directed into the material which is exerting the force per unit area $\mathbf{t}(\mathbf{z},\mathbf{n})$ at the position $\mathbf{z}$ on the surface.

## EXERCISES

**2.2.2-1** Show that the stress tensor is frame indifferent.

**2.2.2-2** In going from Eq. (2-2) to (2-3), prove that

$$n_i = \frac{A_i}{A} \quad i = 1, 2, 3$$

## [2.2] EULER'S LAWS

**2.2.3 Cauchy's first law of motion** For the moment, let us again restrict ourselves to a body in which all quantities are continuous and differentiable as many times as desired. This rules out the appearance of shock waves and phase interfaces (at least if we continue to represent phase interfaces as singular surfaces; see Sec. 1.3.5).

A modification of the transport theorem that takes into account the postulate that mass is conserved (see Exercise 1.3.3-1) may be used to express Euler's first law in Sec. 2.2.1 as

$$\int_{V_{(m)}} \rho \frac{d_{(m)}\mathbf{v}}{dt} dV = \int_{S_{(m)}} \mathbf{t}\, dS + \int_{V_{(m)}} \rho \mathbf{f}\, dV \tag{3-1}$$

If we express the stress vector in terms of the stress tensor as suggested in Sec. 2.2.2, the first term on the right of this equation may be rearranged by an application of Green's transformation (Sec. A.11.2):

$$\int_{S_{(m)}} \mathbf{t}\, dS = \int_{S_{(m)}} \mathbf{T} \cdot \mathbf{n}\, dS = \int_{V_{(m)}} \operatorname{div} \mathbf{T}\, dV \tag{3-2}$$

Substituting Eq. (3-2) into Eq. (3-1), we have

$$\int_{V_{(m)}} \left( \rho \frac{d_{(m)}\mathbf{v}}{dt} - \operatorname{div} \mathbf{T} - \rho \mathbf{f} \right) dV = 0 \tag{3-3}$$

But remember that Eq. (3-1) was written for an arbitrary portion $P$ of a body. No matter whether $P$ is very large or arbitrarily small, Eq. (3-3) remains valid. This implies that the integrand itself must be identically zero:

$$\rho \frac{d_{(m)}\mathbf{v}}{dt} = \operatorname{div} \mathbf{T} + \rho \mathbf{f} \tag{3-4}$$

This equation is known by various names: Cauchy's first law, the differential momentum balance, or the stress equation of motion. It is the balance on momentum which must be satisfied at every point in a body.

### REFERENCES

1. Truesdell, C., and R. A. Toupin: In S. Flügge (ed.), "Handbuch der Physik," vol. 3/1, Springer-Verlag, Berlin, 1960.
2. Slattery, J. C.: *Ind. Eng. Chem., Fundamentals*, **6**:108 (1967); *ibid.*, **7**:672 (1968).
3. Rouse, Hunter, and Simon Ince: "History of Hydraulics," Iowa Institute of Hydraulic Research, State University of Iowa, 1957.

### EXERCISES

**2.2.3-1** *Jump momentum balance* Starting with Euler's first law and neglecting all interfacial effects such as surface tension, derive the *jump momentum balance* for a phase interface represented by a singular surface [1, p. 545; 2]:

$$[\rho \mathbf{v}(\mathbf{v} \cdot \boldsymbol{\xi} - u_{(\xi)}) - \mathbf{T} \cdot \boldsymbol{\xi}] = 0$$

Here $u_{(\xi)}$ is the speed of displacement of the phase interface [1, p. 499]; the boldface bracket notation is defined by Eq. (4-7) in Sec. 1.3.4.

**2.2.3-2** *More on the jump momentum balance* Use the approach outlined in Exercise 1.3.5-2 to derive the jump momentum balance of Exercise 2.2.3-1.

**2.2.3-3** Let us assume that the contact force has the character of a hydrostatic pressure:

$$\mathbf{T} = -p\mathbf{I}$$

and let us assume that there is mass transfer across the phase interface. Prove that under these circumstances the tangential components of velocity are continuous across a phase interface.

**2.2.3-4** Prove the following two theorems of Archimedes [3, p. 17]:

> If a solid lighter than a fluid be forcibly immersed in it, the solid will be driven upwards by a force equal to the difference between its weight and the weight of the fluid displaced.

and

> A solid heavier than a fluid will, if placed in it, descend to the bottom of the fluid, and the solid will, when weighed in the fluid, be lighter than its true weight by the weight of the fluid displaced.

*Hint:* Apply Euler's first law and extend the definition of the fluid stress tensor to the region of space occupied by the solid.

**2.2.3-5** Prove another theorem of Archimedes [3, p. 17]:

> Any solid lighter than a fluid will, if placed in the fluid, be so far immersed that the weight of the solid will be equal to the weight of the fluid displaced.

*Hint:* Proceed as in Exercise 2.2.3-4 making use of the jump momentum balance in Exercise 2.2.3-1.

**2.2.3-6** An irregular solid object of uniform density $\rho_{(s)}$ floats in a liquid whose density $\rho_{(f)}$ varies linearly with depth $y$ according to the relation

$$\rho_{(f)} = a + by$$

The object is completely submerged. Prove that the density of the fluid at the depth $h$ of the object's center of mass is equal to the density of the solid. This implies that

$$h = \frac{\rho_{(s)} - a}{b}$$

The center of mass of a body is that point whose position vector $\mathbf{c}$ is defined as

$$\mathbf{c} \equiv \frac{1}{\mathcal{M}} \int_{V_{(m)}} \rho \mathbf{p} \, dV$$

where $\mathcal{M}$ is the mass of the body. This means that

$$\int_{V_{(m)}} \rho(\mathbf{p} - \mathbf{c}) \, dV = 0$$

or

$$\int_{V_{(m)}} \rho \mathbf{p}' \, dV = 0$$

where $\mathbf{p}'$ is a position vector measured with respect to the center of mass as the origin.

## [2.2] EULER'S LAWS

**2.2.3-7** A barge loaded with steel waits within a closed lock of a canal. The steel accidentally slips off the barge and falls to the bottom of the canal. Does the water level rise or fall as the result of this accident?

**2.2.3-8** *Cauchy's first law in another frame of reference* (a) Use the result of Exercise 1.2.2-2 to write Cauchy's first law for an arbitrary frame of reference:

$$\rho\{\ddot{\mathbf{z}}^* - \ddot{\mathbf{c}} - \dot{\boldsymbol{\omega}} \wedge (\mathbf{z}^* - \mathbf{c}) + \boldsymbol{\omega} \wedge [\boldsymbol{\omega} \wedge (\mathbf{z}^* - \mathbf{c})]$$
$$- 2\boldsymbol{\omega} \wedge (\dot{\mathbf{z}}^* - \dot{\mathbf{c}})\} = \operatorname{div} \mathbf{T}^* + \rho \mathbf{f}^*$$

(b) Determine that Cauchy's first law assumes the form of Eq. (3-4) in every frame of reference which moves at a constant velocity (without rotation) relative to the fixed stars.

### 2.2.4 Cauchy's second law of motion

We again restrict ourselves to a body in which all quantities are smooth functions of position and time as described at the beginning of Sec. 2.2.3.

After an application of the transport theorem and the differential equation of continuity, Euler's second law may be written as

$$\int_{V_{(m)}} \rho \frac{d_{(m)}}{dt} (\mathbf{p} \wedge \mathbf{v}) \, dV = \int_{S_{(m)}} \mathbf{p} \wedge (\mathbf{T} \cdot \mathbf{n}) \, dS + \int_{V_{(m)}} \rho(\mathbf{p} \wedge \mathbf{f}) \, dV \qquad (4\text{-}1)$$

In writing this equation we have also expressed the stress vector in terms of the stress tensor as described in Sec. 2.2.2. Let us consider these expressions individually.

From the left-hand term, we have

$$\frac{d_{(m)}}{dt} (\mathbf{p} \wedge \mathbf{v}) = \frac{d_{(m)}\mathbf{p}}{dt} \wedge \mathbf{v} + \mathbf{p} \wedge \frac{d_{(m)}\mathbf{v}}{dt} \qquad (4\text{-}2)$$

But since we have defined (Sec. 1.1.1)

$$\mathbf{v} \equiv \frac{d_{(m)}\mathbf{p}}{dt} \qquad (4\text{-}3)$$

we are left with

$$\frac{d_{(m)}}{dt} (\mathbf{p} \wedge \mathbf{v}) = \mathbf{p} \wedge \frac{d_{(m)}\mathbf{v}}{dt} \qquad (4\text{-}4)$$

By an application of Green's transformation (Sec. A.11.2), the first term on the right of Eq. (4-1) becomes

$$\int_{S_{(m)}} \mathbf{p} \wedge (\mathbf{T} \cdot \mathbf{n}) \, dS = \int_{V_{(m)}} \frac{\partial}{\partial z_m} (e_{ijk} z_j T_{km}) \, dV \mathbf{e}_i \qquad (4\text{-}5)$$

We find that

$$\frac{\partial}{\partial z_m} (e_{ijk} z_j T_{km}) = e_{ijk} \frac{\partial z_j}{\partial z_m} T_{km} + e_{ijk} z_j \frac{\partial T_{km}}{\partial z_m} \qquad (4\text{-}6)$$

But

$$\frac{\partial z_j}{\partial z_m} = \delta_{jm} \tag{4-7}$$

Equations (4-6) and (4-7) allow us to conclude that

$$\frac{\partial}{\partial z_m}(e_{ijk}z_j T_{km})\mathbf{e}_i = e_{ijk}T_{kj}\mathbf{e}_i + \mathbf{p} \wedge (\text{div } \mathbf{T}) \tag{4-8}$$

Equations (4-4), (4-5), and (4-8) enable us to express Eq. (4-1) in the form

$$\int_{V_{(m)}} \left[ \mathbf{p} \wedge \left( \rho \frac{d_{(m)}\mathbf{v}}{dt} - \text{div } \mathbf{T} - \rho \mathbf{f} \right) - e_{ijk}T_{kj}\mathbf{e}_i \right] dV = 0 \tag{4-9}$$

In view of Cauchy's first law (Sec. 2.2.3), this becomes

$$\int_{V_{(m)}} e_{ijk}T_{kj}\mathbf{e}_i \, dV = 0 \tag{4-10}$$

But Eq. (4-1) was written for a portion of a body with the understanding that the portion considered might be arbitrarily large or small. This implies from Eq. (4-10) that

$$e_{ijk}T_{kj}\mathbf{e}_i = 0 \tag{4-11}$$

Since the rectangular cartesian basis fields are linearly independent, we have

$$e_{ijk}T_{kj} = 0 \tag{4-12}$$

We may also write then (see Exercise A.2.1-2)

$$e_{imn}e_{ijk}T_{kj} = 0$$
$$T_{nm} - T_{mn} = 0 \tag{4-13}$$

Equation (4-13) expresses *Cauchy's second law of motion*:

A necessary and sufficient condition for the balance of moment of momentum in a body where momentum is balanced is that the stress tensor be symmetric:

$$\mathbf{T} = \mathbf{T}^T \tag{4-14}$$

Hereafter, we need not worry about satisfying Euler's second law, the moment of momentum balance, so long as we take the components of the stress tensor to be symmetric.

It is worthwhile to recall at this point that the conclusion that the stress tensor is symmetric is based upon a statement of Euler's law appropriate to the nonpolar case, i.e., all torques acting on the body are the result of forces acting on the body (see Sec. 2.2.1).

## EXERCISE

**2.2.4-1** Show that the jump moment of momentum balance at a phase interface leads to no further restriction on the motion beyond that already implied by the jump momentum balance.

## 2.3 BEHAVIOR

### 2.3.1 Behavior of materials

It should occur to you that we have said nothing as yet about the behavior of materials. The postulate that mass is conserved and the first and second postulates of Euler are made for all materials. Yet our experience tells us that under similar circumstances air and steel respond to forces in drastically different manners. Somewhere in our theoretical structure we must incorporate this information.

To confirm this intuitive feeling, let us consider the mathematical structure we have developed thus far. For simplicity, assume that the material is incompressible so that mass density $\rho$ is a known constant. Let us also assume that the description of the external force field **f** is given; for example, we are commonly concerned with physical situations in an essentially uniform gravitational field. As unknowns in some arbitrary coordinate system we are left with the three components of the velocity vector **v** and the six components of the stress tensor **T**. (We are taking advantage here of the symmetry of the stress tensor proved in Sec. 2.2.4.) As equations, we have the differential equation of continuity, Eq. (3-4) of Sec. 1.3.3, and the three components of Cauchy's first law, Eq. (3-4) of Sec. 2.2.3. This means we have four equations in nine unknowns. This reinforces our intuitive feeling that further information is required.

One way of describing just what it is that we are missing is to say that, while we have assumed the nature of the external force is known, we have said nothing about the character of the force that one portion of a body exerts on its neighboring portion. We indicated in Sec. 2.2.1 that we would neglect mutual forces hereafter. This leaves only the contact force. We must describe how the contact forces in a body depend upon the motion and deformation of the body. More specifically, we must say how the stress tensor **T** varies with the motion and deformation of the body.

Before plunging ahead to relate the stress tensor to motion, let us see if our everyday experience in observing materials in deformations and motions will help us to lay down some rules governing such a relation. For example, it seems obvious that what happens to a body in the future is going to have no influence on the present stress tensor field. This suggests stating a *principle of determinism* [1, p. 6; 2, p. 56]:

> The stress in a body is determined by the history of the motion that the body has undergone.

Our gross experience is that motion in one portion of a body does not *necessarily* have any effect on the state of stress in another portion of the body. For example, if we lay down a bead of caulking compound, we may shape one portion of the bead with a putty knife without disturbing the rest. From a somewhat different point of view, the physical idea of a contact force suggests that the circumstances in the immediate neighborhood of the point in question determine it. We may state this as a *principle of local action* [1, p. 6; 2, p. 56]:

> The motion of the material outside an arbitrarily small neighborhood of the material point $\zeta$ may be ignored in determining the stress at this material point.

Let us consider an experiment in which a series of weights are successively added to one end of a spring, the other end having been attached to the ceiling of a laboratory. Two experimentalists observe this experiment, one standing on the floor of the laboratory near the spring and the other standing across the room on a turntable that rotates with some angular velocity. The frame of reference for the first observer is the walls of the laboratory. The frame of reference for the second observer consists of the turntable's axis and a series of lines painted upon the turntable; to this observer, the spring and weights appear to be revolving in a circle. Yet we expect both observers to come to the same conclusions regarding the behavior of the spring under stress.

Referring to Secs. 1.2.1 and 1.2.2 and to Exercise 2.2.2–1, we can summarize our feeling here with the *principle of material frame indifference*[1] [2, p. 44]:

> Constitutive equations (equations that describe material behavior; an example is an equation that expresses the stress tensor as a function of the motion of the material) must be invariant under changes of frame of reference. If a constitutive equation is satisfied for a process in which the stress tensor and motion are given by

$$\mathbf{T} = \mathbf{T}(\mathbf{z}_\varkappa, t) \quad \text{and} \quad \mathbf{z} = \boldsymbol{\chi}_\varkappa(\mathbf{z}_\varkappa, t) \tag{1-1}$$

> then it must also be satisfied for any equivalent process described with respect to another frame of reference. In particular, the constitutive equation must be satisfied for a process in which the stress tensor and motion are given by

$$\mathbf{T}^* = \mathbf{T}^*(\mathbf{z}^*_\varkappa, t^*) = \mathbf{Q}(t) \cdot \mathbf{T}(\mathbf{z}_\varkappa, t) \cdot \mathbf{Q}^T(t)$$

$$\mathbf{z}^* = \boldsymbol{\chi}^*_\varkappa(\mathbf{z}^*_\varkappa, t^*) = \mathbf{c}(t) + \mathbf{Q}(t) \cdot \boldsymbol{\chi}_\varkappa(\mathbf{z}_\varkappa, t)$$

> and

$$t^* = t - a \tag{1-2}$$

It is possible to make further statements in much the same manner as above [5, p. 700; 2, p. 101]. These principles may then be used to help in the construction

---

[1] In writing this text, I have made an effort to refer the reader to the more lucid reference rather than the historical "first" when a choice has been necessary. A choice was necessary here, because the essential idea of the principle of material frame indifference had been stated by several authors prior to Noll [3]. Oldroyd [4] in particular attracted considerable interest with his viewpoint. Truesdell and Noll [2, p. 45] have gone back to the seventeenth century to trace the development of this idea through the literature.

of particular constitutive equations for the stress tensor. The type of argument involved is illustrated in the following section.

**REFERENCES**

1. Truesdell, C.: "Six Lectures on Modern Natural Philosophy," Springer-Verlag, New York, 1966.
2. Truesdell, C., and W. Noll: In S. Flügge (ed.), "Handbuch der Physik," vol. 3/3, Springer-Verlag, Berlin, 1965.
3. Noll, Walter: *Arch. Rational Mech. Analysis*, **2**:197 (1958).
4. Oldroyd, J. G.: *Proc. Roy. Soc. (London)*, **A200**:523 (1950).
5. Truesdell, C., and R. A. Toupin: In S. Flügge (ed.), "Handbuch der Physik," vol. 3/1, Springer-Verlag, Berlin, 1960.

**2.3.2 A simple constitutive equation for stress** In the last section we discussed three principles which it seemed reasonable to require every constitutive equation for stress to satisfy. Let us propose a simple constitutive equation that is consistent with these principles.

We can satisfy the principle of determinism by requiring the stress to depend only upon a description of the present state of motion in the material. Both the principle of determinism and the principle of local action are satisfied if we assume that the stress at a point is a function of the velocity and velocity gradient at that point:

$$\mathbf{T} = \mathbf{H}(\mathbf{v}, \nabla \mathbf{v}) \tag{2-1}$$

It is understood that stress may also depend upon local thermodynamic state variables, but since this dependence is not of primary concern as yet, it is not shown explicitly.

Every second-order tensor can be written as the sum of a symmetric tensor and a skew-symmetric tensor. For example, the velocity gradient may be expressed as the sum of the rate of deformation tensor (Sec. 1.4.1) $\mathbf{D}$ and the vorticity tensor (Sec. 1.4.2) $\mathbf{W}$:

$$\nabla \mathbf{v} = \tfrac{1}{2}[\nabla \mathbf{v} + (\nabla \mathbf{v})^T] + \tfrac{1}{2}[\nabla \mathbf{v} - (\nabla \mathbf{v})^T]$$
$$= \mathbf{D} + \mathbf{W} \tag{2-2}$$

This allows us to rewrite Eq. (2-1) as

$$\mathbf{T} = \mathbf{H}(\mathbf{v}, \mathbf{D} + \mathbf{W}) \tag{2-3}$$

The principle of material frame indifference discussed in Sec. 2.3.1 requires that

$$\mathbf{T}^* = \mathbf{Q} \cdot \mathbf{T} \cdot \mathbf{Q}^T = \mathbf{H}(\mathbf{v}^*, \mathbf{D}^* + \mathbf{W}^*) \tag{2-4}$$

From Eq. (2-4) in Sec. 1.2.2, Exercise 2.3.2-1, and Eq. (2-3), we find that the function **H** must be such that

$$\mathbf{Q} \cdot \mathbf{H}(\mathbf{v}, \mathbf{D} + \mathbf{W}) \cdot \mathbf{Q}^T = \mathbf{H}(\dot{\mathbf{c}} + \dot{\mathbf{Q}} \cdot \mathbf{z} + \mathbf{Q} \cdot \mathbf{v}, \mathbf{Q} \cdot \mathbf{D} \cdot \mathbf{Q}^T$$
$$+ \mathbf{Q} \cdot \mathbf{W} \cdot \mathbf{Q}^T + \dot{\mathbf{Q}} \cdot \mathbf{Q}^T) \tag{2-5}$$

Let us choose a particular change of frame such that

$$\dot{\mathbf{Q}} = -\mathbf{Q} \cdot \mathbf{W} \tag{2-6}$$

and

$$\dot{\mathbf{c}} = -\dot{\mathbf{Q}} \cdot \mathbf{z} - \mathbf{Q} \cdot \mathbf{v} \tag{2-7}$$

With this change of frame, we have from Eqs. (2-4) and (2-5)

$$\mathbf{T}^* = \mathbf{H}(\mathbf{O}, \mathbf{Q} \cdot \mathbf{D} \cdot \mathbf{Q}^T + \mathbf{O})$$

$$= \mathbf{G}(\mathbf{D}^*) \tag{2-8}$$

Applying the principle of material frame indifference again, we conclude from Eq. (2-8) that

$$\mathbf{T} = \mathbf{G}(\mathbf{D}) \tag{2-9}$$

Equation (2-5) requires that the function $\mathbf{G}$ satisfy

$$\mathbf{Q} \cdot \mathbf{G}(\mathbf{D}) \cdot \mathbf{Q}^T = \mathbf{G}(\mathbf{Q} \cdot \mathbf{D} \cdot \mathbf{Q}^T) \tag{2-10}$$

Let us try to give a physical interpretation to our use of the principle of material frame indifference in the preceding paragraphs. We begin with an observer at the position $\mathbf{z}$ in an arbitrary frame of reference who assumes that the stress tensor depends upon velocity, the rate of deformation tensor, and the vorticity tensor: Eq. (2-3). Equation (2-3) in Sec. 1.2.2 and Eqs. (2-6) and (2-7) describe another observer at the position $\mathbf{z}^*$ in a new frame of reference. The observer at $\mathbf{z}^*$ rotates and translates with the material in such a way that for him the velocity and the vorticity tensor of the material, $\mathbf{v}$ and $\mathbf{W}$, respectively, are zero. He can see no dependence of the stress tensor $\mathbf{T}$ upon $\mathbf{v}$ and $\mathbf{W}$; he sees a dependence of $\mathbf{T}$ upon $\mathbf{D}$ alone, Eq. (2-8). But the principle of material frame indifference requires that all observers come to the same conclusions about the behavior of materials. We consequently conclude that Eq. (2-3) reduces to Eq. (2-9).

The most general form that Eq. (2-9) can take when restricted by Eq. (2-10) is [1, p. 32]

$$\mathbf{T} = \varkappa_0 \mathbf{I} + \varkappa_1 \mathbf{D} + \varkappa_2 \mathbf{D} \cdot \mathbf{D} \tag{2-11}$$

where

$$\varkappa_k = \varkappa_k(\mathrm{I_D}, \mathrm{II_D}, \mathrm{III_D}) \tag{2-12}$$

Here $\mathrm{I_D}$, $\mathrm{II_D}$, and $\mathrm{III_D}$ are the three principal invariants of the rate of deformation tensor, i.e., the coefficients in the equation for the principal values of $\mathbf{D}$:

$$0 = \det(\mathbf{D} - m\mathbf{I}) = -m^3 + \mathrm{I_D}m^2 - \mathrm{II_D}m + \mathrm{III_D} \tag{2-13}$$

[2.3] BEHAVIOR

We leave it as an exercise to show that

$$I_D \equiv \text{tr } \mathbf{D} = \text{div } \mathbf{v} \tag{2-14}$$

$$II_D \equiv \tfrac{1}{2}[(I_D)^2 - \overline{II}_D] \tag{2-15}$$

$$\overline{II}_D \equiv \text{tr } \mathbf{D}^2 = \text{tr } (\mathbf{D} \cdot \mathbf{D}) \tag{2-16}$$

$$III_D \equiv \det \mathbf{D} \tag{2-17}$$

Equation (2-11) was first obtained by Reiner [2] and by Prager [3] for functions $\mathbf{G}(\mathbf{D})$ in the form of a tensor power series [1, p. 33].

Notice that this constitutive equation for stress automatically satisfies Cauchy's second law, since the rate of deformation tensor is symmetric.

It follows immediately from Eq. (2-11) that the most general *linear* relation between the stress tensor and the rate of deformation tensor which is consistent with the principle of material frame indifference is

$$\mathbf{T} = (\alpha + \lambda \text{ div } \mathbf{v})\mathbf{I} + 2\mu \mathbf{D} \tag{2-18}$$

If we take $\alpha \equiv -P$, where $P$ is the thermodynamic pressure (see Sec. 5.1.2), we have a special case of Eq. (2-18):

$$\mathbf{T} = (-P + \lambda \text{ div } \mathbf{v})\mathbf{I} + 2\mu \mathbf{D} \tag{2-19}$$

We refer to this as the *newtonian model* for fluid behavior or Newton's law of viscosity. The coefficient $\mu$ is known as the shear coefficient of viscosity; $\varkappa \equiv \lambda + \tfrac{2}{3}\mu$ is the bulk coefficient of viscosity. Later we use the differential entropy inequality to show that [Exercise 5.5.2-1; 1, p. 357] $\mu \geq 0$ and $\lambda \geq -\tfrac{2}{3}\mu$. It is often stated that the Stokes' relation $\varkappa = \lambda + \tfrac{2}{3}\mu = 0$ has been substantiated for low-density monatomic gases. In fact, this result is implicitly assumed in that theory [4, sec. 61A]. Two reviews of the literature concerned with Eq. (2-14) have been given recently [4, sec. 61A; 5]. They point out that nearly all experimental measurements to date indicate that $\lambda$ is positive and that for many fluids it is orders of magnitude greater than $\mu$.

Another special case of Eq. (2-18) corresponds to an incompressible fluid:

$$\mathbf{T} = -p\mathbf{I} + 2\mu \mathbf{D} \tag{2-20}$$

We refer to this as the *incompressible newtonian model* for fluid behavior or Newton's law of viscosity for an incompressible fluid. Equation (2-20) is sometimes described as a special case of Eq. (2-19), this is objectionable, since $P$ is not defined for an incompressible fluid. The quantity $p$ is known as the *mean pressure*. From Eq. (2-20), we see that we may take as its definition

$$p \equiv -\tfrac{1}{3} \text{tr } \mathbf{T} \tag{2-21}$$

When discussing constitutive equations for the stress tensor, it is common to speak in terms of an extra stress tensor $\mathbf{S}$:

$$\mathbf{S} \equiv \mathbf{T} + P\mathbf{I} \tag{2-22}$$

In this way, the strictly thermodynamic quantity $P$ is separated from those effects arising from deformation. For incompressible fluids, we define the extra stress tensor as

$$\mathbf{S} \equiv \mathbf{T} + p\mathbf{I} \tag{2-23}$$

where $p$ may be taken as any convenient scalar field [6, p. 641]. The usual practice for incompressible fluids is to define $p$ to be the mean pressure, Eq. (2-21). This will be our custom in what follows.

While the newtonian model, Eq. (2-18), has been found to be useful in describing the behavior of many gases and low-molecular-weight liquids, it has not been established that any fluid requires a nonzero value for $\varkappa_2$ or a dependence of $\varkappa_1$ upon $III_D$ in Eq. (2-11). On the other hand, a number of empirical relations based upon limiting forms of Eq. (2-11) have been found to be of some engineering value. A few of these are discussed in the next section.

## REFERENCES

1. Truesdell, C., and W. Noll: In S. Flügge (ed.), "Handbuch der Physik," vol. 3/3, Springer-Verlag, Berlin, 1965.
2. Reiner, Markus: *Am. J. Math.*, **67**:350 (1945).
3. Prager, W.: *J. Appl. Phys.*, **16**:837 (1945).
4. Truesdell, C.: *J. Rational Mech. Analysis*, **1**:125 (1952); corrected reprint, International Science Review Series, vol. 8/1, Gordon and Breach, New York.
5. Karim, S. M., and L. Rosenhead: *Rev. Mod. Phys.*, **24**:108 (1952).
6. Truesdell, C., and R. A. Toupin: In S. Flügge (ed.), "Handbuch der Physik," vol. 3/1, Springer-Verlag, Berlin, 1960.
7. Truesdell, C.: "The Elements of Continuum Mechanics," Springer-Verlag, New York, 1966.

## EXERCISES

**2.3.2-1** [7, p. 25] (a) Let us define the *deformation gradient*

$$\mathbf{F} \equiv \frac{\partial \chi_{\varkappa i}}{\partial z_{\varkappa j}} \mathbf{e}_i \mathbf{e}_j$$

where $\chi_\varkappa$ is the deformation of the body (see Sec. 1.1.1), $\mathbf{z} = \chi_\varkappa(\mathbf{z}_\varkappa, t)$. Show that

$$\dot{\mathbf{F}} \equiv \frac{d_{(m)}\mathbf{F}}{dt} = (\nabla \mathbf{v}) \cdot \mathbf{F}$$

and that

$$\nabla \mathbf{v} = \dot{\mathbf{F}} \cdot \mathbf{F}^{-1}$$

(b) Let $\mathbf{Q}$ be a time-dependent orthogonal transformation associated with a change of frame (Sec. 1.2.1) and let the motions $\chi$ and $\chi^*$ be referred to the same reference configuration. Show that

$$\mathbf{F}^* = \mathbf{Q} \cdot \mathbf{F}$$

(c) Take the material derivative of this equation and show that

$$(\nabla \mathbf{v})^* \cdot \mathbf{F}^* = \mathbf{Q} \cdot (\nabla \mathbf{v}) \cdot \mathbf{Q}^T \cdot \mathbf{F}^* + \dot{\mathbf{Q}} \cdot \mathbf{Q}^T \cdot \mathbf{F}^*$$

[2.3] BEHAVIOR 51

Here the superscript * indicates an association with the new frame.

(d) The decomposition of a second-order tensor into skew-symmetric and symmetric portions is unique. Make use of this fact to show that

$$\mathbf{D}^* = \mathbf{Q} \cdot \mathbf{D} \cdot \mathbf{Q}^T$$

and

$$\mathbf{W}^* = \mathbf{Q} \cdot \mathbf{W} \cdot \mathbf{Q}^T + \mathbf{A}$$

We conclude that the rate of deformation tensor $\mathbf{D}$ is frame indifferent, while the vorticity tensor $\mathbf{W}$ is not. The angular velocity tensor $\mathbf{A}$ is defined by Eq. (2-8) of Sec. 1.2.2.

**2.3.2-2** Derive Eq. (2-13), with the coefficients given by Eqs. (2-14) to (2-17).

**2.3.2-3** Starting with tr $\mathbf{D}$, show that

$$\text{tr } \mathbf{D} = \text{div } \mathbf{v}$$

**2.3.2-4** *The existence of a hydrostatic pressure* Consider a fluid described by Eqs. (2-9) and (2-10). We see from Eq. (2-11) that, when there is no flow,

$$\mathbf{T} = \kappa_0 \mathbf{I}$$

Prove this result directly from Eqs. (2-9) and (2-10) by means of Exercise A.5.3-4.

**2.3.3 Two useful classes of empirical models** At the present time it has not been established that any of the many fluids for which the newtonian model is inadequate can be described by Eq. (2-11). But limited experimental observations for incompressible fluids have shown that empirical models based upon Eq. (2-11) may have some utility, in that they predict some aspects of real fluid behavior.

The most common class of empirical models for incompressible fluids based upon Eq. (2-11) is of the form

$$\mathbf{S} \equiv \mathbf{T} + p\mathbf{I} = 2\eta(\gamma)\mathbf{D} \tag{3-1}$$

where we define

$$\gamma \equiv \sqrt{2\overline{\mathrm{II}}_\mathbf{D}} = \sqrt{2 \text{ tr }(\mathbf{D} \cdot \mathbf{D})} = \sqrt{2 \text{ tr } \mathbf{D}^2} \tag{3-2}$$

We refer to $\mathbf{S}$ as the extra-stress tensor; it is common to call $\eta(\gamma)$ the *apparent-viscosity* function by analogy with Eq. (2-20). From the differential entropy inequality, we can show that (Exercise 5.5.2-5)

$$\eta(\gamma) \geq 0 \tag{3-3}$$

On the basis of experimental observation, we reject the possibility that $\eta(\gamma) = 0$, even in the limit $\gamma \to 0$. We require

$$\eta(\gamma) > 0 \tag{3-4}$$

Let us define

$$\tau \equiv \sqrt{\tfrac{1}{2}\overline{\mathrm{II}}_\mathbf{S}} = \sqrt{\tfrac{1}{2} \text{ tr }(\mathbf{S} \cdot \mathbf{S})} = \sqrt{\tfrac{1}{2} \text{ tr } \mathbf{S}^2} \tag{3-5}$$

From Eq. (3-1),

$$\tau = \tau(\gamma) = \eta(\gamma)\gamma \tag{3-6}$$

We assume that $\eta(\gamma)$ is a differentiable function, which means

$$\left.\frac{d\tau}{d\gamma}\right|_{\gamma=0} = \lim_{\gamma \to 0} \frac{\tau(\gamma) - \tau(0)}{\gamma}$$

$$= \lim_{\gamma \to 0} \frac{\tau(\gamma)}{\gamma} = \eta(0) \tag{3-7}$$

Equations (3-4) and (3-7) require that

$$\left.\frac{d\tau}{d\gamma}\right|_{\gamma=0} > 0 \tag{3-8}$$

If the derivative $d\tau/d\gamma$ is continuous, it must be positive in some neighborhood of $\gamma = 0$. In this neighborhood, $\tau(\gamma)$ will be a strictly increasing function of $\gamma$ and, for this reason, it will have an inverse:

$$\gamma = \lambda(\tau) \tag{3-9}$$

Equation (3-9) follows from Eq. (3-1) for $\gamma$ sufficiently close to zero. It is possible that $\tau(\gamma)$ ceases to increase when $\gamma$ exceeds some value, but such behavior has not been confirmed experimentally.

From Eqs. (3-6) and (3-9),

$$\frac{\gamma}{\tau} = \frac{1}{\eta(\gamma)} = \frac{\lambda(\tau)}{\tau} = \frac{1}{\eta(\lambda(\tau))} \tag{3-10}$$

This suggests that we may write Eq. (3-1) as

$$2\mathbf{D} = \varphi(\tau)\mathbf{S} \tag{3-11}$$

where we define

$$\varphi(\tau) \equiv \frac{1}{\eta(\lambda(\tau))} \tag{3-12}$$

One of the most useful two-parameter generalized newtonian models is the Ostwald-de Waele model or power model [1, p. 243]:

$$\eta(\gamma) = m\gamma^{n-1} \tag{3-13}$$

or

$$\varphi(\tau) = m^{-(1/n)}\tau^{(1-n)/n} \tag{3-14}$$

Here $m$ and $n$ are parameters which are to be determined empirically. When $n = 1$ and $m = \mu$, the power model reduces to the newtonian model for an incompressible fluid, Eq. (2-20). Since this model is relatively simple, it has been used widely in theoretical calculations. Its disadvantage is that it does not reduce to newtonian behavior either in the limit $\gamma \to 0$ ($\tau \to 0$) or in the limit $\gamma \to \infty$ ($\tau \to \infty$) as we currently believe all real fluid does. For most polymers and polymer solutions,

## [2.3] BEHAVIOR

$n$ is less than unity. With this assumption, Eq. (3-13) predicts an infinite viscosity in the limit of zero rate of deformation and a zero viscosity as the rate of deformation becomes unbounded.

The Ellis model [1, p. 246; 2] is one possible superposition of the newtonian model and the power model:

$$\varphi(\tau) = \frac{1}{\eta_0}\left[1 + \left(\frac{\tau}{\tau_{\frac{1}{2}}}\right)^{\alpha-1}\right] \tag{3-15}$$

where $\eta_0$, $\tau_{\frac{1}{2}}$, and $\alpha$ are parameters to be fixed by comparison with experimental data. It includes the power model as a special case corresponding to the limit:

$$\frac{1}{\eta_0} \to 0 \qquad \frac{1}{\eta_0(\tau_{\frac{1}{2}})^{\alpha-1}} \to m^{-1/n} \qquad \alpha \to 1/n$$

For polymers and their solutions, $\alpha$ is usually between 1 and 3 [2], which means that it properly predicts a lower-limiting viscosity $\eta_0$ as $\tau \to 0$. Equation (3-15) is believed to be one of the most useful three-parameter models.

Hermes and Fredrickson [3] proposed a modified Ellis model:

$$\eta(\gamma) = \frac{m\eta_0}{m + \eta_0\gamma^{1-n}} \tag{3-16}$$

where $m$, $\eta_0$, and $n$ are experimentally determined parameters. If we assume that $n$ is less than unity, it predicts a lower-limiting viscosity as $\gamma \to 0$. Equation (3-16) may prove to be more useful than the Ellis model in many cases, since the stress tensor is given as an explicit function of the rate of deformation tensor.

The Sisko model [2, 4] is another superposition of newtonian and power-model behavior:

$$\text{For } \gamma < \gamma_0: \quad \eta(\gamma) = \eta_0\left[1 - \left(\frac{\gamma}{\gamma_0}\right)^{\alpha-1}\right] \tag{3-17}$$

Here $\eta_0$, $\gamma_0$, and $\alpha$ are parameters whose values depend upon the particular material being described. It properly predicts a lower-limiting viscosity $\eta_0$ as $\gamma \to 0$, but cannot be used for $\gamma > \gamma_0$, since $\alpha$ is usually between 1 and 3 [2].

The Bingham plastic model [1, p. 114] is of historical interest, but of limited current practical value. It describes a material which behaves as a rigid solid until the stress has exceeded some critical value:

$$\text{For } \tau > \tau_0: \quad \eta(\gamma) = \eta_0 + \frac{\tau_0}{\gamma} \tag{3-18}$$

$$\text{For } \tau < \tau_0: \quad \mathbf{D} = 0 \tag{3-19}$$

This model contains two parameters: $\eta_0$ and $\tau_0$. It was originally proposed to represent the behavior of paint. The idea of a critical stress $\tau_0$ at which the rigid solid yielded and began to flow probably was postulated on the basis of inadequate data in the limit $\tau \to 0$. Though later work has failed to establish that any materials are true Bingham plastics, the concept is firmly established in the older literature.

Many more models of the form of Eqs. (3-1) and (3-11) have been proposed than those mentioned here [1, 2, 5]. They often have been published in a simplified form within the context of a particular situation. The reader interested in applying to another situation a model that has been presented in this fashion should first express the model in a form consistent with either Eq. (3-1) or Eq. (3-11).

The principal danger with these models appears to be that at the present time we are uncertain as to their limitations. We do know that they can be used as reliable representations for the shear stress in viscometric flows [6] such as steady-state flow through a tube, steady-state tangential flow between concentric cylinders, and steady-state flow between a cone and plate. This is the purpose for which these models were originally developed by experimentalists. We also know that they are incapable of correctly representing the normal stresses [6] in the viscometric flows. The discussion in the next section does not give us much encouragement regarding the appropriateness of these models for more general flows. For this reason, one should be cautious in using these models for other than the shear stress in viscometric flows, although it has been encouraging to see how useful they are in describing steady-state creeping flow past a sphere [7, 8; see also Sec. 4.4.20].

## REFERENCES

1. Reiner, M.: "Deformation, Strain and Flow," 2d ed., Interscience, New York, 1960.
2. Bird, R. B.: *Can. J. Chem. Eng.*, **43**:161 (1965).
3. Hermes, R. A., and A. G. Fredrickson: *A.I.Ch.E.J.*, **13**:253 (1967).
4. Sisko, A. W.: *Ind. Eng. Chem.*, **50**:1789 (1958).
5. Bird, R. B., W. E. Stewart, E. N. Lightfoot, and T. W. Chapman: *A.I.Ch.E. Continuing Education Series* No. 4, 1969.
6. Coleman, B. D., H. Markovitz, and W. Noll: "Viscometric Flows of Non-Newtonian Fluids," Springer-Verlag, New York, 1966.
7. Hopke, S. W., and J. C. Slattery: *A.I.Ch.E.J.*, **16**:224 (1970).
8. Hopke, S. W., and J. C. Slattery: *A.I.Ch.E.J.*, **16**:317 (1970).

## EXERCISES

**2.3.3-1** Show that $p$ in Eq. (3-1) must be the *mean pressure* defined by Eq. (2-21).

**2.3.3-2** Starting from Eq. (3-1), derive Eq. (3-6).

### 2.3.4 The Noll simple fluid

Many commercial processes involve viscoelastic fluids, ranging from polymers and polymer solutions to food products. (Viscoelastic is used here in the sense that the materials obey neither of the classical linear relations, Newton's law of viscosity and Hooke's law of elasticity. Fluids that show a finite relaxation time and fluids that exhibit normal stresses in viscometric flows [1, p. 47] form subclasses of the viscoelastic materials. The term viscoelastic is commonly used in literature in referring to these subclasses.) The behavior of these fluids is generally much more complex than we have suggested in Secs. 2.3.2 and 2.3.3. Sometimes the simple models discussed in Sec. 2.3.3 are adequate for representing the principal aspects of material behavior to be observed in a particular experiment. Many times they are not. Noll [1 to 4] has suggested a constitutive equation for the stress tensor that apparently can be used to explain all aspects of the behavior of viscoelastic liquids that have been observed experimentally.

Before trying to say exactly what is meant by a Noll simple fluid, let us pause for a little background. Let $\xi$ be the place at time $t - s$ ($0 \leq s < \infty$) of the material particle that at time $t$ occupies the place $\mathbf{x}$:

$$\xi = \chi_{(t)}(\mathbf{x}, t - s) \tag{4-1}$$

We call $\chi_{(t)}$ the *relative deformation function*. Equation (4-1) describes the motion that took place in the material at all times $t - s$ prior to the time $t$. The gradient with respect to $\mathbf{x}$ of the relative deformation function is called the *relative deformation gradient* (the more general *deformation gradient* is defined in Exercise 2.3.2-1):

$$\mathbf{F}_{(t)}(t - s) \equiv \nabla \chi_{(t)}(\mathbf{x}, t - s) \tag{4-2}$$

The *relative right Cauchy-Green strain tensor* is defined as

$$\mathbf{C}_{(t)}(t - s) \equiv \mathbf{F}_{(t)}^T(t - s) \cdot \mathbf{F}_{(t)}(t - s) \tag{4-3}$$

Noll [1 to 4] defines an *incompressible simple fluid* as one for which the extra stress $\mathbf{S}$ at the position $\mathbf{x}$ and time $t$ is specified by the history of the relative right Cauchy-Green strain tensor for the material that is within an arbitrarily small neighborhood of $\mathbf{x}$ at time $t$:

$$\mathbf{S} = \frac{\mu_0}{s_0} \underset{\sigma=0}{\overset{\infty}{\mathcal{H}^*}} (\mathbf{C}_{(t)}(t - s_0 \sigma)) \tag{4-4}$$

Here we follow Truesdell's discussion of the dimensional indifference of the definition of a simple material [4, p. 65; 5]. The quantity $\underset{\sigma=0}{\overset{\infty}{\mathcal{H}^*}}$ is a dimensionally invariant tensor-valued functional (by a tensor-valued functional, we mean an operator that maps tensor-valued functions into tensors). The constants $\mu_0$ and $s_0$ are, respectively, a characteristic viscosity and characteristic time or natural time lapse of the fluid. Like any characteristic quantities introduced in defining dimensionless variables, the definitions for $\mu_0$ and $s_0$ are arbitrary. The advantages and disadvantages of particular definitions for $\mu_0$ and $s_0$ have been discussed elsewhere [6].

Equation (4-4) clearly satisfies the principles of determinism and local action (see Sec. 2.3.1). That it also satisfies the principle of material frame indifference is less obvious but, nonetheless, true [13, pp. 39, 58, and 63].

Since the form of the functional $\underset{\sigma=0}{\overset{\infty}{\mathcal{H}^*}}$ is left unspecified, it is clear that the Noll simple fluid incorporates a great deal of flexibility. It is for exactly this reason that many workers currently believe the simple fluid model to be capable of explaining all manifestations of material behavior that have been observed experimentally to date.

The Noll simple fluid should be viewed as representing an entire class of constitutive equations or an entire class of material behaviors.

But the very generality of the Noll simple fluid is also its weakness. Only two classes of flows have been shown to be dynamically possible for *every* simple fluid [1, 3, 4, 7 to 11]. Most flows of engineering interest cannot be analyzed without first specifying a particular form for the functional $\mathop{\mathcal{H}}\limits_{\sigma=0}^{\infty}*$.

But this does not mean that the Noll simple fluid is of no significance to those of us interested in practical problems. It actually is a very simple model for fluid behavior in the sense that it incorporates at most two dimensional parameters: $\mu_0$ and $s_0$. This dimensional simplicity, together with its capacity for representing a wide range of material behavior, makes the Noll simple fluid ideal for use in preparing dimensionless correlations of experimental data. The only limitation upon correlations based on the simple fluid model is that, since material behavior in the form of the functional $\mathop{\mathcal{H}}\limits_{\sigma=0}^{\infty}*$ has not been fully specified, correlations of experimental data can be made for only one fluid at a time. While this is a serious limitation, there is a bright side as well. It is not necessary to have all of the data that would be required in order to describe the behavior of the material under study. For more on scale-ups and data correlations for viscoelastic fluids, see [6, 12].

As you might expect, there have been alternative descriptions for the complex behavior observed in real fluids. Of these, Oldroyd's "generalized elasticoviscous fluid" [14] has attracted perhaps the most interest.

If you wish to learn more about the behavior of real materials and their description, you are fortunate to have several excellent texts available [15 to 17, 1, 4, 13].

## REFERENCES

1. Coleman, B. D., Hershel Markovitz, and Walter Noll: "Viscoelastic Flows of Non-Newtonian Fluids," Springer-Verlag, New York, 1966.
2. Noll, Walter: *Arch. Rational Mech. Analysis*, **2**:197 (1958).
3. Coleman, B. D., and Walter Noll: *Ann. N.Y. Acad. Sci.*, **89**:672 (1961).
4. Truesdell, C., and Walter Noll: In S. Flügge (ed.), "Handbuch der Physik," vol. 3/3, Springer-Verlag, Berlin, 1965.
5. Truesdell, C.: *Phys. Fluids*, **7**:1134 (1964).
6. Slattery, J. C.: *A.I.Ch.E.J.*, **14**:516 (1968).
7. Coleman, B. D., and Walter Noll: *Arch. Rational Mech. Analysis*, **3**:289 (1959).
8. Coleman, B. D.: *Arch. Rational Mech. Analysis*, **9**:273 (1962).
9. Noll, Walter: *Arch. Rational Mech. Analysis*, **11**:97 (1962).
10. Coleman, B. D., and Walter Noll: *Phys. Fluids*, **5**:840 (1962).
11. Slattery, J. C.: *Phys. Fluids*, **7**:1913 (1964).
12. Slattery, J. C.: *A.I.Ch.E.J.*, **11**:831 (1965).
13. Truesdell, C.: "The Elements of Continuum Mechanics," Springer-Verlag, New York, 1966.
14. Oldroyd, J. G.: *Proc. Roy. Soc. (London)*, **A283**:115 (1965).
15. Leigh, D. C.: "Nonlinear Continuum Mechanics," McGraw-Hill, New York, 1968.
16. Lodge, A. S.: "Elastic Liquids," Academic, New York, 1964.
17. Fredrickson, A. G.; "Principles and Applications of Rheology," Prentice-Hall, Englewood Cliffs, N.J., 1964.

### 2.3.5 Surface tension

In this text, we have adopted the widely accepted device of representing a phase interface by a singular surface (see Sec. 1.3.5). Like everything else we do in continuum mechanics, this should be regarded as a model for reality. Our understanding of the phase interface is by no means complete, but there is good experimental evidence that indicates density may be a continuous function of position through the interfacial regions [1, p. 373]. Perhaps all of the intensive variables we are concerned with, including velocity, should more accurately be regarded as continuous functions of position in going from one phase to the next.

There is of course a disadvantage in taking this apparently more realistic view of a continuous interfacial region. Let us consider as an example the phase interface between two newtonian fluids. For reasons that will become obvious in a moment, we are almost certain that the stress tensor cannot be represented by the newtonian model in the interfacial region. For this region, a constitutive equation for the stress tensor might be required to depend upon temperature, density, and density gradient. For one range of temperature and density, it reduces to the newtonian model corresponding to one of the bulk phases; for another range of temperature and density, it reduces to the newtonian model corresponding to the other bulk phase; for an intermediate range of temperature and density, the stress tensor is found to depend upon the density gradient. Constitutive equations of this character have been proposed [2, p. 513], but they are very difficult to use in analyzing dynamic problems. For their proper use, more information is needed regarding the thermodynamic behavior of material in the interfacial region than is currently available.

This accounts for the popularity of the singular-surface model for the phase interface. Yet we cannot ignore experimental effects directly attributable to the behavior of the material in the interfacial region. We attempt to account for these effects in terms of the singular-surface model by introducing sources for momentum, mass, etc., in the singular surface. The most familiar example is in momentum transfer, where we commonly write

$$[\rho \mathbf{v}(\mathbf{v} \cdot \boldsymbol{\xi} - u_{(\xi)}) - \mathbf{T} \cdot \boldsymbol{\xi}] = 2H\sigma\boldsymbol{\xi}$$

$$= (\varkappa_1 + \varkappa_2)\sigma\boldsymbol{\xi} \qquad (5\text{-}1)$$

Here $\sigma$ is surface tension, $H$ is the mean curvature [3, p. 205] of the surface; $\varkappa_1$ and $\varkappa_2$ are the principal curvatures of the surface [3, p. 211]. In arriving at this result, we have assumed that surface tension is independent of position upon the phase interface [4]; experimentally, this corresponds to saying that any surfactant (any surfactive agent such as soap, detergent, etc.) present is uniformly distributed over the phase interface. The term on the right of Eq. (5-1) may be interpreted as the rate of production per unit area of momentum in the singular surface. Other constitutive equations for this surface source of momentum have been proposed that would depend upon the rate of deformation of the phase interface [4], but their usefulness has not as yet been confirmed experimentally.

For a further introduction to interfacial behavior, see [4].

## REFERENCES

1. Hirschfelder, J. O., C. F. Curtiss, and R. B. Bird: "Molecular Theory of Gases and Liquids," Wiley, New York, 1954; corrected with notes added 1964.
2. Truesdell, C., and W. Noll: In S. Flügge (ed.), "Handbuch der Physik," vol. 3/3, Springer-Verlag, Berlin, 1965.
3. McConnell, A. J.: "Application of Tensor Analysis," Dover, New York, 1957.
4. Slattery, J. C.: *Ind. Eng. Chem., Fundamentals*, **6**:108 (1967); *ibid.*, **7**:672 (1968).

## 2.4 NEWTONIAN FLUID

**2.4.1 Cauchy's first law for a newtonian fluid** In Secs. 2.3.1 to 2.3.4, we pointed out the need for information beyond the differential equation of continuity and Cauchy's two laws and we discussed several possible constitutive equations for the stress tensor. One of these was the newtonian model for fluid behavior, often referred to as simply the newtonian fluid, Eq. (2-19) in Sec. 2.3.2.

The divergence of the stress tensor for a newtonian fluid may be written as

$$\text{div } \mathbf{T} = \text{div } (-P\mathbf{I} + \lambda \, [\text{div } \mathbf{v}]\mathbf{I} + 2\mu \mathbf{D}) \tag{1-1}$$

Let us assume that $\lambda$ and $\mu$ are constants with respect to position; Eq. (1-1) becomes

$$\text{div } \mathbf{T} = -\nabla P + \lambda \nabla (\text{div } \mathbf{v}) + 2\mu \text{ div } \mathbf{D} \tag{1-2}$$

Since

$$\text{div } \mathbf{D} = \tfrac{1}{2} \text{div } (\nabla \mathbf{v}) + \tfrac{1}{2} \nabla (\text{div } \mathbf{v}) \tag{1-3}$$

we have

$$\text{div } \mathbf{T} = -\nabla P + (\lambda + \mu) \nabla (\text{div } \mathbf{v}) + \mu \text{ div } (\nabla \mathbf{v}) \tag{1-4}$$

By means of Eq. (1-4), Cauchy's first law, Eq. (3-4) in Sec. 2.2.3, becomes for a newtonian fluid

$$\rho \frac{d_{(m)} \mathbf{v}}{dt} = -\nabla P + (\lambda + \mu) \nabla (\text{div } \mathbf{v}) + \mu \text{ div } (\nabla \mathbf{v}) + \rho \mathbf{f} \tag{1-5}$$

The stress-deformation behavior of an incompressible newtonian fluid is described by Eq. (2-20) in Sec. 2.3.2. Cauchy's first law for this case becomes

$$\rho \frac{d_{(m)} \mathbf{v}}{dt} = -\nabla p + \mu \text{ div } (\nabla \mathbf{v}) + \rho \mathbf{f} \tag{1-6}$$

This is often referred to as the *Navier-Stokes equation*.

## EXERCISE

**2.4.1-1** Starting with div **D**, derive Eq. (1-3).

## 2.5 SUMMARY

**2.5.1 Summary of useful equations** If the reader has studied the sections marked with asterisks and the problems of the appendix, he is in the position of being able to derive whatever equation he needs in any specified coordinate system. While this is desirable for specialists in any area of mechanics in order to comment critically on the work of others, I do not feel it necessary that the majority of readers have this degree of competence. Further, it would be obviously a waste of energy for anyone to rederive fundamental equations at the beginning of each new problem. With this thought in mind, we felt that the reader should be assisted with a set of tables (Tables 2.5.1-1 to 2.5.1-10) presenting some of the most common relationships in rectangular cartesian, cylindrical, and spherical coordinates [1, p. 83].

It is generally more convenient to work with Cauchy's first law, Eq. (3-4) in Sec. 2.2.3, in terms of the extra-stress tensor $\mathbf{S}$ (see Sec. 2.3.2):

$$\rho\left[\frac{\partial \mathbf{v}}{\partial t} + (\nabla \mathbf{v}) \cdot \mathbf{v}\right] = -\nabla P + \text{div } \mathbf{S} + \rho \mathbf{f} \tag{1-1}$$

It is the components of this equation that are presented in Tables 2.5.1-2, 2.5.1-4, and 2.5.1-6.

Commonly, the only external force to be considered is a uniform gravitational field, which we may represent as

$$\mathbf{f} = -\nabla \varphi \tag{1-2}$$

For an incompressible fluid, Eq. (1-2) allows us to express Eq. (1-1) as

$$\rho\left[\frac{\partial \mathbf{v}}{\partial t} + (\nabla \mathbf{v}) \cdot \mathbf{v}\right] = -\nabla \mathscr{P} + \text{div } \mathbf{S} \tag{1-3}$$

where $\mathscr{P}$ is referred to as the modified pressure

$$\mathscr{P} \equiv p + \rho \varphi \tag{1-4}$$

The components of Eq. (1-3) are easily found from Tables 2.5.1-2, 2.5.1-4, and 2.5.1-6 by deleting the components of $\mathbf{f}$ and replacing $P$ with $\mathscr{P}$.

We deal only with physical components of spatial vector fields and second-order tensor fields when discussing curvilinear coordinate systems in the sections and problems without asterisks which form the bulk of this text. For this reason we adopt a somewhat simpler notation for physical components in cylindrical and spherical coordinates than that suggested in the appendix. We denote the physical components of spatial vector fields in cylindrical coordinates as $v_r$, $v_\theta$, and $v_z$ rather than $v_{\langle 1 \rangle}$, $v_{\langle 2 \rangle}$, $v_{\langle 3 \rangle}$; the physical components of second-order tensor fields are indicated as $D_{rr}$, $D_{r\theta}$, $D_{\theta z}$, etc. The notation used in spherical coordinates is very similar.

Because of this change in notation, we do *not* employ the summation convention hereafter with the physical components of spatial vector fields and second-order tensor fields. The quantity $D_{rr}$ is a single physical component of the second-order tensor field **D**; when used in context, there should be no occasion to misinterpret it as the sum of three rectangular cartesian components.

When we have occasion to discuss physical components with respect to other curvilinear coordinate systems, we revert to the notation introduced in the appendix.

**REFERENCE**

1. Bird, R. B., W. E. Stewart, and E. N. Lightfoot: "Transport Phenomena," 7th printing, Wiley, New York, 1960.

**Table 2.5.1-1** The differential equation of continuity, Eq. (3-4) of Sec. 1.3.3, in three coordinate systems

*Rectangular cartesian coordinates* $(z_1, z_2, z_3)$:

$$\frac{\partial \rho}{\partial t} + \frac{\partial}{\partial z_1}(\rho v_1) + \frac{\partial}{\partial z_2}(\rho v_2) + \frac{\partial}{\partial z_3}(\rho v_3) = 0 \quad (A)$$

*Cylindrical coordinates* $(r, \theta, z)$:

$$\frac{\partial \rho}{\partial t} + \frac{1}{r}\frac{\partial}{\partial r}(\rho r v_r) + \frac{1}{r}\frac{\partial}{\partial \theta}(\rho v_\theta) + \frac{\partial}{\partial z}(\rho v_z) = 0 \quad (B)$$

*Spherical coordinates* $(r, \theta, \varphi)$:

$$\frac{\partial \rho}{\partial t} + \frac{1}{r^2}\frac{\partial}{\partial r}(\rho r^2 v_r) + \frac{1}{r \sin \theta}\frac{\partial}{\partial \theta}(\rho v_\theta \sin \theta) + \frac{1}{r \sin \theta}\frac{\partial}{\partial \varphi}(\rho v_\varphi) = 0 \quad (C)$$

**Table 2.5.1-2** Cauchy's first law in rectangular cartesian coordinates

$z_1$ *component*:

$$\rho\left(\frac{\partial v_1}{\partial t} + v_1\frac{\partial v_1}{\partial z_1} + v_2\frac{\partial v_1}{\partial z_2} + v_3\frac{\partial v_1}{\partial z_3}\right) = -\frac{\partial P}{\partial z_1} + \frac{\partial S_{11}}{\partial z_1} + \frac{\partial S_{12}}{\partial z_2} + \frac{\partial S_{13}}{\partial z_3} + \rho f_1 \quad (A)$$

$z_2$ *component*:

$$\rho\left(\frac{\partial v_2}{\partial t} + v_1\frac{\partial v_2}{\partial z_1} + v_2\frac{\partial v_2}{\partial z_2} + v_3\frac{\partial v_2}{\partial z_3}\right) = -\frac{\partial P}{\partial z_2} + \frac{\partial S_{21}}{\partial z_1} + \frac{\partial S_{22}}{\partial z_2} + \frac{\partial S_{23}}{\partial z_3} + \rho f_2 \quad (B)$$

$z_3$ *component*:

$$\rho\left(\frac{\partial v_3}{\partial t} + v_1\frac{\partial v_3}{\partial z_1} + v_2\frac{\partial v_3}{\partial z_2} + v_3\frac{\partial v_3}{\partial z_3}\right) = -\frac{\partial P}{\partial z_3} + \frac{\partial S_{31}}{\partial z_1} + \frac{\partial S_{32}}{\partial z_2} + \frac{\partial S_{33}}{\partial z_3} + \rho f_3 \quad (C)$$

## [2.5] SUMMARY

**Table 2.5.1-3** Cauchy's first law in rectangular cartesian coordinates for a newtonian fluid with constant $\rho$ and $\mu$ (the Navier-Stokes equation), Eq. (1-6) of Sec. 2.4.1

$z_1$ component:

$$\rho\left(\frac{\partial v_1}{\partial t} + v_1\frac{\partial v_1}{\partial z_1} + v_2\frac{\partial v_1}{\partial z_2} + v_3\frac{\partial v_1}{\partial z_3}\right) = -\frac{\partial p}{\partial z_1} + \mu\left(\frac{\partial^2 v_1}{\partial z_1^2} + \frac{\partial^2 v_1}{\partial z_2^2} + \frac{\partial^2 v_1}{\partial z_3^2}\right) + \rho f_1 \quad \text{(A)}$$

$z_2$ component:

$$\rho\left(\frac{\partial v_2}{\partial t} + v_1\frac{\partial v_2}{\partial z_1} + v_2\frac{\partial v_2}{\partial z_2} + v_3\frac{\partial v_2}{\partial z_3}\right) = -\frac{\partial p}{\partial z_2} + \mu\left(\frac{\partial^2 v_2}{\partial z_1^2} + \frac{\partial^2 v_2}{\partial z_2^2} + \frac{\partial^2 v_2}{\partial z_3^2}\right) + \rho f_2 \quad \text{(B)}$$

$z_3$ component:

$$\rho\left(\frac{\partial v_3}{\partial t} + v_1\frac{\partial v_3}{\partial z_1} + v_2\frac{\partial v_3}{\partial z_2} + v_3\frac{\partial v_3}{\partial z_3}\right) = -\frac{\partial p}{\partial z_3} + \mu\left(\frac{\partial^2 v_3}{\partial z_1^2} + \frac{\partial^2 v_3}{\partial z_2^2} + \frac{\partial^2 v_3}{\partial z_3^2}\right) + \rho f_3 \quad \text{(C)}$$

**Table 2.5.1-4** Cauchy's first law in cylindrical coordinates

$r$ component:

$$\rho\left(\frac{\partial v_r}{\partial t} + v_r\frac{\partial v_r}{\partial r} + \frac{v_\theta}{r}\frac{\partial v_r}{\partial \theta} - \frac{v_\theta^2}{r} + v_z\frac{\partial v_r}{\partial z}\right) = -\frac{\partial P}{\partial r} + \frac{1}{r}\frac{\partial}{\partial r}(rS_{rr}) + \frac{1}{r}\frac{\partial}{\partial \theta}(S_{r\theta}) - \frac{S_{\theta\theta}}{r} + \frac{\partial S_{rz}}{\partial z} + \rho f_r \quad \text{(A)}$$

$\theta$ component:

$$\rho\left(\frac{\partial v_\theta}{\partial t} + v_r\frac{\partial v_\theta}{\partial r} + \frac{v_\theta}{r}\frac{\partial v_\theta}{\partial \theta} + \frac{v_r v_\theta}{r} + v_z\frac{\partial v_\theta}{\partial z}\right) = -\frac{1}{r}\frac{\partial P}{\partial \theta} + \frac{1}{r^2}\frac{\partial}{\partial r}(r^2 S_{\theta r}) + \frac{1}{r}\frac{\partial S_{\theta\theta}}{\partial \theta} + \frac{\partial S_{\theta z}}{\partial z} + \rho f_\theta \quad \text{(B)}$$

$z$ component:

$$\rho\left(\frac{\partial v_z}{\partial t} + v_r\frac{\partial v_z}{\partial r} + \frac{v_\theta}{r}\frac{\partial v_z}{\partial \theta} + v_z\frac{\partial v_z}{\partial z}\right) = -\frac{\partial P}{\partial z} + \frac{1}{r}\frac{\partial}{\partial r}(rS_{zr}) + \frac{1}{r}\frac{\partial S_{z\theta}}{\partial \theta} + \frac{\partial S_{zz}}{\partial z} + \rho f_z \quad \text{(C)}$$

**Table 2.5.1-5** Cauchy's first law in cylindrical coordinates for a newtonian fluid with constant $\rho$ and $\mu$ (the Navier-Stokes equation), Eq. (1-6) of Sec. 2.4.1

$r$ component:

$$\rho\left(\frac{\partial v_r}{\partial t} + v_r \frac{\partial v_r}{\partial r} + \frac{v_\theta}{r}\frac{\partial v_r}{\partial \theta} - \frac{v_\theta^2}{r} + v_z \frac{\partial v_r}{\partial z}\right)$$

$$= -\frac{\partial p}{\partial r} + \mu\left\{\frac{\partial}{\partial r}\left(\frac{1}{r}\frac{\partial}{\partial r}(rv_r)\right) + \frac{1}{r^2}\frac{\partial^2 v_r}{\partial \theta^2} - \frac{2}{r^2}\frac{\partial v_\theta}{\partial \theta} + \frac{\partial^2 v_r}{\partial z^2}\right\} + \rho f_r \quad \text{(A)}$$

$\theta$ component:

$$\rho\left(\frac{\partial v_\theta}{\partial t} + v_r \frac{\partial v_\theta}{\partial r} + \frac{v_\theta}{r}\frac{\partial v_\theta}{\partial \theta} + \frac{v_r v_\theta}{r} + v_z \frac{\partial v_\theta}{\partial z}\right)$$

$$= -\frac{1}{r}\frac{\partial p}{\partial \theta} + \mu\left\{\frac{\partial}{\partial r}\left(\frac{1}{r}\frac{\partial}{\partial r}(rv_\theta)\right) + \frac{1}{r^2}\frac{\partial^2 v_\theta}{\partial \theta^2} + \frac{2}{r^2}\frac{\partial v_r}{\partial \theta} + \frac{\partial^2 v_\theta}{\partial z^2}\right\} + \rho f_\theta \quad \text{(B)}$$

$z$ component:

$$\rho\left(\frac{\partial v_z}{\partial t} + v_r \frac{\partial v_z}{\partial r} + \frac{v_\theta}{r}\frac{\partial v_z}{\partial \theta} + v_z \frac{\partial v_z}{\partial z}\right) = -\frac{\partial p}{\partial z} + \mu\left\{\frac{1}{r}\frac{\partial}{\partial r}\left(r\frac{\partial v_z}{\partial r}\right) + \frac{1}{r^2}\frac{\partial^2 v_z}{\partial \theta^2} + \frac{\partial^2 v_z}{\partial z^2}\right\} + \rho f_z \quad \text{(C)}$$

**Table 2.5.1-6** Cauchy's first law in spherical coordinates

$r$ component:

$$\rho\left(\frac{\partial v_r}{\partial t} + v_r \frac{\partial v_r}{\partial r} + \frac{v_\theta}{r}\frac{\partial v_r}{\partial \theta} + \frac{v_\varphi}{r \sin \theta}\frac{\partial v_r}{\partial \varphi} - \frac{v_\theta^2 + v_\varphi^2}{r}\right)$$

$$= -\frac{\partial P}{\partial r} + \frac{1}{r^2}\frac{\partial}{\partial r}(r^2 S_{rr}) + \frac{1}{r \sin \theta}\frac{\partial}{\partial \theta}(S_{r\theta} \sin \theta) + \frac{1}{r \sin \theta}\frac{\partial S_{r\varphi}}{\partial \varphi} - \frac{S_{\theta\theta} + S_{\varphi\varphi}}{r} + \rho f_r \quad \text{(A)}$$

$\theta$ component:

$$\rho\left(\frac{\partial v_\theta}{\partial t} + v_r \frac{\partial v_\theta}{\partial r} + \frac{v_\theta}{r}\frac{\partial v_\theta}{\partial \theta} + \frac{v_\varphi}{r \sin \theta}\frac{\partial v_\theta}{\partial \varphi} + \frac{v_r v_\theta}{r} - \frac{v_\varphi^2 \cot \theta}{r}\right)$$

$$= -\frac{1}{r}\frac{\partial P}{\partial \theta} + \frac{1}{r^3}\frac{\partial}{\partial r}(r^3 S_{\theta r}) + \frac{1}{r \sin \theta}\frac{\partial}{\partial \theta}(S_{\theta\theta} \sin \theta) + \frac{1}{r \sin \theta}\frac{\partial S_{\theta\varphi}}{\partial \varphi} - \frac{\cot \theta}{r} S_{\varphi\varphi} + \rho f_\theta \quad \text{(B)}$$

$\varphi$ component:

$$\rho\left(\frac{\partial v_\varphi}{\partial t} + v_r \frac{\partial v_\varphi}{\partial r} + \frac{v_\theta}{r}\frac{\partial v_\varphi}{\partial \theta} + \frac{v_\varphi}{r \sin \theta}\frac{\partial v_\varphi}{\partial \varphi} + \frac{v_\varphi v_r}{r} + \frac{v_\theta v_\varphi}{r} \cot \theta\right)$$

$$= -\frac{1}{r \sin \theta}\frac{\partial P}{\partial \varphi} + \frac{1}{r^3}\frac{\partial}{\partial r}(r^3 S_{\varphi r}) + \frac{1}{r \sin^2 \theta}\frac{\partial}{\partial \theta}(S_{\varphi\theta} \sin^2 \theta) + \frac{1}{r \sin \theta}\frac{\partial S_{\varphi\varphi}}{\partial \varphi} + \rho f_\varphi \quad \text{(C)}$$

[2.5] SUMMARY

**Table 2.5.1-7** Cauchy's first law in spherical coordinates for a newtonian fluid with constant $\rho$ and $\mu$ (the Navier-Stokes equation), Eq. (1-6) of Sec. 2.4.1

---

$r$ component:

$$\rho\left(\frac{\partial v_r}{\partial t} + v_r \frac{\partial v_r}{\partial r} + \frac{v_\theta}{r}\frac{\partial v_r}{\partial \theta} + \frac{v_\varphi}{r \sin\theta}\frac{\partial v_r}{\partial \varphi} - \frac{v_\theta^2 + v_\varphi^2}{r}\right)$$

$$= -\frac{\partial p}{\partial r} + \mu\left(\mathscr{H} v_r - \frac{2}{r^2}v_r - \frac{2}{r^2}\frac{\partial v_\theta}{\partial \theta} - \frac{2}{r^2}v_\theta \cot\theta - \frac{2}{r^2 \sin\theta}\frac{\partial v_\varphi}{\partial \varphi}\right) + \rho f_r \quad (A)$$

$\theta$ component:

$$\rho\left(\frac{\partial v_\theta}{\partial t} + v_r \frac{\partial v_\theta}{\partial r} + \frac{v_\theta}{r}\frac{\partial v_\theta}{\partial \theta} + \frac{v_\varphi}{r \sin\theta}\frac{\partial v_\theta}{\partial \varphi} + \frac{v_r v_\theta}{r} - \frac{v_\varphi^2 \cot\theta}{r}\right)$$

$$= -\frac{1}{r}\frac{\partial p}{\partial \theta} + \mu\left(\mathscr{H} v_\theta + \frac{2}{r^2}\frac{\partial v_r}{\partial \theta} - \frac{v_\theta}{r^2 \sin^2\theta} - \frac{2\cos\theta}{r^2 \sin^2\theta}\frac{\partial v_\varphi}{\partial \varphi}\right) + \rho f_\theta \quad (B)$$

$\varphi$ component:

$$\rho\left(\frac{\partial v_\varphi}{\partial t} + v_r \frac{\partial v_\varphi}{\partial r} + \frac{v_\theta}{r}\frac{\partial v_\varphi}{\partial \theta} + \frac{v_\varphi}{r \sin\theta}\frac{\partial v_\varphi}{\partial \varphi} + \frac{v_\varphi v_r}{r} + \frac{v_\theta v_\varphi}{r}\cot\theta\right)$$

$$= -\frac{1}{r \sin\theta}\frac{\partial p}{\partial \varphi} + \mu\left(\mathscr{H} v_\varphi - \frac{v_\varphi}{r^2 \sin^2\theta} + \frac{2}{r^2 \sin\theta}\frac{\partial v_r}{\partial \varphi} + \frac{2\cos\theta}{r^2 \sin^2\theta}\frac{\partial v_\theta}{\partial \varphi}\right) + \rho f_\varphi \quad (C)$$

where

$$\mathscr{H} = \frac{1}{r^2}\frac{\partial}{\partial r}\left(r^2 \frac{\partial}{\partial r}\right) + \frac{1}{r^2 \sin\theta}\frac{\partial}{\partial \theta}\left(\sin\theta \frac{\partial}{\partial \theta}\right) + \frac{1}{r^2 \sin^2\theta}\left(\frac{\partial^2}{\partial \varphi^2}\right)$$

---

**Table 2.5.1-8** Components of rate of deformation tensor in rectangular cartesian coordinates

$$D_{11} = \frac{\partial v_1}{\partial z_1} \quad (A)$$

$$D_{22} = \frac{\partial v_2}{\partial z_2} \quad (B)$$

$$D_{33} = \frac{\partial v_3}{\partial z_3} \quad (C)$$

$$D_{12} = D_{21} = \frac{1}{2}\left(\frac{\partial v_1}{\partial z_2} + \frac{\partial v_2}{\partial z_1}\right) \quad (D)$$

$$D_{13} = D_{31} = \frac{1}{2}\left(\frac{\partial v_1}{\partial z_3} + \frac{\partial v_3}{\partial z_1}\right) \quad (E)$$

$$D_{23} = D_{32} = \frac{1}{2}\left(\frac{\partial v_2}{\partial z_3} + \frac{\partial v_3}{\partial z_2}\right) \quad (F)$$

**Table 2.5.1-9** Components of the rate of deformation tensor in cylindrical coordinates

$$D_{rr} = \frac{\partial v_r}{\partial r} \tag{A}$$

$$D_{\theta\theta} = \frac{1}{r}\frac{\partial v_\theta}{\partial \theta} + \frac{v_r}{r} \tag{B}$$

$$D_{zz} = \frac{\partial v_z}{\partial z} \tag{C}$$

$$D_{r\theta} = D_{\theta r} = \frac{1}{2}\left[r\frac{\partial}{\partial r}\left(\frac{v_\theta}{r}\right) + \frac{1}{r}\frac{\partial v_r}{\partial \theta}\right] \tag{D}$$

$$D_{rz} = D_{zr} = \frac{1}{2}\left(\frac{\partial v_r}{\partial z} + \frac{\partial v_z}{\partial r}\right) \tag{E}$$

$$D_{\theta z} = D_{z\theta} = \frac{1}{2}\left(\frac{\partial v_\theta}{\partial z} + \frac{1}{r}\frac{\partial v_z}{\partial \theta}\right) \tag{F}$$

**Table 2.5.1-10** Components of the rate of deformation tensor in spherical coordinates

$$D_{rr} = \frac{\partial v_r}{\partial r} \tag{A}$$

$$D_{\theta\theta} = \frac{1}{r}\frac{\partial v_\theta}{\partial \theta} + \frac{v_r}{r} \tag{B}$$

$$D_{\varphi\varphi} = \frac{1}{r \sin \theta}\frac{\partial v_\varphi}{\partial \varphi} + \frac{v_r}{r} + \frac{v_\theta \cot \theta}{r} \tag{C}$$

$$D_{r\theta} = D_{\theta r} = \frac{1}{2}\left[r\frac{\partial}{\partial r}\left(\frac{v_\theta}{r}\right) + \frac{1}{r}\frac{\partial v_r}{\partial \theta}\right] \tag{D}$$

$$D_{r\varphi} = D_{\varphi r} = \frac{1}{2}\left[\frac{1}{r \sin \theta}\frac{\partial v_r}{\partial \varphi} + r\frac{\partial}{\partial r}\left(\frac{v_\varphi}{r}\right)\right] \tag{E}$$

$$D_{\theta\varphi} = D_{\varphi\theta} = \frac{1}{2}\left[\frac{\sin \theta}{r}\frac{\partial}{\partial \theta}\left(\frac{v_\varphi}{\sin \theta}\right) + \frac{1}{r \sin \theta}\frac{\partial v_\theta}{\partial \varphi}\right] \tag{F}$$

## EXERCISE

**2.5.1-1** In a convenient rectangular cartesian coordinate system, derive an expression for $\varphi$ and show that $\varphi$ is arbitrary to a constant.

## 2.5.2 The Navier-Stokes equation for a two-dimensional flow
In Secs. 1.3.6 and 1.3.7 we expressed the velocity components for a two-dimensional motion of an incompressible fluid in terms of a stream function $\psi$. In this way, the equation of continuity is automatically satisfied. Here we examine the result of the introduction of a stream function upon the equation of motion for an incompressible, newtonian fluid where the external force may be expressed in terms of a potential.

When the external force is representable as the gradient of a scalar potential, we may introduce the modified pressure of Sec. 2.5.1 into the Navier-Stokes equation of Sec. 2.4.1 to obtain

$$\rho \frac{\partial \mathbf{v}}{\partial t} + \rho(\nabla \mathbf{v}) \cdot \mathbf{v} = -\nabla \mathscr{P} + \mu \operatorname{div}(\nabla \mathbf{v}) \tag{2-1}$$

If we take the curl of this equation, modified pressure $\mathscr{P}$ is eliminated to yield

$$\frac{\partial (\operatorname{curl} \mathbf{v})}{\partial t} + \operatorname{curl}([\nabla \mathbf{v}] \cdot \mathbf{v}) = \nu \operatorname{div}(\nabla [\operatorname{curl} \mathbf{v}]) \tag{2-2}$$

Here $\nu$ is the kinematic viscosity:

$$\nu \equiv \frac{\mu}{\rho} \tag{2-3}$$

In any coordinate system for which the velocity vector has only two nonzero components, Eq. (2-2) has only one nonzero component. The nonzero component expressed in terms of the stream function is presented for several situations in Table 2.5.2-1.

Another approach to the differential equations of Table 2.5.2-1 is to recognize that in any two-dimensional flow Eq. (2-1) will have only two nonzero components. Modified pressure may be eliminated between these two equations by recognizing that

$$\frac{\partial^2 \mathscr{P}}{\partial x^i \, \partial x^j} = \frac{\partial^2 \mathscr{P}}{\partial x^j \, \partial x^i} \tag{2-4}$$

The velocity components in the resulting differential equation may be expressed in terms of a stream function.

## REFERENCES

1. Bird, R. B., W. E. Stewart, and E. N. Lightfoot: "Transport Phenomena," 7th printing, Wiley, New York, 1960.
2. Goldstein, S.: "Modern Developments in Fluid Dynamics," Oxford University Press, London, 1938.

**Table 2.5.2-1  The stream function**

| Coordinate system | Assumed form of velocity distribution | Velocity components | Nonzero component of Eq. (2-2)† | Operator |
|---|---|---|---|---|
| Rectangular cartesian | $v_3 = 0$ <br> $v_1 = v_1(z_1, z_2)$ <br> $v_2 = v_2(z_1, z_2)$ | $v_1 = \dfrac{\partial \psi}{\partial z_2}$ <br> $v_2 = -\dfrac{\partial \psi}{\partial z_1}$ | $\dfrac{\partial}{\partial t}(E^2\psi) - \dfrac{\partial(\psi, E^2\psi)}{\partial(z_1, z_2)} = \nu E^4\psi$ | $E^2 = \dfrac{\partial^2}{\partial z_1^2} + \dfrac{\partial^2}{\partial z_2^2}$ <br> $E^4\psi = E^2(E^2\psi) = \left(\dfrac{\partial^4}{\partial z_1^4} + 2\dfrac{\partial^4}{\partial z_1^2 \partial z_2^2} + \dfrac{\partial^4}{\partial z_2^4}\right)\psi$ |
| Cylindrical | $v_z = 0$ <br> $v_r = v_r(r, \theta)$ <br> $v_\theta = v_\theta(r, \theta)$ | $v_r = \dfrac{1}{r}\dfrac{\partial \psi}{\partial \theta}$ <br> $v_\theta = -\dfrac{\partial \psi}{\partial r}$ | $\dfrac{\partial}{\partial t}(E^2\psi) - \dfrac{1}{r}\dfrac{\partial(\psi, E^2\psi)}{\partial(r, \theta)} = \nu E^4\psi$ | $E^2 = \dfrac{\partial^2}{\partial r^2} + \dfrac{1}{r}\dfrac{\partial}{\partial r} + \dfrac{1}{r^2}\dfrac{\partial^2}{\partial \theta^2}$ |
| Cylindrical | $v_\theta = 0$ <br> $v_r = v_r(r, z)$ <br> $v_z = v_z(r, z)$ | $v_r = \dfrac{1}{r}\dfrac{\partial \psi}{\partial z}$ <br> $v_z = -\dfrac{1}{r}\dfrac{\partial \psi}{\partial r}$ | $\dfrac{\partial}{\partial t}(E^2\psi) - \dfrac{1}{r}\dfrac{\partial(\psi, E^2\psi)}{\partial(r, z)} - \dfrac{2}{r^2}\dfrac{\partial \psi}{\partial z} E^2\psi = \nu E^4\psi$ | $E^2 = \dfrac{\partial^2}{\partial r^2} - \dfrac{1}{r}\dfrac{\partial}{\partial r} + \dfrac{\partial^2}{\partial z^2}$ |
| Spherical | $v_\varphi = 0$ <br> $v_r = v_r(r, \theta)$ <br> $v_\theta = v_\theta(r, \theta)$ | $v_r = \dfrac{1}{r^2 \sin\theta}\dfrac{\partial \psi}{\partial \theta}$ <br> $v_\theta = -\dfrac{1}{r \sin\theta}\dfrac{\partial \psi}{\partial r}$ | $\dfrac{\partial}{\partial t}(E^2\psi) - \dfrac{1}{r^2 \sin\theta}\dfrac{\partial(\psi, E^2\psi)}{\partial(r, \theta)} + \dfrac{2E^2\psi}{r^2 \sin^2\theta}\left(\dfrac{\partial \psi}{\partial r}\cos\theta - \dfrac{1}{r}\dfrac{\partial \psi}{\partial \theta}\sin\theta\right) = \nu E^4\psi$ | $E^2 = \dfrac{\partial^2}{\partial r^2} + \dfrac{\sin\theta}{r^2}\dfrac{\partial}{\partial \theta}\left(\dfrac{1}{\sin\theta}\dfrac{\partial}{\partial \theta}\right)$ |

*Source:* This table is taken from Bird, Stewart, and Lightfoot [1, Table 4.2-1] and S. Goldstein [2, p. 114]. Goldstein presents as well relations for axisymmetric flows with a nonzero component of velocity around the axis.

† The Jacobian notation signifies $\dfrac{\partial(f, g)}{\partial(x, y)} = \begin{vmatrix} \dfrac{\partial f}{\partial x} & \dfrac{\partial f}{\partial y} \\ \dfrac{\partial g}{\partial x} & \dfrac{\partial g}{\partial y} \end{vmatrix}$

# 3
# Applications of the Differential Balances to Momentum Transfer

After considerable preparation (Appendix A, Chaps. 1 and 2), we are ready to describe the detailed motions of materials in particular geometries. Try not to be discouraged if some of our initial examples appear to be too simple. The simple problems are there in order to allow you to gain both facility and confidence.

The objection is sometimes raised that studying fluid mechanics, such as we are here, is really useless, since the problems that can be solved are trivial. A statement such as this is normally made as an exaggeration in order to make a point. The point to be made usually is that the interesting problems require numerical solutions or that the interesting problems are susceptible only to approximate solutions.

There is an element of truth here. My feeling is that the best way to be introduced to this material is through the use of problems that can be solved analytically. Concepts and techniques can be polished either in your office or at home, no matter what the time of day may be, since no time is required at the keypunch or computer terminal. Yet these problems are required preparation for those who wish to go on to study more sophisticated problems numerically. It seems unreasonable to attack an involved problem, if the straightforward ones are still giving you trouble. As a further practical matter, analytic solutions for limiting cases are highly desirable in order to check the validity of whatever numerical work is performed.

Now what about the statement that the most interesting problems are susceptible only to approximate solutions? There are at least four classes of approximations that can be easily distinguished.

1. Sometimes the physical problem in which we are primarily interested is too difficult for us to handle. One answer is to replace it by a problem that has most of the important features of our original problem, but which is sufficiently simple for us to analyze. A good example here is provided by flow through a tube. From a practical point of view, we are always concerned with flow through finite tubes. But sometimes the entrance and exit regions are of lesser importance to us and the real problem can be replaced by an idealized one in which entrance and exit effects are negligible: flow through a tube of infinite length.
2. Even after such an idealization of our original physical problem, it may still be too difficult. We may wish to consider a limiting case in which one or more terms in a differential equation are neglected. I have tried to place special emphasis in this text upon the way in which one should argue in order to arrive at such an approximation. Note in particular the discussions of creeping flow, potential flow, and boundary-layer theory.
3. Many times our requirements do not demand detailed solutions of the differential balances. Perhaps we are only interested in some type of integral average. The approach to integral averages is explored in some detail in Chap. 4.
4. Finally, there is the question of mathematical approximations. This is really the problem with which you are primarily concerned in carrying out numerical solutions. A mathematical approximation is applied repeatedly in order to arrive at a solution for a differential equation. The approximate solution presented by the computer can be made to approach the exact solution as closely as desired by using smaller and smaller step sizes. We do not address ourselves to the problems of numerical analysis in this text.

To summarize, this chapter begins by examining those problems for which analytic solutions to the original differential equation can be developed. We then look at three limiting cases which are themselves worthy of textbooks: creeping flow, potential flow, and boundary-layer theory.

## 3.1 PHILOSOPHY

### 3.1.1 The philosophy of solving problems in fluid mechanics
In Chaps. 1 and 2 we developed a structure by means of which we can predict how a given fluid will behave in a particular situation. Our intention here is to discuss how one uses this structure to solve problems.

The first step is to decide just what problem it is that you wish to solve. This means in part that one must choose a particular model by means of which the stress

## [3.1] PHILOSOPHY

deformation behavior of the fluid may be represented; for example, you might wish to study an incompressible newtonian fluid. To complete the specification of the problem, one must indicate the geometry through which the material is to move and the forces which cause the fluid to move. We might wish to study the flow of our incompressible fluid through a horizontal tube in a uniform gravitational field when the stresses indicated by two pressure gauges mounted on the tube at different axial positions are given. (Hereafter, we assume the external force is due to a uniform gravitational field, unless we specifically state otherwise.) The specification of stress at two points on the tube wall is one example of what we term boundary conditions.

By a boundary condition all we generally mean is that one of the variables in the problem is specified or restricted in some fashion at some point in the geometry, usually a point on one of the boundary surfaces. We might have a boundary condition involving the stress tensor as above, or we might say something about the velocity vector. There are several common types of boundary conditions for which one should look in considering an unfamiliar problem.

1. We shall always assume that tangential components of velocity are continuous at a phase interface. This appears to be an excellent assumption, even though we might be dealing with a liquid-solid interface such that the liquid does not "wet" the surface. There is no evidence of slip at the wall when mercury flows through glass tubes. For an excellent discussion of the conditions at the surface of contact of a fluid with a solid body, see Goldstein [1, p. 676].

    For a restricted theoretical argument that supports the concept of adhesion at a phase interface, see Exercise 2.2.3-3. Continuity of the tangential components of velocity at a phase interface is also suggested by visualizing that in a sense local equilibrium is established at a phase boundary. In the limit of a stable equilibrium, we learn in Sec. 8.5.3 that the tangential components of velocity are continuous across an interface.

2. The jump mass balance discussed in Sec. 1.3.5 must be satisfied at every phase interface. This may be used to relate the normal components of velocity in each phase at an interface.

3. The jump momentum balance discussed in Exercise 2.2.3-1 must be satisfied at every phase interface. It is common in solving for the velocity and phase distributions within a fluid to ignore the stress distribution in bounding solid walls. When this is done, we employ the jump momentum balance at fluid-fluid phase interfaces exclusively.

4. The components of the stress tensor are finite at all points in the fluid. This is a special case of a theorem which requires that the components of all second-order tensors be finite [2, p. 140].

5. We assume that the components of velocity remain finite at all points in the fluid.

Having specified what it is that we wish to describe, we are faced with a strictly mathematical problem of solving several partial differential equations simultaneously to find a solution that is consistent with the boundary conditions. This is not easy, even in the relatively simple case of an incompressible newtonian fluid, since no general solution of the Navier-Stokes equation, Eq. (1-6) of Sec. 2.4.1, consistent with the equation of continuity, Eq. (3-3) of Sec. 1.3.3, is known. One must accept the disappointing fact that at the present time we are not capable of solving every fluid mechanics problem. This is true even though high-speed computers are available for help. Relatively easy problems can be solved and many difficult problems can be approximated by simpler problems which in turn can be solved. In the next few sections, we shall look at examples of relatively simple problems.

One point is worth keeping in mind in reading this chapter. We do not make any claims that we are looking for a unique solution. Some uniqueness theorems have been proved [3, 4], but the general case has not been treated as yet. We shall take the approach here of most workers and merely ask for "a" solution. Many times, experiments will suggest that the solution we find is unique, but this is *not always* the case.

## REFERENCES

1. Goldstein, S.: "Modern Developments in Fluid Dynamics," Oxford University Press, London, 1938.
2. Stakgold, Ivar: "Boundary Value Problems of Mathematical Physics," vol. 1, Macmillan, New York, 1967.
3. Ladyzhenskaya, O. A.: "The Mathematical Theory of Viscous Incompressible Flow," Gordon and Breach, New York, 1963.
4. Finn, Robert, and Walter Noll: *Arch. Rational Mech. Analysis*, **1**:97 (1957).

## 3.2 COMPLETE SOLUTIONS

### 3.2.1 Flow of an incompressible newtonian fluid through a tube

As our first example of a problem for which an exact solution can be found, consider the steady-state flow of an incompressible newtonian fluid through a horizontal tube of radius $R$. It is assumed that we are working with a section of the tube which is so removed from the entrance and exit that disturbances originating at the entrance and exit can be neglected. This is equivalent to saying that the tube is infinitely long. In terms of a cylindrical coordinate system whose $z$ axis coincides with the axis of the tube, we have two pressure gauges mounted on the tube which measure the $r$ component of the force per unit area which the fluid exerts on the tube wall, $-T_{rr}$:

$$\text{At } z = 0, r = R, \theta = 0: \quad -T_{rr} = P_0 \tag{1-1}$$

$$\text{At } z = L, r = R, \theta = 0: \quad -T_{rr} = P_L \tag{1-2}$$

As further boundary conditions, we can state that at the tube wall in Fig. 3.2.1-1 the velocity vector must be zero:

$$\text{At } r = R: \quad \mathbf{v} = 0 \tag{1-3}$$

## [3.2] COMPLETE SOLUTIONS

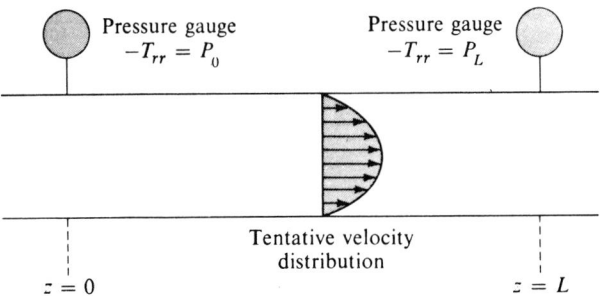

**Fig. 3.2.1-1** Flow through a tube.

As mentioned in Section 3.1.1, we cannot start with a general solution of the Navier-Stokes equation, Eq. (1-6) of Sec. 2.4.1, and look for the particular conditions under which this general solution satisfies the specified boundary conditions. Instead, our approach is to use our intuition to say something about the form of a solution to the problem. The important thing to remember is that we are looking for one solution which satisfies the Navier-Stokes equation, the equation of continuity, and the required boundary conditions. We will not worry about the question of uniqueness.

Usually one's intuition can be assisted markedly by a simple sketch. For example, in Figure 3.2.1-1 we suggest that the flow is in the axial direction and that, if the velocity is to be zero at the tube wall, it must reach a maximum at the centerline of the tube. This suggests that we might assume

$$v_r = v_\theta = 0 \tag{1-4}$$
$$v_z = v_z(r,z) \tag{1-5}$$

and see whether we can find a solution to the Navier-Stokes equation and the equation of continuity which satisfies the boundary conditions in Eqs. (1-1) to (1-3).

Remembering that the fluid is incompressible, we see that the equation of continuity, Eq. (B) in Table 2.5.1-1, requires that

$$\frac{\partial v_z}{\partial z} = 0 \quad \text{or} \quad v_z = v_z(r) \tag{1-6}$$

From Eq. (2-20) of Sec. 2.3.2 for an incompressible newtonian fluid, from the physical components of the rate of deformation tensor for cylindrical coordinates in Table 2.5.1-9, and from Eqs. (1-4) and (1-6), we know that the only nonzero components of the extra stress tensor are $S_{rz} = S_{zr}$:

$$S_{rz} = \mu \frac{dv_z}{dr} \tag{1-7}$$

From Table 2.5.1-4, the three components of Cauchy's first law for a uniform gravitational field, Eq. (1-3) of Sec. 2.5.1, become

$$0 = -\frac{\partial \mathscr{P}}{\partial r} \tag{1-8}$$

$$0 = -\frac{1}{r}\frac{\partial \mathscr{P}}{\partial \theta} \tag{1-9}$$

and

$$\frac{\partial \mathscr{P}}{\partial z} = \frac{1}{r}\frac{d}{dr}(rS_{rz}) \tag{1-10}$$

Here $\mathscr{P}$ is the modified pressure

$$\mathscr{P} = p + \rho\varphi \tag{1-11}$$

and $\varphi$ is the gravitational potential. We showed in Exercise 2.5.1-1 that $\varphi$ was arbitrary to a constant; let us define

$$\text{At } r = R,\ \theta = 0: \qquad \varphi = 0 \tag{1-12}$$

Equations (1-8) and (1-9) imply that $\mathscr{P}$ is a function only of $z$. But the term on the right of Eq. (1-10) is a function only of $r$, while the term on the left is a function only of $z$. This can be true only if

$$\frac{d\mathscr{P}}{dz} = A = \text{constant} \tag{1-13}$$

Integrating this equation using Eqs. (1-1) and (1-2) as boundary conditions [remember that $S_{rr}$ is zero as the result of Eqs. (1-4) and (1-6)], we have

$$-A = -\frac{d\mathscr{P}}{dz} = \frac{P_0 - P_L}{L} \tag{1-14}$$

Having eliminated $d\mathscr{P}/dz$ between Eqs. (1-10) and (1-14), we may integrate:

$$-\frac{(P_0 - P_L)}{L}\int_0^r r\,dr = \int_0^{rS_{rz}} d(rS_{rz})$$

$$-\frac{(P_0 - P_L)}{L}\frac{r}{2} = S_{rz} \tag{1-15}$$

Our only assumption here is that $S_{rz}$ is finite at $r = 0$; this is an example of the type 4 boundary condition discussed in Section 3.1.1. Substituting for $S_{rz}$ from Eq. (1-7), we may again integrate using Eq. (1-3) as the boundary condition:

$$-\frac{(P_0 - P_L)}{2\mu L}\int_R^r r\,dr = \int_0^{v_z} dv_z$$

$$v_z = \frac{(P_0 - P_L)R^2}{4\mu L}\left[1 - \left(\frac{r}{R}\right)^2\right] \tag{1-16}$$

[3.2] COMPLETE SOLUTIONS

We see that this equation confirms the parabolic velocity distribution we sketched in Fig. 3.2.1-1 on the basis of our intuition. This means that our initial assumptions in Eqs. (1-4) and (1-5) were justified and that there is a solution to the problem of the form assumed.

The maximum velocity occurs on the centerline of the tube, where

$$v_{z(\max)} = \frac{(P_0 - P_L)R^2}{4\mu L} \tag{1-17}$$

The volume rate of flow $Q$ is easily computed to be

$$Q = \int_0^{2\pi}\int_0^R v_z r\, dr\, d\theta = 2\pi R^2 \frac{(P_0 - P_L)R^2}{4\mu L} \int_0^1 (1 - x^2)x\, dx$$

$$= \frac{\pi(P_0 - P_L)R^4}{8\mu L} \tag{1-18}$$

Equation (1-18) is often referred to as Poiseuille's law. Its general form was published in 1841 by Jean Louis Poiseuille, a physician interested in experimental physiology. He presented it as an empirical relationship correlating data for the flow of water through glass capillary tubes [1, p. 160].

Notice that, since $S_{rr} = 0$ by reason of Eq. (1-4),

$$-T_{rr} = p - S_{rr}$$
$$= p \tag{1-19}$$

This means that the pressure gauge does indeed measure pressure in a tube, at least under conditions of laminar flow when mounted sufficiently far from either an entrance or an exit.

## REFERENCES

1. Rouse, Hunter, and Simon Ince: "History of Hydraulics," Iowa Institute of Hydraulic Research, State University of Iowa, 1957.
2. Bird, R. B., W. E. Stewart, and E. N. Lightfoot: "Transport Phenomena," 7th printing, Wiley, New York, 1960.
3. Paton, J. B., P. H. Squires, W. H. Darnell, F. M. Cash, and J. F. Carley: Chap. 4 in E. C. Burnhardt (ed.), "Processing of Thermoplastic Materials," Reinhold, New York, 1959.

## EXERCISES

**3.2.1-1** Justify Eq. $(1-19)_1$ (see footnote 1) starting from Eq. (2-20) of Sec. 2.3.2.

**3.2.1-2** Show how the discussion in this section is modified when the tube is inclined at an angle $\alpha$ with respect to the horizon as illustrated in Fig. 3.2.1-2.
*Answer*

$$-\frac{d\mathscr{P}}{dz} = -A = \frac{P_0 - P_L - \rho g L \sin \alpha}{L}$$

---

[1] Subscript numbers after equation numbers refer to the lines in the equation; therefore, subscript 1 here denotes the first line of Eq. (1-19).

74    APPLICATIONS OF THE DIFFERENTIAL BALANCES TO MOMENTUM TRANSFER

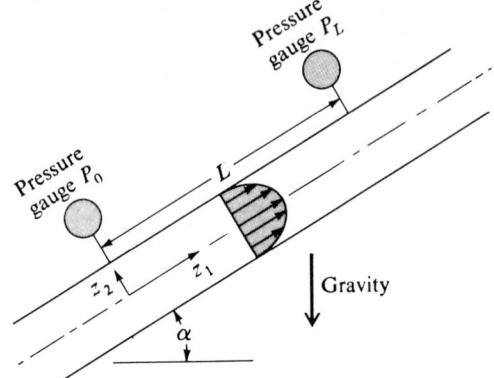

**Fig. 3.2.1-2** Flow through an inclined tube.

**3.2.1-3** *Power-model flow through a tube* Derive the analog of Poiseuille's law for a power-model fluid.

*Answer*

$$v_z = \frac{n}{1+n}\left(\frac{-AR^{1+n}}{2m}\right)^{1/n}\left[1-\left(\frac{r}{R}\right)^{(1+n)/n}\right]$$

$$Q = \frac{\pi n}{(1+3n)}\left(\frac{-AR^{1+3n}}{2m}\right)^{1/n}$$

**3.2.1-4** *Ellis model flow through a tube* Derive the analog of Poiseuille's law for an Ellis model fluid.

*Answer*

$$v_z = \frac{-AR^2}{4\eta_0}\left[1-\left(\frac{r}{R}\right)^2\right] + \left(\frac{-A}{2\tau_{\frac{1}{2}}}\right)^\alpha \frac{\tau_{\frac{1}{2}} R^{\alpha+1}}{\eta_0(\alpha+1)}\left[1-\left(\frac{r}{R}\right)^{\alpha+1}\right]$$

$$Q = \frac{-A\pi R^4}{8\eta_0} + \frac{\pi \tau_{\frac{1}{2}} R^{\alpha+3}}{\eta_0(\alpha+3)}\left(\frac{-A}{2\tau_{\frac{1}{2}}}\right)^\alpha$$

**3.2.1-5** *Bingham plastic flow through a tube* Derive the analog of Poiseuille's law for a Bingham plastic fluid.

*Answer*

For $\dfrac{-AR}{2} < \tau_0$:    no flow

For $\dfrac{-Ar}{2} > \tau_0$:    $v_z = \dfrac{-AR^2}{4\eta_0}\left[1-\left(\dfrac{r}{R}\right)^2\right] - \dfrac{\tau_0 R}{\eta_0}\left(1-\dfrac{r}{R}\right)$

For $\dfrac{-AR}{2} > \tau_0$:    $Q = \dfrac{-A\pi R^4}{8\eta_0}\left[1-\left(\dfrac{2\tau_0}{-AR}\right)^4\right] - \dfrac{\pi R^3 \tau_0}{3\eta_0}\left[1-\left(\dfrac{2\tau_0}{-AR}\right)^3\right]$

**3.2.1-6** *Newtonian flow in a wire-coating die* [2, p. 65; 3] A somewhat simplified picture of a wire-coating die is shown in Fig. 3.2.1-3a. The wire is assumed to be coaxial with the cylindrical die and moving axially with a speed $V$. The reservoir at the left is assumed to be filled with a liquid coating, taken here to be a newtonian fluid. We wish to determine the steady-state velocity distribution, the

## [3.2] COMPLETE SOLUTIONS

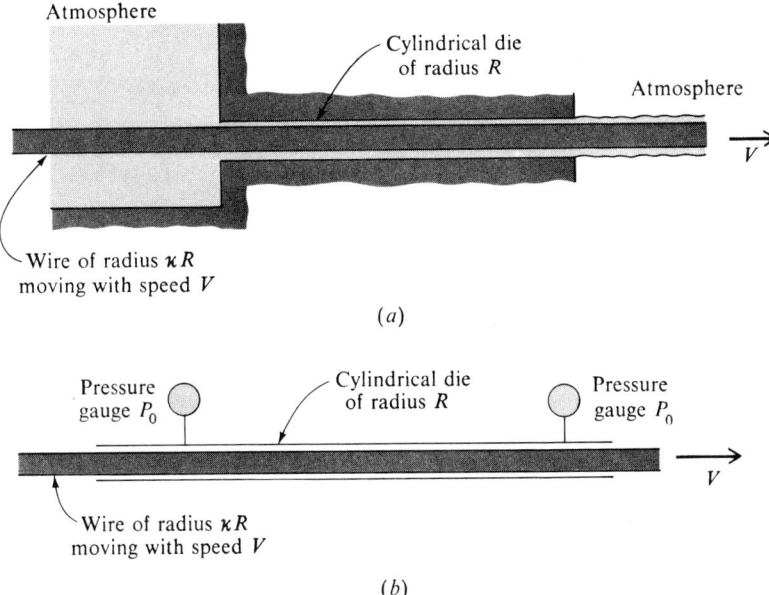

**Fig. 3.2.1-3**  (a) Flow in a wire-coating die; (b) flow in an idealized wire-coating die.

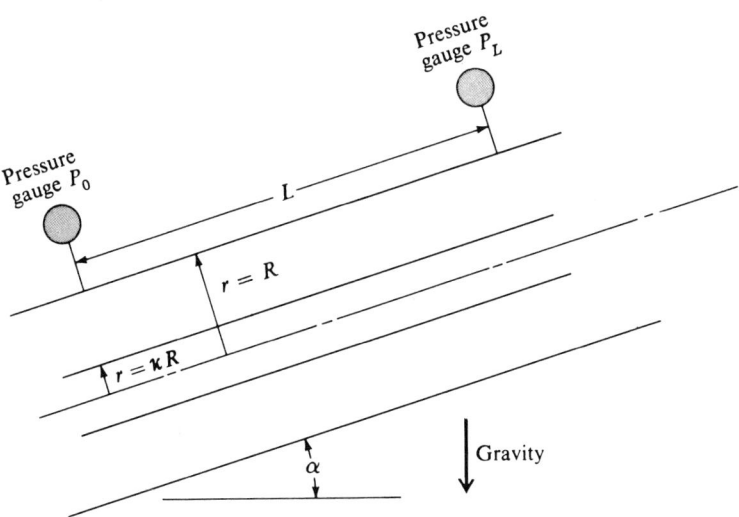

**Fig. 3.2.1-4**  Flow through an inclined annulus.

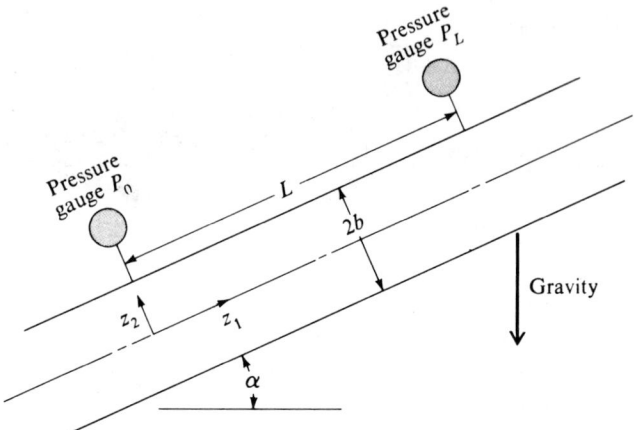

**Fig. 3.2.1-5** Flow through an inclined channel.

volume rate of flow of the fluid in the annular region, and the force per unit length required to pull the wire through the die.

This is a difficult problem to analyze as I have presented it. So let us neglect the "end effects" and determine the quantities requested above for the idealized flow shown in Fig. 3.2.1-3b.

**3.2.1-7** *Newtonian flow through an annulus* An incompressible newtonian fluid flows through the inclined annulus shown in Fig. 3.2.1-4. Determine the velocity distribution in the annular region and the volume rate of flow through the annulus.

Answer

$$v_z = \frac{-AR^2}{4\mu}\left[1 - \left(\frac{r}{R}\right)^2 + \frac{1-\kappa^2}{\ln(1/\kappa)} \ln \frac{r}{R}\right]$$

$$Q = \frac{-A\pi R^4}{8\mu}\left[1 - \kappa^4 - \frac{(1-\kappa^2)^2}{\ln(1/\kappa)}\right]$$

$$-A = \frac{P_0 - P_L - \rho g L \sin \alpha}{L}$$

**3.2.1-8** *Newtonian flow through a channel* An incompressible newtonian fluid flows through the channel of width $2b$ shown in Fig. 3.2.1-5 (two parallel planes of infinite extent separated by a distance $2b$). Determine the velocity distribution and the volume rate of flow per unit width of channel.

Answer

$$v_1 = \frac{-Ab^2}{2\mu}\left[1 - \left(\frac{z_2}{b}\right)^2\right]$$

$$Q = -\frac{2}{3}\frac{Ab^3}{\mu}$$

$$-A \equiv \frac{P_0 - P_L}{L} - \rho g \sin \alpha$$

**3.2.1-9** Repeat Exercise 3.2.1-6 for a power-model fluid.

**3.2.1-10** *Ellis model flow in a wire-coating die* Repeat Exercise 3.2.1-6 for an Ellis model fluid, but assume now that the axial component $F_z$ of the force per unit length required to pull the wire through the die is given and the corresponding speed $V$ of the wire is to be determined.

How does your result simplify for a newtonian fluid (Exercise 3.2.1-6) and for a power-model fluid (Exercise 3.2.1-9)?

*Answer*

$$V = \frac{F_z}{2\pi L \eta_0} \ln \frac{1}{\kappa} + \frac{1}{\eta_0(1-\alpha)} \left(\frac{F_z}{2\pi L}\right)^\alpha (\kappa R \tau_{\frac{1}{2}})^{1-\alpha} \left[\left(\frac{1}{\kappa}\right)^{1-\alpha} - 1\right]$$

## 3.2.2 Flow of an incompressible viscoelastic fluid through a tube

Here we wish to repeat the problem of Sec. 3.2.1 for an incompressible viscoelastic fluid whose material behavior may be described in the simple manner discussed in Sec. 2.3.3. We assume that entrance and exit effects may be neglected, but we take the tube to be inclined with respect to the horizon as shown in Fig. 3.2.1-2.

We begin by looking for a velocity distribution of the form

$$v_r = v_\theta = 0 \qquad v_z = v_z(r) \tag{2-1}$$

As pointed out in the preceding section, the equation of continuity is automatically satisfied in this manner.

From Eqs. (2-1) and Table 2.5.1-9, we see that there is only one nonzero component of the rate of deformation tensor **D** in cylindrical coordinates:

$$D_{rz} = \frac{1}{2} \frac{dv_z}{dr} \tag{2-2}$$

From Sec. 2.3.3 we see that

$$\mathbf{S} = 2\eta(\gamma)\mathbf{D} \tag{2-3}$$

which means that there can be only one nonzero component of the extra-stress tensor **S**: $S_{rz}$. Table 2.5.1-4 indicates that the three components of Cauchy's first law for a uniform gravitational field reduce to the same form as we found in Sec. 3.2.1. This means that $\mathscr{P}$ is only a function of $z$:

$$\frac{d\mathscr{P}}{dz} = A = \text{constant} \tag{2-4}$$

and

$$S_{rz} = Ar/2 \tag{2-5}$$

From Exercise 2.5.1-1 we have that $\varphi$ is arbitrary to a constant. Let us define

At $r = R$, $\theta = 0$, $z = 0$: $\qquad \varphi = 0 \tag{2-6}$

We see from Fig. 3.2.1-2 that, with respect to the rectangular cartesian coordinate system indicated,

$$-\frac{\partial \varphi}{\partial z_2} = f_2 = -g \cos \alpha \tag{2-7}$$

and

$$-\frac{\partial \varphi}{\partial z_1} = f_1 = -g \sin \alpha \qquad (2\text{-}8)$$

The potential energy $\varphi$ may be determined from this information by a line integration:

$$\varphi = \int_0^\varphi d\varphi = \int_R^{z_2} \frac{\partial \varphi}{\partial z_2}\bigg|_{z_1=0} dz_2 + \int_0^{z_1} \frac{\partial \varphi}{\partial z_1} dz_1$$

$$= -(R - z_2)g \cos \alpha + z_1 g \sin \alpha \qquad (2\text{-}9)$$

In terms of cylindrical coordinates, this becomes

$$\varphi = -(R - r \cos \theta)g \cos \alpha + zg \sin \alpha \qquad (2\text{-}10)$$

We are now in a position to integrate Eq. (2-4) in order to determine $A$. Since $S_{rr} = 0$, the boundary conditions on stress, given in Sec. 3.2.1, reduce to

At $z = 0, r = R, \theta = 0$: $\quad p = P_0 \qquad (2\text{-}11)$

and

At $z = L, r = R, \theta = 0$: $\quad p = P_L \qquad (2\text{-}12)$

Equation (2-10) allows us to write these as boundary conditions on modified pressure:

At $z = 0, r = R, \theta = 0$: $\quad \mathscr{P} = P_0 \qquad (2\text{-}13)$

At $z = L, r = R, \theta = 0$: $\quad \mathscr{P} = P_L + \rho g L \sin \alpha \qquad (2\text{-}14)$

Equations (2-13) and (2-14) may be used in integrating Eq. (2-4) to find

$$-A = \frac{P_0 - P_L - \rho g L \sin \alpha}{L} \qquad (2\text{-}15)$$

Compare this with the similar result in Sec. 3.2.1.
From Sec. 2.3.3 we have

$$2\mathbf{D} = \varphi(\tau)\mathbf{S} \qquad (2\text{-}16)$$

For the situation considered,

$$\tau \equiv \sqrt{\tfrac{1}{2} \operatorname{tr} \mathbf{S}^2} = |S_{rz}| \qquad (2\text{-}17)$$

With the help of Eqs. (2-2), (2-5), and (2-17) we may write the only nonzero component of Eq. (2-16) as

$$\frac{dv_z}{dr} = \varphi\left(\frac{|A|r}{2}\right)\frac{Ar}{2} \qquad (2\text{-}18)$$

Using the boundary condition for velocity at the wall of the tube, we may integrate this to obtain for the velocity distribution

$$v_z = \int_r^R \varphi\left(\frac{|A|r}{2}\right)\left[-\frac{Ar}{2}\right] dr \qquad (2\text{-}19)$$

## [3.2] COMPLETE SOLUTIONS

If we require flow to be in the positive $z$ direction as indicated in Fig. 3.2.1-2, it is clear that $-A$ in Eq. (2-15) must be positive, so that we may write

$$v_z = \int_r^R \varphi\left(\frac{-Ar}{2}\right)\left[\frac{-Ar}{2}\right] dr \tag{2-20}$$

Equations (2-4), (2-15), and (2-20) satisfy the equation of continuity, Cauchy's first law, and the constitutive equation for the stress tensor, as well as the required boundary conditions on the velocity vector and stress tensor. This means that we have found a solution to this problem which is consistent with our initial assumptions concerning the velocity distribution.

It is easy to measure experimentally the pressure gradient and the volume rate of flow:

$$Q = 2\pi \int_0^R v_z r \, dr \tag{2-21}$$

It would be useful to be able to determine directly from these measurements the function $\varphi(\tau)$ of Eq. (2-16) or the apparent viscosity function $\eta(\gamma)$ in the equivalent constitutive equation, Eq. (2-3). As we showed in Sec. 2.3.3, we may derive from Eqs. (2-3) and (2-16)

$$\tau = \eta(\gamma)\gamma \tag{2-22}$$

and

$$\gamma = \varphi(\tau)\tau \tag{2-23}$$

Equations (2-5) and (2-15) indicate that we may interpret our pressure measurements in terms of $\tau$ evaluated at the wall of the tube:

$$\tau_R \equiv \tau|_{r=R} = -S_{rz}|_{r=R} = -\frac{AR}{2} \tag{2-24}$$

Our problem is solved if we can measure $\gamma$ at the tube wall. An integration by parts allows us to express Eq. (2-21) as

$$Q = -\pi \int_0^R \frac{dv_z}{dr} r^2 \, dr \tag{2-25}$$

Let us make the change of variable

$$\tau = \frac{r}{R}\tau_R \tag{2-26}$$

in Eq. (2-25) in order that we can write it as

$$Q = -\pi \left(\frac{R}{\tau_R}\right)^3 \int_0^{\tau_R} \frac{dv_z}{dr} \tau^2 \, d\tau \tag{2-27}$$

This in turn may be differentiated with respect to $\tau_R$ to find

$$\frac{d(Q\tau_R^3)}{d\tau_R} = -\pi R^3 \tau_R^2 \left.\frac{dv_z}{dr}\right|_{r=R} \tag{2-28}$$

or

$$\gamma_R \equiv \gamma|_{r=R} = -\left.\frac{dv_z}{dr}\right|_{r=R} = \frac{4Q}{\pi R^3}\left(\frac{3}{4} + \frac{1}{4}\frac{d \ln Q}{d \ln \tau_R}\right) \tag{2-29}$$

From Sec. 3.2.1 we find that

For a newtonian fluid: $\quad \gamma_R = \dfrac{4Q}{\pi R^3}$ \hfill (2-30)

This means that we can interpret the term in parentheses on the right side of Eq. (2-29) as a correction for viscoelastic behavior.

Equations (2-15), (2-24), and (2-29) allow us to prepare a plot of $\tau_R$ as a function of $\gamma_R$ from experimental measurements of pressure drop as a function of volume rate of flow. This plot of $\tau_R$ versus $\gamma_R$ may then be compared with either Eq. (2-22) or (2-23) in order to determine the function $\eta(\gamma)$ or the function $\varphi(\tau)$.

If, as indicated in Fig. 3.2.2-1a, a log-log plot of the experimental data indicates

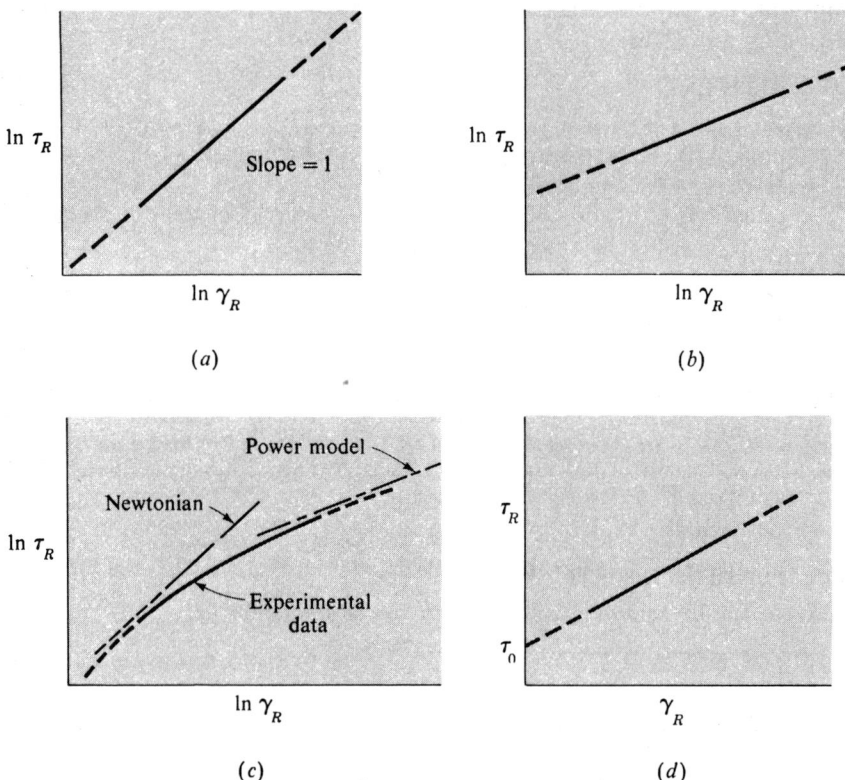

**Fig. 3.2.2-1** (a) Incompressible newtonian model; (b) power model; (c) Ellis model, modified Ellis model, or Sisko model; (d) Bingham plastic model.

a straight line whose slope is unity, the fluid will be described by the incompressible newtonian model. Figure 3.2.2-1b is meant to depict a plot of experimental data that can be represented by a straight line whose slope is other than unity. Such behavior is well described by the power model. A very common situation is illustrated in Fig. 3.2.2-1c. The experimental data appear to be asymptotic to newtonian behavior for small values of $\gamma_R$ and asymptotic to power-model behavior for large values of $\gamma_R$. One should attempt to fit such data with the Ellis model, modified Ellis model, or Sisko model. Ashare, Bird, and Lescarboura [1] have suggested a convenient manner for fitting the three-parameter Ellis model to experimental data. Occasionally a limited range of experimental data will appear to be well described by a linear relationship between $\tau_R$ and $\gamma_R$. Such a case is depicted in Fig. 3.2.2-1d. The Bingham plastic model will describe very well these experimental data, though one should be cautious in extrapolating the results to either larger or smaller values of $\gamma_R$.

The analysis given here can be extended to a more general description of material behavior, the Noll simple fluid introduced in Sec. 2.3.4. For a discussion of the Noll simple fluid and an important class of problems that can be analyzed for this model of material behavior, see an excellent book by Coleman, Markovitz, and Noll [2].

## REFERENCES

1. Ashare, Edward, R. B. Bird, and J. A. Lescarboura: *A.I.Ch.E.J.*, **11**: 910 (1965).
2. Coleman, B. D., H. Markovitz, and W. Noll: "Viscometric Flows of Non-Newtonian Fluids," Springer-Verlag, New York, 1966.

## EXERCISE

**3.2.2-1** *Couette flow* Consider an incompressible fluid whose stress tensor is determined by the most general function of the rate of deformation tensor which is consistent with the principle of material frame indifference (see Sec. 2.3.2):

$$\mathbf{T} = \varkappa_0 \mathbf{I} + \varkappa_1 \mathbf{D} + \varkappa_2 \mathbf{D} \cdot \mathbf{D}$$

Here $\varkappa_0$, $\varkappa_1$, and $\varkappa_2$ are *unknown* functions of $\mathbf{I_D}$, $\mathbf{II_D}$, and $\mathbf{III_D}$. This fluid is contained between the two horizontal parallel plates shown in Fig. 3.2.2-2. The upper plate moves in the $z_1$ direction with a constant speed $V$; the lower plate is stationary.

(a) Assume

$$v_1 = v_1(z_2), \qquad v_2 = v_3 = 0$$

and determine the nonzero components of the rate of deformation tensor.

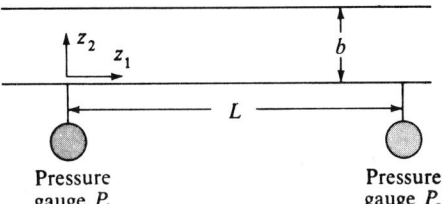

Fig. 3.2.2-2 Couette flow between parallel plates.

(b) Find that the only nonzero components of the stress tensor are $T_{11}$, $T_{22}$, $T_{33}$, $T_{12}$, and $T_{21}$.

(c) Determine that

$$\frac{v_1}{V} = \frac{z_2}{b}$$

(d) If we could study this flow experimentally (unlikely), what measurements would you suggest that we make in order to learn something about the functions $\varkappa_0$, $\varkappa_1$, and $\varkappa_2$?

### 3.2.3 Tangential annular flow of an incompressible Ellis model fluid

Here we consider the flow in an incompressible Ellis model fluid which is trapped between the two concentric cylinders in Fig. 3.2.3-1. The outer cylinder rotates with a constant angular velocity, while the inner cylinder is held stationary. As in the previous problem, we neglect "end effects"; we treat the flow as though the cylinders were infinitely long even though they are a finite length $L$.

This geometry is the basis for one common type of viscometer, an instrument used to study the stress deformation behavior of fluids or to measure the shear viscosity of newtonian fluids. Two measurements are recorded: the angular velocity of the outer cylinder and the axial component of the torque (moment of the force) which the fluid exerts on the inner cylinder. Here we assume that torque is given and that we wish to compute the corresponding angular velocity of the outer cylinder.

Let us begin with the boundary conditions to be satisfied. The inner cylinder is stationary, so that in cylindrical coordinates

At $r = \kappa R$: $\quad \mathbf{v} = 0$ \hfill (3-1)

The moment of the force which the fluid exerts on the inner cylinder, $\mathbf{l}$, may be expressed as

At $r = \kappa R$: $\quad \mathbf{l} = \int_0^{2\pi} \int_0^L \mathbf{p} \wedge (\mathbf{T} \cdot \mathbf{n}) \kappa R \, dz \, d\theta$ \hfill (3-2)

where $\mathbf{p}$ is the position vector with respect to some point on the axis of the cylinders. As pointed out in Exercise A.11.1-1, the integral of a curvilinear component of a vector field generally is not itself a curvilinear component of a vector, although an integral of a rectangular cartesian component of a vector field is a rectangular cartesian component of a vector. Yet here we have a problem where the major portion of our computations are most naturally done in cylindrical coordinates. In terms of a rectangular cartesian coordinate system defined by the relations given in Exercise A.4.1-4, the axial component or $z_3$ component of the torque which the fluid exerts upon the inner cylinder may be expressed as

At $r = \kappa R$: $\quad \bar{l}_3 = \int_0^{2\pi} \int_0^L \bar{a}_3 \kappa R \, dz \, d\theta$ \hfill (3-3)

Here we define

$$\mathbf{a} \equiv \mathbf{p} \wedge (\mathbf{T} \cdot \mathbf{n}) \quad (3\text{-}4)$$

and use overbars to denote rectangular cartesian coordinates. The relation between

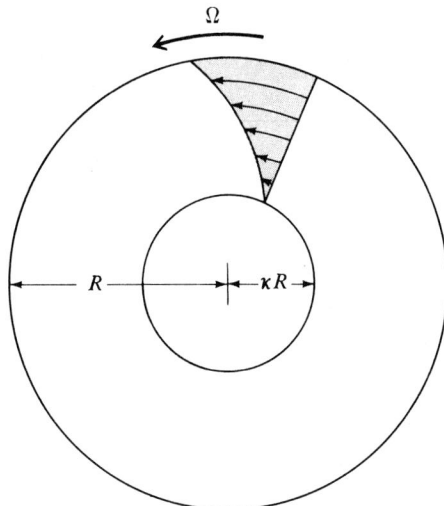

**Fig. 3.2.3-1** Tangential annular flow.

$\bar{a}_3$ and the cylindrical components of **a** is simple (Exercise A.4.1-9):

$$\bar{a}_3 = a_z \tag{3-5}$$

From Exercise A.9.2-1,

$$a_z = a_{\langle 3 \rangle} = e_{3jk} p_{\langle j \rangle} T_{\langle km \rangle} n_{\langle m \rangle} \tag{3-6}$$

Since the only nonzero physical component of **n** in cylindrical coordinates is

$$n_r = n_{\langle 1 \rangle} = 1 \tag{3-7}$$

and since the physical cylindrical components of the position vector are (Exercise A.4.1-11)

$$p_{\langle 1 \rangle} = r \quad p_{\langle 2 \rangle} = 0 \quad p_{\langle 3 \rangle} = z \tag{3-8}$$

Eq. (3-6) reduces to

$$a_z = e_{312} p_{\langle 1 \rangle} T_{\langle 21 \rangle} n_{\langle 1 \rangle} = rT_{\theta r} = rS_{\theta r} \tag{3-9}$$

Equations (3-5) and (3-9) allow us to express Eq. (3-3) as

$$\text{At } r = \kappa R: \quad l_3 = \int_0^{2\pi} \int_0^L S_{\theta r} (\kappa R)^2 \, dz \, d\theta \tag{3-10}$$

Next we must make an assumption about the form of the solution which we are seeking. In Fig. 3.2.3-1 we have sketched what we might imagine the velocity distribution to be like. This suggests assuming in cylindrical coordinates that

$$v_r = v_z = 0 \tag{3-11}$$

and

$$v_\theta = v_\theta(r,\theta) \tag{3-12}$$

Remember that these are assumptions based upon our physical grasp of the problem. We have no guarantee that a solution of this form exists or that, if such a solution exists, it is a unique solution.

The equation of continuity, Eq. (B) in Table 2.5.1-1, indicates that, for an incompressible fluid moving with a velocity distribution of the form suggested by Eqs. (3-11) and (3-12),

$$\frac{\partial v_\theta}{\partial \theta} = 0 \quad \text{or} \quad v_\theta = v_\theta(r) \tag{3-13}$$

From Table 2.5.1-9 and Eqs. (3-11) and (3-13), we see that the only non-zero component of the rate of deformation tensor is

$$D_{r\theta} = \tfrac{1}{2} r \frac{d}{dr}\left(\frac{v_\theta}{r}\right) \tag{3-14}$$

For the Ellis model (see Sec. 2.3.3), we have that

$$2\mathbf{D} = \frac{1}{\eta_0}\left[1 + \left(\frac{\tau}{\tau_{\frac{1}{2}}}\right)^{\alpha-1}\right]\mathbf{S} \tag{3-15}$$

which means that the only nonzero component of the extra-stress tensor is

$$S_{r\theta} = S_{r\theta}(r) \tag{3-16}$$

Accordingly, from Table 2.5.1-4 the three components of Cauchy's first law become

$$\frac{\rho(v_\theta)^2}{r} = \frac{\partial \mathscr{P}}{\partial r} \tag{3-17}$$

$$\frac{\partial \mathscr{P}}{\partial \theta} = \frac{1}{r}\frac{d}{dr}(r^2 S_{r\theta}) \tag{3-18}$$

$$0 = \frac{\partial \mathscr{P}}{\partial z} \tag{3-19}$$

Here we have assumed that the external force per unit mass may be expressed in terms of a potential, and we have introduced the modified pressure $\mathscr{P}$ defined in Sec. 2.5.1.

Equations (3-18) and (3-19) require that

$$\mathscr{P} = \mathscr{P}(r,\theta) = \theta f(r) + g(r) \tag{3-20}$$

Though we did not mention it as a boundary condition, it is clear that all quantities must be periodic functions of $\theta$. In particular,

$$\mathscr{P}(r,\theta) = \mathscr{P}(r, \theta + 2\pi) \tag{3-21}$$

This means that

$$\frac{\partial \mathscr{P}}{\partial \theta} = f(r) = 0 \tag{3-22}$$

[3.2] COMPLETE SOLUTIONS

and Eq. (3-18) reduces to

$$\frac{d}{dr}(r^2 S_{r\theta}) = 0 \tag{3-23}$$

In view of Eq. (3-16), Eq. (3-10) may be simplified to

$$\text{At } r = \kappa R: \quad \bar{I}_3 = 2\pi L (\kappa R)^2 S_{r\theta} \tag{3-24}$$

This is a boundary condition on $S_{r\theta}$ and may be used to integrate Eq. (3-23):

$$\int_{\kappa^2 R^2 S_0}^{r^2 S_{r\theta}} d(r^2 S_{r\theta}) = 0$$

$$S_{r\theta} = \frac{\kappa^2 R^2}{r^2} S_0 \tag{3-25}$$

where

$$S_0 \equiv \frac{\bar{I}_3}{2\pi L \kappa^2 R^2} \tag{3-26}$$

Since the only nonzero component of the extra-stress tensor is $S_{r\theta}$,

$$\tau \equiv \sqrt{\tfrac{1}{2} \operatorname{tr} \mathbf{S}^2} = |S_{r\theta}| = S_{r\theta} \tag{3-27}$$

Here we have made use of Eq. (3-15) to say that $S_{r\theta}$ must have the same sign as $D_{r\theta}$: positive. From Eqs. (3-14), (3-15), (3-25), and (3-27), we arrive at

$$r \frac{d}{dr}\left(\frac{v_\theta}{r}\right) = \frac{1}{\eta_0} \frac{\kappa^2 R^2}{r^2} S_0 + \frac{\tau_{\frac{1}{2}}}{\eta_0}\left(\frac{\kappa^2 R^2 S_0}{r^2 \tau_{\frac{1}{2}}}\right)^\alpha \tag{3-28}$$

Using boundary condition (3-1), we may integrate this last to obtain

$$\frac{v_\theta}{r} = \frac{S_0}{2\eta_0}\left[1 - \left(\frac{\kappa R}{r}\right)^2\right] + \frac{\tau_{\frac{1}{2}}}{2\alpha \eta_0}\left(\frac{S_0}{\tau_{\frac{1}{2}}}\right)^\alpha \left[1 - \left(\frac{\kappa R}{r}\right)^{2\alpha}\right] \tag{3-29}$$

In answer to our original question, the angular velocity of the outer cylinder is, from Eqs. (3-26) and (3-29),

$$\Omega = \frac{\bar{I}_3(1 - \kappa^2)}{4\pi L \kappa^2 R^2 \eta_0} + \frac{\tau_{\frac{1}{2}}}{2\alpha \eta_0}\left(\frac{\bar{I}_3}{2\pi L \kappa^2 R^2 \tau_{\frac{1}{2}}}\right)^\alpha (1 - \kappa^{2\alpha}) \tag{3-30}$$

In a manner somewhat similar to that used in Sec. 3.2.2, it is possible to analyze the tangential annular flow of a viscoelastic fluid of the type discussed in Sec. 2.3.3 without assuming a specific functional form for either $\eta(\gamma)$ or $\varphi(\tau)$ [1, 2]. Without much additional difficulty, tangential annular flow of a Noll simple fluid (Sec. 2.3.4) can be treated; Coleman, Markovitz, and Noll [3] give an excellent discussion of this problem and related topics.

For a survey of the problems that have been solved using the Ellis model, see [4].

## REFERENCES

1. Krieger, I. M., and H. Elrod: *J. Appl. Phys.*, **24**:134 (1953).
2. Pawlowski, J.: *Kolloid-Z.*, **130**:129 (1953).
3. Coleman, B. D., H. Markovitz, and W. Noll: "Viscometric Flows of Non-Newtonian Fluids," Springer-Verlag, New York, 1966.
4. Matsuhisa, Seikichi, and R. B. Bird: *A.I.Ch.E.J.*, **11**:588 (1965).

## EXERCISES

**3.2.3-1** *Tangential annular flow of a power-model fluid* Repeat the problem discussed in this section for an incompressible power-model fluid.

**3.2.3-2** *Tangential annular flow of a newtonian fluid* Repeat this problem for an incompressible newtonian fluid, but assume that the angular velocity of the outer cylinder is given and the axial component of the moment of the force which the fluid exerts on the inner cylinder is to be calculated.

**3.2.3-3** *Radial flow between porous cylindrical shells* Consider radial flow of an incompressible newtonian fluid between concentric porous cylindrical shells. Assume that the volume flow rate is known and solve for the difference in modified pressure between the inner and outer shells. Why does the modified pressure increase in the direction of flow? Is there any special significance to the fact that the velocity distribution does not depend upon viscosity?

**3.2.3-4** *Radial flow between porous spherical shells* Repeat Exercise 3.2.3-3 for radial flow between concentric porous spherical shells.

**3.2.3-5** Starting with Eq. (3-2), derive Eq. (3-3).

**3.2.3-6** *Helical flow of a newtonian fluid* Let us consider the helical flow of an incompressible newtonian fluid through an inclined annulus. Figure 3.2.1-4 again applies with the understanding that the inner cylinder rotates with an angular velocity $\Omega$ while the outer cylinder remains fixed in space. This means that in cylindrical coordinates there are *two* nonzero components of velocity, $v_\theta$ and $v_z$, and the particle paths are helices.

Compute the components of the velocity field in the annulus, the volume rate of flow through the annulus, and the torque that the fluid exerts on the inner cylinder (and that the motor driving the inner cylinder at the angular velocity $\Omega$ must overcome).

*Answer*

$$\frac{v_\theta}{\varkappa R \Omega} = \frac{r/R - R/r}{\varkappa - 1/\varkappa}$$

$$v_z = -\frac{AR^2}{4\mu}\left[1 - \left(\frac{r}{R}\right)^2 + \frac{1-\varkappa^2}{\ln(1/\varkappa)}\ln\frac{r}{R}\right]$$

$$-A = \frac{P_0 - P_L - \rho g L \sin \alpha}{L}$$

$$Q = -\frac{A\pi R^4}{8\mu}\left[1 - \varkappa^4 - \frac{(1-\varkappa^2)^2}{\ln(1/\varkappa)}\right]$$

At $r = \varkappa R$: $\quad l_3 = \dfrac{4\pi\mu L R^2 \Omega}{1 - (1/\varkappa)^2}$

*Hint:* From the $r$ component of Cauchy's first law,

$$\mathscr{P} = h(r) + g(z,\theta)$$

[3.2] COMPLETE SOLUTIONS

From the $\theta$ component,

$$\frac{\partial g}{\partial \theta} = B = \text{a constant}$$

From the $z$ component,

$$\frac{\partial g}{\partial z} = A = \text{a constant}$$

We conclude that

$$\mathscr{P} = h(r) + Az + B\theta + C$$

where $C$ is a constant.

**3.2.3-7** For the problem discussed in the text, show that the other two components of the torque that the fluid exerts on the inner cylinder are zero.

**3.2.3-8** *Tangential annular flow of an incompressible power-model fluid* For the apparatus described in this section, what is the relation between $l_3$ and $\Omega$ for an incompressible power-model fluid?

**3.2.3-9** *Cylinder rotates in an infinite fluid* A long vertical circular cylinder of radius $R$ rotates with an angular velocity $\Omega$ in an infinite incompressible newtonian fluid. Determine the velocity distribution and the torque exerted upon the cylinder by the fluid.

### 3.2.4 A wall which is suddenly set in motion [1, p. 82]

An infinite stationary body of an incompressible newtonian fluid is bounded on one side by the $z_1 z_3$ plane. Gravity acts in the negative $z_2$ direction. We visualize that this body of fluid is actually a deep pool with a horizontal phase interface in contact with the atmosphere. We approximate this experimental situation by saying that, as $z_2$ approaches infinity, pressure is no longer a function of $z_1$ and $z_3$. At time $t = 0$, the wall is suddenly set in motion in the positive $z_1$ direction, with a constant magnitude of velocity $V$. We wish to determine the velocity distribution in the fluid as a function of time.

The boundary conditions on velocity and pressure which must be satisfied are:

For all $t$, as $z_2 \to \infty$: $\quad \dfrac{\partial p}{\partial z_1} = \dfrac{\partial p}{\partial z_3} = 0 \quad$ (4-1)

For all $z_2$, at $t = 0$: $\quad \mathbf{v} = 0 \quad$ (4-2)

For $t > 0$, at $z_2 = 0$: $\quad v_1 = V \quad v_2 = v_3 = 0 \quad$ (4-3)

Some authors feel that as an additional boundary condition one should take the velocity to be zero as $z_2$ approaches infinity for all values of time [1, p. 72; 2, p. 125]. We take the view here that velocity is not directly constrained to be zero at infinity for times greater than zero [3, p. 62]; it is only necessary that velocity be finite as $z_2$ approaches infinity [3, p. 71].

It seems reasonable to ask whether there is a velocity distribution of the form

$$v_1 = v_1(z_2, t) \quad v_2 = v_3 = 0 \quad (4\text{-}4)$$

With this assumption, the rectangular cartesian components of the Navier-Stokes equation in terms of modified pressure, Eq. (1-4) of Sec. 2.5.1, become (see Table 2.5.1-3)

$$\frac{\partial \mathscr{P}}{\partial z_1} = -\rho \frac{\partial v_1}{\partial t} + \mu \frac{\partial^2 v_1}{\partial z_2^2} \qquad (4\text{-}5)$$

and

$$\frac{\partial \mathscr{P}}{\partial z_2} = \frac{\partial \mathscr{P}}{\partial z_3} = 0 \qquad (4\text{-}6)$$

In view of Eq. (4-6) the left side of Eq. (4-5) is a function of $t$ and $z_1$. The right side of this equation can at most be a function of $t$ and $z_2$. This means that each side of this equation is a function of time alone:

$$\frac{\partial \mathscr{P}}{\partial z_1} = A(t) \qquad (4\text{-}7)$$

But this last can satisfy boundary condition (4-1) only if

$$\frac{\partial \mathscr{P}}{\partial z_1} = A(t) = 0 \qquad (4\text{-}8)$$

Our problem has reduced to finding a solution for

$$\frac{\partial v_1}{\partial t} = \frac{\mu}{\rho} \frac{\partial^2 v_1}{\partial z_2^2} \qquad (4\text{-}9)$$

which satisfies

For all $z_2$, at $t = 0$: $\quad v_1 = 0$ (4-10)

and

For $t > 0$, at $z_2 = 0$: $\quad v_1 = V$ (4-11)

Sometimes a partial differential equation such as this may be solved by transforming it into an ordinary differential equation. For example, let us seek a solution of the form

$$\frac{v_1}{V} = g(\eta) \qquad (4\text{-}12)$$

where

$$\eta = a(z_2)^b t^c \qquad (4\text{-}13)$$

It is not difficult to deduce that, if $b = 1$ and $c = -\tfrac{1}{2}$, Eq. (4-9) reduces to an ordinary differential equation. Specifically, if we take

$$\eta = \frac{z_2}{\sqrt{4\mu t/\rho}} \qquad (4\text{-}14)$$

Eq. (4-9) becomes

$$g'' + 2\eta g' = 0 \qquad (4\text{-}15)$$

[3.2] COMPLETE SOLUTIONS

The prime notation is used here to denote differentiation with respect to $\eta$. With this change of variable, boundary condition (4-10) becomes

$$\text{As } \eta \to \infty: \quad g \to 0 \tag{4-16}$$

Boundary condition (4-11) transforms to

$$\text{At } \eta = 0: \quad g = 1 \tag{4-17}$$

Equation (4-15) is a first-order separable differential equation in $g'$, which may be integrated to give

$$g' = C_1 e^{-\eta^2} \tag{4-18}$$

Here $C_1$ is a constant of integration. This in turn may be integrated again using boundary condition (4-17):

$$g - 1 = C_1 \int_0^\eta e^{-\eta^2} \, d\eta \tag{4-19}$$

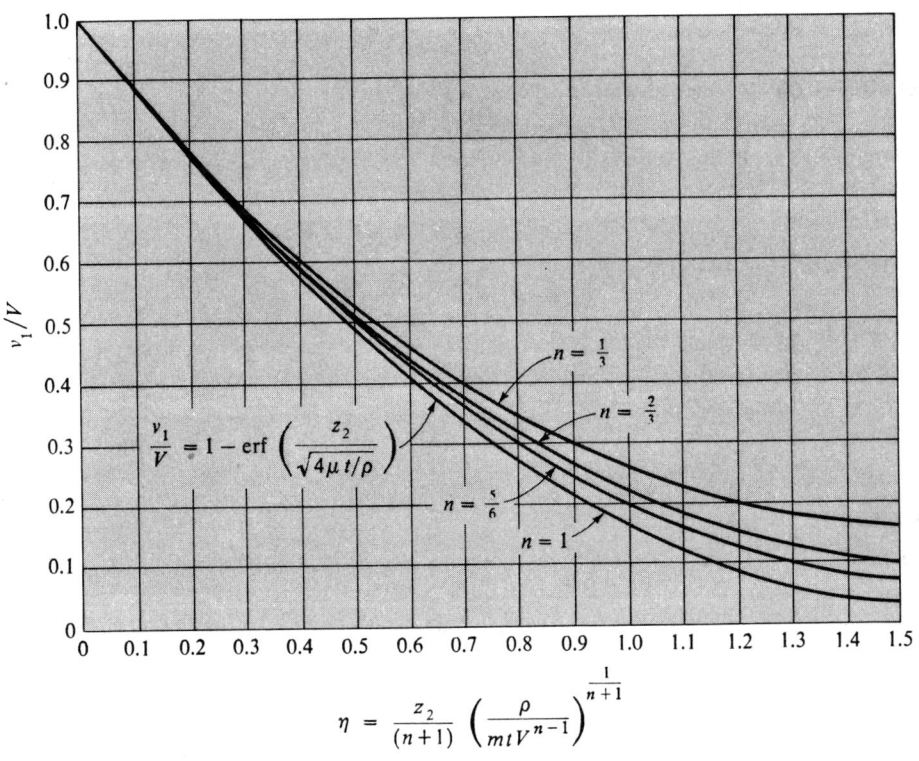

**Fig. 3.2.4-1** Velocity distribution (in dimensionless form) for flow in the neighborhood of a wall suddenly set in motion. The curves are drawn for the power model with parameters $m$ and $n$. For $n = 1$ (newtonian fluid) the variable $\eta$ becomes $z_2/\sqrt{4\mu t/\rho}$. (*From Bird, Stewart, and Lightfoot* [2, fig. 4.1-2].)

Finally, we may determine $C_1$ using Eq. (4-16):

$$-\frac{1}{C_1} = \int_0^\infty e^{-\eta^2}\, d\eta = \frac{\sqrt{\pi}}{2} \qquad (4\text{-}20)$$

As a final result we have

$$\frac{v_1}{V} = g = 1 - \frac{2}{\sqrt{\pi}} \int_0^\eta e^{-\eta^2}\, d\eta \qquad (4\text{-}21)$$

or

$$\frac{v_1}{V} = 1 - \operatorname{erf}\left(\frac{z_2}{\sqrt{4\mu t/\rho}}\right) \qquad (4\text{-}22)$$

We introduce here the error function

$$\operatorname{erf}(x) \equiv \frac{2}{\sqrt{\pi}} \int_0^x e^{-t^2}\, dt \qquad (4\text{-}23)$$

This solution is shown in Fig. 3.2.4-1.

## REFERENCES

1. Schlichting, Hermann: "Boundary-Layer Theory," 6th ed., McGraw-Hill, New York, 1968.
2. Bird, R. B., W. E. Stewart, and E. N. Lightfoot: "Transport Phenomena," 7th printing, Wiley, New York, 1960.
3. Carslaw, H. S., and J. C. Jaeger: "Conduction of Heat in Solids," 2d ed., Oxford University Press, London, 1959.
4. Bird, R. B.: *A.I.Ch.E.J.*, **5**:565 (1959).
5. Bird, R. B.: *Chem. Eng. Progr., Symp.*, (58) **61**:1 (1965).
6. Szymanski, Piotr: *J. Math. Pures Appl.*, (9) **11**:67 (1932).
7. Irving, J., and N. Mullineux: "Mathematics in Physics and Engineering," Academic, New York, 1959.
8. Muller, Wilhelm: *Z. Angew. Math. Mech.*, **16**:227 (1936).
9. Bird, R. B., and C. F. Curtiss: *Chem. Eng. Sci.*, **11**:108 (1959).

## EXERCISES

**3.2.4-1** Starting with Eqs. (4-12) and (4-13), show how one argues to deduce that the choice $b = 1$ and $c = -1/2$ reduces Eq. (4-9) to the simplest possible ordinary differential equation.

**3.2.4-2** *More on the suddenly accelerated wall* [4; 2, fig. 4.1-2] Repeat the problem of this section for a power-model fluid.

(a) Show that Eq. (4-9) is replaced by

$$\frac{\partial v_1}{\partial t} = -\frac{m}{\rho} \frac{\partial}{\partial z_2}\left(-\frac{\partial v_1}{\partial z_2}\right)^n$$

(b) Show that with the change of variable

$$\eta \equiv \frac{z_2}{n+1}\left(\frac{\rho}{mtV^{n-1}}\right)^{1/(n+1)}$$

## [3.2] COMPLETE SOLUTIONS

this partial differential equation reduces to an ordinary differential equation:

$$g''_{(n)}(-g'_{(n)})^{n-1} + \frac{(n+1)^n}{n}\eta g'_{(n)} = 0$$

Here

$$\frac{v_1}{V} = g_{(n)}(\eta)$$

and the prime notation indicates differentiation with respect to $\eta$.

(c) For $n < 1$, show that this differential equation may be integrated twice to give

$$g_{(n)} = \beta_{(n)}^{-[1/(1-n)]} \int_\eta^\infty (B_{(n)} + \eta^2)^{-[1/(1-n)]} d\eta$$

where

$$\beta_{(n)} = \frac{(1+n)^n(1-n)}{2n}$$

and the constant of integration $B_{(n)}$ is to be determined from

$$1 = \beta_{(n)}^{-[1/(1-n)]} \int_0^\infty (B_{(n)} + \eta^2)^{-[1/(1-n)]} d\eta$$

(d) Make the change of variable

$$s^2 = \frac{\eta^2}{B_{(n)} + \eta^2}$$

and show that the solution of part (c) may be expressed as

$$g_{(n)} = \frac{I_{(n)}(s)}{I_{(n)}(0)}$$

where

$$I_{(n)}(s) \equiv \int_s^1 (1-u^2)^{(3n-1)/[2(1-n)]} du$$

and

$$B_{(n)} = \beta_{(n)}^{-[2/(1+n)]} [I_{(n)}(0)]^{[2(1-n)]/(1+n)}$$

This solution has been evaluated for several values of $n$ and the results are presented in Fig. 3.2.4-1.

The results found here may not be a very good representation of the behavior of a real fluid in this geometry. As I indicated in concluding Sec. 2.3.3, one should be cautious in using such simple models for fluid behavior in unsteady-state flows.

**3.2.4-3** Take the Laplace transformation of Eq. (4-9) to obtain a linear second-order differential equation. Solve with the required boundary conditions to find Eq. (4-22).

**3.2.4-4** *An oscillating wall* Consider a situation similar to that described in the text except that a periodic oscillation is superimposed upon the motion of the wall:

At $z_2 = 0$:  $v_1 = V + A \cos(\omega t - \epsilon)$

Let us determine the velocity distribution in the fluid as $t \to \infty$.

After formulating the problem and discussing the pressure distribution, seek a solution of the form

$$v_1^* \equiv \frac{v_1 - V}{A} = u(z_2) \exp[i(\omega t - \epsilon)]$$

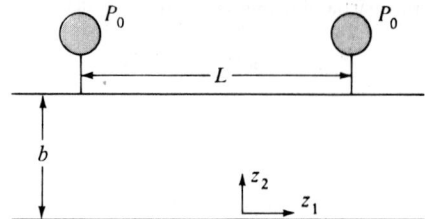

**Fig. 3.2.4-2** Oscillatory flow between two flat plates.

Only the real portion of this solution is of interest to us, but we must allow for the possibility that $u(z_2)$ is a complex function. Determine that

$$v_1^* = \exp(-Kz_2) \cos(\omega t - Kz_2 - \epsilon)$$

where

$$K \equiv \sqrt{\frac{\omega \rho}{2\mu}}$$

*Hint:* Observe that $i = (1+i)^2/2$.

**3.2.4-5** *Oscillatory flow between two flat plates* [5]  An incompressible newtonian fluid is contained between the two parallel plates shown in Fig. 3.2.4-2. The lower plate oscillates periodically:

At $z_2 = 0$:     $v_1 = v_0 \cos \omega t$

while the upper plate is held stationary. Determine the velocity distribution in the fluid.

*Answer*

$$v_1^* \equiv \frac{v_1}{v_0} = \frac{\left\{\begin{bmatrix} \sinh a(1-\xi) \cos a(1-\xi) \sinh a \cos a \\ + \sin a(1-\xi) \cosh a(1-\xi) \sin a \cosh a \end{bmatrix} \cos \omega t \right. }{\sinh^2 a \cos^2 a + \sin^2 a \cosh^2 a}$$

$$\left. + \begin{bmatrix} \sinh a(1-\xi) \cos a(1-\xi) \sin a \cosh a \\ -\sin a(1-\xi) \cosh a(1-\xi) \sinh a \cos a \end{bmatrix} \sin \omega t \right\}$$

where

$$\xi \equiv \frac{z_2}{b} \qquad a \equiv \sqrt{\frac{\omega \rho b^2}{2\mu}}$$

*Hint:* See Exercise 3.2.4-4.

**3.2.4-6** *Unsteady-state flow in a tube*   The inclined tube shown in Fig. 3.2.1-2 is filled with a stationary incompressible newtonian fluid. At time $t = 0$, the pressure gradient indicated in the figure is imposed and the fluid begins to flow. We wish to determine the velocity distribution in the fluid as a function of time.

(a) Argue that the equation of motion in the form

$$\rho \frac{\partial v_z}{\partial t} = -A + \mu \frac{1}{r} \frac{\partial}{\partial r}\left(r \frac{\partial v_z}{\partial r}\right)$$

is to be solved consistent with the boundary conditions

At $t = 0$:     $v_z = 0$

## [3.2] COMPLETE SOLUTIONS

and

At $r = R$:     $v_z = 0$

Here

$$-A = \frac{P_0 - P_L}{L} - \rho g \sin \alpha$$

(b) Let us introduce as dimensionless variables

$$v_z^* = \frac{4\mu}{R^2(-A)} v_z \qquad r^* = \frac{r}{R} \qquad t^* = \frac{\mu t}{\rho R^2}$$

We shall attempt to find a solution of the form

$$v_z^* = v_{z\infty}^* - \Phi(r^*, t^*)$$

where $v_{z\infty}^*$ is the steady-state velocity distribution. Determine that $\Phi(r^*, t^*)$ is a solution to

$$\frac{\partial \Phi}{\partial t^*} = \frac{1}{r^*} \frac{\partial}{\partial r^*} \left( r^* \frac{\partial \Phi}{\partial r^*} \right)$$

that is consistent with the boundary conditions

At $t^* = 0$:     $\Phi = 1 - r^{*2}$

and

At $r^* = 1$:     $\Phi = 0$

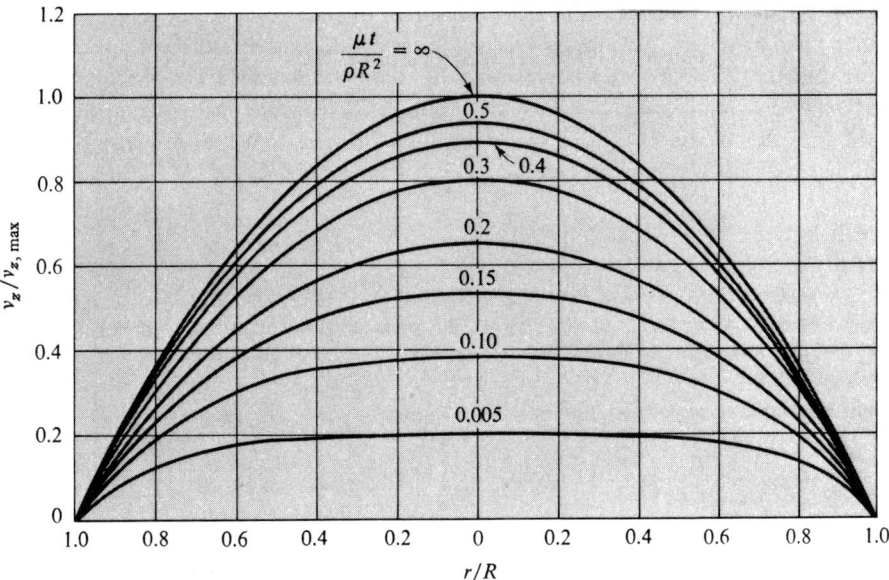

**Fig. 3.2.4-3** Velocity distribution for unsteady-state flow from rest in a circular tube [6].

(c) Solve the problem posed in (b) to find

$$v_z^* = 1 - r^{*2} - \sum_{n=1}^{\infty} \frac{8}{(\alpha_n)^3 J_1(\alpha_n)} J_0(\alpha_n r^*) \exp(-\alpha_n^2 t^*)$$

Here the $\alpha_n$ ($n = 1, 2, \ldots$) are roots of [7, p. 130]

$$J_0(\alpha_n) = 0$$

This solution is shown in Fig. 3.2.4-3.

(d) From a practical point of view, we may be more interested in the volume flow rate $Q$ as a function of time:

$$\frac{8\mu Q}{\pi R^4(-A)} = 1 - \sum_{n=1}^{\infty} \frac{32}{(\alpha_n)^4} \exp(-\alpha_n^2 t^*)$$

This result is plotted in Fig. 4.2.2-1.
For another view of this problem, see Sec. 4.2.2.
Hint: From [7, p. 161] we know

$$\int_0^1 r^*(1 - r^{*2}) J_0(\alpha_m r^*) \, dr^* = \frac{2}{(\alpha_m)^2} J_2(\alpha_m)$$

**3.2.4-7** *Unsteady-state flow in an annulus*  The analysis outlined in Exercise 3.2.4-6 may be used as a guide in treating flow from rest of an incompressible newtonian fluid in a coaxial annulus.

(a) Determine the velocity distribution in unsteady-state axial flow [8].

(b) Determine the velocity distribution in unsteady-state tangential flow [9]. Assume that for $t \sim 0$, the outer cylinder rotates with an angular velocity $\Omega$.

## 3.2.5  A rotating bucket

The right circular cylinder shown in Fig. 3.2.5-1 is partially filled with an incompressible newtonian fluid and rotates with a constant angular velocity $\Omega$. We wish to determine the shape of the gas-liquid phase interface.

In this discussion, we shall neglect viscous effects in the gas phase and assume that the pressure in the gas phase has a uniform value $p_0$. We shall further neglect the influence of surface tension (see Sec. 2.3.5).

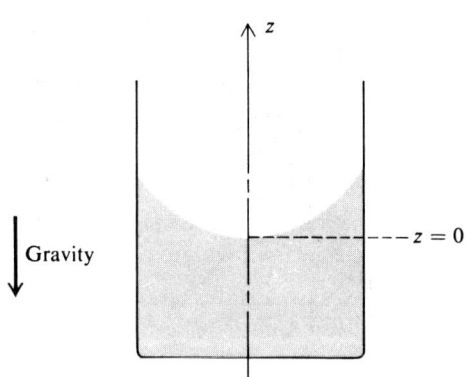

**Fig. 3.2.5-1**  The rotating bucket.

Most readers will immediately recognize that the fluid in the bucket must rotate as a solid body and that, consequently, the velocity distribution in the fluid takes the following form in the cylindrical coordinates (see also Exercise 3.2.5-1):

$$\frac{v_\theta}{r} = \Omega \qquad v_r = v_z = 0 \tag{5-1}$$

For this velocity distribution, the equation of continuity is satisfied identically and the rate of deformation tensor is zero (as it should be since the fluid rotates as a solid body).

This means that the extra-stress tensor **S** is zero as well and the three components of Cauchy's first law reduce to

$$\frac{\partial \mathscr{P}}{\partial r} = \rho r \Omega^2 \tag{5-2}$$

and

$$\frac{\partial \mathscr{P}}{\partial \theta} = \frac{\partial \mathscr{P}}{\partial z} = 0 \tag{5-3}$$

Since there is no mass transfer across the phase interface, the normal component of velocity of the fluid at the phase interface must be the same as the speed of displacement of the phase interface (see Sec. 1.3.4):

$$\mathbf{v} \cdot \boldsymbol{\xi} = u_{(\xi)} \tag{5-4}$$

This assures that the jump mass balance for the phase interface (Sec. 1.3.5) is identically satisfied and the jump momentum balance (Exercise 2.2.3-1) reduces to

$$[\mathbf{T} \cdot \boldsymbol{\xi}] = 0 \tag{5-5}$$

It is easy to show that, in view of Eqs. (5-1), the $\theta$ component of Eq. (5-5) is satisfied identically and the $r$ and $z$ components require only that at the phase interface

$$p = p_0 \tag{5-6}$$

Let us define our coordinate system such that the phase interface passes through the origin ($r = 0$, $z = 0$) and let us define the external force potential $\varphi$ such that

$$\varphi = gz \tag{5-7}$$

In view of Eqs. (5-3), this allows us to say that

At $r = 0$: $\qquad \mathscr{P} = p_0$ (5-8)

We may now integrate Eq. (5-2) using Eq. (5-8) as a boundary condition to determine that the pressure at any point in the fluid is given by

$$p + \rho g z - p_0 = \rho \frac{\Omega^2}{2} r^2 \tag{5-9}$$

The phase interface is that surface on which Eq. (5-6) is satisfied. From Eq.

(5-9) the phase interface is the parabolic surface

$$z = \left(\frac{\Omega^2}{2g}\right) r^2 \qquad (5\text{-}10)$$

In this discussion, we have neglected the effect of surface tension and we have ignored the contact angle boundary condition at the wall. For a more complete discussion, see [1].

## REFERENCES

1. Wasserman, M. L., and J. C. Slattery: *Proc. Phys. Soc.* (*London*), **84**:795 (1964).
2. Whitaker, Stephen: "Introduction to Fluid Mechanics," Prentice-Hall, Englewood Cliffs, N.J., 1968.

## EXERCISES

**3.2.5-1** Repeat the analysis of this section beginning with the assumption that the velocity distribution is of the form

$$\frac{v_\theta}{r} = \omega(r) \qquad v_r = v_z = 0$$

**3.2.5-2** *A bucket filled with two fluids* Repeat the analysis of this section assuming that the bucket is filled with two immiscible, incompressible newtonian fluids as shown in Fig. 3.2.5-2. Determine that the shape of the liquid-liquid phase interface is given by

$$z - z_{(B)} = \frac{\Omega^2 r^2}{2g}$$

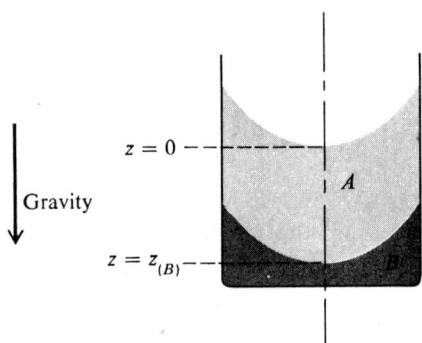

**Fig. 3.2.5-2** A rotating bucket filled with two immiscible, incompressible newtonian fluids.

## [3.2] COMPLETE SOLUTIONS

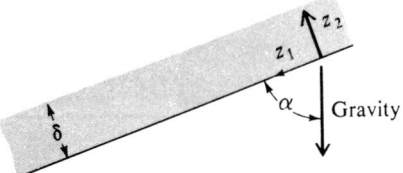

**Fig. 3.2.5-3** Flow down an inclined plane.

**3.2.5-3** *Flow in a film* An incompressible newtonian fluid flows down an inclined plane as shown in Fig. 3.2.5-3. Neglect viscous effects in the gas phase and take the pressure in the gas phase to have a uniform value $p_0$. Attempt to find the velocity distribution in this film assuming that the thickness of the film is a constant $\delta$.

*Answer*

$$v_1 = \frac{\rho g \delta^2 \cos \alpha}{2\mu} \left[ 2 \frac{z_2}{\delta} - \left(\frac{z_2}{\delta}\right)^2 \right]$$

**3.2.5-4** *Flow in a cylindrical film* An incompressible newtonian fluid flows down the inside of a vertical pipe as shown in Fig. 3.2.5-4. Neglect viscous effects in the gas phase and take the pressure in the gas phase to have a uniform value $p_0$. Attempt to find the velocity distribution in this film assuming that the thickness of the film is a constant $\delta$.

*Answer*

$$v_z = \frac{\rho g (R - \delta)^2}{2\mu} \left[ \ln \frac{r}{R} - \frac{1}{2}\left(\frac{r}{R - \delta}\right)^2 + \frac{1}{2}\left(\frac{R}{R - \delta}\right)^2 \right]$$

**3.2.5-5** A tank containing a liquid slides down an inclined plane as shown in Fig. 3.2.5-5. The coefficient of friction between the tank and the plane is $\gamma$. Assuming that the acceleration of the tank is independent of time, determine that the angle $\beta = \tan^{-1} \gamma$.

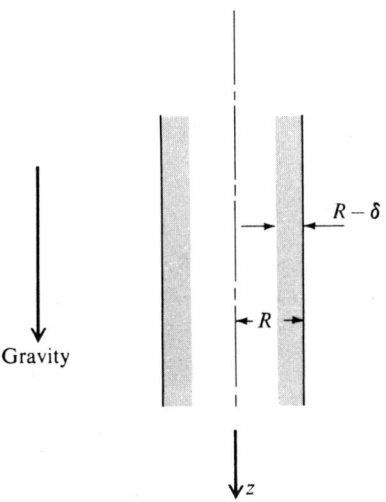

**Fig. 3.2.5-4** Flow in a film on the interior of a vertical pipe.

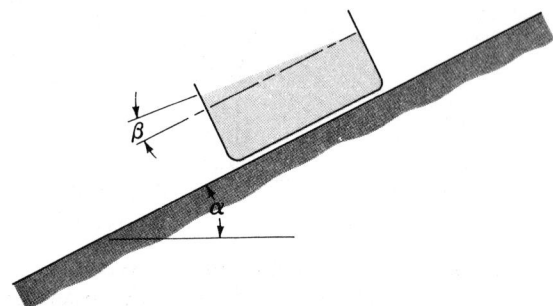

**Fig. 3.2.5-5** Acceleration of tank is constant as it slides down the plane.

**3.2.5-6** [2, p. 183] A film of water flows down a vertical wall as shown in Fig. 3.2.5-6. How large must the air gap $h_{(2)}$ be in order that the effect of the air on the water stream can be neglected?

(a) Assume in both the air and water that

$$v_2 = v_3 = 0 \qquad v_1 = v_1(z_2)$$

Assume that the difference in pressure in the gas phase between $z_1 = 0$ and $z_1 = L$ is that which would exist in a static situation.

Prove that

$$\frac{\partial \mathscr{P}_{(air)}}{\partial z_1} = 0$$

$$\frac{\partial \mathscr{P}_{(H_2O)}}{\partial z_1} = -g(\rho_{(H_2O)} - \rho_{(air)})$$

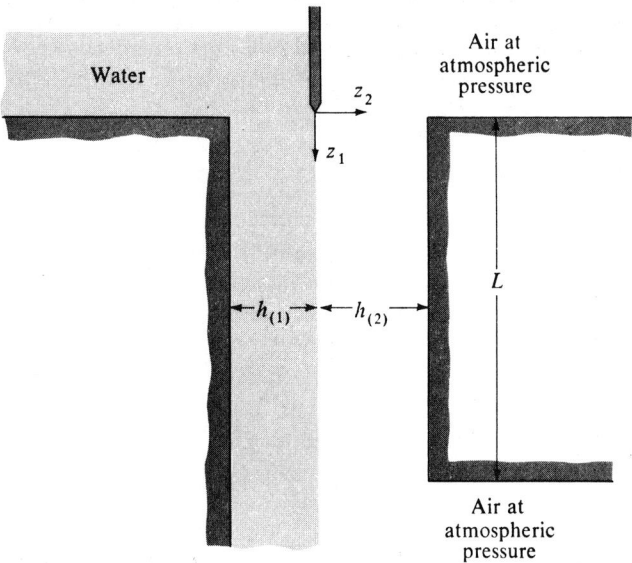

**Fig. 3.2.5-6** The effect of air upon a vertical water film.

[3.3] CREEPING FLOWS 99

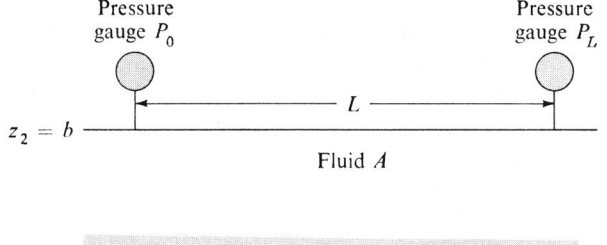

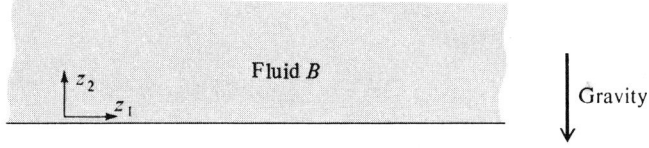

**Fig. 3.2.5-7** Two-phase flow between horizontal parallel plates.

(b) Determine the velocity distribution in the water film to be

$$v_{1(H_2O)} = \frac{g(\rho_{(H_2O)} - \rho_{(air)})h_{(1)}^2}{2\mu_{(H_2O)}}\left[1 - \left(\frac{z_2}{h_1}\right)^2 - \frac{1 + z_2/h_{(1)}}{1 + (\mu_{(H_2O)}h_{(2)}/\mu_{(air)}h_{(1)})}\right]$$

(c) At 68°F,

$\mu_{(air)} = 3.8 \times 10^{-7}$ lb$_f \cdot$ s/ft$^2$

$\mu_{(H_2O)} = 2.1 \times 10^{-5}$ lb$_f \cdot$ s/ft$^2$

Conclude that the air gap has a negligible effect so long as

$$\frac{h_{(2)}}{h_{(1)}} \gg 1.8 \times 10^{-2}$$

**3.2.5-7** *Stratified flow* Fluid A and fluid B are pumped at equal volumetric flow rates between two horizontal parallel plates as indicated in Fig. 3.2.5-7. Determine the location of the phase interface for fully developed laminar flow.

## 3.3 CREEPING FLOWS

**3.3.1 Creeping flow** In the preceding sections we see examples of problems which can be solved exactly. Most problems are not of this nature; after we have made what seem to be reasonable initial assumptions, the partial differential equations with which we are faced may be very difficult. Before beginning an extensive numerical solution, one is advised to consider what might be learned from limiting cases.

Let us look at the Navier-Stokes equation, Eq. (1-6) of Sec. 2.4.1, which is the expression of Cauchy's first law appropriate to an incompressible newtonian fluid with a constant viscosity. We may define the following dimensionless variables:

$$\mathbf{v}^* \equiv \frac{\mathbf{v}}{v_0} \qquad z_i^* \equiv \frac{z_i}{L_0}$$
$$\mathscr{P}^* \equiv \frac{\mathscr{P}}{\mathscr{P}_0} \qquad t^* \equiv \frac{t}{t_0} \qquad (1\text{-}1)$$

where $v_0$ is a characteristic magnitude of velocity, $L_0$ a characteristic length, $\mathcal{P}_0$ a characteristic modified pressure, and $t_0$ a characteristic time. By a characteristic quantity associated with a given problem, we mean one which occurs as a parameter in one of the governing equations or boundary conditions of that problem. We assume that the external force per unit mass is representable by a potential as discussed in Sec. 2.5.1. Under this condition and in terms of these dimensionless variables, Eq. (1-6) of Sec. 2.4.1 may easily be rearranged to the form

$$\frac{1}{N_{\text{St}}} \frac{\partial \mathbf{v}^*}{\partial t^*} + (\nabla \mathbf{v}^*) \cdot \mathbf{v}^* = -\frac{1}{N_{\text{Ru}}} \nabla \mathcal{P}^* + \frac{1}{N_{\text{Re}}} \text{div}(\nabla \mathbf{v}^*) \tag{1-2}$$

Here $N_{\text{St}}$, $N_{\text{Ru}}$, and $N_{\text{Re}}$ are, respectively, the Strouhal number, the Ruark number, and the Reynolds number:

$$N_{\text{St}} \equiv \frac{t_0 v_0}{L_0} \qquad N_{\text{Ru}} \equiv \rho \frac{v_0^2}{\mathcal{P}_0} \qquad N_{\text{Re}} \equiv \frac{L_0 v_0 \rho}{\mu} \tag{1-3}$$

One limiting case is that obtained by letting the Reynolds number $N_{\text{Re}}$ approach zero. In a given geometry with a given fluid, this is accomplished by letting $v_0$ approach zero; hence, the name *creeping flow*. If $N_{\text{St}}$ and $N_{\text{Ru}}$ take arbitrary values as $N_{\text{Re}} \to 0$, we intuitively expect the convective inertial terms $(\nabla \mathbf{v}^*) \cdot \mathbf{v}^*$ to become negligibly small with respect to the viscous terms $1/N_{\text{Re}} \text{div}(\nabla \mathbf{v}^*)$. This suggests that in the limit $N_{\text{Re}} \to 0$, the Navier-Stokes equation reduces to

$$\frac{1}{N_{\text{St}}} \frac{\partial \mathbf{v}^*}{\partial t^*} = -\frac{1}{N_{\text{Ru}}} \nabla \mathcal{P}^* + \frac{1}{N_{\text{Re}}} \text{div}(\nabla \mathbf{v}^*) \tag{1-4}$$

The Strouhal number is usually taken to have a value other than unity only in a periodic flow. For nonperiodic motion we take $t_0 = L_0/v_0$; in the limit $N_{\text{Re}} \to 0$, Eq. (1-2) reduces to

$$0 = -\frac{1}{N_{\text{Ru}}} \nabla \mathcal{P}^* + \frac{1}{N_{\text{Re}}} \text{div}(\nabla \mathbf{v}^*) \tag{1-5}$$

or

$$0 = -\nabla \mathcal{P} + \mu \, \text{div}(\nabla \mathbf{v}) \tag{1-6}$$

It is important to realize that we have used a strictly intuitive argument in the above discussion. There is no mathematical theorem which says that a very small term in a partial differential equation will have a very small effect on the solution of that differential equation. Stokes' paradox [1, p. 33] calls to our attention a situation in which the inertial terms are not negligibly small compared with the viscous terms in Eq. (1-2) as $N_{\text{Re}} \to 0$.

*Stokes' Paradox* No steady creeping flow of an incompressible newtonian fluid past an infinite circular cylinder is possible.

A little later we will see another example of an arbitrarily small term in the Navier-Stokes equation which yields a finite effect. Yet we do not wish to admit as a

# [3.3] CREEPING FLOWS

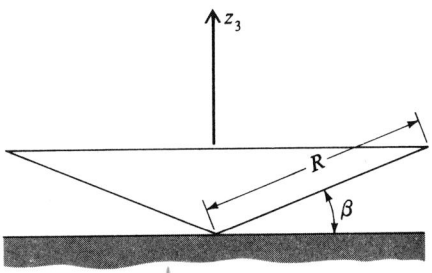

**Fig. 3.3.2-1** The cone-plate viscometer.

universal principle the possibility that small causes lead to large effects. Every experimental observation made is subject to countless small influences; if we were required to account for them all, we would never be able to predict our results. Since many experimental observations can be predicted theoretically, difficulties associated with neglecting small terms must be restricted to certain types of partial differential equations. For a further discussion of this topic, see Birkhoff [1, p. 4].

## REFERENCE

1. Birkhoff, Garrett: "Hydrodynamics, a Study in Logic, Fact and Similitude," Dover, New York, 1955.

## EXERCISE

**3.3.1-1** *Equation for modified pressure* Prove that, for the creeping flow of an incompressible newtonian fluid, the modified pressure satisfies Laplace's equation

$$\text{div}\,(\nabla \mathcal{P}) = 0$$

## 3.3.2 Flow in a cone-plate viscometer of an incompressible power-model fluid

An incompressible power-model fluid is being continuously deformed between the rotating plate and stationary cone shown in Fig. 3.3.2-1. We neglect "edge effects." This is similar to saying that the cone and plate extend to infinity, though we shall see later that there is at least one important difference.

This geometry is the basis for another type of viscometer. Two measurements are made: the angular velocity of the plate and the axial component of the moment of the force that the fluid exerts on the cone. We assume here that the angular velocity of the plate, $\Omega$, is given and that we wish to compute the axial component of the torque (moment of force) which the fluid exerts on the cone.

The boundary conditions to be satisfied in terms of spherical coordinates are:

At $\theta = \dfrac{\pi}{2} - \beta$:   $\mathbf{v} = \mathbf{0}$ \hfill (2-1)

At $\theta = \dfrac{\pi}{2}$:   $v_r = v_\theta = 0$   $v_\varphi = \Omega r \sin \theta$ \hfill (2-2)

Symmetry suggests that we assume our velocity distribution is of the form

$$v_r = v_\theta = 0 \tag{2-3}$$

$$v_\varphi = v_\varphi(r,\theta) \tag{2-4}$$

With a little more reflection, we might guess that the angular velocity of the fluid in the gap was a function only of $\theta$:

$$\frac{v_\varphi}{r \sin \theta} = \omega = \omega(\theta) \tag{2-5}$$

The equation of continuity, Eq. (C) of Table 2.5.1-1, is satisfied identically by Eqs. (2-3) and (2-5).

From Table 2.5.1-10, the only nonzero components of the rate of deformation tensor are

$$D_{\theta\varphi} = D_{\varphi\theta} = \tfrac{1}{2} \sin \theta \, \frac{d\omega}{d\theta} \tag{2-6}$$

From Sec. 2.3.3,

$$\mathbf{S} = 2m\gamma^{n-1}\mathbf{D} \tag{2-7}$$

where

$$\gamma \equiv \sqrt{2 \, \mathrm{tr} \, \mathbf{D}^2} = \sin \theta \left| \frac{d\omega}{d\theta} \right| \tag{2-8}$$

This means that the only nonzero components of the extra-stress tensor are

$$S_{\theta\varphi} = S_{\varphi\theta} = m\left(\sin \theta \, \frac{d\omega}{d\theta}\right)^n \tag{2-9}$$

In writing this last, we have observed that $d\omega/d\theta$ is positive in this situation.

From Table 2.5.1-6, the three components of Cauchy's first law in terms of the modified pressure $\mathscr{P}$ of Sec. 2.5.1 are

$$\rho \frac{(v_\varphi)^2}{r} = \frac{\partial \mathscr{P}}{\partial r} \tag{2-10}$$

$$\rho(v_\varphi)^2 \cot \theta = \frac{\partial \mathscr{P}}{\partial \theta} \tag{2-11}$$

$$\frac{\partial \mathscr{P}}{\partial \varphi} = \frac{1}{\sin \theta} \frac{d}{d\theta}(S_{\theta\varphi} \sin^2 \theta) \tag{2-12}$$

From Eq. (2-12),

$$\mathscr{P} = \mathscr{P}(r,\theta,\varphi) = \varphi f(\theta) + g(r,\theta) \tag{2-13}$$

But we require $\mathscr{P}$ to be a periodic function of $\varphi$ with period $2\pi$:

$$\mathscr{P}(r,\theta,\varphi) = \mathscr{P}(r, \theta, \varphi + 2\pi) \tag{2-14}$$

## [3.3] CREEPING FLOWS

which means that

$$\frac{\partial \mathscr{P}}{\partial \varphi} = f(\theta) = 0 \tag{2-15}$$

This leaves us

$$S_{\theta\varphi} = \frac{A}{\sin^2 \theta} \tag{2-16}$$

where $A$ is a constant of integration. This can be used together with the constitutive equation for the extra-stress tensor to solve for the velocity distribution in the manner of Secs. 3.2.1 to 3.2.3. All we have left to do is to check Eqs. (2-10) and (2-11) for consistency. We require that

$$\frac{\partial^2 \mathscr{P}}{\partial r \, \partial \theta} = \frac{\partial^2 \mathscr{P}}{\partial \theta \, \partial r} \tag{2-17}$$

But this is not satisfied (see Exercise 3.3.2-8).

What does this mean? Our assumptions in Eqs. (2-3) and (2-5) have led us to a contradiction. There is not a steady-state solution of this form to the equation of continuity and Cauchy's first law for an incompressible power-model fluid (or for that matter any model of the type discussed in Sec. 2.3.3). This could mean that the $r$ dependence assumed in Eq. (2-5) is incorrect; this possibility is investigated in Exercise 3.3.2-5 and found to be without merit. This probably means that a solution will have *three* nonzero components of the velocity vector. Suddenly the problem has become very difficult.

At this point we can revise our assumptions in Eqs. (2-3) and (2-5) and attempt a solution of the resulting problem (which is as yet unsolved in the literature even for a newtonian fluid). Or we can stop and ask about limiting cases. Let us take the latter alternative and ask about the creeping flow limit discussed in Sec. 3.3.1.

We may write Cauchy's first law in terms of the modified pressure for an incompressible power-model fluid as

$$\rho \frac{\partial \mathbf{v}}{\partial t} + \rho (\nabla \mathbf{v}) \cdot \mathbf{v} = -\nabla \mathscr{P} + \text{div}\, (2m\gamma^{n-1}\mathbf{D}) \tag{2-18}$$

In terms of the dimensionless variables introduced in Eq. (1-1),

$$\frac{1}{N_{\text{St}}} \frac{\partial \mathbf{v}^*}{\partial t^*} + (\nabla \mathbf{v}^*) \cdot \mathbf{v}^* = -\frac{1}{N_{\text{Ru}}} \nabla \mathscr{P}^* + \frac{1}{N_{\text{Re }PM}} \text{div}\, (2\gamma^{*n-1}\mathbf{D}^*) \tag{2-19}$$

where

$$\gamma^* \equiv \frac{L_0}{v_0} \gamma \qquad \mathbf{D}^* \equiv \frac{L_0}{v_0} \mathbf{D}$$

$$N_{\text{Re }PM} \equiv \frac{L_0^n v_0^{2-n} \rho}{m} \tag{2-20}$$

By analogy with our discussion in Sec. 3.3.1, the creeping flow limit corresponds to the limit $N_{\text{Re }PM} \to 0$. In this limit it appears that inertial effects may be negligibly small with respect to viscous effects in Cauchy's first law, and Eq. (2-19) reduces to

$$\frac{1}{N_{\text{St}}} \frac{\partial \mathbf{v}^*}{\partial t^*} = -\frac{1}{N_{\text{Ru}}} \nabla \mathscr{P}^* + \frac{1}{N_{\text{Re }PM}} \text{div } (2\gamma^{*n-1}\mathbf{D}^*) \qquad (2\text{-}21)$$

or

$$\rho \frac{\partial \mathbf{v}}{\partial t} = -\nabla \mathscr{P} + \text{div } (2m\gamma^{n-1}\mathbf{D}) \qquad (2\text{-}22)$$

If we look at Eqs. (2-10) and (2-11) in the limit of creeping flow, they reduce to

$$0 = \frac{\partial \mathscr{P}}{\partial r} \qquad (2\text{-}23)$$

$$0 = \frac{\partial \mathscr{P}}{\partial \theta} \qquad (2\text{-}24)$$

and condition (2-17) is now satisfied.

We have not yet identified the characteristic length and velocity used in Eqs. (2-20). This is a somewhat unusual problem in that, when we neglect edge effects, we find that no lengths or velocities appear in the boundary conditions, Eqs. (2-1) and (2-2). If we truly wished to analyze flow between an infinite cone and an infinite plate (the mathematical problem to be solved when edge effects are neglected), we would have little choice but to define $v_0 \equiv L_0 \Omega$ and $L_0$ such that $N_{\text{Re}PM} = 1$. This means that, since we could not treat the case $N_{\text{Re }PM} \to 0$, there is no creeping-flow solution for the infinite cone–infinite plate problem.

Our point of view here is that we do not wish to discuss the infinite cone–infinite plate problem, but rather the finite geometry pictured in Fig. 3.3.2-1 under conditions such that edge effects are negligibly small. This suggests that we take

$$v_0 \equiv R\Omega \qquad L_0 \equiv R$$

$$N_{\text{Re }PM} \equiv \frac{R^2 \Omega^{2-n} \rho}{m} \qquad (2\text{-}25)$$

If $m$ is sufficiently large, small values of $N_{\text{Re }PM}$ may be achieved even though $\Omega$ may be large and $0 < n \leq 1$ (the more common situation). (For more general viscoelastic behavior such as that represented by the Noll simple fluid, it is necessary in addition to require $\beta \to 0$ in order that the compatability condition, Eq. (2-17), be satisfied [1, p. 51].)

We may now determine the velocity distribution in the limit of creeping flow. The extra-stress component $S_{\theta\varphi}$ may be eliminated between Eqs. (2-9) and (2-16) to obtain

$$\frac{A}{\sin^2 \theta} = m \left( \sin \theta \frac{d\omega}{d\theta} \right)^n \qquad (2\text{-}26)$$

## [3.3] CREEPING FLOWS

This may be rearranged as

$$\frac{1}{\sin\theta}\left(\frac{A}{m \sin^2\theta}\right)^{1/n} = \frac{d\omega}{d\theta} \tag{2-27}$$

and integrated to satisfy boundary condition (2-2):

$$\left(\frac{A}{m}\right)^{1/n} \int_{\pi/2}^{\theta} \frac{d\theta}{(\sin\theta)^{(n+2)/n}} = \omega - \Omega \tag{2-28}$$

Applying boundary condition (2-1), we have

$$\left(\frac{A}{m}\right)^{1/n} \int_{\pi/2-\beta}^{\pi/2} \frac{d\theta}{(\sin\theta)^{(n+2)/n}} = \Omega \tag{2-29}$$

Together these equations imply that

$$\frac{\Omega - \omega}{\Omega} = \frac{\int_{\theta}^{\pi/2} \frac{d\theta}{(\sin\theta)^{(n+2)/n}}}{\int_{\pi/2-\beta}^{\pi/2} \frac{d\theta}{(\sin\theta)^{(n+2)/n}}} \tag{2-30}$$

These integrals cannot be evaluated analytically, but they pose no numerical difficulty.

Our original object was to compute the axial component of $\mathbf{l}$, the moment of force that the fluid exerts on the stationary cone:

$$\text{At } \theta = \frac{\pi}{2} - \beta: \quad \mathbf{l} = \int_0^{2\pi} \int_0^R \mathbf{p} \wedge (\mathbf{T}\cdot\mathbf{n}) r \sin\left(\frac{\pi}{2} - \beta\right) dr\, d\varphi \tag{2-31}$$

In terms of a rectangular cartesian coordinate system defined by the relations given in Exercise A.4.1-5, the axial component or $z_3$ component of $\mathbf{l}$ may be expressed as

$$\text{At } \theta = \frac{\pi}{2} - \beta: \quad \bar{l}_3 = \int_0^{2\pi} \int_0^R \bar{a}_3 r \sin\left(\frac{\pi}{2} - \beta\right) dr\, d\varphi \tag{2-32}$$

Here we define

$$\mathbf{a} \equiv \mathbf{p} \wedge (\mathbf{T}\cdot\mathbf{n}) \tag{2-33}$$

and use overbars to denote rectangular cartesian coordinates. The relation between $\bar{a}_3$ and the spherical components of $\mathbf{a}$ is (Exercise A.4.1-9)

$$\bar{a}_3 = a_r \cos\theta - a_\theta \sin\theta \tag{2-34}$$

From Exercise A.9.2-1,

$$a_r \equiv a_{\langle 1\rangle} = e_{1jk} p_{\langle j\rangle} T_{\langle km\rangle} n_{\langle m\rangle} \tag{2-35}$$

Since the only nonzero physical component of $\mathbf{n}$ in spherical coordinates is

$$n_\theta \equiv n_{\langle 2\rangle} = 1 \tag{2-36}$$

and since the physical spherical components of the position vector are (Exercise A.4.1-11)

$$p_{\langle 1\rangle} = r \qquad p_{\langle 2\rangle} = p_{\langle 3\rangle} = 0 \qquad (2\text{-}37)$$

Eq. (2-35) reduces to

$$a_r = e_{11k} r T_{\langle k2\rangle} = 0 \qquad (2\text{-}38)$$

In the same manner,

$$a_\theta \equiv a_{\langle 2\rangle} = e_{2jk} p_{\langle j\rangle} T_{\langle km\rangle} n_{\langle m\rangle} = e_{213} r T_{\langle 32\rangle} = -rT_{\varphi\theta} = -rS_{\varphi\theta} \qquad (2\text{-}39)$$

Equations (2-34), (2-38), and (2-39) allow us to express Eq. (2-32) as

$$\text{At } \theta = \frac{\pi}{2} - \beta: \quad \bar{l}_3 = \int_0^{2\pi}\!\!\int_0^R S_{\varphi\theta} r^2 \sin^2\left(\frac{\pi}{2} - \beta\right) dr\, d\varphi \qquad (2\text{-}40)$$

From Eq. (2-16),

$$\bar{l}_3 = \int_0^{2\pi}\!\!\int_0^R A r^2 \, dr\, d\varphi = \frac{2\pi A R^3}{3} \qquad (2\text{-}41)$$

where $A$ is given by Eq. (2-29).

In a manner which is somewhat similar to that used in Sec. 3.2.2, creeping flow between a rotating cone and plate may be analyzed for a viscoelastic fluid of the type discussed in Sec. 2.3.3 without assuming a specific functional form for either $\eta(\gamma)$ or $\varphi(\tau)$ [2]. It is hardly more difficult to treat the flow of a Noll simple fluid (Sec. 2.3.4) in this geometry; see Coleman, Markovitz, and Noll [1] for an excellent discussion.

**REFERENCES**

1. Coleman, B. D., H. Markovitz, and W. Noll: "Viscometric Flows of Non-Newtonian Fluids," Springer-Verlag, New York, 1966.
2. Slattery, J. C.: *J. Colloid Sci.*, **16**:431 (1961).
3. Bird, R. B., W. E. Stewart, and E. N. Lightfoot: "Transport Phenomena," 7th printing, Wiley, New York, 1960.
4. Slattery, J. C.: "Non-Newtonian Flow about a Sphere," doctoral dissertation, Dept. of Chemical Engineering, University of Wisconsin, Madison, 1959.

**EXERCISES**

**3.3.2-1** Determine that the other two components of the torque which the fluid exerts on the cone are zero.

**3.3.2-2** *Ellis model flow in a cone-plate viscometer* Repeat the analysis of this section for an Ellis model fluid, but assume that the axial component of the moment of the force which the fluid exerts on the cone is given and that the angular velocity of the plate is to be calculated.

**3.3.2-3** *Newtonian flow in a cone-plate viscometer* Repeat Exercise 3.3.2-2 for an incompressible newtonian fluid.

## [3.3] CREEPING FLOWS

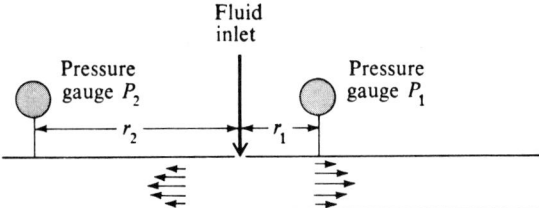

**Fig. 3.3.2-2** Radial flow between parallel disks.

**3.3.2-4** *Radial flow between parallel plates* Consider radial flow of an incompressible newtonian fluid between the two horizontal parallel circular disks shown in Fig. 3.3.2-2. The upper plate is located at $z = +b$ and the lower plate at $z = -b$. A pressure gauge at radius $r_1$ on the upper disk reads $P_1$; one at radius $r_2$ on the upper disk reads $P_2$. Consider only the region $r_1 \leq r \leq r_2$ and determine the velocity distribution in this region. What is the volume rate of flow between the plates? What are the two major assumptions which must be made in solving this problem [3, p. 114]?

**3.3.2-5** *More on newtonian flow in a cone-plate viscometer* [4, p. 175] (*a*) If Eq. (2-4) is assumed rather than Eq. (2-5), does this remove the contradiction found in the text between the $r$ and $\theta$ components of Cauchy's first law?

(*b*) Let us consider the flow of an incompressible newtonian fluid in the cone-plate viscometer described in the text and shown in Fig. 3.3.2-1 and let us assume that the velocity distribution is described by Eqs. (2-3) and (2-4).

(i) Show that Cauchy's first law yields

$$\frac{\partial}{\partial r}\left(r^2 \frac{\partial v_\varphi}{\partial r}\right) + \frac{1}{\sin\theta}\frac{\partial}{\partial \theta}\left(\sin\theta \frac{\partial v_\varphi}{\partial \theta}\right) - \frac{v_\varphi}{\sin^2\theta} = 0$$

for which a solution is to be obtained consistent with the boundary conditions

At $\theta = \frac{\pi}{2} - \beta$: $\quad v_\varphi = 0$

At $\theta = \frac{\pi}{2}$: $\quad v_\varphi = \Omega r$

At $r = 0$: $\quad v_\varphi = 0$

(ii) Look for separable solutions to this equation which have the form

$$v_\varphi = R(r)\Theta(\theta)$$

Find that the differential equations to be solved for $R(r)$ and $\Theta(\theta)$ are

$$\frac{d}{dr}\left(r^2 \frac{dR}{dr}\right) = mR$$

and

$$\frac{d}{dx}\left[(1-x^2)\frac{d\Theta}{dx}\right] + \left(m - \frac{1}{1-x^2}\right)\Theta = 0$$

where

$$x \equiv \cos\theta$$

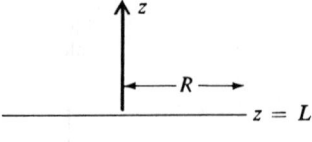

**Fig. 3.3.2-3** Tangential flow between parallel disks.

and $m$ is a constant. If we take

$$m \equiv n(n+1)$$

where $n$ is a positive integer, $\Theta$ is a solution of Legendre's associated equation of the first order and $n$th degree. A solution for the partial differential equation of (i) is, consequently,

$$v_\varphi = \sum_{n=1}^{\infty} \{r^n [A_n P_n^1(\cos\theta) + B_n Q_n^1(\cos\theta)] + r^{-n-1}[C_n P_n^1(\cos\theta) + D_n Q_n^1(\cos\theta)]\}$$

Here $P_n^1(\cos\theta)$ is Legendre's function of the first kind, first order, and $n$th degree; $Q_n^1(\cos\theta)$ is Legendre's function of the second kind, first order, and $n$th degree; $A_n$, $B_n$, $C_n$, $D_n$ are constants to be determined in such a way that this expression for $v_\varphi$ satisfies the required boundary conditions.

(iii) Apply the boundary conditions to find

$$\frac{v_\varphi}{\Omega r \sin\theta} = 1 - \frac{\frac{1}{2}\ln\frac{1+\cos\theta}{1-\cos\theta} + \frac{\cos\theta}{\sin^2\theta}}{\frac{1}{2}\ln\frac{1+\cos(\pi/2-\beta)}{1-\cos(\pi/2-\beta)} + \frac{\cos(\pi/2-\beta)}{\sin^2(\pi/2-\beta)}}$$

**3.3.2-6** *Tangential flow between parallel disks* The gap between the two parallel plates shown in Fig. 3.3.2-3 is filled with an incompressible newtonian fluid. The upper plate rotates with a constant angular velocity $\Omega$. The lower plate is stationary. Determine the velocity distribution in the fluid and the axial component of the torque which the fluid exerts upon the upper plate.

**3.3.2-7** *Nonnewtonian flow in a cone-plate viscometer* Consider an incompressible fluid whose stress tensor is determined by the most general function of the rate of deformation tensor which is consistent with the principle of material frame indifference (see Sec. 2.3.2):

$$\mathbf{T} = \varkappa_0 \mathbf{I} + \varkappa_1 \mathbf{D} + \varkappa_2 \mathbf{D} \cdot \mathbf{D}$$

Here $\varkappa_0$, $\varkappa_1$, and $\varkappa_2$ are unknown functions of $I_D$, $II_D$, and $III_D$. We have available a cone-plate viscometer in which $l_3$, the axial component of the torque which the fluid exerts on the stationary cone, can be measured as a function of $\Omega$, the angular velocity of the plate. If we are willing to neglect inertial effects with respect to viscous effects in the equation of motion and if we are willing to say that $\beta$, the angle between the cone and plate, is very small, what can be learned about the functions $\varkappa_0$, $\varkappa_1$, and $\varkappa_2$?

It is suggested that you approach the answer to this question in the following manner.

(a) Assume

$$v_r = v_\theta = 0 \qquad \frac{v_\varphi}{r\sin\theta} = \omega = \omega(\theta)$$

so that the only nonzero components of the rate of deformation tensor are

$$D_{\theta\varphi} = D_{\varphi\theta} = \tfrac{1}{2}\sin\theta \frac{d\omega}{d\theta}$$

[3.3] CREEPING FLOWS

Determine that the only nonzero components of the stress tensor are

$$T_{rr} = \varkappa_0$$

$$T_{\theta\theta} = T_{\varphi\varphi} = \varkappa_0 + \varkappa_2(D_{\theta\varphi})^2$$

$$T_{\theta\varphi} = T_{\varphi\theta} = \frac{C}{\sin^2\theta} = \varkappa_1 D_{\theta\varphi}$$

where

$$\mathrm{I_D} = 0 \qquad \mathrm{II_D} = -(D_{\theta\varphi})^2 \qquad \mathrm{III_D} = 0$$

and $C$ is a constant.

(b) Using the approach outlined in Exercise 5.5.2-1, it becomes clear that for this flow

$$\varkappa_1(0, -(D_{\theta\varphi})^2, 0) \geq 0$$

where

$$\varkappa_1 = \varkappa_1(\mathrm{I_D}, \mathrm{II_D}, \mathrm{III_D})$$

On the basis of experimental observation (explained more fully below), we reject the possibility $\varkappa_1 = 0$ even in the limit $D_{\theta\varphi} \to 0$ and require

$$\varkappa_1(0, -(D_{\theta\varphi})^2, 0) > 0$$

We assume that $\varkappa_1(0, -(D_{\theta\varphi})^2, 0)$ is a differentiable function and that, consequently, $T(D_{\theta\varphi})$ is a differentiable function. This allows us to conclude that

$$\left.\frac{dT_{\theta\varphi}}{dD_{\theta\varphi}}\right|_{D_{\theta\varphi}=0} = \lim_{D_{\theta\varphi}\to 0} \frac{T_{\theta\varphi}(D_{\theta\varphi}) - T_{\theta\varphi}(0)}{D_{\theta\varphi}}$$

$$= \lim_{D_{\theta\varphi}\to 0} \frac{T_{\theta\varphi}(D_{\theta\varphi})}{D_{\theta\varphi}} = \varkappa_1(0,0,0) > 0$$

If the derivative $dT_{\theta\varphi}/dD_{\theta\varphi}$ is continuous, it must be positive in some neighborhood of $D_{\theta\varphi} = 0$. In this neighborhood, $T_{\theta\varphi} = T_{\theta\varphi}(D_{\theta\varphi})$ will be a strictly increasing function of $D_{\theta\varphi}$, and for this reason it will have an inverse:

$$D_{\theta\varphi} = \lambda(T_{\theta\varphi})$$

(It is possible that $T_{\theta\varphi}$ ceases to increase when $D_{\theta\varphi}$ exceeds some critical value, but this has not been observed experimentally.)

The above discussion allows us to write

$$T_{rr} = T_{rr}(D_{\theta\varphi}) = T_{rr}(\lambda(T_{\theta\varphi})) = \hat{T}_{rr}(T_{\theta\varphi})$$

$$T_{\theta\theta} = \hat{T}_{\theta\theta}(T_{\theta\varphi})$$

$$T_{\varphi\varphi} = \hat{T}_{\varphi\varphi}(T_{\theta\varphi})$$

Use these expressions together with

$$T_{\theta\varphi} = \frac{C}{\sin^2\theta}$$

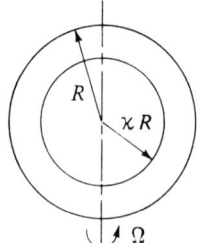

**Fig. 3.3.2-4** Tangential flow between concentric rotating spheres.

to show that the $r$ and $\theta$ components of Cauchy's first law are compatible for this flow if the inertial terms can be neglected and if $\beta$ is sufficiently small.

(c) Determine that

$$\frac{\omega}{\Omega} = \frac{\theta + \beta - \pi/2}{\beta}$$

$$D_{\theta\varphi} = \frac{\Omega}{2\beta}$$

and

$$T_{\theta\varphi} = \frac{3 l_3}{2\pi R^3}$$

We are now in a position to answer our original question. What can be learned about the functions $\varkappa_0$, $\varkappa_1$, and $\varkappa_2$ from measurements of $l_3$ as a function of $\Omega$?

**3.3.2-8** Starting with Eqs. (2-10) and (2-11) and with the realization that the boundary conditions require $\omega$ to be a function of $\theta$, conclude that Eq. (2-17) cannot be satisfied by a velocity distribution described by Eqs. (2-3) and (2-5).

**3.3.2-9** *Tangential flow between concentric rotating spheres* An incompressible newtonian fluid is trapped between the two concentric spheres shown in Fig. 3.3.2-4. The outer sphere rotates with a constant angular velocity $\Omega$, while the inner sphere is held stationary. Determine the torque required to hold the inner sphere stationary.

Neglect any end effects attributable to the axial supports required to hold the inner sphere in position.

(a) The boundary conditions suggest that we assume

$$v_r = v_\theta = 0 \qquad \frac{v_\varphi}{r \sin \theta} = \omega = \omega(r)$$

Determine that this satisfies the equation of continuity.

(b) Write out the components of Cauchy's first law and conclude that the velocity distribution assumed above cannot be a solution to the problem as it has been posed.

(c) Neglect the inertial terms in Cauchy's first law and integrate to find

$$\frac{\omega}{\Omega} = \frac{1/\varkappa^3 - (R/r)^3}{1/\varkappa^3 - 1}$$

(d) Calculate the axial component of torque (in rectangular cartesian coordinates) as

$$l_3 = \frac{-8\pi\mu\Omega R^3}{1/\varkappa^3 - 1}$$

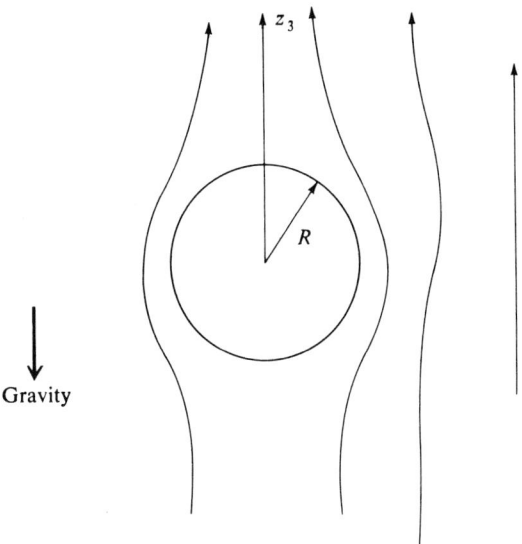

**Fig. 3.3.3-1** Flow past a sphere.

**3.3.2-10** *A sphere rotates in an unbounded fluid* A sphere of radius $R$ rotates with a constant angular velocity in an unbounded incompressible newtonian fluid. Determine the velocity distribution in the fluid as well as the axial component of the torque required to maintain the motion of the sphere.

*Answer*

$$\frac{v_\varphi}{r\Omega \sin\theta} = \left(\frac{R}{r}\right)^3$$

$$I_3 = -8\pi\mu\Omega R^3 \quad \text{exerted by fluid on the sphere}$$

**3.3.3 Flow past a sphere** Consider a sphere that falls at a constant speed $v_\infty$ along the axis of a cylinder filled with an incompressible newtonian fluid. We wish to calculate the force that the fluid exerts on the sphere.

In order to simplify the problem, let us neglect effects attributable to the bounding surfaces of the container and assume that the sphere is moving in the direction of gravity through an infinite expanse of fluid.

This is an unsteady-state problem, since the position of the sphere is changing as a function of time. However, it becomes a steady-state problem if we view the descent of the sphere in a coordinate system that is fixed with respect to the sphere. Specifically, let us adopt a spherical coordinate system whose origin coincides with the center of the sphere. Referring to Fig. 3.3.3-1, we see that the boundary conditions for our problem become

$$\text{At } r = R: \quad \mathbf{v} = 0 \tag{3-1}$$

and

$$\text{As } r \to \infty: \quad \mathbf{v} \to v_\infty \mathbf{e}_3 \tag{3-2}$$

# APPLICATIONS OF THE DIFFERENTIAL BALANCES TO MOMENTUM TRANSFER

It seems reasonable to assume that the velocity distribution in spherical coordinates is of the form

$$v_r = v_r(r,\theta) \qquad v_\theta = v_\theta(r,\theta) \qquad v_\varphi = 0 \tag{3-3}$$

In Secs. 1.3.6 and 1.3.7, we show how the equation of continuity can be satisfied identically in such a two-dimensional motion by the introduction of a stream function $\psi$. In Table 2.5.2-1, we find that we may represent the velocity components in terms of this stream function by

$$v_r = \frac{1}{r^2 \sin\theta} \frac{\partial \psi}{\partial \theta} \qquad v_\theta = -\frac{1}{r \sin\theta} \frac{\partial \psi}{\partial r} \tag{3-4}$$

For a steady-state motion in which the creeping-flow assumption can be made, the Navier-Stokes equation simplifies considerably when written in terms of the stream function (see Table 2.5.2-1):

$$\left[\frac{\partial^2}{\partial r^2} + \frac{\sin\theta}{r^2}\frac{\partial}{\partial \theta}\left(\frac{1}{\sin\theta}\frac{\partial}{\partial \theta}\right)\right]^2 \psi = 0 \tag{3-5}$$

From Eq. (3-1), the corresponding boundary conditions are

$$\text{At } r = R: \qquad \frac{1}{r^2 \sin\theta} \frac{\partial \psi}{\partial \theta} = 0 \tag{3-6}$$

and

$$\text{At } r = R: \qquad \frac{1}{r \sin\theta} \frac{\partial \psi}{\partial r} = 0 \tag{3-7}$$

The boundary condition on $\psi$ implied by Eq. (3-2) must be derived. From Eq. (3-4)$_1$ and Exercise A.4.1-10, we find that

$$\text{As } r \to \infty: \qquad v_r = \frac{1}{r^2 \sin\theta} \frac{\partial \psi}{\partial \theta} \to v_\infty \cos\theta \tag{3-8}$$

The stream function $\psi$ is arbitrary to a constant and we are free to require

$$\text{As } r \to \infty \text{ for } \theta = 0: \qquad \psi \to 0 \tag{3-9}$$

This allows us to say

$$\text{As } r \to \infty: \qquad \int_0^\psi d\psi = \int_0^\theta \frac{\partial \psi}{\partial \theta} d\theta = r^2 v_\infty \int_0^\theta \sin\theta \cos\theta \, d\theta \tag{3-10}$$

or

$$\text{As } r \to \infty: \qquad \psi \to \tfrac{1}{2} r^2 v_\infty \sin^2\theta \tag{3-11}$$

The boundary condition (3-11) suggests that the stream function might be of the form

$$\psi = f(r) \sin^2\theta \tag{3-12}$$

[3.3] CREEPING FLOWS

With this transformation of variable, Eq. (3-5) becomes

$$\frac{d^4f}{dr^4} - \frac{4}{r^2}\frac{d^2f}{dr^2} + \frac{8}{r^3}\frac{df}{dr} - \frac{8f}{r^4} = 0 \tag{3-13}$$

The boundary conditions corresponding to Eqs. (3-6), (3-7), and (3-11) are

$$\text{At } r = R: \quad f = 0 \tag{3-14}$$

$$\text{At } r = R: \quad \frac{df}{dr} = 0 \tag{3-15}$$

and

$$\text{As } r \to \infty: \quad f \to \tfrac{1}{2}r^2 v_\infty \tag{3-16}$$

The fourth-order differential equation (3-13) is linear and homogeneous. One form of solution is

$$f(r) = ar^n \tag{3-17}$$

This implies

$$f(r) = \frac{A}{r} + Br + Cr^2 + Dr^4 \tag{3-18}$$

Boundary conditions (3-14) to (3-16) require

$$A = \tfrac{1}{4}v_\infty R^3 \quad B = -\tfrac{3}{4}v_\infty R$$
$$C = \tfrac{1}{2}v_\infty \quad D = 0 \tag{3-19}$$

To summarize, Eqs. (3-4), (3-12), (3-18), and (3-19) tell us that the components of velocity within the fluid are

$$\frac{v_r}{v_\infty} = \left[1 - \frac{3}{2}\frac{R}{r} + \frac{1}{2}\left(\frac{R}{r}\right)^3\right]\cos\theta \tag{3-20}$$

and

$$\frac{v_\theta}{v_\infty} = -\left[1 - \frac{3}{4}\frac{R}{r} - \frac{1}{4}\left(\frac{R}{r}\right)^3\right]\sin\theta \tag{3-21}$$

Let us next determine the pressure distribution within the fluid. It is most convenient to work in terms of the modified pressure

$$\mathscr{P} = p + \rho\varphi \tag{3-22}$$

where

$$\varphi = gz_3 = gr\cos\theta \tag{3-23}$$

In order to completely specify pressure within the fluid (pressure in an incompressible fluid is determined by the motion only within a constant), we require

$$\text{As } r \to \infty \text{ for } \theta = \frac{\pi}{2}: \quad \mathscr{P} = p_0 \tag{3-24}$$

From the $r$ and $\theta$ components of the Navier-Stokes equation (see Table 2.5.1-7), as well as Eqs. (3-20) and (3-21), we have

$$\frac{\partial \mathscr{P}}{\partial r} = 3\mu v_\infty R \frac{\cos \theta}{r^3} \tag{3-25}$$

and

$$\frac{\partial \mathscr{P}}{\partial \theta} = \tfrac{3}{2}\mu v_\infty R \frac{\sin \theta}{r^2} \tag{3-26}$$

This allows us to determine the pressure distribution by the line integration

$$\mathscr{P} - p_0 = \int_{p_0}^{\mathscr{P}} d\mathscr{P} = \int_\infty^r \frac{\partial \mathscr{P}}{\partial r}\bigg|_{\theta=\pi/2} dr + \int_{\pi/2}^\theta \frac{\partial \mathscr{P}}{\partial \theta} d\theta \tag{3-27}$$

as

$$p = p_0 - \rho g r \cos \theta - \tfrac{3}{2}\mu v_\infty R \frac{\cos \theta}{r^2} \tag{3-28}$$

The $z_3$ component of the force that the fluid exerts upon the sphere can be expressed as (an alternative expression for $f_{(s)3}$ is found in Exercise 4.4.12-5)

$$f_{(s)3} = R^2 \int_0^{2\pi} \int_0^\pi (T_{rr} \cos \theta - T_{\theta r} \sin \theta)\big|_{r=R} \sin \theta \, d\theta \, d\varphi \tag{3-29}$$

In view of Eq. (3-20), we have

At $r = R$:  $\quad S_{rr} = 0$ \hfill (3-30)

and Eq. (3-29) reduces to

$$f_{(s)3} = R^2 \int_0^{2\pi} \int_0^\pi (-p \cos \theta - T_{\theta r} \sin \theta)\big|_{r=R} \sin \theta \, d\theta \, d\varphi \tag{3-31}$$

After carrying out the required integrations we find

$$f_{(s)3} = \tfrac{4}{3}\pi R^3 \rho g + 6\pi \mu R v_\infty \tag{3-32}$$

The first term on the right describes the buoyant force that the fluid exerts upon the sphere; the second term is the result of the motion in the fluid. Equation (3-32) is a classic result known as *Stokes' law*.

It is very common to express the force that a fluid exerts on a submerged object beyond the force attributable to the ambient pressure and to the hydrostatic pressure in terms of a drag coefficient $c_\mathscr{D}$ (see Sec. 4.4.7). For flow past a sphere,

$$c_\mathscr{D} \equiv \frac{f_{(s)3} - \tfrac{4}{3}\pi R^3 \rho g}{\tfrac{1}{2}\rho v_\infty^2 \pi R^2} \tag{3-33}$$

Stokes' law may be expressed in terms of this drag coefficient as

$$c_\mathscr{D} = \frac{6\pi \mu R v_\infty}{\tfrac{1}{2}\rho v_\infty^2 \pi R^2} = \frac{24}{N_{\text{Re}}} \tag{3-34}$$

[3.3] CREEPING FLOWS

where

$$N_{\text{Re}} \equiv \frac{2\rho v_\infty R}{\mu} \tag{3-35}$$

In arriving at Eq. (3-34), we restricted ourselves to the limit as $N_{\text{Re}} \to 0$ in order to justify neglecting the inertial terms with respect to the viscous terms in the equation of motion. A comparison with experimental data [1, p. 17] indicates that Stokes' law is an excellent representation of reality for $N_{\text{Re}} < 0.5$.

For a sphere falling at a constant velocity under the action of gravity, a simple force balance (see Sec. 4.4.5) tells us that the force that the fluid exerts upon the sphere should be equal in magnitude and opposite in direction to the force of gravity upon the sphere:

$$f_{(s)z} = \tfrac{4}{3}\pi R^3 \rho_{(s)} g \tag{3-36}$$

Here $\rho_{(s)}$ indicates the density of the sphere. This gives us an explicit expression for the drag coefficient:

$$c_\mathscr{D} = \frac{8}{3}\frac{Rg}{v_\infty^2}\frac{\rho_{(s)} - \rho}{\rho} \tag{3-37}$$

When Stokes' law is applicable, Eqs. (3-34) and (3-37) represent a useful relationship between the properties of the sphere and the properties of the fluid. If the density of the fluid happens to be larger than the density of the sphere, Eqs. (3-34) and (3-37) are still applicable; it merely means that $v_\infty$ must be negative and the sphere rises rather than falls.

For a discussion of steady-state creeping flow past a sphere of an incompressible Ellis model fluid, see Sec. 4.4.20.

## REFERENCES

1. Schlichting, Hermann: "Boundary-Layer Theory," 6th ed., McGraw-Hill, New York, 1968.
2. Sampson, R. A.: *Phil. Trans. Roy. Soc. London*, **A182**:449 (1891).
3. Proudman, I., and J. R. Pearson: *J. Fluid Mech.*, **2**:237 (1957).
4. Taylor, T. D., and A. Acrivos: *J. Fluid Mech.*, **18**:466 (1964).
5. Taylor, T. D., and A. Acrivos: *Chem. Eng. Sci.*, **19**:445 (1964).
6. Irving, J., and N. Mullineux: "Mathematics in Physics and Engineering," Academic, New York, 1959.
7. Lamb, Horace: "Hydrodynamics," 6th ed., Dover, New York, 1945.
8. Ericksen, J. L.: In S. Flügge (ed.), "Handbuch der Physik," vol. 3/1, Springer-Verlag, Berlin, 1960.
9. Kaplan, Wilfred: "Advanced Calculus," Addison-Wesley, Cambridge, Mass., 1952.
10. Happel, John, and Howard Brenner: "Low Reynolds Number Hydrodynamics," Prentice-Hall, Englewood Cliffs, N.J., 1965.

## EXERCISES

**3.3.3-1** *The pressure distribution in flow past a sphere*  Derive Eqs. (3-25), (3-26), and (3-28).

**3.3.3-2** *The force that the fluid exerts upon the sphere*  (*a*) Starting with

$$f_{(s)3} = R^2 \int_0^{2\pi}\!\!\int_0^{\pi} t_3 \sin\theta\, d\theta\, d\varphi$$

where $t_3$ is the $z_3$ component of the force per unit area that the fluid exerts on the sphere, derive Eq. (3-29).

(b) Starting with Eq. (3-31), arrive at Eq. (3-32).

**3.3.3-3** *A general solution for steady-state axially symmetric creeping flows in spherical coordinates* Let us assume that there are only two nonzero components of velocity in spherical coordinates:

$$v_r = v_r(r,\theta) \qquad v_\theta = v_\theta(r,\theta) \qquad v_\varphi = 0$$

When the Navier-Stokes equation is expressed in terms of the corresponding stream function $\psi$, it reduces for a steady-state creeping flow to Eq. (3-5) (see Table 2.5.2-1).

Sampson [2; see also 3 to 5] presents a general solution to Eq. (3-5):

$$\psi = \sum_{m=1}^{\infty} (A_m r^{m+3} + B_m r^{-m+2} + C_m r^{m+1} + D_m r^{-m}) \bar{P}_m$$

Here

$$\bar{P}_m \equiv \int_{-1}^{\mu} P_m(\mu)\, d\mu$$

$P_m(\mu)$ is the Legendre polynomial of degree $m$ [6, p. 82] and

$$\mu \equiv \cos\theta$$

Use this general solution to determine that the stream function takes the form indicated by Eqs. (3-12), (3-18), and (3-19) for steady-state creeping flow past a sphere as described in the text.

**3.3.3-4** *A general solution for steady-state creeping flow* [7, secs. 335 and 336] For the steady-state creeping flow of an incompressible newtonian fluid, the equation of continuity and the Navier-Stokes equation require

$$\text{div } \mathbf{v} = 0 \qquad (1)$$

and

$$-\nabla \mathscr{P} + \mu \text{ div } \nabla \mathbf{v} = 0 \qquad (2)$$

(a) Determine that Eq. (2) may also be expressed as

$$\nabla \mathscr{P} + \mu \text{ curl (curl } \mathbf{v}) = 0 \qquad (3)$$

Let us begin by confining our efforts to solving the homogeneous equation

$$\text{curl (curl } \mathbf{v}') = 0 \qquad (4)$$

consistent with the equation of continuity

$$\text{div } \mathbf{v}' = 0 \qquad (5)$$

A necessary and sufficient condition that the curl of a vector be zero is that the vector be representable in terms of a scalar potential $\tilde{\chi}$ [8, sec. 33]:

$$\text{curl } \mathbf{v}' = \nabla \tilde{\chi} \qquad (6)$$

That it is a sufficient condition is shown in Exercise A.9.1-1.

(b) Prove that

$$\mathbf{p} \wedge \nabla \tilde{\chi} = \nabla(\mathbf{p} \cdot \mathbf{v}') - \mathbf{v}' - \nabla \mathbf{v}' \cdot \mathbf{p} \qquad (7)$$

(c) Prove that

$$\text{div } \nabla(\mathbf{p} \cdot \mathbf{v}') = 0 \qquad (8)$$

and, therefore, $\mathbf{p} \cdot \mathbf{v}'$ is a solution to Laplace's equation.

## [3.3] CREEPING FLOWS

(d) Prove that

$$\text{div } \nabla \tilde{\chi} = 0 \tag{9}$$

and, therefore, $\tilde{\chi}$ is also a solution to Laplace's equation.

Since each rectangular cartesian component of $\mathbf{v}'$ is a solution to Laplace's equation, $\mathbf{v}'$ can be expanded in a series of spherical harmonics [6, pp. 31 and 194]:

$$\mathbf{v}' = \sum_{n=-\infty}^{\infty} \mathbf{v}'_n \tag{10}$$

where each component of $\mathbf{v}'_n$ is a homogeneous function of degree $n$ [9, p. 90]. Similarly,

$$\tilde{\chi} = \sum_{n=-\infty}^{\infty} \tilde{\chi}_n \tag{11}$$

with the understanding that $\tilde{\chi}_n$ is a spherical harmonic of degree $n$.

(e) Prove that

$$\mathbf{v}'_n = \nabla(\mathbf{p} \cdot \mathbf{v}'_n) - \nabla \mathbf{v}'_n \cdot \mathbf{p} - \mathbf{p} \wedge \nabla \tilde{\chi}_n \tag{12}$$

Use the fact that the spherical harmonics are linearly independent.

(f) Prove that

$$\tilde{\Phi}_{n+1} \equiv \mathbf{p} \cdot \mathbf{v}'_n \tag{13}$$

is a spherical harmonic of degree $n + 1$.

(g) Use Euler's theorem for homogeneous functions [9, p. 90] to say that

$$\nabla \mathbf{v}'_n \cdot \mathbf{p} = n \mathbf{v}'_n \tag{14}$$

(h) Conclude that

$$\mathbf{v}' = \sum_{n=-\infty}^{\infty} (\nabla \Phi_n - \mathbf{p} \wedge \nabla \chi_n) \tag{15}$$

where $\Phi_n$ and $\chi_n$ are spherical harmonics of degree $n$.

Now let us turn our attention to the original problem, the simultaneous solution of Eqs. (1) and (2).

(i) Prove that $\mathscr{P}$ is a solution to Laplace's equation and, therefore, may be expanded in a series of spherical harmonics:

$$\mathscr{P} = \sum_{n=-\infty}^{\infty} \mathscr{P}_n \tag{16}$$

(j) Let us attempt to find a particular solution to

$$\nabla \mathscr{P}_n = \mu \text{ div } \nabla \mathbf{v} \tag{17}$$

of the form

$$\mathbf{v}^{(P)} \equiv A_n r^2 \nabla \mathscr{P}_n + B_n r^{2n+3} \nabla \left( \frac{\mathscr{P}_n}{r^{2n+1}} \right) \tag{18}$$

Here $r$ is the spherical coordinate,

$$r^2 = \mathbf{p} \cdot \mathbf{p} \tag{19}$$

Determine that

$$A_n = \frac{1}{2\mu(2n+1)} \tag{20}$$

(k) Determine that if $\mathbf{v}^{(P)}$ is to satisfy Eq. (1) as well, then

$$B_n = \frac{n}{\mu(n+1)(2n+1)(2n+3)} \tag{21}$$

(l) Conclude that a complete solution for the velocity distribution is

$$\mathbf{v} = \frac{1}{\mu} \sum_{n=-\infty}^{\infty} \left[ \frac{r^2}{2(2n+1)} \nabla \mathscr{P}_n + \frac{n r^{2n+3}}{(n+1)(2n+1)(2n+3)} \nabla \left( \frac{\mathscr{P}_n}{r^{2n+1}} \right) \right]$$

$$+ \sum_{n=-\infty}^{\infty} (\nabla \Phi_n - \mathbf{p} \wedge \nabla \chi_n) \tag{22}$$

(m) Prove that this may also be written as [10, Eq. 3-2.3]

$$\mathbf{v} = \sum_{n=-\infty}^{\infty} \left[ \text{curl}\,(\mathbf{p}\chi_n) + \nabla \Phi_n + \frac{(n+3)r^2}{2\mu(n+1)(2n+3)} \nabla \mathscr{P}_n - \frac{n \mathscr{P}_n \mathbf{p}}{\mu(n+1)(2n+3)} \right] \tag{23}$$

While the development of this solution is dependent upon the use of rectangular cartesian coordinates, Eqs. (22) and (23) are tensorially invariant and valid for all coordinate systems. Although the components of $\mathbf{v}'$ obey Laplace's equation only in rectangular cartesian coordinates, the functions $\mathscr{P}$, $\Phi$, and $\chi$ obey Laplace's equation in all coordinate systems.

**3.3.3-5** *Still another approach to flow past a sphere* We have already seen two approaches to the solution for steady-state creeping flow past a sphere (text and Exercise 3.3.3-3). For a third, use the general solution developed in Exercise 3.3.3-4.

For flow in the region exterior to a sphere, we seek a solution for $(\mathbf{v} - \mathbf{v}_\infty)$ such that

as $r \to \infty$: $\mathbf{v} - \mathbf{v}_\infty \to 0$

The solution may be expressed in the form of Eq. (23) of Exercise 3.3.3-4.

We may write the spherical harmonics that appear in Eq. (23) of Exercise 3.3.3-4 in terms of their corresponding surface harmonics [6, pp. 31 and 194]:

$$\mathscr{P}_n = r^n \mathscr{P}_{(n)}$$

$$\Phi_n = r^n \Phi_{(n)}$$

$$\chi_n = r^n \chi_{(n)}$$

The surface harmonics $\mathscr{P}_{(n)}$, $\Phi_{(n)}$, and $\chi_{(n)}$ are functions only of $\theta$ and $\varphi$ in spherical coordinates [6, p. 31].

(a) Argue that the solution for $(\mathbf{v} - \mathbf{v}_\infty)$ involves only harmonic functions of negative order and

$$\mathbf{v} - \mathbf{v}_\infty = \sum_{n=0}^{\infty} \left[ \text{curl}(\mathbf{p}\chi_{-n-1}) + \nabla \Phi_{-n-1} - \frac{(n-2)r^2}{2\mu n(2n-1)} \nabla \mathscr{P}_{-n-1} \right.$$

$$\left. + \frac{(n+1)}{\mu n(2n-1)} \mathbf{p} \mathscr{P}_{-n-1} \right]$$

(b) Prove that

$$\frac{\mathbf{p} \cdot (\mathbf{v} - \mathbf{v}_\infty)}{|\mathbf{p}|} = \sum_{n=0}^{\infty} \left[ \frac{(n+1)}{2\mu(2n-1)} r^{-n} \mathscr{P}_{(-n-1)} - (n+1) r^{-n-2} \Phi_{(-n-1)} \right]$$

$$\frac{\mathbf{p} \cdot \nabla (\mathbf{v} - \mathbf{v}_\infty) \cdot \mathbf{p}}{|\mathbf{p}|} = \sum_{n=0}^{\infty} \left[ \frac{-n(n+1)}{2\mu(2n-1)} r^{-n} \mathscr{P}_{(-n-1)} \right.$$

$$\left. + (n+1)(n+2) r^{-n-2} \Phi_{(-n-1)} \right]$$

$$\mathbf{p} \cdot \mathrm{curl}(\mathbf{v} - \mathbf{v}_\infty) = \sum_{n=0}^{\infty} n(n+1) r^{-n-1} \chi(-n-1)$$

(c) Let us represent at $r = a$:

$$\frac{\mathbf{p} \cdot (\mathbf{v} - \mathbf{v}_\infty)}{|\mathbf{p}|} = \sum_{n=0}^{\infty} X(n)$$

$$\frac{\mathbf{p} \cdot \nabla(\mathbf{v} - \mathbf{v}_\infty) \cdot \mathbf{p}}{|\mathbf{p}|} = \sum_{n=0}^{\infty} Y(n)$$

$$\mathbf{p} \cdot \mathrm{curl}(\mathbf{v} - \mathbf{v}_\infty) = \sum_{n=0}^{\infty} Z(n)$$

Prove that [10, p. 65]

$$\mathscr{P}(-n-1) = \frac{\mu(2n-1)a^n}{(n+1)} \left[ (n+2) X(n) + Y(n) \right]$$

$$\Phi(-n-1) = \frac{a^{n+2}}{2(n+1)} (n X(n) + Y(n))$$

$$\chi(-n-1) = \frac{Z(n)}{n(n+1)} a^{n+1}$$

(d) For the particular case of flow past a stationary sphere as described in the text, prove that

$X(1) = -v_\infty \cos\theta$, $X(n) = 0$ for $n \geq 0$ and $n \neq 1$
$Y(n) = 0$ for $n \geq 0$
$Z(n) = 0$ for $n \geq 0$

$\mathscr{P}(-2) = -\frac{3}{2} a \mu v_\infty \cos\theta$, $\mathscr{P}(-n-1) = 0$ for $n \geq 0$ and $n \neq 1$

$\Phi(-2) = -\frac{1}{4} a^3 v_\infty \cos\theta$, $\Phi(-n-1) = 0$ for $n \geq 0$ and $n \neq 1$

$\chi(-n-1) = 0$ for $n \geq 0$

and the velocity distribution takes the form of Eqs. (3-20) and (3-21).

## 3.4 NONVISCOUS FLOWS

### 3.4.1 Potential flow

Though it is not entirely necessary, let us confine our attention to an incompressible newtonian fluid.

In Sec. 3.3.1 we examined the limit of creeping flow obtained as the Reynolds number is allowed to approach zero. Let us now examine the limit which is obtained as the Reynolds number goes to infinity. Referring again to the dimensionless form of the Navier-Stokes equation, Eq. (1-2) of Sec. 3.3.1, we have

$$\frac{1}{N_{\mathrm{St}}} \frac{\partial \mathbf{v}^*}{\partial t^*} + (\nabla \mathbf{v}^*) \cdot \mathbf{v}^* = -\frac{1}{N_{\mathrm{Ru}}} \nabla \mathscr{P}^* + \frac{1}{N_{\mathrm{Re}}} \mathrm{div}(\nabla \mathbf{v}^*) \tag{1-1}$$

As the Reynolds number $N_{\mathrm{Re}}$ is allowed to approach infinity, it appears that the viscous terms, $1/N_{Re}$ div $(\nabla \mathbf{v})$, become negligibly small with respect to the convective inertial terms, $(\nabla \mathbf{v}^*) \cdot \mathbf{v}^*$, in this equation. As the Reynolds number becomes very large, it appears that the Navier-Stokes equation reduces to

$$\frac{1}{N_{St}} \frac{\partial \mathbf{v}^*}{\partial t^*} + (\nabla \mathbf{v}^*) \cdot \mathbf{v}^* = -\frac{1}{N_{Ru}} \nabla \mathscr{P}^* \tag{1-2}$$

or

$$\rho \frac{\partial \mathbf{v}}{\partial t} + \rho (\nabla \mathbf{v}) \cdot \mathbf{v} = -\nabla \mathscr{P} \tag{1-3}$$

Our next task is to study the solutions of Eq. (1-3) that are consistent with the equation of continuity. A velocity distribution of the form

$$\mathbf{v} = -\nabla \Phi \tag{1-4}$$

suggests itself, since in this case the Navier-Stokes equation

$$\rho \frac{\partial \mathbf{v}}{\partial t} + \rho (\nabla \mathbf{v}) \cdot \mathbf{v} = -\nabla \mathscr{P} + \mu \, \text{div} \, (\nabla \mathbf{v}) \tag{1-5}$$

reduces to Eq. (1-3) for all values of the Reynolds number:

$$\text{div} \, (\nabla \mathbf{v}) = \frac{\partial^2 v_i}{\partial z_j \, \partial z_j} \mathbf{e}_i = -\frac{\partial^3 \Phi}{\partial z_j \, \partial z_j \, \partial z_i} \mathbf{e}_i$$

$$= \frac{\partial^2 v_j}{\partial z_i \, \partial z_j} \mathbf{e}_i = \nabla (\text{div} \, \mathbf{v}) = 0 \tag{1-6}$$

In view of the equation of continuity for an incompressible fluid, the equation that must be solved for the potential $\Phi$ is Laplace's equation

$$\text{div} \, (\nabla \Phi) = 0 \tag{1-7}$$

Laplace's equation is presented in Table 3.4.1-1 for rectangular cartesian, cylindrical, and spherical coordinates. This equation has received considerable attention in the mathematical literature. See, for example, Churchill [1, chaps. 9 and 10 and app. 2] and Kellogg [2].

Equation (1-3) is, consequently, used only to determine the modified pressure distribution.

When the velocity distribution is given by an equation of the form of Eq. (1-4), we say that the velocity $\mathbf{v}$ is representable by a potential $\Phi$. We refer to such a flow as a *potential flow*.

For a potential flow it is easy to see that the vorticity vector is zero (see Exercise A.9.1-1). As discussed in Sec. 1.4.2, a flow in which the vorticity vector vanishes everywhere is said to be an irrotational flow. The physical picture is that there is no local angular motion; hence, the name.

To summarize, incompressible potential flows (or incompressible irrotational flows) form one class of flows in which the viscous terms in the Navier-Stokes equation are negligibly small with respect to the convective inertial terms. This by no means

## [3.4] NONVISCOUS FLOWS

**Table 3.4.1-1** Laplace's equation div $\nabla \Phi = 0$, in three coordinate systems

*Rectangular cartesian coordinates:*

$$\frac{\partial^2 \Phi}{\partial^2 z_1} + \frac{\partial^2 \Phi}{\partial^2 z_2} + \frac{\partial^2 \Phi}{\partial^2 z_3} = 0 \quad \text{(A)}$$

*Cylindrical coordinates:*

$$\frac{1}{r}\frac{\partial}{\partial r}\left(r \frac{\partial \Phi}{\partial r}\right) + \frac{1}{r^2}\frac{\partial^2 \Phi}{\partial \theta^2} + \frac{\partial^2 \Phi}{\partial z^2} = 0 \quad \text{(B)}$$

*Spherical coordinates:*

$$\frac{\partial}{\partial r}\left(r^2 \frac{\partial \Phi}{\partial r}\right) + \frac{1}{\sin \theta}\frac{\partial}{\partial \theta}\left(\sin \theta \frac{\partial \Phi}{\partial \theta}\right) + \frac{1}{\sin^2 \theta}\frac{\partial^2 \Phi}{\partial \varphi^2} = 0 \quad \text{(C)}$$

should suggest that they are the only flows with this property. It is because of their mathematical simplicity that they have received the most attention.

It is also worth noting that the viscous terms in Cauchy's first law for an incompressible fluid described by Eq. (2-11) of Sec. 2.3.2 or Eq. (3-1) of Sec. 2.3.3 are not automatically zero [3].

In reading Sec. 3.4.3, you may start to wonder whether flows in which viscous effects are neglected with respect to inertial effects have any importance. Their true value will be better appreciated in the context of boundary-layer theory, beginning in Sec. 3.5.1.

## REFERENCES

1. Churchill, R. V.: "Complex Variables and Applications," 2d ed., McGraw-Hill, New York, 1960.
2. Kellogg, O. D.: "Foundations of Potential Theory," Ungar, New York, 1929.
3. Slattery, J. C.: *Chem. Eng. Sci.*, **17**:689 (1962).

## EXERCISE

**3.4.1-1** *Potential flow of a nonnewtonian fluid* [3] Consider the potential flow of an incompressible fluid described by Eq. (2-11) of Sec. 2.3.2. Since the definition of pressure is arbitrary (see Sec. 2.3.2), we take here

$$\mathbf{S} \equiv \mathbf{T} - \varkappa_0 \mathbf{I} = \varkappa_1 \mathbf{D} + \varkappa_2 \mathbf{D} \cdot \mathbf{D}$$

(a) Show that

$$\text{div } \mathbf{S} = \varkappa_1 \text{ div } \mathbf{D} + \mathbf{D} \cdot \nabla \varkappa_1 + \varkappa_2 \text{ div } (\mathbf{D} \cdot \mathbf{D}) + (\mathbf{D} \cdot \mathbf{D}) \cdot \nabla \varkappa_2$$

(b) Show that

$$\nabla \overline{\text{II}}_\mathbf{D} = \tfrac{1}{2}\nabla [\text{div } \nabla (v^2)]$$

(c) Derive an appropriate expression for $VIII_D$.

(d) Deduce the following sufficient conditions under which div S in Cauchy's first law is zero:
1. Newtonian behavior.
2. For an incompressible fluid described by Eq. (3-1) of Sec. 2.3.3: $v^2$ = constant.
3. For an incompressible fluid described by Eq. (2-11) of Sec. 2.3.2: **v** = a constant vector.

### 3.4.2 The Bernoulli equation

In this section let us assume that we are concerned with a compressible fluid and that we have made an intuitive argument to neglect the viscous terms in Cauchy's first law. For example, we might have started with a compressible newtonian fluid and argued that in the limit as both $N_{Re} \to \infty$ and $LV\rho/\lambda \to \infty$ the viscous terms should become negligibly small compared with the convective inertial terms. (See the constitutive equation for a compressible newtonian fluid in Sec. 2.3.2.) As a result, Cauchy's first law is reduced to

$$\rho \frac{\partial \mathbf{v}}{\partial t} + \rho (\nabla \mathbf{v}) \cdot \mathbf{v} = -\nabla P + \rho \mathbf{f} \tag{2-1}$$

Let us further assume that the external force per unit mass is representable by a potential:

$$\mathbf{f} = -\nabla \varphi \tag{2-2}$$

It is easily shown that (see Exercise 3.4.2-1)

$$\mathbf{v} \wedge \mathbf{w} = \nabla(\tfrac{1}{2}v^2) - (\nabla \mathbf{v}) \cdot \mathbf{v} \tag{2-3}$$

where **w** is the vorticity vector (see Sec. 1.4.2):

$$\mathbf{w} \equiv \text{curl } \mathbf{v} \tag{2-4}$$

Equations (2-2) and (2-3) allow us to rewrite Eq. (2-1) as [1, p. 75]

$$\frac{\partial \mathbf{v}}{\partial t} - \mathbf{v} \wedge \mathbf{w} = -\frac{1}{\rho}\nabla P - \nabla(\tfrac{1}{2}v^2) - \nabla\varphi = -\nabla\chi \tag{2-5}$$

where

$$\chi \equiv \int_{P_0}^{P} \frac{dP}{\rho} + \tfrac{1}{2}v^2 + \varphi \tag{2-6}$$

Notice that in arriving at Eq. (2-5) we have assumed that density is a function only of pressure (see Exercise 3.4.2-6).

Let $s$ be a parameter along an arbitrary curve in space. At any point along this arbitrary curve, $d\mathbf{p}/ds$ is a unit tangent vector to the curve. Let us consider the component of Eq. (2-5) in the direction $d\mathbf{p}/ds$:

$$\frac{\partial \mathbf{v}}{\partial t} \cdot \frac{d\mathbf{p}}{ds} - (\mathbf{v} \wedge \mathbf{w}) \cdot \frac{d\mathbf{p}}{ds} = -\frac{d\mathbf{p}}{ds} \cdot \nabla\chi = -\frac{d\chi}{ds} \tag{2-7}$$

## [3.4] NONVISCOUS FLOWS

Since

$$\frac{d}{ds}\int_{s_0}^{s}\left(\frac{\partial \mathbf{v}}{\partial t}\cdot\frac{d\mathbf{p}}{ds}\right)ds = \frac{\partial \mathbf{v}}{\partial t}\cdot\frac{d\mathbf{p}}{ds} \qquad (2\text{-}8)$$

we may express Eq. (2-7) as [2, p. 72]

$$-(\mathbf{v}\wedge\mathbf{w})\cdot\frac{d\mathbf{p}}{ds} = -\frac{dX}{ds} \qquad (2\text{-}9)$$

where

$$X \equiv \int_{P_0}^{P}\frac{dP}{\rho} + \tfrac{1}{2}v^2 + \varphi + \int_{s_0}^{s}\left(\frac{\partial \mathbf{v}}{\partial t}\cdot\frac{d\mathbf{p}}{ds}\right)ds \qquad (2\text{-}10)$$

At any point in time, let $d\mathbf{p}/ds$ be the tangent vector to a streamline. Then

$$-(\mathbf{v}\wedge\mathbf{w})\cdot\frac{d\mathbf{p}}{ds} = 0 = -\frac{dX}{ds} \qquad (2\text{-}11)$$

We conclude that along a streamline $X =$ constant.

If at any point in time, $d\mathbf{p}/ds$ is the tangent vector to a vortex line, Eq. (2-11) is again valid and our conclusion is that along a vortex line $X =$ constant. (The *vortex lines* for time $t$ form that family of curves to which the vorticity field is everywhere tangent at time $t$. The parametric equations for the vortex lines are solutions of the differential equations

$$\frac{d\mathbf{p}}{d\alpha} = \mathbf{w} \qquad (2\text{-}12)$$

where $\alpha$ is an arbitrary parameter measured along the curves and time $t$ is a constant. See the discussion of streamlines in Sec. 1.1.3.)

If we are dealing with an irrotational flow ($\mathbf{w} = 0$) and $d\mathbf{p}/ds$ is the tangent vector to *any* curve in the flow field, Eq. (2-11) still holds. We deduce that in an irrotational flow, $X =$ constant along any curve in the flow field.

A more interesting result for an irrotational flow may be obtained by starting directly with Eq. (2-5). Since $\mathbf{v} = -\nabla\Phi$, we may write

$$\nabla\left(\int_{P_0}^{P}\frac{dP}{\rho} + \tfrac{1}{2}v^2 + \varphi - \frac{\partial\Phi}{\partial t}\right) = 0 \qquad (2\text{-}13)$$

This means that

$$\Psi \equiv \int_{P_0}^{P}\frac{dP}{\rho} + \tfrac{1}{2}v^2 + \varphi - \frac{\partial\Phi}{\partial t} = F(t) \qquad (2\text{-}14)$$

The form of $F(t)$ is without physical significance, for we can define

$$\tilde{\Phi} \equiv \Phi + \int_{0}^{t} F(t)\, dt \qquad (2\text{-}15)$$

such that

$$\int_{P_0}^{P} \frac{dP}{\rho} + \tfrac{1}{2}v^2 + \varphi - \frac{\partial \tilde{\Phi}}{\partial t} = 0 \qquad (2\text{-}16)$$

Yet $\tilde{\Phi}$ retains the important physical significance

$$\nabla\tilde{\Phi} = \nabla\Phi = -\mathbf{v} \qquad (2\text{-}17)$$

For this reason, it is customary to write for an irrotational flow

$$\Psi = \text{constant} \qquad (2\text{-}18)$$

To summarize,

$$X \equiv \int_{P_0}^{P} \frac{dP}{\rho} + \tfrac{1}{2}v^2 + \varphi + \int_{s_0}^{s}\left(\frac{\partial \mathbf{v}}{\partial t} \cdot \frac{d\mathbf{p}}{ds}\right) ds = \text{a constant} \qquad (2\text{-}19)$$

is *Bernoulli's equation* valid along streamlines and vortex lines in any flow such that viscous effects can be neglected in Cauchy's first law. In any irrotational flow such that viscous effects can be neglected in Cauchy's first law, Eq. (2-19) is valid along every curve in the flow field; more important, *Bernoulli's equation for a potential flow* holds through the flow field:

$$\Psi \equiv \int_{P_0}^{P} \frac{dP}{\rho} + \tfrac{1}{2}v^2 + \varphi - \frac{\partial \Phi}{\partial t} = \text{a constant} \qquad (2\text{-}20)$$

## REFERENCES

1. Milne-Thomson, L. M.: "Theoretical Hydrodynamics," 3d ed., Macmillan, New York, 1955.
2. Sabersky, R. H., and A. J. Acosta: "Fluid Flow," Macmillan, New York, 1964.

## EXERCISES

**3.4.2-1** Show that

$$\mathbf{v} \wedge \mathbf{w} = \nabla(\tfrac{1}{2}v^2) - (\nabla \mathbf{v}) \cdot \mathbf{v}$$

where $\mathbf{w}$ is the vorticity vector:

$$\mathbf{w} \equiv \text{curl } \mathbf{v}$$

**3.4.2-2** Derive Eq. (2-8).

**3.4.2-3** Starting with the result that $X =$ a constant along every curve in an irrotational flow field such that viscous effects can be neglected in Cauchy's first law, derive the result that $\Psi =$ a constant throughout the flow field.

**3.4.2-4** *The siphon* In Fig. 3.4.2-1, compute the largest value of $h$ for which flow can be maintained. You may neglect viscous effects and assume that the level of water in the tank is constant as a function of time. The vapor pressure of water at 70°F is 0.36 lb/in².

**3.4.2-5** *More on the siphon* Referring to Exercise 3.4.2-4, determine the value of $h$ for which the volume rate of flow through the siphon is maximized.

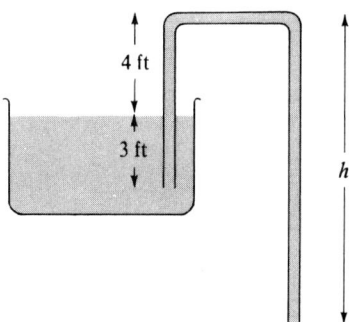

**Fig. 3.4.2-1** Flow through a siphon.

**3.4.2-6** *Restriction on Bernoulli's equation* In arriving at Eq. (2-5), we assumed that the density $\rho$ is a function only of pressure. In this way we restrict ourselves to an isothermal (constant composition) flow. Where was this assumption introduced?

**3.4.2-7** *A simple furnace design* [2, p. 70] We wish to design a chimney for a furnace. The process of which the furnace is a part dictates the energy requirements and, consequently, the rate at which fuel is consumed and the rate at which combustion gases are discharged through the stack. The prevailing winds and a desire to minimize the pollution of the surroundings lead to a decision concerning the height of the chimney. In order to determine the diameter of the chimney, we must estimate the speed of the gas in the chimney. Let us consider this highly simplified problem.

(a) In Fig. 3.4.2-2, air of density $\rho_1$ enters the furnace, where it reacts in part with the fuel. The heated gas leaving the furnace has a somewhat smaller density $\rho_2$ and rises through a chimney. What is the speed of the gas as a function of position in the chimney?

You may assume that the ideal gas law applies, that all viscous effects may be neglected, that the speed of the air entering the furnace is very small, that the gas within the chimney has a uniform temperature $T_2^*$, and that the air surrounding the furnace and chimney has a uniform temperature $T_1$. Indicate any other assumptions that must be made in order to solve the problem. *Hint:* See Exercise 3.4.2-6.

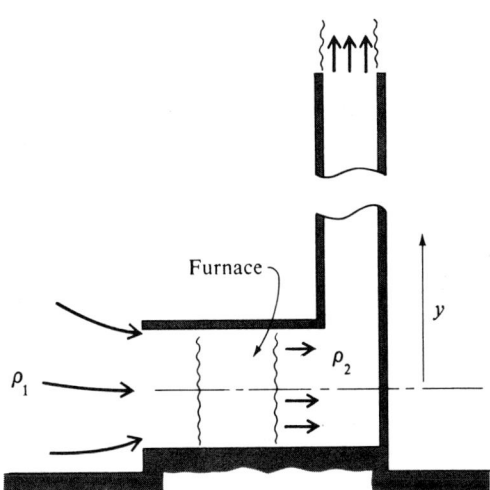

**Fig. 3.4.2-2** A simple chimney design.

*Answer*

$$v = \left[2gy\left(\frac{\rho_1}{\rho_2} - 1\right)\right]^{\frac{1}{2}}$$

(b) How should the cross-sectional area $A$ of the chimney vary as a function of position for a given mass flow rate $\mathcal{M}$ of gas?
*Answer*

$$A = \frac{\mathcal{M}\exp(gy\rho_1/P_1)}{\rho_2\left[2gy\left(\frac{\rho_1}{\rho_2} - 1\right)\right]^{\frac{1}{2}}}$$

**3.4.2-8** A vertical tube of length $L$ is filled with an incompressible fluid and a plate is held over the lower end. At time $t = 0$, the plate is removed and the fluid is allowed to run out of the tube.

Use Bernoulli's equation to derive an expression for the time required to empty the tube. (See also Exercises 3.4.4-4 and 4.4.8-4.)

*Hint:* Don't hesitate to make a reasonable approximation in evaluating the line integral.

**3.4.2-9** Let us now assume that the lower end of the vertical tube in Exercise 3.4.2-8 is finished with a rounded orifice. Use Bernoulli's equation to derive a differential equation that describes (approximately) the height of water in the tube as a function of time. Give the required boundary conditions.

*Answer*

$$f\frac{d^2f}{dt^2} + \frac{1}{2}\left[1 - \left(\frac{A_{(t)}}{A_{(0)}}\right)^2\right]\left(\frac{df}{dt}\right)^2 + gf = 0$$

At $t = 0$: $\quad f = L \quad$ and $\quad \dfrac{df}{dt} = 0$

## 3.4.3 The D'Alembert paradox

Let us consider a steady-state potential flow about a body submerged in an incompressible newtonian fluid which extends to infinity in all directions. Assume that the fluid at infinity is stationary and that the body moves with a constant velocity **U**.

Following our discussion in Sec. 3.4.1, by a potential flow we mean one in which the velocity vector is representable in terms of a scalar potential $\Phi$:

$$\mathbf{v} = -\nabla\Phi \tag{3-1}$$

Since the fluid is incompressible, the scalar potential $\Phi$ satisfies Laplace's equation

$$\text{div}(\nabla\Phi) = 0 \tag{3-2}$$

Let $r$ be a radial spherical coordinate measured from the center of mass of the body and let $S_B$ denote the closed bounding surface of the body. For the problem under consideration, the boundary conditions take the form

As $r \to \infty$: $\quad \nabla\Phi = 0 \tag{3-3}$

and

On $S_B$: $\quad -\nabla\Phi = \mathbf{U} \tag{3-4}$

We define **n** to be the outwardly directed unit normal vector with respect to $S_B$.

At every point on $S_B$ we also define two unit vectors, $\mathbf{t}_{(A)}$ and $\mathbf{t}_{(B)}$, which are mutually orthogonal and which are tangent to the surface. These definitions allow us to write Eq. (3-4) in the general form

On $S_B$: $\quad \nabla\Phi \cdot \mathbf{n}$ (3-5)

is specified and

On $S_B$: $\quad \nabla\Phi \cdot \mathbf{t}_{(A)} \quad$ and $\quad \nabla\Phi \cdot \mathbf{t}_{(B)}$ (3-6)

are specified.

A unique solution of Eq. (3-2) with boundary conditions (3-3) and (3-5) exists with the property [1, pp. 211, 216, 311]

$$\text{As } r \to \infty: \quad r\Phi \quad r^2 \frac{\partial \Phi}{\partial z_1} \quad r^2 \frac{\partial \Phi}{\partial z_2} \quad r^2 \frac{\partial \Phi}{\partial z_3} \quad (3\text{-}7)$$

are bounded in absolute value.

The immediate significance of this is that the specification of the two tangential components of velocity, Eq. (3-6), is not used. Physically this implies that potential flows allow slip at bounding solid surfaces. This is clearly in violation of our discussion of boundary conditions in Sec. 3.1.1 and we must start to worry about the physical significance of potential flows.

Slip at a solid surface suggests that the fluid moves past the solid surface without resistance. Let us see what we can learn about the force $f$ which a fluid in potential flow exerts on a submerged body. We again specifically concern ourselves with the case where the fluid extends to infinity in all directions. The force $f$ may be expressed as

$$f = \int_{S_B} \mathbf{T} \cdot \mathbf{n} \, dS \quad (3\text{-}8)$$

This suggests an application of Green's transformation (Sec. A.11.2), but we must be cautious in applying Green's transformation to unbounded regions. Let $\Sigma_r$ be a spherical surface of radius $r$ which is centered upon the center of mass of the body and which totally encloses the body. Let $V_r$ denote the region of fluid enclosed by $\Sigma_r$ and $S_B$. By Green's transformation

$$f_r = \int_{S_B} \mathbf{T} \cdot \mathbf{n} \, dS = -\int_{V_r} \text{div } \mathbf{T} \, dV + \int_{\Sigma_r} \mathbf{T} \cdot \mathbf{n} \, dS \quad (3\text{-}9)$$

In the last integral on the right, $\mathbf{n}$ is outwardly directed with respect to the closed surface $\Sigma_r$. In the integral on the left, $\mathbf{n}$ is outwardly directed with respect to the closed surface $S_B$; whereas, we need a unit normal which is outwardly directed with respect to the closed bounding surface of $V_r$ for an application of Green's transformation. This accounts for the signs of the integrals.

Let us examine the two terms on the right of Eq. (3-9) individually. By the equation of continuity, we may write

$$\rho\frac{d_{(m)}\mathbf{v}}{dt} = \frac{d_{(m)}(\rho\mathbf{v})}{dt} + \rho\mathbf{v}(\text{div } \mathbf{v}) = \frac{\partial(\rho\mathbf{v})}{\partial t} + \text{div }(\rho\mathbf{vv}) \tag{3-10}$$

This allows us to express Cauchy's first law as

$$\frac{\partial(\rho\mathbf{v})}{\partial t} + \text{div }(\rho\mathbf{vv}) = \text{div } \mathbf{T} + \rho\mathbf{f} \tag{3-11}$$

Since we are considering a steady-state flow, the first term on the right of Eq. (3-9) becomes

$$-\int_{V_r} \text{div } \mathbf{T} \, dV = -\int_{V_r} \text{div }(\rho\mathbf{vv}) \, dV + \int_{V_r} \rho\mathbf{f} \, dV \tag{3-12}$$

By Green's transformation and boundary condition (3-4), the first term on the right of Eq. (3-12) becomes

$$-\int_{V_r} \text{div }(\rho\mathbf{vv}) \, dV = -\int_{\Sigma_r} \rho\mathbf{vv} \cdot \mathbf{n} \, dS + \rho\mathbf{UU} \cdot \int_{S_B} \mathbf{n} \, dS = -\int_{\Sigma_r} \rho\mathbf{vv} \cdot \mathbf{n} \, dS \tag{3-13}$$

If we assume that the external force per unit mass may be represented by a potential as described in Sec. 2.5.1, the second term on the right of Eq. (3-12) may also be subjected to Green's transformation:

$$\int_{V_r} \rho\mathbf{f} \, dV = -\int_{V_r} \nabla(\rho\varphi) \, dV = -\int_{\Sigma_r} \rho\varphi\mathbf{n} \, dS + \int_{S_B} \rho\varphi\mathbf{n} \, dS$$

$$= -\int_{\Sigma_r} \rho\varphi\mathbf{n} \, dS + \int_{V_B} \nabla(\rho\varphi) \, dV = -\int_{\Sigma_r} \rho\varphi\mathbf{n} \, dS - \int_{V_B} \rho\mathbf{f} \, dV \tag{3-14}$$

By $V_B$ we refer to the region occupied by the submerged body.

The Bernoulli equation for a steady-state potential flow of an incompressible newtonian fluid gives

$$P = -\rho\varphi - \tfrac{1}{2}\rho v^2 + P_{(0)} + \rho\varphi_{(0)} + \tfrac{1}{2}\rho U^2 \tag{3-15}$$

Here $P_{(0)}$ and $\varphi_{(0)}$ are evaluated at some reference point on the body. This allows us to write the second term on the right-hand side of Eq. (3-9) as

$$\int_{\Sigma_r} \mathbf{T} \cdot \mathbf{n} \, dS = \int_{\Sigma_r} (\rho\varphi + \tfrac{1}{2}\rho v^2)\mathbf{n} \, dS - (P_{(0)} + \rho\varphi_{(0)} + \tfrac{1}{2}\rho U^2)\int_{\Sigma_r} \mathbf{n} \, dS$$

$$+ \int_{\Sigma_r} \mathbf{S} \cdot \mathbf{n} \, dS = \int_{\Sigma_r} (\rho\varphi + \tfrac{1}{2}\rho v^2)\mathbf{n} \, dS + \mu\int_{\Sigma_r} [\nabla\mathbf{v} + (\nabla\mathbf{v})^T] \cdot \mathbf{n} \, dS \tag{3-16}$$

Equations (3-12), (3-13), (3-14), and (3-16) may be used to obtain from Eq. (3-9)

## [3.4] NONVISCOUS FLOWS

$$f_r = -\int_{\Sigma_r} \rho \mathbf{vv} \cdot \mathbf{n} \, dS + \int_{\Sigma_r} \tfrac{1}{2}\rho v^2 \mathbf{n} \, dS + \mu \int_{\Sigma_r} [\nabla \mathbf{v} + (\nabla \mathbf{v})^T] \cdot \mathbf{n} \, dS - \int_{V_B} \rho \mathbf{f} \, dV \qquad (3\text{-}17)$$

We are ready now to examine the limit as $r \to \infty$. From Eqs. (3-3) and (3-7), we deduce that

$$\text{As } r \to \infty: \quad \mathbf{v} = 0 + O\!\left(\frac{1}{r^2}\right) \quad \nabla \mathbf{v} = 0 + O\!\left(\frac{1}{r^3}\right) \qquad (3\text{-}18)$$

where $O(1/r^2)$, for example, is read "terms of the order $1/r^2$" or "terms which approach zero as rapidly as $1/r^2$." Equations (3-18) allow us to estimate

$$\lim_{r \to \infty} \int_{\Sigma_r} \rho \mathbf{vv} \cdot \mathbf{n} \, dS = \lim_{r \to \infty} \int_{\Sigma_r} O\!\left(\frac{1}{r^4}\right) dS = 0$$

$$\lim_{r \to \infty} \int_{\Sigma_r} \tfrac{1}{2}\rho v^2 \mathbf{n} \, dS = \lim_{r \to \infty} \int_{\Sigma_r} O\!\left(\frac{1}{r^4}\right) dS = 0 \qquad (3\text{-}19)$$

$$\lim_{r \to \infty} \int_{\Sigma_r} [\nabla \mathbf{v} + (\nabla \mathbf{v})^T] \cdot \mathbf{n} \, dS = \lim_{r \to \infty} \int_{\Sigma_r} O\!\left(\frac{1}{r^3}\right) dS = 0$$

From Eqs. (3-17) and (3-19) we conclude that

$$f = \lim_{r \to \infty} f_r = -\int_{V_B} \rho \mathbf{f} \, dV \qquad (3\text{-}20)$$

Many of us prefer to think in terms of the total force on the submerged body rather than only of the force which the fluid exerts on the body. The additional term is the effect of the external force, which is taken here to be gravity. Let us refer to the density of the solid as $\rho_{(s)}$ and the density of the fluid as $\rho_{(f)}$. From Eq. (3-20) the total force exerted on a submerged body in a steady-state potential flow of an incompressible newtonian fluid is

$$\int_{V_B} \rho_{(s)} \mathbf{f} \, dV + \int_S \mathbf{T} \cdot \mathbf{n} \, dS = \int_V (\rho_{(s)} - \rho_{(f)}) \mathbf{f} \, dV \qquad (3\text{-}21)$$

Physically, this means that the total force exerted on the submerged body in the potential flow is the same as that which is exerted on the body when the body is stationary: the buoyant force. Another way of looking at this is to say that no work is required to move the body through the fluid. Clearly this violates our everyday experience and all known experimental measurements.

The result that the force on a submerged body in a potential flow is the buoyant force is known as the D'Alembert paradox, first recognized by Jean D'Alembert (1717–1783) [2, p. 102].

We have found here that a velocity distribution which is representable by a potential cannot satisfy the boundary conditions on the tangential components of velocity required at the surface of a submerged obstacle. Nor can such a velocity distribution be used to determine the force which a fluid in relative motion exerts on a submerged body. A natural reaction by the reader might be to question why we have spent so much time on this subject if it has no practical value.

But it does have practical value. In Sec. 3.5.1 we find that potential flows may often do an excellent job of portraying the movements of real fluids outside the immediate neighborhood of a submerged object. For this reason we consider specific examples of potential flows in the two sections which follow.

**REFERENCES**

1. Kellogg, O. D.: "Foundations of Potential Theory," Ungar, New York, 1929.
2. Rouse, Hunter, and Simon Ince: "History of Hydraulics," State University of Iowa, Iowa City, 1957.

**3.4.4 Potential flow past a sphere** Let us consider steady-state potential flow past a stationary sphere such that a fluid at a very large distance from the sphere moves with a uniform velocity in the positive $z_3$ direction:

$$\text{As } r \to \infty: \quad \mathbf{v} \to V\mathbf{e}_3 \tag{4-1}$$

The origins of the spherical and rectangular cartesian coordinate systems referred to here coincide with the center of the sphere; the relationship between these coordinate systems is that adopted in Exercise A.4.1-5. We wish to determine the velocity potential $\Phi$ for this flow and the corresponding velocity distribution

$$\mathbf{v} = -\nabla\Phi \tag{4-2}$$

Boundary condition (4-1) may be interpreted as a restriction on the velocity potential:

$$\text{As } r \to \infty: \quad \Phi \to -Vz_3 = -Vr\cos\theta \tag{4-3}$$

Our discussion in Sec. 3.4.3 indicates that only the normal component of velocity at the surface of the sphere can be satisfied by the potential flow. If $a$ denotes the radius of a sphere, this means that

$$\text{At } r = a: \quad v_r = -\frac{\partial\Phi}{\partial r} = 0 \tag{4-4}$$

where we have used the physical components of the gradient of a scalar found in Exercise A.4.3-6.

Since a sphere is a body of revolution and since boundary condition (4-3) is axially symmetric, let us look for a velocity distribution of the form

$$v_r = v_r(r,\theta) \qquad v_\theta = v_\theta(r,\theta) \qquad v_\varphi = 0 \tag{4-5}$$

This means that we are seeking a velocity potential which is independent of the spherical coordinate $\varphi$:

$$\Phi = \Phi(r,\theta) \tag{4-6}$$

## [3.4] NONVISCOUS FLOWS

From Table 3.4.1-1, Laplace's equation becomes in spherical coordinates

$$\frac{\partial}{\partial r}\left(r^2 \frac{\partial \Phi}{\partial r}\right) + \frac{1}{\sin \theta} \frac{\partial}{\partial \theta}\left(\sin \theta \frac{\partial \Phi}{\partial \theta}\right) = 0 \tag{4-7}$$

We wish to find a solution to this equation that satisfies boundary conditions (4-3) and (4-4).

If we use the method of separation of the variables, it is easy to show that Eq. (4-7) has a solution of the form [1, p. 30]

$$\Phi = \sum_{n=0}^{\infty} \{r^n[A_n P_n(\cos \theta) + B_n Q_n(\cos \theta)] + r^{-n-1}[C_n P_n(\cos \theta) + D_n Q_n(\cos \theta)]\} \tag{4-8}$$

where $P_n(x)$ is the Legendre function of the first kind, $Q_n(x)$ is the Legendre function of the second kind, and $A_n$, $B_n$, $C_n$, and $D_n$ are constants. Since

$$\text{As } \cos \theta \rightarrow \pm 1: \quad Q_n(\cos \theta) \rightarrow \infty \tag{4-9}$$

we take

$$B_n = D_n = 0 \quad \text{for} \quad n \geq 0 \tag{4-10}$$

Recognizing that

$$P_1(\cos \theta) = \cos \theta \tag{4-11}$$

we see that Eq. (4-8) must reduce to

$$\Phi = -VrP_1(\cos \theta) + \sum_{n=0}^{\infty} r^{-n-1} C_n P_n(\cos \theta) \tag{4-12}$$

in order that it satisfy Eq. (4-3). Applying boundary condition (4-4) to Eq. (4-12), we find that

$$\text{At } r = a: \quad VP_1(\cos \theta) + \sum_{n=0}^{\infty} (n+1)a^{-n-2} C_n P_n(\cos \theta) = 0 \tag{4-13}$$

The Legendre functions of the first kind are orthogonal on the interval $-1 \leq x \leq 1$ [2, p. 116]:

$$\int_{-1}^{+1} P_m(x) P_n(x) \, dx = 0 \quad m \neq n \tag{4-14}$$

$$\int_{-1}^{+1} [P_n(x)]^2 \, dx = \frac{2}{2n+1} \tag{4-15}$$

These orthogonality relations allow us to conclude from Eq. (4-13) that

$$C_1 = -\frac{Va^3}{2} \quad C_0 = 0 \quad C_n = 0 \quad \text{for} \quad n \geq 2 \tag{4-16}$$

To summarize, we find that the velocity potential is given by

$$\Phi = -V\left(r + \frac{a^3}{2r^2}\right)\cos\theta \tag{4-17}$$

Using Exercise A.4.3-6, we learn that the corresponding components of velocity are

$$v_r = -\frac{\partial \Phi}{\partial r} = V\left(1 - \frac{a^3}{r^3}\right)\cos\theta \tag{4-18}$$

$$v_\theta = -\frac{1}{r}\frac{\partial \Phi}{\partial \theta} = -V\left(1 + \frac{a^3}{2r^3}\right)\sin\theta \tag{4-19}$$

and

$$v_\varphi = 0 \tag{4-20}$$

This confirms our initial assumption that there is at least one solution to this problem of the form assumed in Eq. (4-6). It is shown elsewhere that this solution is unique [3, p. 95].

To help in visualizing what we have found here, let us calculate the stream function $\psi$ for this solution. From Table 2.5.2-1,

$$v_r = \frac{1}{r^2 \sin\theta}\frac{\partial \psi}{\partial \theta} \tag{4-21}$$

and

$$v_\theta = -\frac{1}{r \sin\theta}\frac{\partial \psi}{\partial r} \tag{4-22}$$

Equations (4-18) and (4-19) allow us to carry out a line integration to determine the stream function:

$$\psi = \int_0^\psi d\psi = \int_a^r \left.\frac{\partial \psi}{\partial r}\right|_{\theta=0} dr + \int_0^\theta \frac{\partial \psi}{\partial \theta} d\theta$$

$$= \frac{Va^2}{2}\left(\frac{r^2}{a^2} - \frac{a}{r}\right)\sin^2\theta \tag{4-23}$$

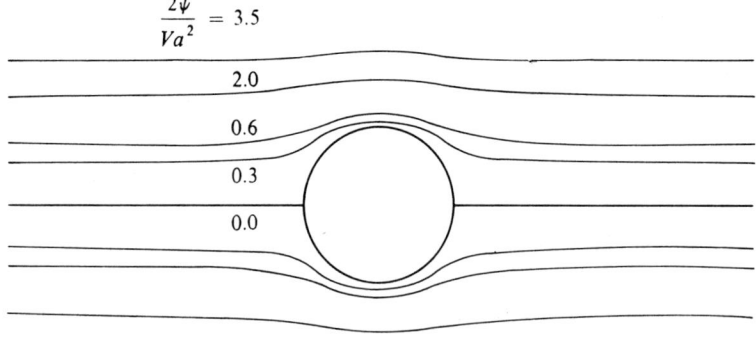

**Fig. 3.4.4-1** Streamlines for potential flow past a sphere.

Since the stream function is arbitrary to a constant, we have chosen in this integration

At $r = a$, $\theta = 0$: $\quad \psi = 0$ (4-24)

Various streamlines corresponding to different values of $2\psi/Va^2$ are shown in Fig. 3.4.4-1.

## REFERENCES

1. Irving, J., and N. Mullineux: "Mathematics in Physics and Engineering," Academic, New York, 1959.
2. Jahnke, Eugene, and Fritz Emde: "Tables of Functions," 4th ed., Dover, New York, 1945.
3. Milne-Thomson, L. M.: "Theoretical Hydrodynamics," 3d ed., Macmillan, New York, 1955.

## EXERCISES

**3.4.4-1** (a) Using the results of this section and Sec. 3.4.2, determine the pressure distribution in the fluid.

(b) Determine that the total force exerted upon the sphere is the buoyant force.

**3.4.4-2** *Potential flow past a cylinder* Consider steady-state, plane potential flow past a stationary cylinder such that

As $r \to \infty$: $\quad \mathbf{v} \to V\mathbf{e}_1$

and such that the viscous terms may be neglected in the equation of motion. By plane flow, we mean that, with respect to a rectangular cartesian coordinate system,

$$v_1 = v_1(z_1, z_2) \quad v_2 = v_2(z_1, z_2) \quad v_3 = 0$$

Determine the velocity potential for this flow and the corresponding velocity components with respect to a cylindrical coordinate system.

Answer

$$\Phi = -V\left(r + \frac{a^2}{r}\right)\cos\theta + \Lambda\theta$$

where $\Lambda$ is an arbitrary constant.

**3.4.4-3** (a) Using the results of Exercise 3.4.4-2 and Sec. 3.4.2, determine the pressure distribution in the fluid.

(b) Calculate both the $z_1$ and $z_2$ components of the force exerted upon the cylinder by the fluid beyond the buoyant force.

**3.4.4-4** A vertical tube of length $L$ is filled with an incompressible fluid and a plate is held over the lower end. At time $t = 0$, the plate is removed and the fluid is allowed to run out of the tube. Derive an expression for the time required to empty the tube.

You may assume that this is a potential flow, neglect viscous effects in Cauchy's first law, and assume that the top surface of the fluid remains a horizontal plane as the tube empties.

See also Exercises 3.4.2-8 and 4.4.8-4.

**3.4.4-5** *Flow in divergent or convergent channels* Consider two plane walls oriented in such a way that they form surfaces of constant $\theta$ in cylindrical coordinates. An incompressible fluid moves in either the positive or negative $r$ direction between these two planes. This is known as flow in either a divergent or a convergent channel. Determine the velocity distribution for the corresponding steady-state potential flow, assuming that the viscous terms may be neglected in Cauchy's first law.

**134** APPLICATIONS OF THE DIFFERENTIAL BALANCES TO MOMENTUM TRANSFER

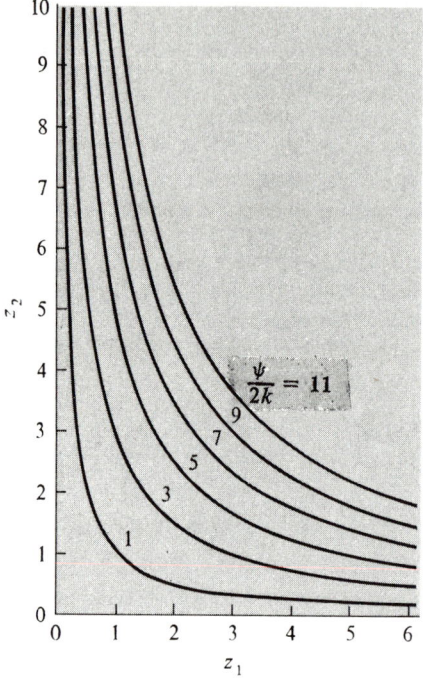

**Fig. 3.4.5-1** Streamlines for plane potential flow in the neighborhood of a corner.

**3.4.5 Plane potential flow in a corner** Consider the plane flow whose streamlines are pictured in Fig. 3.4.5-1. Fluid approaches the plane $z_2 = 0$ from the $z_2$ direction, is turned to one side by the corner, and departs in the $z_1$ direction. By *plane flow*, we mean that

$$v_1 = v_1(z_1, z_2) \qquad v_2 = v_2(z_1, z_2) \qquad v_3 = 0 \tag{5-1}$$

Let us determine the velocity potential and the velocity distribution for the corresponding potential flow.

Rather than trying to describe the fluid movements in this entire semi-infinite region, let us restrict our attention to the immediate neighborhood of the corner where the velocity potential $\Phi$ can be represented by a Taylor expansion about the origin;

$$\Phi = \Phi(z_1, z_2) = \Phi(0,0) + z_1 \frac{\partial \Phi}{\partial z_1}(0,0)$$

$$+ z_2 \frac{\partial \Phi}{\partial z_2}(0,0) + \tfrac{1}{2}(z_1)^2 \frac{\partial^2 \Phi}{\partial z_1^2}(0,0)$$

$$+ z_1 z_2 \frac{\partial^2 \Phi}{\partial z_1 \partial z_2}(0,0)$$

$$+ \tfrac{1}{2}(z_2)^2 \frac{\partial^2 \Phi}{\partial z_2^2}(0,0) + \cdots \tag{5-2}$$

[3.4] NONVISCOUS FLOWS

We require as boundary conditions that

$$\text{At } z_1 = 0: \quad v_1 = -\frac{\partial \Phi}{\partial z_1} = 0 \tag{5-3}$$

and

$$\text{At } z_2 = 0: \quad v_2 = -\frac{\partial \Phi}{\partial z_2} = 0 \tag{5-4}$$

These boundary conditions mean that in Eq. (5-2)

$$\frac{\partial \Phi}{\partial z_1}(0,0) = \frac{\partial \Phi}{\partial z_2}(0,0) = \frac{\partial^2 \Phi}{\partial z_1 \partial z_2}(0,0) = 0 \tag{5-5}$$

Laplace's equation must hold everywhere, in particular at the origin:

$$\frac{\partial^2 \Phi}{\partial z_2^2}(0,0) = -\frac{\partial^2 \Phi}{\partial z_1^2}(0,0) = k = \text{constant} \tag{5-6}$$

Since the velocity potential is arbitrary to a constant, we choose to define

$$\Phi(0,0) = 0 \tag{5-7}$$

Equations (5-5) to (5-7) allow us to express Eq. (5-2) as

$$\Phi = \frac{k}{2}[(z_2)^2 - (z_1)^2] + \cdots \tag{5-8}$$

The corresponding velocity components are

$$v_1 = -\frac{\partial \Phi}{\partial z_1} = kz_1 + \cdots \tag{5-9}$$

and

$$v_2 = -\frac{\partial \Phi}{\partial z_2} = -kz_2 + \cdots \tag{5-10}$$

It is helpful in visualizing these results to compute the stream function and to plot the streamlines. From Table 2.5.2-1,

$$v_1 = \frac{\partial \psi}{\partial z_2} \quad \text{and} \quad v_2 = -\frac{\partial \psi}{\partial z_1} \tag{5-11}$$

Equations (5-9) to (5-11) allow us to carry out a line integration to determine the stream function:

$$\psi = \int_0^\psi d\psi = \int_0^{z_1} \frac{\partial \psi}{\partial z_1}\bigg|_{z_2=0} dz_1 + \int_0^{z_2} \frac{\partial \psi}{\partial z_2} dz_2 = 2kz_1 z_2 + \cdots \tag{5-12}$$

Since the stream function is arbitrary to a constant, we have chosen in this integration

At $z_1 = z_2 = 0$: $\quad \psi = 0$ (5-13)

Various streamlines corresponding to different values of $\psi/k$ are shown in Fig. 3.4.5-1.

It is important to realize that the first terms of the velocity potential given by Eq. (5-8) cannot possibly describe a real flow at large distances from the origin. Equations (5-9) and (5-10) indicate that the magnitude of velocity is proportional to the distance from the origin:

$$|\mathbf{v}| = k\,|\mathbf{z}| \tag{5-14}$$

and, consequently, is unbounded from above in the field of flow. This reinforces the viewpoint suggested when we expanded the velocity potential in a Taylor series earlier in this discussion: the results found here should be thought of as describing a potential flow in the neighborhood of the corner.

Our method of seeking a solution here (expansion of $\Phi$ in a Taylor series about the origin) implies that the first two terms given in Eq. (5-8) cannot possibly represent a unique potential flow for this geometry. But to make the point more plainly, consider [1, p. 211]

$$\Phi = -B e^{(z_1)^2 - (z_2)^2} \cos(2 z_1 z_2) \tag{5-15}$$

This is another solution to Laplace's equation which satisfies boundary conditions (5-3) and (5-4).

As an added benefit, the solution found here for potential flow in the neighborhood of a corner also represents potential flow in the neighborhood of a stagnation point on the plane $z_2 = 0$, the streamlines for which are shown in Fig. 3.4.5-2. Fluid approaching from the positive $z_2$ direction is turned aside by the plane $z_2 = 0$; it subsequently moves in the positive and negative $z_1$ directions. This flow has a plane of symmetry at $z_1 = 0$ along which the $z_1$ component of velocity is zero. Since the velocity is zero at the origin, it is known as a stagnation point. In the neighborhood of this stagnation point, the stream function is again given by Eq. (5-12); the corresponding streamlines are shown in Fig. 3.4.5-2 for various values of $\psi/k$. In Sec. 3.5.5 we give another analysis of plane stagnation flow taking viscous effects into account.

## REFERENCE

1. Churchill, R. V.: "Complex Variables and Applications," 2d ed., McGraw-Hill, New York, 1960.

## EXERCISES

**3.4.5-1** Prove that Eq. (5-15) satisfies Laplace's equation as well as boundary conditions (5-3) and (5-4).

**3.4.5-2** For the flow described in this section, determine the pressure distribution on the plane $z_2 = 0$.

## [3.5] BOUNDARY-LAYER THEORY

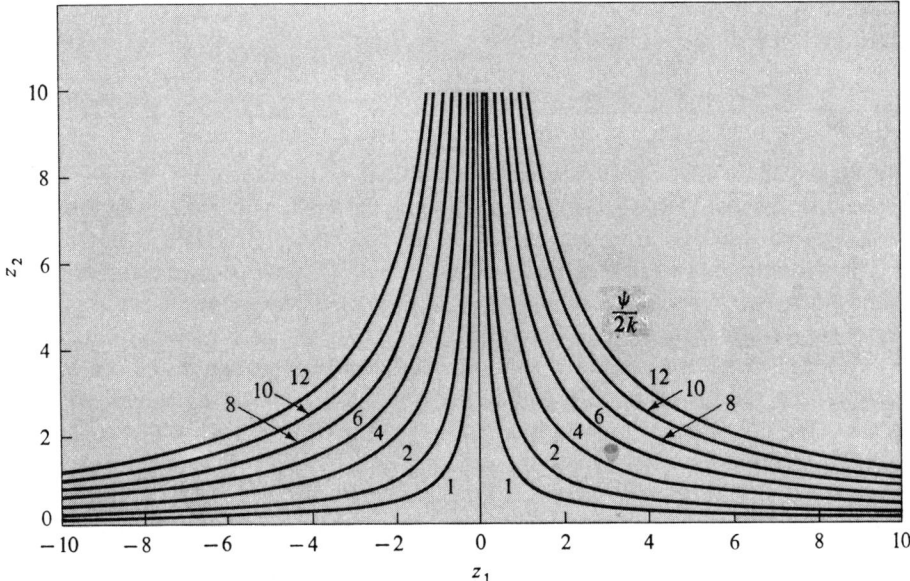

**Fig. 3.4.5-2** Streamlines for plane potential flow in the neighborhood of a stagnation point.

## 3.5 BOUNDARY-LAYER THEORY

**3.5.1 Boundary-layer theory for plane flow past a flat plate** In Sec. 3.3.1 we expressed the Navier-Stokes equation (for an incompressible newtonian fluid under conditions such that the external force vector per unit mass is representable by a potential) in a dimensionless form,

$$\frac{1}{N_{\text{St}}} \frac{\partial \mathbf{v}^*}{\partial t^*} + (\nabla \mathbf{v}^*) \cdot \mathbf{v}^* = -\frac{1}{N_{\text{Ru}}} \nabla \mathscr{P}^* + \frac{1}{N_{\text{Re}}} \text{div } (\nabla \mathbf{v}^*) \tag{1-1}$$

We argued there that for sufficiently small values of the Reynolds number $N_{\text{Re}}$, the convective inertial terms $(\nabla \mathbf{v}^*) \cdot \mathbf{v}^*$ might be neglected with respect to the viscous terms. This intuitive argument seems to work in three-dimensional flows such as flow in a cone-plate viscometer discussed in Sec. 3.3.2, but we pointed out that it fails in plane motions. A velocity field is defined to be *plane* if there is some rectangular cartesian coordinate system such that $v_1 = v_1(z_1, z_2)$, $v_2 = v_2(z_1, z_2)$, and $v_3 = 0$.

In Sec. 3.4.1 we suggested that, for sufficiently large values of the Reynolds number, the viscous terms might be neglected with respect to the convective inertial terms. We investigated one such solution, potential flow, and found in Sec. 3.4.3 that this solution does not satisfy the required boundary conditions on the tangential components of the velocity vector. Further, we discovered that the potential flow solution predicts that an incompressible fluid will exert no force (beyond the buoyant force) on an object submerged in it. This is in obvious disagreement with our

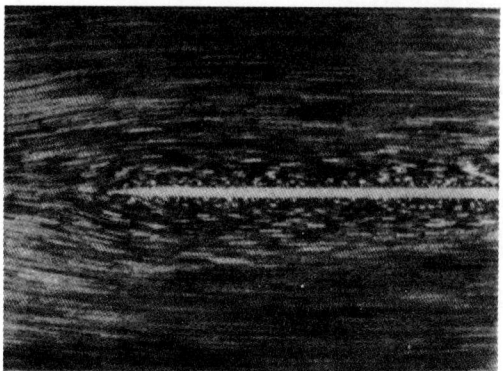

**Fig. 3.5.1-1** Flow past a thin plate of length $L$ for $N_{Re} = v_0 L \rho/\mu = 3$. *(From L. Prandtl and O. G. Tietjens [1, p. 306, plate 27, fig. 68].)*

everyday physical observations, suggesting that we may have been in error when we neglected all the viscous effects with respect to the convective inertial terms in Sec. 3.4.1.

It should not be too surprising that an argument that suggests one term in a differential equation may be neglected with respect to another occasionally fails. There is no mathematical basis for such a step. It is reasonable to neglect one term with respect to another in the *solution* of a differential equation, but entirely another matter in the differential equation used to obtain that solution. Rather than being surprised when an overly simple argument such as this fails, we should be grateful that it is so often helpful.

With these thoughts in mind, let us reexamine the problem of flows at large Reynolds numbers. It helps to be acquainted with what one might observe experimentally. In order to be specific, we restrict our attention to plane flow past a *thin* flat plate. The concepts we develop by looking at this particular flow may be readily extended to other situations.

In the time-lapse photograph shown in Fig. 3.5.1-1, we see particle paths in flow past a flat plate, which were made visible by sprinkling small particles on the water surface. The streaks made by individual particles are proportional to their speed. In a thin region next to the plate the velocity is considerably smaller than at a larger distance, where the flow is essentially undisturbed. In Fig. 3.5.1-2, we illustrate on an exaggerated scale the velocity distribution in the immediate neighborhood of the plate. If we observed this flow as a function of the Reynolds number, we would find that the thickness of this region of retarded velocity decreased as the Reynolds number increased.

It appears on the basis of this limited evidence that outside a very thin region next to a body the flow is essentially that predicted by potential flow. This is the case for relatively thin streamlined bodies, such as airfoils. More generally, this

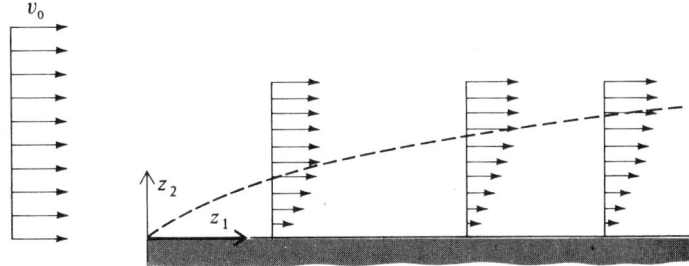

**Fig. 3.5.1-2** The character of the velocity distribution in flow past a flat plate.

thin boundary-layer region, where viscous effects are very important, can appear to "separate" from a body [2, p. 28] and vortices can form at some distance from the body; see, for example, the photograph showing streak lines in flow past a circular cylinder in Fig. 3.5.1-3. Here the flow at some distance from the body is obviously rotational, which means that the flow outside the boundary layer on the leading surface of the cylinder would not be precisely that predicted by the potential flow [2, p. 21]. It is perhaps safer to say that viscous effects may be ignored outside the boundary layer. A potential flow represents one particular type of nonviscous flow for a given set of boundary conditions on velocity, but it does *not* form a *unique solution*.

From what we have observed and suggested above, it seems clear that viscous effects are just as important as inertial effects in the equation of motion as the Reynolds number becomes very large, at least in the boundary layer. The argument we pre-

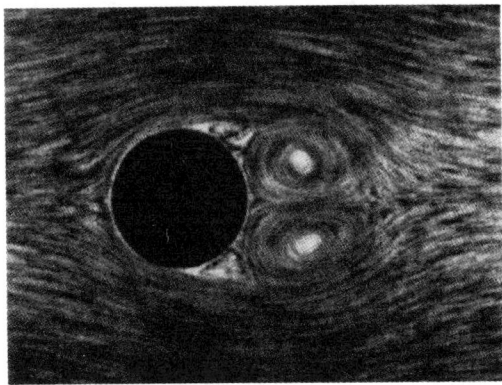

**Fig. 3.5.1-3** Flow from left to right past a circular cylinder; boundary layer has separated from the cylinder surface and two vortices have formed in the cylinder's wake. (*From L. Prandtl and O. G. Tietjens* [*1, p. 279, plate 3, fig. 6*].)

sented in Sec. 3.4.1 is not applicable to this region, since it says that viscous effects are negligibly small compared with inertial effects; apparently, this treatment must be reserved for the flow outside the boundary layer. If an approximation similar to that of Sec. 3.4.1 is to be used for the boundary layer, we must revise the definitions of our dimensionless variables so as to prevent the disappearance of all the viscous terms. Since the thickness of the boundary layer decreases as the Reynolds number increases, it seems reasonable that we should magnify the thickness of the boundary layer by introducing in the discussion of flow past a flat plate

$$z_2^{**} = (N_{Re})^a z_2^* \tag{1-2}$$

while retaining the previous definitions for the other two dimensionless coordinates.

In the context of the definition of the velocity vector as the material derivative of the position vector in Sec. 1.1.1, Eq. (1-2) suggests as a reasonable definition for the dimensionless $z_2$ component of velocity

$$v_2^{**} = (N_{Re})^a v_2^* \tag{1-3}$$

Alternatively, if we require that each term in the equation of continuity have equal weight,

$$\frac{\partial v_1^*}{\partial z_1^*} + (N_{Re})^a \frac{\partial v_2^*}{\partial z_2^{**}} + \frac{\partial v_3^*}{\partial z_3^*} = 0 \tag{1-4}$$

we are again led to Eq. (1-3).

Let us assume that our plate is of infinite width so that we may reasonably take the flow to be plane:

$$\begin{aligned} v_1^* &= v_1^*(z_1^*, z_2^{**}, t^*) \\ v_2^{**} &= v_2^{**}(z_1^*, z_2^{**}, t^*) \\ v_3^* &= 0 \end{aligned} \tag{1-5}$$

Under these conditions, the components of the dimensionless Navier-Stokes equation of Sec. 3.3.1 are

$$\frac{1}{N_{St}} \frac{\partial v_1^*}{\partial t^*} + v_1^* \frac{\partial v_1^*}{\partial z_1^*} + v_2^{**} \frac{\partial v_1^*}{\partial z_2^{**}} = -\frac{1}{N_{Ru}} \frac{\partial \mathscr{P}^*}{\partial z_1^*}$$

$$+ \frac{1}{N_{Re}} \frac{\partial^2 v_1^*}{\partial z_1^{*2}} + (N_{Re})^{2a-1} \frac{\partial^2 v_1^*}{\partial z_2^{**2}} \tag{1-6}$$

$$\frac{1}{N_{St}} \frac{\partial v_2^{**}}{\partial t^*} + v_1^* \frac{\partial v_2^{**}}{\partial z_1^*} + v_2^{**} \frac{\partial v_2^{**}}{\partial z_2^{**}} = -\frac{(N_{Re})^{2a}}{N_{Ru}} \frac{\partial \mathscr{P}^*}{\partial z_2^{**}}$$

$$+ \frac{1}{N_{Re}} \frac{\partial^2 v_2^{**}}{\partial z_1^{*2}} + (N_{Re})^{2a-1} \frac{\partial^2 v_2^{**}}{\partial z_2^{**2}} \tag{1-7}$$

and

$$\frac{\partial \mathscr{P}^*}{\partial z_3^*} = 0 \tag{1-8}$$

## [3.5] BOUNDARY-LAYER THEORY

If any viscous terms are to be the same order of magnitude as the inertial terms as the Reynolds number becomes very large, we must identify

$$a = \tfrac{1}{2} \qquad (1\text{-}9)$$

In the limit, as the Reynolds number becomes unbounded, Eq. (1-6) reduces to

$$\frac{1}{N_{St}} \frac{\partial v_1^*}{\partial t^*} + v_1^* \frac{\partial v_1^*}{\partial z_1^*} + v_2^{**} \frac{\partial v_1^*}{\partial z_2^{**}} = -\frac{1}{N_{Ru}} \frac{\partial \mathscr{P}^*}{\partial z_1^*} + \frac{\partial^2 v_1^*}{\partial z_2^{**2}} \qquad (1\text{-}10)$$

and Eq. (1-7) reduces to

$$\frac{\partial \mathscr{P}^*}{\partial z_2^{**}} = 0 \qquad (1\text{-}11)$$

Equations (1-8) and (1-11) mean that

$$\mathscr{P}^* = \mathscr{P}^*(z_1^*) \qquad (1\text{-}12)$$

To summarize, by the above arguments for flow past a thin flat plate, the equation of continuity and the equation of motion have been reduced, respectively, to

$$\frac{\partial v_1^*}{\partial z_1^*} + \frac{\partial v_2^{**}}{\partial z_2^{**}} = 0 \qquad (1\text{-}13)$$

and

$$\frac{1}{N_{St}} \frac{\partial v_1^*}{\partial t^*} + v_1^* \frac{\partial v_1^*}{\partial z_1^*} + v_2^{**} \frac{\partial v_1^*}{\partial z_2^{**}} = -\frac{1}{N_{Ru}} \frac{d\mathscr{P}^*}{dz_1^*} + \frac{\partial^2 v_1^*}{\partial z_2^{**2}} \qquad (1\text{-}14)$$

These are two equations in three unknowns: $v_1^*$, $v_2^{**}$, and $\mathscr{P}^*$. Fortunately we have some information which we have not yet employed.

Outisde the boundary-layer region, viscous effects can be neglected in the equation of motion. Let

$$v_1^{(e)*} = v_1^{(e)*}(z_1^*, z_2^*) = v_1^{(e)*}\left(z_1^*, \frac{z_2^{**}}{\sqrt{N_{Re}}}\right) \qquad (1\text{-}15)$$

denote the dimensionless $z_1$ component of velocity for the nonviscous external flow. Within the boundary-later region, in the limit as the Reynolds number becomes unbounded,

$$\tilde{v}_1^* \equiv \lim_{\substack{N_{Re} \to \infty \\ z_1^*, z_2^{**} \text{ fixed}}} v_1^{(e)*}\left(z_1^*, \frac{z_2^{**}}{\sqrt{N_{Re}}}\right) = v_1^{(e)*}(z_1^*, 0) \qquad (1\text{-}16)$$

As $z_2^{**} \to \infty$, we must require that the $z_1$ component of velocity from the boundary-layer solution must approach asymptotically the corresponding velocity component for the nonviscous external flow:

$$\text{as } z_2^{**} \to \infty: \quad v_1^* \to \tilde{v}_1^* \qquad (1\text{-}17)$$

In view of Eq. (1-12), we conclude that in Eq. (1-14) we may identify

$$-\frac{1}{N_{\text{Ru}}}\frac{d\mathscr{P}^*}{dz_1^*} = -\frac{1}{N_{\text{Ru}}}\frac{d\tilde{\mathscr{P}}^*}{dz_1^*} \equiv \frac{1}{N_{\text{St}}}\frac{\partial \tilde{v}_1^*}{\partial t^*} + \tilde{v}_1^* \frac{\partial \tilde{v}_1^*}{\partial z_1^*} \quad (1\text{-}18)$$

and write Eq. (1-14) as

$$\frac{1}{N_{\text{St}}}\frac{\partial v_1^*}{\partial t^*} + v_1^* \frac{\partial v_1^*}{\partial z_1^*} + v_2^{**} \frac{\partial v_1^*}{\partial z_2^{**}} = \frac{1}{N_{\text{St}}}\frac{\partial \tilde{v}_1^*}{\partial t^*} + \tilde{v}_1^* \frac{\partial \tilde{v}_1^*}{\partial z_1^*} + \frac{\partial^2 v_1^*}{\partial z_2^{**2}} \quad (1\text{-}19)$$

Since we assume that $\tilde{v}_1^*$ is known a priori, Eqs. (1-13) and (1-17) are to be solved simultaneously for the two unknowns $v_1^*$ and $v_2^{**}$. These equations are commonly referred to as the *boundary-layer equations* for plane flow past a flat plate.

In the next section, we carry through to a solution the ideas developed here for flow past a flat plate.

For an alternative view, consider a perturbation solution for the equation of continuity and Cauchy's first law, in which the perturbation parameter is taken to be $N_{\text{Re}}^{-1}$. This is known as a singular perturbation problem, in the sense that the order of one of the differential equations, Cauchy's first law, changes when the perturbation parameter is set equal to zero. A linearized nonviscous flow arises as the zeroth order term in a perturbation solution that makes no attempt to satisfy the no-slip condition at phase boundaries. We can also think of it as the zeroth order term in an asymptotic expansion for the outer flow (outside the immediate neighborhood of a phase boundary), where viscous effects are negligible. A linearized boundary-layer flow corresponds to the zeroth order term in an asymptotic expansion for the inner flow (inside the immediate neighborhood of a phase boundary), where viscous effects can not be completely neglected. The inner and outer asymptotic expansions are matched by requiring the existance of a region in which both are valid. One advantage of this approach is that the effect of $N_{\text{Re}}$ can be seen in the limit $N_{\text{Re}} \to \infty$. For more on the theory of matched asymptotic expansions, see Cole [3].

## REFERENCES

1. Prandtl, L., and O. G. Tietjens: "Applied Hydro- and Aeromechanics," Dover, New York, 1957.
2. Schlichting, Hermann: "Boundary-Layer Theory," 6th ed., McGraw-Hill, New York, 1968.
3. Cole, J. D.: "Perturbation Methods in Applied Mathematics," Blaisdell, Waltham, Mass., 1968.
4. Schowalter, W. R.: *A.I.Ch.E.J.*, **6**:24 (1960).
5. Acrivos, Andreas, M. J. Shah, and E. E. Petersen: *A.I.Ch.E.J.*, **6**:312 (1960).
6. Yau, Joseph, and Chi Tien, *Can. J. Chem. Eng.*, **41**:139 (1963).
7. Gutfinger, Chaim, and Ruel Shinnar: *A.I.Ch.E.J.*, **10**:631 (1964).
8. White, J. L., and A. B. Metzner: *A.I.Ch.E.J.*, **11**:324 (1965).
9. Acrivos, Andreas, M. J. Shah, and E. E. Petersen: *Chem. Eng. Sci.*, **20**:101 (1965).
10. Hermes, R. A., and A. G. Fredrickson, *A.I.Ch.E.J.*, **13**:253 (1967).

## EXERCISE

**3.5.1-1** *Power-model flow past a flat plate* An incompressible fluid that is described by the power model [Eqs. (3-1) and (3-13) of Sec. 2.3.3] is in plane flow past a flat plate. Construct an argument

## [3.5] BOUNDARY-LAYER THEORY

that suggests that, in the limit as the modified Reynolds number $N_{Re\,PM}$ becomes unbounded, the appropriate boundary-layer equations are Eqs. (1-13) and

$$\frac{1}{N_{St}}\frac{\partial v_1^*}{\partial t^*} + v_1^*\frac{\partial v_1^*}{\partial z_1^*} + v_2^{**}\frac{\partial v_1^*}{\partial z_2^{**}} = \frac{1}{N_{St}}\frac{\partial \tilde{v}_1^*}{\partial t^*} + \tilde{v}_1^*\frac{\partial \tilde{v}_1^*}{\partial z_1^*} + \frac{\partial}{\partial z_2^{**}}\left[\left|\frac{\partial v_1^*}{\partial z_2^{**}}\right|^{n-1}\frac{\partial v_1^*}{\partial z_2^{**}}\right]$$

Here

$$z_2^{**} = (N_{Re\,PM})^{1/(n+1)} z_2^*$$
$$v_2^{**} = (N_{Re\,PM})^{1/(n+1)} v_2^*$$

and

$$N_{Re\,PM} \equiv \frac{\rho v_0^{2-n} L_0^n}{m}$$

As before, $v_0$ and $L_0$ represent, respectively, a magnitude of velocity and a length, which are both characteristic of the flow.

For further reading on boundary-layer flows of viscoelastic fluids, see [4 to 10].

### 3.5.2 More on the boundary layer in plane flow past a flat plate

We wish to discuss here a particular case to which the theory of the previous section is applicable: steady-state plane flow past a flat plate at zero incidence of an incompressible newtonian fluid.

Let us assume that the velocity field outside the boundary layer might be representable in terms of a potential as Figs. 3.5.1-1 and 3.5.1-2 indicate. We neglect any displacement of the potential flow by the boundary layer and by the thin plate. Consequently, the velocity field outside the boundary layer is uniform in magnitude and direction, the positive $z_1$ direction in Fig. 3.5.1-2.

From Sec. 3.5.1, the equations to be solved for the velocity distribution in the boundary layer become

$$\frac{\partial v_1^*}{\partial z_1^*} + \frac{\partial v_2^{**}}{\partial z_2^{**}} = 0 \tag{2-1}$$

and

$$v_1^*\frac{\partial v_1^*}{\partial z_1^*} + v_2^{**}\frac{\partial v_1^*}{\partial z_2^{**}} = \frac{\partial^2 v_1^*}{\partial z_2^{**\,2}} \tag{2-2}$$

The corresponding boundary conditions are

At $z_2^{**} = 0$: $\quad v_1^* = v_2^{**} = 0$ \tag{2-3}

and

As $z_2^{**} \to \infty$: $\quad v_1^* \to 1$ \tag{2-4}

Here

$$v_1^* \equiv \frac{v_1}{v_0} \qquad v_2^{**} \equiv \sqrt{N_{Re}}\,\frac{v_2}{v_0}$$
$$z_1^* \equiv \frac{z_1}{L} \qquad \text{and} \qquad z_2^{**} \equiv \sqrt{N_{Re}}\,\frac{z_2}{L} \tag{2-5}$$

where $L$ is the length of the plate and

$$N_{\text{Re}} \equiv \frac{v_0 L \rho}{\mu} \tag{2-6}$$

We will notice later that the quantity $L$ drops out of the solution, which is consistent with our intuitive view that the solution we obtain should be valid for the semi-infinite flat plate.

The equation of continuity, Eq. (2-1), may be satisfied identically by expressing the velocity components in terms of a stream function as suggested in Secs. 1.3.6 and 1.3.7. Referring to Table 2.5.2-1, we introduce a dimensionless stream function $\psi$ by requiring

$$v_1^* = \frac{\partial \psi}{\partial z_2^{**}} \tag{2-7}$$

and

$$v_2^{**} = -\frac{\partial \psi}{\partial z_1^*} \tag{2-8}$$

Introduction of the stream function into Eq. (2-2) yields

$$\frac{\partial \psi}{\partial z_2^{**}} \frac{\partial^2 \psi}{\partial z_1^* \partial z_2^{**}} - \frac{\partial \psi}{\partial z_1^*} \frac{\partial^2 \psi}{\partial z_2^{**2}} = \frac{\partial^3 \psi}{\partial z_2^{**3}} \tag{2-9}$$

We now must solve one partial differential equation for $\psi$, Eq. (2-9). It sometimes helps in a situation like this to make a change of independent and dependent variables with the aim of transforming the partial differential equation into an ordinary differential equation. For example, if we postulate here that

$$\psi = \sqrt{z_1^*} f(\eta) \tag{2-10}$$

where

$$\eta \equiv \frac{z_2^{**}}{\sqrt{z_1^*}} \tag{2-11}$$

Eq. (2-9) becomes

$$ff'' + 2f''' = 0 \tag{2-12}$$

The primes denote differentiation with respect to $\eta$. Boundary conditions (2-3) and (2-4) may be expressed as

$$\text{At } \eta = 0: \quad f = f' = 0 \tag{2-13}$$

and

$$\text{As } \eta \to \infty: \quad f' \to 1 \tag{2-14}$$

Note that with these changes of variable $L$ drops out of the problem. As we suggested earlier, this is to be expected, since this geometry has no inherent characteristic length.

## [3.5] BOUNDARY-LAYER THEORY

We have arrived at a completely defined mathematical problem, since Eq. (2-12) is a third-order ordinary differential equation to be solved consistent with the three boundary conditions in Eqs. (2-13) and (2-14). Notice that we never said anything about $v_2^{**}$ as $z_2^{**} \to \infty$; it is different from zero as the edge of the boundary layer is approached [1, Fig. 7.8] and determined by the other conditions placed upon the flow.

Equation (2-12) is considerably simpler than Eq. (2-9), but it is still nonlinear. A general solution cannot be given in closed form. The problem was originally solved by Blasius [2; 1, p. 125], though one of the most accurate solutions was given by Howarth [3]. A detailed comparison of Howarth's solution with available experimental data is presented by Schlichting [1, p. 132]; it is perhaps sufficient to say here that the agreement is excellent.

## REFERENCES

1. Schlichting, Hermann: "Boundary-Layer Theory," 6th ed., McGraw-Hill, New York, 1968.
2. Blasius, H.: *Z. Math. Physik*, **56**:1 (1908); English translation, National Advisory Committee for Aeronautics, Tech. Memo. 1256.
3. Howarth, L.: *Proc. Roy. Soc. (London)*, **A164**:547 (1938).

## EXERCISES

**3.5.2-1** Given the change of variables in Eqs. (2-10) and (2-11), derive Eq. (2-12) and boundary conditions (2-13) and (2-14).

**3.5.2-2** Starting with

$$\psi = (z_1^*)^a f(\eta)$$

and

$$\eta = (z_1^*)^b (z_2^{**})^c$$

show how one might argue to arrive at the transformations indicated by Eqs. (2-10) and (2-11).

**3.5.2-3** Show that, when Eqs. (2-12) to (2-14) are expressed in terms of dimensional variables, the characteristic length $L$ does not appear.

### 3.5.3 Boundary-layer theory for plane flow past a curved wall
In Sec. 3.5.1 we discussed plane flow past a flat plate. In the limit, as the Reynolds number becomes unbounded, we found that the equation of motion may be considerably simplified for the fluid in a thin layer (boundary layer) next to the plate. This simplification involved neglecting some (though not all) of the viscous terms with respect to the inertial terms.

In this section we wish to consider the boundary layer formed by an incompressible newtonian fluid in plane flow past a curved wall. Intuitively, it seems clear that the fluid in a sufficiently thin boundary layer behaves in the same manner whether the wall is curved or flat. Our object here is to show in what sense this intuitive feeling is correct.

A portion of this wall is shown in Fig. 3.5.3-1. With respect to the rectangular cartesian coordinate system indicated, the equation of this surface is

$$z_2 = f(z_1) \tag{3-1}$$

In order to better compare flow past a curved wall with flow past a flat plate, let us view this problem in terms of an orthogonal curvilinear coordinate system such that:

$x \equiv x^1$ is defined to be arc length measured along the wall in a plane of constant $z$.
$y \equiv x^2$ is defined to be arc length measured along straight lines which are normal to the wall.
$z \equiv x^3 \equiv z_3$ is the coordinate normal to the plane of flow.

By plane flow we mean here that

$$\begin{aligned} v_x &\equiv v_{\langle 1 \rangle} = v_x(x,y,t) \\ v_y &\equiv v_{\langle 2 \rangle} = v_y(x,y,t) \\ v_z &\equiv v_{\langle 3 \rangle} = 0 \end{aligned} \tag{3-2}$$

With these restrictions, the equation of continuity and the three components of the equation of motion may be written, respectively, as [1, p. 119; see also Exercise 3.5.3-1]

$$\frac{\partial v_x}{\partial x} + (1 + \varkappa y)\frac{\partial v_y}{\partial y} + \varkappa v_y = 0 \tag{3-3}$$

$$(1 + \varkappa y)\frac{\partial v_x}{\partial t} + v_x\frac{\partial v_x}{\partial x} + (1 + \varkappa y)v_y\frac{\partial v_x}{\partial y} + \varkappa v_x v_y$$
$$= -\frac{1}{\rho}\frac{\partial \mathscr{P}}{\partial x} + \frac{\mu}{\rho}\left[\frac{1}{1 + \varkappa y}\frac{\partial^2 v_x}{\partial x^2} + \frac{v_y}{(1 + \varkappa y)^2}\frac{\partial \varkappa}{\partial x} + \frac{2\varkappa}{1 + \varkappa y}\frac{\partial v_y}{\partial x}\right.$$
$$\left. - \frac{y}{(1 + \varkappa y)^2}\frac{d\varkappa}{dx}\frac{\partial v_x}{\partial x} - \frac{\varkappa^2}{1 + \varkappa y}v_x + \varkappa\frac{\partial v_x}{\partial y} + (1 + \varkappa y)\frac{\partial^2 v_x}{\partial y^2}\right] \tag{3-4}$$

$$\frac{\partial v_y}{\partial t} + \frac{v_x}{1 + \varkappa y}\frac{\partial v_y}{\partial x} + v_y\frac{\partial v_y}{\partial y} - \frac{\varkappa}{1 + \varkappa y}v_x^2 = -\frac{1}{\rho}\frac{\partial \mathscr{P}}{\partial y} + \frac{\mu}{\rho}\left[\frac{1}{(1 + \varkappa y)^2}\frac{\partial^2 v_y}{\partial x^2}\right.$$
$$\left. - \frac{y}{(1 + \varkappa y)^3}\frac{d\varkappa}{dx}\frac{\partial v_y}{\partial x} + \frac{\varkappa}{1 + \varkappa y}\frac{\partial v_y}{\partial y} - \frac{2\varkappa}{(1 + \varkappa y)^2}\frac{\partial v_y}{\partial x} - \frac{\varkappa^2}{(1 + \varkappa y)^2}v_y\right.$$
$$\left. - \frac{v_x}{(1 + \varkappa y)^3}\frac{d\varkappa}{dx} + \frac{\partial^2 v_y}{\partial y^2}\right] \tag{3-5}$$

and

$$\frac{\partial \mathscr{P}}{\partial z} = 0 \tag{3-6}$$

## [3.5] BOUNDARY-LAYER THEORY

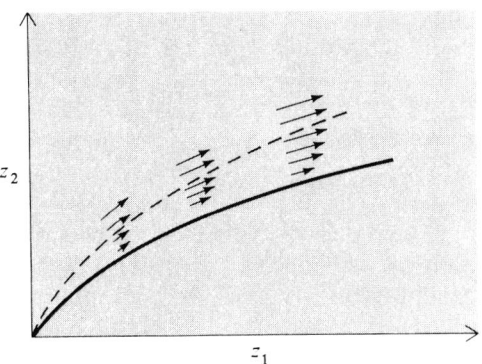

**Fig. 3.5.3-1** Plane flow past a curved wall.

where we define

$$\varkappa \equiv \frac{-f''}{[1 + (f')^2]^{\frac{3}{2}}} = \varkappa(x) \tag{3-7}$$

The primes here indicate differentiation with respect to $z_1$.

In addition to the dimensionless velocity, dimensionless modified pressure, and dimensionless time introduced in Sec. 3.3.1, let us define

$$x^* \equiv \frac{x}{L} \qquad y^* \equiv \frac{y}{L} \qquad \varkappa^* \equiv \varkappa L \tag{3-8}$$

If we extend the arguments in Sec. 3.5.1 to this geometry, we are motivated to express our results in terms of

$$y^{**} \equiv (N_{\text{Re}})^a y^* \qquad v_y^{**} \equiv (N_{\text{Re}})^a v_y^* \tag{3-9}$$

If any viscous terms in the equation of motion are to be of the same order of magnitude as the inertial terms, as the Reynolds number $N_{\text{Re}}$ becomes unbounded, we are again forced to specify that $a = \frac{1}{2}$. With this understanding, as the Reynolds number becomes unbounded, Eqs. (3-3), (3-4), and (3-5) reduce to

$$\frac{\partial v_x^*}{\partial x^*} + (1 + \varkappa^{**} y^{**}) \frac{\partial v_y^{**}}{\partial y^{**}} + \varkappa^{**} v_y^{**} = 0 \tag{3-10}$$

$$\frac{1}{N_{\text{St}}}(1 + \varkappa^{**} y^{**}) \frac{\partial v_x^*}{\partial t^*} + v_x^* \frac{\partial v_x^*}{\partial x^*} + (1 + \varkappa^{**} y^{**}) v_y^{**} \frac{\partial v_x^*}{\partial y^{**}} + \varkappa^{**} v_x^* v_y^{**}$$

$$= -\frac{1}{N_{\text{Ru}}} \frac{\partial \mathscr{P}^*}{\partial x^*} - \frac{\varkappa^{**2}}{1 + \varkappa^{**} y^{**}} v_x^* + \varkappa^{**} \frac{\partial v_x^*}{\partial y^{**}} + (1 + \varkappa^{**} y^{**}) \frac{\partial^2 v_x^*}{\partial y^{**2}} \tag{3-11}$$

and

$$\frac{\varkappa^{**}}{1+\varkappa^{**}y^{**}}\,v_x^{*2} = \frac{1}{N_{\text{Ru}}}\frac{\partial\mathscr{P}^*}{\partial y^{**}} \tag{3-12}$$

where we define

$$\varkappa^{**} \equiv \varkappa^* N_{\text{Re}}^{-\frac{1}{2}} \tag{3-13}$$

For a fixed wall configuration, $\varkappa^{**} \to 0$ in the limit as the Reynolds number becomes unbounded. Equations (3-10), (3-11), and (3-12) simplify under these conditions to

$$\frac{\partial v_x^*}{\partial x^*} + \frac{\partial v_y^{**}}{\partial y^{**}} = 0 \tag{3-14}$$

$$\frac{1}{N_{\text{St}}}\frac{\partial v_x^*}{\partial t^*} + v_x^*\frac{\partial v_x^*}{\partial x^*} + v_y^{**}\frac{\partial v_x^*}{\partial y^{**}} = -\frac{1}{N_{\text{Ru}}}\frac{\partial\mathscr{P}^*}{\partial x^*} + \frac{\partial^2 v_x^*}{\partial y^{**2}} \tag{3-15}$$

and

$$\frac{\partial\mathscr{P}^*}{\partial y^{**}} = 0 \tag{3-16}$$

Equations (3-6) and (3-16) mean that

$$\mathscr{P}^* = \mathscr{P}^*(x^*) \tag{3-17}$$

It can be shown that $-\varkappa = 2H$, where $H$ is the mean curvature of the surface [2, p. 203]. We may alternatively refer to $-\varkappa$ as the normal curvature [2, p. 210] of the surface in the direction $x$; it is also the only nonzero principal curvature [2, p. 211] of the surface.

As we suggested in Sec. 3.5.1,

$$\text{As } y^{**} \to \infty: \quad v_x^* \to \tilde{v}_x^* \tag{3-18}$$

where $\tilde{v}_x^*$ is the dimensionless $x$ component of velocity at the curved wall from the nonviscous flow solution. In view of Eq. (3-17), we conclude that we may identify

$$-\frac{1}{N_{\text{Ru}}}\frac{\partial\mathscr{P}^*}{\partial x^*} = -\frac{1}{N_{\text{Ru}}}\frac{d\tilde{\mathscr{P}}^*}{dx^*} = \tilde{v}_x^*\frac{d\tilde{v}_x^*}{dx^*} \tag{3-19}$$

and write Eq. (3-15) as

$$\frac{1}{N_{\text{St}}}\frac{\partial v_x^*}{\partial t^*} + v_x^*\frac{\partial v_x^*}{\partial x^*} + v_y^{**}\frac{\partial v_x^*}{\partial y^{**}} = \tilde{v}_x^*\frac{d\tilde{v}_x^*}{dx^*} + \frac{\partial^2 v_x^*}{\partial y^{**2}} \tag{3-20}$$

Since we assume that $\tilde{v}_x^*$ is known a priori, Eqs. (3-14) and (3-20) are to be solved simultaneously for the two unknowns $v_x^*$ and $v_y^{**}$. These equations are commonly referred to as the *boundary-layer equations* for plane flow past a curved wall. As our intuition suggested, they have the same form as the boundary-layer equations developed in Sec. 3.5.1 for plane flow past a flat plate.

## [3.5] BOUNDARY-LAYER THEORY

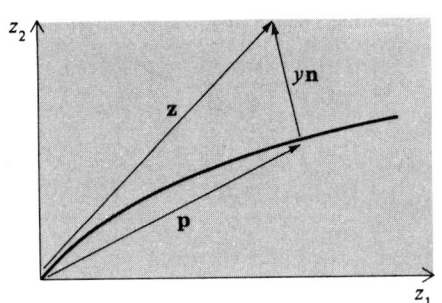

Fig. 3.5.3-2

## REFERENCES

1. Goldstein, S.: "Modern Developments in Fluid Dynamics," Oxford University Press, London, 1938.
2. McConnell, A. J.: "Applications of Tensor Analysis," Dover, New York, 1957.

## EXERCISES

**3.5.3-1** *Derivation of Eqs.* (3-3) *to* (3-6)   (*a*) In Fig. 3.5.3-2, **p** denotes the position vector of any point on the curved wall:

$$\mathbf{p} = z_1 \mathbf{e}_1 + f(z_1)\mathbf{e}_2 + z_3 \mathbf{e}_3$$

Determine that

$$\frac{dz_1}{dx} = [1 + (f')^2]^{-\frac{1}{2}}$$

where the prime denotes differentiation with respect to $z_1$.

(*b*) In Fig. 3.5.3-2, **z** denotes the position vector of any point in the boundary layer:

$$\mathbf{z} = \mathbf{p} + y\mathbf{n}$$

By **n** we mean the unit normal to the wall directed into the fluid. For the curvilinear coordinate system described in the text of this section, find that
   (i)

$$\mathbf{g}_1 \equiv \frac{\partial \mathbf{z}}{\partial x} = \frac{1 + \varkappa y}{[1 + (f')^2]^{\frac{1}{2}}} (\mathbf{e}_1 + f'\mathbf{e}_2)$$

and

$$\mathbf{g}_2 \equiv \frac{\partial \mathbf{z}}{\partial y} = \mathbf{n} = \frac{-f'\mathbf{e}_1 + \mathbf{e}_2}{[1 + (f')^2]^{\frac{1}{2}}}$$

where $\varkappa$ is defined by Eq. (3-7).
   (ii)

$$g_{11} = (1 + \varkappa y)^2$$

and

$$g_{22} = g_{33} = 1$$

*(iii) The only nonzero Christoffel symbols of the second kind are

$$\begin{Bmatrix} 1 \\ 1 \ 1 \end{Bmatrix} = \frac{y(d\varkappa/dx)}{1 + \varkappa y}$$

$$\begin{Bmatrix} 1 \\ 1 \ 2 \end{Bmatrix} = \begin{Bmatrix} 1 \\ 2 \ 1 \end{Bmatrix} = \frac{\varkappa}{1 + \varkappa y}$$

and

$$\begin{Bmatrix} 2 \\ 1 \ 1 \end{Bmatrix} = -\varkappa(1 + \varkappa y)$$

*(c) Derive Eqs. (3-3) through (3-6).

**3.5.3-2** *Derivation of Eqs. (3-10) to (3-12)* (a) Introduce in Eqs. (3-3) to (3-6) the dimensionless velocity, dimensionless modified pressure, and dimensionless time defined in Sec. 3.3.1 as well as those dimensionless variables defined in Eqs. (3-8).

(b) Express these equations in terms of the variables defined in Eqs. (3-9) and construct the reasoning that leads to the definition $a = \frac{1}{2}$ and to Eqs. (3-10) to (3-12).

***3.5.3-3** *Viscoelastic fluids*  Let us restrict our attention to incompressible viscoelastic fluids whose behavior can be represented by one of the simple empirical models discussed in Sec. 2.3.3:

$$\mathbf{S} = 2\eta(\gamma)\mathbf{D}$$

Let us further require $\eta(\gamma)$ to be a homogeneous function of degree $p$. By this I mean that, for a fixed value of $p$,

$$t^p \eta(\gamma) = \eta(t\gamma)$$

no matter what value $t$ assumes.

Argue that, for plane boundary-layer flow past a curved wall, Cauchy's first law requires

$$\frac{1}{N_{St}} \frac{\partial v_x^*}{\partial t^*} + v_x^* \frac{\partial v_x^*}{\partial x^*} + v_y^{**} \frac{\partial v_x^*}{\partial y^{**}} = \frac{1}{N_{St}} \frac{\partial \tilde{v}_x^*}{\partial t^*} + \tilde{v}_x^* \frac{\partial \tilde{v}_x^*}{\partial x^*} + \frac{\partial}{\partial y^{**}}\left[\eta^*\left(\left|\frac{\partial v_x^*}{\partial y^{**}}\right|\right) \frac{\partial v_x^*}{\partial y^{**}}\right] \quad (1)$$

and

$$\frac{\partial \mathscr{P}^*}{\partial y^{**}} = 0$$

Here we have defined

$$\eta^*\left(\left|\frac{\partial v_x^*}{\partial y^{**}}\right|\right) \equiv \frac{1}{\mu_0} (N_{Re})^{(-p)/(p+2)} \eta\left(\left|\frac{\partial v_x}{\partial y}\right|\right)$$

$$y^{**} \equiv y^*(N_{Re})^{1/(p+2)} \qquad v_y^{**} \equiv v_y^*(N_{Re})^{1/(p+2)}$$

and

$$N_{Re} \equiv \frac{L_0 v_0 \rho}{\mu_0}$$

By $\mu_0$, I mean a viscosity characteristic of the fluid.

For a power-model fluid,

$$\eta\left(\left|\frac{\partial v_x}{\partial y}\right|\right) = m \left(\frac{v_0}{L_0}\right)^{n-1} (N_{Re\ PM})^{(n-1)/(n+1)} \left|\frac{\partial v_x^*}{\partial y^{**}}\right|^{n-1}$$

we see that

$$p = n - 1$$

## [3.5] BOUNDARY-LAYER THEORY

and that

$$\eta^*\left(\left|\frac{\partial v_x^*}{\partial y^{**}}\right|\right) = \left|\frac{\partial v_x^*}{\partial y^{**}}\right|^{n-1}$$

If we define

$$\mu_0 \equiv m\left(\frac{v_0}{L_0}\right)^{n-1}$$

then

$$N_{\text{Re }PM} \equiv \rho\, \frac{v_0^{2-n} L_0^n}{m}$$

With these definitions, Eq. (1) above is consistent with the result found in Exercise 3.5.1-1 for flow past a flat plate.

### 3.5.4 Solutions found by combination of variables [1, p. 137]

In Secs. 3.5.1 and 3.5.3 we found that the same differential equations govern plane boundary-layer flow past either a flat wall or a curved wall: Eqs. (3-14) and (3-20). While the latter represents a simplification of the Navier-Stokes equations, Eq. (3-20) is still a nonlinear partial differential equation. Our object here is to determine circumstances in which the solution of these equations reduces to the solution of an ordinary differential equation. We shall focus our attention upon steady-state flows.

As discussed in Secs. 1.3.6 and 1.3.7, Eq. (3-14) may be satisfied identically by expressing the velocity components in terms of a stream function. Referring to Table 2.5.2-1, we introduce a dimensionless stream function $\psi$ by requiring

$$v_x^* = \frac{\partial \psi}{\partial y^{**}} \tag{4-1}$$

and

$$v_y^{**} = -\frac{\partial \psi}{\partial x^*} \tag{4-2}$$

These expressions allow us to write Eq. (3-20) for a steady-state flow as

$$\frac{\partial \psi}{\partial y^{**}} \frac{\partial^2 \psi}{\partial x^* \partial y^{**}} - \frac{\partial \psi}{\partial x^*} \frac{\partial^2 \psi}{\partial y^{**2}} = \tilde{v}_x^* \frac{d\tilde{v}_x^*}{dx^*} + \frac{\partial^3 \psi}{\partial y^{**3}} \tag{4-3}$$

Since it is assumed that $\tilde{v}_x^*$ is known a priori, Eq. (4-3) is a nonlinear partial differential equation which must be solved for $\psi$. In what follows we find that there are several classes of problems for which this equation may be expressed as an ordinary differential equation.

Our approach is to introduce new independent and dependent variables defined, respectively, as

$$\eta \equiv \frac{y^{**}}{g(x^*)} \tag{4-4}$$

and

$$f(\eta) \equiv \frac{\psi}{h(x^*)} \tag{4-5}$$

The functions $g$ and $h$ are to be determined by the requirement that Eq. (4-3) be expressed as an ordinary differential equation for $f$ as a function of $\eta$. In terms of these variables, Eq. (4-3) becomes

$$f''' + g \frac{dh}{dx^*} ff'' - \left(g \frac{dh}{dx^*} - h \frac{dg}{dx^*}\right)(f')^2 + \frac{g^3}{h} \tilde{v}_x^* \frac{d\tilde{v}_x^*}{dx^*} = 0 \tag{4-6}$$

where the prime denotes differentiation with respect to $\eta$. This equation suggests that we define $h$ by requiring

$$g \frac{dh}{dx^*} - h \frac{dg}{dx^*} = \frac{g^3}{h} \tilde{v}_x^* \frac{d\tilde{v}_x^*}{dx^*} \tag{4-7}$$

This may easily be rearranged to read

$$\frac{d}{dx^*}\left(\frac{h^2}{g^2}\right) = \frac{d\tilde{v}_x^{*2}}{dx^*} \tag{4-8}$$

from which we have the definition

$$h \equiv g\tilde{v}_x^* \tag{4-9}$$

This allows us to express Eq. (4-6) as

$$f''' + \alpha ff'' + \beta[1 - (f')^2] = 0 \tag{4-10}$$

where

$$\alpha \equiv g \frac{d}{dx^*}(g\tilde{v}_x^*) \tag{4-11}$$

and

$$\beta \equiv g^2 \frac{d\tilde{v}_x^*}{dx^*} \tag{4-12}$$

In order that $f$ be a function of $x^*$ only through its dependence upon $\eta$, we must require that $\alpha$ and $\beta$ be independent of $x^*$. These two requirements determine the functions $g$ and $\tilde{v}_x^*$. When we determine the function $\tilde{v}_x^*$, we specify those geometries for which Eq. (4-3) may be reduced to an ordinary differential equation by the changes of variable defined in Eqs. (4-4) and (4-5).

Case 1: $2\alpha - \beta \neq 0$ From Eqs. (4-11) and (4-12) we have that

$$2\alpha - \beta = \frac{d}{dx^*}(g^2 \tilde{v}_x^*) \tag{4-13}$$

and

$$\alpha - \beta = g\tilde{v}_x^* \frac{dg}{dx^*} \tag{4-14}$$

This latter equation may be rearranged as

$$\frac{\alpha - \beta}{\tilde{v}_x^*} \frac{d\tilde{v}_x^*}{dx^*} = g \frac{d\tilde{v}_x^*}{dx^*} \frac{dg}{dx^*} = \frac{\beta}{g} \frac{dg}{dx^*} \tag{4-15}$$

## [3.5] BOUNDARY-LAYER THEORY

and integrated to give

$$\tilde{v}_x^{*\alpha-\beta} = Kg^\beta \tag{4-16}$$

where $K$ is a constant of integration. Equation (4-13) may also be integrated:

$$(2\alpha - \beta)x^* = g^2\tilde{v}_x^* + C \tag{4-17}$$

Here $C$ indicates a constant of integration. We may eliminate $g$ between Eqs. (4-16) and (4-17) to obtain

$$(2\alpha - \beta)x^* = \frac{(\tilde{v}_x^*)^{(2\alpha-\beta)/\beta}}{K^{2/\beta}} + C \tag{4-18}$$

If we assume that $2\alpha - \beta \neq 0$, this result may be solved for $\tilde{v}_x^*$:

$$\tilde{v}_x^* = K^{2/(2\alpha-\beta)}[(2\alpha - \beta)x^* - C]^{\beta/(2\alpha-\beta)} \tag{4-19}$$

The corresponding form for $g$ may be obtained by eliminating $\tilde{v}_x^*$ between Eqs. (4-16) and (4-19):

$$g = K^{-[1/(2\alpha-\beta)]}[(2\alpha - \beta)x^* - C]^{(\alpha-\beta)/(2\alpha-\beta)} \tag{4-20}$$

*Case 1a:* $\alpha \neq 0$, $2\alpha - \beta \neq 0$  It is clear from Eq. (4-19) that the result is independent of any common factor of $\alpha$ and $\beta$, since any common factor may be included in $g$. Therefore, we may assume that $\alpha \neq 0$ and put $\alpha = 1$ without loss of generality. Equations (4-19) and (4-20) become as a result

$$\tilde{v}_x^* = K^{2/(2-\beta)}[(2 - \beta)x^* - C]^{\beta/(2-\beta)} \tag{4-21}$$

and

$$g = K^{-[1/(2-\beta)]}[(2 - \beta)x^* - C]^{(1-\beta)/(2-\beta)} \tag{4-22}$$

If we define

$$m \equiv \frac{\beta}{2 - \beta} \tag{4-23}$$

these results take the somewhat simpler forms

$$\tilde{v}_x^* = K^{1+m}\left[\left(\frac{2}{1+m}\right)x^* - C\right]^m \tag{4-24}$$

and

$$g = K^{[-(1+m)]/2}\left[\left(\frac{2}{1+m}\right)x^* - C\right]^{(1-m)/2} \tag{4-25}$$

When $C = 0$, Eq. (4-24) describes the potential flow velocity distribution at the wall near the leading edge of a wedge whose included angle is [1, p. 141]

$$\pi\beta = \frac{2\pi m}{1+m} \tag{4-26}$$

For $\beta = m = 1$, we have plane stagnation flow (see Sec. 3.4.5) which is discussed further in Sec. 3.5.5. For $\beta = m = 0$, we have flow past a flat plate at zero angle of incidence. For a summary of the solutions available describing flow past a wedge, see [1, p. 150].

*Case 1b:* $\alpha = 0$, $2\alpha - \beta \neq 0$  In the event that $\alpha = 0$, Eqs. (4-19) and (4-20) become

$$\tilde{v}_x^* = K^{-(2/\beta)}(-\beta x^* - C)^{-1} \tag{4-27}$$

and

$$g = K^{1/\beta}(-\beta x^* - C) \tag{4-28}$$

It is convenient to take $\beta = \pm 1$, depending upon the sign of $d\tilde{v}_x^*/dx^*$ in Eq. (4-12).

When $C = 0$, Eq. (4-27) may be interpreted as flow in either a convergent or divergent channel with flat walls (see Exercise 3.4.4-5). For a convergent channel, $\tilde{v}_x^*$ is negative and $d\tilde{v}_x^*/dx^*$ is positive; Eqs. (4-12) and (4-27) require that we define $\beta = +1$. For a divergent channel, $\tilde{v}_x^*$ is positive and $d\tilde{v}_x^*/dx^*$ is negative; $\beta = -1$. Flow in a convergent channel is discussed in Sec. 3.5.6.

The full Navier-Stokes equations have been solved for plane flow in convergent and divergent channels with flat walls. Goldstein [2, p. 105] and Schlichting [1, p. 99] give interesting discussions of the available solutions.

*Case 2:* $2\alpha - \beta = 0$  In the event that $2\alpha - \beta = 0$, we have from Eq. (4-13) that

$$\tilde{v}_x^* = \frac{K}{g^2} \tag{4-29}$$

where $K$ is a constant of integration. Equation (4-14), consequently, reduces to

$$-\alpha = \frac{K}{g} \frac{dg}{dx^*} \tag{4-30}$$

which may be integrated to give

$$g = \exp\left(-\frac{\alpha}{K}x^* + C\right) \tag{4-31}$$

Here $C$ denotes another constant of integration. This last allows us to write Eq. (4-29) as

$$\tilde{v}_x^* = \frac{K}{\exp\left(-[2\alpha/K]x^* + 2C\right)} \tag{4-32}$$

## REFERENCES

1. Schlichting, Hermann: "Boundary-Layer Theory," 6th ed., McGraw-Hill, New York, 1968.
2. Goldstein, S.: "Modern Developments in Fluid Dynamics," Oxford University Press, London, 1938.

## [3.5] BOUNDARY-LAYER THEORY

### 3.5.5 Plane stagnation flow

In Sec. 3.5.4, we studied the class of plane flow problems for which solutions to the boundary-layer equations can be found by the method of combination of variables. A particular member of this class is plane flow in the neighborhood of a stagnation point on the plane $z_2 = 0$, potential flow streamlines for which are shown in Fig. 3.4.5-2. Fluid approaching from the positive $z_2$ direction is turned aside by the plane $z_2 = 0$; it subsequently moves in the positive and negative $z_1$ directions. This flow has a plane of symmetry at $z_1 = 0$ on which the $z_1$ component of velocity is zero.

In this section we wish to find a solution to the full Navier-Stokes equation for plane stagnation flow in the neighborhood of the origin. We will find that this solution is identical to the solution found using the boundary-layer equations for plane flow as indicated in Sec. 3.5.4.

We seek then a velocity distribution which satisfies the full Navier-Stokes equation:

$$\rho \frac{d_{(m)}\mathbf{v}}{dt} = -\nabla \mathscr{P} + \mu \, \text{div} \, (\nabla \mathbf{v}) \tag{5-1}$$

as well as the equation of continuity for an incompressible fluid:

$$\text{div } \mathbf{v} = 0 \tag{5-2}$$

Referring to Fig. 3.4.5-2, we require that at a sufficient distance from the plane $z_2 = 0$ this solution predict the same $z_1$ component of velocity found for the corresponding plane potential flow:

$$\text{As } z_2 \to \infty: \quad v_1 \to kz_1 + \cdots \tag{5-3}$$

The velocity distribution must also satisfy the "no-slip" condition

$$\text{At } z_2 = 0: \quad \mathbf{v} = 0 \tag{5-4}$$

These are the same boundary conditions that we impose upon the corresponding boundary-layer flow.

Since we are concerned with the plane flow, there are only two nonzero components of velocity:

$$v_1 = v_1(z_1, z_2) \quad v_2 = v_2(z_1, z_2) \quad v_3 = 0 \tag{5-5}$$

We can automatically satisfy Eq. (5-2) by the introduction of a stream function. From Table 2.5.2-1,

$$v_1 = \frac{\partial \psi}{\partial z_2} \tag{5-6}$$

and

$$v_2 = -\frac{\partial \psi}{\partial z_1} \tag{5-7}$$

The only nonzero component of the curl of Eq. (5-1) may be expressed in terms of this stream function as (see Table 2.5.2-1)

$$-\frac{\partial(\psi, E^2\psi)}{\partial(z_1, z_2)} = \nu E^4 \psi \tag{5-8}$$

Boundary conditions (5-3) and (5-4) become

$$\text{As } z_2 \to \infty: \quad \frac{\partial \psi}{\partial z_2} \to k z_1 + \cdots \tag{5-9}$$

and

$$\text{At } z_2 = 0: \quad \frac{\partial \psi}{\partial z_1} = \frac{\partial \psi}{\partial z_2} = 0 \tag{5-10}$$

Equation (5-9) suggests that we look for a solution of the form

$$\psi = z_1 f(z_2) \tag{5-11}$$

With this assumption, Eq. (5-8) reduces to

$$-ff''' + f'f'' = \nu f^{\text{iv}} \tag{5-12}$$

$$\frac{d}{dz_2}[(f')^2 - ff'' - \nu f'''] = 0$$

where the primes denote differentiation with respect to $z_2$. This may be integrated once consistent with (5-9) to obtain

$$(f')^2 - ff'' - \nu f''' = k^2 \tag{5-13}$$

Equation (5-13) may be simplified by a change of variables. Let us define

$$\varphi \equiv \frac{f}{A} \tag{5-14}$$

and

$$\eta \equiv \alpha z_2 \tag{5-15}$$

where $A$ and $\alpha$ are constants. Equation (5-13) becomes

$$\alpha^2 A^2[(\varphi')^2 - \varphi \varphi''] - \nu A \alpha^3 \varphi''' = k^2 \tag{5-16}$$

The primes now denote differentiation with respect to $\eta$. Let us define $A$ and $\alpha$ such that

$$\alpha^2 A^2 = k^2 \tag{5-17}$$

and

$$\nu A \alpha^3 = k^2 \tag{5-18}$$

## [3.5] BOUNDARY-LAYER THEORY

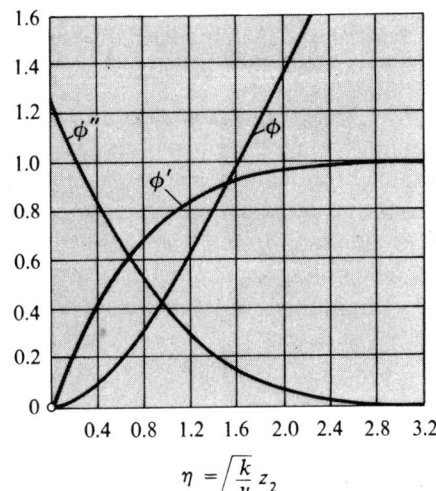

**Fig. 3.5.5-1** Velocity distribution for plane stagnation flow in the neighborhood of the origin. (*From Schlichting [3, fig. 5.10]*.)

$$\eta = \sqrt{\frac{k}{\nu}}\, z_2$$

We have as a result

$$A \equiv k^{\frac{1}{2}} \nu^{\frac{1}{2}} \tag{5-19}$$

$$\alpha \equiv k^{\frac{1}{2}} \nu^{-\frac{1}{2}} \tag{5-20}$$

and Eq. (5-16) becomes

$$\varphi''' + \varphi \varphi'' - (\varphi')^2 + 1 = 0 \tag{5-21}$$

A solution to this last is to be found consistent with boundary conditions (5-9) and (5-10) which may now be expressed as

$$\text{As } \eta \to \infty: \quad \varphi' \to 1 \tag{5-22}$$

and

$$\text{At } \eta = 0: \quad \varphi = \varphi' = 0 \tag{5-23}$$

Hiemenz [1] and Howarth [2] have solved this problem numerically. Howarth's solution is shown in Fig. 3.5.5-1.

In Sec. 3.5.4 we found that the solution to the boundary-layer equations for plane stagnation flow in the neighborhood of the origin was also determined by Eq. (5-21) ($\alpha = \beta = 1$ in Case 1a) with boundary conditions (5-22) and (5-23). This means that the solution to the boundary-layer equations for this particular flow also represents a solution to the full Navier-Stokes equation.

## REFERENCES

1. Hiemenz, K.: *Dinglers Polytechnisches J.*, **326**:321 (1911).

2. Howarth, L.: *Aeron. Res. Council (Great Britain) Reps. and Memoranda*, 1632, 1934.
3. Schlichting, Hermann: "Boundary-Layer Theory," 6th ed., McGraw-Hill, New York, 1968.

### 3.5.6 Flow in a convergent channel

Consider the steady-state flow of an incompressible newtonian fluid in a convergent channel. The convergent channel should be thought of as two plane walls oriented in such a way that they form surfaces of constant $\theta$ in cylindrical coordinates; the fluid moves in the negative $r$ direction. This discussion is only approximately applicable to a real channel of finite length, since we have made no attempt to account for end effects.

In Exercise 3.4.4-5, we analyzed the corresponding potential flow and found

$$v_r^* = -\frac{1}{r^*} \qquad v_\theta^* = v_z^* = 0 \tag{6-1}$$

In this section, we wish to solve the corresponding boundary-layer flow pictured in Fig. 3.5.6-1. One reason for discussing this particular flow is that it is a rare case where an analytic solution to the boundary-layer equations can be determined.

Reverting to the notation of Sec. 3.5.3, we may use the potential flow described by Eq. (6-1) to specify the condition which the corresponding boundary layer must satisfy at infinity:

$$\text{As } y^{**} \to \infty: \quad \tilde{v}_x^* \to -\frac{1}{x^*} \tag{6-2}$$

All of the boundary layers we have considered up to this point have developed downstream of a leading edge, entrance, or stagnation point. This is not true here. The coordinate $x^*$ is measured from the imaginary sink for a convergent channel rather than from the leading edge of a finite channel. But the interpretation of the boundary layer is the same. In a flow for which the Reynolds number is large, it is that region (normally within the immediate neighborhood of a phase interface) in which viscous effects can not be neglected with respect to inertial effects.

We say in Sec. 3.5.4 (Case 1b) that boundary-layer flow in a convergent channel is one of the situations for which the boundary-layer equations may be simplified by the method of combination of variables. Having expressed the dimensionless velocity components in terms of a stream function $\psi$,

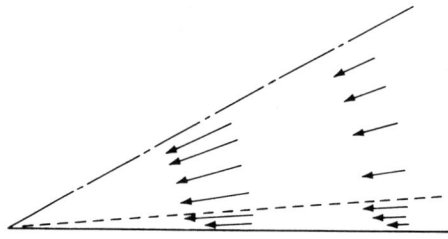

**Fig. 3.5.6-1** Flow in the neighborhood of the wall in a convergent channel.

## [3.5] BOUNDARY-LAYER THEORY

$$v_x^* = \frac{\partial \psi}{\partial y^{**}} \tag{6-3}$$

$$v_y^{**} = -\frac{\partial \psi}{\partial x^*} \tag{6-4}$$

we found that

$$\psi = h(x^*)f(\eta) = g\tilde{v}_x^* f(\eta) = K^{-1}f(\eta) \tag{6-5}$$

where $f = f(\eta)$ is a solution to

$$f''' - (f')^2 + 1 = 0 \tag{6-6}$$

The primes in this last expression indicate differentiation with respect to $\eta$. On comparison of Eq. (6-2) with the expression for $\tilde{v}_x^*$ appropriate to this case in Sec. 3.5.4, we find that the constant $K$ in Eq. (6-5) must be either $+1$ or $-1$. The variable $\eta$ is defined as

$$\eta \equiv \frac{y^{**}}{g} = -\frac{y^{**}}{Kx^*} = \frac{y^{**}}{x^*} \tag{6-7}$$

where we have chosen to define

$$K \equiv -1 \tag{6-8}$$

In addition to Eq. (6-2) the boundary conditions to be satisfied by this boundary-layer flow are

At $y^{**} = 0$:  $\quad v_x^* = v_y^{**} = 0 \tag{6-9}$

From Eqs. (6-3), (6-4), (6-5), (6-7), and (6-8) we find that

$$v_x^* = -\frac{f'}{x^*} \tag{6-10}$$

and

$$v_y^{**} = -\frac{f'\eta}{x^*} \tag{6-11}$$

This means that boundary conditions (6-2) and (6-9) may be expressed as follows:

As $\eta \to \infty$:  $\quad f' \to 1 \tag{6-12}$

and

At $\eta = 0$:  $\quad f' = 0 \tag{6-13}$

If we multiply Eq. (6-6) by $2f''$, we may integrate it once:

$$2f''f''' + 2f''[1 - (f')^2] = 0$$

$$\frac{d}{d\eta}[(f'')^2 - \tfrac{2}{3}(1-f')^2(f'+2)] = 0 \tag{6-14}$$

$$(f'')^2 - \tfrac{2}{3}(1-f')^2(f'+2) = a$$

Here $a$ is a constant of integration. If our primary interest is in the velocity distribution, we may view this last as a separable first-order ordinary differential equation in $f'$ with two boundary conditions, Eqs. (6-12) and (6-13). The problem as it stands is well posed mathematically and there would be no particular difficulty in obtaining a numerical solution for $f'$.

However, with a little insight we may obtain an analytic expression for $f'$. Our discussion of boundary conditions in Secs. 3.5.1 and 3.5.3 suggests that

$$\text{As } y^{**} \to \infty: \quad \frac{\partial v_x^*}{\partial y^{**}} = -\frac{f''}{x^{*2}} \to 0 \tag{6-15}$$

or

$$\text{As } \eta \to \infty: \quad f'' \to 0 \tag{6-16}$$

Equations (6-12) and (6-16) suggest that we take

$$a = 0 \tag{6-17}$$

which allows us to write Eq. (6-14)$_3$ as

$$\frac{f''}{\sqrt{2}(1-f')\sqrt{\tfrac{1}{3}f'+\tfrac{2}{3}}} = 1$$

$$\sqrt{2}\,\frac{d}{d\eta}(\tanh^{-1}\sqrt{\tfrac{1}{3}f'+\tfrac{2}{3}}) = 1 \tag{6-18}$$

This last may be integrated once using boundary condition (6-13) to give

$$\sqrt{2}\int_{f'=0}^{f'} d(\tanh^{-1}\sqrt{\tfrac{1}{3}f'+\tfrac{2}{3}}) = \int_0^\eta d\eta$$

$$\sqrt{2}\,(\tanh^{-1}\sqrt{\tfrac{1}{3}f'+\tfrac{2}{3}} - \tanh^{-1}\sqrt{\tfrac{2}{3}}) = \eta \tag{6-19}$$

This last satisfies boundary condition (6-12) as was originally required.

To repeat, Eq. (6-16) should not be viewed as a boundary condition for this problem. Rather, this is a condition that we suggested as being helpful in seeking an analytic solution to Eq. (6-14)$_3$.

Equation (6-19)$_2$ can be solved for $f'$:

$$\frac{v_x^*}{\bar{v}_x^*} = f' = 3\tanh^2\left(\frac{\eta}{\sqrt{2}} + 1.146\right) - 2 \tag{6-20}$$

## [3.5] BOUNDARY-LAYER THEORY

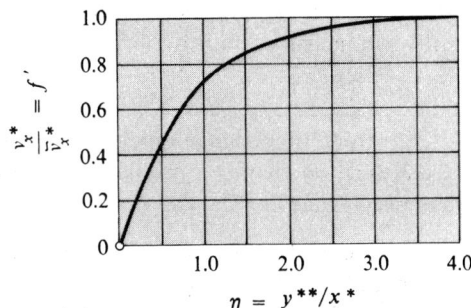

**Fig. 3.5.6-2** Velocity distribution for boundary-layer flow in a convergent channel. (*From Schlichting [1, fig. 9.3].*)

In arriving at this expression, we have noted that $\tanh^{-1}\sqrt{\tfrac{2}{3}} = 1.146$. Equation (6-20) is plotted in Fig. 3.5.6-2.

### REFERENCE

1. Schlichting, Hermann: "Boundary-Layer Theory," 6th ed., McGraw-Hill, New York, 1968.

**3.5.7 Boundary-layer theory for flow past a body of revolution** In Secs. 3.5.1 and 3.5.3, we discussed plane flow past a flat plate and plane flow past a curved wall. We found that the boundary layer could be described by the same set of equations in both cases.

In what follows, we consider the boundary layer formed by an incompressible newtonian fluid flowing past a body of revolution. In some sense we expect the fluid in a sufficiently thin boundary layer to behave in the same manner on a body of revolution as it would on a flat wall. We wish to determine here in what sense this intuitive feeling is correct.

A plane section through the axis of symmetry of the body is shown in Fig. 3.5.7-1.

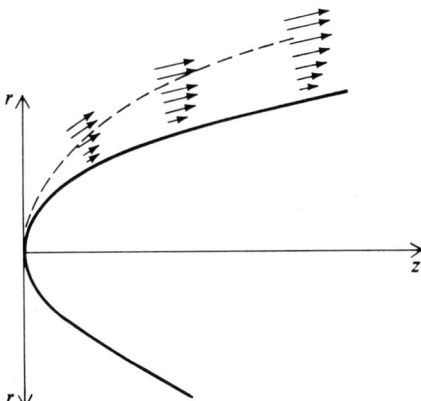

**Fig. 3.5.7-1** Flow past a body of revolution.

## APPLICATIONS OF THE DIFFERENTIAL BALANCES TO MOMENTUM TRANSFER

With respect to the cylindrical coordinate system indicated, the equation of the axially symmetric surface is

$$r = f(z) \tag{7-1}$$

In order to better compare flow past a body of revolution with flow past a flat plate, let us view this problem in terms of an orthogonal curvilinear coordinate system such that

$x \equiv x^3$ is defined to be arc length measured along the wall (in the direction of flow) in a plane of constant $\theta$.

$y \equiv x^1$ is defined to be arc length measured along straight lines that are normal to the wall.

$\theta \equiv x^2$ is the azimuthal cylindrical coordinate (measured around the axis of the body).

The shape of the wall suggests that

$$v_x \equiv v_{\langle 3 \rangle} = v_x(x,y,t)$$
$$v_y \equiv v_{\langle 1 \rangle} = v_y(x,y,t)$$
$$v_\theta \equiv v_{\langle 2 \rangle} = 0 \tag{7-2}$$

With these restrictions, the equation of continuity and the three components of the equation of motion may be written, respectively, as (see Exercise 3.5.7-1)

$$\frac{1}{1+\varkappa y}\frac{\partial v_x}{\partial x} + \frac{1}{1+\varkappa y}\frac{\partial}{\partial y}[(1+\varkappa y)v_y] + \frac{f'}{g}v_x + \frac{1}{g}v_y = 0 \tag{7-3}$$

$$(1+\varkappa y)\frac{\partial v_x}{\partial t} + v_x\frac{\partial v_x}{\partial x} + (1+\varkappa y)v_y\frac{\partial v_x}{\partial y} + \varkappa v_x v_y = -\frac{1}{\rho}\frac{\partial \mathscr{P}}{\partial x}$$

$$+\frac{\mu}{\rho}\left\{\frac{1}{1+\varkappa y}\frac{\partial^2 v_x}{\partial x^2} + \frac{v_y}{(1+\varkappa y)^2}\frac{d\varkappa}{dx} + \frac{2\varkappa}{1+\varkappa y}\frac{\partial v_y}{\partial x} - \frac{y}{(1+\varkappa y)^2}\frac{d\varkappa}{dx}\frac{\partial v_x}{\partial x}\right.$$

$$-\frac{\varkappa^2}{1+\varkappa y}v_x + \varkappa\frac{\partial v_x}{\partial y} + (1+\varkappa y)\frac{\partial^2 v_x}{\partial y^2} - \frac{f'(1+\varkappa y)}{g^2}v_y$$

$$\left.+\frac{\varkappa f'}{g}[1+(f')^2]^{\frac{1}{2}}v_y + \frac{f'}{g}[1+(f')^2]^{\frac{1}{2}}\frac{\partial v_x}{\partial x} + \frac{(1+\varkappa y)}{g}\frac{\partial v_x}{\partial y}\right\} \tag{7-4}$$

$$\frac{\partial v_y}{\partial t} + \frac{v_x}{1+\varkappa y}\frac{\partial v_y}{\partial x} + v_y\frac{\partial v_y}{\partial y} - \frac{\varkappa}{1+\varkappa y}v_x^2 = -\frac{1}{\rho}\frac{\partial \mathscr{P}}{\partial y} + \frac{\mu}{\rho}\left\{\frac{1}{(1+\varkappa y)^2}\frac{\partial^2 v_y}{\partial x^2}\right.$$

$$-\frac{y}{(1+\varkappa y)^3}\frac{d\varkappa}{dx}\frac{\partial v_y}{\partial x} + \frac{\varkappa}{1+\varkappa y}\frac{\partial v_y}{\partial y} - \frac{2\varkappa}{(1+\varkappa y)^2}\frac{\partial v_x}{\partial x}$$

$$\left.-\frac{\varkappa^2}{(1+\varkappa y)^2}v_y - \frac{v_x}{(1+\varkappa y)^3}\frac{d\varkappa}{dx} + \frac{\partial^2 v_y}{\partial y^2} - \frac{v_y}{g^2} + \frac{1}{g}\frac{\partial v_y}{\partial y}\right.$$

[3.5] BOUNDARY-LAYER THEORY

$$+ \frac{f'}{(1+\varkappa y)g}[1+(f')^2]^{\frac{1}{2}}\left(\frac{\partial v_y}{\partial x}-\varkappa v_x\right)\right\} \quad (7\text{-}5)$$

and

$$\frac{\partial \mathscr{P}}{\partial \theta}=0 \quad (7\text{-}6)$$

where we define

$$\varkappa \equiv \frac{-f''}{[1+(f')^2]^{\frac{3}{2}}}=\varkappa(x) \quad (7\text{-}7)$$

and

$$g \equiv f[1+(f')^2]^{\frac{3}{2}}+y \quad (7\text{-}8)$$

The primes here indicate differentiation with respect to the cylindrical coordinate $z$ measured along the axis of revolution.

In addition to the dimensionless velocity, dimensionless modified pressure, and dimensionless time introduced in Sec. 3.3.1, let us define

$$x^* \equiv \frac{x}{L} \quad y^* \equiv \frac{y}{L} \quad f^* \equiv \frac{f}{L} \quad \varkappa^* \equiv \varkappa L \quad (7\text{-}9)$$

If we extend the arguments of Sec. 3.5.1 to this geometry, we are motivated to express our result in terms of

$$y^{**} \equiv (N_{\text{Re}})^a y^* \quad \text{and} \quad v_y^{**} \equiv (N_{\text{Re}})^a v_y^* \quad (7\text{-}10)$$

In order that some viscous terms in the equation of motion have the same order of magnitude as the inertial terms when the Reynolds number $N_{\text{Re}}$ becomes unbounded, we are again forced to specify that $a=\frac{1}{2}$. With this understanding, as the Reynolds number becomes unbounded, Eqs. (3-3) to (3-5) reduce to

$$\frac{1}{1+\varkappa^{**}y^{**}}\frac{\partial v_x^*}{\partial x^*}+\frac{1}{1+\varkappa^{**}y^{**}}\frac{\partial}{\partial y^{**}}[(1+\varkappa^{**}y^{**})v_y^{**}]$$

$$+\frac{f^{*'}}{f^*[1+(f^{*'})^2]^{\frac{1}{2}}}v_x^*=0 \quad (7\text{-}11)$$

$$\frac{1}{N_{\text{St}}}(1+\varkappa^{**}y^{**})\frac{\partial v_x^*}{\partial t^*}+v_x^*\frac{\partial v_x^*}{\partial x^*}+(1+\varkappa^{**}y^{**})v_y^{**}\frac{\partial v_x^*}{\partial y^{**}}+\varkappa^{**}v_x^*v_y^{**}$$

$$=-\frac{1}{N_{\text{Ru}}}\frac{\partial \mathscr{P}^*}{\partial x^*}-\frac{\varkappa^{**2}}{1+\varkappa^{**}y^{**}}v_x^*+\varkappa^{**}\frac{\partial v_x^*}{\partial y^{**}}+(1+\varkappa^{**}y^{**})\frac{\partial^2 v_x^*}{\partial y^{**2}} \quad (7\text{-}12)$$

and

$$\frac{\varkappa^{**}}{1+\varkappa^{**}y^{**}}v_x^{*2} = \frac{1}{N_{\text{Ru}}}\frac{\partial \mathscr{P}^*}{\partial y^{**}} \tag{7-13}$$

where we define

$$\varkappa^{**} \equiv \varkappa^* N_{\text{Re}}^{-\frac{1}{2}} \tag{7-14}$$

The primes in Eq. (7-11) now indicate differentiation with respect to $z^*$.

For a fixed-wall configuration, $\varkappa^{**} \to 0$ in the limit as the Reynolds number becomes unbounded. Equations (7-11) to (7-13) simplify under these conditions to

$$\frac{\partial v_x^*}{\partial x^*} + \frac{\partial v_y^{**}}{\partial y^{**}} + \frac{f^{*\prime}}{f^*[1+(f^{*\prime})^2]^{\frac{1}{2}}}v_x^* = 0 \tag{7-15}$$

$$\frac{1}{N_{\text{St}}}\frac{\partial v_x^*}{\partial t^*} + v_x^*\frac{\partial v_x^*}{\partial x^*} + v_y^{**}\frac{\partial v_x^*}{\partial y^{**}} = -\frac{1}{N_{\text{Ru}}}\frac{\partial \mathscr{P}^*}{\partial x^*} + \frac{\partial^2 v_x^*}{\partial y^{**2}} \tag{7-16}$$

and

$$\frac{\partial \mathscr{P}^*}{\partial y^{**}} = 0 \tag{7-17}$$

Since (see Exercise 3.5.7-1)

$$\frac{dz^*}{dx^*} = [1+(f^{*\prime})^2]^{-\frac{1}{2}} \tag{7-18}$$

Eq. (7-15) may also be written as

$$\frac{\partial(f^*v_x^*)}{\partial x^*} + \frac{\partial(f^*v_y^{**})}{\partial y^{**}} = 0 \tag{7-19}$$

Equations (7-6) and (7-17) mean that

$$\mathscr{P}^* = \mathscr{P}^*(x^*) \tag{7-20}$$

We may refer to $-\varkappa$ as the normal curvature [1, p. 210] of the surface in the direction $x$; it is also one of the principal curvatures [1, p. 211] of the surface.

As we suggested in Sec. 3.5.1,

$$\text{As } y^{**} \to \infty: \quad v_x^* \to \tilde{v}_x^* \tag{7-21}$$

where $\tilde{v}_x^*$ is the dimensionless $x$ component of velocity at the curved wall from the nonviscous flow solution. In view of Eq. (7-20), we conclude that we may identify

$$-\frac{1}{N_{\text{Ru}}}\frac{\partial \mathscr{P}^*}{\partial x^*} = -\frac{1}{N_{\text{Ru}}}\frac{d\tilde{\mathscr{P}}^*}{dx^*} = \tilde{v}_x^*\frac{d\tilde{v}_x^*}{dx^*} \tag{7-22}$$

and write Eq. (7-16) as

$$\frac{1}{N_{\text{St}}}\frac{\partial v_x^*}{\partial t^*} + v_x^*\frac{\partial v_x^*}{\partial x^*} + v_y^{**}\frac{\partial v_x^*}{\partial y^{**}} = \tilde{v}_x^*\frac{d\tilde{v}_x^*}{dx^*} + \frac{\partial^2 v_x^*}{\partial y^{**2}} \tag{7-23}$$

Since we assume that $\tilde{v}_x^*$ is known a priori, Eqs. (7-19) and (7-23) are to be solved simultaneously for the two unknowns $v_x^*$ and $v_y^{**}$. These equations may be referred to as the *boundary-layer equations* for flow past a body of revolution.

Mangler [2, p. 235; see also Exercises 3.5.7-3 and 3.5.7-4] has suggested an interesting change of variables. Define

$$\bar{x}^* \equiv \int_0^{x^*} f^{*2} \, dx^* \tag{7-24}$$

$$\bar{y}^{**} \equiv f^* y^{**} \tag{7-25}$$

$$\bar{t}^* \equiv f^{*2} t^* \tag{7-26}$$

and

$$\bar{v}_y^{**} \equiv \frac{1}{f^*} v_y^{**} + \frac{f^{*\prime}}{f^{*2}[1 + (f^{*\prime})^2]^{\frac{1}{2}}} y^{**} v_x^* \tag{7-27}$$

With this change of variables, Eqs. (7-19) and (7-23) become

$$\frac{\partial v_x^*}{\partial \bar{x}^*} + \frac{\partial \bar{v}_y^{**}}{\partial \bar{y}^{**}} = 0 \tag{7-28}$$

and

$$\frac{1}{N_{St}} \frac{\partial v_x^*}{\partial \bar{t}^*} + v_x^* \frac{\partial v_x^*}{\partial \bar{x}^*} + \bar{v}_y^{**} \frac{\partial v_x^*}{\partial \bar{y}^{**}} = \tilde{v}_x^* \frac{d\tilde{v}_x^*}{d\bar{x}^*} + \frac{\partial^2 v_x^*}{\partial \bar{y}^{**2}} \tag{7-29}$$

These equations have the same form as the boundary-layer equations found in Secs. 3.5.1 and 3.5.3. They may be referred to as *Mangler's boundary-layer equations* for flow past a body of revolution. It is in this sense that our original intuitive feelings are confirmed: the mathematical problems that describe flow past a body of revolution and plane flow past a curved wall have the same form.

## REFERENCES

1. McConnell, A. J.: "Applications of Tensor Analysis," Dover, New York, 1957.
2. Schlichting, Hermann: "Boundary-Layer Theory," 6th ed., McGraw-Hill, New York, 1968.

## EXERCISES

**3.5.7-1** *Derivation of Eqs. (7-3) to (7-6)* (a) In Fig. 3.5.7-2, **p** denotes the position vector of any point on the surface of revolution. With respect to a cylindrical coordinate basis such that $z$ is measured along the axis of the body of revolution,

$$\mathbf{p} = f(z)\mathbf{g}_r + z\mathbf{g}_z$$

With respect to a rectangular cartesian coordinate basis such that $z_3 \equiv z$,

$$\mathbf{p} = [f(z_3)\cos\theta]\mathbf{e}_1 + [f(z_3)\sin\theta]\mathbf{e}_2 + z_3\mathbf{e}_3$$

Determine that

$$\frac{dz_3}{dx} = [1 + (f')^2]^{-\frac{1}{2}}$$

where the prime denotes differentiation with respect to $z_3 (\equiv z)$.

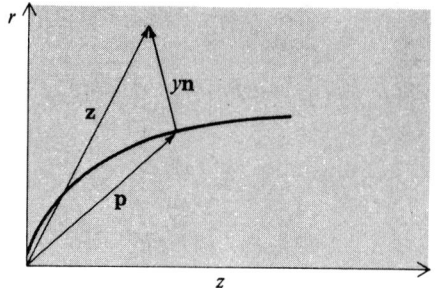

**Fig. 3.5.7-2** Relation between the position vectors for a point on the surface and the corresponding point in the fluid.

(b) In Fig. 3.5.7-2, **z** denotes the position vector of any point in the boundary layer:

$$\mathbf{z} = \mathbf{p} + y\mathbf{n}$$

By **n** we mean the unit normal to the wall directed into the fluid. For the curvilinear coordinate system described in the text of this section, find that

(i)

$$\mathbf{g}_1 \equiv \frac{\partial \mathbf{z}}{\partial y} = \mathbf{n}$$
$$= [1 + (f')^2]^{-\frac{1}{2}}(\cos\theta \mathbf{e}_1 + \sin\theta \mathbf{e}_2 - f'\mathbf{e}_3)$$

$$\mathbf{g}_2 \equiv \frac{\partial \mathbf{z}}{\partial \theta}$$
$$= \{f + y[1 + (f')^2]^{-\frac{1}{2}}\}(-\sin\theta \mathbf{e}_1 + \cos\theta \mathbf{e}_2)$$

and

$$\mathbf{g}_3 \equiv \frac{\partial \mathbf{z}}{\partial x}$$
$$= (1 + \varkappa y)[1 + (f')^2]^{-\frac{1}{2}}(f'\cos\theta \mathbf{e}_1 + f'\sin\theta \mathbf{e}_2 + \mathbf{e}_3)$$

where $\varkappa$ is defined by Eq. (7-7).

(ii)

$$g_{11} = 1$$
$$g_{22} = \{f + y[1 + (f')^2]^{-\frac{1}{2}}\}^2$$

and

$$g_{33} = (1 + \varkappa y)^2$$

*(iii) The only nonzero Christoffel symbols of the second kind are

$$\left\{\begin{matrix} 1 \\ 2\ 2 \end{matrix}\right\} = -\frac{g}{1 + (f')^2}$$

$$\left\{\begin{matrix} 1 \\ 3\ 3 \end{matrix}\right\} = -\varkappa(1 + \varkappa y)$$

$$\left\{\begin{matrix} 2 \\ 1\ 2 \end{matrix}\right\} = \frac{1}{g}$$

## [3.5] BOUNDARY-LAYER THEORY

$$\begin{Bmatrix} 2 \\ 2 \ 3 \end{Bmatrix} = \frac{f'(1+\varkappa y)}{g}$$

$$\begin{Bmatrix} 3 \\ 1 \ 3 \end{Bmatrix} = \frac{\varkappa}{1+\varkappa y}$$

$$\begin{Bmatrix} 3 \\ 2 \ 2 \end{Bmatrix} = -f'g(1+\varkappa y)^{-1}[1+(f')^2]^{-\frac{1}{2}}$$

and

$$\begin{Bmatrix} 3 \\ 3 \ 3 \end{Bmatrix} = \frac{y}{1+\varkappa y}\frac{d\varkappa}{dx}$$

where $g$ is defined by Eq. (7-8).

*(c) Derive Eqs. (7-3) to (7-6).

**3.5.7-2** *Derivation of Eqs. (7-11) to (7-13)* (a) Introduce in Eqs. (7-3) to (7-6) the dimensionless velocity, dimensionless modified pressure, and dimensionless time defined in Sec. 3.3.1, as well as those dimensionless variables defined in Eqs. (7-9).

(b) Express these equations in terms of the variables defined in Eqs. (7-10) and construct the reasoning which leads to the definition $a = \frac{1}{2}$ and to Eqs. (7-11) to (7-13).

**3.5.7-3** *Mangler's transformation* [2, p. 235] (a) If $A$ is any scalar and if $\bar{x}^*$ and $\bar{y}^{**}$ are defined by Eqs. (7-24) and (7-25), prove that

$$\frac{\partial A}{\partial x^*} = f^{*2}\frac{\partial A}{\partial \bar{x}^*} + \frac{f^{*'}}{f^*[1+(f^{*'})^2]^{\frac{1}{2}}}\bar{y}^{**}\frac{\partial A}{\partial \bar{y}^{**}}$$

and

$$\frac{\partial A}{\partial y^{**}} = f^*\frac{\partial A}{\partial \bar{y}^{**}}$$

(b) Starting with Eqs. (7-19) and (7-23), make the change of variables indicated by Eqs. (7-24) to (7-27) to arrive at Eqs. (7-28) and (7-29).

**3.5.7-4** [2, p. 237] Consider a body of revolution such that

$$f^* = f^*(z^*) = x^*$$

and

$$\tilde{v}_x^* = ax^*$$

where $a$ is a dimensionless constant. Show that for this case Mangler's boundary-layer equations, Eqs. (7-28) and (7-29), have the same form as the boundary-layer equations for plane flow past a wedge such that

$$\tilde{v}_x^* = b\bar{x}^{*\frac{1}{3}}$$

Here $b$ is another dimensionless constant.

*3.5.7-5 *Viscoelastic fluids* Let us restrict our attention to incompressible fluids whose behavior can be represented by one of the simple empirical models discussed in Sec. 2.3.3:

$$\mathbf{S} = 2\eta(\gamma)\mathbf{D}$$

We require $\eta(\gamma)$ to be a homogeneous function of degree $p$.

Construct an argument to conclude that, for boundary-layer flow past a body of revolution, Cauchy's first law assumes the same form as that found in Exercise 3.5.3-3.

# 4
# Application of Integral Averaging Techniques to Momentum Transfer

I mentioned in my introduction to Chap. 3 that not every interesting problem should be attacked by directly solving the differential equation of continuity and Cauchy's first law. Some problems are really too difficult to be solved in this manner. In other cases, the amount of effort required for such a solution is not justified when the end purpose for which the solution is being developed is taken into account.

In the majority of fluid-mechanics problems, the quantity of ultimate interest is an integral. Perhaps it is an average velocity, a volume flow rate, or a force on a surface. This suggested that I set aside an entire chapter in order to exploit approaches to problems in which the independent variables are integrals or integral averages.

I begin by approaching turbulence in terms of time-averaged variables. Then I look at some problems that are normally explained in terms of area-averaged variables. The random geometry encountered in flow through porous media suggests the use of a local volume-averaged variable. The chapter concludes with the relatively well-known integral balances for arbitrary systems.

Again I encourage those of you who feel you are primarily interested in energy and mass transfer to pay close attention to this chapter. The ideas developed here

## 4.1 TIME AVERAGING

### 4.1.1 Turbulence

*Turbulence* is defined to be a motion that varies randomly with time over at least a portion of the flow field such that statistically distinct average values can be discerned. More precisely, we should require the motion to vary randomly with time in all possible frames of reference. We certainly would not wish to say that a rigid solid was in a turbulent motion merely because it was subject to random rotations and translations. Any fluid motion that is not turbulent is termed *laminar*. In thinking about turbulence you should carefully distinguish between a complex-appearing laminar flow and a true turbulent flow. A laminar flow may to the eye have a very complex dependence upon space and time, but it is only the turbulent flow that exhibits the random variations with time.

If we think of the fluid in terms of a particulate picture, we expect the molecules to execute a rather random motion. Frictional forces tend to dissipate the small-scale oscillations and to produce in this way a more regular motion. When the Reynolds number is small, we visualize viscous effects as being more important than inertial effects and the flow tends to be laminar. On the other hand, for large values of the Reynolds number, inertial effects tend to dominate the viscous terms in Cauchy's first law and the motion will tend to become random and fluctuating, even when the external boundary conditions are independent of time.

From a practical point of view, turbulent flows are probably more important than laminar flows. Whenever fluid is pumped through a pipe in a commercial process, it is almost certainly in turbulent flow. You might ask why, if turbulent flows are so important, we have waited until now to mention them. The difficulty is that I can tell you nothing about exact solutions for turbulent flow. It is the averaging techniques discussed in Chaps. 4, 7, and 9 that have proved most helpful in our attempts to analyze practical problems.

In reading the literature, it is helpful to have a few commonly accepted definitions in mind. Turbulence that is generated and continuously affected by fixed walls is referred to as *wall turbulence*. It is wall turbulence that we observe in flow through a tube. In the absence of walls, we speak of *free turbulence*. Aircraft encounter free turbulence sometimes in apparently clear skies.

If the turbulence has quantitatively the same structure everywhere in the flow field, it is said to be *homogeneous*. If its statistical features do not depend upon direction, then the turbulence is called *isotropic*. [The word isotropic is overworked. We have isotropic functions [1, p. 22] and isotropic materials [2, p. 60], not to mention isotropic porous media (Sec. 4.3.1). I would personally prefer to talk about *oriented* and *nonoriented* turbulence.] Where the mean velocity shows a gradient, we speak of the turbulence as being nonisotropic or anisotropic. It seems clear that wall turbulence will nearly always be anisotropic.

The approach to turbulence that is outlined in the next few sections is very old.

Everyone recognizes now that it can never by itself lead to a detailed understanding of the phenomena. We recommend this approach, at least for an introduction, because it has been the most fruitful in terms of engineering results. For more detailed studies of turbulence from a statistical point of view see Lin and Reid [3], Corrsin [4], Hinze [5], Batchelor [6], and Townsend [7].

## REFERENCES

1. Truesdell C., and W. Noll: In S. Flügge (ed.), "Handbuch der Physik," vol. 3/3, Springer-Verlag, Berlin, 1965.
2. Truesdell, C.: "The Elements of Continuum Mechanics," Springer-Verlag, New York, 1966.
3. Lin, C. C., and W. H. Reid: In S. Flügge and C. Truesdell (eds.), "Handbuch der Physik," vol. 8/2, Springer-Verlag, Berlin, 1963.
4. Corrsin, Stanley: In S. Flügge and C. Truesdell (eds.), "Handbuch der Physik," vol. 8/2, Springer-Verlag, Berlin, 1963.
5. Hinze, J. O.: "Turbulence," McGraw-Hill, New York, 1959.
6. Batchelor, G. K.: "The Theory of Homogeneous Turbulence," Cambridge, London, 1959.
7. Townsend, A. A.: "The Structure of Turbulent Shear Flow," Cambridge, London, 1956.

**4.1.2 Time averages** Let us consider a constant volumetric flow through a tube. Although the Reynolds number may be in excess of 5000 and we know the flow must be turbulent, a pressure gauge mounted on the wall of the tube will show a reading that is independent of time. If we examine the velocity distribution using a Pitot tube, we will see that only the axial component velocity differs from zero and that it also appears to be independent of time. We appear to have a contradiction because the Reynolds number tells us the flow is turbulent and yet the velocity and pressure distributions appear to be independent of time.

There is, of course, no contradiction. Both the Bourdon-tube pressure gauge and the Pitot tube damp out the high-frequency variations with time. The readings they give us are time averages of pressure and velocity.

The fact that our ordinary instruments measure time averages suggests that, at least for engineering purposes, we might work exclusively in terms of time-averaged variables. Let $B$ be any scalar, vector, or tensor. We will define its time average as

$$\bar{B}(t) \equiv \frac{1}{\Delta t} \int_{t}^{t+\Delta t} B(t') \, dt' \tag{2-1}$$

By $\Delta t$, we mean a finite time interval that is large with respect to the period or time scale of the random fluctuations in time of this variable, but small compared with the period or time scale of any slow variations in the field of flow that we do not wish to regard as belonging to the turbulence. There is a degree of arbitrariness in the choice of the fluctuations that we do not wish to consider. In practice, a choice can usually be made without too much difficulty. Figure 4.1.2-1 is helpful in this regard in that it indicates how a velocity component might vary randomly from its mean value, even though the mean value itself is a function of time.

If we are to talk about the turbulent flows in terms of their time-averaged distributions for velocity, pressure, density, etc., then it seems reasonable that we begin

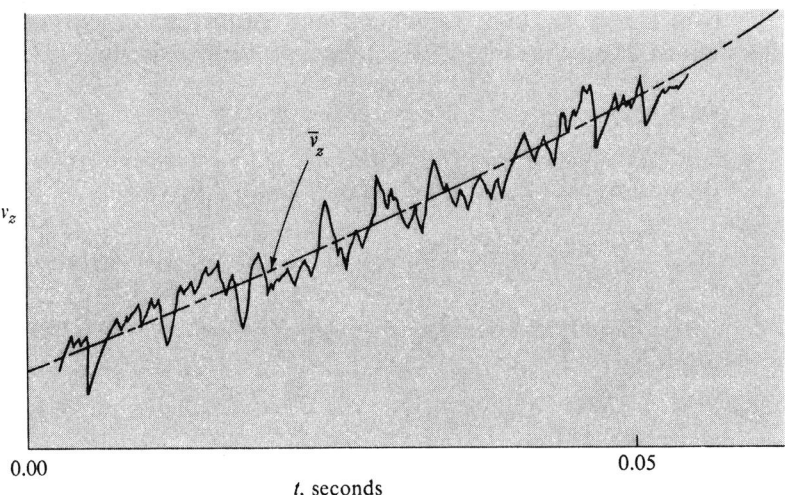

**Fig. 4.1.2-1** Random variation of velocity component from a mean value as might be measured with a hot-wire anemometer.

by taking the time average of the equation of motion and of Cauchy's first law. Let us begin by taking the time average of the equation of continuity:

$$\frac{1}{\Delta t} \int_t^{t+\Delta t} \left[ \frac{\partial \rho}{\partial t'} + \text{div}(\rho \mathbf{v}) \right] dt' = 0 \qquad (2\text{-}2)$$

The first term on the left may be integrated directly and then rearranged using the Leibnitz rule for differentiation of an integral:

$$\frac{1}{\Delta t} \int_t^{t+\Delta t} \frac{\partial \rho}{\partial t'} dt' = \frac{1}{\Delta t} (\rho|_{t+\Delta t} - \rho|_t)$$

$$= \frac{\partial}{\partial t} \left( \frac{1}{\Delta t} \int_t^{t+\Delta t} \rho \, dt' \right)$$

$$= \frac{\partial \bar{\rho}}{\partial t} \qquad (2\text{-}3)$$

In the second term on the left of Eq. (2-2), the divergence operation commutes with time averaging. As a result, Eq. (2-2) becomes

$$\frac{\partial \bar{\rho}}{\partial t} + \text{div}(\overline{\rho \mathbf{v}}) = 0 \qquad (2\text{-}4)$$

This will be referred to as the *time average of the equation of continuity*. For an incompressible fluid, we have the considerably simpler result

$$\text{div } \bar{\mathbf{v}} = 0 \qquad (2\text{-}5)$$

In looking at Eq. (2-4), it is important to realize that in general the time average of a product is not the same as the product of the time averages:

$$\overline{\rho \mathbf{v}} \neq \bar{\rho} \bar{\mathbf{v}} \qquad (2\text{-}6)$$

This point will arise repeatedly throughout our discussion of averaging operations.

Let us now take the time average of Cauchy's first law:

$$\frac{1}{\Delta t} \int_t^{t+\Delta t} \left[ \frac{\partial \rho \mathbf{v}}{\partial t'} + \operatorname{div}(\rho \mathbf{v}\mathbf{v}) + \nabla P - \operatorname{div} \mathbf{S} - \rho \mathbf{f} \right] dt' = 0 \qquad (2\text{-}7)$$

In a manner directly analogous to that used in going from Eq. (2-2) to (2-4), we can express this as

$$\frac{\partial \overline{\rho \mathbf{v}}}{\partial t} + \operatorname{div}(\bar{\rho}\bar{\mathbf{v}}\bar{\mathbf{v}}) = -\nabla \bar{P} + \operatorname{div}(\bar{\mathbf{S}} + \mathbf{T}^{(t)}) + \bar{\rho}\mathbf{f} \qquad (2\text{-}8)$$

Here $\mathbf{T}^{(t)}$ is the *Reynolds stress tensor*:

$$\mathbf{T}^{(t)} \equiv \bar{\rho}\bar{\mathbf{v}}\bar{\mathbf{v}} - \overline{\rho \mathbf{v}\mathbf{v}} \qquad (2\text{-}9)$$

It is introduced in order that the convective inertial terms in Eq. (2-8) can be expressed as the divergence of a product of averages rather than as a divergence of an average of products. For an incompressible fluid, Eqs. (2-8) and (2-9) simplify to

$$\rho \left( \frac{\partial \bar{\mathbf{v}}}{\partial t} + \nabla \bar{\mathbf{v}} \cdot \bar{\mathbf{v}} \right) = -\nabla \bar{p} + \operatorname{div}(\bar{\mathbf{S}} + \mathbf{T}^{(t)}) + \rho \mathbf{f} \qquad (2\text{-}10)$$

and

$$\mathbf{T}^{(t)} = \rho(\bar{\mathbf{v}}\bar{\mathbf{v}} - \overline{\mathbf{v}\mathbf{v}}) \qquad (2\text{-}11)$$

In arriving at Eq. (2-8) and (2-10), we have thought of the external force as gravity and have taken it to be independent of time.

For the sake of simplicity, in the sections that follow we restrict our attention to incompressible fluids. Consequently, we will be interested in solving Eqs. (2-5) and (2-10) consistent with appropriate boundary conditions and descriptions of material behavior. I think it is immediately clear that merely saying we have an incompressible newtonian fluid will not be enough. We have lost a certain amount of detail in carrying out our time average of Cauchy's first law and we must now supply some kind of empirical data correlation for the Reynolds stress tensor $\mathbf{T}^{(t)}$.

## EXERCISES

**4.1.2-1** Fill in the details in going from Eq. (2-7) to Eq. (2-8).

**4.1.2-2** *An incompressible newtonian fluid* Determine that for an incompressible newtonian fluid the time average of Cauchy's first law becomes

$$\rho \left( \frac{\partial \bar{\mathbf{v}}}{\partial t} + \nabla \bar{\mathbf{v}} \cdot \bar{\mathbf{v}} \right) = -\nabla \bar{p} + \operatorname{div}(\mu \nabla \bar{\mathbf{v}} + \mathbf{T}^{(t)}) + \rho \mathbf{f}$$

### 4.1.3 The time average of a time-averaged variable

In the next section and repeatedly in the literature, one is asked to identify the time average of a time-averaged variable:

$$\bar{\bar{B}} \equiv \frac{1}{\Delta t} \int_t^{t+\Delta t} \bar{B}\, dt' \tag{3-1}$$

with simply the time average of that variable. It should be understood that $B$ can be any scalar, vector, or tensor function of time (and usually position as well), $B = B(t)$. This seems intuitively reasonable so long as we are averaging over sufficiently small increments $\Delta t$ in time. Our purpose here is to show in what sense this intuitive feeling is confirmed.

By way of orientation, let us consider a somewhat simpler problem. Given some function $f(x)$, let us ask about the value of

$$B \equiv \frac{1}{R_2 R_1} \int_0^{R_2} \int_0^{R_1} f(x + X)\, dX\, dx \tag{3-2}$$

where the constants $R_1$ and $R_2$ are known. If we expand $f(x + X)$ in a Taylor series, Eq. (3-2) may be expressed as

$$\begin{aligned}
B &= \frac{1}{R_2 R_1} \int_0^{R_2} \int_0^{R_1} \left( f(x) + X \frac{\partial f}{\partial x}(x) + \tfrac{1}{2} X^2 \frac{\partial^2 f}{\partial x^2}(x) + \cdots \right) dX\, dx \\
&= \frac{1}{R_2} \int_0^{R_2} f(x)\, dx + \frac{1}{2} \frac{R_1}{R_2} \int_0^{R_2} \frac{\partial f}{\partial x}(x)\, dx + \cdots \\
&= \frac{1}{R_2} \int_0^{R_2} f(x)\, dx + \frac{1}{2} \frac{R_1}{R_2} [f(R_2) - f(0)] + \cdots
\end{aligned} \tag{3-3}$$

Clearly, this suggests that as $R_1/R_2 \to 0$:

$$\frac{1}{R_2 R_1} \int_0^{R_2} \int_0^{R_1} f(x + X)\, dX\, dx \to \frac{1}{R_2} \int_0^{R_2} f(x)\, dx \tag{3-4}$$

This indicates how we might look at the time average of a time-averaged variable:

$$\begin{aligned}
\bar{\bar{B}} &= \frac{1}{(\Delta t)^2} \int_t^{t+\Delta t} \int_T^{T+\Delta t} B(T')\, dT'\, dT \\
&= \frac{1}{(\Delta t)^2} \int_t^{t+\Delta t} \int_0^{\Delta t} B(T + \tau)\, d\tau\, dT \\
&= \frac{1}{(\Delta t)^2} \int_t^{t+\Delta t} \int_0^{\Delta t} \left[ B(T) + \tau \frac{\partial B}{\partial T} + \tfrac{1}{2} \tau^2 \frac{\partial^2 B}{\partial T^2} + \cdots \right] d\tau\, dT \\
&= \frac{1}{\Delta t} \int_t^{t+\Delta t} \left[ B(T) + \tfrac{1}{2} \Delta t \frac{\partial B}{\partial T} + \frac{(\Delta t)^2}{6} \frac{\partial^2 B}{\partial T^2} + \cdots \right] dT
\end{aligned} \tag{3-5}$$

It is convenient to introduce as a dimensionless variable

$$T^* \equiv \frac{T}{t_0} \qquad (3\text{-}6)$$

where $t_0$ is characteristic of the period of time scale of any slow variations in the field of flow that we do not wish to regard as belonging to the turbulence. In terms of this dimensionless time, Eq. (3-5) becomes

$$\bar{\bar{B}} = \frac{t_0}{\Delta t} \int_{t^*}^{t^* + \Delta t/t_0} \left[ B(T^*) + \frac{1}{2}\frac{\Delta t}{t_0}\frac{\partial B}{\partial T^*} + \frac{1}{6}\left(\frac{\Delta t}{t_0}\right)^2 \frac{\partial^2 B}{\partial T^{*2}} + \cdots \right] dT^* \qquad (3\text{-}7)$$

This motivates our saying that

$$\text{As } \frac{\Delta t}{t_0} \to 0: \quad \bar{\bar{B}} \to \bar{B} \qquad (3\text{-}8)$$

### 4.1.4 Empirical correlations for the Reynolds stress tensor $T^{(t)}$

In this section we use two examples to illustrate how empirical data correlations for the Reynolds stress tensor $T^{(t)}$ can be formulated. We base this discussion on three points.

1. If we limit ourselves to changes of frame such that

$$\bar{Q} \doteq Q \qquad (4\text{-}1)$$

we may use the result of Sec. 4.1.3 to conclude that $T^{(t)}$ is frame indifferent:

$$T^{(t)*} \equiv \rho(\overline{v^* v^*} - \overline{v^*}\,\overline{v^*})$$

$$= \rho\overline{(v^* - \overline{v^*})(v^* - \overline{v^*})}$$

$$= \rho Q \cdot \overline{(\bar{v} - v)(v - \bar{v})} \cdot Q^T$$

$$= Q \cdot T^{(t)} \cdot Q^T \qquad (4\text{-}2)$$

   Here $Q$ is a (possibly) time-dependent orthogonal second-order tensor. We make use of the fact that a velocity difference is frame indifferent (see Exercise 1.2.2-1).
2. We assume that the principle of material frame indifference introduced in Sec. 2.3.1 applies to any empirical correlations developed for $T^{(t)}$ so long as the changes of frame considered satisfy Eq. (4-1).
3. The Buckingham-Pi theorem serves to further limit the form of any expression for $T^{(t)}$.

**EXAMPLE 1 PRANDTL'S MIXING LENGTH THEORY**

Let us attempt to develop an empirical correlation for $T^{(t)}$ appropriate to wall turbulence.

[4.1] TIME AVERAGING

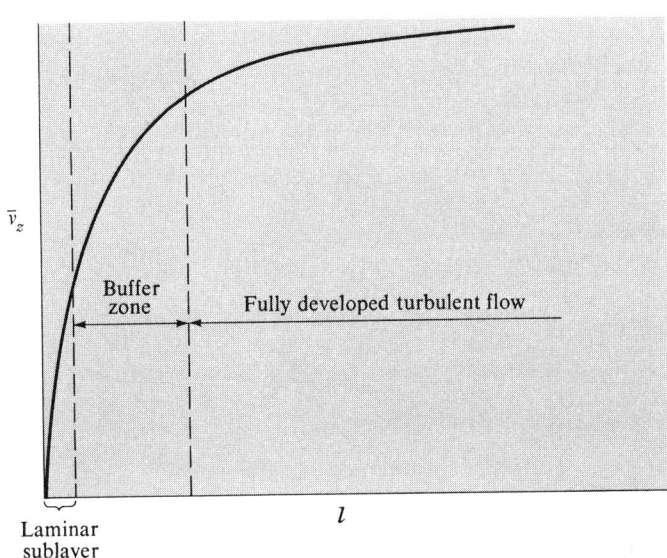

**Fig. 4.1.4-1** Time-averaged velocity as a function of distance measured from the wall.

Before beginning a task such as this, it is advisable to examine available experimental evidence so as to establish the qualitative character of the phenomena to be described. For example, with respect to wall turbulence, most of our experimental evidence concerns flow through a tube.

Though at the center of a tube the direction of the velocity fluctuations is nearly random, in the immediate neighborhood of the tube wall the magnitude of the velocity fluctuation in the axial direction is greater than that in the radial direction. All velocity fluctuations approach zero in the limit of the wall itself. This changing character of the velocity fluctuations suggests that three regimes of wall turbulence might be recognized. Referring to Fig. 4.1.4-1, we can visualize that there is a thin layer next to the wall where the viscous stresses are far more important than the Reynolds stresses; we shall call this the *laminar sublayer*. Outside of this laminar sublayer is an intermediate region in which the viscous stresses are of the same order of magnitude as the Reynolds stresses. This is the *buffer zone*. Beyond this buffer zone, the turbulent flow is said to be *fully developed* and the Reynolds stresses dominate the viscous stresses.

We will fix our attention here on the fully developed flow regime.

We are tempted to think of the Reynolds stress tensor as being in some way similar to the stress tensor **T**. This suggests that we might attack the problem of empirical correlations for $\mathbf{T}^{(t)}$ in the same way that we approached the problem of constitutive equations for **T** in Sec. 2.3.2. For example, we might assume that the Reynolds stress is a function of the density $\rho$ of the fluid, the distance $l$ from the wall, and $\nabla \bar{\mathbf{v}}$:

$$\mathbf{T}^{(t)} = \mathbf{T}^{(t)}(\rho, l, \nabla \bar{\mathbf{v}}) \tag{4-3}$$

We specifically do not include viscosity as an independent variable because we are considering the fully developed turbulent flow regime. Arguing as we did in Sec. 2.3.2., we quickly conclude that the most general form that Eq. (4-3) can take if it is to satisfy our restricted form of the principle of material frame indifference is

$$\mathbf{T}^{(t)} = \eta_0 \mathbf{I} + \eta_1 \bar{\mathbf{D}} + \eta_2 \bar{\mathbf{D}} \cdot \bar{\mathbf{D}} \tag{4-4}$$

where

$$\eta_k = \eta_k(\rho, l, \text{div } \bar{\mathbf{v}}, \text{tr } \bar{\mathbf{D}}^2, \det \bar{\mathbf{D}}) \tag{4-5}$$

Our experience in formulating simple empirical models for $\mathbf{T}$ (see Sec. 2.3.3) suggests that we restrict our attention to the special case

$$\mathbf{T}^{(t)} = \eta_0 \mathbf{I} + \eta_1 \bar{\mathbf{D}} \tag{4-6}$$

Since we are considering only incompressible fluids,

$$-p^{(t)} \equiv \frac{1}{3} \text{tr } \mathbf{T}^{(t)} = \eta_0 \tag{4-7}$$

$$\eta_1 = \eta_1(\rho, l, \text{tr } \bar{\mathbf{D}}^2, \det \bar{\mathbf{D}}) \tag{4-8}$$

and Eq. (4-6) becomes

$$\mathbf{T}^{(t)} + p^{(t)}\mathbf{I} = \eta_1 \bar{\mathbf{D}} \tag{4-9}$$

The coefficient $\eta_1$ is often referred to as the *eddy viscosity coefficient.*

An early proposal by Boussinesq [1] was that

$$\eta_1 = \text{a constant} \tag{4-10}$$

This has not proved to be realistic. Experimentally we find that $\eta_1$ is a function of position in a flow field.

Let us investigate the special case of (4-9) for which

$$\eta_1 = \eta_1(\rho, l, \text{tr } \bar{\mathbf{D}}^2) \tag{4-11}$$

By the Buckingham-Pi theorem [3], we find

$$\frac{\eta_1}{\rho l^2 \sqrt{2 \text{ tr } \bar{\mathbf{D}}^2}} = 2\eta_1^* \tag{4-12}$$

where $\eta_1^*$ is a dimensionless constant. In view of Eqs. (4-7), (4-10), and (4-12), Eq. (4-4) becomes

$$\mathbf{T}^{(t)} + p^{(t)}\mathbf{I} = 2\eta_1^* \rho l^2 \sqrt{2 \text{ tr } \bar{\mathbf{D}}^2} \, \bar{\mathbf{D}} \tag{4-13}$$

This should be viewed as the tensorial form of *Prandtl's mixing length theory* [2, p. 277; 4; 5, p. 545].

It is perhaps worth emphasizing that we should not expect the Prandtl mixing length theory to be appropriate to the laminar sublayer or buffer zone. We assumed at the beginning that we were constructing a representation for the Reynolds stress tensor in the fully developed turbulent flow regime. We see in the next section that it provides an excellent representation of the time-averaged velocity distribution within the fully developed regime for turbulent flow through tubes.

There are also some similarities between this result and Taylor's vorticity transport theory [2, p. 281; 6; 7, p. 209]. Von Karman's similarity hypothesis [5, p. 551] has a considerably different character. His development is in the context of a special two-dimensional flow, but that he expected the Reynolds stress tensor to depend upon second derivatives of $\bar{\mathbf{v}}$ is clear. The tensorial representation of his result given by Bird, Stewart, and Lightfoot [8, p. 161] suggests that his approach may not have been entirely reasonable. If we were to approach his expression for the Reynolds stress tensor in a manner similar to that suggested above, instead of Eq. (4-3) we would start with

$$\mathbf{T}^{(t)} = \mathbf{T}^{(t)}(\rho, l, \bar{\mathbf{D}}, \Omega) \tag{4-14}$$

where

$$\Omega \equiv \nabla \mathbf{w} + (\nabla \mathbf{w})^T \tag{4-15}$$

and $\mathbf{w}$ is the vorticity vector (see Sec. 1.4.2). While it is true that the Bird, Stewart, and Lightfoot tensorial expression for Von Karman's result does obey the principle of material frame indifference, $\Omega$ seems to be an unlikely independent variable in Eq. (4-14), since it is not a frame-indifferent tensor (see for example [9, p. 24]).

**EXAMPLE 2  DEISSLER'S EXPRESSION FOR THE REGION NEAR THE WALL**

If we focus our attention on that portion of the turbulent flow of an incompressible newtonian fluid in the immediate vicinity of a bounding wall, the laminar sublayer and the buffer zone, it might appear reasonable to propose

$$\mathbf{T}^{(t)} = \mathbf{T}^{(t)}(\rho, \mu, l, \bar{\mathbf{v}} - \bar{\mathbf{v}}_{(s)}, \nabla \bar{\mathbf{v}}) \tag{4-16}$$

By $\mathbf{v}_{(s)}$, we mean the velocity of the bounding wall. The most general expression of this form that is consistent both with the principle of material frame indifference and the Buckingham-Pi theorem is exceedingly difficult [10, sec. 7; 11]. Let us instead look at a special case of Eq. (4-16) that satisfies the principle of material frame indifference:

$$\mathbf{T}^{(t)} + p^{(t)}\mathbf{I} = \varkappa(\rho, \mu, l, |\bar{\mathbf{v}} - \mathbf{v}_{(s)}|)\bar{\mathbf{D}} \tag{4-17}$$

After an application of the Buckingham-Pi theorem, we conclude that

$$\mathbf{T}^{(t)} + p^{(t)}\mathbf{I} = 2\eta^* \rho l \, |\bar{\mathbf{v}} - \mathbf{v}_{(s)}| \, \bar{\mathbf{D}} \tag{4-18}$$

Here

$$\eta^* = \eta^*(N) \tag{4-19}$$

and

$$N \equiv \frac{\rho l \, |\bar{\mathbf{v}} - \mathbf{v}_{(s)}|}{\mu} \tag{4-20}$$

Deissler [12] has proposed on empirical grounds that

$$\mathbf{T}^{(t)} + p^{(t)}\mathbf{I} = 2n^2 \rho l \, |\bar{\mathbf{v}} - \mathbf{v}_{(s)}| \, [1 - \exp(-n^2 N)]\bar{\mathbf{D}} \tag{4-21}$$

We see in the next section that this does an excellent job of representing the time-averaged velocity distribution within the laminar sublayer and buffer zone for turbulent flow through tubes.

## REFERENCES

1. Boussinesq, J.: *Mém. prés. par div. savants à l'Acad. Sci. Paris*, **23**:46 (1877).
2. Hinze, J. O.: "Turbulence," McGraw-Hill, New York, 1959.
3. Brand, L.: *Arch. Rational Mech. Analysis*, **1**:35 (1957).
4. Prandtl, L.: *Z. Angew. Math. Mech.*, **5**:136 (1925).
5. Schlichting, Hermann: "Boundary-Layer Theory," 6th ed., McGraw-Hill, New York, 1968.
6. Taylor, G. I.: *Proc. Roy. Soc. (London)*, **A135**:685 (1932).
7. Goldstein, S.: "Modern Developments in Fluid Dynamics," vol. 1, Oxford University Press, London, 1938.
8. Bird, R. B., W. E. Stewart, and E. N. Lightfoot: "Transport Phenomena," 7th printing, Wiley, New York, 1960.
9. Truesdell, C., and W. Noll: In S. Flügge (ed.), "Handbuch der Physik," vol. 3/3, Springer-Verlag, Berlin, 1965.
10. Spencer, A. J.: *Arch. Rational Mech. Analysis*, **4**:214 (1959/60).
11. Smith, G. F.: *Arch. Rational Mech. Analysis*, **18**:282 (1965).
12. Deissler, R. G.: *NACA Rept.* 1210, 1955.

**4.1.5 Turbulent flow through a tube** Our purpose is to discuss the time-averaged velocity distribution for turbulent flow of an incompressible newtonian fluid through a long inclined tube of radius $R$ such as that shown in Fig. 3.2.1-2. For the moment, we will focus our attention on that portion of the flow which can be considered as fully developed. The velocity distribution in the laminar sublayer and buffer zone will be considered later. Since we are primarily concerned with the fully developed flow, we will use the Prandtl mixing length theory for the Reynolds stress tensor (see Sec. 4.1.4, Example 1).

Let us begin by assuming that the time-averaged velocity distribution has the form

$$\bar{v}_z = \bar{v}_z(r) \qquad \bar{v}_r = \bar{v}_\theta = 0 \tag{5-1}$$

This means that the viscous and Reynolds stress tensors have only one nonzero shear component:

$$\bar{S}_{rz} + T_{rz}^{(t)} = \left[\mu + \eta_1^* \rho (R - r)^2 \left|\frac{d\bar{v}_z}{dr}\right|\right] \frac{d\bar{v}_z}{dr} \tag{5-2}$$

The time-averaged equation of continuity for an incompressible fluid is satisfied identically by Eqs. (5-1). The three components of the time-averaged Cauchy's first law reduce to

$$\frac{\partial \mathscr{P}}{\partial r} = \frac{\partial \mathscr{P}}{\partial \theta} = 0 \qquad (5\text{-}3)$$

and

$$\frac{\partial \mathscr{P}}{\partial z} = \frac{1}{r}\frac{d}{dr}[r(\bar{S}_{rz} + T_{rz}^{(t)})] \qquad (5\text{-}4)$$

Arguing in much the same manner as we did in Sec. 3.2.1, we quickly conclude that

$$-\frac{\partial \mathscr{P}}{\partial z} = (P_0 - P_L - \rho g L \sin \alpha)/L \qquad (5\text{-}5)$$

and

$$\bar{S}_{rz} + T_{rz}^{(t)} = -S_0 \frac{r}{R} \qquad (5\text{-}6)$$

where

$$S_0 \equiv \frac{(P_0 - P_L - \rho g L \sin \alpha)R}{2L} \qquad (5\text{-}7)$$

When we recognize that

$$\left|\frac{d\bar{v}_z}{dr}\right| = -\frac{d\bar{v}_z}{dr} \qquad (5\text{-}8)$$

Eqs. (5-2) and (5-6) may be combined to say

$$\left[\mu - \eta_1^* \rho (R-r)^2 \frac{d\bar{v}_z}{dr}\right]\frac{d\bar{v}_z}{dr} = -\frac{S_0 r}{R} \qquad (5\text{-}9)$$

This can be expressed in terms of a dimensionless distance measured from the tube wall

$$s^* \equiv \frac{R-r}{R} = 1 - \frac{r}{R} \qquad (5\text{-}10)$$

and a dimensionless velocity

$$v^* \equiv \frac{\bar{v}_z}{v_0} \qquad (5\text{-}11)$$

as

$$\left(1 + \frac{\rho v_0 R}{\mu}\eta_1^* s^{*2}\frac{dv^*}{ds^*}\right)\frac{dv^*}{ds^*} = \frac{RS_0}{\mu v_0}(1-s^*) \qquad (5\text{-}12)$$

If we identify the characteristic speed as

$$v_0 \equiv \sqrt{\frac{S_0}{\rho}} \tag{5-13}$$

Eq. (5-12) simplifies to

$$\left(1 + N_{(t)}\eta_1^* s^{*2} \frac{dv^*}{ds^*}\right)\frac{dv^*}{ds^*} = N_{(t)}(1 - s^*) \tag{5-14}$$

where we define

$$N_{(t)} \equiv \frac{\rho v_0 R}{\mu} \tag{5-15}$$

In the immediate neighborhood of the wall, $s^* \ll 1$ and Eq. (5-14) reduces to

$$\left(1 + \eta_1^* s^{**2} \frac{dv^*}{ds^{**}}\right)\frac{dv^*}{ds^{**}} = 1 \tag{5-16}$$

The expanded variable

$$s^{**} = N_{(t)} s^* \tag{5-17}$$

has been introduced here for simplicity.

We began by saying that we would confine our attention to that portion of the flow which can be considered to be fully developed. We suggested in Sec. 4.1.4 that in the fully developed portion of the flow the effect of the viscous stresses was negligible compared with that of the Reynolds stresses. Consequently, we will assume that

$$\eta_1^* s^{**2} \frac{dv^*}{ds^{**}} \gg 1 \tag{5-18}$$

and that in the fully developed portion of the flow, Eq. (5-16) reduces to

$$\eta_1^* s^{**2} \left(\frac{dv^*}{ds^{**}}\right)^2 = 1 \tag{5-19}$$

In a moment, we will return to check inequality (5-18).

Equation (5-19) can be integrated to find

$$\text{For } s^{**} \geq s_1^{**}: \qquad v^* - v_1^* = \frac{1}{\sqrt{\eta_1^*}} \ln \frac{s^{**}}{s_1^{**}} \tag{5-20}$$

where it is convenient to interpret $s_1^{**}$ as the outer edge of the buffer zone and $v_1^*$ as the dimensionless velocity at this position.

As a result of the comparison with experimental data shown in Fig. 4.1.5-1, Deissler [1] recommends that we take $\sqrt{\eta_1^*} = 0.36$, $s_1^{**} = 26$, and $v_1^* = 12.85$. With these values, Eq. (5-20) becomes

## [4.1] TIME AVERAGING

For $s^{**} \geq 26$: 
$$v^* = \frac{1}{0.36} \ln s^{**} + 3.8 \qquad (5\text{-}21)$$

This means that

For $s^{**} \geq 26$: 
$$\eta_1^* s^{**2} \frac{dv^*}{ds^{**}} \geq 9.4 \qquad (5\text{-}22)$$

and inequality (5-18) appears to be justified.

Notice in Fig. 4.1.5-1 that $dv^*/ds^{**} \neq 0$ at the center of the tube in contrast with our intuition. But our intuition has not failed us. Rather, as explained previously, we have restricted ourselves to the region near to the wall (although outside the laminar sublayer and buffer zone) in deriving Eq. (5-16) and, consequently, Eq. (5-21). We should not expect it to be valid for the region near the axis of the tube.

Let us now try to describe the velocity distribution in the laminar sublayer and in the buffer zone. Deissler's [1] empirical proposal for describing the Reynolds stress tensor in this region was discussed in Example 2 of Sec. 4.1.4. If we again start with the assumption that the time-averaged velocity distribution has the form indicated in Eq. (5-1), then the viscous and Reynolds stress tensors have only one nonzero component:

$$\bar{S}_{rz} + T_{rz}^{(t)} = \left( \mu + n^2 \rho \bar{v}_z (R - r) \left\{ 1 - \exp\left( \frac{-n^2 \rho \bar{v}_z [R - r]}{\mu} \right) \right\} \right) \frac{d\bar{v}_z}{dr} \qquad (5\text{-}23)$$

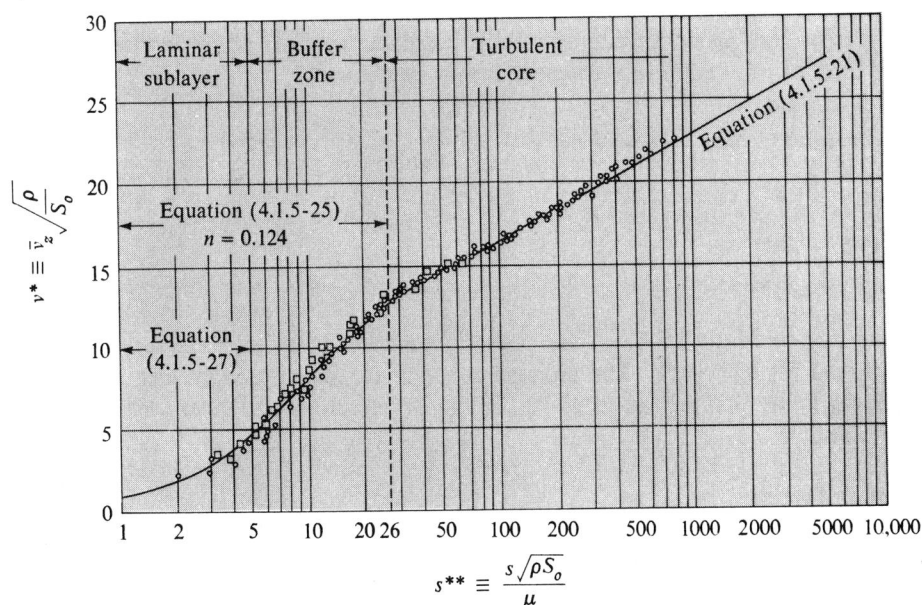

**Fig. 4.1.5-1** Velocity distribution for turbulent isothermal flow in tubes [1, p. 3]. Experimental data are those of Deissler [2] and Laufer [3], circles and squares, respectively.

In view of Eq. (5-6), this means that

$$\{1 + n^2 v^* s^{**}[1 - \exp(-n^2 v^* s^{**})]\} \frac{dv^*}{ds^{**}} = 1 - \frac{s^{**}}{N_{(t)}} \quad (5\text{-}24)$$

where, for convenience, we have introduced the dimensionless variables defined by Eqs. (5-12) and (5-13).

Since we are interested here in the laminar sublayer and buffer zone, we can restrict ourselves to the immediate neighborhood of the wall

$$s^* = \frac{s^{**}}{N_{(t)}} \ll 1:$$

$$\{1 + n^2 v^* s^{**}[1 - \exp(-n^2 v^* s^{**})]\} \frac{dv^*}{ds^{**}} = 1 \quad (5\text{-}25)$$

Deissler [1] has integrated Eq. (5-25) numerically. His result for $n = 0.124$ is shown in Fig. 4.1.5-1. With this value for $n$, Eq. (5-25) does an excellent job in describing the velocity distribution in both the laminar sublayer and in the buffer zone.

If we are primarily interested in the laminar sublayer, then we should examine the limit of Eq. (5-25) as the wall is approached:

$$\text{As } s^{**} \to 0: \quad \frac{dv^*}{ds^{**}} = 1 \quad (5\text{-}26)$$

Integrating, we find that the velocity distribution in the laminar sublayer should have the form

$$v^* = s^{**} \quad (5\text{-}27)$$

Referring to Fig. 4.1.5-1, we see that this relationship provides a very good representation for $s^{**} \leq 5$. The laminar sublayer for flow through very long tubes is defined to be that region in which Eq. (5-27) describes the time-averaged velocity distribution.

**REFERENCES**

1. Deissler, R. G.: *NACA Rept.* 1210, 1955.
2. Deissler, R. G.: *NACA Tech. Note* 2138, 1950.
3. Laufer, John: *NACA Tech. Note* 2954, 1953.

**EXERCISES**

**4.1.5-1** *More on the Prandtl mixing length theory* Integrate Eq. (5-16) directly to find for $s^{**} \geq s_1^{**}$:

$$v^* - v_1^* = \frac{1 - \sqrt{1 + 4n_1^* s^{**2}}}{2\sqrt{n_1^*} s^{**}} - \frac{1 - \sqrt{1 + 4n_1^* s_1^{**2}}}{2\sqrt{n_1^*} s_1^{**}} + \frac{1}{\sqrt{n_1^*}} \ln\left\{ \frac{2\sqrt{n_1^*} s^{**} + \sqrt{1 + 4n_1^* s^{**2}}}{2\sqrt{n_1^*} s_1^{**} + \sqrt{1 + 4n_1^* s_1^{**2}}} \right\}$$

For $2\sqrt{n_1^*} s^{**} \gg 1$, this result reduces to Eq. (5-20). With the values recommended by Deissler [1], $2\sqrt{n_1^*} s_1^{**} = 18.7$.

## [4.2] AREA AVERAGING

**4.1.5-2** *Turbulent flow between two flat plates* Repeat the analysis of this section for turbulent flow through the inclined channel shown in Fig. 3.2.1-5. Determine that Eqs. (5-20) and (5-25) again apply with the understanding that here

$$S_0 \equiv \frac{(P_0 - P_L - \rho g L \sin \alpha)b}{L}$$

and

$$v^* \equiv \frac{\bar{v}_1}{v_0}$$

## 4.2 AREA AVERAGING

**4.2.1 Area averaging** For most engineering purposes complete velocity distributions are not required. We are usually concerned with estimating some gross aspect of a problem such as a volume flow rate or a force on a wall.

When the dependence of the velocity distribution upon the directions normal to the gross flow do not appear to be of prime interest, it may be wise to average the equation of motion over the cross section normal to the flow. This can lead to a sizable simplification. For this reason it is a particularly desirable approach when the original problem posed requires considerable time and money for solution. You must make a judgment in the context of the application with which you are concerned. If you must have an answer accurate within 1 percent, then a solution to the full equation of motion is required. If you are willing to accept as much as a 20 or 25 percent error (and no rigorous error bounds), then an integral averaging technique, such as area averaging, is in order.

With any of the integral averaging techniques, information is lost and you are asked to supply an empiricism or an approximation. In time averaging we found that it was necessary to supply an empirical data correlation for the Reynolds stress tensor $\mathbf{T}^{(t)}$. In area averaging, there are two ways in which this empiricism can be introduced.

In the first class of problems, as illustrated in Sec. 4.2.2, we concern ourselves primarily with an area-averaged variable, perhaps a volume flow rate. Normally, an approximation is made concerning the force per unit area or stress at a bounding wall.

The second and more highly developed class of problems is often referred to as approximate boundary-layer theory. In approximate boundary-layer theory, the form of the velocity distribution is assumed in terms of a function $\delta(x)$ of the arc length $x$. This function is often referred to as the (approximate) boundary-layer thickness. The area-averaged equation of motion yields an ordinary differential equation for $\delta$. Approximate boundary-layer theory is introduced in Sec. 4.2.3.

**4.2.2 Flow from rest in a circular tube** This problem is meant to illustrate the use of area-averaged variables.

In Exercise 3.2.4-6, I outlined an exact solution for flow from rest in an inclined tube. I assumed there that I was dealing with an incompressible newtonian fluid. But what happens if you wish to treat the same problem for an incompressible power-model fluid? You are faced with the nasty job of solving a nonlinear partial differential equation. This is certainly not an impossible task, but the use to which the solution is to be put may not justify the time and expense required.

I would like to suggest an alternative approach based on the premise that we are primarily concerned with determining the volume flow rate $Q$ through the tube as a function of time. In other words, we would like to find the area-averaged axial component of velocity as a function of time. We see in Exercise 4.2.2-1 that numerical work is still required for the case of the power-model fluid, but only a relatively straightforward numerical integration.

But what about accuracy? I suggest we begin by again looking at an incompressible newtonian fluid, so that we may compare the results obtained here with the exact solution outlined in Exercise 3.2.4-6.

Having assumed in Exercise 3.3.4-6 that there was only one nonzero component of velocity:

$$v_r = v_\theta = 0 \qquad v_z = v_z(t,r) \tag{2-1}$$

we found that Cauchy's first law implied for an incompressible newtonian fluid

$$\frac{\partial v_z}{\partial t} = -A + \frac{1}{r}\frac{\partial}{\partial r}(rS_{rz}) \tag{2-2}$$

where

$$-A \equiv \frac{P_0 - P_L}{L} - \rho g \sin \alpha \tag{2-3}$$

Equation (2-2) and the appropriate constitutive equation for $S_{rz}$ were solved simultaneously, consistent with the boundary conditions

$$\text{At } t = 0: \qquad v_z = 0 \tag{2-4}$$

and

$$\text{At } r = R: \qquad v_z = 0 \tag{2-5}$$

But if we are primarily interested in the area-averaged velocity

$$\bar{v}_z \equiv \frac{1}{\pi R^2} \int_0^{2\pi} \int_0^R v_z r \, dr \, d\theta \tag{2-6}$$

then we might consider averaging Eq. (2-2) over the cross section of the tube normal to flow:

$$\rho \frac{d\bar{v}_z}{dt} = -A + \frac{1}{\pi R^2} \int_0^{2\pi} \int_0^R \frac{\partial}{\partial r}(rS_{rz}) \, dr \, d\theta$$

## [4.2] AREA AVERAGING

$$= -A + \frac{2}{R} S_{rz}\big|_{r=R} \tag{2-7}$$

There is a problem here. We do not know a priori the shear stress at the wall of the tube. We have lost some information in averaging, just as we did in Sec. 4.1.2. We need either an empirical data correlation or an approximation, in order to evaluate the wall shear stress. Perhaps the simplest thing to do is to say that the relationship between the wall shear stress and the average velocity in the tube is that found for the Poiseuille flow in Sec. 3.2.1:

$$S_{rz}\big|_{r=R} = -\frac{4\mu}{R}\bar{v}_z \tag{2-8}$$

In applying this approximation, we find that Eq. (2-7) yields an ordinary differential equation for the area-averaged velocity:

$$\rho \frac{d\bar{v}_z}{dt} = -A - \frac{8\mu}{R^2}\bar{v}_z \tag{2-9}$$

This can easily be integrated consistent with boundary condition (2-4) in the form of

At $t = 0$: $\quad \bar{v}_z = 0$ \hfill (2-10)

to find

$$\frac{8\mu Q}{\pi R^4(-A)} = 1 - \exp(-8t^*) \tag{2-11}$$

In Fig. 4.2.2-1, we compare this last with the exact solution, Eq. (1) of Exercise 3.2.4-6. The error introduced in using Eq. (2-11) is less than 20 percent for $t^* > 0.05$.

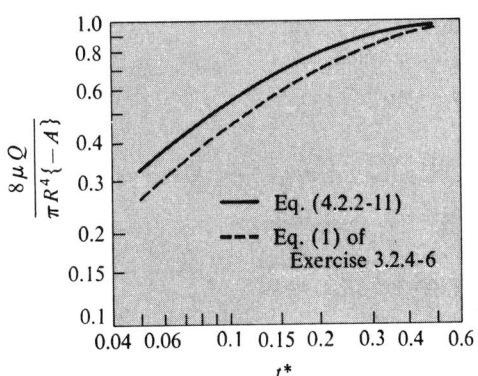

**Fig. 4.2.2-1** Comparison of area-averaged analysis with exact solution for flow from rest in a circular tube.

## 186 APPLICATION OF INTEGRAL AVERAGING TECHNIQUES TO MOMENTUM TRANSFER

For many purposes, this is a small price to pay when the simplicity of the calculation is taken into account.

But our primary purpose here is to suggest how we might attack the problem of flow from rest in a tube of a nonnewtonian fluid. For more on this, see Exercise 4.2.2-1.

### EXERCISE

**4.2.2-1** *Flow from rest of a power-model fluid* Repeat the problem discussed in the text for an incompressible power-model fluid. Determine that the area-averaged velocity is a solution to

$$\rho \frac{d\bar{v}_z}{dt} = -A - \frac{2m}{R^{n+1}}\left(\frac{1+3n}{n}\right)^n \bar{v}_z{}^n$$

consistent with boundary condition (2-10). The constant $A$ is defined again by Eq. (2-3).

The simplicity of this differential equation is sharply in contrast with the nonlinear, second-order partial differential equation that must be solved in the exact analysis (the exact analysis would parallel that outlined in Exercise 3.2.4-6).

### 4.2.3 Approximate boundary-layer theory for plane flow past a curved wall

[1, p. 192] Boundary-layer theory, discussed in Secs. 3.5.1 to 3.5.7, is developed by keeping our attention upon the limit as $N_{Re} \to \infty$. The boundary-layer equation of motion is more amenable to solution than the Navier-Stokes equation, but some problems do not warrant the time and effort required for solution of a partial differential equation. A less detailed and more approximate solution seems in order.

Our success in Sec. 4.2.2 suggests that we might integrate the boundary-layer equations over the cross section normal to the flow. We shall do this, but in contrast with the previous section we shall not work in terms of area-averaged variables. Instead, we shall take an entirely different approach and estimate some of the integrals involved, using an approximate velocity distribution. I think you will find the results are surprisingly accurate when compared with the exact solutions of the boundary-layer equations, particularly considering the dramatic reduction in effort required to find a solution.

Let us confine our attention to steady-state plane flow past a curved wall of an incompressible newtonian fluid. In Sec. 3.5.3 we found that, under these circumstances, the equation of continuity and the equation of motion reduce to

$$\frac{\partial v_x^*}{\partial x^*} + \frac{\partial v_y^{**}}{\partial y^{**}} = 0 \tag{3-1}$$

and

$$v_y^* \frac{\partial v_x^*}{\partial x^*} + v_y^{**} \frac{\partial v_x^*}{\partial y^{**}} = \tilde{v}_x^* \frac{d\tilde{v}_x^*}{dx^*} + \frac{\partial^2 v_x^*}{\partial y^{**2}} \tag{3-2}$$

where

$$y^{**} \equiv (N_{Re})^{\frac{1}{2}} y^* \qquad v_y^{**} \equiv (N_{Re})^{\frac{1}{2}} v_y^* \tag{3-3}$$

## [4.2] AREA AVERAGING

and $\tilde{v}_x^*$ is the dimensionless velocity distribution at the curved wall from the nonviscous flow solution. For plane boundary-layer flow past a curved wall, we seek a solution to Eqs. (3-1) and (3-2) that is consistent with the conditions that

$$\text{At } y^{**} = 0: \quad v_x^* = v_y^{**} = 0 \tag{3-4}$$

and

$$\text{As } y^{**} \to \infty: \quad v_x^* \to \tilde{v}_x^* \tag{3-5}$$

Let us begin by integrating Eq. (3-2) over the boundary layer:

$$\int_0^\infty \left( v_x^* \frac{\partial v_x^*}{\partial x^*} + v_y^{**} \frac{\partial v_x^*}{\partial y^{**}} - \tilde{v}_x^* \frac{d\tilde{v}_x^*}{dx^*} - \frac{\partial^2 v_x^*}{\partial y^{**2}} \right) dy^{**} = 0 \tag{3-6}$$

Some rearrangement is called for.

We can integrate Eq. (3-1) to find

$$\int_0^{y^{**}} \left( \frac{\partial v_x^*}{\partial x^*} + \frac{\partial v_y^{**}}{\partial y^{**}} \right) dy^{**} = 0 \tag{3-7}$$

or

$$v_y^{**} = -\int_0^{y^{**}} \frac{\partial v_x^*}{\partial x^*} dy^{**} \tag{3-8}$$

This together with an integration by parts may be used to state the second term on the left of Eq. (3-6) as

$$\int_0^\infty v_y^{**} \frac{\partial v_x^*}{\partial y^{**}} dy^{**} = -\int_0^\infty \frac{\partial v_x^*}{\partial y^{**}} \left( \int_0^{y^{**}} \frac{\partial v_x^*}{\partial x^*} dy^{**} \right) dy^{**}$$

$$= -\tilde{v}_x^* \int_0^\infty \frac{\partial v_x^*}{\partial x^*} dy^{**} + \int_0^\infty v_x^* \frac{\partial v_x^*}{\partial x^*} dy^{**} \tag{3-9}$$

The fourth term on the left of Eq. (3-6) may be integrated directly as

$$\int_0^\infty \frac{\partial^2 v_x^*}{\partial y^{**2}} dy^{**} = -\left. \frac{\partial v_x^*}{\partial y^{**}} \right|_{y^{**}=0} \tag{3-10}$$

so long as we are willing to require that

$$\text{As } y^{**} \to \infty: \quad \frac{\partial v_x^*}{\partial y^{**}} \to 0 \tag{3-11}$$

Equations (3-9) and (3-10) allow us to rewrite Eq. (3-6) as

$$\frac{d}{dx^*} \int_0^\infty v_x^*(\tilde{v}_x^* - v_x^*) \, dy^{**} + \frac{d\tilde{v}_x^*}{dx^*} \int_0^\infty (\tilde{v}_x^* - v_x^*) \, dy^{**} = \left. \frac{\partial v_x^*}{\partial y^{**}} \right|_{y^{**}=0} \tag{3-12}$$

I propose next that we express this equation in terms of an approximate velocity distribution

$$\frac{v_x^*}{\tilde{v}_x^*} = h(x^*, \eta) \tag{3-13}$$

where we define

$$\eta = \frac{y^{**}}{\delta^{**}(x^*)} = \frac{y}{\delta(x)} \tag{3-14}$$

The function $\delta(x)$ does not have a carefully defined physical meaning, though it is usually loosely thought of as the approximate thickness of the boundary layer. With this thought in mind, we require the function $h(x^*,\eta)$ to be such that

$$\text{For } \eta \geq 1: \quad h = 1 \tag{3-15}$$

This allows us to evaluate in Eq. (3-12)

$$\int_0^\infty v_x^*(\tilde{v}_x^* - v_x^*)\, dy^{**} = (\tilde{v}_x^*)^2 \delta^{**} \alpha_1 \tag{3-16}$$

$$\int_0^\infty (\tilde{v}_x^* - v_x^*)\, dy^{**} = \tilde{v}_x^* \delta^{**} \alpha_2 \tag{3-17}$$

and

$$\left.\frac{\partial v_x^*}{\partial y^{**}}\right|_{y^{**}=0} = \frac{\tilde{v}_x^*}{\delta^{**}} \beta_1 \tag{3-18}$$

with the understanding that

$$\alpha_1 \equiv \int_0^1 h(1-h)\, d\eta \tag{3-19}$$

$$\alpha_2 \equiv \int_0^1 (1-h)\, d\eta \tag{3-20}$$

and

$$\beta_1 \equiv \left.\frac{\partial h}{\partial \eta}\right|_{\eta=0} \tag{3-21}$$

Equations (3-16) to (3-18) enable us to write Eq. (3-12) as

$$\tilde{v}_x^* \delta^{**} \alpha_1 \frac{d}{dx^*}(\delta^{**}\alpha_1) + \left(2 + \frac{\alpha_2}{\alpha_1}\right)(\delta^{**}\alpha_1)^2 \frac{d\tilde{v}_x^*}{dx^*} = \alpha_1 \beta_1 \tag{3-22}$$

Let us choose a particular approximate velocity distribution:

$$\frac{v_x^*}{\tilde{v}_x^*} = h(x^*,\eta) = a + b\eta + c\eta^2 + d\eta^3 + e\eta^4 \tag{3-23}$$

We evaluate the five functions of $x^*$ in this velocity distribution by requiring Eq.

## [4.2] AREA AVERAGING

(3-23) to satisfy Eqs. (3-4), (3-5), and (3-11), as well as

$$\text{At } \eta^{**} = 0: \quad \tilde{v}_x^* \frac{d\tilde{v}_x^*}{dx^*} + \frac{\partial^2 v_x^*}{\partial y^{**2}} = 0 \tag{3-24}$$

and

$$\text{As } \eta^{**} \to 1: \quad \frac{\partial^2 v_x^*}{\partial y^{**2}} \to 0 \tag{3-25}$$

As a result, we find

$$h = 1 - (1-\eta)^3(1+\eta) + \frac{\Lambda}{6}\eta(1-\eta)^3 \tag{3-26}$$

We introduce here as a definition

$$\Lambda \equiv \delta^{**2} \frac{d\tilde{v}_x^*}{dx^*} = \frac{\rho}{\mu} \delta^2 \frac{d\tilde{v}_x}{dx} \tag{3-27}$$

This enables us to obtain explicit expressions for $\alpha_1$, $\alpha_2$, and $\beta_1$ in Eqs. (3-19) to (3-21):

$$\alpha_1 = \frac{1}{63}\left(\frac{37}{5} - \frac{\Lambda}{15} - \frac{\Lambda^2}{144}\right) \tag{3-28}$$

$$\alpha_2 = \frac{3}{10} - \frac{\Lambda}{120} \tag{3-29}$$

and

$$\beta_1 = 2 + \frac{\Lambda}{6} \tag{3-30}$$

At this point it seems convenient to write Eq. (3-22) in the more compact form

$$\frac{dZ}{dx^*} = \frac{F(\Lambda)}{\tilde{v}_x^*} \tag{3-31}$$

Here

$$Z \equiv (\delta^{**}\alpha_1)^2 \tag{3-32}$$

and

$$F(\Lambda) \equiv 2\alpha_1[\beta_1 - (2\alpha_1 + \alpha_2)\Lambda]$$
$$= 2\left(\frac{37}{315} - \frac{\Lambda}{945} - \frac{\Lambda^2}{9072}\right)$$
$$\times \left[2 - \frac{116}{315}\Lambda + \left(\frac{2}{945} + \frac{1}{120}\right)\Lambda^2 + \frac{2}{9072}\Lambda^3\right] \tag{3-33}$$

Equation (3-31) is to be solved simultaneously with

$$Z\frac{d\tilde{v}_x^*}{dx^*} = (\alpha_1)^2 \Lambda \tag{3-34}$$

in order to determine $Z$ and $\Lambda$ as functions of the dimensionless arc length $x^*$.

To be more explicit, the computations required may be summarized as follows:

1. We assume that we are given $\tilde{v}_x^* = \tilde{v}_x^*(x^*)$, the dimensionless velocity distribution at the curved wall from the nonviscous flow solution.
2. Equations (3-31) and (3-34) are to be solved simultaneously for $Z$ and $\Lambda$ as functions of the dimensionless arc length $x^*$. The boundary condition required in carrying out this integration is that at the stagnation point

at $x^* = 0$: $\tilde{v}_x = 0$, $\Lambda = \Lambda_0 = 7.052$ (3-35)

where from Eq. (3-31)

$$F(\Lambda_0) = 0 \tag{3-36}$$

3. The point at which the boundary layer separates from the wall is defined as the limit between forward and reverse flow in the immediate neighborhood of the wall [1, p. 122]:

At point of separation: $\left.\dfrac{\partial v_x^*}{\partial y^{**}}\right|_{y^{**}=0} = 0$ (3-37)

From Eqs. (3-18) and (3-30), it is clear that

At point of separation: $\Lambda = -12$ (3-38)

4. The approximate boundary-layer thickness $\delta$ as a function of $x$ follows from Eq. (3-27).
5. Finally, the approximate velocity distribution is found from Eq. (3-26).

Schlichting [1, p. 201] has an excellent comparison between the results obtained using the approximate boundary-layer theory as described here and the exact boundary-layer theory as described in Sec. 3.5.3. For many purposes, the approximate boundary-layer theory does surprisingly well. It is also interesting to know that the results appear to be relatively insensitive to the form of the approximate velocity distribution chosen in Eq. (3-23) [1, p. 192].

### REFERENCES

1. Schlichting, Hermann: "Boundary-Layer Theory," 6th ed., McGraw-Hill, New York, 1968.
2. Bizzell, G. D.: *Chem. Eng. Sci.*, **17**:777 (1962); *ibid.*, **20**:364 (1965).

### EXERCISES

**4.2.3-1** *Flow past a flat plate* As a particular example, let us consider flow past a flat plate at zero

[4.2] AREA AVERAGING

incidence, for which $\tilde{v}_x^* = 1$. If we define the drag coefficient as (we discuss the concept of drag coefficients in Sec. 4.4.7)

$$c \equiv \frac{2\int_0^L \mu \left.\frac{\partial v_x}{\partial y}\right|_{y=0} dx}{\frac{1}{2}\rho(v_\infty)^2 2L}$$

use the discussion in the text to estimate that

$$c = \frac{1.37}{\sqrt{N_{Re}}} \qquad N_{Re} \equiv \rho\frac{v_\infty L}{\mu}$$

Here $L$ is the length of the plate.

For comparison, the exact solution tells us that [1, p. 128]

$$c = \frac{1.33}{\sqrt{N_{Re}}}$$

Approximate boundary-layer theory leads to a 3 percent error, but far less time and trouble are required to obtain the result.

**4.2.3-2** *Steady-state flow past a body of revolution* Let us follow the example of the text in developing approximate boundary-layer theory for steady-state flow of an incompressible newtonian fluid past a body of revolution.

(a) Starting with the boundary-layer equations derived in Sec. 3.5.7, determine that

$$\tilde{v}_x^* \delta^{**}\alpha_1 \frac{d}{dx^*}(\delta^{**}\alpha_1) + \left(2 + \frac{\alpha_2}{\alpha_1}\right)(\delta^{**}\alpha_1)^2 \frac{d\tilde{v}_x^*}{dx^*} + \frac{1}{f^*}\frac{df^*}{dx^*}(\tilde{v}_x^* \delta^{**2}\alpha_1^2) = \alpha_1\beta_1$$

where $\alpha_1$, $\alpha_2$, and $\beta_1$ are again defined by Eqs. (3-19) through (3-21).

(b) Introduce the approximate velocity distribution of Eq. (3-23) and conclude that $\alpha_1$, $\alpha_2$, and $\beta_1$ are still represented by Eqs. (3-28) to (3-30), with $\Lambda$ defined by Eq. (3-27).

(c) Conclude that the approximate boundary-layer equations, which are to be solved simultaneously as explained in the text, are

$$\frac{\tilde{v}_x^*}{f^{*2}}\frac{d}{dx^*}(f^{*2}Z) = F(\Lambda)$$

and Eq. (3-34). The functions $Z$ and $F(\Lambda)$ are again defined by Eqs. (3-32) and (3-33). The approximate boundary-layer equations are to be solved consistent with the boundary condition that at the stagnation point

at $x^* = 0$: $\tilde{v}_x^* = 0$, $\Lambda = \Lambda_0$

where [1, p. 230]

$$F(\Lambda_0) - 2(\alpha_1)^2 \Lambda_0 \left\{\lim_{x^* \to 0}\left[\frac{1}{f^*}\frac{df^*}{dx^*}\tilde{v}_x^*\left(\frac{d\tilde{v}_x^*}{dx^*}\right)^{-1}\right]\right\} = 0$$

**4.2.3-3** *Steady-state flow past a body of revolution of an incompressible power-model fluid* [2] Let us repeat Exercise 4.2.3-2 for an incompressible power-model fluid.

(a) Starting with the boundary-layer equations derived in Exercise 3.5.7-3, determine that

## 192 APPLICATION OF INTEGRAL AVERAGING TECHNIQUES TO MOMENTUM TRANSFER

$$\tilde{v}_x^{*2-n} \delta^{**n} \alpha_1^n \frac{d}{dx^*}(\delta^{**}\alpha_1) + \left(2 + \frac{\alpha_2}{\alpha_1}\right)(\delta^{**}\alpha_1)^{n+1}\tilde{v}_x^{*1-n} \frac{d\tilde{v}_x^*}{dx^*}$$

$$+ \frac{1}{f^*}\frac{df^*}{dx^*}(\tilde{v}_x^{*2-n} \delta^{**n+1}\alpha_1^{n+1}) = \alpha_1^n \beta_n$$

where $\alpha_1$ and $\alpha_2$ are defined by Eqs. (3-19) and (3-20) and

$$\beta_n \equiv \left(\frac{\partial h}{\partial \eta}\right)^n \bigg|_{\eta=0}$$

(b) Introduce the approximate velocity distribution of Eq. (3-23) and conclude that

$$\frac{v_x^*}{\tilde{v}_x^*} = 1 - (1 - \eta)^3(1 + \eta) + \frac{\Lambda_n}{6}\eta(1 - \eta)^3$$

$$\alpha_1 = \frac{1}{63}\left[\frac{37}{5} - \frac{\Lambda_n}{15} - \frac{\Lambda_n^2}{144}\right]$$

$$\alpha_2 = \frac{3}{10} - \frac{\Lambda_n}{120}$$

and

$$\beta_n = \left(2 + \frac{\Lambda_n}{6}\right)^n$$

Here $\Lambda_n$ is the solution to

$$\Lambda_n = \frac{1}{n} \delta^{**n+1}\tilde{v}_x^{*1-n} \frac{d\tilde{v}_x^*}{dx^*}\left(2 + \frac{\Lambda_n}{6}\right)^{1-n}$$

(c) Conclude that the approximate boundary-layer equations, which are to be solved simultaneously as explained in the text, are

$$\frac{\tilde{v}_x^*}{(f^*)^{n+1}}\frac{d}{dx^*}(f^{*n+1}Z_n) = F(\Lambda_n)$$

and

$$(\alpha_1)^{n+1}\Lambda_n = \frac{Z_n}{n}\frac{d\tilde{v}_x^*}{dx^*}\left(2 + \frac{\Lambda_n}{6}\right)^{1-n}$$

We define here

$$Z_n \equiv (\delta^{**}\alpha_1)^{n+1}\tilde{v}_x^{*1-n}$$

and

$$F(\Lambda_n) \equiv (n+1)\alpha_1^n\left[\beta_n - \frac{n\Lambda_n(2\alpha_1 + \alpha_2)}{(2 + \Lambda_n/6)^{1-n}} + \frac{n\Lambda_n\alpha_1(1-n)}{(1+n)(2 + \Lambda_n/6)^{1-n}}\right]$$

These are to be satisfied consistent with the boundary condition that at the stagnation point

at $x^* = 0$: $\tilde{v}_x^* = 0$, $\Lambda_n = \Lambda_{n0}$

where

$$F(\Lambda_{n0}) - n(n+1)(\alpha_1)^{n+1}\Lambda_{n0}\left(2 + \frac{\Lambda_{n0}}{6}\right)^{n-1}\left\{\lim_{x^* \to 0}\left[\frac{1}{f^*}\frac{df^*}{dx^*}\tilde{v}_x^*\left(\frac{d\tilde{v}_x^*}{dx^*}\right)^{-1}\right]\right\} = 0$$

[4.3] LOCAL VOLUME AVERAGING

The results obtained here include three interesting special cases.

1. We have the appropriate equations for newtonian flow past an axially symmetric body (Exercise 4.2.3-2) when we set $n = 1$.
2. We have the results for newtonian flow past a curved wall (text) when we set $n = 1$ and $f^* = a$ constant.
3. We have the results for the flow of a power-model fluid past a curved wall when we set $f^* = a$ constant.

This development has been used to estimate the point of separation in the flow of an incompressible power-model fluid past a sphere [2].

## 4.3 LOCAL VOLUME AVERAGING

**4.3.1 Flow through porous media** The movement of gases and liquids through porous media is common to many industrial processes. Distillation and absorption columns are often filled with beads or packing in a variety of shapes. A chemical reactor may be filled with porous pellets impregnated with a catalyst. Filters are employed in most chemical processes. The movements of water and oil through porous strata are important in water conservation and oil exploration.

In 1856 Darcy [1] made the first serious study of this problem. Among other things, as a correlation of experimental data for water moving axially with a volume flow rate $Q$ through a cylindrical packed bed of cross-sectional area $A$ and length $L$ under the influence of a pressure difference $\Delta P$, he proposed [2, p. 634]

$$\frac{\Delta P}{L} = b \frac{Q}{A} \tag{1-1}$$

It was later observed that $b$ is proportional to the coefficient of viscosity for incompressible newtonian fluids:

$$\frac{\Delta P}{L} = \frac{\mu}{k} \frac{Q}{A} \tag{1-2}$$

This last is usually referred to as *Darcy's law* and the coefficient $k$ is referred to as the *permeability* of the bed.

But there are at least three major questions that remain unanswered by Eq. (1-2).

1. What should be done about flow in other geometries or under the influence of other boundary conditions?

    The standard answer has been to say that a differential equation, inspired by Eq. (1-2), describes the flow at each point in the porous medium [2, p. 634]:

$$\nabla P + \frac{\mu}{k} \mathbf{w} = 0 \tag{1-3}$$

A major difficulty with this equation is that, since it has not been derived, the average pressure $P$ and the average velocity $\mathbf{w}$ are not defined in terms of the local pressure distribution and local velocity distribution in the pores.
2. What should be done about oriented (the term "anisotropic" is in common use) porous media?

By an *oriented* porous structure, I mean one that has a direction or a set of directions intrinsically associated with the pore geometry. For example, in a naturally occurring stratified rock there is often a gradient in "particle diameter" in the direction of gravity or in what was originally the direction of gravity. A *nonoriented* porous structure (often referred to as *isotropic*) has no such direction intrinsically associated with it.

Equation (1-3) is usually said to describe flow through a nonoriented porous medium. For an oriented porous medium, the scalar permeability $k$ is replaced by a second-order permeability tensor $\mathbf{K}$:

$$\nabla P + \mu \mathbf{K}^{-1} \cdot \mathbf{w} = 0 \tag{1-4}$$

The tensor $\mathbf{K}$ is usually said to be both symmetric and invertible, but the justification of these properties has often left room for doubt.
3. How should one describe the movements of a viscoelastic fluid through a porous medium?

We will attempt to answer these questions in the sections that follow.

Throughout this discussion, we assume that the porous structure is rigid and stationary.

I owe special thanks to Prof. Stephen Whitaker (University of California, Davis), with whom I have spent many hours discussing the mechanics of flow through porous media.

Most of the developments of the next few sections are taken from two papers that discuss single-phase flow through porous media [3, 4]. We have also discussed multiphase flow through porous media [5-7], but we will not take up this topic here.

**REFERENCES**

1. Darcy, H. P.: "Les fontaines publiques de la ville de Dijon," Dalmont, Paris, 1856.
2. Scheidegger, A. E.: In S. Flügge and C. Truesdell (eds.), "Handbuch der Physik," vol. 8/2, Springer-Verlag, Berlin, 1963.
3. Slattery, J. C.: *A.I.Ch.E.J.*, **13**:1066 (1967).
4. Slattery, J. C.: *A.I.Ch.E.J.*, **15**:866 (1969).
5. Slattery, J. C.: *A.I.Ch.E.J.*, **14**:50 (1968).
6. Slattery, J. C.: *A.I.Ch.E.J.*, **16**:345 (1970).
7. Patel, J. G., M. G. Hegde, and J. C. Slattery: *A.I.Ch.E.J.*, **18**:1062 (1972).

**4.3.2 Local volume averaging** Our initial objective is to associate with every point in a porous medium a local volume average of the differential equation of continuity:

$$\frac{\partial \rho}{\partial t} + \text{div}(\rho \mathbf{v}) = 0 \tag{2-1}$$

When I say every point in the porous medium, I include the solid phase as well as the fluid phase and the solid-fluid phase interface.

Referring to Fig. 4.3.2-1, let us begin by thinking of a particular point z in the porous medium. It makes no difference whether this point is in fact located in the solid phase, the fluid phase, or on the solid-fluid phase interface; the argument remains unchanged. Let us associate with this point a closed surface $S$. I happen to have chosen a sphere in Fig. 4.3.2-1.

We will associate this averaging surface $S$ with every point in the porous medium by a simple translation of $S$ without rotation. If $S$ is a unit sphere whose center coincides with the point initially considered, we center upon each point in the porous medium a unit sphere. The diameter of $S$ should be sufficiently large that averages over the pore space enclosed by $S$ vary smoothly with position. Whenever possible, the diameter of $S$ should be so small as to be negligible with respect to a characteristic dimension of the macroscopic porous body. Yet it should not be so small that $S$ encloses only solid or only fluid at many points in the porous structure. The minimum size of $S$ will be discussed shortly.

Let $V$ be the volume enclosed by $S$. Let $V_{(f)}$ denote the pores that contain fluid in the interior of $S$; the volume and shape of $V_{(f)}$ in general will change from point to point in the porous medium. The closed boundary surface $S_{(f)}$ of $V_{(f)}$ is the sum of $S_e$ and $S_w$ : $S_e$ coincides with $S$ and $S_w$ coincides with the pore walls. We may think of $S_e$ as the entrance and exit surfaces of $V_{(f)}$ through which fluid passes in and out of $V_{(f)}$.

Let us write a mass balance for the fluid contained within this closed surface $S$. The most convenient way of doing this is to integrate the differential equation of

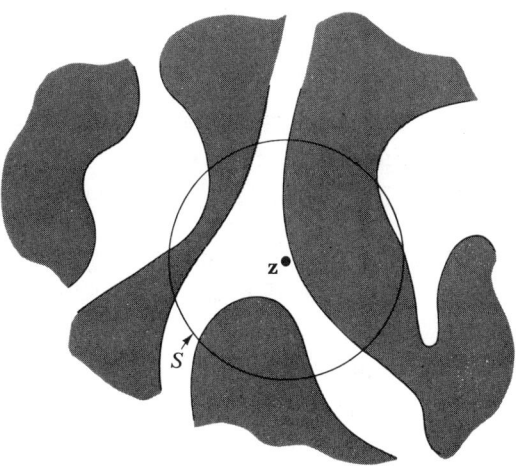

**Fig. 4.3.2-1** The averaging surface $S$ to be associated with every point z in the porous medium.

continuity, Eq. (2-1), over $V_{(f)}$, the region of space occupied by the fluid within $S$:

$$\int_{V_{(f)}} \left[\frac{\partial \rho}{\partial t} + \text{div}(\rho \mathbf{v})\right] dV = 0 \tag{2-2}$$

We can immediately interchange the operations of volume integration and differentiation with respect to time in the first term on the left to find

$$\frac{\partial \bar{\rho}}{\partial t} + \frac{1}{V} \int_{V_{(f)}} \text{div}(\rho \mathbf{v}) \, dV = 0 \tag{2-3}$$

where

$$\bar{\rho} \equiv \frac{1}{V} \int_{V_{(f)}} \rho \, dV \tag{2-4}$$

Assume that $B$ is some quantity associated with the fluid. We will have occasion to speak of two averages: the local volume average of $B$ (the mean value of $B$ in $V$)

$$\bar{B} \equiv \frac{1}{V} \int_{V_{(f)}} B \, dV \tag{2-5}$$

and the intrinsic volume average of $B$ (the mean value of $B$ in $V_{(f)}$)

$$\langle B \rangle \equiv \frac{1}{V_{(f)}} \int_{V_{(f)}} B \, dV \tag{2-6}$$

Let $L_0$ be a characteristic dimension of $S$. The minimum acceptable size of $S$ or the minimum acceptable value of $L_0$ is such that $\bar{B}$ is nearly independent of position over distances of the same order of magnitude. This implies that

$$\langle \bar{B} \rangle = \bar{B} \tag{2-7}$$

$$\langle \langle B \rangle \rangle = \langle B \rangle \tag{2-8}$$

It would be nice if we could interchange the volume integration with the divergence operation in the second term on the left of Eq. (2-3). But the limits on this volume integration depend upon the pore geometry enclosed by $S$ and must be functions of position $z$. The next section explores this problem in more detail.

**4.3.3 Theorem for the local volume average of a gradient** Let $B$ be any scalar, spatial vector, or second-order tensor associated with the fluid. Given

$$\overline{\nabla B} \equiv \frac{1}{V} \int_{V_{(f)}} \nabla B \, dV \tag{3-1}$$

let us ask in what sense we might interchange the volume average with the gradient operation to obtain

$$\nabla \bar{B} \equiv \nabla \left(\frac{1}{V} \int_{V_{(f)}} B \, dV\right) \tag{3-2}$$

# [4.3] LOCAL VOLUME AVERAGING

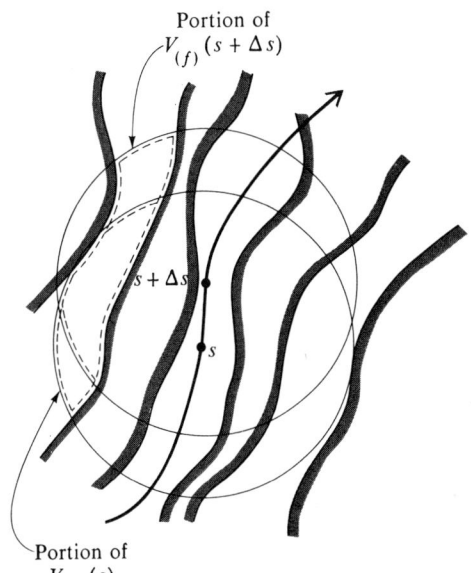

**Fig. 4.3.3-1** An arbitrary curve running through the porous medium, where $s$ is a parameter measured along this curve.

Let us now associate the averaging surface $S$, which was introduced in Sec. 4.3.2, with every point in the porous medium. We will do this by a simple translation of $S$ without rotation. As an example, if $S$ is a unit sphere the center of which coincides with the point initially considered, we center upon each point in the porous medium a unit sphere. If $S$ is small compared with the average pore diameter, it may enclose only solid or only fluid at many points; if it is large, many pores may intersect $S$, the intersections serving as entrances and exits to the fluid enclosed by $S$.

Consider any arbitrary curve running through the porous medium as shown in Fig. 4.3.3-1. Let $s$ be a parameter such as arc length measured along this curve. We can identify with each point along this curve a system denoted by $V_{(f)}$, composed of the pores containing fluid enclosed by surface $S$. We may think of $V_{(f)}$ as a function of the parameter $s$ along this curve. If we simply replace the parameter time by $s$ in the generalized transport theorem of Sec. 1.3.2, we have

$$\frac{d}{ds}\int_{V_{(f)}} B\, dV = \int_{V_{(f)}} \frac{\partial B}{\partial s}\, dV + \int_{S_{(f)}} B\, \frac{d\mathbf{p}}{ds}\cdot \mathbf{n}\, dS \tag{3-3}$$

Here $\mathbf{p}$ is the position vector field.

Let us further restrict ourselves to quantities $B$ that are explicit functions of position (and time) only:

$$\frac{\partial B}{\partial s} = 0 \tag{3-4}$$

(By $\partial B/\partial s$, we mean a derivative with respect to $s$ holding position and time fixed.)

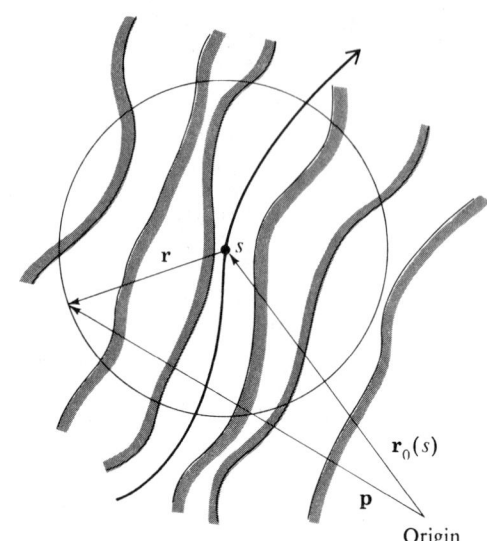

**Fig. 4.3.3-2** The vector $\mathbf{r}_0(s)$ denotes the position of the point $s$ along the curve; $\mathbf{r}$ denotes position on $S_{(f)}$ with respect to the point $s$.

As a function of $s$, the artificial system particles clearly move tangent to the fixed pore walls on $S_w$:

$$\text{On } S_w: \quad \frac{d\mathbf{p}}{ds} \cdot \mathbf{n} = 0 \tag{3-5}$$

Let $\mathbf{r}_0(s)$ be the position vector locating the point $s$ on the arbitrary curve and let $\mathbf{r}(s)$ be the position vector locating points on $S_{(f)}$ relative to this point $s$ (which is at the center of the sphere for the case illustrated in Figs. 4.3.3-1 and 4.3.3-2). Provided $S$ is translated *without rotation* along this arbitrary curve in identifying it with every point in the porous medium, then

$$\text{On } S_e: \quad \frac{d\mathbf{p}}{ds} = \frac{d\mathbf{r}_0}{ds} + \frac{d\mathbf{r}}{ds} = \frac{d\mathbf{r}_0}{ds} \tag{3-6}$$

Equations (3-4) to (3-6) allow us to rewrite Eq. (3-3) as

$$\frac{d}{ds}\int_{V_{(f)}} B\, dV = \frac{d\mathbf{r}_0}{ds} \cdot \nabla \int_{V_{(f)}} B\, dV = \int_{S_e} B \frac{d\mathbf{r}_0}{ds} \cdot \mathbf{n}\, dS \tag{3-7}$$

or, since $d\mathbf{r}_0/ds$ is independent of position on $S_e$,

$$\frac{d\mathbf{r}_0}{ds} \cdot \nabla \int_{V_{(f)}} B\, dV = \left(\int_{S_e} B\mathbf{n}\, dS\right) \cdot \frac{d\mathbf{r}_0}{ds} \tag{3-8}$$

Because we have been concerned with any arbitrary curve running through the porous medium, this implies

## [4.3] LOCAL VOLUME AVERAGING

$$\nabla \int_{V_{(f)}} B \, dV = \int_{S_e} B\mathbf{n} \, dS \tag{3-9}$$

An application of Green's transformation allows us to express Eq. (3-1) as

$$\frac{1}{V} \int_{V_{(f)}} \nabla B \, dV = \frac{1}{V} \int_{S_e + S_w} B\mathbf{n} \, dS \tag{3-10}$$

In view of Eq. (3-9), we have the useful result

$$\overline{\nabla B} \equiv \frac{1}{V} \int_{V_{(f)}} \nabla B \, dV = \nabla \left( \frac{1}{V} \int_{V_{(f)}} B \, dV \right) + \frac{1}{V} \int_{S_w} B\mathbf{n} \, dS$$

$$= \nabla \bar{B} + \frac{1}{V} \int_{S_w} B\mathbf{n} \, dS \tag{3-11}$$

We may refer to this as the *theorem for the volume average of a gradient*.

A special case of Eq. (3-11) is

$$\overline{\mathrm{div}\,\mathbf{B}} \equiv \frac{1}{V} \int_{V_{(f)}} \mathrm{div}\,\mathbf{B} \, dV$$

$$= \mathrm{div}\,\bar{\mathbf{B}} + \frac{1}{V} \int_{S_w} \mathbf{B} \cdot \mathbf{n} \, dS \tag{3-12}$$

Here **B** should be interpreted as a spatial vector field or second-order tensor field. Equation (3-12) may be referred to as the *theorem for the volume average of a divergence*.

Let us now apply these results in obtaining the local volume average of the equation of continuity.

### 4.3.4 The local volume average of the equation of continuity

In Sec. 4.3.2, we found that the local volume average of the equation of continuity could be written as

$$\frac{\partial \bar{\rho}}{\partial t} + \frac{1}{V} \int_{V_{(f)}} \mathrm{div}\,(\rho \mathbf{v}) \, dV = 0 \tag{4-1}$$

An application of the theorem of Sec. 4.3.3 allows us to express this as

$$\frac{\partial \bar{\rho}}{\partial t} + \mathrm{div}\,(\overline{\rho \mathbf{v}}) = 0 \tag{4-2}$$

In arriving at this result, we have observed that the velocity of the fluid is zero on the pore walls $S_w$. It is unfortunate that the local volume average $\overline{\rho v}$ occurs rather than the product of the local volume averages $\bar{\rho}\bar{v}$.

For the special case of an incompressible fluid, we have the simpler result

$$\mathrm{div}\,\bar{\mathbf{v}} = 0 \tag{4-3}$$

It is for this reason that incompressible fluids are easier to work with in considering flows through porous media.

### EXERCISE

**4.3.4-1** Equation (4-3) can be obtained immediately by taking a local volume average of the equation of continuity for an incompressible fluid. How does one obtain the same result from Eq. (4-2)?
*Hint:* Note that

$$\tilde{\rho} = \rho \frac{V_{(f)}}{V}$$

## 4.3.5 The local volume average of Cauchy's first law

We can start as we did in Sec. 4.3.2, where we began considering the local volume average of the equation of continuity. Let us think of a particular point **z** in the porous medium and let us integrate Cauchy's first law over $V_{(f)}$, the region of space occupied by the fluid within $S$ associated with **z**:

$$\frac{1}{V} \int_{V_{(f)}} \left[ \frac{\partial (\rho \mathbf{v})}{\partial t} + \text{div}\,(\rho \mathbf{v v}) - \text{div}\,\mathbf{T} - \rho \mathbf{f} \right] dV = 0 \tag{5-1}$$

The operations of volume integration and differentiation with respect to time may be interchanged in the first term on the left:

$$\frac{1}{V} \int_{V_{(f)}} \frac{\partial (\rho \mathbf{v})}{\partial t} dV = \frac{\partial \overline{\rho \mathbf{v}}}{\partial t} \tag{5-2}$$

The theorem of Sec. 4.3.3 can be used to express the second and third terms on the left of Eq. (5-1) as

$$\frac{1}{V} \int_{V_{(f)}} \text{div}\,(\rho \mathbf{v v}) \, dV = \text{div}\,(\overline{\rho \mathbf{v v}}) \tag{5-3}$$

and

$$\frac{1}{V} \int_{V_{(f)}} \text{div}\,\mathbf{T} \, dV = \text{div}\,\overline{\mathbf{T}} + \frac{1}{V} \int_{S_w} \mathbf{T} \cdot \mathbf{n} \, dS \tag{5-4}$$

In arriving at Eq. (5-3), we have observed that the velocity vector must be zero at the fluid-solid phase interface $S_w$. In view of Eqs. (5-2) to (5-4), Eq. (5-1) becomes

$$\frac{\partial \overline{\rho \mathbf{v}}}{\partial t} + \text{div}\,(\overline{\rho \mathbf{v v}}) = \text{div}\,\overline{\mathbf{T}} + \overline{\rho \mathbf{f}} + \frac{1}{V} \int_{S_w} \mathbf{T} \cdot \mathbf{n} \, dS \tag{5-5}$$

Throughout the remainder of our discussion of flow through porous media, we will restrict ourselves to incompressible fluids and we will assume that all inertial effects may be neglected in the local volume average of Cauchy's first law. We shall also find it convenient to assume that the external force per unit mass **f** may be represented by a scalar potential $\varphi$:

## [4.3] LOCAL VOLUME AVERAGING

$$\mathbf{f} = -\nabla \varphi \tag{5-6}$$

With these restrictions, Eq. (5-5) simplifies to

$$\operatorname{div}(\overline{\mathbf{T}} - \overline{\rho \varphi} \mathbf{I}) + \frac{1}{V} \int_{S_w} (\mathbf{T} - \rho \varphi \mathbf{I}) \cdot \mathbf{n} \, dS = 0 \tag{5-7}$$

or

$$\nabla(\overline{\mathscr{P} - p_0}) - \operatorname{div} \overline{\mathbf{S}} + \mathbf{g} = 0 \tag{5-8}$$

Here $\mathscr{P}$ is the modified pressure

$$\mathscr{P} \equiv p + \rho \varphi \tag{5-9}$$

and $\mathbf{S}$ is the extra-stress tensor. A constant reference or ambient pressure $p_0$ is introduced here in order that we may identify

$$\mathbf{g} = -\frac{1}{V} \int_{S_w} [\mathbf{T} + (p_0 - \rho \varphi) \mathbf{I}] \cdot \mathbf{n} \, dS \tag{5-10}$$

as the force per unit volume that the fluid exerts upon the pore walls contained within $S$ beyond the hydrostatic force and beyond any force attributable to the ambient pressure. This force per unit volume $\mathbf{g}$ is entirely assignable to the motion of the fluid.

For an incompressible newtonian fluid,

$$\overline{\mathbf{S}} = \mu[\overline{\nabla \mathbf{v}} + \overline{(\nabla \mathbf{v})^T}] \tag{5-11}$$

Green's transformation, the theorem of Sec. 4.3.3, and the fact that the velocity of the fluid is zero at the pore walls allow us to say

$$\overline{\nabla \mathbf{v}} \equiv \frac{1}{V} \int_{V_{(f)}} \nabla \mathbf{v} \, dV$$

$$= \frac{1}{V} \int_{S_e + S_w} \mathbf{v} \mathbf{n} \, dS$$

$$= \nabla \overline{\mathbf{v}} \tag{5-12}$$

Exactly the same argument may be used to show

$$\overline{(\nabla \mathbf{v})^T} = (\nabla \overline{\mathbf{v}})^T \tag{5-13}$$

Consequently,

$$\overline{\mathbf{S}} = \mu[\nabla \overline{\mathbf{v}} + (\nabla \overline{\mathbf{v}})^T] \tag{5-14}$$

and

$$\operatorname{div} \overline{\mathbf{S}} = \mu \operatorname{div}(\nabla \overline{\mathbf{v}}) \tag{5-15}$$

To summarize, when we neglect all inertial effects and assume that the external force per unit mass can be represented by the gradient of a scalar potential, the local volume average of Cauchy's first law for an incompressible newtonian fluid can be written as

$$\nabla \overline{(\mathscr{P} - p_0)} - \mu \operatorname{div}(\nabla \bar{\mathbf{v}}) + \mathbf{g} = 0 \qquad (5\text{-}16)$$

In the next section we discuss the preparation of empirical correlations for **g**.

### 4.3.6 Empirical correlations for g

In this section, we use three examples to indicate how experimental data can be used to prepare correlations for **g**, introduced in Sec. 4.3.5. We base this discussion upon four points:

1. The force per unit volume **g** is frame indifferent:

$$\mathbf{g}^* = -\frac{1}{V} \int_{S_w} [\mathbf{T}^* + (p_0 - \rho\varphi)\mathbf{I}] \cdot \mathbf{n}^* \, dS$$

$$= -\frac{1}{V} \int_{S_w} \mathbf{Q} \cdot [\mathbf{T} + (p_0 - \rho\varphi)\mathbf{I}] \cdot \mathbf{n} \, dS$$

$$= \mathbf{Q} \cdot \mathbf{g} \qquad (6\text{-}1)$$

Here **Q** is a (possibly) time-dependent, orthogonal, second-order tensor.

2. We assume that the principle of material frame indifference introduced in Sec. 2.3.1 applies to any empirical correlation developed for **g**.
3. The Buckingham-Pi theorem serves to further restrict the form of any expression for **g**.
4. The averaging surface $S$ is sufficiently large that **g** may be assumed *not* to be an *explicit* function of position in the porous structure, though it very well may be an *implicit* function of position as the result of its dependence upon other quantities.

**EXAMPLE I  FLOW OF A NEWTONIAN FLUID THROUGH A NONORIENTED MEDIUM**

For the moment, let us assume that **g** is a function of the difference between the intrinsic volume average velocity of the fluid $\langle \mathbf{v} \rangle$ and the intrinsic volume average velocity of the solid

$$\langle \mathbf{u} \rangle^{(s)} \equiv \frac{1}{V_{(s)}} \int_{V_{(s)}} \mathbf{u} \, dV \qquad (6\text{-}2)$$

Here **u** is the velocity distribution within the solid (which may be undergoing a rigid body rotation and translation); $V_{(s)}$ is the volume occupied by the solid within the averaging surface $S$. Equivalently, we can assume **g** is a function of the difference between the local volume average velocity of the fluid $\bar{\mathbf{v}}$ and $\Psi \langle \mathbf{u} \rangle^{(s)}$

$$\mathbf{g} = \hat{\mathbf{g}}(\bar{\mathbf{v}} - \Psi \langle \mathbf{u} \rangle^{(s)}) \qquad (6\text{-}3)$$

where

$$\Psi \equiv \frac{V_{(f)}}{V} \qquad (6\text{-}4)$$

denotes the local porosity of the structure (assuming all of the pores are filled with fluid).

By the principle of material frame indifference, the functional relationship between these variables should be the same in every frame of reference. This means that

$$\mathbf{g}^* = \mathbf{Q} \cdot \mathbf{g} = \mathbf{Q} \cdot \hat{\mathbf{g}}(\bar{\mathbf{v}} - \Psi \langle \mathbf{u} \rangle^{(s)})$$
$$= \hat{\mathbf{g}}[\mathbf{Q} \cdot (\bar{\mathbf{v}} - \Psi \langle \mathbf{u} \rangle^{(s)})] \tag{6-4}$$

or **g** is an isotropic function [1, p. 22]:

$$\hat{\mathbf{g}}(\bar{\mathbf{v}} - \Psi \langle \mathbf{u} \rangle^{(s)}) = \mathbf{Q}^T \cdot \hat{\mathbf{g}}[\mathbf{Q} \cdot (\bar{\mathbf{v}} - \Psi \langle \mathbf{u} \rangle^{(s)})] \tag{6-5}$$

By a representation theorem for a vector-valued isotropic function of one vector [1, p. 35], we may write

$$\mathbf{g} = \hat{\mathbf{g}}(\bar{\mathbf{v}} - \Psi \langle \mathbf{u} \rangle^{(s)}) = R\,[\bar{\mathbf{v}} - \Psi \langle \mathbf{u} \rangle^{(s)}] \tag{6-6}$$

It is to be understood here that the resistance coefficient $R$ is a function of the magnitude of the local volume-averaged velocity of the fluid relative to the local volume-averaged velocity of the solid $|\bar{\mathbf{v}} - \Psi \langle \mathbf{u} \rangle^{(s)}|$, a function of the viscosity of the fluid $\mu$, the porosity $\Psi$, as well as a characteristic length $l_0$ of the porous medium:

$$R = R(|\bar{\mathbf{v}} - \Psi \langle \mathbf{u} \rangle^{(s)}|, \mu, \Psi, l_0) \tag{6-7}$$

We have not considered the fluid density here, since it does not appear in the local volume-averaged Cauchy first law, Eq. (5-8), and since it does not enter the local volume-averaged equation of continuity for an incompressible fluid, Eq. (4-3). By the Buckingham-Pi theorem [2], Eq. (6-7) can be written in terms of a dimensionless permeability $k_0^*$ which is a function of $\Psi$ only.

$$R = \frac{\Psi \mu}{l_0^2 k_0^*} \tag{6-8}$$

To summarize, Eqs. (6-6) and (6-8) may be used to describe the force per unit volume that an incompressible newtonian fluid exerts on a nonoriented porous structure (beyond the hydrostatic force and the force attributable to the ambient pressure).

**EXAMPLE 2  FLOW OF A VISCOELASTIC FLUID THROUGH A NONORIENTED MEDIUM**

Let us repeat Example 1 for an incompressible viscoelastic fluid. In order that our results have a wide range of applicability, let us assume that the behavior of this viscoelastic fluid can be described by the Noll simple fluid discussed in Sec. 2.3.4.

The initial argument given under Example 1 and concluding with Eq. (6-6) is again applicable here. The only modification necessary is to say that the resistance coefficient $R$ is a function of a characteristic viscosity $\mu_0$ and characteristic time $s_0$ of the fluid as well as $|\bar{\mathbf{v}} - \Psi \langle \mathbf{u} \rangle^{(s)}|$, the porosity $\Psi$, and the characteristic length $l_0$:

$$R = R(|\bar{\mathbf{v}} - \Psi \langle \mathbf{u} \rangle^{(s)}|, \mu_0, s_0, \Psi, l_0) \tag{6-9}$$

The Buckingham-Pi theorem [2] allows us to conclude that

$$R = \frac{\Psi \mu_0}{l_0^2 k^*} \tag{6-10}$$

where $k^*$ is a function of the local Weissenberg number $N_{Wi}$ and $\Psi$:

$$k^* = k^*(N_{Wi}, \Psi) \tag{6-11}$$

$$N_{Wi} \equiv \frac{s_0 |\bar{v} - \Psi \langle u \rangle^{(s)}|}{l_0} \tag{6-12}$$

Anticipating the result of Sec. 4.3.10, we can postulate as an alternative to Eq. (6-9)

$$R = R(|\nabla(\overline{\mathscr{P} - p_0})|, \mu_0, s_0, \Psi, l_0) \tag{6-13}$$

By the Buckingham-Pi theorem [2], we conclude that

$$R = \frac{\Psi \mu_0}{l_0^2 \kappa^*} \tag{6-14}$$

where

$$\kappa^* = \kappa^* \left( \frac{|\nabla(\overline{\mathscr{P} - p_0})| l_0 s_0}{\mu_0}, \Psi \right) \tag{6-15}$$

In summary, Eqs. (6-6) and (6-10) describe the force per unit volume that an incompressible viscoelastic fluid exerts upon a nonoriented porous structure, so long as the behavior of the fluid is representable by the Noll simple fluid. The dimensionless permeability $k^*$ must now be considered to be a function of the Weissenberg number $N_{Wi}$. As mentioned in Sec. 2.3.4 (see also [3] and [4]), an empirical correlation of this type can be prepared for only one viscoelastic fluid at a time, since the functional $\mathscr{H}^*_{\sigma=0}^{\infty}$ in the simple fluid model has not been fully specified. But if we are interested in only a single fluid, then it is really not necessary to have particular values for the characteristic viscosity $\mu_0$ and the characteristic time $s_0$. As far as this empirical correlation is concerned, we can avoid an extensive (and perhaps somewhat indeterminate) study of material behavior in a set of viscometers.

Experimental data [5; see also Sec. 4.3.11] suggest that, over a limited range of $|\nabla(\overline{\mathscr{P} - p_0})|$, a useful relation in the form of Eq. (6-14) is

$$\frac{1}{R} = m(l_0)^{n+1} |\nabla(\overline{\mathscr{P} - p_0})|^{n-1} \tag{6-16}$$

In order to account for newtonian behavior at very low values of $|\nabla(\overline{\mathscr{P} - p_0})|$, we might say instead that

## [4.3] LOCAL VOLUME AVERAGING

$$\frac{1}{R} = \alpha(l_0)^2[1 + (\beta l_0 |\nabla(\overline{\mathscr{P} - p_0})|)^{\gamma-1}] \tag{6-17}$$

Equations (6-16) and (6-17) are analogous to the power and Ellis models of Sec. 2.3.3. But one should not necessarily expect any relation between, for example, the parameters appearing in Eq. (6-17) and those appearing in the Ellis model. Equations (6-16) and (6-17) are empiricisms that may be helpful in correlating data for the flow of Noll simple fluids in porous media.

**EXAMPLE 3 FLOW OF A NEWTONIAN FLUID THROUGH AN ORIENTED MEDIUM**

One should not expect Eq. (6-6) to be applicable to the flow of an incompressible newtonian fluid through a porous structure in which particle diameter $l$ is a function of position. For such a structure, Eq. (6-2) must be altered to include a dependence upon additional vector and possibly tensor quantities. For example, one might postulate a dependence of **g** upon the local gradient in particle diameter as well as $\bar{\mathbf{v}} - \Psi\langle\mathbf{u}\rangle^{(s)}$:

$$\mathbf{g} = \hat{\mathbf{g}}(\bar{\mathbf{v}} - \Psi\langle\mathbf{u}\rangle^{(s)}, \nabla l) \tag{6-18}$$

For the moment, we leave the additional dependence of **g** upon $\mu$ and $l$ understood.

The principle of material frame indifference again requires $\hat{\mathbf{g}}$ to be an isotropic function:

$$\hat{\mathbf{g}}(\bar{\mathbf{v}} - \Psi\langle\mathbf{u}\rangle^{(s)}, \nabla l) = \mathbf{Q}^T \cdot \hat{\mathbf{g}}[\mathbf{Q} \cdot (\bar{\mathbf{v}} - \Psi\langle\mathbf{u}\rangle^{(s)}), \mathbf{Q} \cdot \nabla l] \tag{6-19}$$

By representation theorems of Spencer and Rivlin [6, sec. 7] and of Smith [7], the most general polynomial isotropic vector function of two vectors has the form

$$\mathbf{g} = \varphi_{(1)}(\bar{\mathbf{v}} - \Psi\langle\mathbf{u}\rangle^{(s)}) + \varphi_{(2)}\nabla l \tag{6-20}$$

Here $\varphi_{(1)}$ and $\varphi_{(2)}$ are scalar-valued polynomials in $|\bar{\mathbf{v}} - \Psi\langle\mathbf{u}\rangle^{(s)}|$, $|\nabla l|$, and $[(\bar{\mathbf{v}} - \Psi\langle\mathbf{u}\rangle^{(s)}) \cdot \nabla l]$:

For $i = 1, 2$: $\quad \varphi_{(i)} = \varphi_{(i)}[|\bar{\mathbf{v}} - \Psi\langle\mathbf{u}\rangle^{(s)}|, |\nabla l|, (\bar{\mathbf{v}} - \Psi\langle\mathbf{u}\rangle^{(s)}) \cdot \nabla l, \mu, \Psi, l]$ (6-21)

(In applying the theorem of Spencer and Rivlin, we identify a vector **b** that has covariant components $b_i$ with the skew-symmetric tensor that has contravariant components $\epsilon^{ijk}b_i$. Their theorem requires an additional term in Eq. (6-20) proportional to the vector product $[(\bar{\mathbf{v}} - \Psi\langle\mathbf{u}\rangle^{(s)}) \wedge \nabla l]$. This term is not consistent with the requirement that **g** be isotropic [1, p. 24] and consequently is dropped.)

An application of the Buckingham-Pi theorem [2] allows us to conclude that

$$\varphi_{(1)} = \frac{\Psi\mu}{l^2 k^*_{(1)}} \tag{6-22}$$

and

$$\varphi_{(2)} = \frac{\mu|\bar{\mathbf{v}} - \Psi\langle\mathbf{u}\rangle^{(s)}|}{l^2 k^*_{(2)}} \tag{6-23}$$

where

For $i = 1, 2$:  $\quad k_{(i)}^* = k_{(i)}^* \left( |\nabla l|, \dfrac{\overline{\mathbf{v}} - \Psi \langle \mathbf{u} \rangle^{(s)}}{|\overline{\mathbf{v}} - \Psi \langle \mathbf{u} \rangle^{(s)}|} \cdot \nabla l, \Psi \right)$  (6-24)

As we would expect, $\varphi_{(2)} = 0$ for $|\overline{\mathbf{v}} - \Psi \langle \mathbf{u} \rangle^{(s)}| = 0$, in order that $\mathbf{g} = 0$ in this limit.

In summary, Eqs. (6-20) and (6-22) to (6-24) may be used to represent the force per unit volume that an incompressible newtonian fluid exerts upon an oriented structure such that the orientation of the structure is fully described by the local gradient of particle diameter. The resulting expression for $\mathbf{g}$ is somewhat more complicated than that which we found for a nonoriented structure in Example 1.

## REFERENCES

1. Truesdell, C., and W. Noll: In S. Flügge (ed.), "Handbuch der Physik," vol. 3/3, Springer-Verlag, Berlin, 1965.
2. Brand, L.: *Arch. Rational Mech. Analysis*, **1**:35 (1957).
3. Slattery, J. C.: *A.I.Ch.E.J.*, **11**:831 (1965).
4. Slattery, J. C.: *A.I.Ch.E.J.*, **14**:516 (1968).
5. Slattery, J. C.: *A.I.Ch.E.J.*, **15**:866 (1969).
6. Spencer A. J., and R. S. Rivlin: *Arch. Rational Mech. Analysis*, **4**:214 (1959/60).
7. Smith, G. F.: *Arch. Rational Mech. Analysis*, **18**:282 (1965).

**4.3.7 Summary of results for an incompressible newtonian fluid**  In Sec. 4.3.4, we found that the local volume average of the equation of continuity requires

$$\text{div } \overline{\mathbf{v}} = 0 \tag{7-1}$$

Under conditions such that inertial effects can be neglected and the external force per unit mass can be represented as the gradient of a scalar potential, Secs. 4.3.5 and 4.3.6 indicate that the local volume average of Cauchy's first law for an incompressible newtonian fluid flowing through a nonoriented porous medium has the form

$$\nabla(\overline{\mathscr{P} - p_0}) - \mu \text{ div }(\nabla \overline{\mathbf{v}}) + \dfrac{\Psi \mu}{l_0^2 k_0^*} \overline{\mathbf{v}} = 0 \tag{7-2}$$

Here $p_0$ is a reference or ambient pressure, $l_0$ is a characteristic length of the porous medium, and $k_0^*$ is a dimensionless function of porosity. In writing Eq. (7-2), we have recognized that, in the frame of reference for which Eq. (5-16) is appropriate, the velocity of the proous medium is assumed to be zero. Interestingly, Brinkman [1] proposed a similar relationship without derivation.

In Sec. 4.3.6, we considered the flow of an incompressible newtonian fluid through a porous medium such that the orientation of the gradient of the "local pore diameter" $l$ is significant. Under conditions such that inertial effects can be neglected and the external force per unit mass can be represented in terms of a scalar potential, Secs. 4.3.5 and 4.3.6 suggest that the local volume average of Cauchy's first law is

$$\nabla(\overline{\mathscr{P} - p_0}) - \mu \text{ div }(\nabla \overline{\mathbf{v}}) + \dfrac{\Psi \mu}{l^2 k_{(1)}^*} \overline{\mathbf{v}} + \dfrac{\mu |\overline{\mathbf{v}}|}{l^2 k_{(2)}^*} \nabla l = 0 \tag{7-3}$$

## [4.3] LOCAL VOLUME AVERAGING

where, for $i = 1, 2$,

$$k^*_{(i)} = k^*_{(i)}\left(|\nabla l|, \frac{\bar{\mathbf{v}}}{|\bar{\mathbf{v}}|} \cdot \nabla l, \Psi\right) \tag{7-4}$$

In stating these results in the form of Eqs. (7-3) and (7-4), we have recognized that, in the frame of reference being considered, the porous structure is stationary.

With the exception of the second term on the left, Eq. (7-2) has the same form as the extended Darcy law for a nonoriented porous structure. Equation (7-3) might also be written as [2]

$$\nabla\overline{(\mathscr{P} - p_0)} - \mu \operatorname{div}(\nabla\bar{\mathbf{v}}) + \frac{\Psi\mu}{l^2} \mathbf{R}^* \cdot \bar{\mathbf{v}} = 0 \tag{7-5}$$

where we define

$$\mathbf{R}^* \equiv \frac{1}{k^*_{(1)}}\mathbf{I} + \frac{|\bar{\mathbf{v}}|}{k^*_{(2)}(\bar{\mathbf{v}} \cdot \nabla l)} \nabla l \nabla l \tag{7-6}$$

Again, with the exception of the second term on the left, Eq. (7-5) is clearly similar to the extended form of Darcy's law for flow through an oriented porous structure, Eq. (1-4). In Sec. 4.3.10, we further explore these similarities.

Before illustrating how we can use these equations to analyze flows through porous media, let us stop and ask how local volume-averaged variables can be used to calculate some of the gross area and volume averages of practical interest.

### REFERENCES

1. Brinkman, H. C.: *Appl. Sci. Res.*, **A1**:27, 81 (1949).
2. Whitaker, Stephen: *Ind. Eng. Chem.*, **61**(12):14 (1969).

**4.3.8 Averages of volume-averaged variables** One commonly used type of packed bed is prepared by filling a cylindrical tube with small particles (sand, glass beads, catalyst pellets, etc.). In the next section, we analyze the flow through such a bed to determine the local volume-averaged velocity distribution. From a practical point of view, we are more interested in the average of velocity over the cross section of the tube. Intuitively, we feel that the average of velocity over the cross section is equal to the average over the cross section of the local volume-averaged velocity distribution, so long as the pores in the porous structure are sufficiently small compared with the diameter of the tube. The purpose of this section is to confirm these intuitive feelings.

More generally, we would like to discuss under what circumstances

$$\frac{1}{\mathscr{A}}\int_{\mathscr{S}}\bar{f}\,dA = \frac{1}{\mathscr{A}}\int_{\mathscr{S}}f\,dA \tag{8-1}$$

where $f$ is some quantity (scalar, vector, or second-order tensor field) associated with the fluid, $\mathscr{S}$ is some macroscopic surface, and $\mathscr{A}$ is the area of $\mathscr{S}$.

Let us rearrange (8-1) as

$$\frac{1}{\mathscr{A}}\int_{\mathscr{S}}\bar{f}\,dA = \frac{1}{\mathscr{A}}\int_{\mathscr{S}}[f+(\bar{f}-f)]\,dA \tag{8-2}$$

and fix our attention upon the last two terms on the right, writing

$$\frac{1}{\mathscr{A}}\int_{\mathscr{S}}(\bar{f}-f)\,dA = \frac{1}{\mathscr{A}}\sum_{n}\int_{\mathscr{S}_n}(\bar{f}-f)\,dA \tag{8-3}$$

The understanding here is that

$$\mathscr{S} = \sum_{n}\mathscr{S}_n \tag{8-4}$$

The only limitation placed upon the sub-surface $\mathscr{S}_n$ is that the characteristic dimension of each be of the order $L_0$, the characteristic dimension of the averaging surface $S$ (see Sec. 4.3.2). With this limitation, it seems reasonable to approximate the mean value of $f$ on $\mathscr{S}_n$ by

$$\bar{f} \doteq \frac{1}{\mathscr{A}_n}\int_{\mathscr{S}_n}f\,dA \tag{8-5}$$

and to say, since $\bar{f}$ is nearly independent of position over distances of order $L_0$ (Sec. 4.3.2),

$$\int_{\mathscr{S}_n}(\bar{f}-f)\,dA \doteq 0 \tag{8-6}$$

By $\mathscr{A}_n$ I mean the area of $\mathscr{S}_n$.

The desired result (8-1) follows from (8-2), (8-3), and (8-6). The limitations on (8-1) are these:

1. The characteristic dimension $L_0$ of the averaging surface $S$ is chosen such that $\bar{f}$ is nearly independent of position over distances of the same order.
2. The characteristic dimension of $\mathscr{S}$ must be greater than or equal to $L_0$.

### EXERCISE

**4.3.8-1** Let $R$ be some macroscopic region in space whose volume is $\mathscr{V}$. Construct an argument similar to that given in the text to conclude that

$$\frac{1}{\mathscr{V}}\int_{\mathscr{R}} \bar{f}\,dv = \frac{1}{\mathscr{V}}\int_{\mathscr{R}} f\,dv$$

Assumption (1) of the text again applies as well as

2' The characteristic dimension of $\mathscr{R}$ must be greater than or equal to $L_0$.

### 4.3.9 Flow through a packed tube
An incompressible newtonian fluid flows through a nonoriented permeable structure of uniform porosity

$$\Psi \equiv \frac{V_{(f)}}{V} \tag{9-1}$$

bounded by a cylindrical tube of radius $r_0$. We wish to determine the local volume-averaged velocity distribution for the fluid as well as the corresponding volume rate of flow through the tube.

Referring to Fig. 4.3.9-1, we say that

At $r = r_0$, $\theta = 0$, $z = 0$: $\quad \bar{p} = \Psi P_0 \tag{9-2}$

and

At $r = r_0$, $\theta = 0$, $z = L$: $\quad \bar{p} = \Psi P_L \tag{9-3}$

Here we express our belief that experimentalists measure more nearly (see Exercise 4.3.9-1)

$$\langle p \rangle \equiv \frac{1}{V_{(f)}} \int_{\mathscr{V}_{(f)}} p\,dV = \frac{\bar{p}}{\Psi} \tag{9-4}$$

Because of the finite size of the averaging surface $S$, the local volume-averaged

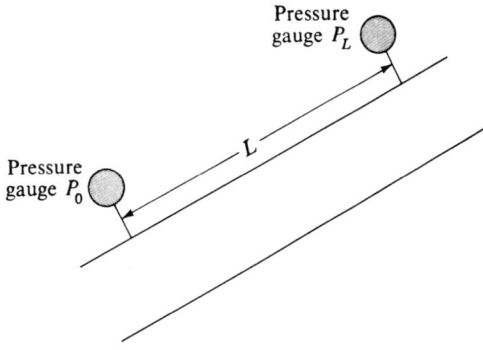

Fig. 4.3.9-1 Flow through a packed tube.

velocity $\bar{\mathbf{v}}$ will not be zero at the tube wall. But as we advance into the pipe wall $\bar{\mathbf{v}}$ decreases, since less of the averaging surface $S$ intercepts the porous structure that contains the fluid. At some distance $\epsilon$ inside the impermeable wall, $\bar{\mathbf{v}}$ goes to zero (this distance $\epsilon$ will depend upon the averaging surface $S$ chosen):

At $r = r_0 + \epsilon$: $\quad \bar{\mathbf{v}} = 0$ \hfill (9-5)

A solution of the form

$$\bar{v}_z = \bar{v}_\theta = 0 \qquad \bar{v}_z = \bar{v}_z(r) \tag{9-6}$$

satisfies the local volume-averaged equation of continuity for an incompressible fluid, Eq. (4-3). The three components of Eq. (7-2) reduce to

$$\frac{\partial(\overline{\mathscr{P} - p_0})}{\partial r} = \frac{\partial(\overline{\mathscr{P} - p_0})}{\partial \theta} = 0 \tag{9-7}$$

and

$$\frac{\partial(\overline{\mathscr{P} - p_0})}{\partial z} = \frac{\mu}{r}\frac{d}{dr}\left(r\frac{d\bar{v}_z}{dr}\right) - \frac{\Psi\mu}{l_0^2 k_0^*}\bar{v}_z \tag{9-8}$$

These equations imply that

$$C = \frac{d(\overline{\mathscr{P} - p_0})}{dz} = \frac{\mu}{r}\frac{d}{dr}\left(r\frac{d\bar{v}_z}{dr}\right) - \frac{\Psi\mu}{l_0^2 k_0^*}\bar{v}_z$$

$$= \text{a constant} \tag{9-9}$$

We can integrate Eq. (9-9)$_1$ with boundary conditions (9-2) and (9-3) to find

$$C = -\frac{1}{L}\{\Psi(P_0 - P_L) + \rho[\bar{\varphi}(r_0,0,0) - \bar{\varphi}(r_0,0,L)]\} \tag{9-10}$$

where $\varphi = \varphi(r,\theta,z)$. If in Eq. (9-9) we introduce

$$u^* \equiv -\frac{\bar{v}_z \Psi \mu}{C l_0^2 k_0^*} - 1 \tag{9-11}$$

and

$$r^* = \frac{r}{r_0 + \epsilon} \tag{9-12}$$

we obtain a form of Bessel's equation,

$$\frac{d^2 u^*}{dr^{*2}} + \frac{1}{r^*}\frac{du^*}{dr^*} - N^2 u^* = 0 \tag{9-13}$$

Here

## [4.3] LOCAL VOLUME AVERAGING

$$N \equiv \left[\frac{\Psi(r_0 + \epsilon)^2}{l_0^2 k_0^*}\right]^{1/2} \tag{9-14}$$

The boundary conditions for Eq. (9-13) are that $u^*$ remains finite at $r^* = 0$ and

At $r^* = 1$: $\quad u^* = -1$ (9-15)

The required solution is

$$u^* = -\frac{I_0(Nr^*)}{I_0(N)} \tag{9-16}$$

or

$$\bar{v}_z = -\frac{Cl_0^2 k_0^*}{\Psi\mu}\left[1 - \frac{I_0(Nr^*)}{I_0(N)}\right] \tag{9-17}$$

By $I_0$, we mean the zero-order modified Bessel function of the first kind [1, p. 143].

The volume rate of flow through the packed tube can be calculated using Eq. (9-17):

$$Q = 2\pi(r_0 + \epsilon)^2 \int_0^{\frac{R}{R+\epsilon}} \bar{v}_z r^* \, dr^*$$

$$= -\frac{\pi r_0^2 l_0^2 k_0^* C}{\Psi\mu}\left[1 - \frac{2(R + \epsilon)}{NR} \frac{I_1(N\frac{R}{R+\epsilon})}{I_0(N)}\right] \tag{9-18}$$

For sufficiently large values of $N$, Eq. (9-17) tells us that $\bar{v}_z$ is essentially constant over the cross section, except in the immediate vicinity of the wall where it approaches zero. Equation (9-18) indicates that as $N \to \infty$ the volume rate of flow through the tube is essentially that found by multiplying the centerline velocity by the cross-sectional area.

In order to get a better feeling for the magnitude of the wall effect, let us consider an example. Say that we have a tube, 2 cm in diameter, packed with spherical particles of diameter 2 mm such that the void fraction or porosity $\Psi = 0.3$. Comparison of Eq. (7-2) with the extended form of Darcy's law, Eq. (1-3), suggests that

$$l_0^2 k_0^* = k \tag{9-19}$$

where $k$ is the permeability of the packed bed. We may use the Blake-Kozeny equation [2, p. 199] to estimate $k = 1.5 \times 10^{-5}$ cm². We conclude

$$N = 141(1 + \epsilon) \tag{9-20}$$

In this case, there is a small wall effect in (9-18).

## REFERENCES

1. Irving, J., and N. Mullineux: "Mathematics in Physics and Engineering," Academic, New York, 1959.

2. Bird, R. B., W. E. Stewart, and E. N. Lightfoot: "Transport Phenomena," 7th printing, Wiley, New York, 1960.
3. Whitaker, Stephen: *Ind. Eng. Chem.*, **61**(12):14 (1969).
4. Slattery, J. C.: *A.I.Ch.E.J.*, **15**:866 (1969).

## EXERCISES

**4.3.9-1** *The pressure-gauge measurement* Figure 4.3.9-2 shows the detail of a somewhat idealized pressure-gauge probe. Let $A_{(f)}$ indicate the area at which the fluid acts upon the probe; $A$ denotes the total cross-sectional area of the probe.

(a) If $P_0$ is the pressure-gauge reading, argue that

$$P_0 = \frac{A}{A_{(f)}} \bar{p}$$

(b) Let us define the void-volume distribution function

$$\alpha(z) = \begin{cases} 1 & \text{if z lies in the fluid} \\ 0 & \text{if z lies in the solid} \end{cases}$$

Estimate that [3]

$$\frac{A_{(f)}}{A} = \frac{V_{(f)}}{V} = \Psi$$

This allows us to conclude

$$P_0 = \Psi^{-1} \bar{p}$$

**4.3.9-2** *Channel flow* [4] Let us now use Fig. 4.3.9-1 to describe flow of an incompressible newtonian fluid through a nonoriented permeable structure of uniform porosity bounded by two infinite parallel planes. The distance between the planes is $2b$. Determine the local volume-averaged velocity distribution corresponding to the fluid as well as the volume rate of flow of the fluid through the channel.

**4.3.9-3** *Radial flow* [4] A problem closely related to radial flow to or from a well bore may be described in cylindrical coordinates by a local volume-averaged velocity distribution of the form

$$\bar{v}_r = \bar{v}_r(r) \qquad \bar{v}_\theta = \bar{v}_z = 0$$

with an associated local volume-averaged pressure distribution that satisfies

At $r = r_1$, $\theta = 0$, $z = 0$: $\quad \bar{p} = \Psi P_1$

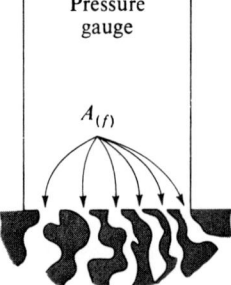

**Fig. 4.3.9-2** Detail of a somewhat idealized pressure-gauge probe.

## [4.3] LOCAL VOLUME AVERAGING

and

At $r = r_2$, $\theta = 0$, $z = 0$: $\quad \bar{p} = \Psi P_2$

Assume that the cylindrical coordinate system is oriented with the $z$ axis in the direction opposite to the action of gravity. Determine the radial component of the local volume-averaged velocity distribution as well as a volume rate of flow through the cylindrical surface $r = r_1$.

### 4.3.10 Neglecting the divergence of the local volume-averaged extra stress

Our discussion of flow through a packed tube in Sec. 4.3.9 suggests that under many circumstances we may be able to neglect the effect of the divergence of the local volume-averaged extra stress in the equation of motion for a porous medium, Eq. (5-8). In this section, we justify neglecting this term for the flow of incompressible newtonian and incompressible viscoelastic fluids through nonoriented porous media. The extension to oriented permeable structures follows along the same lines (see Exercise 4.3.10-1).

Equation (7-2) describes the flow of an incompressible newtonian fluid through a nonoriented porous structure. In terms of dimensionless variables, it says

$$\frac{P_0 l_0^2 k_0^*}{L_0 \mu v_0} \nabla(\overline{\mathscr{P} - p_0})^* - \frac{l_0^2 k_0^*}{L_0^2} \operatorname{div}(\nabla \bar{\mathbf{v}}^*) + \bar{\mathbf{v}}^* = 0 \qquad (10\text{-}1)$$

Here $P_0$ is a characteristic pressure, $v_0$ is a characteristic magnitude of velocity, and $L_0$ is a length characteristic of the gross geometry. If, as we suggested in Sec. 4.3.9, $l_0^2 k_0$ is of the same order of magnitude as the permeability $k$ in Darcy's law, Eq. (1-3), then $l_0^2 k_0^* / L_0^2$ would be a very small number for common porous media problems. This suggests that the second term on the left of Eq. (10-1) be neglected with respect to the third term to obtain

$$\nabla(\overline{\mathscr{P} - p_0}) + \frac{\mu}{l_0^2 k_0^*} \bar{\mathbf{v}} = 0 \qquad (10\text{-}2)$$

But the reader is cautioned that this is an intuitive argument that need not be true in every situation. There is no theorem that says that, because a term of a differential equation is multiplied by a small parameter, the term can be neglected.

This same argument can be repeated for a viscoelastic fluid as represented by the Noll simple fluid model discussed in Sec. 2.3.4. For flow through a nonoriented porous structure, Eqs. (5-8), (6-6), and (6-10) tell us that

$$\nabla(\overline{\mathscr{P} - p_0}) - \operatorname{div} \mathbf{S} + \frac{\mu_0}{l_0^2 k^*} \bar{\mathbf{v}} = 0 \qquad (10\text{-}3)$$

where $k^*$ is a function of the Weissenberg number $N_{\text{Wi}}$:

$$N_{\text{Wi}} \equiv \frac{s_0 |\bar{\mathbf{v}}|}{l_0} \qquad (10\text{-}4)$$

Let us introduce the same dimensionless variables as we did above, with the additional definition

$$\bar{\bar{S}}^* = \frac{s_0 \bar{\bar{S}}}{\mu_0} \tag{10-5}$$

This allows us to write Eq. (10-3) in a dimensionless form as

$$\frac{P_0 l_0^2 k^*}{L_0 \mu_0 v_0} \nabla \overline{(\mathscr{P} - p_0)}^* - \frac{l_0^2 k^*}{s_0 L_0 v_0} \operatorname{div} \bar{\bar{S}}^* + \bar{v}^* = 0 \tag{10-6}$$

In order to get a feeling for the magnitude of the parameter multiplying the second term on the left of Eq. (10-6), let us consider an example using some typical values for the various characteristic quantities appropriate to flow in an oil-bearing rock structure:

$l_0^2 k^* \approx 250$ millidarcys or $2.5 \times 10^{-9}$ cm²

$s_0 = 10^{-2}$ s

$v_0 = 1$ ft/day or $3.5 \times 10^{-4}$ cm/s

$$\frac{l_0^2 k^*}{s_0 L_0 v_0} \approx \frac{7 \times 10^{-4}}{L_0} \tag{10-7}$$

A characteristic time of $10^{-2}$ s appears to be reasonable for some viscoelastic fluids [1, 2]. If we remember that $L_0$ is a length characteristic of the gross geometry, then this suggests that the second term on the left of Eq. (10-6) may be neglected with respect to the third term to obtain

$$\nabla \overline{(\mathscr{P} - p_0)} + \frac{\mu_0}{l_0^2 k^*} \bar{v} = 0 \tag{10-8}$$

In dropping out the divergence of the local volume-averaged extra stress to obtain Eqs. (10-2) and (10-8), there is a corresponding reduction in order of the differential equation. This is most obvious in comparing Eq. (10-1) with (10-2) for an incompressible newtonian fluid. With this reduction in order, we lose our ability to satisfy boundary conditions on the tangential components of the local volume-averaged velocity vector. This is similar to the problem we ran into in discussing potential flow in Sec. 3.4.3. So long as we are not very interested in the velocity distribution in the immediate neighborhood of an impermeable boundary to a porous medium, the approximation suggested here should be entirely satisfactory.

As an example of the type of problem where one might get into trouble by neglecting the divergence of the local volume-averaged extra stress, consider flow through a porous-walled tube. The description of the local volume-averaged velocity distribution in the immediate neighborhood of the boundary of the porous medium would appear to be important in determining the proper boundary conditions for the fluid flowing through the tube. Since the tangential component of velocity would not necessarily go to zero at the tube wall, one might expect to see in the experimental data an apparent slip at the wall.

To summarize, so long as we can neglect inertial effects and represent the external force vector in terms of a scalar potential, we will almost always be justified in writing the local volume average of Cauchy's first law as

$$\nabla(\overline{\mathscr{P} - p_0}) + \mathbf{g} = 0 \tag{10-9}$$

where the force per unit volume $\mathbf{g}$ that the fluid exerts upon the porous structure must be determined from an empirical data correlation of the form suggested in Sec. 4.3.6.

What about the relation of these results to the force balances that have been in use in the literature for some time? For the flow of an incompressible newtonian fluid through a nonoriented porous structure, Eq. (10-9) takes the form of Eq. (10-2). Upon comparison of Eq. (10-2) with Eq. (1-3), it becomes clear that we have derived here the extended form of Darcy's law with a clear interpretation for the variables being used. For the flow of an incompressible newtonian fluid through an oriented porous structure of the type discussed in Sec. 4.3.6, Eq. (10-9) becomes

$$\nabla(\overline{\mathscr{P} - p_0}) + \frac{\mu}{l^2}\mathbf{R}^* \cdot \bar{\mathbf{v}} = 0 \tag{10-10}$$

where

$$\mathbf{R}^* \equiv \frac{1}{k_{(1)}^*}\mathbf{I} + \frac{|\bar{\mathbf{v}}|}{k_{(2)}^*(\mathbf{v}\cdot\nabla l)}\nabla l \nabla l \tag{10-11}$$

and

$$k_{(i)}^* = k_{(i)}^*\left(|\nabla l|, \frac{\bar{\mathbf{v}}}{|\bar{\mathbf{v}}|}\cdot\nabla l\right) \quad \text{for } i = 1, 2 \tag{10-12}$$

Comparison with Eq. (1-4) indicates that Eq. (10-11) should be interpreted as an expression for $l^2\mathbf{K}^{-1}$. From this point of view, we see that $\mathbf{K}^{-1}$ must be symmetric, although it is not so clear that it is always invertible (that is, given $\mathbf{K}^{-1}$, $\mathbf{K}$ exists).

### REFERENCES

1. Shertzer, C. R., and A. B. Metzner: *Proc. 4th Intern. Congr. Rheol.*, pt. 2, p. 603, 1965.
2. Ginn, R. F., and A. B. Metzner: *Proc. 4th Intern. Congr. Rheol.*, pt. 2, p. 583, 1965.

### EXERCISE

**4.3.10-1** *Flow of an incompressible newtonian fluid through an oriented porous medium* Justify the use of Eq. (10-10) to describe the flow of an incompressible newtonian fluid through an oriented porous structure.

**4.3.11 Flow of a viscoelastic fluid through a tube** As an example of how one might use Eq. (10-9), let us consider the flow of a viscoelastic fluid through the packed tube shown in Fig. 4.3.9-1. We will assume that the porous structure is nonoriented and that the behavior of the fluid can be represented by the Noll simple fluid of Sec. 2.3.4. To be more specific, we assume that **g**, the force per unit volume that the fluid exerts on the porous structure, is representable by Eqs. (6-6) and (6-17).

As boundary conditions we have

At $r = r_0$, $\theta = 0$, $z = 0$: $\quad \bar{p} = \Psi P_0$ (11-1)

At $r = r_0$, $\theta = 0$, $z = L$: $\quad \bar{p} = \Psi P_L$ (11-2)

and

At $r = r_0 + \epsilon$: $\quad \bar{v}_r = 0$ (11-3)

The local volume-averaged equation of continuity, Eq. (4-3), as well as boundary condition (11-3) are satisfied by a velocity distribution of the form

$$\bar{v}_r = \bar{v}_\theta = 0 \qquad \bar{v}_z = \bar{v}_z(r) \tag{11-4}$$

With this assumption and Eq. (6-6), we find that the three components of the local volume average of Cauchy's first law become

$$\frac{1}{R} \frac{\partial (\overline{\mathscr{P} - p_0})}{\partial z} + \bar{v}_z = 0 \tag{11-5}$$

and

$$\frac{\partial (\overline{\mathscr{P} - p_0})}{\partial r} = \frac{\partial (\overline{\mathscr{P} - p_0})}{\partial \theta} = 0 \tag{11-6}$$

With the help of Eq. (6-17), Eq. (11-5) can be written as

$$\bar{v}_z = -\frac{d(\overline{\mathscr{P} - p_0})}{dz} \alpha l_0^2 \left\{ 1 + \left[ -\beta l_0 \frac{d(\overline{\mathscr{P} - p_0})}{dz} \right]^{\gamma-1} \right\} \tag{11-7}$$

This last equation implies that

$$\frac{d(\overline{\mathscr{P} - p_0})}{dz} = C = \text{a constant} \tag{11-8}$$

Upon integration and application of boundary conditions (11-1) and (11-2), we learn that

$$C = -\frac{1}{L} \{\Psi(P_0 - P_L) + \rho[\bar{\varphi}(r_0,0,0) - \bar{\varphi}(r_0,0,L)]\} \tag{11-9}$$

where the local volume average of the scalar force potential $\bar{\varphi} = \bar{\varphi}(r,\theta,z)$.

For the same problem, if the resistance coefficient is described by Eq. (6-16), we obtain

$$\bar{v}_z = m l_0^{n+1} \left[ -\frac{d(\overline{\mathscr{P} - p_0})}{dz} \right]^n \tag{11-10}$$

Equations (11-8) and (11-9) are again applicable.

## [4.3] LOCAL VOLUME AVERAGING

Using the method suggested by Ashare, Bird, and Lescarboura [1], we have fitted Eq. (11-7) to the data of Christopher [2] for a 1.25 percent carboxymethyl cellulose solution in water flowing through a bed 1.76 cm long. Figure 4.3.11-1 indicates excellent agreement with the experimental data. This figure also illustrates that Eq. (11-10) represents very well the data of Sadowski [3, 4] for an 18.5 percent Carbowax (polyethylene glycol, viscosity average molecular weight equal to 20,000) solution in water. In this way we confirm the potential utility of Eqs. (6-16) and (6-17).

It is perhaps worth emphasizing that the primary use of Eqs. (11-7) and (11-10)

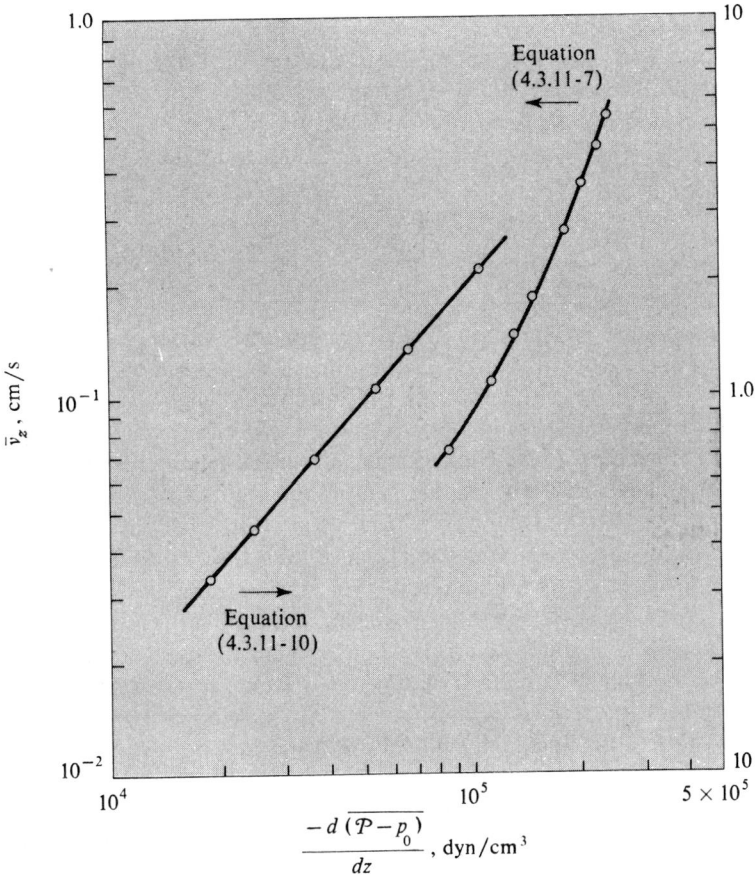

**Fig. 4.3.11-1** Comparison of Eq. (11-7) with data for 1.25 percent carboxymethyl cellulose solution in water [2] ($\alpha l_0^2 = 7.14 \times 10^{-7}$ cm/s.; $\beta l_0 = 5.49 \times 10^{-6}$ cm$^2$/dyn; $\gamma = 3.5$). Comparison of Eq. (11-10) with data for 18.5 percent Carbowax (polyethylene glycol, viscosity average molecular weight equal to 20,000) solution in water [3] ($m l_0^{n+1} = 8.90 \times 10^{-6}$; $n = 1.07$). Taken from [5].

is not in the correlation and extrapolation of data for a particular fluid in a particular porous structure. They are primarily useful to the extent that they allow one to predict on the basis of data for a single packed bed how a given fluid will behave in a wide class of geometrically similar porous beds corresponding to different mean pore diameters $l_0$. By *geometrically similar* porous media, we mean porous structures that were formed in roughly the same manner. For example, one might hope to predict how a 1 percent carboxymethyl cellulose solution in water would behave in most limestone formations on the basis of data for a single limestone. The practicality of such an extrapolation has been demonstrated by Savins [6].

## REFERENCES

1. Ashare, Edward, R. B. Bird, and J. A. Lescarboura: *A.I.Ch.E.J.*, **11**:910 (1965).
2. Christopher, R. H.: M.S. thesis, Dept. of Chem. Engng., University of Rochester, Rochester, N.Y., 1965.
3. Sadowski, T. J.: doctoral dissertation, Dept. of Chem. Engng., University of Wisconsin, Madison, Wis. 1963.
4. Sadowski, T. J.: *Trans. Soc. Rheol.*, **9**(2):251 (1965).
5. Slattery, J. C.: *A.I.Ch.E.J.*, **15**:866 (1969).
6. Savins, J. G.: *Ind. Eng. Chem.*, **61**(10):18 (1969).

## 4.4 INTEGRAL BALANCES

**4.4.1 The integral balances** By an integral balance I mean an equation that describes accumulation in terms of influx and outflow for any quantity. The implication is that the system over which the balance is made need not be a collection of material particles. The system may be the fluid in a surge tank or the fuel in a jet aircraft which is moving along an arbitrary curve in space. In these two examples the system is not a material one, since fluid may be entering or leaving the surge tank and fuel is being continuously consumed by the jet engine.

We frequently use integral balances in situations where we wish to make a statement about the system as a whole without worrying about a detailed description of the motions of the fluids within the system. For example, in relating the thrust developed by a rocket engine to the average velocity of its exhaust gases, it is not important to do a detailed study of the atomization of the liquid fuel. This atomization process may be important to the subsequent combustion and in this way may directly affect the rocket's performance. But once we have been given the average velocity of the rocket's exhaust, we will find we have sufficient information to estimate the thrust developed by the engine.

Integral balances are one of the most commonly used techniques in engineering, since they result in algebraic equations or relatively simple differential equations. Their importance cannot be overstressed. Yet their simplicity is misleading in that one is often forced to make a series of approximations based upon intuitive judgments or related experimental knowledge. It becomes more difficult to say whether a particular analysis will describe an experimental observation within a prescribed error. One often hears remarks to the effect that "a good engineer develops a facility for making intuitive judgments." This certainly is true of the engineer who is successful in applying integral balances.

Many of the ideas associated with integral balances that we present here are due to Bird [1].

**REFERENCE**

1. Bird, R. B.: *Chem. Eng. Sci.*, **6**:123 (1957).

**4.4.2 The integral mass balance** The key to the development of the integral balances is the generalized transport theorem discussed in Sec. 1.3.2 for an arbitrary system:

$$\frac{d}{dt}\int_{V_{(s)}} \Psi\, dV = \int_{V_{(s)}} \frac{\partial \Psi}{\partial t}\, dV + \int_{S_{(s)}} \Psi(\mathbf{v}_{(s)} \cdot \mathbf{n})\, dS \tag{2-1}$$

Here the system has a volume $V_{(s)}$ and closed bounding surface $S_{(s)}$; $\mathbf{v}_{(s)}$ is the velocity of the closed bounding surface, which may be a function of position on the surface; $\Psi$ is any quantity, though in our discussion it will have the dimensions of something per unit volume.

Our initial object is to write a mass balance for an arbitrary system; we want an equation that describes the time rate of change of mass in an arbitrary system. This suggests taking $\Psi = \rho$ in Eq. (2-1) and writing

$$\frac{d}{dt}\int_{V_{(s)}} \rho\, dV = \int_{V_{(s)}} \frac{\partial \rho}{\partial t}\, dV + \int_{S_{(s)}} \rho(\mathbf{v}_{(s)} \cdot \mathbf{n})\, dS \tag{2-2}$$

This equation, while true, is not useful in itself, because in general we will have no idea how $\partial \rho/\partial t$ varies with position in the system. We could determine $\rho$ as a function of position and time by solving the equation of continuity and Cauchy's first law, but often, when we apply the integral balances, we are trying to avoid this. Yet, if we are trying to formulate a mass balance for a large system, it seems only reasonable to see what the equation of continuity, which expresses the concept of conservation of mass at each point in the material, implies.

Let us integrate the differential equation of continuity over our system:

$$\int_{V_{(s)}} \frac{\partial \rho}{\partial t}\, dV + \int_{V_{(s)}} \text{div}\,(\rho \mathbf{v})\, dV = 0 \tag{2-3}$$

After an application of Green's transformation to the second integral on the left, we obtain

$$\int_{V_{(s)}} \frac{\partial \rho}{\partial t} dV = -\int_{S_{(s)}} \rho(\mathbf{v} \cdot \mathbf{n}) \, dS \tag{2-4}$$

This allows us to rewrite Eq. (2-2) in a more useful form:

$$\frac{d}{dt} \int_{V_{(s)}} \rho \, dV = \int_{S_{(s)}} \rho(\mathbf{v} - \mathbf{v}_{(s)}) \cdot (-\mathbf{n}) \, dS \tag{2-5}$$

In words, this equation tells us that the time rate of change of mass in the system is equal to the net rate at which mass enters the system. Notice how naturally the velocity of the fluid relative to the boundary of the system, $(\mathbf{v} - \mathbf{v}_{(s)})$, enters. Since $(\mathbf{v} - \mathbf{v}_{(s)})$ is different from zero only on the entrance and exit portions of $S_{(s)}$, $S_{(\text{ent ex})}$, we may write Eq. (2-5) as

$$\frac{d}{dt} \int_{V_{(s)}} \rho \, dV = \int_{S_{(\text{ent ex})}} \rho(\mathbf{v} - \mathbf{v}_{(s)}) \cdot (-\mathbf{n}) \, dS \tag{2-6}$$

This is the *integral mass balance* for a single-phase system.

It is important to realize that as Eq. (2-6) has been derived it applies only to a single-phase system. More often than not in practice we are concerned with multiphase systems. For example, if we should wish to talk about the mass of an accelerating rocket as a function of time, the most natural choice of systems would be the rocket and all its contents. This system would consist of millions of distinct phases—gases, liquids, and solids. Let us derive for such a system a relation analogous to Eq. (2-6).

For simplicity, let us assume that our arbitrary system consists of only two phases. Each phase may itself be regarded as a system to which Eq. (2-6) applies. Referring to Fig. 4.4.2-1, if we denote the region occupied by phase $i$ in the system as $V_{(i)}$, then the region occupied by the entire system may be indicated as

$$V_{(s)} = V_{(s1)} + V_{(s2)}$$

The entrance and exit surfaces of $V_{(si)}$ are $S_{(\text{ent ex } i)}$. Some of these entrance and exit surfaces may coincide with a portion of the entrance and exit surface for the system as a whole, $S_{(\text{ent ex})}$; others will coincide with the singular surfaces (phase interfaces) $S_{(\text{sing})}$ at which the two phases meet, since we wish to allow for the possibility of interphase mass transfer. Adding the statements of Eq. (2-6) made for each phase, we have

$$\frac{d}{dt} \int_{V_{(s)}} \rho \, dV = \frac{d}{dt} \int_{V_{(s1)}} \rho \, dV + \frac{d}{dt} \int_{V_{(s2)}} \rho \, dV$$

$$= \int_{S_{(\text{ent ex 1})}} \rho(\mathbf{v} - \mathbf{v}_{(s)}) \cdot (-\mathbf{n}) \, dS$$

$$+ \int_{S_{(\text{ent ex 2})}} \rho(\mathbf{v} - \mathbf{v}_{(s)}) \cdot (-\mathbf{n}) \, dS$$

## [4.4] INTEGRAL BALANCES

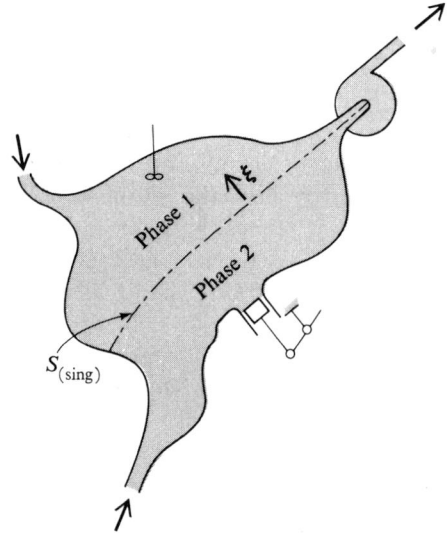

**Fig. 4.4.2-1** A system consisting of two phases.

or

$$\frac{d}{dt} \int_{V_{(s)}} \rho \, dV = \int_{S_{(\text{ent ex})}} \rho(\mathbf{v} - \mathbf{v}_{(s)}) \cdot (-\mathbf{n}) \, dS$$
$$+ \int_{S_{(\text{sing})}} [\rho(\mathbf{v} \cdot \boldsymbol{\xi} - u_{(\xi)})] \, dS \quad (2\text{-}7)$$

Here we have employed the notation first introduced in Sec. 1.3.3. If, for any quantity $\Psi$ and for a particular position on $S_{(\text{sing})}$, $\Psi_{(i)}$ denotes the value of $\Psi$ obtained in the limit as this position is approached from the interior of phase $i$, then

$$\text{At } S_{(\text{sing})}: \quad [\Psi] = \Psi_{(1)} - \Psi_{(2)} \quad (2\text{-}8)$$

The unit normal to $S_{(\text{sing})}$ directed from phase 2 into phase 1 is denoted by $\boldsymbol{\xi}$; $u_{(\xi)}$ is the speed of displacement of $S_{(\text{sing})}$ [1, p. 499]. This argument for two-phase systems is readily extended to systems containing any number of phases. Equation (2-7) is the *general integral mass balance* appropriate to a multiphase system.

We will usually be willing to say that the jump mass balance of Sec. 1.3.5 is applicable. In this case, Eq. (2-7) reduces to (2-6). The integral mass balance (2-6) applies equally well to single-phase and multiphase systems so long as the jump mass balance of Sec. 1.3.5 is valid for all phase interfaces involved.

## REFERENCE

1. Truesdell, C., and R. A. Toupin: In S. Flügge (ed.), "Handbuch der Physik," vol. 3/1, Springer-Verlag, Berlin, 1960.

### 4.4.3 The integral mass balance for turbulent flows

Some of the most important and interesting applications of the integral balances are to systems, portions of which are in turbulent flow.

For example, we may wish to determine the pressure drop between pumping stations in a pipeline that transports natural gas from the southwest to Chicago. It is convenient to choose the natural gas in the pipeline between the pumping stations as our system in analyzing this problem. Our first step is to determine the restrictions placed upon this system by the integral mass balance. To do this, we must make some statement about the time rate of change of the mass of the system. Is this a steady-state problem or an unsteady-state problem? Since the gas is in turbulent flow, this must by definition be an unsteady-state situation. Yet, when the pipeline superintendent speaks about the volume rate of flow past a particular cross section in the pipeline, he ignores any random fluctuations caused by turbulence. He has little choice, since whatever instruments he uses to infer the volume flow rate report only time-averaged readings. The integral mass balance derived in Sec. 4.4.2 is not directly applicable, because it relates the instantaneous mass of the system to the instantaneous mass flow rates through the entrances and exits.

An integral mass balance that is more appropriate to systems in turbulent flow may be obtained. Let us limit ourselves to single-phase or multiphase systems that do not involve fluid-fluid phase interfaces. Starting with the time-averaged differential equation of continuity found in Sec. 4.1.2,

$$\frac{\partial \bar{\rho}}{\partial t} + \text{div}\,(\overline{\rho \mathbf{v}}) = 0 \tag{3-1}$$

the derivation of Sec. 4.4.2 may be repeated to find

$$\frac{d}{dt}\int_{V_{(s)}} \bar{\rho}\, dV = \int_{S_{(\text{ent ex})}} (\overline{\rho \mathbf{v}} - \bar{\rho}\mathbf{v}_{(s)}) \cdot (-\mathbf{n})\, dS$$

$$+ \int_{S_{(\text{sing})}} \left[\overline{\rho \mathbf{v}} \cdot \boldsymbol{\xi} - \bar{\rho} u_{(\xi)}\right] dS \tag{3-2}$$

The time-averaged jump mass balance of Exercise 4.4.3-1 simplifies this to

$$\frac{d}{dt}\int_{V_{(s)}} \bar{\rho}\, dV = \int_{S_{(\text{ent ex})}} (\overline{\rho \mathbf{v}} - \bar{\rho}\mathbf{v}_{(s)}) \cdot (-\mathbf{n})\, dS \tag{3-3}$$

This says that the time rate of change of the time-averaged mass of the system equals the net time-averaged rate at which mass is brought into the system through the entrances and exits. Equation (3-3) is the *integral mass balance* appropriate to a turbulent single-phase or multiphase system that does not involve fluid-fluid phase interfaces.

I do not recommend this approach when discussing systems that contain or are bounded by fluid-fluid phase interfaces, since the positions of these phase interfaces will in general be random functions of time. Even after a definition for a time-averaged phase interface is agreed upon, we find that it is not the time average of the

[4.4] INTEGRAL BALANCES 223

Sec. 1.3.4 jump mass balance that applies. There are important interfacial effects that are directly attributable to turbulence.

For single-phase or multiphase systems that include one or more fluid-fluid phase interfaces, I recommend time averaging the integral mass balance obtained in Sec. 4.4.2:

$$\frac{d}{dt}\int_{V_{(s)}} \rho \, dV = \int_{S_{(\text{ent ex})}} \rho(\mathbf{v} - \mathbf{v}_{(s)}) \cdot (-\mathbf{n}) \, dS \tag{3-4}$$

to find

$$\overline{\frac{d}{dt}\int_{V_{(s)}} \rho \, dV} = \overline{\int_{S_{(\text{ent ex})}} \rho(\mathbf{v} - \mathbf{v}_{(s)}) \cdot (-\mathbf{n}) \, dS} \tag{3-5}$$

Here we have observed that

$$\overline{\frac{d}{dt}\int_{V_{(s)}} \rho \, dV} \equiv \frac{1}{\Delta t} \int_t^{t+\Delta t} \left(\frac{d}{dt'}\int_{V_{(s)}} \rho \, dV\right) dt'$$

$$= \frac{1}{\Delta t}\left(\int_{V_{(s)}}\rho \, dV\Big|_{t+\Delta t} - \int_{V_{(s)}}\rho \, dV\Big|_t\right) = \frac{d}{dt}\left(\frac{1}{\Delta t}\int_t^{t+\Delta}\int_{V_{(s)}} \rho \, dV \, dt'\right)$$

$$= \frac{d}{dt}\overline{\int_{V_{(s)}} \rho \, dV} \tag{3-6}$$

Equation (3-5) tells us that the time rate of change of the time-averaged mass of the system is equal to the sum of the time-averaged mass flow rates through the entrances of the system minus the time-averaged mass flow rates through the exits of the system.

As we discussed in Sec. 4.1.1, the interval $\Delta t$ for time averaging is chosen to be large with respect to a time characteristic of a random fluctuation in turbulent flow but small with respect to a time characteristic of the gross process being studied. This suggests that, if we restrict ourselves to single-phase or multiphase systems that do not involve fluid-fluid phase interfaces, for any quantity $\Psi$

$$\overline{\int_{V_{(s)}} \Psi \, dV} = \int_{V_{(s)}} \overline{\Psi} \, dV \tag{3-7}$$

$$\overline{\int_{S_{(\text{ent ex})}} \Psi \, dS} = \int_{S_{(\text{ent ex})}} \overline{\Psi} \, dS \tag{3-8}$$

and

$$\overline{\mathbf{v}_{(s)} \cdot \mathbf{n}} = \overline{\mathbf{v}_{(s)}} \cdot \mathbf{n} \tag{3-9}$$

Under these circumstances, Eq. (3-5) reduces to Eq. (3-3).

### EXERCISE

**4.4.3-1** *Time-averaged jump mass balance* Determine that the time-averaged jump mass balance applicable to solid-fluid phase interfaces that bound turbulent flows is identical to the balance found in Sec. 1.3.4:

$$[\overline{\rho \mathbf{v} \cdot \boldsymbol{\xi} - \bar{\rho} u_{(\xi)}}] = [\overline{\rho(\mathbf{v} \cdot \boldsymbol{\xi} - u_{(\xi)})}] = 0$$

**4.4.4 Integral mass balance: An example** A local reservoir for a community is supplied by a pipeline from a series of wells at a steady volume flow rate $q_1$. The volume flow rate demanded by the community during an average 24-h day may be approximately represented by

$$q_2 = A + B \cos \frac{\pi t}{12} \tag{4-1}$$

(In this representation time is measured in hours and $t = 0$ denotes the time of peak demand rather than midnight.) In order to allow for the possibility of a pump failure, we require that the reservoir never contain less than a 3-day supply of water. Our object is to determine $q_1$ and the minimum reservoir size.

Let us choose the water in the reservoir to be our system. Since one of the bounding surfaces of the system is an air-water phase interface and since it is likely that a portion of the system is in turbulent flow (near the entrance and exit), we use the more general time-averaged integral mass balance Eq. (3-5). [Equation (3-3) could probably be used just as well, because it is likely that the air-water interface is undisturbed by turbulence.] If we denote the volume of this system by $V$ and if we recognize that water is an incompressible fluid, the integral mass balance tells us that

$$\frac{dV}{dt} = q_1 - q_2 \tag{4-2}$$

In writing this equation, we have recognized that

$$q_1 = \overline{\int_{S_{(\text{ent})}} (\mathbf{v} - \mathbf{v}_{(s)}) \cdot (-\mathbf{n}) \, dS} \tag{4-3}$$

and

$$-q_2 = \overline{\int_{S_{(\text{ex})}} (\mathbf{v} - \mathbf{v}_{(s)}) \cdot (-\mathbf{n}) \, dS} \tag{4-4}$$

Having eliminated $q_2$ by means of Eq. (4-1), we may integrate Eq. (4-2) to find

$$V = V(t) = V(0) + (q_1 - A)t - \frac{12B}{\pi} \sin \frac{\pi t}{12} \tag{4-5}$$

Here $V(0)$ denotes the volume of water in the reservoir at $t = 0$.
Since we require

$$V(0) = V(24) \tag{4-6}$$

it is clear from Eq. (4-5) that

$$q_1 = A \tag{4-7}$$

## [4.4] INTEGRAL BALANCES

By examining

$$\frac{dV}{dt} = -B\cos\frac{\pi t}{12} = 0 \qquad (4\text{-}8)$$

and

$$\frac{d^2V}{dt^2} = \frac{\pi B}{12}\sin\frac{\pi t}{12} \qquad (4\text{-}9)$$

we find that $V$ is minimized at $t = 6$ and maximized at $t = 18$. Since we require the reservoir to contain at least 3 days supply of water at all times, we have from Eq. (4-5)

$$72A = V(0) - \frac{12B}{\pi}\sin\frac{\pi}{2} \qquad (4\text{-}10)$$

or

$$V(0) = 72A + \frac{12B}{\pi} \qquad (4\text{-}11)$$

It follows that the minimum reservoir size must be the maximum amount of water contained in the reservoir during any 24-h period:

$$V_{(\max)} = V(0) - \frac{12B}{\pi}\sin\frac{3\pi}{2}$$

$$= 72A + \frac{24B}{\pi} \qquad (4\text{-}12)$$

**4.4.5 The integral momentum balance** Our object here is to write a momentum balance that describes the time rate of change of the momentum associated with an arbitrary system.

Let us take the same approach that we used in arriving at the integral mass balance. In the generalized transport theorem, let us identify $\Psi = \rho\mathbf{v}$, momentum per unit volume, to obtain

$$\frac{d}{dt}\int_{V_{(s)}} \rho\mathbf{v}\,dV = \int_{V_{(s)}} \frac{\partial(\rho\mathbf{v})}{\partial t}\,dV + \int_{V_{(s)}} \rho\mathbf{v}(\mathbf{v}_{(s)}\cdot\mathbf{n})\,dS \qquad (5\text{-}1)$$

While this is a correct statement, the value of the first integral on the right will not be immediately obvious in any given problem. Nor have we made use of the fact that Cauchy's first law must be satisfied at every point in the system.

By means of the differential equation of continuity, Cauchy's first law may be written as

$$\frac{\partial(\rho\mathbf{v})}{\partial t} + \text{div}\,(\rho\mathbf{v}\mathbf{v}) = \text{div}\,\mathbf{T} + \rho\mathbf{f} \qquad (5\text{-}2)$$

Integrating this over the volume of our arbitrary system, we have

$$\int_{V_{(s)}} \left[ \frac{\partial(\rho \mathbf{v})}{\partial t} + \text{div}\,(\rho \mathbf{v}\mathbf{v}) - \text{div}\,\mathbf{T} - \rho \mathbf{f} \right] dV = 0 \qquad (5\text{-}3)$$

The first integral on the left is just the one needed in Eq. (5-1). The fourth integral usually has clear significance as it stands. The second and third integrals have clearer physical meaning after an application of Green's transformation:

$$\int_{V_{(s)}} \text{div}\,(\rho \mathbf{v}\mathbf{v})\,dV = \int_{S_{(s)}} \rho \mathbf{v}(\mathbf{v}\cdot\mathbf{n})\,dS \qquad (5\text{-}4)$$

$$\int_{V_{(s)}} \text{div}\,\mathbf{T}\,dV = \int_{S_{(s)}} \mathbf{T}\cdot\mathbf{n}\,dS \qquad (5\text{-}5)$$

Equations (5-3), (5-4), and (5-5) allow us to rewrite Eq. (5-1) as

$$\frac{d}{dt}\int_{V_{(s)}} \rho \mathbf{v}\,dV = \int_{S_{(s)}} \rho \mathbf{v}(\mathbf{v} - \mathbf{v}_{(s)})\cdot(-\mathbf{n})\,dS$$

$$- \int_{S_{(s)}} \mathbf{T}\cdot(-\mathbf{n})\,dS + \int_{V_{(s)}} \rho \mathbf{f}\,dV \qquad (5\text{-}6)$$

The first term on the right is the net rate at which momentum is brought into the system with whatever material is crossing the boundaries. The second term is the force that the material in the system exerts on the bounding surfaces of the system. This is the total force, whereas we usually speak in terms of the force in excess of the force of atmospheric pressure. In order to correct for the effect of atmospheric pressure $p_0$, it is easiest to return to Cauchy's first law and note that

$$\text{div}\,\mathbf{T} = \text{div}\,(\mathbf{T} + p_0\mathbf{I}) \qquad (5\text{-}7)$$

This means that Eq. (5-6) may be replaced by

$$\frac{d}{dt}\int_{V_{(s)}} \rho \mathbf{v}\,dV = \int_{S_{(s)}} \rho \mathbf{v}(\mathbf{v} - \mathbf{v}_{(s)})\cdot(-\mathbf{n})\,dS$$

$$- \int_{S_{(s)}} (\mathbf{T} + p_0\mathbf{I})\cdot(-\mathbf{n})\,dS + \int_{V_{(s)}} \rho \mathbf{f}\,dV \qquad (5\text{-}8)$$

If we recognize that $(\mathbf{v} - \mathbf{v}_{(s)})\cdot\mathbf{n}$ is different from zero only on the exit and entrance surfaces $S_{(\text{ent ex})}$, we may rewrite Eq. (5-8) as

$$\frac{d}{dt}\int_{V_{(s)}} \rho \mathbf{v}\,dV = \int_{S_{(\text{ent ex})}} \rho \mathbf{v}(\mathbf{v} - \mathbf{v}_{(s)})\cdot(-\mathbf{n})\,dS$$

$$- \int_{S_{(\text{ent ex})}} (\mathbf{T} + p_0\mathbf{I})\cdot(-\mathbf{n})\,dS - \mathscr{F} + \int_{V_{(s)}} \rho \mathbf{f}\,dV \qquad (5\text{-}9)$$

where we define

$$\mathscr{F} \equiv \int_{S_{(s)} - S_{(\text{ent ex})}} (\mathbf{T} + p_0\mathbf{I})\cdot(-\mathbf{n})\,dS \qquad (5\text{-}10)$$

Physically $\mathscr{F}$ denotes the force that the system exerts upon the impermeable portion

## [4.4] INTEGRAL BALANCES

of its bounding surface beyond the force attributable to the ambient pressure $p_0$. Equation (5-9) is the *general integral momentum balance* applicable to a single-phase system.

Commonly, one chooses a system for an application of the integral momentum balance in such a way that viscous forces on the entrance and exit surfaces $S_{(\text{ent ex})}$ may be neglected. This might be done by selecting entrance and exit surfaces as cross sections normal to flow in long straight pipes and by thinking of the flow in these straight pipes as being approximately represented by Poiseuille flow (Sec. 3.2.1). In most problems, one is forced to do this for lack of sufficient information to do otherwise. Under these conditions Eq. (5-9) becomes

$$\frac{d}{dt}\int_{V_{(s)}} \rho\mathbf{v}\, dV = \int_{S_{(\text{ent ex})}} \rho\mathbf{v}(\mathbf{v}-\mathbf{v}_{(s)})\cdot(-\mathbf{n})\, dS$$

$$- \int_{S_{(\text{ent ex})}} (P-p_0)\mathbf{n}\, dS - \mathscr{F} + \int_{V_{(s)}} \rho\mathbf{f}\, dV \qquad (5\text{-}11)$$

This form of the integral momentum balance is more commonly employed than Eq. (5-9).

When dealing with incompressible fluids under circumstances such that $\mathbf{f}$ is representable by a potential $\varphi$,

$$\mathbf{f} = -\nabla\varphi \qquad (5\text{-}12)$$

it may be more convenient to express Eq. (5-11) as

$$\frac{d}{dt}\int_{V_{(s)}} \rho\mathbf{v}\, dV = \int_{S_{(\text{ent ex})}} \rho\mathbf{v}(\mathbf{v}-\mathbf{v}_{(s)})\cdot(-\mathbf{n})\, dS$$

$$- \int_{S_{(\text{ent ex})}} (\mathscr{P}-p_0)\mathbf{n}\, dS - \mathscr{G} \qquad (5\text{-}13)$$

where we define

$$\mathscr{G} = \int_{S_{(s)}-S_{(\text{ent ex})}} [\mathbf{T} + (p_0 - \rho\varphi)\mathbf{I}]\cdot(-\mathbf{n})\, dS \qquad (5\text{-}14)$$

By $\mathscr{G}$, we mean the force that the system exerts upon the impermeable portion of its bounding surface beyond the force attributable to the ambient pressure $p_0$ and to the hydrostatic pressure.

As Eqs. (5-11) and (5-13) have been derived, they are applicable only to a single-phase system. As we pointed out in Sec. 4.4.2, we are more commonly concerned with multiphase systems. Using the approach and notation of Sec. 4.4.2 and making no additional assumptions, we find that the *general* integral momentum balance appropriate to a multiphase system is

$$\frac{d}{dt}\int_{V_{(s)}} \rho \mathbf{v}\, dV = \int_{S_{(\text{ent ex})}} \rho \mathbf{v}(\mathbf{v} - \mathbf{v}_{(s)}) \cdot (-\mathbf{n})\, dS$$

$$- \int_{S_{(\text{ent ex})}} (\mathbf{T} + p_0 \mathbf{I}) \cdot (-\mathbf{n})\, dS - \mathscr{F} + \int_{V_{(s)}} \rho \mathbf{f}\, dV$$

$$+ \int_{S_{(\text{sing})}} [\rho \mathbf{v}(\mathbf{v} \cdot \boldsymbol{\xi} - u_{(\xi)}) - \mathbf{T} \cdot \boldsymbol{\xi}]\, dS \qquad (5\text{-}15)$$

For a multiphase system composed of incompressible materials, this may be written as

$$\frac{d}{dt}\int_{V_{(s)}} \rho \mathbf{v}\, dV = \int_{S_{(\text{ent ex})}} \rho \mathbf{v}(\mathbf{v} - \mathbf{v}_{(s)}) \cdot (-\mathbf{n})\, dS$$

$$- \int_{S_{(\text{ent ex})}} [\mathbf{T} + (p_0 - \rho\varphi)\mathbf{I}] \cdot (-\mathbf{n})\, dS - \mathscr{G}$$

$$+ \int_{S_{(\text{sing})}} [\rho \mathbf{v}(\mathbf{v} \cdot \boldsymbol{\xi} - u_{(\xi)}) - (\mathbf{T} - \rho\varphi\mathbf{I}) \cdot \boldsymbol{\xi}]\, dS \qquad (5\text{-}16)$$

If we assume that the jump momentum balance of Exercise 2.2.3-1 is applicable and if we neglect viscous forces on the entrance and exit surfaces, we find Eq. (5-15) reduces to the equivalent result for a single-phase system, Eq. (5-11), but Eq. (5-16) simplifies to

$$\frac{d}{dt}\int_{V_{(s)}} \rho \mathbf{v}\, dV = \int_{S_{(\text{ent ex})}} \rho \mathbf{v}(\mathbf{v} - \mathbf{v}_{(s)}) \cdot (-\mathbf{n})\, dS$$

$$- \int_{S_{(\text{ent ex})}} (\mathscr{P} - p_0)\mathbf{n}\, dS - \mathscr{G} + \int_{S_{(\text{sing})}} [\rho]\varphi \boldsymbol{\xi}\, dS \qquad (5\text{-}17)$$

There are three common types of problems in which the integral momentum balance is applied: the force $\mathscr{F}$ ($\mathscr{G}$) may be neglected, it may be the unknown to be determined, or it may be known from previous experimental data. In this last case, one employs an empirical correlation of data for $\mathscr{F}$ ($\mathscr{G}$). In Sec. 4.4.7, we discuss the form that these empirical correlations should take.

**EXERCISE**

**4.4.5-1** Derive Eqs. (5-15) and (5-16).

### 4.4.6 The integral momentum balance for turbulent flows

As I pointed out in Sec. 4.4.3, some of the most important applications of the integral balances are to systems, portions of which are in turbulent flow. The remarks made there in the context of the integral mass balance are equally applicable here.

## [4.4] INTEGRAL BALANCES

An integral momentum balance that is more appropriate to systems in turbulent flow may be obtained. Let us limit ourselves (for reasons explained in Sec. 4.4.3) to single-phase or multiphase systems that do not involve fluid-fluid phase interfaces. Starting with the time-averaged version of Cauchy's first law found in Sec. 4.1.3,

$$\frac{\partial(\bar{\rho}\bar{\mathbf{v}})}{\partial t} + \text{div}\,(\bar{\rho}\bar{\mathbf{v}}\bar{\mathbf{v}}) = -\nabla\bar{P} + \text{div}\,(\bar{\mathbf{S}} + \mathbf{T}^{(t)}) + \bar{\rho}\mathbf{f} \tag{6-1}$$

the derivation of Sec. 4.4.5 may be repeated to find

$$\frac{d}{dt}\int_{V_{(s)}} \bar{\rho}\bar{\mathbf{v}}\, dV = \int_{S_{(\text{ent ex})}} (\bar{\rho}\bar{\mathbf{v}}\bar{\mathbf{v}} - \overline{\rho\mathbf{v}\mathbf{v}_{(s)}}) \cdot (-\mathbf{n})\, dS$$

$$- \int_{S_{(\text{ent ex})}} (\bar{P} - p_0)\mathbf{n}\, dS - \mathscr{F} + \int_{V_{(s)}} \bar{\rho}\mathbf{f}\, dV$$

$$+ \int_{S_{(\text{sing})}} [\bar{\rho}\bar{\mathbf{v}}\bar{\mathbf{v}} \cdot \boldsymbol{\xi} - \overline{\rho\mathbf{v}u_{(\xi)}} - \mathbf{T} \cdot \boldsymbol{\xi}]\, dS \tag{6-2}$$

where

$$\mathscr{F} \equiv \int_{S_{(s)} - S_{(\text{ent ex})}} (\mathbf{T} + p_0\mathbf{I}) \cdot (-\mathbf{n})\, dS$$

$$= \int_{S_{(s)} - S_{(\text{ent ex})}} (\bar{\mathbf{T}} + p_0\mathbf{I}) \cdot (-\mathbf{n})\, dS \tag{6-3}$$

Here we have noted that, for this case, $S_{(s)} - S_{(\text{ent ex})}$ and $S_{(\text{sing})}$ must be composed of fluid-solid and solid-solid phase interfaces on which the turbulent fluctuations are identically zero. The time-averaged jump momentum balance of Exercise 4.4.6-1 simplifies Eq. (6-2) to

$$\frac{d}{dt}\int_{V_{(s)}} \bar{\rho}\bar{\mathbf{v}}\, dV = \int_{S_{(\text{ent ex})}} (\bar{\rho}\bar{\mathbf{v}}\bar{\mathbf{v}} - \overline{\rho\mathbf{v}\mathbf{v}_{(s)}}) \cdot (-\mathbf{n})\, dS$$

$$- \int_{S_{(\text{ent ex})}} (\bar{P} - p_0)\mathbf{n}\, dS - \mathscr{F} + \int_{V_{(s)}} \bar{\rho}\mathbf{f}\, dV \tag{6-4}$$

This is the form of the *integral momentum balance* recommended for turbulent single-phase or multiphase systems that do not involve fluid-fluid phase interfaces. Note that in arriving at this result we have neglected any viscous or turbulent forces acting at the entrances and exits to the system.

For single-phase or multiphase systems that do not involve fluid-fluid phase interfaces and that are composed only of incompressible materials, we find in a similar manner

$$\frac{d}{dt}\int_{V_{(s)}} \rho\bar{\mathbf{v}}\, dV = \int_{S_{(\text{ent ex})}} \rho(\bar{\mathbf{v}}\bar{\mathbf{v}} - \bar{\mathbf{v}}\mathbf{v}_{(s)}) \cdot (-\mathbf{n})\, dS$$

$$- \int_{S_{(\text{ent ex})}} (\bar{\mathscr{P}} - p_0)\mathbf{n}\, dS - \mathscr{G} + \int_{S_{(\text{sing})}} [\rho]\varphi\boldsymbol{\xi}\, dS \tag{6-5}$$

Here we have

$$\mathcal{G} = \int_{S_{(s)}-S_{(ent\,ex)}} [\mathbf{T} + (p_0 - \rho\varphi)\mathbf{I}] \cdot (-\mathbf{n})\, dS$$

$$= \int_{S_{(s)}-S_{(ent\,ex)}} \overline{[\mathbf{T} + (p_0 - \rho\varphi)\mathbf{I}] \cdot (-\mathbf{n})}\, dS \tag{6-6}$$

For single-phase or multiphase systems that include one or more fluid-fluid phase interfaces, I recommend time averaging the integral momentum balances of Sec. 4.4.5 to find

$$\frac{d}{dt}\overline{\int_{V_{(s)}} \rho\mathbf{v}\,dV} = \overline{\int_{S_{(ent\,ex)}} \rho\mathbf{v}(\mathbf{v}-\mathbf{v}_{(s)})\cdot(-\mathbf{n})\,dS}$$

$$- \overline{\int_{S_{(ent\,ex)}} (P - p_0)\mathbf{n}\,dS} - \overline{\mathscr{F}} + \overline{\int_{V_{(s)}} \rho\mathbf{f}\,dV} \tag{6-7}$$

and, for incompressible materials,

$$\frac{d}{dt}\overline{\int_{V_{(s)}} \rho\mathbf{v}\,dV} = \overline{\int_{S_{(ent\,ex)}} \rho\mathbf{v}(\mathbf{v}-\mathbf{v}_{(s)})\cdot(-\mathbf{n})\,dS}$$

$$- \overline{\int_{S_{(ent\,ex)}} (\mathscr{P} - p_0)\mathbf{n}\,dS} - \overline{\mathscr{G}} \tag{6-8}$$

In these results,

$$\overline{\mathscr{F}} \equiv \int_{S_{(s)}-S_{(ent\,ex)}} \overline{(\mathbf{T} + p_0\mathbf{I})\cdot(-\mathbf{n})}\,dS \tag{6-9}$$

and

$$\overline{\mathscr{G}} \equiv \int_{S_{(s)}-S_{(ent\,ex)}} \overline{[\mathbf{T} + (p_0 - \rho\varphi)\mathbf{I}]\cdot(-\mathbf{n})}\,dS \tag{6-10}$$

Again, Eqs. (6-7) and (6-8) assume that the time-averaged viscous forces acting at the entrances and exits to the system may be neglected.

When we restrict ourselves to single-phase or multiphase systems that do not involve fluid-fluid phase interfaces and when we neglect any turbulent forces acting at the entrances and exits to the system, we can use Eqs. (3-7), (3-8), and (3-9) to show that Eqs. (6-7) to (6-10) reduce to Eqs. (6-4), (6-5), (6-3), and (6-6), respectively.

## EXERCISE

**4.4.6-1** *Time-averaged jump momentum balance* Determine that the time-averaged jump momentum balance applicable to solid-fluid phase interfaces that bound turbulent flows is identical to the balance found in Exercise 2.2.3-1:

$$[\bar{\rho}\overline{\mathbf{v}\mathbf{v}}\cdot\boldsymbol{\xi} - \bar{\rho}\overline{\mathbf{v}}u_{(\xi)} - \mathbf{T}\cdot\boldsymbol{\xi}] = [\rho\mathbf{v}(\mathbf{v}\cdot\boldsymbol{\xi} - u_{(\xi)}) - \mathbf{T}\cdot\boldsymbol{\xi}] = 0$$

## [4.4] INTEGRAL BALANCES

**4.4.7 Empirical correlations for $\mathscr{F}$ and $\mathscr{G}$** In this section, we use two examples to illustrate how empirical data correlations for $\mathscr{F}$ and $\mathscr{G}$ (or, when dealing with turbulent flows, $\overline{\mathscr{F}}$ and $\overline{\mathscr{G}}$), introduced in Sec. 4.4.5, can be formulated. We base this discussion on three points:

1. The forces $\mathscr{F}$ and $\mathscr{G}$ are frame indifferent. For example,

$$\mathscr{G}^* = \int_{S_{(s)}-S_{(\text{ent ex})}} [\mathbf{T}^* + (p_0 - p\varphi)\mathbf{I}] \cdot (-\mathbf{n}^*) \, dS$$

$$= \int_{S_{(s)}-S_{(\text{ent ex})}} \mathbf{Q} \cdot [\mathbf{T} + (p_0 - p\varphi)\mathbf{I}] \cdot (-\mathbf{n}) \, dS$$

$$= \mathbf{Q} \cdot \mathscr{G} \qquad (7\text{-}1)$$

where $\mathbf{Q}$ is a (possibly) time-dependent orthogonal second-order tensor.

2. We assume that the principle of material frame indifference introduced in Sec. 2.3.1 applies to any empirical correlation developed for either $\mathscr{F}$ or $\mathscr{G}$.
3. The Buckingham-Pi theorem serves to further restrict the form of any expression for $\mathscr{F}$ or $\mathscr{G}$.

**EXAMPLE I  FLOW PAST A SPHERE**

As our first example, let us consider a sphere of radius $a$ that is in relative motion with respect to a large body of an incompressible newtonian fluid. In a frame of reference that is fixed with respect to earth, we observe that the sphere translates without rotation at a constant velocity $\mathbf{v}_0$ and that at a very large distance from the sphere the fluid moves with a uniform velocity $\mathbf{v}_\infty$. It seems reasonable to say that $\mathscr{G}$ should be a function of the fluid's density $\rho$ and viscosity $\mu$, the sphere's radius $a$ and the velocity difference $\mathbf{v}_\infty - \mathbf{v}_0$:

$$\mathscr{G} = \mathbf{h}(\rho, \mu, a, \mathbf{v}_\infty - \mathbf{v}_0) \qquad (7\text{-}2)$$

Our reason for choosing $\mathbf{v}_\infty - \mathbf{v}_0$ as an independent variable in this expression, rather than $\mathbf{v}_0$ and $\mathbf{v}_\infty$ separately, is that velocity is not frame indifferent, whereas a velocity difference is (see Exercise 1.2.2-1).

For the moment, let the dependence of $\mathscr{G}$ upon $\rho$, $\mu$, and $a$ be understood and let us concentrate our attention upon $(\mathbf{v}_\infty - \mathbf{v}_0)$:

$$\mathscr{G} = \hat{\mathbf{h}}(\mathbf{v}_\infty - \mathbf{v}_0) \qquad (7\text{-}3)$$

By the principle of material frame indifference, the functional relationship between these variables should be the same in every frame of reference. This means that

$$\mathscr{G}^* = \mathbf{Q} \cdot \mathscr{G} = \mathbf{Q} \cdot \hat{\mathbf{h}}(\mathbf{v}_\infty - \mathbf{v}_0)$$

$$= \hat{\mathbf{h}}[\mathbf{Q} \cdot (\mathbf{v}_\infty - \mathbf{v}_0)] \qquad (7\text{-}4)$$

or $\hat{\mathbf{h}}$ is an isotropic function [1, p. 22]:

$$\hat{\mathbf{h}}(\mathbf{v}_\infty - \mathbf{v}_0) = \mathbf{Q}^T \cdot \hat{\mathbf{h}}[\mathbf{Q} \cdot (\mathbf{v}_\infty - \mathbf{v}_0)] \qquad (7\text{-}5)$$

By a representation theorem for a vector-valued isotropic function of one vector [1, p. 35], we may write

$$\mathscr{G} = \hat{\mathbf{h}}(\mathbf{v}_\infty - \mathbf{v}_0) = \mathscr{G} \frac{\mathbf{v}_\infty - \mathbf{v}_0}{|\mathbf{v}_\infty - \mathbf{v}_0|} \tag{7-6}$$

where it is to be understood that $\mathscr{G} \equiv |\mathscr{G}|$ is a function of the magnitude of the undisturbed fluid relative to the sphere, $|\mathbf{v}_\infty - \mathbf{v}_0|$, as well as a function of $\rho$, $\mu$, and $a$:

$$\mathscr{G} = \mathscr{G}(\rho, \mu, a, |\mathbf{v}_\infty - \mathbf{v}_0|) \tag{7-7}$$

It is customary to express $\mathscr{G}$ in terms of a drag coefficient or friction factor $c$. The drag coefficient or friction factor is introduced in many contexts but it is almost always defined as the ratio of $\mathscr{G}$ to an area $A$ that is characteristic of $S_{(s)} - S_{(\text{ent ex})}$ and to a kinetic energy per unit volume that is characteristic of the flow:

General:

$$c \equiv \frac{\mathscr{G}}{\frac{1}{2}\rho u^2 A} \tag{7-8}$$

For this situation we define $A = \pi a^2$ and $u = |\mathbf{v}_\infty - \mathbf{v}_0|$ so that

Sphere:

$$c \equiv \frac{\mathscr{G}}{(\frac{1}{2}\rho |\mathbf{v}_\infty - \mathbf{v}_0|^2)\pi a^2} \tag{7-9}$$

The drag coefficient for a sphere is, consequently, a dimensionless function

$$c = c(\rho, \mu, a, |\mathbf{v}_\infty - \mathbf{v}_0|) \tag{7-10}$$

By the Buckingham-Pi theorem [2] we find that $c$ is a function of the Reynolds number,

$$c = \hat{c}(N_{\text{Re}}) \tag{7-11}$$

where we define

$$N_{\text{Re}} \equiv \frac{a |\mathbf{v}_\infty - \mathbf{v}_0| \rho}{\mu} \tag{7-12}$$

To summarize, we find for this situation

$$\mathscr{G} = c(\frac{1}{2}\rho |\mathbf{v}_\infty - \mathbf{v}_0|^2)\pi a^2 \frac{\mathbf{v}_\infty - \mathbf{v}_0}{|\mathbf{v}_\infty - \mathbf{v}_0|} \tag{7-13}$$

where the drag coefficient $c$ is to be determined from a correlation of experimental data in the form of Eq. (7-11) [3, p. 17].

If we had wished to develop an empirical correlation for $\mathscr{F}$ instead, we would have had to include the additional dependence of $\mathscr{F}$ upon the external force vector $\mathbf{f}$ (see Exercise 4.4.7-1). Because of this additional complication with constitutive equations for $\mathscr{F}$, it is not common to find them being used in the literature.

## [4.4] INTEGRAL BALANCES

### EXAMPLE 2 PLANE FLOW PAST A CYLINDRICAL BODY

An infinitely long cylindrical body (the surface of which is traced by a straight line moving parallel to a fixed straight line and intersecting a fixed closed curve) is in relative motion with respect to a large body of an incompressible newtonian fluid. In a frame of reference that is fixed with respect to the earth, the cylindrical body translates without rotation at a constant velocity $\mathbf{v}_0$ and the fluid at a very large distance from the body moves with a uniform velocity $\mathbf{v}_\infty$. The vectors $\mathbf{v}_0$ and $\mathbf{v}_\infty$ are normal to the axis of the cylinder so that we may expect that the fluid moves in a plane flow. One unit vector $\boldsymbol{\alpha}$ is sufficient to describe the orientation of the cylinder. Following the discussion of the previous example, we postulate that $\mathscr{G}$ is a function of $\rho$, $\mu$, a length $L$ that is characteristic of the cylinder's cross section, $\mathbf{v}_\infty - \mathbf{v}_0$, and $\boldsymbol{\alpha}$:

$$\mathscr{G} = \mathbf{h}(\rho, \mu, L, \mathbf{v}_\infty - \mathbf{v}_0, \boldsymbol{\alpha}) \tag{7-14}$$

Let us concentrate our attention upon the independent variables $\mathbf{v}_\infty - \mathbf{v}_0$ and $\boldsymbol{\alpha}$:

$$\mathscr{G} = \hat{\mathbf{h}}(\mathbf{v}_\infty - \mathbf{v}_0, \boldsymbol{\alpha}) \tag{7-15}$$

By the principle of material frame indifference, we conclude that $\hat{\mathbf{h}}$ is a vector-valued isotropic function of two vectors:

$$\hat{\mathbf{h}}(\mathbf{v}_\infty - \mathbf{v}_0, \boldsymbol{\alpha}) = \mathbf{Q}^T \cdot \hat{\mathbf{h}}(\mathbf{Q} \cdot [\mathbf{v}_\infty - \mathbf{v}_0], \mathbf{Q} \cdot \boldsymbol{\alpha}) \tag{7-16}$$

Again, $\mathbf{Q}$ is an orthogonal second-order tensor. The representation theorems of Spencer and Rivlin [4, sec. 7] and of Smith [5] tell us that the most general polynomial isotropic vector function of two vectors has the form

$$\mathscr{G} = \hat{\mathbf{h}}(\mathbf{v}_\infty - \mathbf{v}_0, \boldsymbol{\alpha}) = \varphi_{(1)} \frac{\mathbf{v}_\infty - \mathbf{v}_0}{|\mathbf{v}_\infty - \mathbf{v}_0|} + \varphi_{(2)} \boldsymbol{\alpha} \tag{7-17}$$

where $\varphi_{(1)}$ and $\varphi_{(2)}$ are scalar-valued polynomials in $|\mathbf{v}_\infty - \mathbf{v}_0|$ and $(\mathbf{v}_\infty - \mathbf{v}_0) \cdot \boldsymbol{\alpha}$ as well as $\rho$, $\mu$, and $L$. (In applying the theorem of Spencer and Rivlin, we identify a vector $\mathbf{b}$ that has covariant components $b_i$ with the skew-symmetric tensor that has contravariant components $\epsilon^{ijk} b_i$. Their theorem requires an additional term in Eq. (7-17) proportional to the vector product $(\mathbf{v}_\infty - \mathbf{v}_0) \wedge \boldsymbol{\alpha}$. This term is not consistent with the requirement that $\hat{\mathbf{h}}$ be isotropic [1, p. 24] and, consequently, is dropped.) We expect $\varphi_{(2)} = 0$ for $(\mathbf{v}_\infty - \mathbf{v}_0) = 0$ in order that $\mathscr{G} = 0$ in this limit.

Equation (7-17), as such, is not commonly seen in the literature. Rather, $\mathscr{G}$ is expressed as a linear combination of the direction of the relative motion $(\mathbf{v}_\infty - \mathbf{v}_0)/(|\mathbf{v}_\infty - \mathbf{v}_0|)$ and the direction orthogonal to the relative motion $\boldsymbol{\lambda}$:

$$\mathscr{G} = \mathscr{D}\left(\frac{\mathbf{v}_\infty - \mathbf{v}_0}{|\mathbf{v}_\infty - \mathbf{v}_0|}\right) + \mathscr{L}\boldsymbol{\lambda} \tag{7-18}$$

We refer to $\mathscr{D}$ as the *drag* component of $\mathscr{G}$; $\mathscr{L}$ is the *lift* component.

Current practice is to express $\mathscr{D}$ and $\mathscr{L}$ in terms of drag and lift coefficients defined in the manner of Eq. (7-8):

$$c_\mathscr{D} \equiv \frac{\mathscr{D}}{\tfrac{1}{2}\rho |\mathbf{v}_\infty - \mathbf{v}_0|^2 A_\mathscr{D}} \tag{7-19}$$

and

$$c_{\mathscr{L}} = \frac{\mathscr{L}}{\frac{1}{2}\rho \, |\mathbf{v}_\infty - \mathbf{v}_0|^2 \, A_{\mathscr{L}}} \qquad (7\text{-}20)$$

The characteristic areas for drag and lift are not necessarily the same. The area of the body projected on a plane normal to the direction of flow is usually chosen for $A_{\mathscr{D}}$. In the case of airfoil sections, $A_{\mathscr{L}}$ is taken to be the product of the chord length and wing length.

The Buckingham-Pi theorem [2] tells us finally that the dimensionless drag and lift coefficients are functions of two Reynolds numbers:

$$c_{\mathscr{D}} = c_{\mathscr{L}}(N_{\text{Re}(1)}, N_{\text{Re}(2)}) \qquad (7\text{-}21)$$

and

$$c_{\mathscr{L}} = c_{\mathscr{L}}(N_{\text{Re}(1)}, N_{\text{Re}(2)}) \qquad (7\text{-}22)$$

Here

$$N_{\text{Re}(1)} \equiv \frac{L \, |\mathbf{v}_\infty - \mathbf{v}_0| \, \rho}{\mu} \qquad (7\text{-}23)$$

and

$$N_{\text{Re}(2)} \equiv \frac{L[(\mathbf{v}_\infty - \mathbf{v}_0) \cdot \boldsymbol{\alpha}]\rho}{\mu} \qquad (7\text{-}24)$$

To summarize,

$$\mathscr{G} = c_{\mathscr{D}} A_{\mathscr{D}}(\tfrac{1}{2}\rho \, |\mathbf{v}_\infty - \mathbf{v}_0|^2) \frac{|\mathbf{v}_\infty - \mathbf{v}_0|}{\mathbf{v}_\infty - \mathbf{v}_0} + c_{\mathscr{L}} A_{\mathscr{L}}(\tfrac{1}{2}\rho \, |\mathbf{v}_\infty - \mathbf{v}_0|^2)\boldsymbol{\lambda} \qquad (7\text{-}25)$$

where $c_{\mathscr{D}}$ and $c_{\mathscr{L}}$ are to be determined from empirical correlations of experimental data. These empirical correlations should be expected to have the form of Eqs. (7-21) and (7-22) or their equivalent [3, p. 22].

## REFERENCES

1. Truesdell, C., and W. Noll: In S. Flügge (ed.), "Handbuch der Physik," vol. 3/3, Springer-Verlag, Berlin, 1965.
2. Brand, L.: *Arch. Rational Mech. Analysis*, **1**:35 (1957).
3. Schlichting, Hermann: "Boundary-Layer Theory," 6th ed., McGraw-Hill, New York, 1968.
4. Spencer, A. J., and R. S. Rivlin: *Arch. Rational Mech. Analysis*, **4**:214 (1959/60).
5. Smith, G. F.: *Arch. Rational Mech. Analysis*, **18**:282 (1965).

## EXERCISES

**4.4.7-1** Repeat Example 1 to obtain a constitutive equation for $\mathscr{F}$.

**4.4.7-2** Repeat Example 1 for a power-model fluid (see Sec. 2.3.3).

**4.4.7-3** Repeat Example 1 for any body of revolution.

**4.4.7-4** Repeat Example 1 for a spinning sphere. Assume that the orientation of the axis of rotation may be a function of time.

### 4.4.8 The integral momentum balance: An example

A liquid-fueled rocket engine is operating at steady-state conditions on a stationary test stand. Oxidizer and fuel react at a high pressure in the combustion chamber to form gases that exit through the nozzle at a time-averaged pressure $\bar{P}_e$, which is in the neighborhood of atmospheric pressure $p_0$, and at a supersonic time-averaged velocity $\bar{\mathbf{v}}_e$. We wish to calculate the thrust developed by the rocket engine, if the mass flow of oxidizer and fuel to the engine is $q$.

Referring to Fig. 4.4.8-1, let us choose our system to be the self-contained rocket engine and the exhaust gases in the exit gas stream out to the cross section where the velocity $\bar{\mathbf{v}}_e$ is measured. We assume that at this exit cross section the velocity of the gas is uniform in magnitude and direction. By the thrust of the rocket engine we mean the magnitude of the force $\mathscr{F}$ that the rocket engine exerts upon the test stand beyond simply the weight of the engine and its fuel tank (and of course beyond the force attributable to atmospheric pressure). Our object is to compute the rectangular cartesian component $\mathscr{F}_x$.

Since a portion of this system is in turbulent flow, let us employ the time-averaged integral momentum balance that is applicable to any multiphase system. Note that, because $S_{(s)} - S_{(\text{ent ex})}$ is composed of entirely fluid-solid and solid-solid phase interfaces,

$$\mathscr{F}_x = \overline{\mathscr{F}_x} \qquad (8\text{-}1)$$

If we are willing to neglect the effect of the turbulent Reynolds stress at the exit, the $x$ component of Eq. (6-7) reduces to

$$\mathscr{F}_x = -\overline{\int_{S_{(\text{ex})}} \rho v_x (\mathbf{v} \cdot \mathbf{n})\, dS} - \overline{\int_{S_{(\text{ex})}} (P - p_0) n_x\, dS}$$

$$\doteq -\bar{v}_e \int_{S_{(\text{ex})}} \overline{\rho \mathbf{v}} \cdot \mathbf{n}\, dS - (\bar{P}_e - p_0) S_{(\text{ex})}$$

$$\doteq -\bar{v}_e q - (\bar{P}_e - p_0) S_{(\text{ex})} \qquad (8\text{-}2)$$

where we have identified

$$q \equiv \overline{\int_{S_{(\text{ex})}} \rho \mathbf{v} \cdot \mathbf{n}\, dS} = \int_{S_{(\text{ex})}} \overline{\rho \mathbf{v}} \cdot \mathbf{n}\, dS \qquad (8\text{-}3)$$

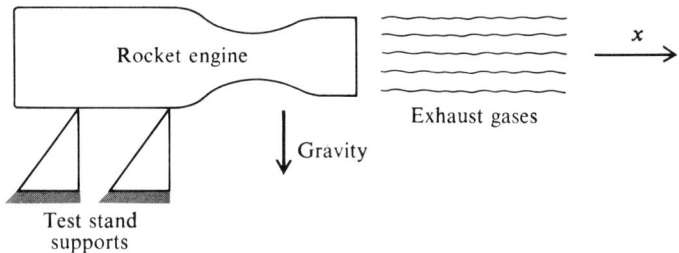

**Fig. 4.4.8-1** Static test of rocket engine.

**236** APPLICATION OF INTEGRAL AVERAGING TECHNIQUES TO MOMENTUM TRANSFER

Often the second term on the right of Eq. (8-2) may be neglected with respect to the first to find

$$\mathscr{F}_x \doteq -\bar{v}_e q \tag{8-4}$$

### EXERCISES

**4.4.8-1** *The Borda mouthpiece*  A tank has an orifice as shown in Fig. 4.4.8-2. With such an orifice, there is relatively little movement of the fluid in the neighborhood of the tank walls. Neglecting any effects of friction and assuming the fluid is incompressible, find the contraction coefficient for this nozzle $A_j/A_0$. This orifice is referred to as a *Borda mouthpiece*.

*Hint:* The Bernoulli equation is useful here.

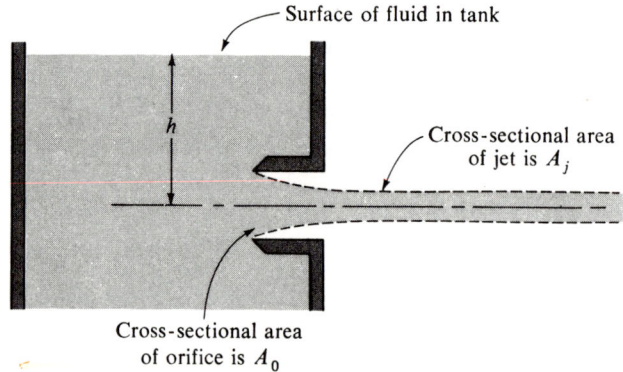

**Fig. 4.4.8-2**  The Borda mouthpiece.

**4.4.8-2** *Deflection of a two-dimensional jet*  (a) A two-dimensional jet of incompressible fluid strikes a plane surface. Referring to Fig. 4.4.8-3, let $v_i$ ($i = 1, 2, 3$) denote the average magnitude of the velocity of stream $i$ ($i = 1, 2, 3$) and $b_i$ the corresponding width of the stream. Neglecting viscous

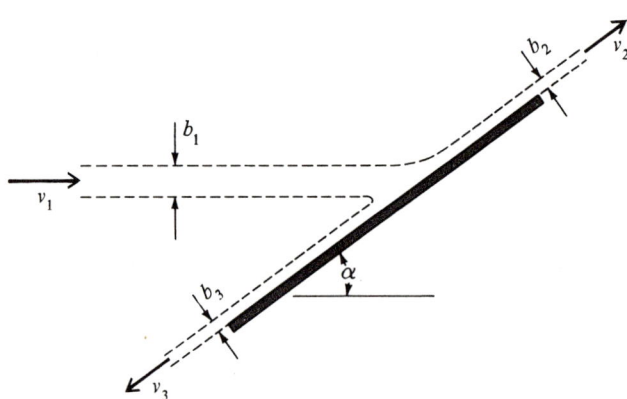

**Fig. 4.4.8-3**  Deflection of a two-dimensional jet by a flat plate.

## [4.4] INTEGRAL BALANCES

effects and the effect of gravity, find $v_2$, $v_3$, $b_2$, and $b_3$ in terms of the density of the fluid $\rho$, $v_1$, $b_1$, and $\alpha$.

(b) Find the direction and magnitude of the force per unit width required to hold the plate stationary.

**4.4.8-3** *An accelerating rocket* At time $t = 0$ a rocket, the initial mass of which is $M_0$, is ignited. It accelerates along a straight line in the opposite direction to the action of gravity. The rocket consumes propellant at a constant rate $G$ (mass per unit time). The speed of the exhaust gas with respect to the accelerating rocket is $v_e$. Determine the speed of the rocket as a function of time.

**4.4.8-4** *More on emptying a vertical tube* Do Exercise 3.4.4-4 using the integral momentum balance.

**4.4.8-5** *The force on a dam* Referring to Fig. 4.4.8-4, estimate the $x$ component of the force that the fluid exerts upon the dam per unit width of the dam. You may neglect viscous effects.

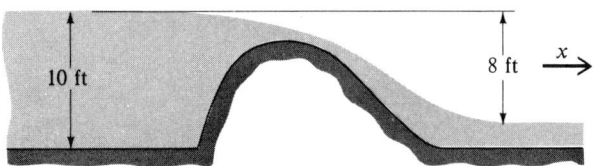

**Fig. 4.4.8-4** The force on a dam.

**4.4.8-6** *A moving frame of reference* (a) Determine that the form of the integral mass balance is independent of the frame of reference.

(b) Prove that the form of the integral momentum balance is unchanged for any frame of reference that moves at a constant velocity with respect to the fixed stars.

(c) Prove that Bernoulli's equation is unchanged for any frame of reference that moves at a constant velocity with respect to the fixed stars.

**4.4.8-7** *Force on a moving plate* [1, p. 276] Figure 4.4.8-5 shows a cylindrical jet of water issuing from a 1-in-ID nozzle at a speed of 30 ft/s. It strikes a flat plate that is moving away from the jet at a speed of 17 ft/s. What force does the jet exert on the plate?

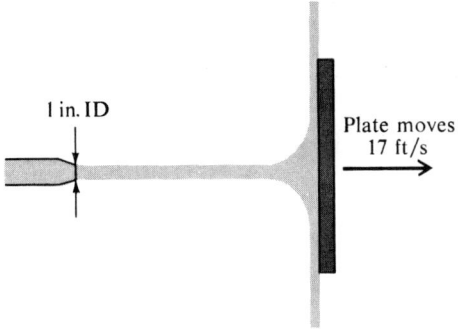

**Fig. 4.4.8-5** Force on a moving plate.

*Answer:* 1.78 lb$_f$.

*Hint:* Choose a system and a frame of reference that are fixed with respect to the plate. See Exercise 4.4.8-6.

**4.4.8-8** *More about the force on a moving plate* Repeat Exercise 4.4.8-2 assuming now that the plate moves to the right at a constant speed $v_p$.

**238**  APPLICATION OF INTEGRAL AVERAGING TECHNIQUES TO MOMENTUM TRANSFER

In carrying out this calculation, let $v_1$, $v_2$, and $v_3$ denote speeds measured with respect to a stationary frame of references and let $\tilde{v}_1$, $\tilde{v}_2$, and $\tilde{v}_3$ signify speeds measured with respect to the moving frame of reference.

**4.4.8-9** *A fire hose* Does the nozzle on a fire hose place the hose in tension or compression? Derive an expression for the force that the fluid exerts upon the nozzle that confirms your intuitive opinion.

### 4.4.9 The mechanical energy balance for incompressible materials

In Sec. 3.4.2, we developed the Bernoulli equation by means of which we were able to relate differences in pressure between two points in a fluid to differences in kinetic and potential energy. But the use of the Bernoulli equation is restricted to situations such that the viscous terms may be neglected in Cauchy's first law. Let us try to develop an equation that tells us something about the kinetic energy of an arbitrary system without making any assumption about viscous effects.

While we will find that the practical results from this section are for incompressible fluids, I suggest that we refrain from imposing this restriction until we are nearly finished. In this way, our intermediate results will be directly applicable to a further discussion of the integral mechanical energy balance in Sec. 7.4.6.

We can begin with the same approach that we used in arriving at the integral mass and momentum balances. In the generalized transport theorem of Sec. 1.3.2, let us identify $\Psi = \frac{1}{2}\rho v^2$, kinetic energy per unit volume, to obtain

$$\frac{d}{dt}\int_{V_{(s)}} \tfrac{1}{2}\rho v^2 \, dV = \int_{V_{(s)}} \frac{\partial}{\partial t}(\tfrac{1}{2}\rho v^2) \, dV + \int_{S_{(s)}} \tfrac{1}{2}\rho v^2 (\mathbf{v}_{(s)} \cdot \mathbf{n}) \, dS \qquad (9\text{-}1)$$

But in any given problem the value of the first integral on the right will not be immediately obvious. The difficulty is essentially the same as those we encountered in Secs. 4.4.2 and 4.4.5.

Let us take the scalar product of Cauchy's first law with the velocity vector

$$\mathbf{v} \cdot \left(\rho \frac{d_{(m)}\mathbf{v}}{dt} - \operatorname{div} \mathbf{T} - \rho \mathbf{f}\right) = 0 \qquad (9\text{-}2)$$

and rearrange the first term that appears on the left:

$$\rho \mathbf{v} \cdot \frac{d_{(m)}\mathbf{v}}{dt} = \rho \frac{d_{(m)}}{dt}(\tfrac{1}{2}v^2)$$

$$= \frac{d_{(m)}}{dt}(\tfrac{1}{2}\rho v^2) - \tfrac{1}{2}v^2 \frac{d_{(m)}\rho}{dt}$$

$$= \frac{d_{(m)}}{dt}(\tfrac{1}{2}\rho v^2) + \tfrac{1}{2}\rho v^2 \operatorname{div} \mathbf{v}$$

$$= \frac{\partial}{\partial t}(\tfrac{1}{2}\rho v^2) + \operatorname{div}(\tfrac{1}{2}\rho v^2 \mathbf{v}) \qquad (9\text{-}3)$$

## [4.4] INTEGRAL BALANCES

This result, together with Green's transformation, may be used to express the integral over $V_{(s)}$ of Eq. (9-2) as

$$\int_{V_{(s)}} \frac{\partial}{\partial t}(\tfrac{1}{2}\rho v^2)\, dV = -\int_{S_{(s)}} \tfrac{1}{2}\rho v^2 (\mathbf{v}\cdot\mathbf{n})\, dS + \int_{V_{(s)}} (\mathbf{v}\cdot\operatorname{div}\mathbf{T} + \rho\mathbf{v}\cdot\mathbf{f})\, dV \quad (9\text{-}4)$$

which may be used to eliminate the first term on the right of Eq. (9-1). Before doing so, let us rearrange the last two integrals on the right of Eq. (9-4).

The second integral on the right of Eq. (9-4) has no direct physical significance as it stands, which suggests that we rearrange it by an application of Green's transformation. In terms of a rectangular cartesian coordinate system, we have

$$\int_{V_{(s)}} \mathbf{v}\cdot\operatorname{div}\mathbf{T}\, dV = \int_{V_{(s)}} v_i \frac{\partial T_{ij}}{\partial z_j}\, dV$$

$$= \int_{V_{(s)}} \left[\frac{\partial}{\partial z_j}(v_i T_{ij}) - T_{ij}\frac{\partial v_i}{\partial z_j}\right] dV$$

$$= \int_{S_{(s)}} v_i T_{ij} n_j\, dS - \int_{V_{(s)}} \operatorname{tr}(\mathbf{T}^T\cdot\nabla\mathbf{v})\, dV$$

$$= \int_{S_{(s)}} \mathbf{v}\cdot(\mathbf{T}\cdot\mathbf{n})\, dS - \int_{V_{(s)}} \operatorname{tr}(\mathbf{T}\cdot\nabla\mathbf{v})\, dV \quad (9\text{-}5)$$

Note that we use the symmetry of the stress tensor in arriving at this last expression. We find it convenient to observe further that

$$\int_{V_{(s)}} \operatorname{tr}(\mathbf{T}\cdot\nabla\mathbf{v})\, dV = \int_{V_{(s)}} \operatorname{tr}([-P\mathbf{I}+\mathbf{S}]\cdot\nabla\mathbf{v})\, dV$$

$$= -\int_{V_{(s)}} P\operatorname{div}\mathbf{v}\, dV + \int_{V_{(s)}} \operatorname{tr}(\mathbf{S}\cdot\nabla\mathbf{v})\, dV \quad (9\text{-}6)$$

This allows us to express Eq. (9-5)$_4$ as

$$\int_{V_{(s)}} \mathbf{v}\cdot\operatorname{div}\mathbf{T}\, dV = \int_{S_{(s)}} \mathbf{v}\cdot(\mathbf{T}\cdot\mathbf{n})\, dS + \int_{V_{(s)}} P\operatorname{div}\mathbf{v}\, dV$$

$$-\int_{V_{(s)}} \operatorname{tr}(\mathbf{S}\cdot\nabla\mathbf{v})\, dV \quad (9\text{-}7)$$

The third term on the right of Eq. (9-4) may be left unchanged for many purposes. A somewhat more familiar form is obtained if the external force can be expressed in terms of a potential energy per unit mass $\varphi$:

$$\mathbf{f} = -\nabla\varphi \quad (9\text{-}8)$$

This means that

$$\int_{V_{(s)}} \rho \mathbf{v} \cdot \mathbf{f}\, dV = -\int_{V_{(s)}} \rho \mathbf{v} \cdot \nabla \varphi\, dV$$

$$= -\int_{V_{(s)}} \operatorname{div}(\rho \varphi \mathbf{v})\, dV + \int_{V_{(s)}} \varphi \operatorname{div}(\rho \mathbf{v})\, dV$$

$$= -\int_{S_{(s)}} \rho \varphi (\mathbf{v} \cdot \mathbf{n})\, dS + \int_{V_{(s)}} \varphi \operatorname{div}(\rho \mathbf{v})\, dV \qquad (9\text{-}9)$$

From the equation of continuity,

$$\int_{V_{(s)}} \varphi \operatorname{div}(\rho \mathbf{v})\, dV = -\int_{V_{(s)}} \varphi \frac{\partial \rho}{\partial t}\, dV \qquad (9\text{-}10)$$

Let us *restrict* ourselves to an external force potential that is independent of time at a fixed position. For most situations, including the case where gravity is the only external force, this is completely satisfactory. We are now in position for an application of the generalized transport theorem:

$$\int_{V_{(s)}} \varphi \operatorname{div}(\rho \mathbf{v})\, dV = -\int_{V_{(s)}} \frac{\partial (\rho \varphi)}{\partial t}\, dV$$

$$= -\frac{d}{dt}\int_{V_{(s)}} \rho \varphi\, dV + \int_{S_{(s)}} \rho \varphi (\mathbf{v}_{(s)} \cdot \mathbf{n})\, dS \qquad (9\text{-}11)$$

Equations (9-9) and (9-11) give

$$\int_{V_{(s)}} \rho \mathbf{v} \cdot \mathbf{f}\, dV = -\frac{d}{dt}\int_{V_{(s)}} \rho \varphi\, dV - \int_{S_{(s)}} \rho \varphi (\mathbf{v} - \mathbf{v}_{(s)}) \cdot \mathbf{n}\, dS \qquad (9\text{-}12)$$

We may use Eqs. (9-4), (9-7), and (9-12) to express Eq. (9-1) as

$$\frac{d}{dt}\int_{V_{(s)}} \rho(\tfrac{1}{2}v^2 + \varphi)\, dV = \int_{S_{(s)}} \rho(\tfrac{1}{2}v^2 + \varphi)(\mathbf{v} - \mathbf{v}_{(s)}) \cdot (-\mathbf{n})\, dS$$

$$+ \int_{V_{(s)}} P \operatorname{div} \mathbf{v}\, dV - \int_{S_{(s)}} \mathbf{v} \cdot [\mathbf{T} \cdot (-\mathbf{n})]\, dS - \mathscr{E} \qquad (9\text{-}13)$$

where

$$\mathscr{E} \equiv \int_{V_{(s)}} \operatorname{tr}(\mathbf{S} \cdot \nabla \mathbf{v})\, dV \qquad (9\text{-}14)$$

The third term on the right of Eq. (9-13) is to be interpreted as the rate at which work is done by the system on the surroundings. Intuitively, we realize that there should be no work due to the action of the ambient pressure $p_0$. As an obvious

## [4.4] INTEGRAL BALANCES

extension of our discussion in Sec. 4.4.5, we may write Eq. (9-13) as

$$\frac{d}{dt}\int_{V_{(s)}} \rho(\tfrac{1}{2}v^2 + \varphi)\, dV = \int_{S_{(ent\ ex)}} \rho(\tfrac{1}{2}v^2 + \varphi)(\mathbf{v} - \mathbf{v}_{(s)}) \cdot (-\mathbf{n})\, dS$$

$$+ \int_{V_{(s)}} (P - p_0)\, \mathrm{div}\, \mathbf{v}\, dV - \int_{S_{(s)}} \mathbf{v} \cdot [(\mathbf{T} + p_0\mathbf{I}) \cdot (-\mathbf{n})]\, dS - \mathscr{E} \quad (9\text{-}15)$$

or

$$\frac{d}{dt}\int_{V_{(s)}} \rho(\tfrac{1}{2}v^2 + \varphi)\, dV = \int_{S_{(ent\ ex)}} \rho\left(\tfrac{1}{2}v^2 + \varphi + \frac{P - p_0}{\rho}\right)(\mathbf{v} - \mathbf{v}_{(s)}) \cdot (-\mathbf{n})\, dS$$

$$+ \int_{V_{(s)}} (P - p_0)\, \mathrm{div}\, \mathbf{v}\, dV - \mathscr{W} - \mathscr{E}$$

$$+ \int_{S_{(ent\ ex)}} [-(P - p_0)\mathbf{v}_{(s)} \cdot \mathbf{n} + \mathbf{v} \cdot (\mathbf{S} \cdot \mathbf{n})]\, dS \quad (9\text{-}16)$$

where we define

$$\mathscr{W} \equiv \int_{S_{(s)} - S_{(ent\ ex)}} \mathbf{v} \cdot [(\mathbf{T} + p_0\mathbf{I}) \cdot (-\mathbf{n})]\, dS \quad (9\text{-}17)$$

We should think of $\mathscr{W}$ as the rate at which work is done by the system on the surroundings at the moving impermeable surfaces of the system (beyond any work done on these surfaces by the ambient pressure $p_0$). Equation (9-16) can be thought of as one *general form of the mechanical energy balance*.

The difficulty with Eq. (9-16) is that the value of the second integral on the right is not immediately obvious for most situations. More will be said on this point in Sec. 7.4.6.

For the moment, we are primarily interested in the special case of an incompressible fluid. By the equation of continuity,

$$\int_{V_{(s)}} (P - p_0)\, \mathrm{div}\, \mathbf{v}\, dV = 0 \quad (9\text{-}18)$$

and Eq. (9-16) reduces to

$$\frac{d}{dt}\int_{V_{(s)}} \rho(\tfrac{1}{2}v^2 + \varphi)\, dV = \int_{S_{(ent\ ex)}} \rho\left(\tfrac{1}{2}v^2 + \varphi + \frac{P - p_0}{\rho}\right)(\mathbf{v} - \mathbf{v}_{(s)}) \cdot (-\mathbf{n})\, dS$$

$$- \mathscr{W} - \mathscr{E} + \int_{S_{(ent\ ex)}} [-(p - p_0)\mathbf{v}_{(s)} \cdot \mathbf{n} + \mathbf{v} \cdot (\mathbf{S} \cdot \mathbf{n})]\, dS \quad (9\text{-}19)$$

Notice that in writing this last, we recognize that the pressure being used can no longer be the thermodynamic pressure $P$ (see Sec. 2.3.2). We shall refer to this equation as a general form of the *mechanical energy balance* for incompressible materials.

In words, the term on the left of Eq. (9-19) denotes the time rate of change of the kinetic and potential energy (often referred to as the *mechanical energy*) associated with the system. On the right of this equation,

$$\int_{S_{(\text{ent ex})}} \rho(\tfrac{1}{2}v^2 + \varphi)(\mathbf{v} - \mathbf{v}_{(s)}) \cdot (-\mathbf{n})\, dS$$

is the net rate at which kinetic and potential energy is brought into the system with any material that moves across the boundary. By

$$\int_{S_{(\text{ent ex})}} \rho\left(\frac{p - p_0}{\rho}\right)(\mathbf{v} - \mathbf{v}_{(s)}) \cdot (-\mathbf{n})\, dS + \int_{S_{(\text{ent ex})}} -(p - p_0)\mathbf{v}_{(s)} \cdot \mathbf{n}\, dS$$

$$= \int_{S_{(\text{ent ex})}} (p - p_0)(-\mathbf{v} \cdot \mathbf{n})\, dS$$

we mean the net rate at which pressure forces (beyond the reference or ambient pressure $p_0$) do work on the system at the entrances and exists. We have already defined $\mathscr{W}$ to be the rate at which work is done by the system on the surroundings at the moving impermeable surfaces of the system (beyond any work done on the surfaces by the ambient pressure $p_0$); see Eq. (9-17). The rate at which mechanical energy is dissipated by the action of viscous forces is denoted by $\mathscr{E}$, defined by Eq. (9-14). We will see in our discussion of the energy balance in Sec. 5.3.2 that we may also interpret $\mathscr{E}$ as the rate of production of internal energy by the action of viscous forces. The last term on the right

$$\int_{S_{(\text{ent ex})}} \mathbf{v} \cdot (\mathbf{S} \cdot \mathbf{n})\, dS$$

represents the rate at which work is done on the system at the entrances and exits by the viscous forces.

One usually attempts to choose a system for an application of the mechanical energy balance in such a way that work done by viscous forces at entrances and exits may be neglected. In this way, we may write Eq. (9-19) as

$$\frac{d}{dt}\int_{V_{(s)}} \rho(\tfrac{1}{2}v^2 + \varphi)\, dV = \int_{S_{(\text{ent ex})}} \rho\left[\tfrac{1}{2}v^2 + \varphi + \frac{(p - p_0)}{\rho}\right](\mathbf{v} - \mathbf{v}_{(s)}) \cdot (-\mathbf{n})\, dS$$

$$- \mathscr{W} - \mathscr{E} - \int_{S_{(\text{ent ex})}} (p - p_0)(\mathbf{v}_{(s)} \cdot \mathbf{n})\, dS \qquad (9\text{-}20)$$

This would be rigorous if the entrances and exits could be chosen as normal to the motion in very long straight pipes in which the flow is laminar (see Exercise 4.4.9-3).

As Eq. (9-20) has been derived, it is applicable only to a single-phase system. We are more commonly concerned with multiphase systems. Using the approach and notation of Sec. 4.4.2 and neglecting the work done by viscous forces at entrances and exits, we find that the *mechanical energy balance* for a multiphase system of

## [4.4] INTEGRAL BALANCES

incompressible materials is

$$\frac{d}{dt}\int_{V_{(s)}} \rho(\tfrac{1}{2}v^2 + \varphi)\,dV$$

$$= \int_{S_{(\text{ent ex})}} \rho\left[\tfrac{1}{2}v^2 + \varphi + \frac{(p-p_0)}{\rho}\right](\mathbf{v}-\mathbf{v}_{(s)})\cdot(-\mathbf{n})\,dS$$

$$-\mathscr{W} - \mathscr{E} - \int_{S_{(\text{ent ex})}} (p-p_0)(\mathbf{v}_{(s)}\cdot\mathbf{n})\,dS$$

$$+ \int_{S_{(\text{sing})}} [\rho(\tfrac{1}{2}v^2 + \varphi)(\mathbf{v}\cdot\boldsymbol{\xi} - u_{(\xi)}) - \mathbf{v}\cdot[(\mathbf{T}+p_0\mathbf{I})\cdot\boldsymbol{\xi}]]\,dS \quad (9\text{-}21)$$

When there is no mass transfer across internal phase interfaces and when the jump mass and momentum balances of Sec. 1.3.5 and Exercise 2.2.3-1 apply, this reduces to Eq. (9-20). For this reason, it is Eq. (9-20) with which we will be primarily concerned.

There are three common categories of problems in which the integral mechanical energy balances apply: $\mathscr{E}$ may be neglected, it may be the unknown to be determined, or it may be known from previous experimental data. In this last case, one employs an empirical correlation of data for $\mathscr{E}$. In Sec. 4.4.11 we discuss the form that these empirical data correlations should take.

For more about the mechanical energy balance, see Sec. 7.4.6. An extensive compilation of alternative forms that it can take is given in Tables 7.4.6-1 to 7.4.6-3.

### EXERCISES

**4.4.9-1** Derive Eq. (9-18).

**4.4.9-2** *Relation to Bernoulli's equation* Show that for a steady-state flow through a system with one entrance and one exit, Eq. (9-16) reduces to a form that is similar to Bernoulli's equation.

**4.4.9-3** Prove that, if the entrances and exits are located in very long straight pipes in which the flow is laminar,

$$\int_{S_{(\text{ent ex})}} \mathbf{v}\cdot(\mathbf{S}\cdot\mathbf{n})\,dS = 0$$

### 4.4.10 The mechanical energy balance for incompressible materials in turbulent flow

The remarks made in Sec. 4.4.3 when developing the integral mass balance for turbulent flows are equally applicable here.

A mechanical energy balance that is appropriate to systems in turbulent flow can be derived using much the same approach as we used in Sec. 4.4.9. Let us limit ourselves (for reasons explained in Sec. 4.4.3) to single-phase or multiphase systems that do not involve fluid-fluid phase interfaces. We shall continue to assume that the external force can be expressed in terms of a potential energy per unit mass $\varphi$:

$$\mathbf{f} = -\nabla\varphi \qquad (10\text{-}1)$$

## 244 APPLICATION OF INTEGRAL AVERAGING TECHNIQUES TO MOMENTUM TRANSFER

Under these conditions, the scalar product of the time-averaged velocity vector with the time average of Cauchy's first law (see Sec. 4.1.2),

$$\bar{\mathbf{v}} \cdot \left[ \frac{\partial(\rho \bar{\mathbf{v}})}{\partial t} + \text{div} \, (\rho \bar{\mathbf{v}} \bar{\mathbf{v}}) - \text{div} \, (\bar{\mathbf{T}} + \mathbf{T}^{(t)}) + \rho \, \nabla \varphi \right] = 0 \qquad (10\text{-}2)$$

can be rearranged to read

$$\frac{\partial}{\partial t} \left[ \rho(\tfrac{1}{2}\bar{v}^2 + \varphi) \right] + \text{div} \, (\rho [\tfrac{1}{2}\bar{v}^2 + \varphi] \bar{\mathbf{v}})$$
$$- \text{div} \, ([\bar{\mathbf{T}} + \mathbf{T}^{(t)} + p_0 \mathbf{I}] \cdot \bar{\mathbf{v}}) + \text{tr} \, ([\bar{\mathbf{S}} + \mathbf{T}^{(t)}] \cdot \nabla \bar{\mathbf{v}}) = 0 \qquad (10\text{-}3)$$

Here $\mathbf{T}^{(t)}$ is the Reynolds stress tensor, which for an incompressible fluid takes the form

$$\mathbf{T}^{(t)} = \rho(\bar{\mathbf{v}}\bar{\mathbf{v}} - \overline{\mathbf{v}\mathbf{v}}) \qquad (10\text{-}4)$$

Equation (10-3) may be integrated over the region occupied by the system to obtain

$$\frac{d}{dt} \int_{V_{(s)}} \rho(\tfrac{1}{2}\bar{v}^2 + \varphi) \, dV$$
$$= \int_{S_{(\text{ent ex})}} \rho \left[ \tfrac{1}{2}\bar{v}^2 + \varphi + \frac{(\bar{p} - p_0)}{\rho} \right] (\bar{\mathbf{v}} - \mathbf{v}_{(s)}) \cdot (-\mathbf{n}) \, dS$$
$$- \int_{S_{(\text{ent ex})}} (\bar{p} - p_0)(\mathbf{v}_{(s)} \cdot \mathbf{n}) \, dS - \mathscr{W} - \mathscr{E}^{(t)}$$
$$+ \int_{S_{(\text{sing})}} [\rho(\tfrac{1}{2}\bar{v}^2 + \varphi)(\bar{\mathbf{v}} \cdot \boldsymbol{\xi} - u_{(\xi)}) - \bar{\mathbf{v}} \cdot [(\bar{\mathbf{T}} + p_0 \mathbf{I}) \cdot \boldsymbol{\xi}]] \, dS \qquad (10\text{-}5)$$

where

$$\mathscr{W} \equiv \int_{S_{(s)} - S_{(\text{ent ex})}} \mathbf{v} \cdot [(\mathbf{T} + p_0 \mathbf{I}) \cdot (-\mathbf{n})] \, dS$$
$$= \int_{S_{(s)} - S_{(\text{ent ex})}} \bar{\mathbf{v}} \cdot [(\bar{\mathbf{T}} + p_0 \mathbf{I}) \cdot (-\mathbf{n})] \, dS \qquad (10\text{-}6)$$

and

$$\mathscr{E}^{(t)} \equiv \int_{V_{(s)}} \text{tr}[(\bar{\mathbf{S}} + \mathbf{T}^{(t)}) \cdot \nabla \bar{\mathbf{v}}] \, dV \qquad (10\text{-}7)$$

In arriving at this result, we have neglected the rate at which work is done by viscous and turbulent forces on the entrances and exits of the system and we have noted that for this case $S_{(s)} - S_{(\text{ent ex})}$ and $S_{(\text{sing})}$ must be composed of fluid-solid and solid-solid phase interfaces on which the turbulent fluctuations are identically zero. When there is no mass transfer across internal phase interfaces, the jump mass and

## [4.4] INTEGRAL BALANCES

momentum balances of Exercises 4.4.3-1 and 4.4.6-1 allow Eq. (10-5) to be written as

$$\frac{d}{dt}\int_{V_{(s)}} \rho(\tfrac{1}{2}\bar{v}^2 + \varphi)\, dV$$

$$= \int_{S_{(ent\ ex)}} \rho\left[\tfrac{1}{2}\bar{v}^2 + \varphi + \frac{(\bar{p} - p_0)}{\rho}\right](\bar{\mathbf{v}} - \mathbf{v}_{(s)}) \cdot (-\mathbf{n})\, dS$$

$$- \int_{S_{(ent\ ex)}} (\bar{p} - p_0)(\mathbf{v}_{(s)} \cdot \mathbf{n})\, dS - \mathscr{W} - \mathscr{E}^{(t)} \quad (10\text{-}8)$$

For single-phase or multiphase systems that include one or more fluid-fluid phase interfaces, I recommend time averaging the mechanical energy balance of Sec. 4.4.9 to find

$$\frac{d}{dt}\int_{V_{(s)}} \rho(\tfrac{1}{2}v^2 + \varphi)\, dV$$

$$= \overline{\int_{S_{(ent\ ex)}} \rho\left[\tfrac{1}{2}v^2 + \varphi + \frac{(p - p_0)}{\rho}\right](\mathbf{v} - \mathbf{v}_{(s)}) \cdot (-\mathbf{n})\, dS}$$

$$- \overline{\int_{S_{(ent\ ex)}} (p - p_0)(\mathbf{v}_{(s)} \cdot \mathbf{n})\, dS} - \overline{\mathscr{W}} - \overline{\mathscr{E}}$$

$$+ \overline{\int_{S_{(sing)}} [\rho(\tfrac{1}{2}v^2 + \varphi)(\mathbf{v} \cdot \boldsymbol{\xi} - u_{(\xi)}) - \mathbf{v} \cdot [(\mathbf{T} + p_0\mathbf{I}) \cdot \boldsymbol{\xi}]]\, dS} \quad (10\text{-}9)$$

where

$$\mathscr{W} \equiv \int_{S_{(s)} - S_{(ent\ ex)}} \mathbf{v} \cdot [(\mathbf{T} + p_0\mathbf{I}) \cdot (-\mathbf{n})]\, dS \quad (10\text{-}10)$$

and

$$\mathscr{E} \equiv \int_{V_{(s)}} \text{tr}\, (\mathbf{S} \cdot \nabla \mathbf{v})\, dV \quad (10\text{-}11)$$

When there is no mass transfer across internal phase interfaces and when the jump mass and momentum balances of Sec. 1.3.5 and Exercise 2.2.3-1 apply, this reduces to

$$\frac{d}{dt}\int_{V_{(s)}} \rho(\tfrac{1}{2}v^2 + \varphi)\, dV$$

$$= \overline{\int_{S_{(ent\ ex)}} \rho\left[\tfrac{1}{2}v^2 + \varphi + \frac{(p - p_0)}{\rho}\right](\mathbf{v} - \mathbf{v}_{(s)}) \cdot (-\mathbf{n})\, dS}$$

$$- \overline{\int_{S_{(ent\ ex)}} (p - p_0)(\mathbf{v}_{(s)} \cdot \mathbf{n})\, dS} - \overline{\mathscr{W}} - \overline{\mathscr{E}} \quad (10\text{-}12)$$

When we restrict ourselves to single-phase or multiphase systems that do not involve fluid-fluid phase interfaces and when we neglect the rate at which work is done by turbulent forces on the entrances and exits of the system, we can use Eqs. (3-7), (3-8), and (3-9) to show that Eqs. (10-9) to (10-12) reduce to Eqs. (10-5) to (10-8), respectively.

**4.4.11 Empirical correlations for $\mathscr{E}$** In formulating empirical data correlations for $\mathscr{E}$ ($\overline{\mathscr{E}}$ when concerned with turbulent flows), one proceeds in much the same manner as we described in Sec. 4.4.5, where we discussed correlations for $\mathscr{F}$ and $\mathscr{G}$. Three principal thoughts should be kept in mind.

1. The total rate of dissipation of mechanical energy is frame indifferent:

$$\mathscr{E}^* \equiv \int_{V_{(s)}} \text{tr}\,(\mathbf{S}^* \cdot \nabla \mathbf{v}^*)\, dV$$
$$= \int_{V_{(s)}} \text{tr}\,(\mathbf{S} \cdot \nabla \mathbf{v})\, dV$$
$$= \mathscr{E} \qquad (11\text{-}1)$$

2. We assume that the principle of material frame indifference introduced in Sec. 2.3.1 applies to any empirical correlation developed for $\mathscr{E}$.
3. The form of any expression for $\mathscr{E}$ must satisfy the Buckingham-Pi theorem [1].

The most common class of problems where we employ correlations for $\mathscr{E}$ includes flow through a conduit and flow through a wide range of pipe or tubing fittings (valves, elbows, tees, nozzles, etc.). As an illustration of the general approach, let us consider the flow of an incompressible newtonian fluid through a valve (of a specified design) mounted in a run of tubing on a jet aircraft.

Let us say that we have already described the contribution to $\mathscr{E}$ as a result of viscous dissipation in the tubing. We now wish to consider the additional contribution to $\mathscr{E}$ resulting from the presence of the valve. Let us define

$$u \equiv \frac{4G}{\rho \pi D^2} \qquad (11\text{-}2)$$

where $D$ denotes the inside diameter of the tubing, $\rho$ the density of the fluid, and $G$ the mass flow rate of fluid through the tube. (Note that $G$, and therefore $u$ as well, is a frame-indifferent scalar, since it is based upon a velocity measured with respect to the tubing, which is a difference of velocities. See Exercise 1.2.2-1.) If $\mu$ denotes the viscosity of the fluid, it seems reasonable to assume that

$$\mathscr{E} = \mathscr{E}(\rho, \mu, D, u) \qquad (11\text{-}3)$$

which automatically satisfies the principle of material frame indifference.

The Buckingham-Pi theorem [1] requires that this last be of the form

$$e = e(N_{\text{Re}}) \qquad (11\text{-}4)$$

where $e$ is known as the *energy loss coefficient*:

$$e \equiv \frac{\mathscr{E}}{\frac{1}{2}u^2 G} \tag{11-5}$$

and $N_{\text{Re}}$ is a Reynolds number:

$$N_{\text{Re}} \equiv \frac{Du\rho}{\mu} \tag{11-6}$$

The details of this discussion apply equally well to any pipeline fitting, in either laminar or turbulent flow (for a turbulent flow, we merely replace $\mathscr{E}$ by $\bar{\mathscr{E}}$). The dependence of the friction loss coefficient $e$ for a given fitting upon a Reynolds number has been confirmed experimentally, at least for laminar flows [2]. For turbulent flows, $e$ appears to be approximately a constant [3; 4, p. 217].

**REFERENCES**

1. Brand, Louis: *Arch. Rational Mech. Analysis*, **1**:35 (1957).
2. Kittredge, C. P., and D. S. Rowley: *Trans. ASME*, **79**:1759 (1957).
3. Lapple, C. E.: *Chem. Eng.*, **56**(5):96 (1949).
4. Bird, R. B., W. E. Stewart, E. N. Lightfoot: "Transport Phenomena," 7th printing, Wiley, New York, 1960.

**4.4.12 The mechanical energy balance for incompressible materials: An example** Our intention here is not only to illustrate how the mechanical energy balance can be used but also to point out that sometimes the same problem can be analyzed in more than one way. Since in each case somewhat different approximations may be made, it should not be surprising when differing answers are obtained.

To illustrate these points, let us estimate for an incompressible fluid the change in pressure across the sudden expansion in a pipeline pictured in Fig. 4.4.12-1. More specifically, we seek the change in pressure between cross sections 0 and 2. It is assumed that the velocity distributions at these cross sections are essentially unaffected by the presence of the sudden expansion, although the pressure difference is

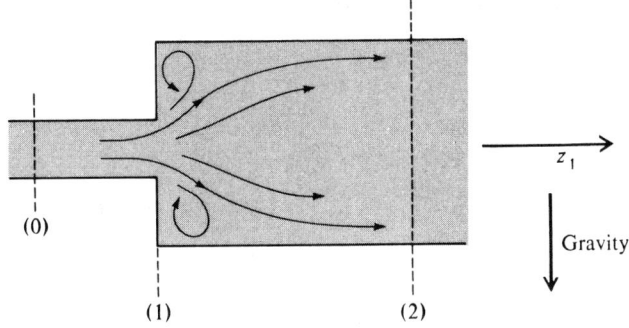

**Fig. 4.4.12-1** Flow through a sudden expansion.

**248** APPLICATION OF INTEGRAL AVERAGING TECHNIQUES TO MOMENTUM TRANSFER

primarily attributable to the change in cross-sectional area rather than pipe friction. We contrast results obtained using the integral momentum and mechanical energy balances. We leave it to the reader in Exercise 4.4.12-1 to analyze this same problem using the Bernoulli equation.

**THE INTEGRAL MOMENTUM BALANCE**

Let us choose as our system all the fluid in the pipe between cross sections 0 and 2.

Let us understand that $\langle \bar{v}_1 \rangle_{(0)}$ denotes the area-averaged $z_1$ component of the time-averaged velocity at cross section 0:

$$\langle \bar{v}_1 \rangle_{(0)} \equiv \frac{1}{S_{(0)}} \int_{S_{(0)}} \bar{v}_1 \, dS \tag{12-1}$$

The time-averaged integral mass balance tells us that

$$\langle \bar{v}_1 \rangle_{(2)} = \langle \bar{v}_1 \rangle_{(0)} \left( \frac{D_{(0)}}{D_{(2)}} \right)^2 \tag{12-2}$$

where $D_{(i)}$ denotes the diameter of the pipe at cross section $i$.

For this system we may use the integral momentum balance in the form of Eq. (6-4), the $z_1$ component of which tells us that

$$\rho \langle (\bar{v}_1)^2 \rangle_{(0)} \frac{\pi D_{(0)}^2}{4} - \rho \langle (\bar{v}_1)^2 \rangle_{(2)} \frac{\pi D_{(2)}^2}{4}$$

$$+ (\langle \bar{p} \rangle_{(0)} - p_0) \frac{\pi D_{(0)}^2}{4} - (\langle \bar{p} \rangle_{(2)} - p_0) \frac{\pi D_{(2)}^2}{4}$$

$$+ (\langle p \rangle_{(1)} - p_0) \left( \frac{\pi D_{(2)}^2}{4} - \frac{\pi D_{(0)}^2}{4} \right) = 0 \tag{12-3}$$

In writing this last we have neglected the effect of the turbulent Reynolds stresses at the entrances and exits and we have neglected the viscous contributions to $\mathscr{F}$ on the bounding metal surfaces. Let us assume that $\bar{v}_1$ is nearly independent of position at cross sections 0 and 2 (see Exercise 4.4.12-3), so that we may write

$$\langle (\bar{v}_1)^2 \rangle_{(0)} \doteq \langle \bar{v}_1 \rangle_{(0)}^2 \quad \text{and} \quad \langle (\bar{v}_1)^2 \rangle_{(2)} \doteq \langle \bar{v}_1 \rangle_{(2)}^2 \tag{12-4}$$

We estimate that

$$\langle p \rangle_{(1)} \doteq \langle \bar{p} \rangle_{(0)} \tag{12-5}$$

Equations (12-2), (12-4), and (12-5) allow us to rearrange Eq. (12-3) in the form of the desired result:

$$\frac{\langle \bar{p} \rangle_{(2)} - \langle \bar{p} \rangle_{(0)}}{\frac{1}{2}\rho \langle \bar{v}_1 \rangle_{(2)}^2} = 2 \left[ \left( \frac{D_{(2)}}{D_{(0)}} \right)^2 - 1 \right] \tag{12-6}$$

## [4.4] INTEGRAL BALANCES

**THE MECHANICAL ENERGY BALANCE**

The mechanical energy balance in the form of Eq. (10-8) may be applied to this system to find

$$\tfrac{1}{2}\rho\langle(\bar{v}_1)^3\rangle_{(0)}\frac{\pi D_{(0)}^2}{4} + (\langle\bar{p}\bar{v}_1\rangle_{(0)} - p_0\langle\bar{v}_1\rangle_{(0)})\frac{\pi D_{(0)}^2}{4}$$
$$- \tfrac{1}{2}\rho\langle(\bar{v}_1)^3\rangle_{(2)}\frac{\pi D_{(2)}^2}{4} - (\langle\bar{p}\bar{v}_1\rangle_{(2)} - p_0\langle\bar{v}_1\rangle_{(2)})\frac{\pi D_{(2)}^2}{4} = 0 \quad (12\text{-}7)$$

In arriving at this result, we have neglected all effects of turbulence at the entrances and exits as well as the time-averaged rate of dissipation of mechanical energy $\bar{\mathscr{E}}^{(t)}$. Again we assume that $\bar{v}_1$ is sufficiently uniform with respect to position at the entrance and exit that we may approximate (see Exercise 4.4.12-3):

$$\langle(\bar{v}_1)^3\rangle_{(0)} \doteq \langle\bar{v}_1\rangle_{(0)}^3 \qquad \langle(\bar{v}_1)^3\rangle_{(2)} \doteq \langle\bar{v}_1\rangle_{(2)}^3$$

$$\langle\bar{p}\bar{v}_1\rangle_{(0)} \doteq \langle\bar{p}\rangle_{(0)}\langle\bar{v}_1\rangle_{(0)}$$

and

$$\langle\bar{p}\bar{v}_1\rangle_{(2)} \doteq \langle\bar{p}\rangle_{(2)}\langle\bar{v}_1\rangle_{(2)} \quad (12\text{-}8)$$

Equations (12-2) and (12-8) allow us to express (12-7) as

$$\frac{\langle\bar{p}\rangle_{(2)} - \langle\bar{p}\rangle_{(0)}}{\tfrac{1}{2}\rho\langle\bar{v}_1\rangle_{(2)}^2} = \left[\left(\frac{D_{(2)}}{D_{(0)}}\right)^4 - 1\right] \quad (12\text{-}9)$$

**DISCUSSION**

Whitaker [1, p. 242] makes an interesting comparison of Eqs. (12-6) and (12-9) with experimental data. He finds that the result from the integral momentum balance, Eq. (12-6), is in reasonable agreement with the data, whereas the result from the mechanical energy balance, Eq. (12-9), consistently gives values that are too large. If we go back to the mechanical energy balance and include the rate of dissipation of mechanical energy by the action of viscous and turbulent forces, we find

$$\frac{\langle\bar{p}\rangle_{(2)} - \langle\bar{p}\rangle_{(0)}}{\tfrac{1}{2}\rho\langle\bar{v}_1\rangle_{(2)}^2} = \left[\left(\frac{D_{(2)}}{D_{(0)}}\right)^4 - 1\right] - \frac{8\bar{\mathscr{E}}^{(t)}}{\pi D_{(2)}^2 \rho\langle\bar{v}_1\rangle_{(2)}^3} \quad (12\text{-}10)$$

While it is clear that $\bar{\mathscr{E}} \geq 0$ for an isothermal flow (see Sec. 7.4.8), a similar statement has not been proved for $\bar{\mathscr{E}}^{(t)}$. However, the experimental data quoted by Whitaker suggest that, at least for this situation, $\bar{\mathscr{E}}^{(t)} \geq 0$.

Before the results of these comparisons with experimental data were described to you, it may not have been obvious that the error involved in neglecting $\bar{\mathscr{E}}^{(t)}$ in the mechanical energy balance was any more serious than the error incurred in neglecting the viscous contribution to $\mathscr{F}$ in the integral momentum balance. I can only tell you that there is no substitute for experience, otherwise known as a well-formed "engineering judgment." This is a good illustration of why one must be wary in applying the integral balances. It is usually necessary to make approximations in order to obtain answers that are in a useful form, but these approximations are made in such a way that it is rarely possible to place error bounds on the results. It is only after a successful comparison with experimental data that most of us can have confidence in using these analyses.

Because Eq. (12-6) does do a reasonable job of representing available experimental data, Eqs. (12-6) and (12-10) are often used to estimate the friction loss coefficient (introduced in Sec. 4.4.11) for this flow as

$$\bar{e} \equiv \frac{8\mathscr{E}^{(t)}}{\pi D_{(2)}^2 \rho \langle \bar{v}_1 \rangle_{(2)}^3} = \left[\left(\frac{D_{(2)}}{D_{(0)}}\right)^2 - 1\right]^2 \tag{12-11}$$

## REFERENCES

1. Whitaker, Stephen: "Introduction to Fluid Mechanics," Prentice-Hall, Englewood Cliffs, N.J., 1968.
2. Bird, R. B., W. E. Stewart, and E. N. Lightfoot: "Transport Phenomena," 7th printing, Wiley, New York, 1960.
3. Shames, I. H.: "Mechanics of Fluids," McGraw-Hill, New York, 1962.
4. Schlichting, Hermann: "Boundary-Layer Theory," 6th ed., McGraw-Hill, New York, 1968.
5. Rouse, Hunter, and Simon Ince: "History of Hydraulics," Iowa Institute of Hydraulic Research, State University of Iowa, 1957.

## EXERCISES

**4.4.12-1** *Pipe flow* [2, p. 216] (*a*) Consider flow down a vertical pipe under steady-state conditions (either laminar flow or turbulent flow such that the time-averaged variables are independent of time at fixed positions). If $\bar{v}$ is the area-averaged axial component of velocity in the pipe, show that

$$\mathscr{E}^{(t)} = \bar{v}\mathscr{F}$$

where $\mathscr{F}$ is the axial component of the force that the fluid exerts on the walls of the pipe in the direction of flow.

(*b*) Conclude that

$$e = \frac{4cL}{D}$$

where $c$ is the drag coefficient defined in the manner indicated in Sec. 4.4.7 and $D$ is the diameter of the pipe.

**4.4.12-2** [3, p. 172] Water in a large tank shown in Fig. 4.4.12-2 is under a pressure of 10 lb$_f$/in$^2$ gauge at the free surface. It is pumped out of the tank and through a nozzle to form a jet. What is the horsepower required of the pump? Neglect any frictional losses.

## [4.4] INTEGRAL BALANCES

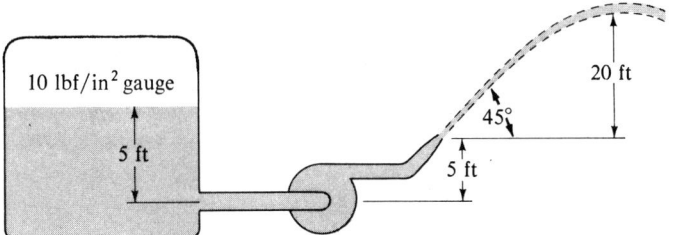

**Fig. 4.4.12-2**

**4.4.12-3** *More on pipe flow* [1, p. 242]  In Sec. 3.2.1, we found that the steady-state velocity distribution of an incompressible newtonian fluid flowing through a very long tube of radius $R$ is given by

$$\frac{v_z}{v_{z,\max}} = 1 - \left(\frac{r}{R}\right)^2$$

The time-averaged velocity distribution for turbulent flow through a pipe is found to be approximately represented by [4, p. 563]

$$\frac{\bar{v}_z}{\bar{v}_{z,\max}} = \left(1 - \frac{r}{R}\right)^{1/n}$$

where $n$ is a function of the Reynolds number as indicated in Table 4.4.12-1.
  Determine that for turbulent flow through a pipe

$$\frac{\langle \bar{v}_z \rangle}{\bar{v}_{z,\max}} = \frac{2n^2}{(1+n)(1+2n)}$$

$$\frac{\langle (\bar{v}_z)^2 \rangle}{\langle \bar{v}_z \rangle^2} = \frac{(1+n)(1+2n)^2}{4n^2(2+n)}$$

and

$$\frac{\langle (\bar{v}_z)^3 \rangle}{\langle \bar{v}_z \rangle^3} = \frac{(1+n)^3(1+2n)^3}{4n^4(3+n)(3+2n)}$$

These results are evaluated for selected values of $n$ in Table 4.4.12-2.

**Table 4.4.12-1**  Dependence of $n$ upon $N_{\text{Re}}$ [4, p. 563]

| $n =$ | 6 | 7 | 10 |
|---|---|---|---|
| $N_{\text{Re}} =$ | $4 \times 10^3$ | $1 \times 10^5$ | $3 \times 10^6$ |

**Table 4.4.12-2 Deviations from the flat velocity profile† [1, p. 243]**

|  | Turbulent | | | | | Laminar |
|---|---|---|---|---|---|---|
| $n$ | 6 | 7 | 8 | 9 | 10 | Parabolic profile |
| $\dfrac{\langle \bar{v}_z \rangle}{\bar{v}_{z,\max}}$ | 0.76 | 0.82 | 0.84 | 0.85 | 0.86 | 0.50 |
| $\dfrac{\langle (\bar{v}_z)^2 \rangle}{\langle \bar{v}_z \rangle^2}$ | 1.03 | 1.01 | 1.01 | 1.02 | 1.03 | 1.33 |
| $\dfrac{\langle (\bar{v}_z)^3 \rangle}{\langle \bar{v}_z \rangle^3}$ | 1.08 | 1.05 | 1.03 | 1.05 | 1.06 | 2 |

† Whitaker's Table 7.5-3 is in error. For laminar flow, 3.20 should be $\tfrac{4}{3}$ and 7.32 should be 2.

**4.4.12-4** *The Egyptian water clock* [1, p. 277] The Egyptians used water clocks similar to that illustrated in Fig. 4.4.12-3 [5, p. 7]. The radius $r_0$ of the circular bowl is a function of $z$, the distance from the bottom of the bowl. Determine the functional dependence of $r_0$ on $z$ required in order that the depth of liquid be a linear function of time.

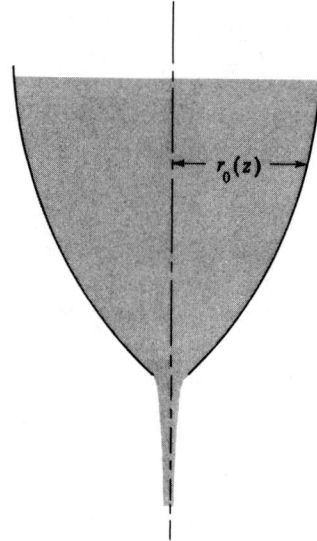

**Fig. 4.4.12-3** Egyptian water clock.

(a) Use the integral mass and mechanical energy balances. Do not neglect the time rate of change of potential energy in the mechanical energy balance.

(b) Use the integral mass balance and the steady-state Bernoulli equation. This is known as a quasi-steady-state analysis.

*Answer for both cases*

$$r_0(z) = (2g)^{\frac{1}{4}} \left[ \frac{S_{\text{ex}}}{-\pi (dz/dt)} \right]^{\frac{1}{2}} z^{\frac{1}{4}}$$

## [4.4] INTEGRAL BALANCES

**4.4.12-5** *An alternative expression for the force on a sphere*  Consider an infinite expanse of fluid that moves past a sphere fixed in space. Very far away from the sphere, the fluid moves with a constant velocity $v_\infty$. (Mathematically, this is the same as considering a sphere that moves with a constant velocity $-v_\infty$ through an infinite expanse of fluid that is stationary very far away from the sphere.)

(a) Use the integral momentum and mechanical energy balances to find
$$\mathbf{v}_\infty \cdot \mathcal{G} = \int_0^{2\pi} \int_0^\pi \int_0^R \text{tr}\,(\mathbf{S} \cdot \nabla \mathbf{v}) r^2 \sin\theta\, dr\, d\theta\, d\varphi$$

(b) For flow of an incompressible newtonian fluid past a sphere as discussed in Sec. 3.3.3, determine that
$$v_\infty f_{(s)3} - \tfrac{4}{3}\pi R^3 \rho g v_\infty = 2\pi \mu \int_R^\infty \int_0^\pi \left\{ 2\left(\frac{\partial v_r}{\partial r}\right)^2 + 2\left(\frac{1}{r}\frac{\partial v_\theta}{\partial \theta} + \frac{v_r}{r}\right)^2 \right.$$
$$\left. + 2\left(\frac{v_r}{r} + \frac{v_\theta \cot\theta}{r}\right)^2 + \left[r\frac{\partial}{\partial r}\left(\frac{v_\theta}{r}\right) + \frac{1}{r}\frac{\partial v_r}{\partial \theta}\right]^2 \right\} r^2 \sin\theta\, dr\, d\theta$$

(c) Use the expression obtained in (b) as well as the velocity distribution from Sec. 3.3.3 to arrive at Stokes' law. Compared with the expression used in Sec. 3.3.3, the result from (b) has the advantage of not requiring the pressure distribution in the fluid.

**4.4.12-6** *The hydraulic ram* [1, p. 275]  A simple hydraulic brake consists of a cylindrical ram that displaces fluid from a slightly larger cylinder as shown in Fig. 4.4.12-4. The speed of the ram is $v_0$. We wish to determine the magnitude $F$ of the force required to maintain this motion.

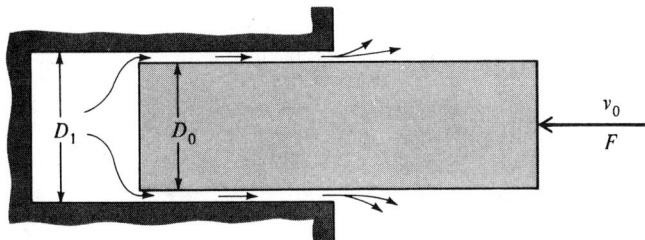

**Fig. 4.4.12-4**  The hydraulic ram.

(a) Use the integral mass balance to estimate the area-averaged $z_1$ component of velocity in the annular space between the ram and cylinder as
$$\langle v_1 \rangle_{\text{ex}} = \frac{v_0}{(D_1/D_0)^2 - 1}$$

(b) Use the integral momentum balance to suggest
$$F = \frac{\rho(v_0)^2[(\pi/4)(D_1)^2]}{[(D_1/D_0)^2 - 1]^2}$$
Neglect viscous effects in this analysis.

(c) Use the integral mechanical energy balance to say
$$F = \frac{\rho(v_0)^2[(\pi/4)(D_1)^2]}{2[(D_1/D_0)^2 - 1]^2}$$
Viscous effects should be neglected in this analysis as well.

(d) The integral mechanical energy balance probably gives a better answer here. Why?

**4.4.12-7** *Ejector pump*  Figure 4.4.12-5 shows a very simple device that can be used to pump fluids, the ejector pump.  Let us assume that both fluids are the same.
   (a) Derive an expression for the pressure rise between cross section $A$ and cross section $B$.
   (b) Derive an equation or set of equations to be solved for $Q_1$ as a function of $Q_0$, $A_0$, and $A_1$.

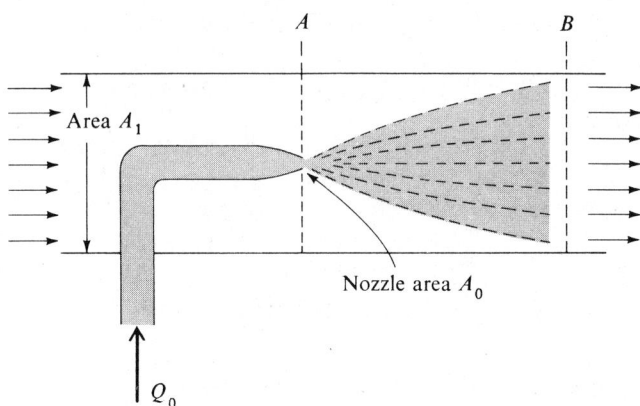

**Fig. 4.4.12-5**  A simple ejector pump.

**4.4.12-8** *Drag on an arbitrary body*  Consider an infinite expanse of fluid that moves past an arbitrary body that is fixed in space.  Very far away from the body, the fluid moves with a constant velocity $\mathbf{v}_\infty$.
   (a) Use the integral momentum and mechanical energy balances to find

$$\mathbf{v}_\infty \cdot \mathscr{G} = \mathscr{E} \equiv \int_{V_{\text{(fluid)}}} \text{tr}\,(\mathbf{S} \cdot \nabla \mathbf{v})\, dV$$

   (b) Determine that

$$\mathbf{v}_\infty \cdot \mathscr{G} = v_\infty \cdot \mathscr{F} - \mathscr{M}(\mathbf{v}_\infty \cdot f)$$

where $\mathscr{M}$ is the mass of fluid displaced by the body.
   (c) For plane flow past a cylindrical body, find that (see Sec. 4.4.7, Example 2)

$$c_\mathscr{D} = \frac{\mathscr{E}}{A_\mathscr{D} \tfrac{1}{2}\rho\, |\mathbf{v}_\infty|^3}$$

where $c_\mathscr{D}$ is the dimensionless drag coefficient.

### 4.4.13  The integral moment of momentum balance

Corresponding to the equation of continuity and Cauchy's first law, we have written the integral mass and momentum balances.  This suggests that we develop an integral balance corresponding to Cauchy's second law, or an integral moment of momentum balance.  From a different point of view, it could be useful to have an equation that would allow us to estimate the torque exerted on a solid surface as the result of a flow.

## [4.4] INTEGRAL BALANCES

Let us take the same approach that we have used in deriving our other integral balances. In the generalized transport theorem, let us identify

$$\Psi = \mathbf{p} \wedge \rho \mathbf{v} \tag{13-1}$$

moment of momentum per unit volume, to obtain

$$\frac{d}{dt}\int_{V_{(s)}} \mathbf{p} \wedge \rho\mathbf{v}\, dV = \int_{V_{(s)}} \frac{\partial}{\partial t}(\mathbf{p}\wedge \rho\mathbf{v})\, dV + \int_{S_{(s)}} (\mathbf{p}\wedge \rho\mathbf{v})(\mathbf{v}_{(s)}\cdot \mathbf{n})\, dS \tag{13-2}$$

Here $\mathbf{p}$ is the position vector field measured with respect to a conveniently chosen origin. Let us concentrate on finding a more easily evaluated expression for the second term on the right of Eq. (13-2).

Cauchy's second law is not of direct assistance to us, since it states merely that the stress tensor field $\mathbf{T}$ must be symmetric. Considering our intention of developing an integral moment of momentum balance, we are prompted to consider the integral over an arbitrary system of the vector product of the position vector field $\mathbf{p}$ and Cauchy's first law:

$$\int_{V_{(s)}} \mathbf{p} \wedge \left(\rho \frac{d_{(m)}\mathbf{v}}{dt} - \operatorname{div} \mathbf{T} - \rho \mathbf{f}\right) dV = 0 \tag{13-3}$$

Let us consider the terms in this equation individually.

The integral of the first term in Eq. (13-3) may be expressed as

$$\mathbf{p} \wedge \left(\rho \frac{d_{(m)}\mathbf{v}}{dt}\right) = \mathbf{p} \wedge \left[\frac{\partial(\rho\mathbf{v})}{\partial t} + \operatorname{div}(\rho\mathbf{vv})\right] \tag{13-4}$$

in terms of rectangular cartesian components. The first term on the right of this last becomes

$$\mathbf{p} \wedge \frac{\partial(\rho\mathbf{v})}{\partial t} = e_{ijk} z_j \frac{\partial(\rho v_k)}{\partial t}\mathbf{e}_i$$

$$= \frac{\partial}{\partial t}(\mathbf{p}\wedge \rho\mathbf{v}) \tag{13-5}$$

The second term on the right of Eq. (13-4) may be expressed as

$$\mathbf{p} \wedge [\operatorname{div}(\rho\mathbf{vv})] = e_{ijk} z_j \frac{\partial}{\partial z_m}(\rho v_k v_m)\mathbf{e}_i$$

$$= \frac{\partial}{\partial z_m}(e_{ijk} z_j \rho v_k v_m)\mathbf{e}_i - e_{ijk}\delta_{jm}\rho v_k v_m \mathbf{e}_i$$

$$= \frac{\partial}{\partial z_m}(e_{ijk} z_j \rho v_k v_m)\mathbf{e}_i$$

$$= \operatorname{div}([\mathbf{p}\wedge \rho\mathbf{v}]\mathbf{v}) \tag{13-6}$$

Equations (13-5) and (13-6) allow us to express the first term on the left of (13-3) as

$$\int_{V_{(s)}} \mathbf{p} \wedge \left(\rho \frac{d_{(m)}\mathbf{v}}{dt}\right) dV = \int_{V_{(s)}} \left\{\frac{\partial}{\partial t}(\mathbf{p} \wedge \rho\mathbf{v}) + \text{div}\left[(\mathbf{p} \wedge \rho\mathbf{v})\mathbf{v}\right]\right\} dV \qquad (13\text{-}7)$$

or, after an application of Green's transformation,

$$\int_{V_{(s)}} \mathbf{p} \wedge \left(\frac{\rho\, d_{(m)}\mathbf{v}}{dt}\right) dV = \int_{V_{(s)}} \frac{\partial}{\partial t}(\mathbf{p} \wedge \rho\mathbf{v})\, dV + \int_{S_{(s)}} (\mathbf{p} \wedge \rho\mathbf{v})(\mathbf{v} \cdot \mathbf{n})\, dS \qquad (13\text{-}8)$$

The first term on the right is exactly the one needed in Eq. (13-2).

Rearrangement of the second term on the left of Eq. (13-3) is again best seen in terms of rectangular cartesian components:

$$-\int_{V_{(s)}} \mathbf{p} \wedge (\text{div } \mathbf{T})\, dV = -\int_{V_{(s)}} e_{ijk} z_j \frac{\partial T_{km}}{\partial z_m} \mathbf{e}_i\, dV$$

$$= -\int_{V_{(s)}} \frac{\partial}{\partial z_m} (e_{ijk} z_j T_{km}) \mathbf{e}_i\, dV + \int_{V_{(s)}} e_{ijk} \delta_{jm} T_{km} \mathbf{e}_i\, dV$$

$$= -\int_{V_{(s)}} \frac{\partial}{\partial z_m} (e_{ijk} z_j T_{km}) \mathbf{e}_i\, dV$$

$$= -\int_{V_{(s)}} \text{div } (\mathbf{p} \wedge \mathbf{T})\, dV \qquad (13\text{-}9)$$

We have taken advantage of Cauchy's second law in declaring the second term on the right of Eq. (13-9)$_2$ to be zero. After an application of Green's transformation, we are left with

$$-\int_{V_{(s)}} \mathbf{p} \wedge (\text{div } \mathbf{T})\, dV = -\int_{S_{(s)}} e_{ijk} z_j T_{km} n_m \mathbf{e}_i\, dS$$

$$= -\int_{S_{(s)}} \mathbf{p} \wedge (\mathbf{T} \cdot \mathbf{n})\, dS \qquad (13\text{-}10)$$

Equations (13-3), (13-8), and (13-10) allow us to express Eq. (13-2) as

$$\frac{d}{dt}\int_{V_{(s)}} \mathbf{p} \wedge \rho\mathbf{v}\, dV = \int_{S_{(s)}} (\mathbf{p} \wedge \rho\mathbf{v})(\mathbf{v} - \mathbf{v}_{(s)}) \cdot (-\mathbf{n})\, dS$$

$$- \int_{S_{(s)}} \mathbf{p} \wedge [\mathbf{T} \cdot (-\mathbf{n})]\, dS + \int_{V_{(s)}} \mathbf{p} \wedge \rho\mathbf{f}\, dV \qquad (13\text{-}11)$$

The first term on the right of this equation is the net rate at which moment of momentum is brought into the system with whatever material is crossing the boundaries. The second term is the torque or moment of the force that the material in the system exerts on the bounding surfaces of the system. This is the total torque, whereas we usually speak in terms of the torque in excess of that attributable to an ambient or atmospheric pressure $p_0$. Repeating the argument we used in Sec. 4.4.5 to account for $p_0$ and neglecting viscous effects at the entrances and exits of the system, we can finally write [1]

$$\frac{d}{dt}\int_{V_{(s)}} \mathbf{p} \wedge \rho\mathbf{v}\, dV = \int_{S_{(\text{ent ex})}} (\mathbf{p} \wedge \rho\mathbf{v})(\mathbf{v} - \mathbf{v}_{(s)}) \cdot (-\mathbf{n})\, dS$$

$$-\int_{S_{(\text{ent ex})}} \mathbf{p} \wedge [(\mathbf{T} + p_0\mathbf{I}) \cdot (-\mathbf{n})]\, dS - \mathcal{T} + \int_{V_{(s)}} \mathbf{p} \wedge \rho\mathbf{f}\, dV \quad (13\text{-}12)$$

where we define

$$\mathcal{T} \equiv \int_{S_{(s)} - S_{(\text{ent ex})}} \mathbf{p} \wedge [(\mathbf{T} + p_0\mathbf{I}) \cdot (-\mathbf{n})]\, dS \quad (13\text{-}13)$$

Physically, $\mathcal{T}$ denotes the torque that the system exerts upon the impermeable portion of its bounding surface beyond the torque attributable to the ambient pressure $p_0$. Equation (13-12) is the *general integral moment of momentum balance* applicable to a single-phase system.

We generally try to choose a system for application of the integral moment of momentum balance in such a way that the torque attributable to viscous forces at the entrance and exit surfaces may be neglected. This might be done by selecting entrance and exit surfaces as cross sections normal to the flow in long straight pipes and by thinking of the flow in these straight pipes as being approximately represented by Poiseuille flow (Sec. 3.2.1). In most problems, one is forced to do this for lack of sufficient information to do otherwise. Under these conditions, Eq. (13-12) reduces to

$$\frac{d}{dt}\int_{V_{(s)}} \mathbf{p} \wedge \rho\mathbf{v}\, dV = \int_{S_{(\text{ent ex})}} (\mathbf{p} \wedge \rho\mathbf{v})(\mathbf{v} - \mathbf{v}_{(s)}) \cdot (-\mathbf{n})\, dS$$

$$-\int_{S_{(\text{ent ex})}} (P - p_0)(\mathbf{p} \wedge \mathbf{n})\, dS - \mathcal{T} + \int_{V_{(s)}} \mathbf{p} \wedge \rho\mathbf{f}\, dV \quad (13\text{-}14)$$

This form of the integral moment of momentum balance is more commonly employed than Eq. (13-12).

As Eqs. (13-12) and (13-14) have been derived, they are applicable only to a single-phase system. As we pointed out in Sec. 4.4.2, we are more often concerned with multiphase systems. Using the approach and notation of Sec. 4.4.2 and making no additional assumptions, we find that the *general integral moment of momentum balance* appropriate to a multiphase system is

$$\frac{d}{dt}\int_{V_{(s)}} \mathbf{p} \wedge \rho\mathbf{v}\, dV = \int_{S_{(\text{ent ex})}} (\mathbf{p} \wedge \rho\mathbf{v})(\mathbf{v} - \mathbf{v}_{(s)}) \cdot (-\mathbf{n})\, dS$$

$$-\int_{S_{(\text{ent ex})}} \mathbf{p} \wedge [(\mathbf{T} + p_0\mathbf{I}) \cdot (-\mathbf{n})]\, dS - \mathcal{T} + \int_{V_{(s)}} \mathbf{p} \wedge \rho\mathbf{f}\, dV$$

$$+\int_{S_{(\text{sing})}} \mathbf{p} \wedge [\rho\mathbf{v}(\mathbf{v} \cdot \boldsymbol{\xi} - u_{(\xi)}) - \mathbf{T} \cdot \boldsymbol{\xi}]\, dS \quad (13\text{-}15)$$

**258** APPLICATION OF INTEGRAL AVERAGING TECHNIQUES TO MOMENTUM TRANSFER

If we assume that the jump momentum balance of Exercise 2.2.3-1 is applicable and if we neglect viscous effects at the entrance and exit surfaces, we find that Eq. (13-15) reduces to the equivalent result for a single-phase system, Eq. (13-14).

We can visualize the three types of problems in which the integral moment of momentum balance might be applied: the torque $\mathcal{T}$ might be neglected, it might be the unknown to be determined, or it might be known from previous experimental data. Perhaps because the integral moment of momentum balance has not seen as much use in the literature as the integral momentum balance, or more likely, because the need has not generally arisen, very little in the way of empirical data correlations for $\mathcal{T}$ are available. For this reason, we will not devote a special section to empirical correlation for $\mathcal{T}$. For anyone interested, the approach should be very much the same as that taken for $\mathcal{F}$ and $\mathcal{G}$ in Sec. 4.4.7.

### REFERENCE

1. Slattery, J. C., and R. A. Gaggioli: *Chem. Eng. Sci.*, **17**:893 (1962).

### EXERCISE

**4.4.13-1** Give the detailed argument required in going from Eqs. (13-6)$_2$ to (13-6)$_3$ and from (13-9)$_2$ to (13-9)$_3$.

### 4.4.14 The integral moment of momentum balance for turbulent flows

Perhaps the most important applications of the integral balances are to systems, portions of which are in turbulent flow.

We can take two approaches in constructing an integral moment of momentum balance more appropriate to systems in turbulent flow. In the first approach, let us limit ourselves (for reasons explained in Sec. 4.4.3) to single-phase or multiphase systems that do not involve fluid-fluid phase interfaces. Starting with the time-averaged version of Cauchy's first law found in Sec. 4.1.2,

$$\frac{\partial(\bar{\rho}\bar{\mathbf{v}})}{\partial t} + \text{div}\,(\bar{\rho}\bar{\mathbf{v}}\bar{\mathbf{v}}) = -\nabla\bar{P} + \text{div}\,(\bar{\mathbf{S}} + \mathbf{T}^{(t)}) + \bar{\rho}\mathbf{f} \tag{14-1}$$

The derivation of Sec. 4.4.13 may be repeated to find

$$\frac{d}{dt}\int_{V_{(s)}} \mathbf{p} \wedge \overline{\rho\mathbf{v}}\, dV = \int_{S_{(\text{ent ex})}} \mathbf{p} \wedge (\overline{\rho\mathbf{v}\mathbf{v}} - \overline{\rho\mathbf{v}}\mathbf{v}_{(s)}) \cdot (-\mathbf{n})\, dS$$

$$- \int_{S_{(\text{ent ex})}} (\bar{P} - p_0)(\mathbf{p} \wedge \mathbf{n})\, dS - \mathcal{T} + \int_{V_{(s)}} \mathbf{p} \wedge \bar{\rho}\mathbf{f}\, dV$$

$$+ \int_{S_{(\text{sing})}} \mathbf{p} \wedge [\bar{\rho}\bar{\mathbf{v}}\bar{\mathbf{v}} \cdot \boldsymbol{\xi} - \overline{\rho\mathbf{v}}u_{(\xi)} - \bar{\mathbf{T}} \cdot \boldsymbol{\xi}]\, dS \tag{14-2}$$

where

[4.4]  INTEGRAL BALANCES   259

$$\mathscr{T} = \int_{S_{(s)}-S_{(\text{ent ex})}} \mathbf{p} \wedge [(\mathbf{T}+p_0\mathbf{I})\cdot(-\mathbf{n})]\,dS$$

$$= \int_{S_{(s)}-S_{(\text{ent ex})}} \mathbf{p} \wedge [(\overline{\mathbf{T}}+p_0\mathbf{I})\cdot(-\mathbf{n})]\,dS \qquad (14\text{-}3)$$

In arriving at this result, we have observed that for this case $S_{(s)} - S_{(\text{ent ex})}$ and $S_{(\text{sing})}$ must be composed of fluid-solid and solid-solid phase interfaces on which the turbulent fluctuations are identically zero. The time-averaged jump momentum balance of Exercise 4.4.6-1 simplifies Eq. (14-2) to

$$\frac{d}{dt}\int_{V_{(s)}} \mathbf{p} \wedge \overline{\rho\mathbf{v}}\,dV = \int_{S_{(\text{ent ex})}} \mathbf{p} \wedge (\overline{\rho\tilde{\mathbf{v}}\tilde{\mathbf{v}}} - \overline{\rho\mathbf{v}}\,\mathbf{v}_{(s)}) \cdot (-\mathbf{n})\,dS$$

$$- \int_{S_{(\text{ent ex})}} (\overline{P}-p_0)(\mathbf{p}\wedge\mathbf{n})\,dS - \mathscr{T} + \int_{V_{(s)}} \mathbf{p}\wedge\bar{\rho}\mathbf{f}\,dV \qquad (14\text{-}4)$$

This is the form of the *integral moment of momentum balance* recommended for turbulent single-phase or multiphase systems that do not involve fluid-fluid phase interfaces. Note that in arriving at this result we have neglected any viscous or turbulent forces acting at the entrances and exits of the system.

For a single-phase or multiphase system that includes one or more fluid-fluid phase interfaces, I recommend time averaging the integral moment of momentum balance of Sec. 4.4.13 to find

$$\frac{d}{dt}\overline{\int_{V_{(s)}} \mathbf{p}\wedge\rho\mathbf{v}\,dV} = \overline{\int_{S_{(\text{ent ex})}} (\mathbf{p}\wedge\rho\mathbf{v})(\mathbf{v}-\mathbf{v}_{(s)})\cdot(-\mathbf{n})\,dS}$$

$$- \overline{\int_{S_{(\text{ent ex})}} (P-p_0)(\mathbf{p}\wedge\mathbf{n})\,dS} - \overline{\mathscr{T}} + \overline{\int_{V_{(s)}} \mathbf{p}\wedge\rho\mathbf{f}\,dV} \qquad (14\text{-}5)$$

Here we define

$$\overline{\mathscr{T}} \equiv \overline{\int_{S_{(s)}-S_{(\text{ent ex})}} \mathbf{p}\wedge[(\mathbf{T}+p_0\mathbf{I})\cdot(-\mathbf{n})]\,dS} \qquad (14\text{-}6)$$

Again Eq. (14-5) assumes that the jump momentum balance of Exercise 2.2.3-1 is applicable and that the time-averaged viscous forces acting at the entrances and exits of the system may be neglected.

When we restrict ourselves to single-phase or multiphase systems that do not involve fluid-fluid phase interfaces and when we neglect any turbulent forces acting at the entrances and exits to the system, we can use Eqs. (3-7) to (3-9) to show that Eqs. (14-5) and (14-6) reduce to Eqs. (14-4) and (14-3), respectively.

### 4.4.15 The integral moment of momentum balance: An example[1]   The Catherine wheel is a toy wheel around which is wound a tube filled with black

---
[1] This problem was suggested by Sabersky and Acosta [1, p. 135], though their answer appears to be incorrect.

gunpowder. When the gunpowder is ignited, gases formed by the combustion flow rapidly out of the open end of the tube and the wheel is forced to rotate about its axis. Let us determine the angular speed $\Omega$ of the wheel as a function of time after the powder is ignited.

For simplicity, let us assume that the mass of the wheel $\mathcal{M}$ can be considered to be located at a distance $R$ from the axis of rotation. The speed of the ejecting gases is $V_e$ relative to a point on the rotating wheel at a radius $R$. The rate at which the gases are ejected is a constant $G$ slugs/s. The initial mass of power is $\mathcal{M}_0$; it may also be assumed to be located at the same radius $R$.

This problem is most naturally analyzed in terms of a cylindrical coordinate system, the axis of which coincides with the axis of rotation of the wheel. Let us choose as our system the wheel and the gunpowder. For simplicity, let us ignore any effects attributable to turbulence and ask about the implications of the axial component (z component) of the angular momentum balance in the form of Eq. (13-14):

$$\frac{d}{dt}\int_{V_{(s)}} e_{3jk} p_{\langle j \rangle} \rho v_{\langle k \rangle}\, dV = \int_{S_{(ex)}} e_{3jk} p_{\langle j \rangle} \rho v_{\langle k \rangle} (\mathbf{v} - \mathbf{v}_{(s)}) \cdot (-\mathbf{n})\, dS$$
$$+ \int_{V_{(s)}} e_{3jk} p_{\langle j \rangle} f_{\langle k \rangle}\, dV \quad (15\text{-}1)$$

In writing this equation we neglect any torque resulting from friction between the wheel and any axial support. We also assume, for lack of further information, that the pressure of exhaust gases at the exit of the system is essentially the ambient pressure $p_0$. The convention of Exercise A.4.1-4 is understood in numbering the coordinates.

In Exercise A.4.1-11 we see that there are only two nonzero physical components of the position vector field in cylindrical coordinates:

$$p_{\langle 1 \rangle} = r \qquad p_{\langle 2 \rangle} = 0 \qquad p_{\langle 3 \rangle} = z \quad (15\text{-}2)$$

This means that, without any further assumptions, Eq. (15-1) simplifies to

$$\frac{d}{dt}\int_{V_{(s)}} p_{\langle 1 \rangle} \rho v_{\langle 2 \rangle}\, dV = \int_{S_{(ex)}} p_{\langle 1 \rangle} \rho v_{\langle 2 \rangle} (\mathbf{v} - \mathbf{v}_{(s)}) \cdot (-\mathbf{n})\, dS + \int_{V_{(s)}} p_{\langle 1 \rangle} f_{\langle 2 \rangle}\, dV$$

or $\qquad\qquad\qquad\qquad\qquad\qquad\qquad\qquad\qquad\qquad\qquad\qquad\qquad\qquad\quad (15\text{-}3)$

$$\frac{d}{dt}\int_{V_{(s)}} r \rho v_\theta\, dV = \int_{S_{(ex)}} r \rho v_\theta (\mathbf{v} - \mathbf{v}_{(s)}) \cdot (-\mathbf{n})\, dS \quad (15\text{-}4)$$

In going from Eq. (15-4) to Eq. (15-5), we have chosen to interpret the problem statement as having been made for a horizontally rotating wheel:

$$f_{\langle 2 \rangle} = f_\theta = 0 \quad (15\text{-}5)$$

Recognizing that the system rotates as a solid body whose mass is located at a radius $R$ and that

At $S_{(ex)}$: $\qquad r v_\theta = R(R\Omega - V_e) \quad (15\text{-}6)$

## [4.4] INTEGRAL BALANCES

we can further simplify Eq. (15-4) to

$$\frac{d}{dt}\left(R^2\Omega \int_{V_{(s)}} \rho \, dV\right) = -R(R\Omega - V_e)G \tag{15-7}$$

The integral mass balance

$$\frac{d}{dt}\int_{V_{(s)}} \rho \, dV = \int_{S_{(ex)}} \rho(\mathbf{v} - \mathbf{v}_{(s)}) \cdot (-\mathbf{n}) \, dS$$

$$= -G \tag{15-8}$$

may then be integrated to find

$$\int_{V_{(s)}} \rho \, dV = \mathcal{M} + \mathcal{M}_0 - Gt \tag{15-9}$$

Substituting this into Eq. (15-7), we find the following differential equation determines $\Omega$ as a function of time:

$$\frac{d\Omega}{dt} = \frac{V_e G}{R(\mathcal{M} + \mathcal{M}_0 - Gt)} \tag{15-10}$$

This in turn is easily integrated to yield the desired result:

$$\Omega = \frac{V_e}{R} \ln \frac{\mathcal{M} + \mathcal{M}_0}{\mathcal{M} + \mathcal{M}_0 - Gt} \tag{15-11}$$

### REFERENCE

1. Sabersky, R. H., and A. J. Acosta: "Fluid Flow," Macmillan, New York, 1964.

### EXERCISES

**4.4.15-1** *A stationary pump impeller* [1, p. 133] Consider a two-dimensional, frictionless, steady, incompressible flow in Fig. 4.4.15-1 from $r = R_1$ to $r = R_2$ such that in cylindrical coordinates

At $r = R_1$:    $v_r = v_{r(1)}$    and    $v_\theta = v_{\theta(1)}$

The guide vanes of the impeller align the flow such that

At $r = R_2$:    $v_r = v_{r(2)} = -v_{\theta(2)} \tan \alpha$

Assuming that the flow is axially symmetric and neglecting any disturbance to the flow due to the thickness of the guide vanes, determine the torque exerted on the guide vanes per unit width.

**4.4.15-2** *A rotating pump impeller* [1, p. 135] The impeller in Exercise 4.4.15-1 rotates at a constant angular velocity $\Omega$ in the direction of increasing $\theta$. This imples that

$r^* = r$

$\theta^* = \theta + \Omega t$

$z^* = z$

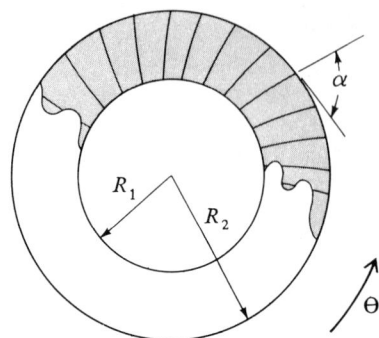

**Fig. 4.4.15-1** Cutaway pump impeller.

where $(r^*, \theta^*, z^*)$ is a set of cylindrical coordinates in a frame fixed with respect to the stars and $(r, \theta, z)$ is a set of cylindrical coordinates in a frame fixed with respect to the guide vanes on the rotating impeller. These two frames share a common origin on the axis of rotation of the impeller. Consider the velocity components introduced in Exercise 4.4.15-1 to be measured in the cylindrical coordinate system $(r, \theta, z)$.

(a) Determine the torque required to turn the impeller for a volume rate of flow $Q$.

(b) What is the rate at which work is done by the impeller on the fluid?

$$\mathcal{W} = \int_{S_{(\text{imp})}} \mathbf{v} \cdot (\mathbf{T} + p_0 \mathbf{I}) \cdot \mathbf{n} \, dS$$

(In writing this expression, our understanding is that $\mathbf{n}$ is the unit normal to the impeller surface directed from the fluid into the metal.)

### 4.4.16 Two extremum principles

In Sec. 4.4.11, I suggested how we could develop empirical correlations for $\mathscr{E}$, the rate at which mechanical energy is dissipated by the action of viscous forces. In the next few sections, I indicate how upper and lower bounds to $\mathscr{E}$ can be calculated, at least for some simple classes of material behavior.

The extremum principles presented in Secs. 4.4.17 and 4.4.18 have been developed in the literature from two different points of view. Johnson [1, 2] used the calculus of variations (the velocity extremum principle was previously presented by Pawlowski [3] and by Bird [4]). In a much more limited context, Sani [5] has indicated how these bounding principles can be viewed from function space. I present here the earlier developments by Hill [6] and by Hill and Power [7].

The developments here are limited to incompressible fluids whose behavior can be described by the two classes of empirical models described in Sec. 2.3.3. It further has been necessary to neglect all the inertial terms in Cauchy's first law and to assume that the external force can be represented as the gradient of the scalar potential energy per unit mass:

$$\mathbf{f} = -\nabla \varphi \tag{16-1}$$

Under these circumstances, Cauchy's first law may be written as

$$\text{div}\,(\mathbf{T} - \rho\varphi\mathbf{I}) = 0 \tag{16-2}$$

After developing the velocity and stress extremum principles in Secs. 4.4.17 and 4.4.18, in Sec. 4.4.19 I indicate for what type of fluid behavior these extremum principles bound the rate of dissipation of energy in a system. The use of these two principles is illustrated in Sec. 4.4.20 for a problem that otherwise would be difficult to handle.

## REFERENCES

1. Johnson, M. W., Jr.: *Phys. Fluids*, **3**:871 (1960).
2. Johnson, M. W., Jr.: *Trans. Soc. Rheol.*, **5**:9 (1961).
3. Pawlowski, J.: *Kolloid-Z.*, **138**:6 (1954).
4. Bird, R. B.: *Phys. Fluids*, **3**:539 (1960).
5. Sani, R. L.: *A.I.Ch.E.J.*, **9**:277 (1963).
6. Hill, R.: *J. Mech. Phys. Solids*, **5**:66 (1956).
7. Hill, R., and G. Power: *Quart. J. Mech. Appl. Math.*, **9**:313 (1956).

**4.4.17 The velocity extremum principle** The space $\mathscr{L}$ of all second-order tensors is nine dimensional (see Secs. A.5.1 and A.5.2). The symmetric second-order tensors form a six-dimensional subspace $\mathscr{S}$. Let $\epsilon = \epsilon(\mathbf{A})$ be a scalar-valued function of one second-order tensor $\mathbf{A}$ in either $\mathscr{L}$ or $\mathscr{S}$. Let $A_{ij}$ be the rectangular cartesian components of $\mathbf{A}$. Then $\epsilon(\mathbf{A})$ may also be viewed as a real-valued function of the nine real variables $A_{ij}$, if $\mathbf{A}$ belongs to $\mathscr{L}$, or of the six real variables $A_{ij}$ ($i \leq j$), if $\mathbf{A}$ belongs to $\mathscr{S}$.

If $\mathbf{A}$ belongs to $\mathscr{L}$, we define the gradient of $\epsilon$ as [1, p. 24]

$$\frac{\partial \epsilon}{\partial \mathbf{A}} \equiv \frac{\partial \epsilon}{\partial A_{ij}} \mathbf{e}_i \mathbf{e}_j = \frac{\partial \epsilon}{\partial \bar{A}_{ij}} \bar{\mathbf{g}}_i \bar{\mathbf{g}}_j = \frac{\partial \epsilon}{\partial \bar{A}_{\langle ij \rangle}} \bar{\mathbf{g}}_{\langle i \rangle} \bar{\mathbf{g}}_{\langle j \rangle} \tag{17-1}$$

Here $\bar{A}_{ij}$ and $\bar{A}_{\langle ij \rangle}$ are, respectively, the covariant and physical components of $\mathbf{A}$ with respect to some curvilinear coordinate system. If $\mathbf{A}$ belongs to $\mathscr{S}$, then the domain of definition for $\epsilon$ must be extended to $\mathscr{L}$ by setting $\tilde{\epsilon}(\mathbf{A}) \equiv \epsilon[\frac{1}{2}(\mathbf{A} + \mathbf{A}^T)]$. The derivatives of $\tilde{\epsilon}$ are taken with respect to the nine components of $\mathbf{A}$ as described in Eq. (17-1), after which $\mathbf{A}$ must again be restricted to $\mathscr{S}$.[1]

In addition to restricting ourselves to creeping flows of incompressible fluids as mentioned in Sec. 4.4.16, we require that the constitutive equation for the extra stress $\mathbf{S}$ be expressible in terms of a scalar-valued function $E$ of the rate of deformation

---

[1] If $\mathbf{A}$ belongs to $\mathscr{S}$ and $\epsilon(\mathbf{A}) \equiv A_{12}$, then

$$\frac{\partial \epsilon}{\partial A_{12}} = \frac{\partial \epsilon}{\partial A_{21}} = \frac{1}{2}$$

and

$$\frac{\partial \epsilon}{\partial \mathbf{A}} = \tfrac{1}{2}(\mathbf{e}_1 \mathbf{e}_2 + \mathbf{e}_2 \mathbf{e}_1)$$

tensor:

$$\mathbf{S} = \frac{\partial E}{\partial \mathbf{D}} \tag{17-2}$$

In Sec. 2.3.3 we discussed a common class of empirical models for incompressible fluids of the form

$$\mathbf{S} = 2\eta(D^2)\mathbf{D} \tag{17-3}$$

where we find it convenient to introduce

$$D \equiv \sqrt{\text{tr } \mathbf{D}^2} \tag{17-4}$$

For this class of models,

$$E = E(D) = \int_0^{D^2} \eta(D^2) \, dD^2 \tag{17-5}$$

In order to confirm that Eq. (17-3) has the required form of Eq. (17-2), differentiate Eq. (17-5) to find

$$\frac{\partial E}{\partial \mathbf{D}} = 2\eta(D^2)\mathbf{D} \tag{17-6}$$

Let us now visualize a particular flow for which $\mathbf{v}$ is the actual velocity distribution. By this I mean $\mathbf{v}$ satisfies the equation of motion, equation of continuity, and all the required boundary conditions. Now let $\mathbf{v}^*$ be some trial or approximate velocity distribution, the properties of which we shall specify later. Let us expand $E(D)$ in a truncated Taylor series with a remainder term:

$$E(D^*) - E(D) = \frac{\partial E}{\partial D_{ij}}\bigg|_D (D_{ij}^* - D_{ij})$$

$$+ \frac{1}{2} \frac{\partial^2 E}{\partial D_{ij} \, \partial D_{mn}}\bigg|_{\bar{D}} (D_{ij}^* - D_{ij})(D_{mn}^* - D_{mn}) \tag{17-7}$$

For the sake of simplicity, we speak in terms of a rectangular cartesian coordinate system; the $D_{ij}$ are the rectangular cartesian components of $\mathbf{D}$. Here $\bar{\mathbf{D}}$ is a suitable value between $\mathbf{D}$ and $\mathbf{D}^*$. Recognizing that $E$ is a function of $D$, we can calculate

$$\frac{\partial E}{\partial D_{ij}} = \frac{1}{D} \frac{dE}{dD} D_{ij} \tag{17-8}$$

and

$$\frac{\partial^2 E}{\partial D_{ij} \, \partial D_{mn}} = \left( \frac{1}{D^2} \frac{d^2 E}{dD^2} - \frac{1}{D^3} \frac{dE}{dD} \right) D_{mn} D_{ij} + \frac{1}{D} \frac{dE}{dD} \delta_{im} \delta_{jn} \tag{17-9}$$

## [4.4] INTEGRAL BALANCES

In view of Eqs. (17-8) and (17-9), we can express Eq. (17-7) as

$$E(D^*) - E(D) = \frac{1}{\bar{D}} \frac{dE}{dD}\bigg|_{\bar{D}} \bar{D}_{ij}(D^*_{ij} - D_{ij})$$

$$+ \frac{1}{2}\left(\frac{1}{\bar{D}^2}\frac{d^2E}{dD^2}\bigg|_{\bar{D}} - \frac{1}{\bar{D}^3}\frac{dE}{dD}\bigg|_{\bar{D}}\right)[\bar{D}_{ij}(D^*_{ij} - D_{ij})]^2$$

$$+ \frac{1}{2}\frac{1}{\bar{D}}\frac{dE}{dD}\bigg|_{\bar{D}}(D^*_{ij} - D_{ij})(D^*_{ij} - D_{ij}) \quad (17\text{-}10)$$

The Schwarz inequality [2, p. 125] requires

$$[\bar{D}_{ij}(D^*_{ij} - D_{ij})]^2 \leq \bar{D}_{mn}\bar{D}_{mn}[(D^*_{ij} - D_{ij})(D^*_{ij} - D_{ij})]$$

$$= \bar{D}^2(D^*_{ij} - D_{ij})(D^*_{ij} - D_{ij}) \quad (17\text{-}11)$$

Assuming for the moment that

$$\frac{1}{\bar{D}}\frac{dE}{dD}\bigg|_{\bar{D}} \geq 0 \quad (17\text{-}12)$$

we can use inequality (17-11) to learn

$$E(D^*) - E(D) \geq \frac{1}{\bar{D}}\frac{dE}{dD}\bigg|_{\bar{D}}\bar{D}_{ij}(D^*_{ij} - D_{ij})$$

$$+ \frac{1}{2}\frac{1}{\bar{D}^2}\frac{d^2E}{dD^2}\bigg|_{\bar{D}}[\bar{D}_{ij}(D^*_{ij} - D_{ij})]^2 \quad (17\text{-}13)$$

Differentiating Eq. (17-5), we find

$$\frac{1}{D}\frac{dE}{dD} = 2\eta \quad (17\text{-}14)$$

This justifies inequality (17-12); in Sec. 2.3.3 we pointed out that all known experimental data indicate that $\eta$ should be positive. Equation (17-14) further allows us to say

$$S \equiv \sqrt{\operatorname{tr} \mathbf{S}^2} = \frac{1}{D}\frac{dE}{dD}\sqrt{\operatorname{tr} \mathbf{D}^2}$$

$$= \frac{dE}{dD} \quad (17\text{-}15)$$

Remembering that in Sec. 2.3.3 we argued that

$$\frac{dS}{dD} = \frac{d^2E}{dD^2} \geq 0 \quad (17\text{-}16)$$

we can finally write inequality (17-13) as

$$E(D^*) - E(D) \geq \frac{1}{\bar{D}}\frac{dE}{dD}\bigg|_{\bar{D}}\bar{D}_{ij}(D^*_{ij} - D_{ij}) \quad (17\text{-}17)$$

Equations (17-3) and (17-14) allow us to express this in a more useful alternate form:

$$E(D^*) - E(D) \geq \text{tr}(\mathbf{S} \cdot [\mathbf{D}^* - \mathbf{D}]) \tag{17-18}$$

Our primary interest is in the integral of inequality (17-18) over an arbitrary system:

$$\int_{V_{(s)}} [E(D^*) - E(D) - \text{tr}(\mathbf{S} \cdot \mathbf{D}^*) + \text{tr}(\mathbf{S} \cdot \mathbf{D})] \, dV \geq 0 \tag{17-19}$$

Looking at the third term on the left, we can reason that

$$\begin{aligned}
\int_{V_{(s)}} \text{tr}(\mathbf{S} \cdot \mathbf{D}) \, dV &= \int_{V_{(s)}} \text{tr}(\mathbf{S} \cdot \nabla \mathbf{v}) \, dV \\
&= \int_{V_{(s)}} \text{tr}([\mathbf{T} - \rho\varphi\mathbf{I}] \cdot \nabla \mathbf{v}) \, dV \\
&= \int_{V_{(s)}} \{\text{div}([\mathbf{T} - \rho\varphi\mathbf{I}] \cdot \mathbf{v}) - \mathbf{v} \cdot \text{div}(\mathbf{T} - \rho\varphi\mathbf{I})\} \, dV \\
&= \int_{V_{(s)}} \text{div}([\mathbf{T} - \rho\varphi\mathbf{I}] \cdot \mathbf{v}) \, dV \\
&= \int_{S_{(s)}} \mathbf{v} \cdot [(\mathbf{T} - \rho\varphi\mathbf{I}) \cdot \mathbf{n}] \, dS
\end{aligned} \tag{17-20}$$

In the first line we have used symmetry of the extra-stress tensor; in the second, the equation of continuity for incompressible fluid; the third is an integration by parts; at the fourth, we have applied Cauchy's first law for creeping motion; last we have used Green's transformation. Let us now *define* $\mathbf{v}^*$ to be an approximate or trial velocity distribution that satisfies the equation of continuity for an incompressible fluid as well as any explicit boundary conditions on velocity. This allows us to calculate in the same way that

$$\int_{V_{(s)}} \text{tr}(\mathbf{S} \cdot \mathbf{D}^*) \, dV = \int_{S_{(s)}} \mathbf{v}^* \cdot [(\mathbf{T} - \rho\varphi\mathbf{I}) \cdot \mathbf{n}] \, dS \tag{17-21}$$

Equations (17-19) to (17-21) enable us to conclude that

$$\int_{V_{(s)}} E(D^*) \, dV - \int_{S_{(s)} - S_v} (\mathbf{v}^* - \mathbf{v}) \cdot ([\mathbf{T} - \rho\varphi\mathbf{I}] \cdot \mathbf{n}) \, dS$$

$$\geq \int_{V_{(s)}} E(D) \, dV \tag{17-22}$$

Here $S_v$ is that portion of the closed bounding surface for the system upon which velocity is specified explicitly; $S_{(s)} - S_v$ is that portion of the closed bounding surface upon which velocity is not explicitly specified. Inequality (17-22) may be referred to as the *velocity extremum principle*.

The physical significance of the velocity extremum principle will be developed in Sec. 4.4.19.

## REFERENCES

1. Truesdell, C., and W. Noll: In S. Flügge (ed.), "Handbuch der Physik," vol. 3/3, Springer-Verlag, Berlin, 1965.
2. Halmos, P. R.: "Finite-dimensional Vector Spaces," 2d ed., Van Nostrand, Princeton, N.J., 1958.

**4.4.18 The stress extremum principle** The development in this section parallels that presented in Sec. 4.4.17.

We again restrict ourselves to creeping motions of incompressible fluids. In addition, we require that the constitutive equation for the extra stress be expressible in the form

$$\mathbf{D} = \frac{\partial E_c}{\partial \mathbf{S}} \tag{18-1}$$

The derivative operation is the same as that defined in Sec. 4.4.17.

In Sec. 2.3.3, we talk about a simple class of constitutive equations that have the general form

$$2\mathbf{D} = \varphi(S^2)\mathbf{S} \tag{18-2}$$

where we now find it convenient to speak in terms of

$$S = \sqrt{\operatorname{tr} \mathbf{S}^2} \tag{18-3}$$

For this class of constitutive equations,

$$E_c = E_c(S) = \int_0^{S^2} \tfrac{1}{4}\varphi(S^2)\,dS^2 \tag{18-4}$$

as can be readily verified by differentiating to find

$$\frac{\partial E_c}{\partial \mathbf{S}} = \tfrac{1}{2}\varphi(S^2)\mathbf{S} \tag{18-5}$$

As suggested in Sec. 2.3.3, all experimental observations to this point have been consistent with saying

$$\varphi(S^2) \geq 0 \tag{18-6}$$

and

$$\frac{dD}{dS} \geq 0 \tag{18-7}$$

This enables us to follow by analogy each step in the derivation of Eq. (17-18) to conclude that

$$E_c(S^*) - E_c(S) \geq \operatorname{tr}\left(\mathbf{D} \cdot [\mathbf{S}^* - \mathbf{S}]\right) \tag{18-8}$$

In arriving at this result, we should think of $\mathbf{S}^*$ as being an approximate or trial stress distribution, the precise restrictions on which will be indicated shortly.

We are primarily interested in the volume integral of Eq. (18-8) over the region occupied by an arbitrary system:

$$\int_{V_{(s)}} [E_c(S^*) - E_c(S) - \text{tr}(\mathbf{D} \cdot \mathbf{S}^*) + \text{tr}(\mathbf{D} \cdot \mathbf{S})] \, dV \geq 0 \tag{18-9}$$

Let us identify $\mathbf{T}^*$ as a trial or approximate stress distribution that satisfies Cauchy's first law for creeping motion:

$$\text{div}(\mathbf{T}^* - \rho\varphi\mathbf{I}) = 0 \tag{18-10}$$

but which does not necessarily satisfy all of the required boundary conditions on the stress tensor. Let us further define the trial extra-stress tensor $\mathbf{S}^*$ as

$$\mathbf{S}^* \equiv \mathbf{T}^* - \tfrac{1}{3}(\text{tr } \mathbf{T}^*)\mathbf{I} \tag{18-11}$$

This amounts to defining $p^*$ as the mean pressure (see Sec. 2.3.2). Because of Eq. (18-10), we can calculate in a manner analogous to the derivation of Eq. (17-20) that

$$\int_{V_{(s)}} \text{tr}(\mathbf{D} \cdot \mathbf{S}^*) \, dV = \int_{S_{(s)}} \mathbf{v} \cdot [(\mathbf{T}^* - \rho\varphi\mathbf{I}) \cdot \mathbf{n}] \, dS \tag{18-12}$$

This, together with Eq. (17-20), allows inequality (18-9) to be expressed as

$$\int_{V_{(s)}} E_c(S^*) \, dV - \int_{S_{(s)}} \mathbf{v} \cdot [(\mathbf{T}^* - \mathbf{T}) \cdot \mathbf{n}] \, dS \geq \int_{V_{(s)}} E_c(S) \, dV \tag{18-13}$$

From Eqs. (18-2) and (18-4), we know that

$$\frac{dE_c}{dS} = \frac{\varphi S}{2} = D \tag{18-14}$$

In a similar manner, from Eqs. (17-3) and (17-5) we have

$$\frac{dE}{dD} = 2\eta D = S \tag{18-15}$$

These expressions can be used to reason that

$$\int_0^S \frac{dE_c}{dS} \, dS + \int_0^D \frac{dE}{dD} \, dD = \int_0^S D \, dS + \int_0^D S \, dD = \int_0^{SD} d(SD) \tag{18-16}$$

or

$$E_c + E = SD = 2\eta D^2 = \text{tr}(\mathbf{S} \cdot \mathbf{D}) \tag{18-17}$$

Integrating this over the region occupied by our arbitrary system, we find with the help of Eq. (17-20) that

$$\int_{V_{(s)}} E_c(S) \, dV = -\int_{V_{(s)}} E(D) \, dV + \int_{V_{(s)}} \text{tr}(\mathbf{S} \cdot \mathbf{D}) \, dV$$

$$= -\int_{V_{(s)}} E(D) \, dV + \int_{S_{(s)}} \mathbf{v} \cdot [(\mathbf{T} - \rho\varphi\mathbf{I}) \cdot \mathbf{n}] \, dS \tag{18-18}$$

Equation (18-18) is the key to obtaining an interesting relationship, because it allows us to conclude from Eq. (18-13) that

$$\int_{V_{(s)}} E(D)\, dV \geq -\int_{V_{(s)}} E_c(S^*)\, dV + \int_{S_{(s)}} \mathbf{v} \cdot [(\mathbf{T}^* - \rho\varphi\mathbf{I}) \cdot \mathbf{n}]\, dS \tag{18-19}$$

We will hereafter refer to this inequality as the *stress extremum principle*.

Notice that the velocity and stress extremum principles give us upper and lower bounds to the same quantity:

$$\int_{V_{(s)}} E(D)\, dV$$

In the next section we shall examine the physical significance of this integral.

### EXERCISE

**4.4.18-1**  Derive Eq. (18-12).

### 4.4.19 Physical interpretation of the velocity and stress extremum principles

If for a fixed value of $p$,

$$t^p E(D) = E(tD) \tag{19-1}$$

no matter what value $t$ assumes, then $E$ is called a homogeneous function of degree $p$. Euler's theorem on homogeneous functions says that [1, p. 90]

$$pE = D\frac{dE}{dD} \tag{19-2}$$

Using Eqs. (17-2) and (17-8), we arrive at an interesting result:

$$\begin{aligned}
E &= \frac{1}{p} D_{ij} \frac{\partial E}{\partial D_{ij}} \\
&= \frac{1}{p} D_{ij} S_{ij} \\
&= \frac{1}{p} \operatorname{tr}(\mathbf{S} \cdot \nabla\mathbf{v})
\end{aligned} \tag{19-3}$$

The function $E$ is proportional to the rate of dissipation of energy in the fluid per unit volume.

The practicality of this expression is that $E$ is homogeneous for some simple models of fluid behavior, such as the newtonian fluid ($p = 2$) and the power model ($p = n + 1$). But not all of the simple models for fluid behavior discussed in Sec. 2.3.3 correspond to homogeneous functions $E$. The Ellis model must be handled in a different manner in order to relate $E$ to the rate of dissipation of energy in the system.

When $E$ is a homogeneous function of degree $p$, the velocity and stress extremum principles discussed in Secs. 4.4.17 and 4.4.18, respectively, require

$$\int_{V_{(s)}} E(D^*)\, dV - \int_{S_{(s)}-S_v} (\mathbf{v}^* - \mathbf{v}) \cdot [(\mathbf{T} - \rho\varphi\mathbf{I}) \cdot \mathbf{n}]\, dS \geq \frac{1}{p} \int_{V_{(s)}} \mathrm{tr}(\mathbf{S} \cdot \nabla \mathbf{v})\, dV \quad (19\text{-}4)$$

and

$$\frac{1}{p} \int_{V_{(s)}} \mathrm{tr}\,(\mathbf{S} \cdot \nabla \mathbf{v})\, dV \geq -\int_{V_{(s)}} E_c(S^*)\, dV + \int_{S_{(s)}} \mathbf{v} \cdot [(\mathbf{T}^* - \rho\varphi\mathbf{I}) \cdot \mathbf{n}]\, dS \quad (19\text{-}5)$$

For this restricted class of fluids, the extremum principles can be used to bound the rate of dissipation of energy in an arbitrary system.

The integral mechanical energy balance appears to be the key to the application of these extremum principles, since it alone of the integral balances discussed so far involves the rate of dissipation of energy. This will become clearer upon reading the next section.

## REFERENCES

1. Kaplan, Wilfred: "Advanced Calculus," Addison-Wesley, Cambridge, Mass., 1952.
2. Hopke, S. W., and J. C. Slattery: *A.I.Ch.E.J.*, **16**:224 (1970).
3. Bird, R. B.: *Can. J. Chem. Eng.*, **8**:161 (1965).

## EXERCISES

**4.4.19-1** *The newtonian and power models* (a) Prove that for the newtonian fluid $E$ is a homogeneous function of degree 2.

(b) Prove that for the power model $E$ is a homogeneous function of degree $n + 1$.

**4.4.19-2** *The Ellis model* [2] The use of these bounding principles for an Ellis model fluid poses a special problem as will become obvious shortly.

(a) Determine that for an Ellis model fluid

$$E_c = \frac{1}{4\eta_0}\left[1 + \frac{2}{\alpha+1}\left(\frac{S}{\sqrt{2}\,\tau_{\frac{1}{2}}}\right)^{\alpha-1}\right] S^2$$

$$\mathrm{tr}\,(\mathbf{S} \cdot \nabla \mathbf{v}) = \frac{1}{2\eta_0}\left[1 + \left(\frac{S}{\sqrt{2}\,\tau_{\frac{1}{2}}}\right)^{\alpha-1}\right] S^2$$

and

$$E = \frac{1}{4\eta_0}\left[1 + \frac{2\alpha}{\alpha+1}\left(\frac{S}{\sqrt{2}\,\tau_{\frac{1}{2}}}\right)^{\alpha-1}\right] S^2$$

Notice that $E$ is not a homogeneous function and the discussion in the text is not applicable.

(b) For the majority of fluids, $\alpha \geq 1$ [3]. If we restrict ourselves to $\alpha \geq 1$, determine that

$$E \geq \tfrac{1}{2}\,\mathrm{tr}\,(\mathbf{S} \cdot \nabla \mathbf{v})$$

and

$$E \leq \frac{\alpha}{\alpha+1}\,\mathrm{tr}\,(\mathbf{S} \cdot \nabla \mathbf{v})$$

## [4.4] INTEGRAL BALANCES

These inequalities together with the velocity and stress extremum principles allow us to obtain upper and lower bounds to the rate of dissipation of energy in an arbitrary system. Notice however that by their very nature these bounds cannot be as close as those that can be obtained for newtonian and power-model fluids.

### 4.4.20 Upper and lower bounds to the drag coefficient for a sphere [1, 2]

An incompressible Ellis model fluid moves slowly past a stationary sphere of radius $R$. The fluid very far away from the sphere moves with a uniform velocity $\mathbf{v}_\infty$. We wish to determine the drag coefficient for this sphere.

Turning back to Sec. 3.3.3, you will see that this is basically the same problem, except for the change in the description of fluid behavior. This is a major change. When we considered a newtonian fluid in Sec. 3.3.3, the mathematical problem developed was linear. Now Cauchy's first law leads to a nonlinear differential equation for the stream function. It does not appear that this differential equation has an analytic solution. Nor does it appear to be particularly attractive to grind out a solution to this partial differential equation numerically on a computer. Having made these two decisions, we can begin casting about for an approximate solution or for a way of estimating the drag coefficient.

Exercise 4.4.19-2 indicates how the velocity and stress extremum principles can be used to calculate upper and lower bounds to the rate of dissipation of energy in the fluid. Exercise 4.4.12-8 develops the relationship between the rate of the dissipation of energy in an unbounded fluid and the component of force in the direction of flow on an arbitrary submerged body. Putting these two ideas together, it becomes clear that we should be able to bound the drag coefficient for a sphere moving slowly through an Ellis model fluid.

I think it would be clearer if we consider the bounding principles independently.

#### THE VELOCITY EXTREMUM PRINCIPLE

It may be best to begin by looking at what it is we wish to obtain. We know that our extremum principles give us upper and lower bounds to the rate of dissipation of energy in the fluid. From Exercise 4.4.12-8 we see that

$$\mathbf{v}_\infty \cdot \mathscr{G} = \int_{V_{(s)}} \operatorname{tr}(\mathbf{S} \cdot \nabla \mathbf{v}) \, dV \tag{20-1}$$

where $\mathscr{G}$ is the force that the fluid exerts upon the sphere over and above the buoyancy force. In Example 1 of Sec. 4.4.7, I defined the drag coefficient for a sphere. For the case considered here, the drag coefficient can be expressed as

$$c = \frac{\mathbf{v}_\infty \cdot \mathscr{G}}{\frac{1}{2} \rho v_\infty^3 \pi R^2} \tag{20-2}$$

These relationships indicate that the desired drag coefficient is directly proportional to the rate of dissipation of energy in the fluid.

In order to use the velocity extremum principle, it is first necessary that we propose a trial velocity distribution $\mathbf{v}^*$ which satisfies the equation of continuity as well as all the boundary conditions on velocity. Looking at the way that we approached the problem for a newtonian fluid in Sec. 3.3.3, we are motivated to assume that in spherical coordinates this trial velocity distribution has only two nonzero components, $v_r^*$ and $v_\theta^*$. With this understanding, the equation of continuity for an incompressible fluid can be satisfied identically be expressing these two velocity components in terms of a stream function:

$$v_r^* = \frac{1}{r^2 \sin\theta} \frac{\partial \psi^*}{\partial \theta} \qquad v_\theta^* = -\frac{1}{r \sin\theta} \frac{\partial \psi^*}{\partial r} \tag{20-3}$$

The boundary conditions that we must require this trial stream function to satisfy are developed in Sec. 3.3.3:

$$\text{At } r = R: \qquad \frac{1}{r^2 \sin\theta} \frac{\partial \psi^*}{\partial \theta} = 0 \tag{20-4}$$

$$\text{At } r = R: \qquad \frac{1}{r \sin\theta} \frac{\partial \psi^*}{\partial r} = 0 \tag{20-5}$$

and

$$\text{As } r \to \infty: \qquad \psi^* \to \tfrac{1}{2} r^2 v_\infty \sin^2\theta \tag{20-6}$$

A simple function that satisfies these three conditions is

$$\psi^* = \tfrac{1}{2} v_\infty r^2 \sin^2\theta \left[ 1 - \left(\frac{R}{r}\right)^a \right]^2 \tag{20-7}$$

Here $a$ is an as yet undetermined parameter.

For this trial stream function, we find that

$$\tilde{D}^{*2} = 6a^2 x^{2(a+1)} \{(1 - x^a)^2 \cos^2\theta + \tfrac{1}{12}[a - 1 - (2a - 1)x^a]^2 \sin^2\theta\} \tag{20-8}$$

It has been convenient here to introduce as new dimensionless variables:

$$\tilde{D} \equiv \frac{RD}{v_\infty} \qquad x \equiv \frac{R}{r} \tag{20-9}$$

Exercise 4.4.19-2 presents an expression for $E$ in terms of $S$. Unfortunately, the Ellis model is such that an explicit expression for $S$ in terms of $D$ cannot be obtained. This suggests that we define a quantity $S^{**}$ in terms of $D^*$ by means of

$$\tilde{D}^{*2} = \tfrac{1}{4}[1 + (N_1 \tilde{S}^{**})^{\alpha-1}]^2 \tilde{S}^{**2} \tag{20-10}$$

We have introduced here

$$\tilde{S} \equiv \frac{RS}{\eta_0 v_\infty} \qquad N_1 \equiv \frac{\eta_0 v_\infty}{\sqrt{2}\, R \tau_{\frac{1}{2}}} \tag{20-11}$$

## [4.4] INTEGRAL BALANCES

This allows us to write

$$E(D^*) = \frac{1}{4\eta_0}\left[1 + \frac{2\alpha}{\alpha+1}\left(\frac{S^{**}}{\sqrt{2\,\tau_{\frac{1}{2}}}}\right)^{\alpha-1}\right]S^{**2} \tag{20-12}$$

where $\tilde{S}^{**}$ is to be found from Eq. (20-10).

Since velocity is specified on the entire bounding surface of the system, which is the fluid, Eq. (20-12) can be used to write the velocity extremum principle of Sec. 4.4.17 as

$$\frac{4}{Rv_\infty^2\eta_0}\int_{V_{(s)}} E\,dV \leq \frac{1}{R^3}\int_{V_{(s)}}\left[1 + \frac{2\alpha}{\alpha+1}(N_1\tilde{S}^{**})^{\alpha-1}\right]\tilde{S}^{**2}\,dV \tag{20-13}$$

But in Exercise 4.4.19-2 we proved that

$$E \geq \tfrac{1}{2}\,\text{tr}\,(\mathbf{S}\cdot\nabla\mathbf{v}) \tag{20-14}$$

This, together with Eqs. (20-1) and (20-2), enables us to conclude from Eq. (20-13) that

$$cN_2 \leq 4\int_0^\pi\int_0^1\left[1 + \frac{2\alpha}{\alpha+1}(N_1\tilde{S}^{**})^{\alpha-1}\right]\tilde{S}^{**2}x^{-4}\sin\theta\,dx\,d\theta \tag{20-15}$$

By definition here,

$$N_2 \equiv \frac{2Rv_\infty\rho}{\eta_0} \tag{20-16}$$

The two integrals on the right side of inequality (20-15) were evaluated using Simpson's rule in two dimensions. At each point and for each value of $a$, $\tilde{S}^{**}$ was calculated from Eqs. (20-8) and (20-10) by a Newton-Raphson iteration. The optimum value of $a$, corresponding to a minimum upper bound in inequality (20-15), was found by a Fibonacci search [3, p. 24]. The results of these computations are shown in Fig. 4.4.20-1.

### THE STRESS EXTREMUM PRINCIPLE

In order to apply the stress extremum principle, we require a trial stress distribution $\mathbf{T}^*$ that satisfies Cauchy's first law for a creeping motion. One possibility is to say that

$$S_{r\theta}^* = -B\left(\frac{\eta_0 v_\infty}{R}\right)x^4\sin\theta \tag{20-17}$$

$$S_{\theta\theta}^* = S_{\varphi\varphi}^* = -\tfrac{1}{2}S_{rr}^*$$

$$= -B\left(\frac{\eta_0 v_\infty}{R}\right)(x^2 - x^4)\cos\theta \tag{20-18}$$

and

$$p^* + \rho\varphi \equiv p_0 - B\frac{\eta_0 v_\infty}{R}x^2\cos\theta \tag{20-19}$$

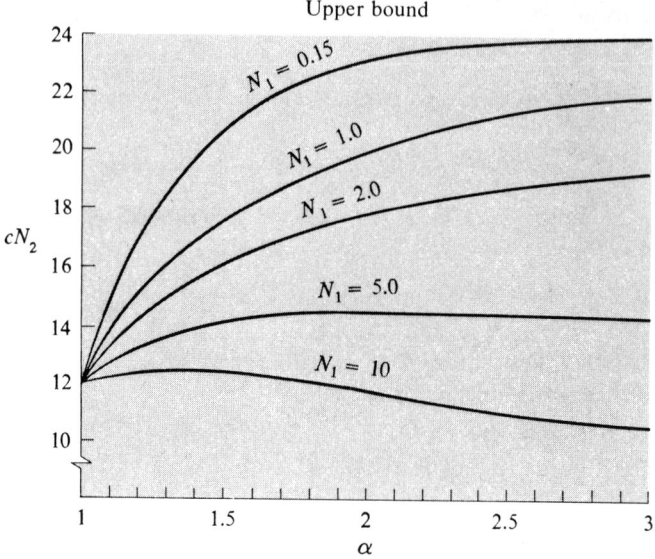
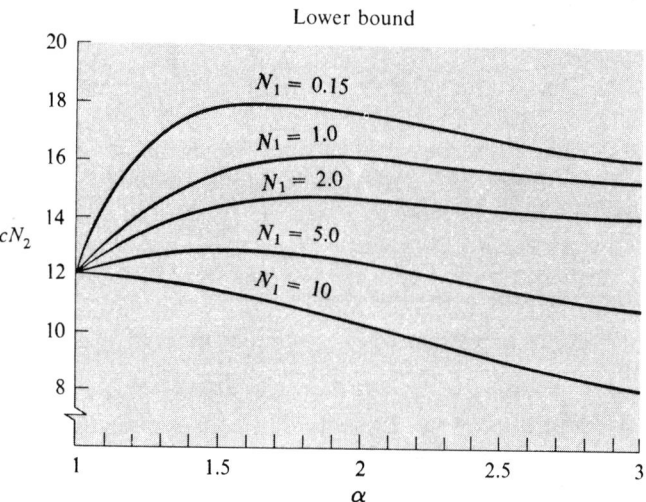

**Fig. 4.4.20-1** Upper and lower bounds to $cN_2$ as functions of $\alpha$ and $N_1$.

## [4.4] INTEGRAL BALANCES

The parameter $B$ is as yet undetermined.

In Exercise 4.4.19-2, we found

$$E_c(S^*) = \frac{1}{4\eta_0}\left[1 + \frac{2}{\alpha+1}\left(\frac{S^*}{\sqrt{2}\,\tau_{\frac{1}{2}}}\right)^{\alpha-1}\right]S^{*2} \qquad (20\text{-}20)$$

Equations (20-17) to (20-19) allow us to compute

$$\tilde{S}^{*2} = 2B^2 x^4[x^4 \sin^2\theta + 3(1-x^2)^2\cos^2\theta] \qquad (20\text{-}21)$$

and

$$\int_{S_{(s)}} \mathbf{v}\cdot[(\mathbf{T}^* - \rho\varphi\mathbf{I})\cdot\mathbf{n}]\,dS = 4\pi\eta_0 R v_\infty^2 B \qquad (20\text{-}22)$$

We may now employ Eqs. (20-20) and (20-22) to write the stress extremum principle of Sec. 4.4.18 in the form

$$\frac{4}{Rv_\infty^2\eta_0}\int_{V_{(s)}} E\,dV \geq 16\pi B - \frac{1}{R^3}\int_{V_{(s)}}\left[1 + \frac{2}{\alpha+1}(N_1\tilde{S}^*)^{\alpha-1}\right]\tilde{S}^{*2}\,dV \qquad (20\text{-}23)$$

the first integral on the right of which may be evaluated as

$$\frac{1}{R^3}\int_V \tilde{S}^{*2}\,dV = 2\pi \int_0^\pi \int_0^1 \tilde{S}^{*2} x^{-4}\sin\theta\,dx\,d\theta$$

$$= \tfrac{16}{3}\pi B^2 \qquad (20\text{-}24)$$

But in Exercise 4.4.19-2, we prove that

$$\frac{\alpha+1}{\alpha}E \leq \operatorname{tr}(\mathbf{S}\cdot\nabla\mathbf{v}) \qquad (20\text{-}25)$$

This, together with Eqs. (20-1), (20-2), and (20-24), allows us to conclude from inequality (20-23) that

$$cN_2 \geq \frac{16B(\alpha+1)}{\alpha} - \frac{16(\alpha+1)}{3}\frac{B^2}{\alpha}$$

$$- \frac{4B^{\alpha+1}}{\alpha}\int_0^\pi\int_0^1 N_1^{\alpha-1}\left(\frac{\tilde{S}^*}{B}\right)^{\alpha+1} x^{-4}\sin\theta\,dx\,d\theta \qquad (20\text{-}26)$$

The integral in inequality (20-26) was evaluated by applying Simpson's rule in two dimensions. The computational error is estimated to be 1 percent; in several cases, the number of panels was doubled and redoubled with less than 1 percent change in the result. The parameter $B$ in Eqs. (20-17) to (20-19) was determined by requiring the lower bound in inequality (20-26) to be a maximum. Figure 4.4.20-1 shows the result of these computations.

**COMPARISON WITH EXPERIMENT**

Besides experimental error, there are five reasons that may explain any differences between the results predicted here and experimentally observed drag coefficients.

1. The Ellis model is an approximation for the actual stress deformation relationship as observed for flow through a tube (see Sec. 3.2.2) or in some other viscometer.
2. There are good theoretical reasons which suggest that, no matter how well the Ellis model describes the stress rate of deformation relationship for flow through a tube, it may not be able to describe the behavior exhibited by a fluid in flowing past a sphere [4].
3. The developments for the two extremum principles assume from the beginning that inertial effects may be neglected with respect to viscous effects in Cauchy's first law. Any criterion for creeping flow must be the result of experimental observations. Our best current estimate is that the analysis presented here should be applicable for [2]

$$N_2 < 0.1 \tag{20-27}$$

4. It should be clear from inequalities (20-14) and (20-25) that as $(\alpha - 1) \to \infty$ our upper and lower bounds to $cN_2$ must diverge, no matter how close they are for $E$. Best results are obtained for values of $\alpha$ not too far removed from unity.
5. Finally, our upper and lower bounds will diverge to the extent that $\mathbf{v}^*$ and $\mathbf{T}^*$ are not reasonable representations for the actual velocity and stress distributions.

A detailed comparison of available experimental data with the average of the calculated upper and lower bounds to $c$ predicted here has been presented elsewhere [2]. It is perhaps sufficient to say that the average error in predicting 24 experiments observed by Turian [5] is 16 percent; for 19 experiments by Dallon [6] the average error is 10.2 percent. The agreement appears reasonable in view of the problems mentioned above.

For additional applications, see Refs. 4 and 7 to 13.

## REFERENCES

1. Hopke, S. W., and J. C. Slattery: *A.I.Ch.E.J.*, **16**:224 (1970).
2. Hopke, S. W., and J. C. Slattery: *A.I.Ch.E.J.*, **16**:317 (1970).
3. Wilde, D. J.: "Optimum Seeking Methods," Prentice-Hall, Englewood Cliffs, N.J., 1964.
4. Wasserman, M. L., and J. C. Slattery: *A.I.Ch.E.J.*, **10**:383 (1964).
5. Turian, R. M.: doctoral dissertation, Dept. of Chem. Engng., University of Wisconsin, Madison, 1963.
6. Dallon, D. S.: doctoral dissertation, Dept. of Chem. Engng., University of Utah, Salt Lake City, 1967.
7. Hill, R., and G. Power: *Quart. J. Mech. Appl. Math.*, **9**:313 (1956).
8. Kearsley, E. A.: *Arch. Rational Mech. Analysis*, **5**:347 (1960).
9. Stewart, W. E.: *A.I.Ch.E.J.*, **8**:425 (1962).
10. Bird, R. B., and R. M. Turian: *Chem. Eng. Sci.*, **17**:331 (1962).
11. Slattery, J. C.: *A.I.Ch.E.J.*, **8**:663 (1962).
12. Sani, R. L.: *A.I.Ch.E.J.*, **9**:277 (1963).
13. Ehrlich, Robert, and J. C. Slattery: *Ind. Eng. Chem., Fundamentals*, **7**:239 (1968).
14. Lamb, Horace: "Hydrodynamics," 6th ed., Dover, New York, 1945.

[4.4] INTEGRAL BALANCES  277

**EXERCISES**

**4.4.20-1** *Comparing the drags on two bodies* [7]  An infinite expanse of fluid moves past an arbitrary body $A$ that is fixed in space.  Very far away from the body, the fluid moves with a constant velocity $\mathbf{v}_\infty$.  An infinite expanse of the same fluid also moves past another arbitrary body $B$ with the same boundary condition at infinity.  The shape or orientation of body $B$ may be entirely different from that of $A$, but body $B$ could be completely contained within body $A$.

Restricting ourselves to incompressible materials for which $E$ exists and is a homogeneous function of degree $n$, use the velocity extremum principle to prove that

$$\mathscr{G}_{(A)} \cdot \frac{\mathbf{v}_\infty}{|\mathbf{v}_\infty|} \geq \mathscr{G}_{(B)} \cdot \frac{\mathbf{v}_\infty}{|\mathbf{v}_\infty|}$$

This says that for creeping motions the drag (beyond any effect attributable to the ambient pressure or buoyancy force) on body $B$ is less than or equal to the drag on body $A$.

*Hint:* See Exercise 4.4.12-8.

**4.4.20-2** *A closed torus* [7]  The closed torus is a surface of revolution that is obtained by revolving a circle of diameter $a$ about a tangent.  In spherical coordinates, its surface is $r = a \sin\theta$.

Consider creeping flow of an unbounded incompressible newtonian fluid past a stationary closed torus.  Very far away from the torus, the fluid moves with a uniform velocity $\mathbf{v}_\infty$.  The body is oriented in such a way that its axis of revolution coincides with $\mathbf{v}_\infty$.

Use Exercise 4.4.20-1, the drag on a disk [14, p. 605], and Stokes' law for spheres to conclude

$$\frac{10.19}{N_{\text{Re}}} \leq c \leq \frac{12}{N_{\text{Re}}}$$

where

$$c = \frac{\mathbf{v}_\infty \cdot \mathscr{G}}{\frac{1}{2}\rho v_\infty^3 \pi a^3}$$

and

$$N_{\text{Re}} \equiv \frac{a v_\infty \rho}{\mu}$$

This means that the drag coefficient for a torus is

$$c = \frac{11.09}{N_{\text{Re}}}$$

with an error not greater than 0.91 in the numerical coefficient.

Hill and Power [7] suggest a lower upper bound, using the known drag on a spheroid.

# 5
# Foundations for Energy Transfer

This chapter is concerned with the foundations for energy transfer. And yet perhaps that is not sufficiently descriptive of what we are actually going to do here, because the foundations for energy transfer are really the foundations for the subject we normally think of as thermodynamics.

Now by thermodynamics I do not mean precisely the subject presented to us by Gibbs, since he was concerned with materials at equilibrium. We are very definitely concerned here with nonequilibrium situations in which momentum and energy are being transferred. (Don't make the mistake of confusing the terms *steady state* and *equilibrium*.)

There are three things that I find give students trouble in their initial introduction to thermodynamics.

1. The caloric equation of state (sometimes referred to as the fundamental equation) is not a fundamental postulate in the same sense as Euler's first and second laws. It is rather a statement of material behavior of the same character as Newton's law of viscosity.

2. Initially there is considerable concern over the physical meaning of specific entropy. I personally find that it is best to look upon entropy as a parameter in the caloric equation of state which is restricted only later with the introduction of the entropy inequality. In Sec. 8.5.1, where I finally attach a precise meaning to the term *equilibrium*, we discover that the entropy of an isolated body is maximized as equilibrium is approached.
3. In the (nonequilibrium) thermodynamics presented here, there are two forms for the entropy inequality (or second law of thermodynamics). There is the differential inequality, the sole purpose of which is to place restrictions upon constitutive equations that describe the behavior of materials. And then there is the integral inequality, which will be developed in Sec. 7.4.7. This will be more familiar to most readers, since it is generally the only form mentioned in undergraduate texts on thermodynamics.

In addition to the ideas introduced in Chaps. 1 and 2, the foundations for thermodynamics and energy transfer are laid upon two postulates: the energy balance and the entropy inequality. The purpose of this chapter is to develop the implications of these two postulates as complemented by two statements of material behavior: the caloric equation of state and Fourier's law.

## 5.1 THERMODYNAMICS

### 5.1.1 Classical thermostatics[1]

The introduction and development of classical thermostatic principles normally follow the methods of Gibbs [1], in which the differential equation (often called the Gibbs equation)

$$dU = T^* \, dS - P^* \, dV + \sum_{A=1}^{N} \mu^*_{(A)} \, dm_{(A)} \qquad (1\text{-}1)$$

its integrated form (sometimes referred to as the Euler form)

$$U = T^* S - P^* V + \sum_{A=1}^{N} \mu^*_{(A)} m_{(A)} \qquad (1\text{-}2)$$

and the Gibbs-Duhem equation [as derived from Eqs. (1-1) and (1-2)]

$$S \, dT^* - V \, dP^* + \sum_{A=1}^{N} m_{(A)} \, d\mu^*_{(A)} = 0 \qquad (1\text{-}3)$$

are applied to a static single-phase system throughout which all intensive or local properties are uniform. Such a system is hereafter described as being *uniform*. (The term *homogeneous* is often used to describe a system throughout which all local properties are uniform. We use the term *thermodynamically homogeneous* [2, p. 620] in another context in Sec. 5.1.2.) Here $U$, $T^*$, $S$, $P^*$, and $V$ are the internal

---

[1] Sections 5.1.1 and 5.1.2 were coauthored with Prof. G. M. Brown, Department of Chemical Engineering, Northwestern University, Evanston, Ill. 60201.

energy, temperature, entropy, thermodynamic pressure, and volume of the system; $m_{(A)}$ is the mass of species $A$ in the system, and $\mu^*_{(A)}$ is the chemical potential of species $A$. With this interpretation, these equations apply rigorously only to those single-phase systems that are not influenced by body forces (such as gravity), by interfacial phenomena, or by other effects which might induce nonuniformity.

An alternative development is to postulate for a static, uniform system the fundamental relation (see, for example, Callen [3, p. 26])

$$U = U(S, V, m_{(1)}, m_{(2)}, \ldots, m_{(N)}) \tag{1-4}$$

and define temperature, thermodynamic pressure, and chemical potential, respectively, as

$$T^* \equiv \frac{\partial U}{\partial S} \qquad P^* \equiv -\frac{\partial U}{\partial V} \qquad \mu^*_{(A)} \equiv \frac{\partial U}{\partial m_{(A)}} \tag{1-5}$$

Equations (1-1), (1-2), and (1-3) may be easily derived from Eqs. (1-4) and (1-5).

We wish to emphasize that all the variables in Eqs. (1-1) to (1-5) refer to properties of the static uniform system. The quantities $T^*$, $P^*$, and $\mu^*_{(A)}$ as applied to the system have no meaning if the system is not static and uniform.

In many cases, it is more convenient to use specific (per unit mass, mole, or volume) properties in place of system properties. Let us denote by $\wedge$ over a symbol that we are dealing with a quantity per unit mass; $\vee$ over a symbol indicates that the quantity is per unit volume. Let us substitute in Eqs. (1-2) and (1-3)

$$\hat{U} \equiv \frac{U}{m} \qquad \hat{S} \equiv \frac{S}{m} \qquad \omega_{(A)} \equiv \frac{m_{(A)}}{m} \tag{1-6}$$

and divide by $m$ $\left(\text{noting that } \sum_{A=1}^{N} \omega_{(A)} = 1\right)$ to obtain

$$\hat{U} = T^*\hat{S} - P^*\hat{V} + \sum_{A=1}^{N} \mu^*_{(A)} \omega_{(A)}$$

$$= T^*\hat{S} - P^*\hat{V} + \sum_{A=1}^{N-1} (\mu^*_{(A)} - \mu^*_{(N)})\omega_{(A)} + \mu^*_{(N)} \tag{1-7}$$

and

$$\hat{S}\,dT^* - \hat{V}\,dP^* + \sum_{A=1}^{N} \omega_{(A)}\,d\mu^*_{(A)} = 0 \tag{1-8}$$

On substituting Eq. (1-6) into Eq. (1-1), we have

$$m\,d\hat{U} + \hat{U}\,dm = mT^*\,d\hat{S} - mP^*\,d\hat{V}$$
$$+ m\sum_{A=1}^{N} \mu^*_{(A)}\,d\omega_{(A)} + \left(T^*\hat{S} - P^*\hat{V} + \sum_{A=1}^{N} \mu^*_{(A)}\omega_{(A)}\right) dm \tag{1-9}$$

After eliminating $\hat{U}$ by Eq. (1-7) and dividing by $m$, we are left with

$$d\hat{U} = T^*\,d\hat{S} - P^*\,d\hat{V} + \sum_{A=1}^{N} \mu^*_{(A)}\,d\omega_{(A)}$$

$$= T^*\,d\hat{S} - P^*\,d\hat{V} + \sum_{A=1}^{N-1} (\mu^*_{(A)} - \mu^*_{(N)})\,d\omega_{(A)} \tag{1-10}$$

[5.1] THERMODYNAMICS

Since Eqs. (1-7), (1-8), and (1-10) have been derived from Eqs. (1-1), (1-2), and (1-3), they would appear to be subject to the same restrictions (a static, uniform system in the absence of gravitational field, etc.). In practice, these equations often are successfully applied without adequate justification to relate the properties at a point in a nonequilibrium system. In the next section we see why this might be expected.

## REFERENCES

1. Gibbs, J. W.: *Trans. Conn. Acad.*, **2**:382 (1873); **3**:108 (1875); **3**:343 (1877); see also "Collected Works," vol. 1, Longmans, New York, 1928.
2. Truesdell, C., and R. A. Toupin: In S. Flügge (ed.), "Handbuch der Physik," vol. 3/1, Springer-Verlag, Berlin, 1960.
3. Callen, H. B.: "Thermodynamics," Wiley, New York, 1963.

## EXERCISES

**5.1.1-1** *The Maxwell relations* Let us define

$$A \equiv U - T^*S$$

$$H \equiv U + P^*V$$

$$G \equiv H - T^*S$$

We refer to $A$ as the Helmholtz free energy, $H$ as enthalpy, and $G$ as the Gibbs free energy. Determine that

(a)
$$\left(\frac{\partial T^*}{\partial V}\right)_{S, m_{(B)}} = -\left(\frac{\partial P^*}{\partial S}\right)_{V, m_{(B)}}$$

$$\left(\frac{\partial T^*}{\partial m_{(A)}}\right)_{S, V, m_{(B)}(B \neq A)} = \left(\frac{\partial \mu^*_{(A)}}{\partial S}\right)_{V, m_{(B)}}$$

$$-\left(\frac{\partial P^*}{\partial m_{(A)}}\right)_{S, V, m_{(B)}(B \neq A)} = \left(\frac{\partial \mu^*_{(A)}}{\partial V}\right)_{S, m_{(B)}}$$

$$\left(\frac{\partial \mu^*_{(A)}}{\partial m_{(B)}}\right)_{S, V, m_{(C)}(C \neq B)} = \left(\frac{\partial \mu^*_{(B)}}{\partial m_{(A)}}\right)_{S, V, m_{(C)}(C \neq A)}$$

(b)
$$\left(\frac{\partial S}{\partial V}\right)_{T^*, m_{(B)}} = \left(\frac{\partial P^*}{\partial T^*}\right)_{V, m_{(B)}}$$

$$-\left(\frac{\partial S}{\partial m_{(A)}}\right)_{T^*, V, m_{(B)}(B \neq A)} = \left(\frac{\partial \mu^*_{(A)}}{\partial T^*}\right)_{V, m_{(B)}}$$

$$-\left(\frac{\partial P^*}{\partial m_{(A)}}\right)_{T^*, V, m_{(B)}(B \neq A)} = \left(\frac{\partial \mu^*_{(A)}}{\partial V}\right)_{T^*, m_{(B)}}$$

$$\left(\frac{\partial \mu^*_{(A)}}{\partial m_{(B)}}\right)_{T^*, V, m_{(C)}(C \neq B)} = \left(\frac{\partial \mu^*_{(B)}}{\partial m_{(A)}}\right)_{T^*, V, m_{(C)}(C \neq A)}$$

(c)
$$\left(\frac{\partial T^*}{\partial P^*}\right)_{S,m_{(B)}} = \left(\frac{\partial V}{\partial S}\right)_{P^*,m_{(B)}}$$

$$\left(\frac{\partial T^*}{\partial m_{(A)}}\right)_{S,P^*,m_{(B)}(B\neq A)} = \left(\frac{\partial \mu_{(A)}^*}{\partial S}\right)_{P^*,m_{(B)}}$$

$$\left(\frac{\partial V}{\partial m_{(A)}}\right)_{S,P^*,m_{(B)}(B\neq A)} = \left(\frac{\partial \mu_{(A)}^*}{\partial P^*}\right)_{S,m_{(B)}}$$

$$\left(\frac{\partial \mu_{(A)}^*}{\partial m_{(A)}}\right)_{S,P^*,m_{(C)}(C\neq B)} = \left(\frac{\partial \mu_{(B)}^*}{\partial m_{(A)}}\right)_{S,P^*,m_{(C)}(C\neq A)}$$

(d)
$$-\left(\frac{\partial S}{\partial P^*}\right)_{T^*,m_{(B)}} = \left(\frac{\partial V}{\partial T^*}\right)_{P^*,m_{(B)}}$$

$$-\left(\frac{\partial S}{\partial m_{(A)}}\right)_{T^*,P^*,m_{(B)}(B\neq A)} = \left(\frac{\partial \mu_{(A)}^*}{\partial T^*}\right)_{P^*,m_{(B)}}$$

$$\left(\frac{\partial V}{\partial m_{(A)}}\right)_{T^*,P^*,m_{(B)}(B\neq A)} = \left(\frac{\partial \mu_{(A)}^*}{\partial P^*}\right)_{T^*,m_{(B)}}$$

$$\left(\frac{\partial \mu_{(A)}^*}{\partial m_{(B)}}\right)_{T^*,P^*,m_{(C)}(C\neq B)} = \left(\frac{\partial \mu_{(B)}^*}{\partial m_{(A)}}\right)_{T^*,P^*,m_{(C)}(C\neq A)}$$

**5.1.1-2** *Partial mass variables* Let $\Phi$ be any extensive variable such as internal energy, enthalpy, or volume. The *partial mass* variable $\overline{\Phi}_{(A)}$ is defined as

$$\overline{\Phi}_{(A)} \equiv \left(\frac{\partial \Phi}{\partial m_{(A)}}\right)_{T^*,P^*,m_{(B)}(B\neq A)}$$

Determine that

(a) $\overline{S}_{(A)} = -\left(\dfrac{\partial \mu_{(A)}^*}{\partial T^*}\right)_{P^*,m_{(B)}}$

(b) $\overline{V}_{(A)} = \left(\dfrac{\partial \mu_{(A)}^*}{\partial P^*}\right)_{T^*,m_{(B)}}$

**5.1.1-3** *Partial molal variables* Let $\Phi$ be any extensive variable such as internal energy, enthalpy, or volume. The *partial molal* variable $\overline{\Phi}_{(A)}^{(m)}$ is defined as

$$\overline{\Phi}_{(A)}^{(m)} \equiv \left(\frac{\partial \Phi}{\partial n_{(A)}}\right)_{T^*,P^*,n_{(B)}(B\neq A)}$$

Here $n_{(A)}$ denotes the number of moles of species $A$. Determine that

$$\overline{\Phi}_{(A)}^{(m)} = M_{(A)}\overline{\Phi}_{(A)}$$

where $M_{(A)}$ is the molecular weight of species $A$.

### 5.1.2 Thermodynamics

As we saw in Sec. 5.1.1, classical thermostatics deals with static systems whose local properties are uniform throughout.

As engineers we are required to deal with and comment upon systems which only an elastic imagination could describe as either uniform or *quasistatic*. We successfully use thermostatics in describing these processes. We are not troubled by discussing how the pressure and temperature may vary from point to point within a material; we do not wait until we have a uniform system to measure temperature and pressure. Nor do we shy away from an analysis of a rapid expansion of a gas through a valve—surely not a quasistatic process.

In practice, we extend the field concepts, which we successfully used in discussing mass and momentum balances in Chaps. 2 to 4, to our study of entropy in dynamic systems: *thermodynamics*. Using a continuum model to represent a real material we introduced there a velocity field $\mathbf{v}$, and a mass density field $\rho$, which allowed us to discuss the local velocity and mass density of the material. In developing thermodynamics, we introduce further fields such as temperature and thermodynamic pressure; in this way we are able to assign to each point in the material a local temperature and thermodynamic pressure. But perhaps we are getting ahead of our story. There are two fundamental quantities which we must introduce first: internal energy and entropy.

If we think in terms of a particulate model for a real material, the electrons, neutrons, and protons have kinetic energy associated with them beyond the kinetic energy of the macroscopic material of which they are a small part. There is also a great deal of energy stored in the material by the action of all the various internal forces acting upon the individual particles. We must surely account for all of this energy in our continuum picture; we call it the *internal energy* of the material. At each point in the continuum model for the material, we say that there is a certain amount of *internal energy per unit mass*, $\hat{U}$. The scalar field $\hat{U}$ is fundamental to our discussion of energy transfer and thermodynamics.

Let $\rho_{(A)}$ denote the mass of species $A$ per unit volume. For an $N$-component system, it seems reasonable that the internal energy per unit mass $\hat{U}$ should be a function of $\rho_{(1)}, \rho_{(2)}, \ldots, \rho_{(N)}$ and position within the material (as specified by a set of material coordinates). But is this sufficient? If we heat a pan of water on the stove, we sense a distinct difference in the water with our fingers. Certainly the internal energy of the water has increased as the result of adding energy in the form of heat. Yet, if we checked the density of the water, we would find little change. This suggests that, if we are to describe the functional dependence of $\hat{U}$, we must say that it depends upon more than mass density and position in the material. For this reason, in our continuum picture we introduce another scalar field $\hat{S}$, called the *entropy per unit mass*. The properties of entropy result from the entropy inequality postulate in Sec. 5.5.1 as well as the definition of equilibrium in Sec. 8.5.1. The physical dimension of $\hat{S}$ is left unnamed, but the dimension of $\hat{U}/\hat{S}$ is referred to as the *dimension of temperature*.

It is perhaps helpful to justify the introduction of $\hat{S}$ from another point of view. Let us think about heating water on the stove in terms of a particulate picture. We visualize that, as we heat the water, more energy is associated with the neutrons, protons, and electrons as the result of their increased motion. When the water is heated sufficiently, the molecules are in such violent motion with respect to one another that they tend to separate and the water is vaporized. As we heat the water, there is greater and greater disorder in our particulate picture of the water until finally the liquid water is transformed into a gas. In our continuum picture, we can think of $\hat{S}$ as being a local measure of disorder in the material.

Let us refer to any variables, other than $\hat{S}$, that we a priori regard as influencing $\hat{U}$ (such as $\rho_{(1)}, \ldots, \rho_{(N)}$ and quantities which describe the deformation of the material from some reference configuration) as the substate variables. The basic assumption of thermodynamics is [1, p. 619]:

> The substate variables $\rho_{(1)}, \ldots, \rho_{(N)}$, etc., plus a single further dimensionally independent scalar parameter $\hat{S}$, are sufficient to determine $\hat{U}$ for the material particle $\mathbf{z}_\varkappa$ independently of time, place, motion, and stress. That is,

$$\hat{U} = \hat{U}(\hat{S}, \rho_{(1)}, \rho_{(2)}, \ldots, \rho_{(N)}, \ldots, \mathbf{z}_\varkappa) \tag{2-1}$$

The basic assumption is really a statement about the behavior of a material—its thermodynamic behavior. For this reason we might refer to Eq. (2-1), the *caloric equation of state* or *fundamental equation*, as being a constitutive equation for internal energy.

We realize that the assumption of the caloric equation of state is not always justified; it does appear to be justified for materials which do not exhibit memory effects, for materials undergoing very rapid deformation, for materials undergoing very slow deformations, and, of course, for materials at equilibrium [2, 3].

When $\mathbf{z}_\varkappa$ does not appear in Eq. (2-1), the material is said to be *thermodynamically homogeneous* [1, p. 620]. The following two equivalent statements represent special cases of Eq. (2-1) for a thermodynamically homogeneous material:

$$\check{U} = \check{U}(\check{S}, \rho_{(1)}, \rho_{(2)}, \ldots, \rho_{(N)}) \tag{2-2}$$

and

$$\hat{U} = \hat{U}(\hat{S}, \hat{V}, \omega_{(1)}, \omega_{(2)}, \ldots, \omega_{(N-1)}) \tag{2-3}$$

Here we introduce the volume per unit mass:

$$\hat{V} \equiv \frac{1}{\rho} \equiv \frac{1}{\sum_{A=1}^{N} \rho_{(A)}} \tag{2-4}$$

and the mass fraction of species $A$:

$$\omega_{(A)} \equiv \frac{\rho_{(A)}}{\rho} \tag{2-5}$$

We denote by ∧ over a symbol that we are dealing with a quantity per unit mass; ∨ over a symbol indicates that the quantity is per unit volume. We will assume that all materials we discuss are thermodynamically homogeneous and may be described by either Eq. (2-2) or (2-3).

[5.1] THERMODYNAMICS

We make a further assumption of regularity: all of the above thermodynamic functions are differentiable as many times as needed and are invertible to yield any one variable as a function of the others [1, p. 622]. We define temperature, thermodynamic pressure, and chemical potential of species $A$, respectively, as

$$T \equiv \left(\frac{\partial \hat{U}}{\partial \hat{S}}\right)_{\hat{V},\omega_{(A)}} = \left(\frac{\partial \check{U}}{\partial \check{S}}\right)_{\rho_{(A)}} \tag{2-6}$$

$$P \equiv -\left(\frac{\partial \hat{U}}{\partial \hat{V}}\right)_{\hat{S},\omega_{(A)}} \tag{2-7}$$

and

$$\mu_{(A)} \equiv \left(\frac{\partial \check{U}}{\partial \rho_{(A)}}\right)_{\check{S},\rho_{(B)}(B \neq A)} \tag{2-8}$$

(Note that a chemical potential on a molar basis is defined in Exercise 5.1.2-1.) It is clear that $T$, $P$, and $\mu_{(A)}$ become $T^*$, $P^*$, and $\mu^*_{(A)}$ as defined in Sec. 5.1.1 when discussing a system whose local properties are uniform throughout. Although thermostatics and thermodynamics employ parallel concepts on different scales, one should not expect to read off results in thermodynamics by analogy with similar results in thermostatics any more than fluid mechanics can be read off from mass-point mechanics [1, p. 616].

In view of Eqs. (2-6), (2-7), and (2-8), differentials of Eqs. (2-2) and (2-3) are

$$d\check{U} = T\,d\check{S} + \sum_{A=1}^{N} \mu_{(A)}\,d\rho_{(A)} \tag{2-9}$$

and

$$d\hat{U} = T\,d\hat{S} - P\,d\hat{V} + \sum_{A=1}^{N-1} \left(\frac{\partial \hat{U}}{\partial \omega_{(A)}}\right)_{\hat{S},\hat{V},\omega_{(B)}(B \neq A,N)} d\omega_{(A)} \tag{2-10}$$

Equation (2-9) can be written in terms of unit mass variables as

$$d\frac{\check{U}}{\check{V}} = T\,d\frac{\hat{S}}{\check{V}} + \sum_{A=1}^{N} \mu_{(A)}\,d\frac{\omega_{(A)}}{\check{V}} \tag{2-11}$$

or, upon rearrangement, we have

$$d\hat{U} = T\,d\hat{S} - \left(-\hat{U} + T\hat{S} + \sum_{A=1}^{N} \mu_{(A)}\omega_{(A)}\right)\frac{d\hat{V}}{\hat{V}}$$
$$+ \sum_{A=1}^{N-1} (\mu_{(A)} - \mu_{(N)})\,d\omega_{(A)} \tag{2-12}$$

Comparison of coefficients in Eqs. (2-10) and (2-12) gives us an extension of Eq. (1-7), Euler's equation,

$$\hat{U} = T\hat{S} - P\hat{V} + \sum_{A=1}^{N} \mu_{(A)}\omega_{(A)} \tag{2-13}$$

as well as

$$\left(\frac{\partial \hat{U}}{\partial \omega_{(A)}}\right)_{\hat{S}, \hat{V}, \omega_B(B \neq A, N)} = \mu_{(A)} - \mu_{(N)} \qquad (2\text{-}14)$$

Equations (2-10) and (2-14) yield the extension of Eq. (1-10), the Gibbs equation:

$$d\hat{U} = T\,d\hat{S} - P\,d\hat{V} + \sum_{A=1}^{N-1} (\mu_{(A)} - \mu_{(N)})\,d\omega_{(A)} \qquad (2\text{-}15)$$

The extension of Eq. (1-8), the Gibbs-Duhem equation, follows immediately by subtracting Eq. (2-15) from the differential of Eq. (2-13):

$$\hat{S}\,dT - \hat{V}\,dP + \sum_{A=1}^{N} \omega_{(A)}\,d\mu_{(A)} = 0 \qquad (2\text{-}16)$$

Note that Eqs. (2-13), (2-15), and (2-16) are applicable at each point in a material described by the caloric equation of state assumed above, either Eq. (2-2) or (2-3). It is not necessary that all local properties be uniform throughout the material or that the material be at equilibrium.

It should now be clear how one might go about developing from this point of view all of the usual relations found in thermodynamic texts. An excellent outline of thermodynamics is presented by Truesdell and Toupin [1, p. 615]; we recommend it for anyone seriously interested in this area.

We base our discussion of energy transfer upon a limited form of Eq. (2-3):

$$\hat{U} = \hat{U}(\hat{S}, \rho) \qquad (2\text{-}17)$$

This is appropriate to a single-component material (or a material for which composition is not a function of position in the material), whose thermodynamic behavior does not vary from one point in the material to the next (thermodynamically homogeneous). We also assume here that the deformation of the material is not an important factor in determining the internal energy of the material. This last is probably an excellent assumption for gases and many liquids, but inappropriate for elastic solids such as rubber or steel. For the same reason, it may also be in error when applied to highly viscoelastic liquids such as polymer melts and some polymer solutions.

## REFERENCES

1. Truesdell, C., and R. A. Toupin: In S. Flügge (ed.), "Handbuch der Physik," vol. 3/1, Springer-Verlag, Berlin, 1960.
2. Coleman, B. D.: *Arch. Rational Mech. Analysis*, **17**:1 (1964).
3. Coleman, B. D.: *Arch. Rational Mech. Analysis*, **17**:230 (1964).

## EXERCISES

**5.1.2-1** *Specific variables per unit mole* Let $c_{(A)}$ denote moles of species $A$ per unit volume. Denote by $\sim$ over a symbol that we are dealing with a quantity per unit mole.

[5.1] THERMODYNAMICS

(a) Show that alternate expressions for temperature and thermodynamic pressure are

$$T = \left(\frac{\partial \tilde{U}}{\partial \tilde{S}}\right)_{\tilde{V}, x_{(A)}} = \left(\frac{\partial \check{U}}{\partial \check{S}}\right)_{c_{(A)}}$$

and

$$P = -\left(\frac{\partial \tilde{U}}{\partial \tilde{V}}\right)_{\tilde{S}, x_{(A)}}$$

Here we introduce the mole fraction $x_{(A)}$:

$$x_{(A)} \equiv \frac{c_{(A)}}{c}$$

and the total mole density:

$$c = \frac{1}{\tilde{V}}$$

(b) The above definitions for temperature and thermodynamic pressure suggest that we define another chemical potential on a mole basis:

$$\mu_{(A)}^{(m)} \equiv \left(\frac{\partial \check{U}}{\partial c_{(A)}}\right)_{\check{S}, c_{(B)}(B \neq A)}$$

Derive the analogs to Eqs. (2-13) to (2-16) in terms of this molar chemical potential.

**5.1.2-2 The Maxwell relations** Let us define

$$\hat{A} \equiv \hat{U} - T\hat{S}$$

$$\hat{H} \equiv \hat{U} + P\hat{V}$$

$$\hat{G} \equiv \hat{H} - T\hat{S}$$

We refer to $\hat{A}$ as *Helmholtz free energy per unit mass*, $\hat{H}$ as *enthalpy per unit mass*, and $\hat{G}$ as *Gibbs free energy per unit mass*. (Truesdell and Toupin [1, p. 627] introduce an alternate definition for specific enthalpy.) Determine that

(a)
$$\left(\frac{\partial T}{\partial \hat{V}}\right)_{\hat{S}, \omega_{(B)}} = -\left(\frac{\partial P}{\partial \hat{S}}\right)_{\hat{V}, \omega_{(B)}}$$

$$\left(\frac{\partial T}{\partial \omega_{(A)}}\right)_{\hat{S}, \hat{V}, \omega_{(B)}(B \neq A, N)} = \left(\frac{\partial (\mu_{(A)} - \mu_{(N)})}{\partial \hat{S}}\right)_{\hat{V}, \omega_{(B)}}$$

$$-\left(\frac{\partial P}{\partial \omega_{(A)}}\right)_{\hat{S}, \hat{V}, \omega_{(B)}(B \neq A, N)} = \left(\frac{\partial (\mu_{(A)} - \mu_{(N)})}{\partial \hat{V}}\right)_{\hat{S}, \omega_{(B)}}$$

$$\left(\frac{\partial (\mu_{(A)} - \mu_{(N)})}{\partial \omega_{(B)}}\right)_{\hat{S}, \hat{V}, \omega_{(C)}(C \neq B, N)} = \left(\frac{\partial (\mu_{(B)} - \mu_{(N)})}{\partial \omega_{(A)}}\right)_{\hat{S}, \hat{V}, \omega_{(C)}(C \neq A, N)}$$

(b)
$$\left(\frac{\partial \hat{S}}{\partial \hat{V}}\right)_{T,\omega_{(B)}} = \left(\frac{\partial P}{\partial T}\right)_{\hat{V},\omega_{(B)}}$$

$$-\left(\frac{\partial \hat{S}}{\partial \omega_{(A)}}\right)_{T,\hat{V},\omega_{(B)}(B \neq A,N)} = \left(\frac{\partial (\mu_{(A)} - \mu_{(N)})}{\partial T}\right)_{\hat{V},\omega_{(B)}}$$

$$-\left(\frac{\partial P}{\partial \omega_{(A)}}\right)_{T,\hat{V},\omega_{(B)}(B \neq A,N)} = \left(\frac{\partial (\mu_{(A)} - \mu_{(N)})}{\partial \hat{V}}\right)_{T,\omega_{(B)}}$$

$$\left(\frac{\partial (\mu_{(A)} - \mu_{(N)})}{\partial \omega_{(B)}}\right)_{T,\hat{V},\omega_{(C)}(C \neq B,N)} = \left(\frac{\partial (\mu_{(B)} - \mu_{(N)})}{\partial \omega_{(A)}}\right)_{T,\hat{V},\omega_{(C)}(C \neq A,N)}$$

(c)
$$\left(\frac{\partial T}{\partial P}\right)_{\hat{S},\omega_{(B)}} = \left(\frac{\partial \hat{V}}{\partial \hat{S}}\right)_{P,\omega_{(B)}}$$

$$\left(\frac{\partial T}{\partial \omega_{(A)}}\right)_{\hat{S},P,\omega_{(B)}(B \neq A,N)} = \left(\frac{\partial (\mu_{(A)} - \mu_{(N)})}{\partial \hat{S}}\right)_{P,\omega_{(B)}}$$

$$\left(\frac{\partial \hat{V}}{\partial \omega_{(A)}}\right)_{\hat{S},P,\omega_{(B)}(B \neq A,N)} = \left(\frac{\partial (\mu_{(A)} - \mu_{(N)})}{\partial P}\right)_{\hat{S},\omega_{(B)}}$$

$$\left(\frac{\partial (\mu_{(A)} - \mu_{(N)})}{\partial \omega_{(B)}}\right)_{\hat{S},P,\omega_{(C)}(C \neq B,N)} = \left(\frac{\partial (\mu_{(B)} - \mu_{(N)})}{\partial \omega_{(A)}}\right)_{\hat{S},P,\omega_{(C)}(C \neq A,N)}$$

(d)
$$-\left(\frac{\partial \hat{S}}{\partial P}\right)_{T,\omega_{(B)}} = \left(\frac{\partial \hat{V}}{\partial T}\right)_{P,\omega_{(B)}}$$

$$-\left(\frac{\partial \hat{S}}{\partial \omega_{(A)}}\right)_{T,P,\omega_{(B)}(B \neq A,N)} = \left(\frac{\partial (\mu_{(A)} - \mu_{(N)})}{\partial T}\right)_{P,\omega_{(B)}}$$

$$\left(\frac{\partial \hat{V}}{\partial \omega_{(A)}}\right)_{T,P,\omega_{(B)}(B \neq A,N)} = \left(\frac{\partial (\mu_{(A)} - \mu_{(N)})}{\partial P}\right)_{T,\omega_{(B)}}$$

$$\left(\frac{\partial (\mu_{(A)} - \mu_{(N)})}{\partial \omega_{(B)}}\right)_{T,P,\omega_{(C)}(C \neq B,N)} = \left(\frac{\partial (\mu_{(B)} - \mu_{(N)})}{\partial \omega_{(A)}}\right)_{T,P,\omega_{(C)}(C \neq A,N)}$$

**5.1.2-3** *More Maxwell relations* Following the definitions introduced in Exercise 5.1.2-2, we have that

$$\check{A} = \check{U} - T\check{S}$$
$$\check{H} = \check{U} + P$$
$$\check{G} = \check{H} - T\check{S}$$

Determine that

(a)
$$\left(\frac{\partial T}{\partial \rho_{(A)}}\right)_{\check{S},\rho_{(B)}(B \neq A)} = \left(\frac{\partial \mu_{(A)}}{\partial \check{S}}\right)_{\rho_{(B)}}$$

$$\left(\frac{\partial \mu_{(A)}}{\partial \rho_{(B)}}\right)_{\check{S},\rho_{(C)}(C \neq B)} = \left(\frac{\partial \mu_{(B)}}{\partial \rho_{(A)}}\right)_{\check{S},\rho_{(C)}(C \neq A)}$$

[5.1] THERMODYNAMICS

(b) $\quad -\left(\dfrac{\partial \check{S}}{\partial \rho_{(A)}}\right)_{T,\rho_{(B)}(B\neq A)} = \left(\dfrac{\partial \mu_{(A)}}{\partial T}\right)_{\rho_{(B)}}$

$\left(\dfrac{\partial \mu_{(A)}}{\partial \rho_{(B)}}\right)_{T,\rho_{(C)}(C\neq B)} = \left(\dfrac{\partial \mu_{(B)}}{\partial \rho_{(A)}}\right)_{T,\rho_{(C)}(C\neq A)}$

**5.1.2.-4** *More on partial mass variables* (a) In the context of a uniform system, let $\Phi$ be any extensive variable

$$\Phi = \Phi(T^*, P^*, m_{(1)}, m_{(2)}, \ldots, m_{(N)})$$

Let us define

$$\hat{\Phi} = \dfrac{\Phi}{m}$$

where

$$\hat{\Phi} = \hat{\Phi}(T^*, P^*, \omega_{(1)}, \omega_{(2)}, \ldots, \omega_{(N-1)})$$

Taking roughly the same approach as was used in deriving Eqs.(5.1.2-13) and (5.1.2-14), determine that

$$\overline{\Phi}_{(A)} - \overline{\Phi}_{(N)} = \left(\dfrac{\partial \hat{\Phi}}{\partial \omega_{(A)}}\right)_{T^*, P^*, \omega_{(B)}(B\neq A, N)}$$

and

$$\hat{\Phi} = \sum_{A=1}^{N} \overline{\Phi}_{(A)} \omega_{(A)}$$

(b) The results of (a) suggest that we *define* $\overline{\Phi}_{(A)}$ ($A = 1, 2, \ldots, N$) for a *nonequilibrium* system by

$$\overline{\Phi}_{(A)} - \overline{\Phi}_{(N)} = \left(\dfrac{\partial \hat{\Phi}}{\partial \omega_{(A)}}\right)_{T, P, \omega_{(B)}(B\neq A, N)}$$

and

$$\hat{\Phi} = \sum_{A=1}^{N} \overline{\Phi}_{(A)} \omega_{(A)}$$

In this way it is more obvious that partial mass variables are intensive variables that have meaning in a discussion of nonequilibrium thermodynamics.

Show from these definitions that

$$\sum_{A=1}^{N} \left(\dfrac{\partial \overline{\Phi}_{(A)}}{\partial \omega_{(B)}}\right)_{T, P, \omega_{(C)}(C\neq B, N)} \omega_{(A)} = 0$$

**5.1.2-5** *More on partial molal variables* Derive results for *partial molal* variables (see Exercise 5.1.1-3) that parallel those obtained in Exercise 5.1.2-4 for partial mass variables.

**5.1.2-6** Determine that

(a) $\quad \overline{S}_{(A)} = -\left(\dfrac{\partial \mu_{(A)}}{\partial T}\right)_{P, \omega_{(B)}}$

(b) $\quad \overline{V}_{(A)} = \left(\dfrac{\partial \mu_{(A)}}{\partial P}\right)_{T, \omega_{(B)}}$

*Hint:* See Gibbs–Duhem equation and Exercise 5.1.2-2.

**5.1.2-7** *Heat capacities* We define the heat capacity per unit mass at constant pressure $\hat{c}_P$ and the heat capacity per unit mass at constant specific volume $\hat{c}_V$ as

$$\hat{c}_P \equiv T\left(\frac{\partial \hat{S}}{\partial T}\right)_{P,\omega_{(B)}}$$

and

$$\hat{c}_V \equiv T\left(\frac{\partial \hat{S}}{\partial T}\right)_{\hat{V},\omega_{(B)}}$$

(a) Determine that

$$\hat{c}_P = \left(\frac{\partial \hat{H}}{\partial T}\right)_{P,\omega_{(B)}}$$

and

$$\hat{c}_V = \left(\frac{\partial \hat{U}}{\partial T}\right)_{\hat{V},\omega_{(B)}}$$

(b) Prove that

$$\rho\hat{c}_P - \left(\frac{\partial P}{\partial T}\right)_{\hat{V},\omega_{(B)}} \left(\frac{\partial \ln \hat{V}}{\partial \ln T}\right)_{P,\omega_{(B)}} = \rho\hat{c}_V$$

(c) For an ideal gas, conclude that

$$\hat{c}_P = \hat{c}_V + \frac{R}{M}$$

where $M$ is the average molecular weight

$$M \equiv \frac{\rho}{c}$$

## 5.2 ENERGY TRANSMISSION

**5.2.1 Energy transmission** The rate at which energy is transmitted is entirely analogous to force. Let us recall our discussion in Sec. 2.1.1.

Corresponding to each body $B$, there is a distinct body (or set of bodies) $\mathring{B}$ such that the mass of the union of these two bodies is the mass of the universe. We call $\mathring{B}$ the exterior or the surroundings of $B$.

A system of energy transmission rates is a scalar-valued function $Q(B,C)$ of pairs of bodies. The value of $Q(B,C)$ is called the rate of energy transmission from body $C$ to body $B$. A system of energy transmission rates is restricted by the following two properties or axioms.

1. For a specified body $B$, $Q(C,\mathring{B})$ is an additive function defined over the subbodies $C$ of $B$.
2. Conversely, for a specified body $B$, $Q(B,C)$ is an additive function defined over the subbodies $C$ of $\mathring{B}$.

## [5.2] ENERGY TRANSMISSION

This says that parts of a body receive energy independently of each other from the various parts of the environment.

There are three types of energy transmission with which we may be concerned.

1. *External energy transmission.* Energy is transmitted from outside the body to the various material particles of which the body is composed. One example is radiation from the sun to the gas that composes the earth's atmosphere. Another example is induction heating in which energy is transferred to the polar molecules of a body by means of an alternating magnetic field. Let $P$ indicate a portion of a body $B$ as illustrated in Fig. 2.1.1-1. Taking $Q_e$ to be the external energy transmission rate per unit mass from the surroundings $\mathring{B}$ to the body $B$, we write the total rate of external energy transmission to $P$ in terms of an integral over the volume of $P$: $\int_{V_P} \rho Q_e \, dV$.
2. *Mutual energy transmission.* By mutual energy transmission we mean that energy is transmitted between pairs of material particles which are part of the same body. Radiation within a hot gas stream is an example. If $Q_m$ is the mutual energy transmission rate per unit mass from $B - P$ to $P$, the total rate of mutual energy transmission to $P$ may be expressed as an integral over the volume of $P$: $\int_{V_P} \rho Q_m \, dV$. (We recognize here that the sum of mutual energy transmissions between any two parts of $P$ must be zero; the proof is much the same as that for mutual forces [1, p. 533].)
3. *Contact energy transmission.* This is energy transmission which is not assignable as a function of position, but which is to be imagined as energy transmission through the bounding surface of a material in such a way as to be equivalent to the energy transmission from one portion of a material to another beyond that accounted for by mutual energy transmission. As an example, when we press our hands to a hot metal surface, there is contact energy transfer with the result that our hands may be burned. Let $h(\mathbf{z},P)$ represent the rate of energy transmission per unit area from $B - P$ to the boundary of $P$ at the position $\mathbf{z}$. This rate of energy transmission per unit area may be referred to as the *contact energy flux*. The total rate of contact energy transfer from $B - P$ to $P$ may be written as an integral over the bounding surface of $P$: $\int_{S_P} h(\mathbf{z},P) \, dS$.

The energy flux principle specifies the nature of the contact energy transmission.

*Energy flux principle* There is a scalar-valued function $h(\mathbf{z},\mathbf{n})$ defined at all points $\mathbf{z}$ in a body $B$ and for all unit vectors $\mathbf{n}$ such that the rate of contact energy transmission per unit area to any portion $P$ of $B$ may be given by

$$h(\mathbf{z},P) = h(\mathbf{z},\mathbf{n}) \tag{1-1}$$

Here $\mathbf{n}$ is the unit normal which is outwardly directed with respect to the closed bounding surface of $P$. The scalar $h = h(\mathbf{z},\mathbf{n})$ is referred to as the contact energy flux at the position $\mathbf{z}$ across the oriented surface element with normal $\mathbf{n}$; $\mathbf{n}$ points into the material from which the contact energy flux to the surface element is $h$.

In any particular problem, we regard the rates of energy transmission to a body as being given a priori to all observers; all observers would assume a priori the same set of energy transmission rates in a given problem. In prescribing these energy transmission rates, we specify a particular dynamic problem. We consequently assume that all energy transmission rates are independent of the observer or are frame indifferent (see Sec. 1.2.1):

$$Q_e^* = Q_e \qquad Q_m^* = Q_m \qquad (1\text{-}2)$$

and

$$h^* = h \qquad (1\text{-}3)$$

## REFERENCE

1. Truesdell, C., and R. A. Toupin: In S. Flügge (ed.), "Handbuch der Physik," vol. 3/1, Springer-Verlag, Berlin, 1960.

## 5.3 ENERGY BALANCE

**5.3.1 The energy balance** In Secs. 1.3.1 and 2.2.1 we introduced three postulates of mechanics: conservation of mass, Euler's first law, and Euler's second law. We are now ready for a fourth postulate, the *energy balance:*

> The time rate of change of the internal and kinetic energy of a body relative to the fixed stars is equal to the rate at which work is done on the body by the forces which act upon the body plus the rate of energy transmission to the body.

Let the volume and closed bounding surface of a body or any portion of a body be denoted, respectively, as $V_{(m)}$ and $S_{(m)}$. In a frame of reference that is stationary with respect to the fixed stars, the energy balance says

$$\frac{d}{dt} \int_{V_{(m)}} \rho(\hat{U} + \tfrac{1}{2}v^2)\, dV = \int_{S_{(m)}} \mathbf{v} \cdot (\mathbf{T} \cdot \mathbf{n})\, dS$$

$$+ \int_{V_{(m)}} \rho(\mathbf{v} \cdot \mathbf{f})\, dV + \int_{S_{(m)}} h\, dS + \int_{V_{(m)}} \rho Q\, dV \qquad (1\text{-}1)$$

The first term on the right describes the rate at which work is done on the body by the contact forces which the surroundings exert on its bounding surface. The second term indicates the rate at which work is done on the body by the external and mutual forces. The third term represents the rate of contact energy transmission from the surrounding to the body through its bounding surface. The fourth term stands for the rate of external and mutual energy transmission to the body; $Q$ is the scalar field which represents the sum of the external and mutual energy transmission rates per unit mass.

[5.3] ENERGY BALANCE

**EXERCISES**

**5.3.1-1** Consider two neighboring portions of a continuous body. Apply the energy balance to each portion and to their union. Deduce that on their common boundary

$$h(z,n) = -h(z, -n)$$

This says that the contact energy fluxes upon opposite sides of the same surface at a given point are equal in magnitude and opposite in sign.

**5.3.1-2** *The energy flux vector* By a development which parallels that given in Sec. 2.2.2, show that the contact energy flux may be expressed as

$$h(z,n) = -\mathbf{q} \cdot \mathbf{n}$$

where **q** is known as the *energy flux vector*.

**5.3.1-3** Show that the energy flux vector is frame indifferent.

**5.3.2 The differential energy balance** Let us restrict ourselves to a body in which all quantities are smooth functions of position and time as described at the beginning of Sec. 2.2.3.

The energy balance postulated in the previous section may be used to imply a differential equation which expresses a balance of energy at every point within a material. The steps are very similar to those used to obtain Cauchy's first law from Euler's first law in Sec. 2.2.3.

The transport theorem of Sec. 1.3.2 and the equation of continuity of Sec. 1.3.3 allow us to write the left of this equation as (see Exercise 1.3.3-1)

$$\frac{d}{dt}\int_{V_{(m)}} \rho(\hat{U} + \tfrac{1}{2}v^2)\, dV = \int_{V_{(m)}} \rho \frac{d_{(m)}}{dt}(\hat{U} + \tfrac{1}{2}v^2)\, dV \tag{2-1}$$

The first term on the right of Eq. (1-1) may be expressed in terms of a volume integral by means of Green's transformation (see Sec. A.11.2):

$$\int_{S_{(m)}} \mathbf{v} \cdot (\mathbf{T} \cdot \mathbf{n})\, dS = \int_{V_{(m)}} \operatorname{div}(\mathbf{T} \cdot \mathbf{v})\, dV \tag{2-2}$$

In writing this expression, we take advantage of the symmetry of the stress tensor.

The third term on the right of Eq. (1-1) may be rewritten in terms of a volume integral only if we can take advantage of Green's transformation. By Exercise 5.3.1-2, we may express the contact energy flux in terms of the *energy flux vector* **q**:

$$h = h(z,n) = -\mathbf{q} \cdot \mathbf{n} \tag{2-3}$$

We now may write

$$\int_{S_{(m)}} h\, dS = -\int_{S_{(m)}} \mathbf{q} \cdot \mathbf{n}\, dS$$

$$= -\int_{V_{(m)}} \operatorname{div} \mathbf{q}\, dV \tag{2-4}$$

By means of Eqs. (2-1), (2-2), and (2-4), Eq. (1-1) becomes

$$\int_{V_{(m)}} \left[ \rho \frac{d_{(m)}}{dt} (\hat{U} + \tfrac{1}{2}v^2) - \text{div} (\mathbf{T} \cdot \mathbf{v}) - \rho(\mathbf{v} \cdot \mathbf{f}) + \text{div } \mathbf{q} - \rho Q \right] dV = 0 \quad (2\text{-}5)$$

Since the size of our body is arbitrary, we conclude that the integrand must be identically zero at every point in the material:

$$\rho \frac{d_{(m)}}{dt} (\hat{U} + \tfrac{1}{2}v^2) = -\text{div } \mathbf{q} + \text{div} (\mathbf{T} \cdot \mathbf{v}) + \rho(\mathbf{v} \cdot \mathbf{f}) + \rho Q \quad (2\text{-}6)$$

This differential equation expresses the requirement of a local balance of energy at every point in the material. This is one form of what we shall refer to as the *differential energy balance*.

Equation (2-6) may be considerably simplified if we take advantage of Cauchy's first law. From the scalar product of the velocity vector, with Cauchy's first law

$$\mathbf{v} \cdot \left( \rho \frac{d_{(m)} \mathbf{v}}{dt} - \text{div } \mathbf{T} - \rho \mathbf{f} \right) = 0 \quad (2\text{-}7)$$

we have

$$\rho \frac{d_{(m)}}{dt} (\tfrac{1}{2}v^2) = \text{div} (\mathbf{T} \cdot \mathbf{v}) - \text{tr} (\mathbf{T} \cdot \nabla \mathbf{v}) + \rho(\mathbf{v} \cdot \mathbf{f}) \quad (2\text{-}8)$$

Subtracting this last from Eq. (2-6), we are left with another form of the differential energy balance:

$$\rho \frac{d_{(m)} \hat{U}}{dt} = -\text{div } \mathbf{q} + \text{tr} (\mathbf{T} \cdot \nabla \mathbf{v}) + \rho Q \quad (2\text{-}9)$$

The second term on the right is commonly written in terms of the thermodynamic pressure $P$ (see Sec. 5.1.2) and an extra-stress tensor $\mathbf{S}$,

$$\mathbf{S} \equiv \mathbf{T} + P\mathbf{I} \quad (2\text{-}10)$$

to obtain

$$\rho \frac{d_{(m)} \hat{U}}{dt} = -\text{div } \mathbf{q} - P \text{ div } \mathbf{v} + \text{tr} (\mathbf{S} \cdot \nabla \mathbf{v}) + \rho Q \quad (2\text{-}11)$$

If we are dealing with an incompressible fluid, an extra-stress tensor is defined as

$$\mathbf{S} \equiv \mathbf{T} + p\mathbf{I} \quad (2\text{-}12)$$

where the pressure $p$ may be taken to be any convenient scalar [1, p. 647]. For an incompressible fluid, Eq. (2-9) reduces to

$$\rho \frac{d_{(m)} \hat{U}}{dt} = -\text{div } \mathbf{q} + \text{tr} (\mathbf{S} \cdot \nabla \mathbf{v}) + \rho Q \quad (2\text{-}13)$$

## [5.3] ENERGY BALANCE

In either case we may interpret tr $(\mathbf{S} \cdot \nabla \mathbf{v})$ as a volume source of internal energy (rate of production of internal energy per unit volume) by action of viscous forces.

We are more often concerned about variations in temperature within a body than variations in internal energy density. We would consequently rather have a differential equation to solve for a temperature field. We know that (see Sec. 5.1.2 for a single component material)

$$d\hat{U} = T\,d\hat{S} - P\,d\hat{V}$$

$$= T\left(\frac{\partial \hat{S}}{\partial T}\right)_{\hat{V}} dT + \left[T\left(\frac{\partial \hat{S}}{\partial \hat{V}}\right)_T - P\right] d\hat{V}$$

$$= \hat{c}_V\,dT + \left[T\left(\frac{\partial P}{\partial T}\right)_{\hat{V}} - P\right] d\hat{V} \qquad (2\text{-}14)$$

Here $\hat{c}_V$ is the material's heat capacity per unit mass at constant specific volume (see Exercise 5.1.2-7):

$$\hat{c}_V \equiv T\left(\frac{\partial \hat{S}}{\partial T}\right)_{\hat{V}} \qquad (2\text{-}15)$$

By the same reasoning used in Eq. (2-14), we have that

$$\rho \frac{d_{(m)}\hat{U}}{dt} = \rho \hat{c}_V \frac{d_{(m)}T}{dt} + \left[T\left(\frac{\partial P}{\partial T}\right)_{\hat{V}} - P\right]\rho \frac{d_{(m)}\hat{V}}{dt} \qquad (2\text{-}16)$$

From the equation of continuity,

$$\rho \frac{d_{(m)}\hat{V}}{dt} = -\frac{1}{\rho} \frac{d_{(m)}\rho}{dt} = \operatorname{div} \mathbf{v} \qquad (2\text{-}17)$$

By Eqs. (2-16) and (2-17), Eq. (2-11) yields

$$\rho \hat{c}_V \frac{d_{(m)}T}{dt} = -\operatorname{div} \mathbf{q} - T\left(\frac{\partial P}{\partial T}\right)_{\hat{V}} \operatorname{div} \mathbf{v} + \operatorname{tr}(\mathbf{S} \cdot \nabla \mathbf{v}) + \rho Q \qquad (2\text{-}18)$$

Several different forms of the differential energy balance which are in common use are given in Table 5.6.-1.

## REFERENCES

1. Truesdell, C., and R. A. Toupin: In S. Flügge (ed.), "Handbuch der Physik," vol. 3/1, Springer-Verlag, Berlin, 1960.
2. Slattery, J. C.: *Ind. Eng. Chem., Fundamentals,* **6**:108 (1967).

## EXERCISES

**5.3.2-1** *The jump energy balance* Starting with the postulated energy balance of Sec. 5.3.1 and neglecting all interfacial effects, derive the *jump energy balance* for a phase interface represented by a singular surface [1, p. 610; 2]:

$$[\rho(\hat{U} + \tfrac{1}{2}v^2)(\mathbf{v}\cdot\boldsymbol{\xi} - u_{(\xi)}) + \mathbf{q}\cdot\boldsymbol{\xi} - \mathbf{v}\cdot(\mathbf{T}\cdot\boldsymbol{\xi})] = 0$$

Here $u_{(\xi)}$ is the speed of displacement of the phase interface [1, p. 499]; the boldface bracket notation is defined by Eq. (4-7) of Sec. 1.3.4.

**5.3.2-2** Use the approach outlined in Exercise 1.3.5-2 to derive the jump energy balance of Exercise 5.3.2-1.

**5.3.2-3** If the external force per unit mass may be expressed in terms of a potential energy per unit mass $\varphi$:

$$\mathbf{f} = -\nabla \varphi$$

show that Eq. (2-6) may be written as

$$\rho \frac{d_{(m)}}{dt}(\hat{U} + \tfrac{1}{2}v^2 + \varphi) = -\operatorname{div}\mathbf{q} + \operatorname{div}(\mathbf{T} \cdot \mathbf{v}) + \rho Q$$

**5.3.2-4** Derive the other forms of the differential energy balance presented in Table 5.6.1-1.

**5.3.2-5** Discuss how Eq. (2-18) simplifies when dealing with (*a*) ideal gas, (*b*) incompressible materials, and (*c*) solids. Obtain a differential equation for temperature applicable to a material in which thermodynamic pressure is a constant.

**5.3.2-6** Show that for a rigid body tumbling in space (see Exercise 2.3.2-1)

$$\operatorname{tr}(\mathbf{T} \cdot \nabla \mathbf{v}) = \operatorname{tr}(\mathbf{S} \cdot \nabla \mathbf{v}) = 0$$

## 5.4 BEHAVIOR

**5.4.1 More on the behavior of materials** In Sec. 2.3.1 we saw that our common experience suggested three principles which we subsequently found to be useful in constructing constitutive equations for the stress tensor. The essential elements of these principles also apply to representations for the energy flux vector.

Whatever is to happen in the future should have no influence on the present energy flux vector. This implies a *principle of determinism:*

> The energy flux vector is determined by the thermal history of the body and by the history of the motion which the body has undergone.

It seems reasonable to say that the temperature distribution and motion in one portion of a body should not necessarily affect the energy flux vector in another portion of the body. We may state this as a *principle of local action:*

> The temperature distribution and motion of the material outside an arbitrarily small neighborhood of the material particle $\zeta$ may be ignored in determining the energy flux vector at this material particle.

In view of Exercise 5.3.1-3, we may extend the *principle of material frame indifference:*

> Constitutive equations must be invariant under changes of frame of reference. If a constitutive equation is satisfied for a process in which the energy flux

vector, stress tensor, and motion are given by

$$\mathbf{q} = \mathbf{q}(\mathbf{z}_\varkappa,t) \qquad \mathbf{T} = \mathbf{T}(\mathbf{z}_\varkappa,t)$$

and

$$\mathbf{z} = \boldsymbol{\chi}_\varkappa(\mathbf{z}_\varkappa,t) \tag{1-1}$$

then it must also be satisfied for any equivalent process described with respect to another frame of reference. In particular, the constitutive equation must be satisfied for a process in which the energy flux vector, stress tensor, and motion are given by

$$\mathbf{q}^* = \mathbf{q}^*(\mathbf{z}_\varkappa^*,t^*) = \mathbf{Q}(t) \cdot \mathbf{q}(\mathbf{z}_\varkappa,t)$$

$$\mathbf{T}^* = \mathbf{T}^*(\mathbf{z}_\varkappa^*,t^*) = \mathbf{Q}(t) \cdot \mathbf{T}(\mathbf{z}_\varkappa,t) \cdot \mathbf{Q}(t)^T$$

$$\mathbf{z}^* = \boldsymbol{\chi}_\varkappa^*(\mathbf{z}_\varkappa^*,t^*) = \mathbf{c}(t) + \mathbf{Q}(t) \cdot \boldsymbol{\chi}_\varkappa(\mathbf{z}_\varkappa,t)$$

and

$$t^* = t - a \tag{1-2}$$

In the next section we illustrate the use of these principles in constructing a constitutive equation for the energy flux vector.

**5.4.2 Fourier's law** Let us propose a simple constitutive equation for the energy flux vector which satisfies the three principles discussed in the last section.

We can satisfy the principle of determinism by requiring the energy flux vector to depend only upon a description of the present temperature distribution in the material. Both the principle of determinism and the principle of local action are satisfied if we assume that the energy flux at a point is a function of the temperature and temperature gradient at that point:

$$\mathbf{q} = \mathbf{h}(T, \nabla T) \tag{2-1}$$

The principle of material frame indifference requires that (we require temperature to be a frame-indifferent scalar field)

$$\mathbf{q}^* = \mathbf{Q} \cdot \mathbf{q} = \mathbf{h}(T,(\nabla T)^*) \tag{2-2}$$

From Exercise 1.2.1-1 and Eq. (2-1), we find that the function $\mathbf{h}$ must be a vector-valued isotropic function [1, p. 22] of $T$ and $\nabla T$:

$$\mathbf{Q} \cdot \mathbf{h}(T, \nabla T) = \mathbf{h}(T, \mathbf{Q} \cdot \nabla T) \tag{2-3}$$

Equation (2-3) requires that [1, p. 35]

$$\mathbf{q} = \mathbf{h}(T, \nabla T) = \tilde{k}(T,|\nabla T|) \nabla T \tag{2-4}$$

*Fourier's law* is a special case of Eq. (2-4):

$$\mathbf{q} = -k(T) \nabla T \tag{2-5}$$

The scalar $k = k(T)$ is referred to as the *thermal conductivity*.

There has been speculation as to the validity of Fourier's law for materials subject to highly nonuniform temperature fields [2, p. 239; 3, p. 366]. This speculation is prompted by the analogy with momentum transfer, where there is considerable evidence of nonlinearities associated with the flow and deformation of viscoelastic fluids. Very little is known about the nature and importance of nonlinear energy transfer. It has been only in the past few years that definite suggestions have been made in this area. For an excellent review see Truesdell and Noll [1, sec. 96]; Bowen [4] and Müller [5] consider the problems associated with a multicomponent system.

A recent experimental study of solids [6, 7] was not able to produce clear evidence of deviations from Fourier's law for gradients as large as 300°C/cm in two high-temperature cements. Chapman and Cowling [8, p. 271] indicate that deviations from Fourier's law are negligible for dilute gases at pressures greater than $10^{-6}$ atm. Another kinetic theory computation [7] confirms that Eq. (2-5) is a linearization, but suggests that for dilute gases higher-order terms are unimportant for gradients less than $10^{6}$°C/cm.

In our remaining discussion of energy transfer, we will assume that Fourier's law, Eq. (2-5), is an accurate description of material behavior.

## REFERENCES

1. Truesdell, C., and W. Noll: In S. Flügge (ed.), "Handbuch der Physik," vol. 3/3, Springer-Verlag, Berlin, 1965.
2. Serrin, J.: In S. Flügge and C. Truesdell (eds.), "Handbuch der Physik," vol. 8/1, Springer-Verlag, Berlin, 1959.
3. Truesdell, C.: Principles of Continuum Mechanics, *Colloquium Lectures in Pure and Applied Science*, No. 5, Mobil Research and Development Corp., Field Research Laboratory, Dallas, Texas, 1960.
4. Bowen, R. M.: *Arch. Rational Mech. Analysis*, **24**:370 (1967).
5. Müller, Ingo: *Arch. Rational Mech. Analysis*, **28**:1 (1968).
6. Flumerfelt, R. W., and J. C. Slattery: *A.I.Ch.E.J.*, **15**:291 (1969).
7. Flumerfelt, R. W.: doctoral dissertation, Dept. of Chem. Engng., Northwestern University, Evanston, Ill., 1965.
8. Chapman, Sydney, and T. G. Cowling: "The Mathematical Theory of Non-Uniform Gases," Cambridge, London, 1952.
9. Spencer, A. J., and R. S. Rivlin: *Arch. Rational Mech. Analysis*, **4**:214 (1959/60).
10. Smith, G. F.: *Arch. Rational Mech. Analysis*, **18**:282 (1965).
11. Truesdell, C.: "The Elements of Continuum Mechanics," Springer-Verlag, New York, 1966.

## EXERCISE

**5.4.2-1** *An oriented or anisotropic solid [6, app.]* One would not expect either Eq. (2-4) or Eq. (2-5) to be valid for a solid such as wood or stratified limestone, in which density is a function of position. For such materials, Eq. (2-1) must be altered to include a dependence upon additional vector and possibly tensor quantities. For example, one might postulate a dependence of the energy flux vector upon the local density gradient, as well as upon temperature and temperature gradient:

$$\mathbf{q} = \mathbf{h}(T, \nabla T, \nabla \rho) \tag{1}$$

[5.5] ENTROPY INEQUALITY

(a) Show that the principle of material frame indifference requires **h** to be an isotropic function:

$$\mathbf{Q} \cdot \mathbf{h}(T, \nabla T, \nabla \rho) = \mathbf{h}(T, \mathbf{Q} \cdot \nabla T, \mathbf{Q} \cdot \nabla \rho) \tag{2}$$

By the representation theorems of Spencer and Rivlin [9, sec. 7] and of Smith [10], the most general polynomial vector function of two vectors is of the form

$$\mathbf{q} = \varkappa_{(1)} \nabla T + \varkappa_{(2)} \nabla \rho + \varkappa_{(3)} \nabla T \wedge \nabla \rho \tag{3}$$

where $\varkappa_{(1)}$, $\varkappa_{(2)}$, and $\varkappa_{(3)}$ are scalar-valued polynomials in $|\nabla T|$, $|\nabla \rho|$, and $(\nabla T \cdot \nabla \rho)$. (In applying the theorem of Spencer and Rivlin, we identify a vector **b**, which has covariant components $b_i$, with the skew-symmetric tensor, which has contravariant components $\epsilon^{ijk} b_i$.)

(b) Show that

$$(\mathbf{Q} \cdot \nabla T) \wedge (\mathbf{Q} \cdot \nabla \rho) = \det \mathbf{Q} [\mathbf{Q} \cdot (\nabla T \wedge \nabla \rho)]$$

which means that $\nabla T \wedge \nabla \rho$ is not frame indifferent.

(c) Show that the third term on the right of Eq. (3) is not consistent with the requirement that **h** be an isotropic function, Eq. (2).

Consequently, Eq. (3) reduces to

$$\mathbf{q} = \varkappa_{(1)} \nabla T + \varkappa_{(2)} \nabla \rho$$

We expect $\varkappa_{(2)} = 0$ for $|\nabla T| = 0$, in order that $\mathbf{q} = 0$ in this limit.

Commonly, an *anisotropic* solid is one for which dynamic response in a process depends upon a direction (or a set of directions) intrinsically associated with the material, such as $\nabla \rho$ in Eq. (1). This seems to be an unfortunate use of the word anisotropic, since we see in Eq. (2) that **h** is an *isotropic function* by the principle of material frame indifference. Furthermore, since we do not allow the energy flux vector **q** to depend upon deformation in Eqs. (2-1) and (1), we have no need to make a statement of material symmetry [11, p. 56] by means of which we might identify a solid as belonging to one of the classical crystallographic groups. For this reason, we suggest referring to a material described by Eq. (2-1) as *nonoriented;* we prefer to say a material described by Eq. (1) is *oriented*.

## 5.5 ENTROPY INEQUALITY

### 5.5.1 The entropy inequality

The differential energy balance of Sec. 5.3.2

$$\rho \frac{d_{(m)} \hat{U}}{dt} = -\operatorname{div} \mathbf{q} - P \operatorname{div} \mathbf{v} + \operatorname{tr}(\mathbf{S} \cdot \nabla \mathbf{v}) + \rho Q \tag{1-1}$$

may be used to gain a differential equation which describes the entropy distribution within a material. The constitutive equation for internal energy which is being used here

$$\hat{U} = \hat{U}(\hat{S}, \hat{V}) \tag{1-2}$$

implies that (see Sec. 5.1.2)

$$\rho \frac{d_{(m)} \hat{U}}{dt} = T \rho \frac{d_{(m)} \hat{S}}{dt} - P \rho \frac{d_{(m)} \hat{V}}{dt} = T \rho \frac{d_{(m)} \hat{S}}{dt} - P \operatorname{div} \mathbf{v} \tag{1-3}$$

This allows us to write Eq. (1-1) as

$$\rho \frac{d_{(m)}\hat{S}}{dt} = -\frac{1}{T}\operatorname{div} \mathbf{q} + \frac{1}{T}\operatorname{tr}(\mathbf{S}\cdot\nabla\mathbf{v}) + \frac{\rho Q}{T}$$

$$= -\operatorname{div}\left(\frac{1}{T}\mathbf{q}\right) - \frac{1}{T^2}\mathbf{q}\cdot\nabla T + \frac{1}{T}\operatorname{tr}(\mathbf{S}\cdot\nabla\mathbf{v}) + \rho\frac{Q}{T} \quad (1\text{-}4)$$

Let us see what this equation implies for a body which contains no surfaces of discontinuity. If $V_{(m)}$ is the volume of the body and $S_{(m)}$ is its closed bounding surface, we may integrate Eq. (1-4) over $V_{(m)}$ to obtain

$$\int_{V_{(m)}} \left[\rho\frac{d_{(m)}\hat{S}}{dt} + \operatorname{div}\left(\frac{1}{T}\mathbf{q}\right) + \frac{1}{T^2}\mathbf{q}\cdot\nabla T - \frac{1}{T}\operatorname{tr}(\mathbf{S}\cdot\nabla\mathbf{v}) - \frac{\rho Q}{T}\right]dV = 0 \quad (1\text{-}5)$$

The transport theorem and the equation of continuity allow us to write the first term as

$$\int_{V_{(m)}} \rho\frac{d_{(m)}\hat{S}}{dt}\, dV = \frac{d}{dt}\int_{V_{(m)}} \rho\hat{S}\, dV \quad (1\text{-}6)$$

The generalized Green's transformation may be applied to the second term with the result

$$\int_{V_{(m)}} \operatorname{div}\left(\frac{1}{T}\mathbf{q}\right) dV = \int_{S_{(m)}} \frac{1}{T}\mathbf{q}\cdot\mathbf{n}\, dS \quad (1\text{-}7)$$

Equations (1-6) and (1-7) permit our rewriting Eq. (1-5) as

$$\frac{d}{dt}\int_{V_{(m)}} \rho\hat{S}\, dV = -\int_{S_{(m)}} \frac{1}{T}\mathbf{q}\cdot\mathbf{n}\, dS + \int_{V_{(m)}}\left[-\frac{1}{T^2}\mathbf{q}\cdot\nabla T\right.$$

$$\left.+\frac{1}{T}\operatorname{tr}(\mathbf{S}\cdot\nabla\mathbf{v}) + \frac{\rho Q}{T}\right]dV \quad (1\text{-}8)$$

This equation is in the form of an entropy balance for continuous bodies. We interpret $(1/T)\mathbf{q}$ as the *entropy flux vector* (rate of contact entropy transmission per unit area) and

$$-\frac{1}{T^2}\mathbf{q}\cdot\nabla T + \frac{1}{T}\operatorname{tr}(\mathbf{S}\cdot\nabla\mathbf{v}) + \frac{\rho Q}{T}$$

as the local rate of production of entropy per unit volume. Equation (1-8) says that the time rate of change of the entropy associated with a continuous body is equal to the rate at which entropy is transferred to the body through the bounding surfaces of the body by contact energy transmission plus the rate at which entropy is generated at each point in the body by thermal dissipation, by viscous dissipation, and by external and mutual energy transmission. Note that Eq. (1-8) is derived only for a continuous body and not for a body which contains surfaces of discontinuity (such as shock surfaces or, perhaps, phase interfaces).

## [5.5] ENTROPY INEQUALITY

Let us examine the experimental evidence available concerning the signs of the second and third terms on the right of Eq. (1-4). Everyday experience indicates that contact energy transmission is always in the opposite direction to that of the temperature gradient. This suggests that

$$-\frac{1}{T^2}\mathbf{q}\cdot\nabla T \geq 0 \qquad (1\text{-}9)$$

If one opens a paper clip and repeatedly twists the ends with respect to one another until the metal breaks, there is a noticeable temperature rise in the region of the deformation. In terms of Eq. (2-18) in Sec. 5.3.2, at any point in the region of the deformation the left side is positive. Since there is no external energy transfer and since the density of the metal is essentially constant, we conclude that

$$\text{tr}\,(\mathbf{S}\cdot\nabla\mathbf{v}) \geq 0 \qquad (1\text{-}10)$$

If one takes the trouble to check, the grease-packed front-wheel bearings on an automobile (with power transmitted to the rear axle) experience a considerable temperature rise as the result of a highway trip. Applying the energy balance to each point in the grease, we arrive at inequality (1-10). At each point in the deforming metal of the paper clip and at each point in the wheel-bearing grease, mechanical energy is transformed into internal energy. In our common experience, it appears that inequality (1-10) is satisfied.

This experimental evidence suggests a fifth fundamental postulate applicable to all bodies, the *entropy inequality*:

The minimum rate of production of entropy in a body is proportional to the rate of energy transmission to the body. Comparison with Eq. (1-5) suggests that

$$\frac{d}{dt}\int_{V_{(m)}} \rho \hat{S}\,dV + \int_{S_{(m)}} \frac{1}{T}\mathbf{q}\cdot\mathbf{n}\,dS - \int_{V_{(m)}} \rho\frac{Q}{T}\,dV \geq 0 \qquad (1\text{-}11)$$

The entropy inequality or one of its implications is usually referred to as the second law of thermodynamics, though statements of the second law are often obscured by references to perpetual motion machines [1, p. 64].

Although a particular constitutive equation for specific internal energy as a function of entropy, Eq. (1-2), has been employed in *motivating* inequality (1-11), the entropy inequality is stated as a fundamental postulate like mass conservation. It should, consequently, be satisfied by all bodies composed of a single species, no matter what constitutive equations are used to describe their behavior.

**REFERENCE**

1. Keenan, J. H.: "Thermodynamics," Wiley, New York, 1941.

## 5.5.2 The differential entropy inequality

Let us again restrict ourselves to a body in which all quantities are smooth functions of position and time as described at the beginning of Sec. 2.2.3.

The entropy inequality, together with Eq. (1-4), implies that at every point in a continuous body

$$\rho \frac{d_{(m)}\hat{S}}{dt} + \text{div}\left(\frac{1}{T}\mathbf{q}\right) - \frac{\rho Q}{T} = -\frac{1}{T^2}\mathbf{q}\cdot\nabla T + \frac{1}{T}\text{tr}\,(\mathbf{S}\cdot\nabla\mathbf{v}) \geq 0 \qquad (2\text{-}1)$$

We refer to this as the *differential entropy inequality*.

The function of the differential entropy inequality is not unlike that of Cauchy's second law. Cauchy's second law places a restriction on the stress tensor: it must be symmetric. The entropy inequality can be viewed as placing restrictions both on the energy flux vector $\mathbf{q}$ and the stress tensor. For a further development of this idea see Exercise 5.5.2-1.

Notice that inequalities (1-9) and (1-10) do not *necessarily* follow from inequality (2-1). One must first state that $\mathbf{q}\cdot\nabla T$ is not a function of the rate of deformation tensor $\mathbf{D}$ (or $\dot{\mathbf{F}}$; see Exercise 2.3.2-1) and that tr $(\mathbf{S}\cdot\mathbf{D})$ is not a function of $\nabla T$ [1, p. 362]. In recent discussions of thermodynamics, some workers have been unwilling to make such statements a priori [2].

### REFERENCES

1. Truesdell, C., and W. Noll: In S. Flügge (ed.), "Handbuch der Physik," vol. 3/3, Springer-Verlag, Berlin, 1965.
2. Coleman, B. D.: *Arch. Rational Mech. Analysis*, **17**:1 (1964).
3. Truesdell, C., and R. A. Toupin: In S. Flügge (ed.), "Handbuch der Physik," vol. 3/1, Springer-Verlag, Berlin, 1960.

### EXERCISES

**5.5.2-1** *The newtonian fluid* Assume that a fluid is described by the newtonian model for the stress tensor:

$$\mathbf{T} = (-P + \lambda\,\text{div}\,\mathbf{v})\mathbf{I} + 2\mu\mathbf{D}$$

and by Fourier's law for the energy flux vector:

$$\mathbf{q} = -k\,\nabla T$$

(a) Under these circumstances, show that the differential entropy inequality may be written as

$$\frac{k}{T^2}(\nabla T)^2 + \frac{\lambda}{T}(\text{tr}\,\mathbf{D})^2 + \frac{2\mu}{T}\,\text{tr}\,(\mathbf{D}^2) \geq 0$$

(b) Show that the differential entropy inequality requires that [1, p. 357; see also Sec. 2.3.2]

$$k \geq 0 \qquad \mu \geq 0 \qquad \text{and} \qquad \lambda \geq -\tfrac{2}{3}\mu$$

## [5.5] ENTROPY INEQUALITY

*Hint:* In order to show that $\lambda \geq -\frac{2}{3}\mu$, realize that $\nabla T$ and the six components of the rate of deformation tensor may be chosen independently. Set $\nabla T$ and the nondiagonal components of **D** equal to zero. Rearrange the differential entropy inequality to read

$$(3\lambda + 2\mu)[(D_{11})^2 + (D_{22})^2 + (D_{33})^2] - \lambda[(D_{11} - D_{22})^2 + (D_{11} - D_{33})^2 + (D_{22} - D_{33})^2] \geq 0$$

Require that $D_{11} = D_{22} = D_{33}$ to conclude that $\lambda \geq -\frac{2}{3}\mu$.

**5.5.2-2** *The jump entropy inequality* Use the approach outlined in Exercise 1.3.5-2 to derive the *jump entropy inequality* for a phase interface represented by a singular surface [3, p. 645]:

$$\left[\rho\hat{S}(\mathbf{v}\cdot\boldsymbol{\xi} - u_{(\xi)}) + \frac{1}{T}\mathbf{q}\cdot\boldsymbol{\xi}\right] \geq 0$$

Here $u_{(\xi)}$ is the speed of displacement of the phase interface [3, p. 499]; the boldface bracket notation is defined by Eq. (4-7) of Sec. 1.3.4.

**5.5.2-3** Discuss the difficulty encountered in deriving the jump entropy inequality of Exercise 5.5.2-2 when the approach of Sec. 1.3.5 is used.

**5.5.2-4** Show that, if we neglect interphase mass transfer and if we assume that temperature is continuous across a phase interface, the jump entropy inequality reduces to

$$[\mathbf{q}\cdot\boldsymbol{\xi}] \geq 0$$

or

$$[\mathbf{T}\cdot\boldsymbol{\xi}] \geq 0$$

We show later (Sec. 8.5.3) that temperature is continuous across a singular surface or phase interface in stable equilibrium. This may explain the apparent contradictions between the results obtained here and those obtained previously for the jump energy balance (Exercise 5.3.2-1) and for the jump momentum balance (Exercise 2.2.3-1).

**5.5.2-5** *Simple nonnewtonian behavior* (a) Assume that the extra-stress tensor of an incompressible fluid is described by (see Sec. 2.3.3)

$$\mathbf{S} = 2\eta(\gamma)\mathbf{D}$$

and that the energy flux vector is represented by Fourier's law. Here

$$\gamma \equiv \sqrt{2\,\mathrm{tr}\,\mathbf{D}^2}$$

Use the approach outlined in Exercise 5.5.2-1 to prove that

$$k \geq 0 \quad \text{and} \quad \eta(\gamma) \geq 0$$

(b) Consider an incompressible fluid for which the energy flux vector is described by Fourier's law and for which the extra-stress tensor and rate of deformation tensor are related by

$$2\mathbf{D} = \varphi(\tau)\mathbf{S}$$

where

$$\tau \equiv \sqrt{\tfrac{1}{2}\,\mathrm{tr}\,\mathbf{S}^2}$$

Prove that

$$k \geq 0 \quad \text{and} \quad \varphi(\tau) \geq 0$$

**5.5.2-6** *More on the newtonian fluid* Repeat Exercise 5.5.2-1, assuming that the fluid is described by

$$\mathbf{T} = (-P + \alpha + \lambda\,\mathrm{div}\,\mathbf{v})\mathbf{I} + 2\mu\mathbf{D}$$

Prove that $\alpha = 0$.

*Hint:* Set $\nabla T$ and the nondiagonal components of **D** equal to zero, require $D_{11} = D_{22} = D_{33}$, and let $D_{11}$ assume both positive and negative values.

## 5.6 SUMMARY

### 5.6.1 Summary of useful equations

In Table 5.6.1-1 we present seven equivalent forms of the differential energy balance discussed in Sec. 5.3.2. Often one form will have a particular advantage in any given problem.

Commonly, the only external force to be considered is a uniform gravitational field and we may represent it as (see Sec. 2.5.1 and Exercise 2.5.1-1)

$$\mathbf{f} = -\nabla \varphi \tag{1-1}$$

We will hereafter refer to $\varphi$ as *potential energy per unit mass*.

Equation (E) of Table 5.6.1-1 may be rewritten as

$$\rho \hat{c}_V \frac{d_{(m)}T}{dt} = -\text{div } \mathbf{q} - T\left(\frac{\partial P}{\partial T}\right)_{\hat{V}} \text{div } \mathbf{v} + \text{tr}(\mathbf{S} \cdot \mathbf{D}) + \rho Q \tag{1-2}$$

This equation is shown for rectangular cartesian, cylindrical, and spherical coordinates in Table 5.6.1-2. It should not be difficult to use these as guides in immediately writing the corresponding expressions for the other six forms of the differential energy

**Table 5.6.1-1** Various forms of the differential energy balance [1, p. 322]

$$\rho \frac{d_{(m)}}{dt}(\hat{U} + \tfrac{1}{2}v^2 + \varphi) = -\text{div } \mathbf{q} + \text{div}(\mathbf{T} \cdot \mathbf{v}) + \rho Q \tag{A}\dagger$$

$$\rho \frac{d_{(m)}}{dt}(\hat{U} + \tfrac{1}{2}v^2) = -\text{div } \mathbf{q} + \text{div}(\mathbf{T} \cdot \mathbf{v}) + \rho(\mathbf{v} \cdot \mathbf{f}) + \rho Q \tag{B}$$

$$\rho \frac{d_{(m)}\hat{U}}{dt} = -\text{div } \mathbf{q} - P \text{ div } \mathbf{v} + \text{tr}(\mathbf{S} \cdot \nabla \mathbf{v}) + \rho Q \tag{C}$$

$$\rho \frac{d_{(m)}\hat{H}}{dt} = -\text{div } \mathbf{q} + \frac{d_{(m)}P}{dt} + \text{tr}(\mathbf{S} \cdot \nabla \mathbf{v}) + \rho Q \tag{D}$$

$$\rho \hat{c}_V \frac{d_{(m)}T}{dt} = -\text{div } \mathbf{q} - T\left(\frac{\partial P}{\partial T}\right)_{\hat{V}} \text{div } \mathbf{v} + \text{tr}(\mathbf{S} \cdot \nabla \mathbf{v}) + \rho Q \tag{E}$$

$$\rho \hat{c}_P \frac{d_{(m)}T}{dt} = -\text{div } \mathbf{q} + \left(\frac{\partial \ln \hat{V}}{\partial \ln T}\right)_P \frac{d_{(m)}P}{dt} + \text{tr}(\mathbf{S} \cdot \nabla \mathbf{v}) + \rho Q \tag{F}$$

$$\rho \frac{d_{(m)}\hat{S}}{dt} = -\text{div}\left(\frac{1}{T}\mathbf{q}\right) - \frac{1}{T^2}\mathbf{q} \cdot \nabla T + \frac{1}{T}\text{tr}(\mathbf{S} \cdot \nabla \mathbf{v}) + \frac{1}{T}\rho Q \tag{G}$$

† We assume that $\partial \varphi / \partial t = 0$.

## [5.6] SUMMARY

**Table 5.6.1-2 The differential energy balance [Eq. (1-2)] in several coordinate systems**

*Rectangular cartesian coordinates:*

$$\rho \hat{c}_V \left( \frac{\partial T}{\partial t} + v_1 \frac{\partial T}{\partial z_1} + v_2 \frac{\partial T}{\partial z_2} + v_3 \frac{\partial T}{\partial z_3} \right) = -\left( \frac{\partial q_1}{\partial z_1} + \frac{\partial q_2}{\partial z_2} + \frac{\partial q_3}{\partial z_3} \right) - T\left(\frac{\partial P}{\partial T}\right)_{\hat{V}} \left( \frac{\partial v_1}{\partial z_1} + \frac{\partial v_2}{\partial z_2} + \frac{\partial v_3}{\partial z_3} \right)$$

$$+ S_{11} \frac{\partial v_1}{\partial z_1} + S_{22} \frac{\partial v_2}{\partial z_2} + S_{33} \frac{\partial v_3}{\partial z_3} + S_{12}\left( \frac{\partial v_1}{\partial z_2} + \frac{\partial v_2}{\partial z_1} \right)$$

$$+ S_{13}\left( \frac{\partial v_1}{\partial z_3} + \frac{\partial v_3}{\partial z_1} \right) + S_{23}\left( \frac{\partial v_2}{\partial z_3} + \frac{\partial v_3}{\partial z_2} \right) + \rho Q \quad \text{(A)}$$

*Cylindrical coordinates:*

$$\rho \hat{c}_V \left( \frac{\partial T}{\partial t} + v_r \frac{\partial T}{\partial r} + \frac{v_\theta}{r} \frac{\partial T}{\partial \theta} + v_z \frac{\partial T}{\partial z} \right) = -\left[ \frac{1}{r} \frac{\partial}{\partial r}(r q_r) + \frac{1}{r} \frac{\partial q_\theta}{\partial \theta} + \frac{\partial q_z}{\partial z} \right]$$

$$- T\left(\frac{\partial P}{\partial T}\right)_{\hat{V}} \left[ \frac{1}{r} \frac{\partial}{\partial r}(r v_r) + \frac{1}{r} \frac{\partial v_\theta}{\partial \theta} + \frac{\partial v_z}{\partial z} \right] + S_{rr} \frac{\partial v_r}{\partial r} + S_{\theta\theta} \frac{1}{r}\left( \frac{\partial v_\theta}{\partial \theta} + v_r \right) + S_{zz} \frac{\partial v_z}{\partial z}$$

$$+ S_{r\theta}\left[ r \frac{\partial}{\partial r}\left(\frac{v_\theta}{r}\right) + \frac{1}{r} \frac{\partial v_r}{\partial \theta} \right] + S_{rz}\left( \frac{\partial v_z}{\partial r} + \frac{\partial v_r}{\partial z} \right) + S_{\theta z}\left[ \frac{1}{r} \frac{\partial v_z}{\partial \theta} + \frac{\partial v_\theta}{\partial z} \right] + \rho Q \quad \text{(B)}$$

*Spherical coordinates:*

$$\rho \hat{c}_V \left( \frac{\partial T}{\partial t} + v_r \frac{\partial T}{\partial r} + \frac{v_\theta}{r} \frac{\partial T}{\partial \theta} + \frac{v_\varphi}{r \sin \theta} \frac{\partial T}{\partial \varphi} \right) = -\left[ \frac{1}{r^2} \frac{\partial}{\partial r}(r^2 q_r) + \frac{1}{r \sin \theta} \frac{\partial}{\partial \theta}(q_\theta \sin \theta) + \frac{1}{r \sin \theta} \frac{\partial q_\varphi}{\partial \varphi} \right]$$

$$- T\left(\frac{\partial P}{\partial T}\right)_{\hat{V}} \left[ \frac{1}{r^2} \frac{\partial}{\partial r}(r^2 v_r) + \frac{1}{r \sin \theta} \frac{\partial}{\partial \theta}(v_\theta \sin \theta) + \frac{1}{r \sin \theta} \frac{\partial v_\varphi}{\partial \varphi} \right]$$

$$+ S_{rr} \frac{\partial v_r}{\partial r} + S_{\theta\theta}\left( \frac{1}{r} \frac{\partial v_\theta}{\partial \theta} + \frac{v_r}{r} \right) + S_{\varphi\varphi}\left( \frac{1}{r \sin \theta} \frac{\partial v_\varphi}{\partial \varphi} + \frac{v_r}{r} + \frac{v_\theta \cot \theta}{r} \right)$$

$$+ S_{r\theta}\left( \frac{\partial v_\theta}{\partial r} + \frac{1}{r} \frac{\partial v_r}{\partial \theta} - \frac{v_\theta}{r} \right) + S_{r\varphi}\left( \frac{\partial v_\varphi}{\partial r} + \frac{1}{r \sin \theta} \frac{\partial v_r}{\partial \varphi} - \frac{v_\varphi}{r} \right)$$

$$+ S_{\theta\varphi}\left( \frac{1}{r} \frac{\partial v_\varphi}{\partial \theta} + \frac{1}{r \sin \theta} \frac{\partial v_\theta}{\partial \varphi} - \frac{\cot \theta}{r} v_\varphi \right) + \rho Q \quad \text{(C)}$$

balance given in Table 5.6.1-1. You will also find useful the rectangular cartesian, cylindrical, and spherical components of Fourier's law shown in Table 5.6.1-4.

For an incompressible newtonian fluid with constant viscosity and constant thermal conductivity, Eqs. (E) and (F) of Table 5.6.1-1 reduce to

$$\rho \hat{c} \frac{d_{(m)} T}{dt} = k \text{ div } \nabla T + 2\mu \text{ tr } (\mathbf{D} \cdot \mathbf{D}) + \rho Q \quad (1\text{-}3)$$

**Table 5.6.1-3** The differential energy balance for newtonian fluids with constant $\rho$, and $k$ [see Eq. (1-3)]

*Rectangular cartesian coordinates:*

$$\rho\hat{c}\left(\frac{\partial T}{\partial t} + v_1\frac{\partial T}{\partial z_1} + v_2\frac{\partial T}{\partial z_2} + v_3\frac{\partial T}{\partial z_3}\right) = k\left(\frac{\partial^2 T}{\partial z_1^2} + \frac{\partial^2 T}{\partial z_2^2} + \frac{\partial^2 T}{\partial z_3^2}\right) + 2\mu\left[\left(\frac{\partial v_1}{\partial z_1}\right)^2 + \left(\frac{\partial v_2}{\partial z_2}\right)^2 + \left(\frac{\partial v_3}{\partial z_3}\right)^2\right]$$

$$+ \mu\left[\left(\frac{\partial v_1}{\partial z_2} + \frac{\partial v_2}{\partial z_1}\right)^2 + \left(\frac{\partial v_1}{\partial z_3} + \frac{\partial v_3}{\partial z_1}\right)^2 + \left(\frac{\partial v_2}{\partial z_3} + \frac{\partial v_3}{\partial z_2}\right)^2\right] + \rho Q \quad \text{(A)}$$

*Cylindrical coordinates:*

$$\rho\hat{c}\left(\frac{\partial T}{\partial t} + v_r\frac{\partial T}{\partial r} + \frac{v_\theta}{r}\frac{\partial T}{\partial \theta} + v_z\frac{\partial T}{\partial z}\right) = k\left[\frac{1}{r}\frac{\partial}{\partial r}\left(r\frac{\partial T}{\partial r}\right) + \frac{1}{r^2}\frac{\partial^2 T}{\partial \theta^2} + \frac{\partial^2 T}{\partial z^2}\right]$$

$$+ 2\mu\left\{\left(\frac{\partial v_r}{\partial r}\right)^2 + \left[\frac{1}{r}\left(\frac{\partial v_\theta}{\partial \theta} + v_r\right)\right]^2 + \left(\frac{\partial v_z}{\partial z}\right)^2\right\} + \mu\left\{\left(\frac{\partial v_\theta}{\partial z} + \frac{1}{r}\frac{\partial v_z}{\partial \theta}\right)^2 + \left(\frac{\partial v_z}{\partial r} + \frac{\partial v_r}{\partial z}\right)^2\right.$$

$$\left. + \left[\frac{1}{r}\frac{\partial v_r}{\partial \theta} + r\frac{\partial}{\partial r}\left(\frac{v_\theta}{r}\right)\right]^2\right\} + \rho Q \quad \text{(B)}$$

*Spherical coordinates:*

$$\rho\hat{c}\left(\frac{\partial T}{\partial t} + v_r\frac{\partial T}{\partial r} + \frac{v_\theta}{r}\frac{\partial T}{\partial \theta} + \frac{v_\varphi}{r\sin\theta}\frac{\partial T}{\partial \varphi}\right) = k\left[\frac{1}{r^2}\frac{\partial}{\partial r}\left(r^2\frac{\partial T}{\partial r}\right) + \frac{1}{r^2\sin\theta}\frac{\partial}{\partial \theta}\left(\sin\theta\frac{\partial T}{\partial \theta}\right) + \frac{1}{r^2\sin^2\theta}\frac{\partial^2 T}{\partial \varphi^2}\right]$$

$$+ 2\mu\left[\left(\frac{\partial v_r}{\partial r}\right)^2 + \left(\frac{1}{r}\frac{\partial v_\theta}{\partial \theta} + \frac{v_r}{r}\right)^2 + \left(\frac{1}{r\sin\theta}\frac{\partial v_\varphi}{\partial \varphi} + \frac{v_r}{r} + \frac{v_\theta\cot\theta}{r}\right)^2\right]$$

$$+ \mu\left\{\left[r\frac{\partial}{\partial r}\left(\frac{v_\theta}{r}\right) + \frac{1}{r}\frac{\partial v_r}{\partial \theta}\right]^2 + \left[\frac{1}{r\sin\theta}\frac{\partial v_r}{\partial \varphi} + r\frac{\partial}{\partial r}\left(\frac{v_\varphi}{r}\right)\right]^2\right.$$

$$\left. + \left[\frac{\sin\theta}{r}\frac{\partial}{\partial \theta}\left(\frac{v_\varphi}{\sin\theta}\right) + \frac{1}{r\sin\theta}\frac{\partial v_\theta}{\partial \varphi}\right]^2\right\} + \rho Q \quad \text{(C)}$$

Here we make the identification for an incompressible fluid:

$$\hat{c} \equiv \hat{c}_V \equiv T\left(\frac{\partial \hat{S}}{\partial T}\right)_{\hat{V}} = \hat{c}_P \equiv T\left(\frac{\partial \hat{S}}{\partial T}\right)_P \quad (1\text{-}4)$$

The rectangular cartesian, cylindrical, and spherical components of this equation are displayed in Table 5.6.1-3. The reader may use these as guides in immediately writing the corresponding expressions for the various forms of the differential energy balance given in Table 5.6.1-1.

## [5.6] SUMMARY

**Table 5.6.1-4 Components of the energy flux vector as represented by Fourier's law, Eq. (2-5) of Sec. 5.4.2**

*Rectangular cartesian coordinates:*

$$q_1 = -k \frac{\partial T}{\partial z_1} \quad \text{(A)}$$

$$q_2 = -k \frac{\partial T}{\partial z_2} \quad \text{(B)}$$

$$q_3 = -k \frac{\partial T}{\partial z_3} \quad \text{(C)}$$

*Cylindrical coordinates:*

$$q_r = -k \frac{\partial T}{\partial r} \quad \text{(D)}$$

$$q_\theta = -k \frac{1}{r} \frac{\partial T}{\partial \theta} \quad \text{(E)}$$

$$q_z = -k \frac{\partial T}{\partial z} \quad \text{(F)}$$

*Spherical coordinates:*

$$q_r = -k \frac{\partial T}{\partial r} \quad \text{(G)}$$

$$q_\theta = -k \frac{1}{r} \frac{\partial T}{\partial \theta} \quad \text{(H)}$$

$$q_\varphi = -k \frac{1}{r \sin \theta} \frac{\partial T}{\partial \varphi} \quad \text{(I)}$$

In Sec. 2.5.1, we discussed the notation to be used for vector and tensor components in cylindrical and spherical coordinate systems. These comments continue to apply.

## REFERENCE

1. Bird, R. B., W. E. Stewart, and E. N. Lightfoot: "Transport Phenomena," 7th printing, Wiley, New York, 1960.

# 6
# Applications of the Differential Balances to Energy Transfer

In this chapter, we are primarily concerned with the formulation and solution of boundary-value problems involving the differential energy balance. You should look upon this chapter as paralleling Chap. 3, where we focused our attention upon solutions of Cauchy's first law.

There is one point that may be worth emphasizing. You will perhaps notice as you go through this chapter that none of the problems directly involve satisfying the differential entropy inequality. This inequality has been satisfied identically by placing certain restrictions upon the constitutive equations for the stress tensor and the energy flux vector. We saw, for example, in Exercise 5.5.2-1 that, for a newtonian fluid which obeys Fourier's law,

$$k \geq 0 \quad \mu \geq 0 \quad \text{and} \quad \lambda \geq -\tfrac{2}{3}\mu \tag{0-1}$$

This situation is entirely analogous to the way in which Cauchy's second law was satisfied identically when approaching problems in Chap. 3.

## 6.1 PHILOSOPHY

### 6.1.1 The philosophy of solving problems involving the energy balance

There is an essential complication in the problems we consider here as compared with those we treated in Chap. 3. There we were concerned with simultaneous solutions of the equation of continuity and Cauchy's first law for some assumed constitutive equation for the stress tensor. Here we analyze problems that require the simultaneous solution of the equation of continuity, Cauchy's first law, and the differential energy balance, with particular constitutive equations for both the stress tensor and the energy flux vector.

As we described in Sec. 3.1.1, the first step is to decide what the problem is. In part this means that constitutive equations for the stress tensor and energy flux vector must be chosen. Though a variety of representations for the stress tensor are used in this chapter, only Fourier's law is employed for the energy flux vector. Alternatives to Fourier's law have found little support in experimental data (see Sec. 5.4.2).

To complete the specification of a particular problem, we must describe the geometry of the material or the geometry through which the material moves, the forces that cause the material to move, and any energy transmission to the material. As in Chap. 3, every problem requires a statement of boundary conditions in its formulation. Beyond those indicated in Sec. 3.1.1, there are several common types of boundary conditions that we suggest should be looked for in an unfamiliar physical situation.

1. We shall assume temperature to be continuous at a phase interface. This is suggested by anticipating that in a sense local equilibrium is established at the phase boundary. In the limit of a stable equilibrium, we learn in Sec. 8.5.3 that temperature is continuous across an interface.
2. The jump energy balance discussed in Exercise 5.3.2-1 must be satisfied at every phase interface.
3. We assume that temperatures and energy fluxes remain finite at all points in a material.

The advice we gave in Sec. 3.1.1 to launch the discussion of solutions for specific fluid mechanics problems is again applicable. Either we are not going to be able to solve every problem we formulate, or we are going to find many solutions to be very expensive considering present costs for computer time. Sometimes, it is more worthwhile to approximate a realistic but difficult problem by a somewhat simpler problem for which an analytic solution is available. This may be all that is needed. At worst, it should prove helpful as a limiting-case check on whatever numerical work is to be done.

The problems considered in this chapter are ones that can be solved exactly. Concepts and principles can be introduced in this way in the minimum amount of time. Further, I feel that relatively simple problems must be understood well before tackling a long numerical study.

As in our discussion of solutions for Cauchy's first law, we do not say that the solutions we find here are unique, though some uniqueness and existence theorems for Laplace's equation are available [1, pp. 211 and 277]. We are most interested in finding *a* solution. Sometimes experimental evidence will suggest that the solutions we seek are unique, but this may not always be so clear.

**REFERENCE**

1. Kellogg, O. D.: "Foundations of Potential Theory," Ungar, New York, 1929.

## 6.2 CONDUCTION

**6.2.1 Conduction in solids** One of the simplest classes of problems involving energy transmission is conduction in a stationary solid. Since the velocity vector is identically zero, the differential energy balance reduces to

$$\rho \hat{c} \frac{\partial T}{\partial t} = \text{div}\,(k\,\nabla T) + \rho Q \tag{1-1}$$

In writing this equation, we recognize that for a solid there is no need to distinguish between heat capacities at constant specific volume and heat capacities at constant thermodynamic pressure:

$$\hat{c} \equiv \hat{c}_V = \hat{c}_P \tag{1-2}$$

We often are not concerned with the stress distribution in the heated solid and, since the velocity distribution is known, Cauchy's first law and the equation of continuity need not be considered.

One commonly considered case is a solid with a thermal conductivity independent of temperature and, consequently, independent of position within the solid. For a steady-state temperature distribution with no external or mutual energy transmission ($Q = 0$), the differential energy balance reduces to Laplace's equation:

$$\text{div}\,(\nabla T) = 0 \tag{1-3}$$

The most attractive feature of this limiting case is that much attention has been given to Laplace's equation in the literature [1].

Remember that Laplace's equation occurred in our treatment of incompressible potential flow (of newtonian fluids). The mathematical problems that we saw there occur here with a different physical significance. For example, a boundary condition in which velocity is specified for a potential flow corresponds here to one in which the energy flux vector is designated.

For a discussion of many more solutions of Eq. (1-1), see Carslaw and Jaeger [2].

## REFERENCES

1. Kellogg, O. D.: "Foundations of Potential Theory," Ungar, New York, 1929.
2. Carslaw, H. S., and J. C. Jaeger: "Conduction of Heat in Solids," 2d ed., Oxford University Press, London, 1959.

### 6.2.2 Cooling of a semi-infinite slab: Constant surface temperature [1, p. 353]

Let us consider a quenching operation in which a hot body of metal with a uniform temperature $T_0$ is suddenly plunged into a cooling bath whose average temperature is controlled automatically by a refrigeration system. We wish to leave the body in the bath until the maximum temperature anywhere in the body is no greater than some specified value. Further, in order to achieve the maximum rate of production with the available equipment, we wish to remove the body from the quenching bath as soon as possible. This means that we must determine the temperature distribution in the body as a function of time.

The real problem is to solve for the temperature distribution both in the metal and in the surrounding quenching oil. Energy cannot flow from the solid into the oil unless there is a temperature gradient in the oil near the surface of the body. Our boundary conditions at the bounding surface of the metal should be that temperature is continuous across the phase interface and that the jump energy balance is satisfied. For the sake of simplicity, we shall say that the surface temperature of the metal is the same as the average temperature of the quenching oil and we shall ignore the temperature distribution in the oil.

In order that this problem be as simple as possible, let us also replace our finite body with a semi-infinite solid which occupies all space corresponding to $z_2 \geq 0$. The initial and boundary conditions become

$$\text{At } t = 0, \text{ for all } z_2 > 0: \quad T = T_0 \tag{2-1}$$

and

$$\text{At } z_2 = 0, \text{ for all } t > 0: \quad T = T_1 \tag{2-2}$$

Some authors feel that as an additional boundary condition one should take the temperature to be $T_0$ as $z_2$ approaches infinity for all values of time [1, p. 353; 2, p. 82]. We take the view here that temperature is not directly constrained to be $T_0$ at infinity for times greater than zero [3, p. 62]; it is only necessary that temperature be finite as $z_2$ approaches infinity [3, p. 71].

We shall further assume that the thermal conductivity of the metal may be taken to be a constant. Since there is no external or mutual energy transmission, Eq. (1-1) reduces to

$$\rho \hat{c} \frac{\partial T}{\partial t} = k \operatorname{div}(\nabla T) \tag{2-3}$$

We assume that

$$T = T(t, z_2) \tag{2-4}$$

From Eq. (A) of Table 5.6.1-3, we see that Eq. (2-3) reduces to

$$\rho \hat{c} \frac{\partial T}{\partial t} = k \frac{\partial^2 T}{\partial z_2^2} \tag{2-5}$$

A solution consistent with boundary conditions (2-1) and (2-2) is sought for this equation.

Our search for a solution can be somewhat simplified if we introduce a dimensionless temperature:

$$\theta = \frac{T - T_0}{T_1 - T_0} \tag{2-6}$$

Equation (2-5) becomes

$$\frac{\partial \theta}{\partial t} = \alpha \frac{\partial^2 \theta}{\partial z_2^2} \tag{2-7}$$

where

$$\alpha = \frac{k}{\rho \hat{c}} \tag{2-8}$$

Equations (2-1) and (2-2) reduce to

At $t = 0$, for all $z_2 > 0$: $\quad \theta = 0 \tag{2-9}$

and

At $z_2 = 0$, for $t > 0$: $\quad \theta = 1 \tag{2-10}$

This problem is mathematically of the same form as the one that resulted when we analyzed the flow in a fluid bounded by a wall that is suddenly set in motion (Sec. 3.2.4). Consequently, the solution is

$$\theta = \frac{T - T_0}{T_1 - T_0} = 1 - \operatorname{erf}\left(\frac{z_2}{\sqrt{4\alpha t}}\right) \tag{2-11}$$

Because erf (2) = 0.995, the temperature of the slab remains essentially unchanged outside a region of thickness $\delta_T$:

$$\delta_T \doteq 4\sqrt{\alpha t} \tag{2-12}$$

Consequently, Eq. (2-11) may be used as an approximate temperature distribution for a slab whose thickness is large compared with $\delta_T$.

## REFERENCES

1. Bird, R. B., W. E. Stewart, and E. N. Lightfoot: "Transport Phenomena," 7th printing, Wiley, New York, 1960.
2. Schlichting, Hermann: "Boundary-Layer Theory," 6th ed., McGraw-Hill, New York, 1968.
3. Carslaw, H. S., and J. C. Jaeger: "Conduction of Heat in Solids," 2d ed., Oxford University Press, London, 1959.

## EXERCISES

**6.2.2-1** *An infinitely long cylinder* [3, p. 199]  An infinitely long solid circular cylinder is initially at a uniform temperature $T_0$.  For time greater than zero, the surface temperature is constrained to be $T_1$.  Determine the temperature distribution in the rod whose radius is $R$.

*Answer*

$$\frac{T - T_1}{T_0 - T_1} = \sum_{n=1}^{\infty} \frac{2 J_0[\lambda_n(r/R)]}{\lambda_n J_1(\lambda_n)} \exp\left(\frac{-\lambda_n^2 k t}{\rho \hat{c} R^2}\right)$$

where the $\lambda_n$ ($n = 1, 2, \ldots$) are the positive roots of $J_0(\lambda_n) = 0$.

**6.2.2-2** *Two large blocks brought into contact*  A large block of metal A at a uniform temperature $T_{0(A)}$ is brought into contact with a large block of metal B at a uniform temperature $T_{0(B)}$ along two plane faces of the blocks.  Estimate the temperature of the interface as a function of time.

*Answer:* At the phase interface,

$$\theta = \left(1 + \sqrt{\frac{\rho_{(A)} c_{(A)} k_{(A)}}{\rho_{(B)} c_{(B)} k_{(B)}}}\right)^{-1}$$

*Hint:* Pose the problem in terms of

$$\theta = \frac{T - T_{0(A)}}{T_{0(B)} - T_{0(A)}}$$

Take the Laplace transform and solve for the temperature distribution in each phase.

**6.2.2-3** *Periodic surface temperature* [3, p. 65]  Conduction of energy in solids with periodic surface temperature is of considerable practical importance.  Problems of this type arise in designing automatic temperature control systems, in estimating the periodic temperatures (and, therefore, the periodic thermal stresses) in the cylinder walls of steam and internal combustion engines, and in studying the periodic heating of the earth's crust by the sun [3, p. 81].

Perhaps the simplest problem of this type is to consider a semi-infinite solid ($z_2 > 0$) whose surface is subjected to a periodic temperature variation:

At $z_2 = 0$: $\quad T = T_0 + A \cos(\omega t - \epsilon)$

Determine that

$$T^* \equiv \frac{T - T_0}{A} = \exp(-K z_2) \cos(\omega t - K z_2 - \epsilon)$$

where

$$K = \sqrt{\frac{\omega \rho \hat{c}}{2 k}}$$

*Hint:* See Exercise 3.2.4-4.

### 6.2.3 Cooling a semi-infinite slab: Newton's law of cooling [1, p. 70]

In order to solve the real quenching problem discussed in the introduction to Sec. 6.2.2, we saw that it would be necessary to determine the temperature distribution both in the metal and in the oil.  At the metal-oil phase interface we would require that temperature be continuous and that the jump energy balance be satisfied.  In the name of simplicity, we made the approximation there that the temperature of the metal surface was equal to the average temperature of the oil and we ignored the temperature distribution in the oil.

A somewhat better approximation would be to employ *Newton's law of cooling*, an empirical observation that:

> The energy flux across a fluid-solid phase interface is roughly proportional to the temperature difference between the surface and the surrounding medium.

In equation form, we write

$$\mathbf{q} \cdot \mathbf{n} = h(T_{surf} - T_{surr}) \tag{3-1}$$

with the understanding that $\mathbf{n}$ is the unit normal to the phase interface that is directed into the surroundings. The coefficient $h$ is referred to as the *heat-transfer coefficient*. Experimentally, $h$ is usually not found to be a constant, although it is often assumed to be a constant in constructing a mathematical model of a real situation. For another use of the heat-transfer coefficient concept, see Sec. 7.4.4.

Let us solve for the temperature distribution in the semi-infinite slab that was described in Sec. 6.2.2. We make only one change in the statement of the problem. Instead of requiring the temperature of the phase interface to be the average temperature of the oil, we specify that the energy transfer across the phase interface be described by Newton's law of cooling with a constant heat-transfer coefficient $h$:

$$\text{At } z_2 = 0, \text{ for all } t > 0: \quad k\frac{\partial T}{\partial z_2} = h(T - T_1) \tag{3-2}$$

With the change of variable

$$\theta = \frac{T - T_1}{T_0 - T_1} \tag{3-3}$$

Eq. (2-5) becomes

$$\frac{\partial \theta}{\partial t} = \alpha \frac{\partial^2 \theta}{\partial z_2^2} \tag{3-4}$$

The initial condition, Eq. (2-1), and the boundary condition, Eq. (3-2), may be written as

$$\text{At } t = 0, \text{ for all } z_2 > 0: \quad \theta = 1 \tag{3-5}$$

and

$$\text{At } z_2 = 0, \text{ for all } t > 0: \quad \frac{\partial \theta}{\partial z_2} = \frac{h}{k}\theta \tag{3-6}$$

Let us define the function

$$A = A(t, z_2) \equiv \theta - \frac{k}{h}\frac{\partial \theta}{\partial z_2} \tag{3-7}$$

[6.2] CONDUCTION

In terms of $A$, Eqs. (3-4), (3-5), and (3-6) become

$$\frac{\partial A}{\partial t} = \alpha \frac{\partial^2 A}{\partial z_2^2} \tag{3-8}$$

At $t = 0$, for all $z_2 > 0$:  $A = 1$ (3-9)

and

At $z_2 = 0$, for all $t > 0$:  $A = 0$ (3-10)

Taking the approach used in Sec. 3.2.4, we find that

$$A = \text{erf}\left(\frac{z_2}{\sqrt{4\alpha t}}\right) \tag{3-11}$$

Equations (3-7) and (3-11) combine to yield a first-order ordinary differential equation for $\theta$:

$$\frac{\partial \theta}{\partial z_2} - \frac{h}{k}\theta = -\frac{h}{k}\text{erf}\left(\frac{z_2}{\sqrt{4\alpha t}}\right) \tag{3-12}$$

We require that $\theta$ remain finite as $z_2$ approaches infinity. The solution to Eq. (3-12) which satisfies this condition is

$$\theta = -\frac{h}{k}\exp\left(\frac{hz_2}{k}\right)\int_\infty^{z_2}\text{erf}\left(\frac{\zeta}{\sqrt{4\alpha t}}\right)\exp\left(\frac{-h\zeta}{k}\right)d\zeta \tag{3-13}$$

A change of variable

$$\eta \equiv \zeta - z_2 \tag{3-14}$$

allows this solution to be expressed as

$$\theta = \frac{2h}{\sqrt{\pi}\,k}\int_0^\infty \exp\left(-\frac{h\eta}{k}\right)\left[\int_0^{(z_2+\eta)/\sqrt{4\alpha t}}\exp(-u^2)\,du\right]d\eta \tag{3-15}$$

Upon an integration by parts, we have

$$\theta = \text{erf}\left(\frac{z_2}{\sqrt{4\alpha t}}\right) + \frac{1}{\sqrt{\pi\alpha t}}\int_0^\infty \exp\left(-\frac{h\eta}{k} - \frac{[z_2+\eta]^2}{4\alpha t}\right)d\eta \tag{3-16}$$

or

$$\theta = \text{erf}\left(\frac{z_2}{\sqrt{4\alpha t}}\right) + \frac{1}{\sqrt{\pi\alpha t}}\exp\left(\frac{hz_2}{k} + \frac{h^2\alpha t}{k^2}\right)\int_0^\infty \exp\left(-\frac{[z_2+\eta+2h\alpha t/k]^2}{4\alpha t}\right)d\eta \tag{3-17}$$

This may be simplified by defining

$$u \equiv \frac{z_2 + \eta + 2h\alpha t/k}{\sqrt{4\alpha t}} \tag{3-18}$$

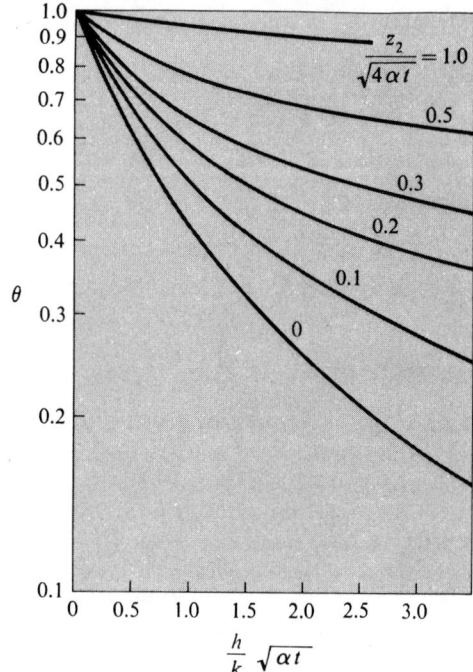

**Fig. 6.2.3-1** Cooling a semi-finite slab according to Eq. (3-20).

with the result

$$\theta = \text{erf}\left(\frac{z_2}{\sqrt{4\alpha t}}\right) + \frac{2}{\sqrt{\pi}} \exp\left(\frac{hz_2}{k} + \frac{h^2\alpha t}{k^2}\right) \int_{\frac{(z_2+2h\alpha t/k)}{\sqrt{4\alpha t}}}^{\infty} \exp(-u^2)\, du \qquad (3\text{-}19)$$

or

$$\theta = \text{erf}\left(\frac{z_2}{\sqrt{4\alpha t}}\right) + \exp\left(\frac{hz_2}{k} + \frac{h^2\alpha t}{k^2}\right)\left[1 - \text{erf}\left(\frac{z_2}{\sqrt{4\alpha t}} + \frac{h}{k}\sqrt{\alpha t}\right)\right] \qquad (3\text{-}20)$$

Equation (3-20) is plotted in Fig. 6.2.3-1.

The value of $\theta$ at the surface is

$$\theta = \exp\frac{h^2\alpha t}{k^2}\left[1 - \text{erf}\left(\frac{h}{k}\sqrt{\alpha t}\right)\right] \qquad (3\text{-}21)$$

This may be expressed in the form of a power series as [1, p. 483]

$$\theta = \frac{k}{h\sqrt{\pi\alpha t}}\left(1 - \frac{k^2}{2h^2\alpha t} + \frac{3k^4}{4h^4\alpha^2 t} - \cdots\right) \qquad (3\text{-}22)$$

Consequently, after a very long time the surface temperature may be approximated by

$$\theta = \frac{T - T_1}{T_0 - T_1} = \frac{k}{h\sqrt{\pi\alpha t}} \qquad (3\text{-}23)$$

The error involved is less than

$$\frac{k^3}{2h^3(\pi\alpha^3 t^3)^{\frac{1}{2}}}$$

## REFERENCE

1. Carslaw, H. S., and J. C. Jaeger: "Conduction of Heat in Solids," 2d ed., Oxford University Press, London, 1959.

## EXERCISES

**6.2.3-1** (a) Starting with the transformation defined by Eq. (3-7), derive Eq. (3-8).

(b) Starting with Eqs. (3-8), (3-9), and (3-10), determine the solution given by Eq. (3-11).

(c) Starting with Eq. (3-12) and the requirement that $\theta$ remain finite as $z_2$ approaches infinity, determine the solution given by Eq. (3-13).

**6.2.3-2** *Energy transfer from a pipe* [1, p. 189] A pipe is used to transport a fluid whose average temperature is $T_0$. The pipe is mounted in an airstream, the average temperature of which is $T_1$.

You may assume that the temperature of the interior surface of the pipe ($r = \varkappa R$) is $T_0$. But assume that Newton's law of cooling applies on the exterior surface ($r = R$).

(a) Show that the temperature distribution in the wall of the pipe is given by

$$\frac{T - T_1}{T_0 - T_1} = 1 - \frac{hR/k \ln(r/\varkappa R)}{1 + hR/k \ln(1/\varkappa)}$$

(b) Show that, if $\varkappa Rh/k > 1$, the magnitude of the energy transfer between the exterior wall of the pipe and the surroundings continuously decreases as $R$ increases for a fixed value of $\varkappa R$. Show that, if $\varkappa Rh/k < 1$, the magnitude of this energy transfer is a maximum at $R = k/h$.

## 6.2.4 Cooling a flat sheet: Constant surface temperature (1, pp. 93 to 97)

Let us replace the semi-infinite slab of Sec. 6.2.2 with a stationary infinite flat sheet which occupies all space between $z_2 = -a$ and $z_2 = +a$. We require the material to initially have a uniform temperature:

At $t = 0$, for $-a < z_2 < a$:  $T = T_0$  (4-1)

and we approximate the plunge of the hot sheet into a quenching bath by stating that

At $z_2 = \pm a$, for $t > 0$:  $T = T_1$  (4-2)

We shall take thermal conductivity to be a constant.

We assume that

$$T = T(t, z_2) \qquad (4\text{-}3)$$

The search for a solution can be somewhat simplified if we introduce an appropriate set of dimensionless variables. Let

$$\theta = \frac{T - T_1}{T_0 - T_1} \tag{4-4}$$

$$\tau = \frac{tk}{\rho \hat{c} a^2} \tag{4-5}$$

and

$$\xi = \frac{z_2}{a} \tag{4-6}$$

The differential energy balance, Eq. (1-1), becomes

$$\frac{\partial \theta}{\partial \tau} = \frac{\partial^2 \theta}{\partial \xi^2} \tag{4-7}$$

with the boundary conditions:

At $\tau = 0$, for $-1 < \xi < 1$: $\quad \theta = 1$ (4-8)

and

At $\xi = \pm 1$, for $\tau > 0$: $\quad \theta = 0$ (4-9)

Let us examine solutions to Eq. (4-7), which are of the form

$$\theta = \mathscr{T}(\tau) X(\xi) \tag{4-10}$$

This implies that

$$\frac{\mathscr{T}'}{\mathscr{T}} = \frac{X''}{X} = -\lambda^2 \tag{4-11}$$

where $\lambda^2$ is a constant.

The solution for

$$\mathscr{T}' + \lambda^2 \mathscr{T} = 0 \tag{4-12}$$

is of the form

$$\mathscr{T} = A e^{-\lambda^2 \tau} \tag{4-13}$$

The solution for

$$X'' + \lambda^2 X = 0 \tag{4-14}$$

is of the form

$$X = B \sin \lambda \xi + C \cos \lambda \xi \tag{4-15}$$

The boundary conditions (4-9) imply that

$$\lambda = \frac{(2n + 1)\pi}{2} \quad n = 0, 1, 2, \ldots \tag{4-16}$$

[6.2] CONDUCTION

and

$$B = 0 \tag{4-17}$$

A linear combination of all the solutions to Eq. (4-7) indicated by Eqs. (4-10), (4-13), and (4-15) to (4-17) yields

$$\theta = \sum_{n=0}^{\infty} D_n \exp\left(\frac{-(2n+1)^2\pi^2\tau}{4}\right) \cos\frac{(2n+1)\pi\xi}{2} \tag{4-18}$$

Boundary condition (4-8) requires that

$$1 = \sum_{n=0}^{\infty} D_n \cos\frac{(2n+1)\pi\xi}{2} \tag{4-19}$$

This implies that

$$\int_{-1}^{1} \cos\left(\frac{(2m+1)\pi\xi}{2}\right) d\xi = \sum_{n=0}^{\infty} D_n \int_{-1}^{1} \cos\left(\frac{(2m+1)\pi\xi}{2}\right) \cos\left(\frac{(2n+1)\pi\xi}{2}\right) d\xi$$

$$= \begin{cases} 0 & \text{if } m \neq n \\ D_m & \text{if } m = n \end{cases} \tag{4-20}$$

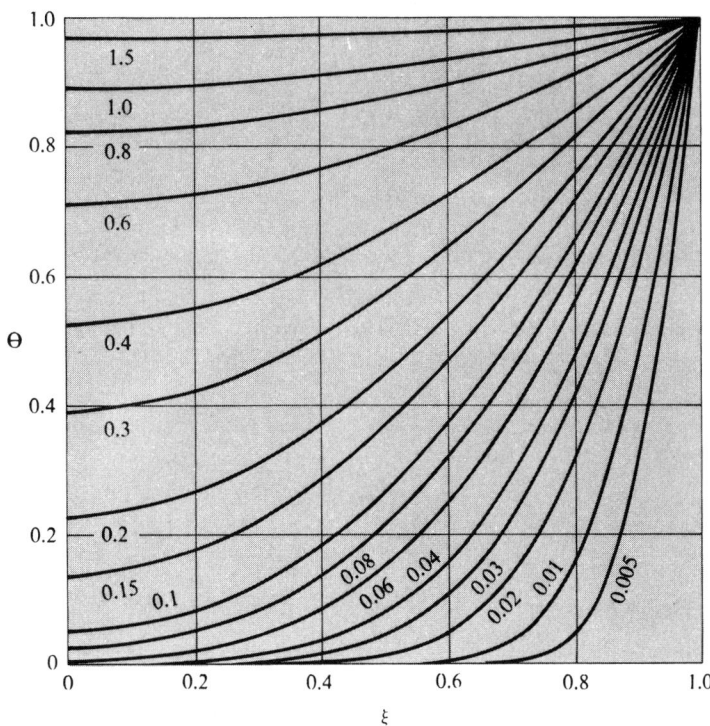

**Fig. 6.2.4-1** Cooling a flat sheet according to Eq. (4-22). (*From Carslaw and Jaeger* [1, p. 101, fig. 11].)

or

$$D_n = \frac{4(-1)^n}{(2n+1)\pi} \tag{4-21}$$

The final result for temperature has the form [1, p. 100]

$$\theta = \sum_{n=0}^{\infty} \frac{4(-1)^n}{(2n+1)\pi} \exp\left(\frac{-(2n+1)^2\pi^2\tau}{4}\right) \cos\frac{(2n+1)\pi\xi}{2} \tag{4-22}$$

or

$$\frac{T - T_1}{T_0 - T_1} = \sum_{n=0}^{\infty} \frac{4(-1)^n}{(2n+1)\pi} \exp\left(\frac{-(2n+1)^2\pi^2 kt}{4\rho\hat{c}a^2}\right) \cos\frac{(2n+1)\pi z_2}{2a} \tag{4-23}$$

This solution is shown in Fig. 6.2.4-1.

## REFERENCE

1. Carslaw, H. S., and J. C. Jaeger: "Conduction of Heat in Solids," 2d ed., Oxford University Press, London, 1959.

## EXERCISES

**6.2.4-1** *A flat sheet* [1, p. 122] We wish to determine the temperature distribution in a stationary infinite flat sheet that occupies all space between $z_2 = -a$ and $z_2 = +a$ and that initially has a uniform temperature $T_0$. For time greater than zero, Newton's law of cooling with a constant-heat-transfer coefficient $h$ is used to describe the loss of energy from the sheet to a fluid with an average temperature $T_1$.

(a) Define

$$\theta \equiv \frac{T - T_1}{T_0 - T_1} \qquad \tau \equiv \frac{tk}{\rho\hat{c}a^2} \qquad \xi \equiv \frac{z_2}{a}$$

and write the differential energy balance and the required boundary conditions in the following dimensionless form:

$$\frac{\partial\theta}{\partial\tau} = \frac{\partial^2\theta}{\partial\xi^2}$$

At $\xi = 1$: $\quad -\dfrac{\partial\theta}{\partial\xi} = H\theta$

At $\xi = -1$: $\quad \dfrac{\partial\theta}{\partial\xi} = H\theta$

At $\tau = 0$: $\quad \theta = 1$

Here we define

$$H \equiv \frac{ha}{k}$$

[6.2] CONDUCTION

(b) Look for a solution by the method of separation of variables. Satisfying all but the initial condition, determine that it has the form

$$\theta = \sum_{n=1}^{\infty} B_n \cos(\lambda_n \xi) \exp(-\lambda_n^2 \tau)$$

where the $\lambda_n$ ($n = 1, 2, \ldots$) denote the positive roots of

$$\lambda \tan \lambda = H \tag{1}$$

Carslaw and Jaeger [1, p. 491] tabulate the first six roots.
What additional condition is required to reach this form of a solution?

(c) As shorthand, denote

$$X_n \equiv \cos(\lambda_n \xi)$$

In the process of answering (b) above, we find that $X_n$ satisfies

$$X_n'' + \lambda_n^2 X_n = 0$$

in such a manner that

At $\xi = +1$: $\quad -X_n' = HX_n \tag{2}$

At $\xi = -1$: $\quad X_n' = HX_n \tag{3}$

Primes denote differentiation with respect to $\xi$.

Assume that $\lambda_n \neq \lambda_m$. Integrate by parts to prove that

$$(\lambda_n^2 - \lambda_m^2) \int_{-1}^{1} X_m X_n \, d\xi = \int_{-1}^{1} (-X_m X_n'' + X_n X_m'') \, d\xi$$

$$= [-X_m X_n' + X_n X_m']_{\xi=-1}^{\xi=1}$$

$$= 0$$

to conclude that [1, p. 116]

$$\int_{-1}^{1} X_m X_n \, d\xi = 0$$

(d) Use the definition of $X_n$ given in (c) to prove that

$$(\lambda_n X_n)^2 + (X_n')^2 = \lambda_n^2 \tag{4}$$

and, consequently, that

$$\lambda_n^2 \int_{-1}^{1} X_n^2 \, d\xi + \int_{-1}^{1} (X_n')^2 \, d\xi = 2\lambda_n^2 \tag{5}$$

Integrate by parts as in (c) to find

$$\lambda_n^2 \int_{-1}^{1} X_n^2 \, d\xi - \int_{-1}^{1} (X_n')^2 \, d\xi = -[X_n X_n']_{\xi=-1}^{\xi=1} \tag{6}$$

Adding Eqs. (5) and (6), we have

$$2\lambda_n^2 \int_{-1}^{1} X_n^2 \, d\xi = 2\lambda_n^2 - [X_n X_n']_{\xi=-1}^{\xi=1} \tag{7}$$

In view of Eq. (4), Eqs. (2) and (3) can be written as

$$\text{At } \xi = 1: \quad X_n X_n' = \frac{-H(\lambda_n)^2}{(\lambda_n)^2 + H^2} \tag{8}$$

$$\text{At } \xi = -1: \quad X_n X_n' = \frac{H(\lambda_n)^2}{(\lambda_n)^2 + H^2} \tag{9}$$

Substituting Eqs. (8) and (9) into Eq. (7), we conclude that [1, p. 116]

$$\int_{-1}^{1} X_n^2 \, d\xi = 1 + \frac{H}{(\lambda_n)^2 + H^2}$$

(e) Use the results of (c) and (d) in determining the coefficients $B_n$. Show that the final expression for the temperature distribution is

$$\theta = \sum_{n=1}^{\infty} \frac{2H \cos(\lambda_n \xi) \sec \lambda_n}{H(H+1) + \lambda_n^2} \exp(-\lambda_n^2 \tau)$$

where the $\lambda_n$ are the positive roots of Eq. (1).

**6.2.4-2** *A bar* [1, p. 173] Consider the cooling of a semi-infinite bar of metal with a uniform cross section $-a \leq z_2 \leq a$, $-b \leq z_1 \leq b$. The initial temperature of the bar is $T_0$ and the temperature of the surface of the bar is maintained at $T_1$. If we denote the solution given in Eq. (4-23) as $\varphi(z_2, a)$, show that the solution to this problem is

$$\frac{T - T_1}{T_0 - T_1} = \varphi(z_2, a) \varphi(z_1, b)$$

**6.2.4-3** *More on a bar* (1, p. 173) Consider the problem of Exercise 6.2.4-2 but use Newton's law of cooling with a constant heat transfer coefficient to describe the loss of energy from the bar to a fluid whose average temperature is $T_1$. If we denote the solution given in Exercise 6.2.4-1 as $\psi(z_2, a, h)$, show that the solution to this problem is

$$\frac{T - T_1}{T_0 - T_1} = \psi(z_2, a, h) \psi(z_1, b, h)$$

**6.2.4-4** *A sphere* [1, p. 233] A sphere is initially at a uniform temperature $T_0$. For time greater than zero, the surface temperature is constrained to be $T_1$. Determine the temperature distribution in the sphere whose radius is $R$.
*Answer*

$$\frac{T - T_1}{T_0 - T_1} = \frac{2R}{\pi} \sum_{n=1}^{\infty} \frac{(-1)^{n+1}}{nr} \sin\left(n\pi \frac{r}{R}\right) \exp\left(\frac{-n^2 \pi^2 kt}{\rho \hat{c} R^2}\right)$$

*Hint:* Introduce a transformation of the general form $u = rT$.

**6.2.4-5** *More on a flat sheet* [1, p. 100] A stationary flat sheet occupies all space between $z_2 = 0$ and $z_2 = b$ and initially has a uniform temperature $T_0$. For time greater than zero the wall at $z_2 = b$ is held at a fixed temperature $T_1$ while the wall at $z_2 = 0$ is insulated. Determine the temperature distribution in this sheet as a function of time and position.
*Answer*

$$\frac{T - T_1}{T_0 - T_1} = \sum_{m=0}^{\infty} \frac{4(-1)^m}{(2m+1)\pi} \cos\left(\frac{(2m+1)\pi z_2}{2 b}\right) \exp\left(\frac{-[2m+1]^2 \pi^2 kt}{4 \rho \hat{c} b^2}\right)$$

## 6.3 MORE COMPLETE SOLUTIONS

**6.3.1 Conduction, convection, and viscous dissipation in fluids** In the succeeding sections, there are two general types of problems with which we will be concerned: either the fluids will be required to be incompressible or the motion will be found to be isochoric.

For both of these cases, the differential energy balance in the form of Eq. (E) of Table 5.6.1-1 reduces to

$$\rho \hat{c}_V \frac{d_{(m)}T}{dt} = -\text{div } \mathbf{q} + \text{tr } (\mathbf{S} \cdot \nabla \mathbf{v}) + \rho Q \qquad (1\text{-}1)$$

We will assume that Fourier's law applies, since alternatives have found little support in experimental data (see Sec. 5.4.2). We will not be concerned here with external or mutual energy transmission ($Q = 0$), and we will generally neglect the temperature dependence of thermal conductivity for the sake of mathematical simplicity. Under these conditions, Eq. (1-1) further simplifies to

$$\rho \hat{c}_V \frac{d_{(m)}T}{dt} = k \text{ div } (\nabla T) + \text{tr } (\mathbf{S} \cdot \nabla \mathbf{v}) \qquad (1\text{-}2)$$

In the sections that follow, we shall generally be concerned with simultaneous solutions of Eq. (1-2), the equation of continuity

$$\text{div } \mathbf{v} = 0 \qquad (1\text{-}3)$$

Cauchy's first law

$$\rho \frac{d_{(m)}\mathbf{v}}{dt} = \text{div } \mathbf{T} + \rho \mathbf{f} \qquad (1\text{-}4)$$

and an appropriate constitutive equation for the stress tensor.

We are motivated to restrict ourselves to incompressible fluids or isochoric motions not primarily for the simplification gained in the differential energy balance. When the density is independent of temperature, Eq. (F) of Table 5.6.1-1 also reduces to Eq. (1-1) (see Exercise 5.1.2-7).[1] Rather, all analytic solutions to Cauchy's first law known to us predict isochoric motions.

**6.3.2 Couette flow of a compressible newtonian fluid** In Fig. 6.3.2-1 a compressible newtonian fluid is trapped between two parallel planes that are separated by a distance $b$. The lower plane is stationary and has a fixed uniform temperature $T_0$:

$$\text{At } z_2 = 0: \quad \mathbf{v} = 0 \quad T = T_0 \qquad (2\text{-}1)$$

---

[1] No similar simplification appears to exist for fluids whose densities are nearly independent of pressure but are functions of temperature.

324  APPLICATIONS OF THE DIFFERENTIAL BALANCES TO ENERGY TRANSFER

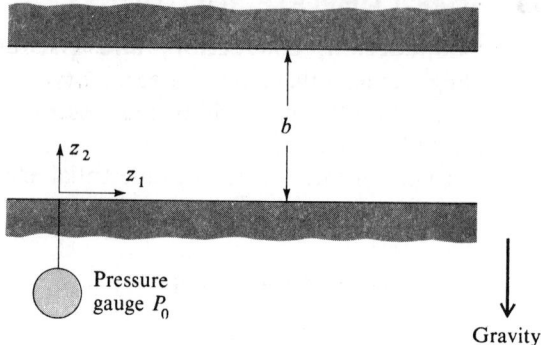

Fig. 6.3.2-1 Couette flow.

The upper plane moves with a constant speed $V$ in the $z_1$ direction and it has a uniform temperature $T_1$:

At $z_2 = b$:  $v_1 = V$  $v_2 = v_3 = 0$  $T = T_1$ (2-2)

A pressure gauge mounted on the lower plate reads $P_0$:

At $z_2 = z_1 = z_3 = 0$:  $-T_{22} = P_0$ (2-3)

Let us determine the velocity and temperature distributions within the fluid.

For lack of further information we will assume that the planes are very large and that edge effects may be neglected. We shall also take all physical properties (other than density) to be constants.

The boundary conditions (2-1) and (2-2) suggest that we assume

$$v_1 = v_1(z_2) \quad v_2 = v_3 = 0 \tag{2-4}$$

This form of velocity distribution satisfies the condition for an isochoric motion:

$$\text{div } \mathbf{v} = 0 \tag{2-5}$$

In view of Eqs. (2-4) and (2-5), the equation of continuity tells us that

$$\frac{d_{(m)}\rho}{dt} = \frac{\partial \rho}{\partial z_1} v_1 = 0 \tag{2-6}$$

or

$$\frac{\partial \rho}{\partial z_1} = 0 \tag{2-7}$$

This implies that density may be a function of $z_2$ and $z_3$:

$$\rho = \rho(z_2, z_3) \tag{2-8}$$

## [6.3] MORE COMPLETE SOLUTIONS

The components of Cauchy's first law simplify to

$$0 = -\frac{\partial P}{\partial z_1} + \mu \frac{\partial^2 v_1}{\partial z_2{}^2} \tag{2-9}$$

$$0 = -\frac{\partial P}{\partial z_2} - \rho g \tag{2-10}$$

and

$$0 = -\frac{\partial P}{\partial z_3} \tag{2-11}$$

Equations (2-10) and (2-11) indicate that

$$\frac{\partial^2 P}{\partial z_3 \, \partial z_2} = -\frac{\partial \rho}{\partial z_3} g = 0 \tag{2-12}$$

This means that density is not a function of $z_3$ and Eq. (2-8) reduces to

$$\rho = \rho(z_2) \tag{2-13}$$

From Eqs. (2-10) and (2-11), we have that

$$P = -g \int_0^{z_2} \rho \, dz_2 + h(z_1) \tag{2-14}$$

Equation (2-9) allows us to say

$$\frac{\partial P}{\partial z_1} = \frac{dh}{dz_1} = \mu \frac{d^2 v_1}{dz_2{}^2} = A = \text{a constant} \tag{2-15}$$

or

$$P = -g \int_0^{z_2} \rho \, dz_2 + A z_1 + B \tag{2-16}$$

In view of Eqs. (2-3) and (2-4), we reason that

$$B = P_0 \tag{2-17}$$

We must wait in order to determine the constant $A$.
Integration of Eq. (2-15) yields

$$v_1 = \frac{A}{2\mu}(z_2)^2 + C_1 z_2 + C_2 \tag{2-18}$$

The boundary conditions on velocity given by Eqs. (2-1) and (2-2) require

$$C_1 = \frac{V}{b} - \frac{A}{2\mu} b \tag{2-19}$$

and

$$C_2 = 0 \tag{2-20}$$

In view of the boundary conditions on temperature given by Eqs. (2-1) and (2-2), it seems reasonable to look for a temperature distribution in the form of

$$T = T(z_2) \tag{2-21}$$

Let us assume that our compressible fluid is such that pressure can be expressed as a function of density and temperature:

$$P = P(\rho, T) \tag{2-22}$$

Then Eqs. (2-13), (2-21), and (2-22) require that in Eq. (2-16)

$$A = 0 \tag{2-23}$$

The differential energy balance in the form of Eq. (A) of Table 5.6.1-2 reduces for the situation to

$$0 = k \frac{d^2 T}{dz_2^2} + \mu \left(\frac{dv_1}{dz_2}\right)^2 \tag{2-24}$$

or, to be consistent with Eqs. (2-18), (2-19), (2-20), and (2-23),

$$\frac{d^2 T}{dz_2^2} = -\frac{\mu}{k}\left(\frac{V}{b}\right)^2 \tag{2-25}$$

It is helpful to write this in a dimensionless form as

$$\frac{d^2 \theta}{dz_2^{*2}} = -N_{\text{Br}} \tag{2-26}$$

where

$$\theta \equiv \frac{T - T_0}{T_1 - T_0} \qquad z_2^* \equiv \frac{z_2}{b} \tag{2-27}$$

and where the Brinkman number is defined as

$$N_{\text{Br}} \equiv \frac{\mu V^2}{k(T_1 - T_0)} \tag{2-28}$$

From Eqs. (2-1) and (2-2), the boundary conditions that must be satisfied by Eq. (2-26) are

At $z_2^* = 0$: $\theta = 0$

At $z_2^* = 1$: $\theta = 1$ \qquad (2-29)

Equation (2-26) is easily integrated to satisfy Eq. (2-29). We find

$$\theta = -\frac{N_{\text{Br}}}{2}[z_2^{*2} - z_2^*] + z_2^* \tag{2-30}$$

## [6.3] MORE COMPLETE SOLUTIONS

To summarize, the velocity, pressure, and temperature distributions in the fluid are, respectively,

$$\frac{v_1}{V} = \frac{z_2}{b} \tag{2-31}$$

$$P = -g \int_0^{z_2} \rho \, dz_2 + P_0 \tag{2-32}$$

and

$$\frac{T - T_0}{T_1 - T_0} = \frac{N_{Br}}{2} \left[ \frac{z_2}{b} - \left(\frac{z_2}{b}\right)^2 \right] + \frac{z_2}{b} \tag{2-33}$$

Upon comparison of Eq. (2-26) with Eq. (2-24), it becomes obvious that the Brinkman number $N_{Br}$ is indicative of the rate at which energy is dissipated by viscous forces within the fluid. If we neglect this dissipation of energy, we see that the temperature distribution is linear:

$$\text{As } N_{Br} \to 0: \quad \frac{T - T_0}{T_1 - T_0} \to \frac{z_2}{b} \tag{2-34}$$

The next section takes up a similar problem, where not only density but the other physical properties as well are allowed to depend upon temperature.

## REFERENCE

1. Bird, R. B.: *Chem. Eng. Progr. Symp. Ser.*, (58)**61**:1 (1965).

## EXERCISES

**6.3.2-1** How is the temperature distribution of the problem discussed in the text altered if the temperature of the upper wall is changed to $T_0$?
*Answer*

$$\frac{T - T_0}{T_0} = \frac{1}{2} N_{Br} \left[ \frac{z_2}{b} - \left(\frac{z_2}{b}\right)^2 \right]$$

**6.3.2-2** Redo the problem of the text, now taking the lower wall to be insulated.
*Answer*

$$\frac{T - T_1}{T_1} = \frac{1}{2} N_{Br} \left[ 1 - \left(\frac{z_2}{b}\right)^2 \right]$$

**6.3.2-3** For the problem of the text, when is the upper wall cooled and when is it heated?

**6.3.2-4** How is the analysis of the text altered if the fluid is taken to be incompressible?

**6.3.2-5** What happens to the problem of the text when both walls are said to be insulated?

**328**  APPLICATIONS OF THE DIFFERENTIAL BALANCES TO ENERGY TRANSFER

**6.3.2-6** Rework the problem of the text, assuming that the gap between the planes is filled with an incompressible power-model fluid. Show that the temperature distribution has the same form with $N_{Br}$ replaced by

$$N_{Br(PM)} \equiv \frac{mV^{n+1}}{kb^{n-1}(T_1 - T_0)}$$

For this problem, use Fig. 6.3.2-2 in place of Fig. 6.3.2-1.

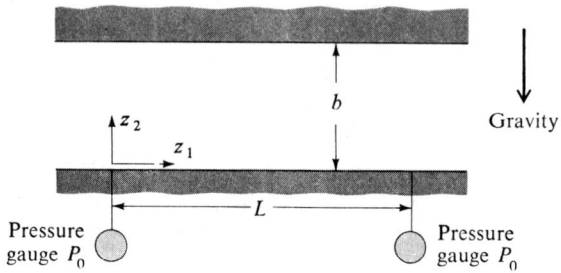

**Fig. 6.3.2-2** Couette flow of an incompressible fluid.

**6.3.2-7** *Flow through a channel*  An incompressible newtonian fluid flows through the channel of width $2b$ shown in Fig. 3.2.1-5 (see Exercise 3.2.1-8). Both planes are maintained at constant temperature $T_0$. Determine the temperature distribution in the fluid. The temperature dependence of viscosity and thermal conductivity may be neglected.

*Answer*

$$T - T_0 = \frac{\mu(v_{1(max)})^2}{3k}\left[1 - \left(\frac{z_2}{b}\right)^4\right]$$

**6.3.2-8** Repeat Exercise 6.3.2-7, assuming that the lower wall ($z_2 = -b$) is insulated.

**6.3.2-9** *Tangential annular flow of an incompressible newtonian fluid*  As explained in Sec. 3.2.3, tangential annular flow is the basis for one common type of viscometer. As the result of viscous dissipation, the temperature distribution within the fluid can be appreciably different from the known temperature of the walls. Apparent non-newtonian behavior in viscous materials' may in fact be attributable to the temperature dependence of a newtonian viscosity.

Referring to Fig. 3.2.3-1, let us assume that the inner wall is stationary, that the angular velocity $\Omega$ of the outer wall is specified, and that the temperature of both walls is a constant $T_0$. For simplicity, we will take the viscosity and thermal conductivity of the fluid to be independent of temperature.

Solve for the velocity and temperature distributions in the fluid and determine the position at which temperature is maximized.

*Answer*

$$\frac{T - T_0}{T_0} = N\left[1 - \left(\frac{R}{r}\right)^2\right] + N\left(\frac{1}{\varkappa^2} - 1\right)\frac{\ln(r/R)}{\ln \varkappa}$$

where

$$N \equiv N_{Br}\frac{\varkappa^4}{(1 - \varkappa^2)^2} \qquad N_{Br} \equiv \frac{\mu\Omega^2 R^2}{kT_0}$$

[6.3] MORE COMPLETE SOLUTIONS 329

**6.3.2-10** *More on tangential angular flow of an incompressible newtonian fluid* Repeat Exercise 6.3.2-9, assuming that the inner wall is insulated.
*Answer*

$$\frac{T-T_0}{T_0} = N\left[1 - \left(\frac{R}{r}\right)^2\right] - \frac{2N}{\varkappa^2}\ln\frac{r}{R}$$

**6.3.2-11** *Tangential annular flow of an incompressible power-model fluid* Repeat Exercise 6.3.2-9 for an incompressible power-model fluid.
*Answer*

$$\frac{T-T_0}{T_0} = \left(\frac{n}{2}\right)^2 N_{\text{Br}(PM)} \left[\frac{2\varkappa^{2/n}}{n(1-\varkappa^{2/n})}\right]^{n+1}\left[\frac{1}{\varkappa^{2/n}} - \left(\frac{R}{r}\right)^{2/n} - \frac{\ln(r/\varkappa R)}{\ln(1/\varkappa)}\left(\frac{1}{\varkappa^{2/n}}-1\right)\right]$$

**6.3.2-12** Show that result of Exercise 6.3.2-9 reduces to that for Exercise 6.3.2-1 as $\varkappa$ approaches unity.

**6.3.2-13** *Viscous heating in an oscillatory flow* [1] For the flow described in Exercise 3.2.4-5, determine the temperature distribution in the fluid in the limit as $\omega \to \infty$. You may assume that both walls are maintained at a constant and uniform temperature $T_0$.

In the limit as $a \to \infty$, it is not really the instantaneous temperature distribution that is of practical interest. We are more concerned with the distribution of the average temperature over one period of the oscillation:

$$\bar{T} \equiv \frac{\omega}{2\pi}\int_t^{t+2\pi/\omega} T\,dt'$$

It is reasonable to begin this problem by assuming

$$T = T(t, z_2)$$

for which the differential energy balance takes the form

$$\rho\hat{c}\frac{\partial T}{\partial t} = k\frac{\partial^2 T}{\partial z_2^2} + \mu\left(\frac{\partial v_1}{\partial z_2}\right)^2$$

It is the time average of this equation with which we shall be concerned:

$$\frac{\omega}{2\pi}\int_t^{t+2\pi/\omega}\left[\rho\hat{c}\frac{\partial T}{\partial t} - k\frac{\partial^2 T}{\partial z_2^2} - \mu\left(\frac{\partial v_1}{\partial z_2}\right)^2\right]dt' = 0$$

Referring to Sec. 4.1.2, we see that this reduces to

$$\rho\hat{c}\frac{\partial \bar{T}}{\partial t} = k\frac{\partial^2 \bar{T}}{\partial z_2^2} + \mu\overline{\left(\frac{\partial v_1}{\partial z_2}\right)^2}$$

But the time-averaged temperature defined here should be independent of time:

$$\bar{T} = \bar{T}(z_2)$$

and the time-averaged differential energy balance reduces to

$$0 = k\frac{\partial^2 \bar{T}}{\partial z_2^2} + \mu\overline{\left(\frac{\partial v_1}{\partial z_2}\right)^2}$$

(a) Determine that

$$\frac{1}{a^2}\left(\frac{b}{v_0}\right)^2\overline{\left(\frac{\partial v_1}{\partial z_2}\right)^2} = \frac{1}{a^2}\overline{\left(\frac{\partial v_1^*}{\partial \xi}\right)^2} = \frac{\sinh^2 a(1-\xi)\sin^2 a(1-\xi) + \cosh^2 a(1-\xi)\cos^2 a(1-\xi)}{\sinh^2 a\cos^2 a + \sin^2 a\cosh^2 a}$$

where

$$a = \sqrt{\frac{\omega \rho b^2}{2\mu}}$$

(b) We are concerned here with the limit as $\omega \to \infty$. Find that

$$\lim_{a \to \infty} \frac{1}{a^2} \left( \frac{\overline{\partial v_1^*}}{\partial \xi} \right)^2 = \exp(-2a\xi)$$

(c) Calculate the time-averaged temperature distribution in the fluid to be

$$\overline{T} - T_0 = \frac{\mu v_0^2}{4k} [1 - e^{-2a\xi} - (1 - e^{-2a})\xi]$$

Conclude that in the limit as $a \to \infty$, the maximum temperature occurs at the lower wall:

$$\overline{T}_{max} - T_0 = \frac{\mu v_0^2}{4k}$$

### 6.3.3 Couette flow of a compressible newtonian fluid with variable viscosity and thermal conductivity

Let us repeat the analysis of Sec. 6.3.2 (more precisely, Exercise 6.3.2-1) for a compressible newtonian fluid in laminar flow between the two planes shown in Fig. 6.3.2-1. The principal changes from Sec. 6.3.2 are that we shall now assume that both planes are maintained at constant temperature $T_0$ and that both viscosity and thermal conductivity are functions of temperature. For the moment, we shall leave the functional dependence of $\mu$ and $k$ unspecified. Our object is to determine the pressure, velocity, and temperature distributions within the fluid.

The boundary conditions to be satisfied are

At $z_2 = 0$:   $\mathbf{v} = 0$   $T = T_0$ (3-1)

At $z_2 = b$:   $v_1 = V$   $v_2 = v_3 = 0$   $T = T_0$ (3-2)

At $z_2 = z_1 = z_3 = 0$:   $-T_{22} = P_0$ (3-3)

We shall again assume that the fluid is in isochoric motion,

$$v_1 = v_1(z_2) \qquad v_2 = v_3 = 0 \tag{3-4}$$

and reason as we did in Sec. 6.3.2 to conclude that

$$\rho = \rho(z_2, z_3) \tag{3-5}$$

The three components of Cauchy's first law take a somewhat different form here:

$$0 = -\frac{\partial P}{\partial z_1} + \frac{\partial}{\partial z_2}\left(\mu \frac{dv_1}{dz_2}\right) \tag{3-6}$$

$$0 = -\frac{\partial P}{\partial z_2} - \rho g \tag{3-7}$$

$$0 = -\frac{\partial P}{\partial z_3} \tag{3-8}$$

[6.3] MORE COMPLETE SOLUTIONS

From Eqs. (3-7) and (3-8), we are able to argue as before that

$$\frac{\partial^2 P}{\partial z_3 \, \partial z_2} = -\frac{\partial \rho}{\partial z_3} g = 0 \tag{3-9}$$

and, consequently, that

$$\rho = \rho(z_2) \tag{3-10}$$

By analogy with our discussion in Sec. 6.3.2, we shall assume that

$$T = T(z_2) \tag{3-11}$$

Since viscosity is a function only of temperature, Eq. (3-11) allows us to rewrite the second term on the right of Eq. (3-6) using only the ordinary total derivative:

$$0 = -\frac{\partial P}{\partial z_1} + \frac{d}{dz_2}\left(\mu \frac{dv_1}{dz_2}\right) \tag{3-12}$$

Equations. (3-7) and (3-12) indicate that the thermodynamic pressure distribution is of the form

$$P = -g \int_0^{z_2} \rho \, dz_2 + h(z_1) \tag{3-13}$$

From Eqs. (3-12) and (3-13), we see

$$\frac{\partial P}{\partial z_1} = \frac{dh}{dz_1} = \frac{d}{dz_2}\left(\mu \frac{dv_1}{dz_2}\right) = A = \text{a constant} \tag{3-14}$$

The pressure distribution, consequently, reduces to

$$P = -g \int_0^{z_2} \rho \, dz_2 + A z_1 + P_0 \tag{3-15}$$

which satisfies boundary condition (3-3). If we assume that thermodynamic pressure is a function only of density and temperature,

$$P = P(\rho, T) \tag{3-16}$$

then Eqs. (3-10) and (3-11) tell us that

$$A = 0 \tag{3-17}$$

At this point, we see from Eq. (3-14) that

$$\frac{d}{dz_2}\left(\mu \frac{dv_1}{dz_2}\right) = 0 \tag{3-18}$$

must be solved simultaneously with the differential energy balance:

$$\frac{d}{dz_2}\left(k \frac{dT}{dz_2}\right) + \mu \left(\frac{dv_1}{dz_2}\right)^2 = 0 \tag{3-19}$$

for the velocity and temperature distributions. Let us introduce as dimensionless variables

$$\varphi \equiv \frac{v_1}{V} \qquad \theta \equiv \frac{T - T_0}{T_0} \qquad \xi \equiv \frac{z_2}{b} \tag{3-20}$$

in terms of which Eqs. (3-18) and (3-19) become

$$\frac{d\varphi}{d\xi} = \frac{\mu_0}{\mu} C_1 \tag{3-21}$$

and

$$\frac{\mu_0}{\mu} \frac{d}{d\xi}\left(\frac{k}{k_0} \frac{d\theta}{d\xi}\right) + N_{\mathrm{Br}} \left(\frac{d\varphi}{d\xi}\right)^2 = 0 \tag{3-22}$$

Here $\mu_0$ and $k_0$ are, respectively, a characteristic viscosity and characteristic thermal conductivity; the Brinkman number is defined here as

$$N_{\mathrm{Br}} \equiv \frac{\mu_0 V^2}{T_0 k_0} \tag{3-23}$$

The corresponding boundary conditions are

$$\text{At } \xi = 0: \qquad \varphi = \theta = 0 \tag{3-24}$$

and

$$\text{At } \xi = 1: \qquad \varphi = 1 \qquad \theta = 0 \tag{3-25}$$

Equations (3-21) and (3-22) suggest that we expand $\mu_0/\mu$ and $k/k_0$ in Taylor series as functions of the dimensionless temperature $\theta$:

$$\frac{k}{k_0} = 1 + \alpha_1\theta + \alpha_2\theta^2 + \cdots \tag{3-26}$$

$$\frac{\mu_0}{\mu} = 1 + \beta_1\theta + \beta_2\theta^2 + \cdots \tag{3-27}$$

In these series expansions we have identified $\mu_0$ and $k_0$ as the viscosity and the thermal conductivity of the fluid at $T = T_0$. We will assume that the parameters $\alpha_i$ and $\beta_i$ ($i = 1, 2, 3, \ldots$) are known from available experimental data.

With Eqs. (3-26) and (3-27), we see that Eqs. (3-21) and (3-22) are highly nonlinear equations. There appears to be little hope of obtaining an analytic solution. We certainly could solve the problem numerically. But even if we should decide to do so, it would still be nice to have an analytic solution for some limiting case that could be used to check the validity of our numerical work.

Whenever one is faced with the prospect of solving an apparently intractable nonlinear equation or system of nonlinear equations, one should ask whether the analysis of some limiting case might be almost as interesting. For example, here it would be interesting to ask about the effects of viscous dissipation in the limit of very small values of the Brinkman number $N_{Br}$. If we did have a solution to the complete problem, we could visualize that $\theta$, $\varphi$, and the constant of integration $C_1$ [in Eq. (3-21)] could all be expanded in Taylor series as functions of the Brinkman number:

$$\theta = \theta_0 + \theta_1 N_{Br} + \theta_2 N_{Br}^2 + \cdots \qquad (3\text{-}28)$$

$$\varphi = \varphi_0 + \varphi_1 N_{Br} + \varphi_2 N_{Br}^2 + \cdots \qquad (3\text{-}29)$$

$$C_1 = C_{10} + C_{11} N_{Br} + C_{12} N_{Br}^2 + \cdots \qquad (3\text{-}30)$$

The quantities $\theta_0$, $\varphi_0$, and $C_{10}$ are known, respectively, as the zeroth perturbations (with respect to $N_{Br} = 0$) of $\theta$, $\varphi$, and $C_1$; $\theta_1$, $\varphi_1$, and $C_{11}$ are known as the first perturbations (with respect to $N_{Br} = 0$).

Substituting Eqs. (3-27) to (3-30) into Eq. (3-21) we find

$$\frac{d\varphi_0}{d\xi} + N_{Br} \frac{d\varphi_1}{d\xi} + \cdots$$

$$= [1 + \beta_1(\theta_0 + \theta_1 N_{Br} + \cdots) + \beta_2(\theta_0 + \theta_1 N_{Br} + \cdots)^2 + \cdots]$$

$$(C_{10} + C_{11} N_{Br} + \cdots) \qquad (3\text{-}31)$$

or

$$\left[\frac{d\varphi_0}{d\xi} - C_{10}(1 + \beta_1 \theta_0 + \beta_2 \theta_0^2 + \cdots)\right]$$

$$+ N_{Br}\left[\frac{d\varphi_1}{d\xi} - C_{10}(\beta_1 \theta_1 + 2\beta_2 \theta_0 \theta_1 + \cdots)\right.$$

$$\left. - C_{11}(1 + \beta_1 \theta_0 + \beta_2 \theta_0^2 + \cdots)\right] + \cdots = 0 \qquad (3\text{-}32)$$

Since $N_{Br}^0$, $N_{Br}^1$, $N_{Br}^2$, ... are linearly independent, we conclude that the coefficients of these quantities in Eq. (3-32) must individually be zero. Looking at the first two coefficients in Eq. (3-32), we have

$$\frac{d\varphi_0}{d\xi} - C_{10}(1 + \beta_1 \theta_0 + \beta_2 \theta_0^2 + \cdots) = 0 \qquad (3\text{-}33)$$

and

$$\frac{d\varphi_1}{d\xi} - C_{10}(\beta_1 \theta_1 + 2\beta_2 \theta_0 \theta_1 + \cdots) - C_{11}(1 + \beta_1 \theta_0 + \beta_2 \theta_0^2 + \cdots) = 0 \qquad (3\text{-}34)$$

Going through the same argument with Eq. (3-22), we find that the coefficient of $N_{Br}^0$ is

$$(1 + \beta_1 \theta_0 + \beta_2 \theta_0^2 + \cdots) \frac{d}{d\xi}\left[(1 + \alpha_1 \theta_0 + \alpha_2 \theta_0^2 + \cdots) \frac{d\theta_0}{d\xi}\right] = 0 \qquad (3\text{-}35)$$

The coefficients of $N_{Br}{}^0$ in Eqs. $(3\text{-}24)_2$ and $(3\text{-}25)_2$ tell us that

At $\xi = 0$:  $\quad \theta_0 = 0$ \hfill (3-36)

and

At $\xi = 1$:  $\quad \theta_0 = 0$ \hfill (3-37)

Clearly, Eqs. (3-35) to (3-37) are satisfied by

$$\theta_0 = 0 \tag{3-38}$$

Equation (3-33), consequently, simplifies to

$$\frac{d\varphi_0}{d\xi} - C_{10} = 0 \tag{3-39}$$

The corresponding boundary conditions are found by looking at the coefficients $N_{Br}{}^0$ in Eqs. $(3\text{-}24)_1$ and $(3\text{-}25)_1$:

At $\xi = 0$:  $\quad \varphi_0 = 0$ \hfill (3-40)

At $\xi = 1$:  $\quad \varphi_0 = 1$ \hfill (3-41)

Equation (3-39) is readily integrated with boundary conditions (3-40) and (3-41) to find

$$\varphi_0 = \xi \tag{3-42}$$

and

$$C_{10} = 1 \tag{3-43}$$

The coefficient of $N_{Br}$ in Eq. (3-22) is

$$\frac{d^2\theta_1}{d\xi^2} + 1 = 0 \tag{3-44}$$

The boundary conditions for this equation are found from the coefficients of $N_{Br}$ in Eqs. $(3\text{-}24)_2$ and $(3\text{-}25)_2$:

At $\xi = 0$:  $\quad \theta_1 = 0$ \hfill (3-45)

At $\xi = 1$:  $\quad \theta_1 = 0$ \hfill (3-46)

The solution to Eq. (3-44) that satisfies Eqs. (3-45) and (3-46) is

$$\theta_1 = \tfrac{1}{2}(\xi - \xi^2) \tag{3-47}$$

From Eqs. (3-34), (3-38), (3-43), and (3-47), the coefficient of $N_{Br}$ in Eq. (3-21) yields

$$\frac{d\varphi_1}{d\xi} - \tfrac{1}{2}\beta_1(\xi - \xi^2) - C_{11} = 0 \tag{3-48}$$

The boundary conditions for this equation are determined by the coefficients $N_{Br}$ in

## [6.3] MORE COMPLETE SOLUTIONS

Eqs. $(3\text{-}24)_1$ and $(3\text{-}25)_1$:

At $\xi = 0$: $\quad \varphi_1 = 0$ \hfill (3-49)

At $\xi = 1$: $\quad \varphi_1 = 0$ \hfill (3-50)

Equations (3-48) to (3-50) are satisfied by

$$\varphi_1 = \frac{-\beta_1}{12}(\xi - 3\xi^2 + 2\xi^3) \tag{3-51}$$

and

$$C_{11} = -\frac{\beta_1}{12} \tag{3-52}$$

Looking at the coefficient of $N_{Br}^2$ in Eq. (3-22), we have

$$\frac{d^2\theta_2}{d\xi^2} + \alpha_1(\tfrac{1}{4} - \tfrac{3}{2}\xi + \tfrac{3}{2}\xi^2) + \frac{\beta_1}{2}(-\tfrac{1}{3} + \xi - \xi^2) = 0 \tag{3-53}$$

The boundary conditions that the solutions to this equation must satisfy are found by examining the coefficients of $N_{Br}^2$ in Eqs. $(3\text{-}24)_2$ and $(3\text{-}25)_2$:

At $\xi = 0$: $\quad \theta_2 = 0$ \hfill (3-54)

At $\xi = 1$: $\quad \theta_2 = 0$ \hfill (3-55)

Equations (3-53) to (3-55) are satisfied by

$$\theta_2 = -\frac{\alpha_1}{8}(\xi^2 - 2\xi^3 + \xi^4) - \frac{\beta_1}{24}(\xi - 2\xi^2 + 2\xi^3 - \xi^4) \tag{3-56}$$

To summarize, Eqs. (3-28), (3-29), (3-38), (3-42), (3-47), (3-51), and (3-56) describe the dimensionless velocity and temperature profiles as

$$\frac{v_1}{V} = \varphi = \xi - N_{Br}\frac{\beta_1}{12}(\xi - 3\xi^2 + 2\xi^3) + \cdots \tag{3-57}$$

and

$$\frac{T - T_0}{T_0} = \theta = N_{Br}\tfrac{1}{2}(\xi - \xi^2) - N_{Br}^2 \frac{\alpha_1}{8}(\xi^2 - 2\xi^3 + \xi^4)$$

$$- N_{Br}^2 \frac{\beta_1}{24}(\xi - 2\xi^2 + 2\xi^3 - \xi^4) + \cdots \tag{3-58}$$

This is an example of how one carries out a *perturbation solution* of one or more nonlinear differential equations. The *perturbation parameter* was chosen to be the Brinkman number $N_{Br}$. Equations (3-57) and (3-58), consequently, describe the velocity and temperature distributions for sufficiently small values of the perturbation parameter. There are two drawbacks to a perturbation solution.

1. One has no firm guarantee that the series solution developed converges.
2. Assuming that the series converges, there is no firm estimate of the error involved in truncating the series. The best that can be said is that the error should be less than the magnitude of the last term retained.

The solution developed here is based upon a similar solution suggested by Bird, Stewart, and Lightfoot [1, p. 306] for an incompressible newtonian fluid. Other approaches to the same general problem have been suggested [2 to 5].

A somewhat similar approach has been taken by Turian and Bird [6] in their discussion of viscous heating in a cone-plate viscometer for an incompressible newtonian fluid. Their comparison with experimental data is particularly helpful in indicating how important it is to consider viscous heating in that geometry.

## REFERENCES

1. Bird, R. B., W. E. Stewart, and E. N. Lightfoot: "Transport Phenomena," 7th printing, Wiley, New York, 1960.
2. Illingworth, C. R.: *Proc. Cambridge Phil. Soc.*, **46**:469 (1950).
3. DeGroff, H. M.: *J. Aeron. Sci.*, **23**:395 (1956).
4. DeGroff, H. M.: *J. Aeron. Sci.*, **23**:978 (1956).
5. Morgan, A. J. A.: *J. Aeron. Sci.*, **24**:315 (1957).
6. Turian, R. M., and R. B. Bird: *Chem. Eng. Sci.*, **18**:689 (1963).

## EXERCISES

**6.3.3-1** Fill in the missing steps in the text between Eqs. (3-35) and (3-56).

**6.3.3-2** *Tangential annular flow of an incompressible newtonian fluid* Repeat the problem described in the text for an incompressible newtonian fluid undergoing the tangential annular flow described in Fig. 3.2.3-1. The inner wall is stationary, the outer wall rotates at a constant angular velocity $\Omega$, and both walls are maintained at a fixed temperature $T_0$.

## 6.4 NO DISSIPATION

**6.4.1 When viscous dissipation is neglected** As we have mentioned previously, many problems are difficult to solve in full generality. Some ingenuity in the use of numerical analysis may be required as well as considerable effort on the part of the programmer, not to mention the cost of computer time. This dollar outlay may be justified easily in certain instances, but before undertaking such an expenditure it is usually worthwhile to consider some pertinent limiting cases. It may turn out that a limiting case is all that a situation demands. At worst, the solution for the limiting case should provide a necessary check on whatever numerical work is required.

Limiting cases are most easily delineated in terms of dimensionless forms of the equation of continuity, Cauchy's first law, and the differential energy balance. To be specific, let the material be a compressible newtonian fluid with variable coefficients

## [6.4] NO DISSIPATION

of viscosity, thermal conductivity, and heat capacity. We define the following dimensionless variables:

$$\rho^* = \frac{\rho}{\rho_0} \quad \mathbf{v}^* = \frac{\mathbf{v}}{v_0} \quad P^* = \frac{P}{P_0} \quad T^* = \frac{T}{T_0}$$

$$t^* = \frac{t}{t_0} \quad \mathbf{D}^* = \frac{L_0}{v_0}\mathbf{D} \quad \mathbf{S}^* = \frac{L_0}{\mu_0 v_0}\mathbf{S} \quad z_i^* = \frac{z_i}{L_0}$$

$$\mu^* = \frac{\mu}{\mu_0} \quad \lambda^* = \frac{\lambda}{\mu_0} \quad k^* = \frac{k}{k_0} \quad \hat{c}_V^* = \frac{\hat{c}_V}{\hat{c}_0}$$

$$\mathbf{f}^* = \frac{\mathbf{f}}{f_0}$$

(1-1)

Here the quantities distinguished by a subscript 0 are reference (or characteristic) quantities having the same dimensions as the variables being made dimensionless. For example, $v_0$ is a characteristic magnitude of velocity and $f_0$ might be the acceleration of gravity. The choice of the reference quantities in any particular problem is arbitrary; it is usually made so as to simplify the equations and boundary conditions as much as possible.

In terms of these dimensionless variables, the equation of continuity has the form

$$\frac{1}{N_{St}}\frac{\partial \rho^*}{\partial t^*} + \text{div }(\rho^*\mathbf{v}^*) = 0 \tag{1-2}$$

Cauchy's first law for a compressible newtonian fluid becomes

$$\frac{1}{N_{St}}\rho^*\frac{\partial \mathbf{v}^*}{\partial t^*} + \rho^*(\nabla\mathbf{v}^*)\cdot\mathbf{v}^* = -\frac{1}{N_{Ru}}\nabla P^*$$

$$+ \frac{1}{N_{Re}}\text{div }(2\mu^*\mathbf{D}^* + \lambda^*[\text{div }\mathbf{v}^*]\mathbf{I}) + \frac{1}{N_{Fr}}\rho^*\mathbf{f}^* \tag{1-3}$$

In a similar way for the differential energy balance in the form of Eq. (E) of Table 5.6.1-1, assuming there is no external or mutual energy transmission, we have

$$\frac{1}{N_{St}}\rho^*\hat{c}_V^*\frac{\partial T^*}{\partial t^*} + \rho^*\hat{c}_V^*(\nabla T^*)\cdot\mathbf{v}^* = \frac{1}{N_{Pr}N_{Re}}\text{div }(k^*\nabla T^*)$$

$$- \frac{N_{Br}}{N_{Ru}N_{Pr}} T^*\left(\frac{\partial P^*}{\partial T^*}\right)_{\hat{V}*} \text{div }\mathbf{v}^* + \frac{N_{Br}}{N_{Pr}N_{Re}}\text{tr }(\mathbf{S}^*\cdot\nabla\mathbf{v}^*) \tag{1-4}$$

In these equations, we have defined the Strouhal, Ruark, Reynolds, Froude, Prandtl, and Brinkman numbers, respectively, as

$$N_{St} \equiv \frac{t_0 v_0}{L_0} \qquad N_{Ru} \equiv \frac{\rho_0 v_0^2}{P_0}$$

$$N_{Re} \equiv \frac{\rho_0 v_0 L_0}{\mu_0} \qquad N_{Fr} \equiv \frac{v_0^2}{f_0 L_0} \tag{1-5}$$

$$N_{Pr} \equiv \frac{\hat{c}_0 \mu_0}{k_0} \qquad N_{Br} \equiv \frac{\mu_0 v_0^2}{k_0 T_0}$$

(The product of the Prandtl and Reynolds numbers is often referred to as the Peclet number. The ratio of the Brinkman to the Prandtl number is known as the Eckert number.)

One common limiting case is obtained by saying that, for sufficiently small values of $N_{Br}/N_{Pr}N_{Re}$, one should be able to neglect the effects of viscous dissipation with respect to convection in Eq. (1-4). In this way, Eq. (1-4) reduces to

$$\frac{1}{N_{St}} \rho^* \hat{c}_V^* \frac{\partial T^*}{\partial t^*} + \rho^* \hat{c}_V^* (\nabla T^*) \cdot \mathbf{v}^*$$

$$= \frac{1}{N_{Pr} N_{Re}} \operatorname{div}(k^* \nabla T^*) - \frac{N_{Br}}{N_{Ru} N_{Pr}} T^* \left(\frac{\partial P^*}{\partial T^*}\right)_{\hat{V}^*} \operatorname{div} \mathbf{v}^* \tag{1-6}$$

It is important to keep in mind that this is an intuitive argument and we consequently have no guarantee that it will always work. Our experience in treating creeping flows in Secs. 3.3.1 to 3.3.3 and in examining boundary-layer flows in Secs. 3.5.1 to 3.5.7 both encourages us and warns us of possible difficulties.

**6.4.2 Natural convection between vertical heated plates** We consider here steady-state natural convection between the two parallel plates shown in Fig. 6.4.2-1. We take the plates to be infinitely long; another way of putting this is that we do not consider the circulation patterns near the end of the channel. If we say that $T_2 > T_1$, we visualize the warmer fluid on the left rising and the cooler fluid on the right descending. We recognize that the circulation arises because the density of the warmer fluid is less than that of the cooler fluid. We must explicitly recognize the fluid to be compressible.

Before doing this, let us specify the problem more completely. The fluid is a compressible newtonian liquid. For simplicity, we take the coefficients of viscosity and thermal conductivity to be constants. As a liquid, the density may be assumed to be a function only of temperature. We assume that the acceleration of gravity is a constant.

We are given two boundary conditions on velocity:

At $z_2 = \pm b$: $\quad \mathbf{v} = 0$ \hfill (2-1)

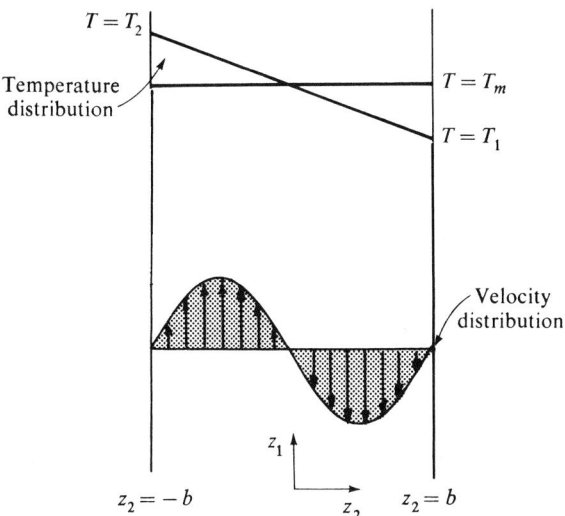

**Fig. 6.4.2-1** Natural convection between vertical heated plates.

and two boundary conditions on temperature:

At $z_2 = b$: $\quad T = T_1$ (2-2)

At $z_2 = -b$: $\quad T = T_2$ (2-3)

If we visualize this very long channel between the two plates to be closed by plates at either end, there is no net flow either up or down in the channel. This gives us a further condition to be met:

For all $z_1$: $\quad \int_{-b}^{b} v_1 \, dz_2 = 0$ (2-4)

The boundary conditions on temperature and the velocity distribution intuitively sketched in Fig. 6.4.2-1 suggest that there is only one nonzero component of velocity, $v_1$, and that both $v_1$ and $T$ are functions only of $z_2$:

$v_2 = v_3 = 0$

$v_1 = v_1(z_2)$ (2-5)

$T = T(z_2)$

This form of a velocity distribution obeys the condition for an isochoric motion:

div **v** = 0 (2-6)

Consequently, the equation of continuity requires only that

$$\frac{d_{(m)}\rho}{dt} = \frac{\partial \rho}{\partial t} + (\nabla \rho) \cdot \mathbf{v}$$

$$= \frac{d\rho}{dT}\left[\frac{\partial T}{\partial t} + (\nabla T) \cdot \mathbf{v}\right]$$

$$= 0 \quad (2\text{-}7)$$

which is satisfied identically by the assumed forms for the velocity and temperature distributions.

The three components of Cauchy's first law for a newtonian fluid reduce to

$$0 = -\frac{\partial P}{\partial z_1} + \mu \frac{d^2 v_1}{dz_2^2} - \rho g \quad (2\text{-}8)$$

and

$$\frac{\partial P}{\partial z_2} = \frac{\partial P}{\partial z_3} = 0 \quad (2\text{-}9)$$

The differential energy balance in the form of Eq. (E) of Table 5.6.1-1 becomes

$$0 = k\frac{d^2 T}{dz_2^2} + \mu \left(\frac{dv_1}{dz_2}\right)^2 \quad (2\text{-}10)$$

As the statement of the problem implied, we assume that there is no external or mutual energy transmission.

Our problem is clarified by the introduction of dimensionless variables in the manner indicated in Sec. 6.4.1:

$$\frac{\partial P^*}{\partial z_2^*} = \frac{\partial P^*}{\partial z_3^*} = 0 \quad (2\text{-}11)$$

$$0 = -\frac{\partial P^*}{\partial z_1^*} + \frac{1}{N_{\text{Re}}}\frac{d^2 v_1^*}{dz_2^{*2}} - \frac{1}{N_{\text{Fr}}}\rho^* \quad (2\text{-}12)$$

$$0 = \frac{d^2 T^*}{dz_2^{*2}} + N_{\text{Br}}\left(\frac{dv_1^*}{dz_2^*}\right)^2 \quad (2\text{-}13)$$

Here, the dimensionless variables are defined as

$$T^* \equiv \frac{T}{T_0} \qquad v_z^* \equiv \frac{v_z}{v_0} \qquad P^* \equiv \frac{P}{\rho_0 v_0^2}$$

$$z_i^* \equiv \frac{z_i}{b} \qquad \rho^* \equiv \frac{\rho}{\rho_0} \quad (2\text{-}14)$$

and the dimensionless groups are

$$N_{\text{Re}} \equiv \frac{\rho_0 v_0 b}{\mu} \qquad N_{\text{Fr}} \equiv \frac{v_0^2}{bg} \qquad N_{\text{Br}} \equiv \frac{\mu v_0^2}{kT_0} \quad (2\text{-}15)$$

## [6.4] NO DISSIPATION

We have not as yet defined $T_0$, $v_0$, and $\rho_0$. The quantity $v_0$ in particular is troublesome, since no magnitude of velocity arises in the boundary conditions. Since we are free to pick $v_0$ in any manner we desire, we choose it so as to make $N_{\text{Re}} = 1$:

$$v_0 \equiv \frac{\mu}{\rho_0 b} \tag{2-16}$$

We shall delay for the moment defining $T_0$ and $\rho_0$.

In view of Eq. (2-16), we have that

$$N_{\text{Br}} \equiv \frac{\mu^3}{k T_0 \rho_0^2 b^2} \tag{2-17}$$

For sufficiently small values of $N_{\text{Br}}$, it appears that the viscous dissipation terms in the energy balance may be neglected. We shall make this assumption and comment further on the magnitude of $N_{\text{Br}}$ at the completion of our discussion. Equation (2-13) consequently reduces to

$$\frac{d^2 T^*}{d z_2^{*2}} = 0 \tag{2-18}$$

Upon integrating this equation and applying the boundary conditions of Eqs. (2-2) and (2-3), we have

$$T^* = \frac{T_1 - T_2}{2 T_0} z_2^* + \frac{T_1 + T_2}{2 T_0} \tag{2-19}$$

This suggests that we define

$$T_0 \equiv \Delta T \equiv T_2 - T_1 \tag{2-20}$$

in order to write Eq. (2-19) in the somewhat simpler form

$$\theta \equiv \frac{T - T_m}{\Delta T} = T^* - \frac{T_1 + T_2}{2 T_0} = -\frac{z_2^*}{2} \tag{2-21}$$

We define here

$$T_m \equiv \frac{T_1 + T_2}{2} \tag{2-22}$$

We now must solve Eq. (2-12) for the velocity distribution. In view of Eqs. (2-11), we recognize that

$$\frac{dP^*}{dz_1^*} = \frac{d^2 v_1^*}{dz_2^{*2}} - \frac{1}{N_{\text{Fr}}} \rho^* = \text{a constant} \tag{2-23}$$

Before proceeding, we must specify the dependence of $\rho^*$ upon temperature. Let us write $\rho$ as a Taylor series about $T = T_m$:

$$\rho = \rho|_{T=T_m} + \left(\frac{d\rho}{dT}\right)_{T=T_m} (T - T_m) + \cdots \tag{2-24}$$

We know that

$$\frac{d\rho}{dT} = \frac{d}{dT}\left(\frac{1}{\hat{V}}\right) = -\frac{1}{\hat{V}^2}\frac{d\hat{V}}{dT} = -\rho\beta \tag{2-25}$$

where the quantity

$$\beta \equiv \frac{1}{\hat{V}}\frac{d\hat{V}}{dT} \tag{2-26}$$

is known as the coefficient of volume expansion. This suggests defining

$$\rho_0 \equiv \rho|_{T=T_m} \qquad \beta_0 \equiv \beta|_{T=T_m} \tag{2-27}$$

and rearranging Eq. (2-24) as

$$\rho^* = 1 - (\beta_0 \Delta T)\theta + \cdots \tag{2-28}$$

For sufficiently small values of $\beta_0 \Delta T$, Eq. (2-23) becomes

$$\frac{d^2 v_1^*}{dz_2^{*2}} - \frac{dP^*}{dz_1^*} - \frac{1}{N_{Fr}} + \frac{\beta_0 \Delta T}{N_{Fr}}\theta = 0 \tag{2-29}$$

or, from Eq. (2-21),

$$\frac{d^2 v_1^*}{dz_2^{*2}} - \frac{dP^*}{dz_1^*} - \frac{1}{N_{Fr}} - \frac{\beta_0 \Delta T}{2N_{Fr}} z_2^* = 0 \tag{2-30}$$

This last equation may be integrated to obtain, after application of the two boundary conditions in Eq. (2-1),

$$v_1^* = \frac{1}{2}\left(\frac{dP^*}{dz_1^*} + \frac{1}{N_{Fr}}\right)(z_2^{*2} - 1) + \frac{\beta_0 \Delta T}{12 N_{Fr}}(z_2^{*3} - z_2^*) \tag{2-31}$$

The unknown constant $dP^*/dz_1^*$ may be determined by Eq. (2-4) as

$$\frac{dP^*}{dz_1^*} = -\frac{1}{N_{Fr}} \tag{2-32}$$

As a final result, we obtain

$$v_1^* = \frac{\beta_0 \Delta T}{12 N_{Fr}}(z_2^{*3} - z_2^*) \tag{2-33}$$

or [1, p. 300]

$$v_1 = \frac{\rho_0 \beta_0 g b^2 \Delta T}{12 \mu}(z_2^{*3} - z_2^*) \tag{2-34}$$

This result is restricted to small values of $\beta_0 \Delta T$ by the approximation introduced in taking only the first two terms on the right of Eq. (2-28). For a wide range of liquids [2, p. 1639], $\beta_0 \approx 10^{-3}°C^{-1}$. This means that $\Delta T$ probably could be as large as $10^2 °C$ without seriously affecting the accuracy of this result.

[6.4] NO DISSIPATION

Returning to Eqs. (2-13) and (2-17), we recall that the application of Eq. (2-34) is also restricted to small values of

$$N_{\text{Br}} \equiv \frac{\mu^3}{k\,\Delta T \rho_0^2 b^2} \tag{2-35}$$

There seems to be no particular difficulty in satisfying this requirement. Water, for example, has $\mu^3/k \approx 10^{-10}$ in cgs units.

### REFERENCES

1. Bird, R. B., W. E. Stewart, and E. N. Lightfoot: "Transport Phenomena," 7th printing, Wiley, New York, 1960.
2. Lange, N. A.: "Handbook of Chemistry," 7th ed., Handbook Publishers, Sandusky, Ohio, 1949.

### EXERCISES

**6.4.2-1** *Natural convection between concentric vertical heated cylinders* Repeat the problem discussed in the text for natural convection between concentric vertical cylinders as described in Fig. 6.4.2-2. The temperature of the inner wall (at $r = \varkappa R$) is $T_1$; the temperature of the outer wall (at

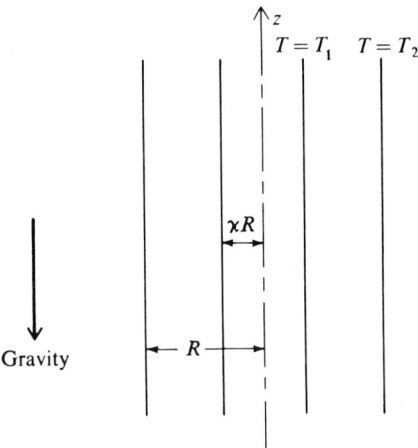

**Fig. 6.4.2-2** Natural convection between concentric vertical heated cylinders.

$r = R$) is $T_2$. Again, assume that the fluid is a newtonian liquid whose density is a function only of temperature, but whose viscosity and thermal conductivity may be assumed to be constants.

*Answer*

$$\frac{v_z \rho_0 R}{\mu} = \left[\left(\frac{r}{R}\right)^2 - 1\right](A - B) - \frac{\ln(r/R)}{\ln \varkappa}[A(\varkappa^2 - 1) + B(\varkappa^2 \ln \varkappa - \varkappa^2 + 1)] + B\left(\frac{r}{R}\right)^2 \ln \frac{r}{R}$$

where

$$A \equiv \frac{1}{4}\left(\frac{R^3 \rho_0}{\mu^2}\frac{dP}{dz} + \frac{R^3 \rho_0^2 g}{\mu^2}\right)$$

$$= B \frac{[(\tfrac{7}{4}\varkappa^2 + \tfrac{3}{4})(\varkappa^2 - 1) - \varkappa^4 \ln \varkappa - (\varkappa^2 - 1)^2/\ln \varkappa]}{[\varkappa^4 - 1 - (\varkappa^2 - 1)^2/\ln \varkappa]}$$

and

$$B \equiv \frac{1}{4} \frac{\beta_0(T_2 - T_1)R^3 \rho_0^2 g}{\mu^2 \ln \varkappa}$$

The parameters $\rho_0$ and $\beta_0$ are evaluated at $T_2$.

**6.4.2-2** *A more general approach to natural convection* Most natural convection problems cannot be approached in precisely the same manner that was used in the text. The difference is that most natural convection flows are not isochoric motions. In this exercise I outline a more general discussion of natural convection that is required for the majority of problems of practical interest.

Let us restrict ourselves to situations where density may be considered to be a function only of temperature and not of pressure. We shall again assume that the natural convection arises as the result of the difference between two characteristic temperatures $T_1$ and $T_2$.

(a) Following Eqs. (2-24) through (2-28), let us expand the density in a Taylor series about $T = T_0$ to find

$$\rho^* \equiv \frac{\rho}{\rho_0} = 1 - \beta_0 \Delta T \theta + \cdots \tag{1}$$

Here

$$\theta \equiv \frac{T - T_0}{\Delta T} \qquad \Delta T \equiv T_1 - T_2 \tag{2}$$

$$\rho_0 \equiv \rho|_{T=T_0} \qquad \beta_0 \equiv \beta|_{T=T_0} \tag{3}$$

and $T_0$ is some reference temperature. Substitute this expression for $\rho^*$ in Eqs. (1-2) through (1-4). In particular, arrange Eq. (1-3) in the form

$$[1 - \beta_0 \Delta T \theta + \cdots] \left[ \frac{1}{N_{\text{St}}} \frac{\partial \mathbf{v}^*}{\partial t^*} + (\nabla \mathbf{v}^*) \cdot \mathbf{v}^* \right] = -\frac{1}{N_{\text{Ru}}} \nabla \mathscr{P}^*$$

$$+ \frac{1}{N_{\text{Re}}} \operatorname{div} (2\mu^* \mathbf{D}^* + \lambda^*[\operatorname{div} \mathbf{v}^*]\mathbf{I}) - \frac{N_{\text{Gr}}}{N_{\text{Re}}^2} \theta \mathbf{f}^* + \cdots \tag{4}$$

where

$$\mathscr{P} \equiv P + \rho_0 \varphi \tag{5}$$

and

$$\mathbf{f}^* \equiv \frac{\mathbf{f}}{g} = -\frac{1}{g} \nabla \varphi \tag{6}$$

Here $N_{\text{Gr}}$ is the Grashof number

$$N_{\text{Gr}} \equiv \frac{\beta_0 \Delta T g L_0^3 \rho_0^2}{\mu_0^2} \tag{7}$$

and $g$ is the acceleration of gravity.

(b) As suggested by the discussion in the text, our discussion of natural convection problems will be confined to the limit as $\beta_0 \Delta T \to 0$. But before we can take this limit in the equations found in (a), we must settle two points. First, we must insure that, in this limit, the last term on the right of Eq. (2) is retained, since this is the driving force for the flow. Second, we must define $v_0$, because a characteristic speed will normally not arise in the boundary conditions for a natural convection problem. Both of these difficulties are resolved if we define $v_0$ by requiring

$$N_{\text{Re}} = N_{\text{Gr}}^{\frac{1}{2}} \tag{8}$$

With Eq. (8) in mind, determine that the dimensionless equation of continuity, the equation of motion for a newtonian fluid, and the differential energy balance reduce to

$$\text{div } \mathbf{v}^* = 0 \tag{9}$$

$$\frac{1}{N_{\text{St}}} \frac{\partial \mathbf{v}^*}{\partial t^*} + (\nabla \mathbf{v}^*) \cdot \mathbf{v}^* = -\frac{1}{N_{\text{Ru}}} \nabla \mathscr{P}^* + \frac{1}{N_{\text{Gr}}^{\frac{1}{2}}} \text{div } (2\mu^* \mathbf{D}^*) - \theta \mathbf{f}^* \tag{10}$$

and

$$\frac{1}{N_{\text{St}}} \hat{c}_V^* \frac{\partial \theta}{\partial t^*} + \hat{c}_V^* (\nabla \theta) \cdot \mathbf{v}^* = \frac{1}{N_{\text{Pr}} N_{\text{Gr}}^{\frac{1}{2}}} \text{div } (k^* \nabla \theta) \tag{11}$$

in the limit as both

$$\beta_0 \Delta T \to 0$$

and

$$\frac{N_{\text{Br}}}{N_{\text{Pr}} N_{\text{Gr}}^{\frac{1}{2}}} = \frac{\beta_0 g L_0}{\hat{c}_0 N_{\text{Gr}}^{\frac{1}{2}}} \to 0$$

(c) Reanalyze the problem described in this section using as your starting point Eqs. (9) to (11).

For more on natural convection, see Exercises 6.7.1-1 and 6.7.6-8.

## 6.5 NO CONVECTION

### 6.5.1 When convection is neglected (creeping flow)
Let us restrict ourselves to incompressible fluids and in this way rule out the possibility of natural convection (Sec. 6.4.2).

In Sec. 3.3.1, we introduced the concept of a *creeping flow*, obtained in the limit as a Reynolds number $N_{\text{Re}} \to 0$. For a creeping flow, the convective inertial terms are neglected with respect to the viscous terms in Cauchy's first law.

Let us visualize that we are working with a particular material. In this way, we may regard the Prandtl number $N_{\text{Pr}}$ as having a fixed value.

Referring to the dimensionless form of the energy balance displayed in Sec. 6.4.1,

$$\frac{1}{N_{\text{St}}} \hat{c}^* \frac{\partial T^*}{\partial t^*} + \rho^* \hat{c}^* (\nabla T^*) \cdot \mathbf{v}^* = \frac{1}{N_{\text{Pr}} N_{\text{Re}}} \text{div } (k^* \nabla T^*)$$

$$+ \frac{N_{\text{Br}}}{N_{\text{Pr}} N_{\text{Re}}} \text{tr } (\mathbf{S}^* \cdot \nabla \mathbf{v}^*) \quad (1\text{-}1)$$

we see that in the limit, as $N_{\text{Re}} \to 0$ for a fixed value of the Prandtl number, it is intuitively appealing to neglect convection (as represented by the second term on the left) with respect to conduction (described by the first term on the right). For this limiting case the energy balance reduces to

$$\frac{1}{N_{St}} \hat{c}^* \frac{\partial T^*}{\partial t^*} = \frac{1}{N_{Pr}N_{Re}} \text{div}(k^* \nabla T^*) + \frac{N_{Br}}{N_{Pr}N_{Re}} \text{tr}(\mathbf{S}^* \cdot \nabla \mathbf{v}^*) \qquad (1\text{-}2)$$

When neglecting convection with respect to conduction, it is also common to neglect viscous dissipation. For the limiting case, as both $N_{Re} \to 0$ and $N_{Br}/N_{Re} \to 0$ for a fixed value of the Prandtl number, it seems reasonable to say that the energy balance reduces to

$$\frac{1}{N_{St}} \hat{c}^* \frac{\partial T^*}{\partial t^*} = \frac{1}{N_{Pr}N_{Re}} \text{div}(k^* \nabla T^*) \qquad (1\text{-}3)$$

The interesting aspect of this limiting case is that the energy balance has the same form as that used to describe conduction in solids. Consequently, all of the problems discussed in the context of conduction in solids (Secs. 6.2.1 to 6.2.4) can be understood to be applicable in this limit to fluids as well.

It is important to realize that our discussion here has a certain degree of artificiality associated with it. We began by assuming that we were dealing only with incompressible fluids. In effect this meant that we were ruling out the possibility that density might depend upon temperature. The densities of all real fluids are dependent upon temperature to some degree.

Let us think a little further about the type of problem discussed in Exercise 6.4.2-2. There is no forced convection and yet there is motion in the fluid (natural convection) as the result of the temperature dependence of the fluid's density. Since there is no forced convection, the characteristic velocity of the problem is chosen by setting $N_{Re} = N_{Gr}^{1/2}$. Assuming that density is a function of temperature and that $\beta_0 \Delta T \to 0$, we can consider the limiting case as $N_{Pr} N_{Gr}^{1/2} \to 0$. Under these circumstances the dimensionless energy balance of Exercise 6.4.2-2 reduces to

$$\frac{1}{N_{St}} \hat{c} \frac{\partial T^*}{\partial t^*} = \frac{1}{N_{Pr}N_{Gr}^{1/2}} \text{div}(k^* \Delta T^*) \qquad (1\text{-}4)$$

### REFERENCE

1. Bird, R. B., W. E. Stewart, and E. N. Lightfoot: "Transport Phenomena," 7th printing, Wiley, New York, 1960.

### EXERCISES

**6.5.1-1** *Conduction from a sphere to a stagnant fluid* [1, p. 303] A heated sphere of diameter $D$ is suspended in a large body of an incompressible fluid (the temperature dependence of density is neglected). The temperature of the sphere is maintained at $T_0$; the temperature of the fluid at the large distance from the sphere is known to be $T_\infty$. The thermal conductivity $k$ of the fluid may be assumed constant.

(a) Determine the temperature distribution in the fluid.

(b) Use this temperature distribution to obtain an expression for the energy flux at the surface of the sphere. Equate this result to the expression for the energy flux written in terms of Newton's law of cooling. Show that the Nusselt number

$$N_{\mathrm{Nu}} \equiv \frac{hD}{k} = 2$$

This is a well-known result in agreement with experimental data in the limit as the Reynolds number approaches zero [1, p. 409].

## 6.6 NO CONDUCTION

### 6.6.1 When conduction is neglected

In Sec. 6.5.1, we consider the limit as the Reynolds number goes to zero for a fixed value of the Prandtl number (that is, for a particular material). Now let us go to the other extreme and study the limit as the Reynolds number becomes unbounded, again for a fixed value of the Prandtl number.

In terms of the dimensionless form of the energy balance presented in Sec. 6.4.1:

$$\frac{1}{N_{\mathrm{St}}} \rho^* \hat{c}_V^* \frac{\partial T^*}{\partial t^*} + \rho^* \hat{c}_V^*(\nabla T^*) \cdot \mathbf{v}^* = \frac{1}{N_{\mathrm{Pr}} N_{\mathrm{Re}}} \operatorname{div}(k^* \nabla T^*)$$

$$- \frac{N_{\mathrm{Br}}}{N_{\mathrm{Ru}} N_{\mathrm{Pr}}} T^* \left(\frac{\partial P^*}{\partial T^*}\right)_{\hat{V}^*} \operatorname{div} \mathbf{v}^* + \frac{N_{\mathrm{Br}}}{N_{\mathrm{Pr}} N_{\mathrm{Re}}} \operatorname{tr}(\mathbf{S}^* \cdot \nabla \mathbf{v}^*) \quad (1\text{-}1)$$

This limit suggests that the first term on the right, representing conduction, be neglected with respect to the second term on the left, representing convection. As a result of this intuitive argument, the energy balance is simplified to

$$\frac{1}{N_{\mathrm{St}}} \rho^* \hat{c}_V^* \frac{\partial T^*}{\partial t^*} + \rho^* \hat{c}_V^*(\nabla T^*) \cdot \mathbf{v}^*$$

$$= - \frac{N_{\mathrm{Br}}}{N_{\mathrm{Ru}} N_{\mathrm{Pr}}} T^* \left(\frac{\partial P^*}{\partial T^*}\right)_{\hat{V}^*} \operatorname{div} \mathbf{v}^* + \frac{N_{\mathrm{Br}}}{N_{\mathrm{Pr}} N_{\mathrm{Re}}} \operatorname{tr}(\mathbf{S}^* \cdot \nabla \mathbf{v}^*) \quad (1\text{-}2)$$

This neglect of conduction with respect to convection should remind you of our introduction to potential flow in Sec. 3.4.1, where the viscous terms in Cauchy's first law were neglected with respect to the convective inertial terms. In Sec. 3.4.3 we found that this limiting case was not capable of representing important aspects of real fluid behavior, at least in the neighborhood of boundary surfaces. This was not altogether surprising, since all the second derivative terms representing viscous effects had been dropped with a resulting reduction in the order of the differential equations to be solved for the velocity distribution.

The situation is not so different here when conduction is entirely neglected with respect to convection to arrive at Eq. (1-2). All the second derivatives of temperature are dropped from the differential energy balance with a resulting reduction in the order of the differential equation. On the basis of our discussion in Secs. 3.4.1 to 3.4.5, we can anticipate that Eq. (1-2) will not allow for a realistic description of the temperature distribution in a fluid near its bounding surfaces.

By analogy with the applications found for potential flow, we might expect Eq. (1-2) to be used to describe the temperature distribution in fluids outside the immediate neighborhood of their bounding surfaces. While very little direct use of Eq. (1-2) has been made in the literature, it is common to neglect both viscous and conduction effects in discussions of sound waves (see next section) and shock waves (surfaces of discontinuity with respect to the normal component of velocity). Serrin [1, secs. 51 to 57] gives an excellent introduction to these topics.

**REFERENCE**

1. Serrin, James: In S. Flügge and C. Truesdell (eds.), "Handbuch der Physik," vol. 8/1, Springer-Verlag, Berlin, 1959.

### 6.6.2 Speed of propagation of sound waves

[1, p. 179; 2, p. 245] For the moment let us define the speed of sound to be the speed of propagation of pressure waves resulting from a small amplitude disturbance in a compressible fluid.

If $a$ is the dimensionless amplitude of the disturbance, then we wish to determine the speed of the propagation of pressure waves in the limit as $a \to 0$. We will assume in this analysis that, when $a = 0$, the temperature, pressure, and density assume uniform values throughout the fluid:

$$\text{At } a = 0: \quad T = T_0 \quad P = P_0 \quad \rho = \rho_0 \quad \mathbf{v} = 0 \tag{2-1}$$

Our analysis must be based upon the equation of continuity, Cauchy's first law, and the differential energy balance. In terms of dimensionless variables, these three equations become

$$\frac{1}{N_{\text{St}}} \frac{\partial \rho^*}{\partial t^*} + \text{div}\,(\rho^* \mathbf{v}^*) = 0 \tag{2-2}$$

$$\frac{1}{N_{\text{St}}} \rho^* \frac{\partial \mathbf{v}^*}{\partial t^*} + \rho^* \nabla \mathbf{v}^* \cdot \mathbf{v}^* = -\frac{1}{N_{\text{Ru}}} \nabla P^* + \frac{1}{N_{\text{Re}}} \text{div}\,\mathbf{S}^* + \frac{1}{N_{\text{Fr}}} \rho^* \mathbf{f}^* \tag{2-3}$$

and

$$\frac{1}{N_{\text{St}}} \rho^* \hat{c}_P^* \frac{\partial T^*}{\partial t^*} + \rho^* \hat{c}_P^* \nabla T^* \cdot \mathbf{v}^*$$

$$= \frac{1}{N_{\text{Pr}} N_{\text{Re}}} \text{div}\,(k^* \nabla T^*) + \frac{N_{\text{Br}}}{N_{\text{St}} N_{\text{Ru}} N_{\text{Pr}}} \left(\frac{\partial \ln \hat{V}^*}{\partial \ln T^*}\right)_{P*} \frac{\partial P^*}{\partial t^*}$$

$$+ \frac{N_{\text{Br}}}{N_{\text{Ru}} N_{\text{Pr}}} \left(\frac{\partial \ln \hat{V}^*}{\partial \ln T^*}\right)_{P*} \nabla P^* \cdot \mathbf{v}^* + \frac{N_{\text{Br}}}{N_{\text{Pr}} N_{\text{Re}}} \text{tr}\,(\mathbf{S}^* \cdot \nabla \mathbf{v}^*) \tag{2-4}$$

## [6.7] BOUNDARY-LAYER THEORY

If our characteristic length is chosen to be representative of the gross system and if our characteristic velocity is the speed of sound, it seems reasonable to confine our attention to the limit as $N_{\text{Re}} \to \infty$, $N_{\text{Fr}} \to \infty$, and $N_{\text{Br}}/N_{\text{Re}} \to 0$. In this limit, Eqs. (2-2) to (2-4) become in dimensional form

$$\frac{d_{(m)}\rho}{dt} + \rho \operatorname{div} \mathbf{v} = 0 \tag{2-5}$$

$$\rho \frac{d_{(m)}\mathbf{v}}{dt} = -\nabla P \tag{2-6}$$

and

$$\rho \hat{c}_P \frac{d_{(m)}T}{dt} = \left(\frac{\partial \ln \hat{V}}{\partial \ln T}\right)_P \frac{d_{(m)}P}{dt} \tag{2-7}$$

We wish to obtain a solution to these equations valid in the limit as $a \to 0$. This suggests that we carry out a perturbation analysis as we did in Sec. 6.3.3. We begin by expressing $T$, $P$, $\rho$, and $\mathbf{v}$ as Taylor series:

$$T = T_0 + aT_1 + a^2 T_2 + \cdots \tag{2-8}$$

$$P = P_0 + aP_1 + a^2 P_2 + \cdots \tag{2-9}$$

$$\rho = \rho_0 + a\rho_1 + a^2 \rho_2 + \cdots \tag{2-10}$$

and

$$\mathbf{v} = a\mathbf{v}_1 + a^2 \mathbf{v}_2 + \cdots \tag{2-11}$$

Substituting these series into Eq. (2-7) and looking only at the coefficient of $a$, we find

$$\rho_0 \hat{c}_{P_0} \frac{\partial T_1}{\partial t} = \left(\frac{\partial \ln \hat{V}}{\partial \ln T}\right)_{P_0} \frac{\partial P_1}{\partial t} \tag{2-12}$$

where we define

$$\left(\frac{\partial \ln \hat{V}}{\partial \ln T}\right)_{P_0} \equiv \left(\frac{\partial \ln \hat{V}}{\partial \ln T}\right)_P \bigg|_{\substack{T=T_0 \\ P=P_0}} \tag{2-13}$$

and

$$\hat{c}_{P_0} \equiv \hat{c}_P \big|_{\substack{T=T_0 \\ P=P_0}} \tag{2-14}$$

We also know that

$$\frac{\partial P}{\partial t} = \left(\frac{\partial P}{\partial \rho}\right)_T \frac{\partial \rho}{\partial t} + \left(\frac{\partial P}{\partial T}\right)_\rho \frac{\partial T}{\partial t} \tag{2-15}$$

the first perturbation of which says

$$\frac{\partial P_1}{\partial t}\left[1 - \left(\frac{\partial P}{\partial T}\right)_{\rho_0} \frac{\partial T_1/\partial t}{\partial P_1/\partial t}\right] = \left(\frac{\partial P}{\partial \rho}\right)_{T_0} \frac{\partial \rho_1}{\partial t} \tag{2-16}$$

Equations (2-12) and (2-16) may now be combined to tell us

$$\frac{\partial P_1/\partial t}{\partial \rho_1/\partial t} = \left(\frac{\partial P}{\partial \rho}\right)_{T_0} \frac{\rho_0 \hat{c}_{P_0}}{\rho_0 \hat{c}_{P_0} - (\partial P/\partial T)_{\rho_0}(\partial \ln \hat{V}/\partial \ln T)_{P_0}} \quad (2\text{-}17)$$

Since (see Exercise 5.1.2-7)

$$\rho \hat{c}_P - \left(\frac{\partial P}{\partial T}\right)_\rho \left(\frac{\partial \ln \hat{V}}{\partial \ln T}\right)_P = \rho \hat{c}_V \quad (2\text{-}18)$$

Eq. (2-17) may finally be rearranged in what will prove to be a more interesting form:

$$\frac{\partial P_1/\partial t}{\partial \rho_1/\partial t} = \gamma_0 \left(\frac{\partial P}{\partial \rho}\right)_{T_0} = (v_s)^2 \quad (2\text{-}19)$$

We have introduced as definitions here

$$\gamma_0 \equiv \frac{\hat{c}_{P_0}}{\hat{c}_{V_0}} \quad (2\text{-}20)$$

and

$$v_s \equiv \sqrt{\gamma_0 \left(\frac{\partial P}{\partial \rho}\right)_{T_0}} \quad (2\text{-}21)$$

The first perturbation of the equation of continuity, Eq. (2-5), yields

$$\frac{\partial \rho_1}{\partial t} + \rho_0 \operatorname{div} \mathbf{v}_1 = 0 \quad (2\text{-}22)$$

We can use Eq. (2-19) to write this as

$$\frac{\partial P_1}{\partial t} + \rho_0 (v_s)^2 \operatorname{div} \mathbf{v}_1 = 0 \quad (2\text{-}23)$$

Upon differentiating with respect to time, we find that this becomes

$$\frac{\partial^2 P_1}{\partial t^2} + \rho_0 (v_s)^2 \operatorname{div} \frac{\partial \mathbf{v}_1}{\partial t} = 0 \quad (2\text{-}24)$$

The second term in this last equation suggests that we look at the first perturbation of Cauchy's first law, Eq. (2-6):

$$\rho_0 \frac{\partial \mathbf{v}_1}{\partial t} = -\nabla P_1 \quad (2\text{-}25)$$

Equations (2-24) and (2-25) imply that the first perturbation in pressure is a solution of the wave equation:

$$\frac{\partial^2 P_1}{\partial t^2} - (v_s)^2 \operatorname{div} \nabla P_1 = 0 \quad (2\text{-}26)$$

The discussion at this point is clarified if we restrict ourselves to a one-dimensional pressure wave in rectangular cartesian coordinates:

## [6.6] NO CONDUCTION

$$P_1 = P_1(t, z_1) \tag{2-27}$$

For this case, Eq. (2-26) reduces to

$$\frac{\partial^2 P_1}{\partial t^2} - (v_s)^2 \frac{\partial^2 P_1}{\partial z_1^2} = 0 \tag{2-28}$$

If we introduce as changes of variable

$$\xi \equiv z_1 - v_s t \tag{2-29}$$

and

$$\eta \equiv z_1 + v_s t \tag{2-30}$$

Eq. (2-28) takes the simpler form

$$\frac{\partial^2 P_1}{\partial \xi \, \partial \eta} = 0 \tag{2-31}$$

This equation can be integrated immediately to find

$$P_1 = F(\xi) + G(\eta) \tag{2-32}$$

where $F(\xi)$ and $G(\eta)$ are arbitrary functions.

The physical interpretation of this result is more obvious if we set $G(\eta) = 0$. In this event

$$P = P_0 + aP_1 = P_0 + aF(\xi) \tag{2-33}$$

Taking the derivative of this equation with respect to time, while holding $P$ constant, we find

$$0 = \left[ \left( \frac{\partial z_1}{\partial t} \right)_P - v_s \right] \frac{dF}{d\xi} \tag{2-34}$$

Since $F(\xi)$ is an arbitrary function, we must conclude that

$$\text{Speed of sound:} \quad \left( \frac{\partial z_1}{\partial t} \right)_P = v_s \equiv \sqrt{\gamma_0 \left( \frac{\partial P}{\partial \rho} \right)_{T_0}} \tag{2-35}$$

The function $F(\xi)$ describes a pressure wave traveling in the positive $z_1$ direction. The speed of propagation of a surface on which pressure is a constant is $v_s$. But this, of course, was our definition for the speed of sound.

We often think in terms of ideal gases for which

$$\left( \frac{\partial P}{\partial \rho} \right)_{T_0} = \frac{P_0}{\rho_0} \tag{2-36}$$

and

$$\text{Ideal gas:} \quad v_s = \sqrt{\gamma_0 \frac{P_0}{\rho_0}} \tag{2-37}$$

The discussion we have presented here is simple, but it rests upon a definition for the speed of sound that is not unequivocal when applied to an entirely arbitrary fluid motion. Serrin [1, p. 212] presents a more satisfying treatment based upon the conception of sound waves as surfaces of discontinuity with respect to the pressure gradient.

## REFERENCES
1. Serrin, James: In S. Flügge and C. Truesdell (eds.), "Handbuch der Physik," vol. 8/1, Springer-Verlag, Berlin, 1959.
2. Landau, L. D., and E. M. Lifshitz: "Fluid Mechanics," Addison-Wesley, Reading, Mass., 1959.

## EXERCISE

**6.6.2-1** *When conduction and viscous dissipation can be neglected in an ideal gas* In the limit as $N_{Re} \to \infty$ and $N_{Br}/N_{Pr}N_{Re} \to 0$, we have seen that the differential energy balance appears to simplify to Eq. (2-7). Conclude that for an ideal gas undergoing such a process

$$P^{(\gamma-1)/\gamma} T^{-1} = \text{a constant}$$

or

$$P \rho^{-\gamma} = \text{a constant}$$

## 6.7 BOUNDARY-LAYER THEORY

### 6.7.1 Thermal boundary-layer theory for plane flow past a flat plate

In Sec. 6.6.1 we explored the possibility of neglecting all the conduction terms of the energy balance in the limit as the Reynolds number goes to infinity for a fixed value of the Prandtl number (that is, for a particular material). We anticipated that elimination of all of the conduction terms would be inconsistent with a realistic description of the temperature distribution near its bounding surfaces.

This problem is really quite similar to the one that we encountered in our discussion of potential flow in Secs. 3.4.1 to 3.4.5. Our answer there was to develop boundary-layer theory in Sec. 3.5.1. In boundary-layer theory, we reason that a portion of the viscous terms might be neglected as $N_{Re} \to \infty$, with a resulting considerable simplification in the components of Cauchy's first law.

Our intention here is to extend the boundary-layer concept to the energy balance. Unfortunately, we are missing direct experimental evidence, such as the temperature distribution in the immediate neighborhood of a flat plate, which might be used to suggest our next step. We shall rely heavily upon our previous success with boundary-layer theory in Secs. 3.5.1 to 3.5.7 and argue here by analogy.

Let us begin by considering in some detail the same class of flows that we used to introduce boundary-layer theory in Sec. 3.5.1: plane flow past a flat plate. With reference to Fig. 6.7.1-1, the temperature of the fluid as it approaches the plate is known to be $T_\infty$ at $z_1 = 0$ for all values of $z_2$. For the moment we will say no more about the external flow and the thermal boundary condition at the plate.

We shall find it convenient to work in terms of the following dimensionless variables:

# [6.7] BOUNDARY-LAYER THEORY

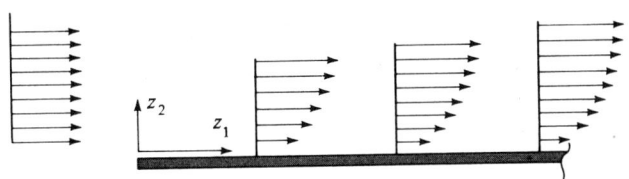

**Fig. 6.7.1-1** Flow past a flat plate.

For $i = 1, 2$:
$$v_i^* \equiv \frac{v_i}{v_0} \qquad z_i^* \equiv \frac{z_i}{L} \qquad T^* \equiv \frac{T - T_\infty}{T_0 - T_\infty} \tag{1-1}$$

Here $v_0$ is a magnitude of velocity characteristic of the plane nonviscous flow outside the boundary layer, $L$ is the length of the plate, and $T_0$ is characteristic of the temperature distribution on the plate. For this plane flow, it seems reasonable to assume that

$$T^* = T^*(z_1^*, z_2^*) \tag{1-2}$$

For simplicity, we limit ourselves to an incompressible newtonian fluid whose viscosity and thermal conductivity are constants independent of temperature.

The velocity distribution appropriate to this flow is precisely that described in Sec. 3.5.1, since both density and viscosity are independent of temperature. We can consequently concentrate our attention upon the temperature distribution.

If we were able to make very careful measurements of temperature in the fluid, we would intuitively expect to find the temperature to be $T_\infty$ nearly everywhere in the fluid, the only exception being a very thin region next to the plate in which the temperature undergoes a rapid change in order to satisfy whatever condition is required of temperature at the plate. As illustrated in Fig. 6.7.1-2 for an isothermal plate, we would anticipate that the thickness of this region of nonuniform temperature increases as we go downstream (for increasing values of $z_1$). We should also expect to find that at any given value of $z_1$ the thickness of this region decreases as the flow past the plate increases.

If we go back to the argument of Sec. 6.6.1 and neglect conduction with respect to convection in high-speed flows, for this situation we predict the temperature to be uniform everywhere at $T_\infty$.

This picture suggests that conduction is just as important as convection in the energy balance as the Reynolds number becomes very large, at least in the boundary

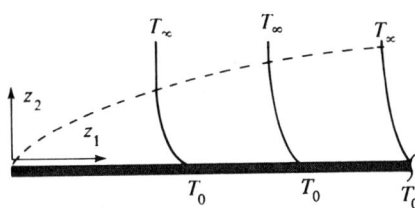

**Fig. 6.7.1-2** Temperature distribution in flow past an isothermal flat plate.

layer. The reasoning we presented in Sec. 6.6.1 is not applicable to this region, since it says that conduction is negligibly small compared with convection; apparently, this treatment must be reserved for the temperature distribution outside the boundary layer. In order to use an approximation similar to that of Sec. 6.6.1 for the boundary layer, we must revise the definitions of our dimensionless variables so as to prevent the disappearance of all the conduction terms. If, as we indicated above, the thickness of the thermal boundary layer decreases as the Reynolds number increases, we should magnify the thickness of the boundary layer by introducing as we did in Sec. 3.5.1

$$z_2^{**} \equiv (N_{Re})^{\frac{1}{2}} z_2^* \tag{1-3}$$

where

$$N_{Re} \equiv \frac{v_0 L \rho}{\mu} \tag{1-4}$$

As explained in Sec. 3.5.1, the equation of continuity requires that we also introduce a magnified $z_2$ component of velocity:

$$v_2^{**} \equiv (N_{Re})^{\frac{1}{2}} v_2^* \tag{1-5}$$

In view of Eq. (1-2), the dimensionless differential energy balance of Sec. 6.4.1 reduces for this plane flow to

$$\frac{1}{N_{St}} \frac{\partial T^*}{\partial t^*} + \frac{\partial T^*}{\partial z_1^*} v_1^* + \frac{\partial T^*}{\partial z_2^*} v_2^* = \frac{1}{N_{Pr} N_{Re}} \left( \frac{\partial^2 T^*}{\partial z_1^{*2}} + \frac{\partial^2 T^*}{\partial z_2^{*2}} \right)$$
$$+ \frac{4 N_{Br}}{N_{Pr} N_{Re}} \left( \frac{\partial v_1^*}{\partial z_1^*} \right)^2 + \frac{N_{Br}}{N_{Pr} N_{Re}} \left( \frac{\partial v_1^*}{\partial z_2^*} + \frac{\partial v_2^*}{\partial z_1^*} \right)^2 \tag{1-6}$$

or, in terms of $z_2^{**}$ and $v_2^{**}$,

$$\frac{1}{N_{St}} \frac{\partial T^*}{\partial t^*} + \frac{\partial T^*}{\partial z_1^*} v_1^* + \frac{\partial T^*}{\partial z_2^{**}} v_2^{**} = \frac{1}{N_{Pr}} \left( \frac{1}{N_{Re}} \frac{\partial^2 T^*}{\partial z_1^{*2}} + \frac{\partial^2 T^*}{\partial z_2^{**2}} \right)$$
$$+ \frac{4 N_{Br}}{N_{Pr} N_{Re}} \left( \frac{\partial v_1^*}{\partial z_1^*} \right)^2 + \frac{N_{Br}}{N_{Pr}} \left( \frac{\partial v_1^*}{\partial z_2^*} + \frac{1}{N_{Re}} \frac{\partial v_2^{**}}{\partial z_1^*} \right)^2$$
$$\tag{1-7}$$

Here

$$N_{Pr} \equiv \frac{\hat{c} \mu}{k} \qquad N_{Br} \equiv \frac{\mu v_0^2}{k(T_0 - T_\infty)} \tag{1-8}$$

Equation (1-7) suggests that as $N_{Re} \to \infty$ the dimensionless differential energy balance may be simplified to

$$\frac{1}{N_{St}} \frac{\partial T^*}{\partial t^*} + \frac{\partial T^*}{\partial z_1^*} v_1^* + \frac{\partial T^*}{\partial z_2^*} v_2^{**} = \frac{1}{N_{Pr}} \frac{\partial^2 T^*}{\partial z_2^{**2}} + \frac{N_{Br}}{N_{Pr}} \left( \frac{\partial v_1^*}{\partial z_2^{**}} \right)^2 \tag{1-9}$$

## [6.7] BOUNDARY-LAYER THEORY

Since the velocity distribution for this flow may be presumed to be already known following the discussion of Sec. 3.5.1, Eq. (1-9) represents the differential equation to be solved for the dimensionless temperature distribution.

Outside the boundary-layer region, conduction effects can be neglected in the equation of motion. Let

$$T^{(e)*} = T^{(e)*}(z_1^*, z_2^*) = T^{(e)*}\left(z_1^*, \frac{z_2^{**}}{\sqrt{N_{Re}}}\right) \qquad (1\text{-}10)$$

denote the dimensionless temperature distribution for the nonconducting, nonviscous external flow. Within the boundary-layer region, in the limit as the Reynolds number becomes unbounded,

$$\tilde{T}^* \equiv \lim_{\substack{N_{Re} \to \infty \\ z_1^*, z_2^{**} \text{ fixed}}} T^{(e)*}\left(z_1^*, \frac{z_2^{**}}{\sqrt{N_{Re}}}\right) = T^{(e)*}(z_1^*, 0) \qquad (1\text{-}11)$$

As $z_2^{**} \to \infty$, we must require that $T^*$ from the boundary-layer solution must approach asymptotically the corresponding temperature from the nonconducting, nonviscous flow:

$$\text{as } z_2^{**} \to \infty: T^* \to \tilde{T}^*$$

## EXERCISE

**6.7.1-1 The boundary-layer equations for natural convection** In Exercise 6.4.2-2, we derived the dimensionless forms of the equation of continuity, the equation of motion, and the differential energy balance appropriate to natural convection in a newtonian fluid.

Let us now assume that $\hat{c}_V$, $k$, and $\mu$ are all constants.

(a) Extend the discussion of Sec. 3.5.1 for the velocity boundary layer on a flat plate. Find that in the limit as $N_{Gr} \to \infty$ the equation of continuity and the equation of motion imply

$$\frac{\partial v_1^*}{\partial z_1^*} + \frac{\partial v_2^{**}}{\partial z_2^{**}} = 0$$

and

$$\frac{1}{N_{St}} \frac{\partial v_1^*}{\partial t^*} + v_1^* \frac{\partial v_1^*}{\partial z_1^*} + v_2^{**} \frac{\partial v_1^*}{\partial z_2^{**}} = -\frac{1}{N_{Ru}} \frac{d\mathcal{P}^*}{dz_1^*} + \frac{\partial^2 v_1^*}{\partial z_2^{**2}} - T^* f_1^*$$

where

$$-\frac{1}{N_{Ru}} \frac{d\mathcal{P}^*}{dz_1^*} = -\frac{1}{N_{Ru}} \frac{d\tilde{\mathcal{P}}^*}{dz_1^*} = \frac{1}{N_{St}} \frac{\partial \tilde{v}_1^*}{\partial t^*} + \tilde{v}_1^* \frac{\partial \tilde{v}_1^*}{\partial z_1^*} + \tilde{T}^* f_1^*$$

Here $\tilde{\mathcal{P}}^*$, $\tilde{v}_1^*$, and $\tilde{T}^*$ are the dimensionless modified pressure, velocity, and temperature distributions at the plate as determined by the nonviscous, nonconducting flow outside the boundary layer.

(b) Determine that the corresponding differential energy balance appropriate to the thermal boundary layer is

$$\frac{1}{N_{\text{St}}}\frac{\partial T^*}{\partial t^*} + \frac{\partial T^*}{\partial z_1^*}v_1^* + \frac{\partial T^*}{\partial z_2^{**}}v_2^{**} = \frac{1}{N_{\text{Pr}}}\frac{\partial^2 T^*}{\partial z_2^{**2}}$$

### 6.7.2 More on the thermal boundary layer in plane flow past a flat plate

Here we take up a particular case to which the theory in the previous section is applicable: steady-state flow past a flat plate at zero incidence of an incompressible newtonian fluid. The temperature of the plate is maintained constant at $T_0$; the temperature of the fluid has a uniform value $T_\infty$ on the plane $z_1 = 0$ in Fig. 6.7.1-1. For simplicity, we take the viscosity and thermal conductivity of the fluid to be independent of temperature.

In Sec 6.7.1 let us now define the characteristic magnitude of velocity of the fluid to be $\tilde{v}_1$, the $z_1$ component of velocity in the nonviscous, nonconducting fluid outside the boundary layer; the characteristic length is $L$, the length of the plate; the characteristic temperature is taken to be the temperature difference $T_0 - T_\infty$. In terms of these characteristic quantities we define the dimensionless velocity components, coordinates, and temperature as

$$\text{For } i = 1, 2: \quad v_i^* \equiv \frac{v_i}{v_0} \quad z_i^* \equiv \frac{z_i}{L} \quad T^* \equiv \frac{T - T_\infty}{T_0 - T_\infty} \tag{2-1}$$

The velocity distribution for this flow is unchanged from that we found in Sec. 3.5.2, since both viscosity and density are taken to be independent of temperature. Just to review, we found that

$$v_1^* \equiv \frac{v_1}{v_0} = f' \quad v_2^{**} \equiv \sqrt{N_{\text{Re}}}\,\frac{v_2}{v_0} = \frac{1}{2\sqrt{z_1^*}}(\eta f' - f) \tag{2-2}$$

where $f = f(\eta)$ and

$$\eta \equiv \frac{z_2^{**}}{\sqrt{z_1^*}} = \frac{\sqrt{N_{\text{Re}}}\,z_2^*}{\sqrt{z_1^*}} \tag{2-3}$$

The prime is used to denote differentiation with respect to $\eta$. The function $f$ is a solution of

$$ff'' + 2f''' = 0 \tag{2-4}$$

that satisfies the boundary conditions

$$\text{At } \eta = 0: \quad f = f' = 0 \tag{2-5}$$

and

$$\text{As } \eta \to \infty: \quad f' \to 1 \tag{2-6}$$

By the Reynolds number we mean here

## [6.7] BOUNDARY-LAYER THEORY

$$N_{\text{Re}} \equiv \frac{\tilde{v}_1 L \rho}{\mu} \tag{2-7}$$

For the nonviscous, nonconducting flow at the outer edge of the boundary layer, we know that

$$\text{At } z_1^* = 0: \quad T^{(e)*} = 0 \tag{2-8}$$

Since

$$v_1^{(e)*} = 1 \quad v_2^{(e)*} = v_3^{(e)*} = 0 \tag{2-9}$$

the differential energy balance applicable to this flow simplifies to

$$\frac{\partial T^{(e)*}}{\partial z_1^*} v_1^{(e)} = \frac{\partial T^{(e)*}}{\partial z_1^*} = 0 \tag{2-10}$$

We conclude from Eqs. (2-8) and (2-10) that for all $z_1^*$

$$\tilde{T}^* \equiv T^{(e)*}(z_1^*, 0) = 0 \tag{2-11}$$

From Sec. 6.7.1, the differential energy balance applicable to the boundary layer is

$$\frac{\partial T^*}{\partial z_1^*} v_1^* + \frac{\partial T^*}{\partial z_2^{**}} v_2^{**} = \frac{1}{N_{\text{Pr}}} \frac{\partial^2 T^*}{\partial z_2^{**2}} + \frac{N_{\text{Br}}}{N_{\text{Pr}}} \left( \frac{\partial v_1^*}{\partial z_2^{**}} \right)^2 \tag{2-12}$$

Here

$$N_{\text{Pr}} \equiv \frac{\hat{c}\mu}{k} \quad N_{\text{Br}} \equiv \frac{\mu(v_0)^2}{k(T_0 - T_\infty)} \tag{2-13}$$

Equation (2-12) can be regarded as the differential equation to be solved for the dimensionless temperature distribution in the boundary layer, since the velocity distribution in the boundary layer is already known from Sec. 3.5.2. The boundary conditions to be satisfied by the desired solution to Eq. (2-12) are that

$$\text{At } z_2^{**} = 0: \quad T^* = 1 \tag{2-14}$$

and

$$\text{As } z_2^{**} \to \infty: \quad T^* \to \tilde{T}^* = 0 \tag{2-15}$$

By analogy with our analysis for the boundary-layer velocity distribution in Sec. 3.5.2, we anticipate that we might be able to find a solution to Eq. (2-12) by combining the two independent variables in such a way as to transform this equation into an ordinary differential equation. Specifically, if we anticipate a solution of the form

$$T^* = T^*(\eta) \tag{2-16}$$

with the help of Eqs. (2-2), Eq. (2-12) becomes

$$T^{*\prime\prime} + \tfrac{1}{2}N_{\text{Pr}}fT^{*\prime} = -N_{\text{Br}}(f'')^2 \tag{2-17}$$

The corresponding boundary conditions are

At $\eta = 0$: $\quad T^* = 1$ (2-18)

and

As $\eta \to \infty$: $\quad T^* \to 0$ (2-19)

Again primes denote differentiation with respect to $\eta$.

It is convenient to seek the solution to Eq. (2-17) as a linear combination of the solutions to two somewhat simpler problems [1, p. 280],

$$T^* = A\Theta_1 + \Theta_2 \tag{2-20}$$

We can think of $\Theta_1$ as being the dimensionless temperature distribution corresponding to boundary-layer flow past an isothermal plate when viscous dissipation, the non-homogeneous term on the right of Eq. (2-17), is neglected:

$$\Theta_1'' + \tfrac{1}{2}N_{\text{Pr}}f\Theta_1' = 0 \tag{2-21}$$

At $\eta = 0$: $\quad \Theta_1 = 1$ (2-22)

As $\eta \to \infty$: $\quad \Theta_1 \to 0$ (2-23)

The dimensionless temperature distribution $\Theta_2$ corresponds to boundary-layer flow past an adiabatic (insulated) flat plate:

$$\Theta_2'' + \tfrac{1}{2}N_{\text{Pr}}f\Theta_2' = -N_{\text{Br}}(f'')^2 \tag{2-24}$$

At $\eta = 0$: $\quad \Theta_2' = 0$ (2-25)

As $\eta \to \infty$: $\quad \Theta_2 \to 0$ (2-26)

The coefficient $A$ in Eq. (2-20) is required in order that boundary condition (2-18) be satisfied.

Let us look at the solutions for $\Theta_1$ and $\Theta_2$ individually.

### ISOTHERMAL WALL BUT NO VISCOUS DISSIPATION

Since we know from Eq. (2-4) that

$$f = -\frac{2f'''}{f''} \tag{2-27}$$

Eq. (2-21) may be rearranged in a separable form,

$$\frac{d}{d\eta}\ln\Theta_1' = N_{\text{Pr}}\frac{d}{d\eta}\ln f'' \tag{2-28}$$

and integrated once to find

$$\Theta_1' = C_1(f'')^{N_{\text{Pr}}} \tag{2-29}$$

The coefficient $C_1$ is an as yet undetermined constant of integration. After another integration and after boundary conditions (2-22) and (2-23) have been satisfied, we find

$$\Theta_1 = \frac{\int_\eta^\infty (f'')^{N_{\rm Pr}}\, d\eta}{\int_0^\infty (f'')^{N_{\rm Pr}}\, d\eta} \qquad (2\text{-}30)$$

For the special case of $N_{\rm Pr} = 1$, Eq. (2-30) reduces to

At $N_{\rm Pr} = 1$: $\qquad \Theta_1 = 1 - f' \qquad (2\text{-}31)$

**ADIABATIC WALL**

Equation (2-27) may again be employed, this time to allow rearrangement of Eq. (2-24) into

$$\frac{d}{d\eta}\left[\frac{1}{(f'')^{N_{\rm Pr}}}\Theta_2'\right] = -N_{\rm Br}(f'')^{2-N_{\rm Pr}} \qquad (2\text{-}32)$$

After one integration using Eq. (2-25), we have

$$\frac{1}{(f'')^{N_{\rm Pr}}}\Theta_2' = -N_{\rm Br}\int_0^\eta (f'')^{2-N_{\rm Pr}}\, d\tau \qquad (2\text{-}33)$$

Another integration satisfying boundary condition (2-26) results in

$$\Theta_2 = N_{\rm Br}\int_\eta^\infty (f'')^{N_{\rm Pr}}\left[\int_0^\xi (f'')^{2-N_{\rm Pr}}\, d\tau\right] d\xi \qquad (2\text{-}34)$$

For the special case $N_{\rm Pr} = 1$, Eq. (2-34) may be expressed as

For $N_{\rm Pr} = 1$: $\qquad \Theta_2 = \tfrac{1}{2}N_{\rm Br}[1 - (f')^2] \qquad (2\text{-}35)$

**SUMMARY**

In order that Eq. (2-20) satisfy boundary condition (2-18), we must require

$$A = 1 - \Theta_2(\eta = 0) \qquad (2\text{-}36)$$

Our final result for the dimensionless temperature distribution in the boundary layer is, consequently,

$$T^* = [1 - \Theta_2(\eta = 0)]\Theta_1 + \Theta_2 \qquad (2\text{-}37)$$

where $\Theta_1$ and $\Theta_2$ are given by Eqs. (2-30) and (2-34), respectively. Equation (2-37) is displayed in Fig. 6.7.2-1 for the important case $N_{\rm Pr} = 0.7$ (air).

Note that Eq. (2-37) does not depend upon the characteristic length $L$. This is in agreement with our intuitive feelings that the solution should be valid for a semi-infinite flat plate.

The $z_2$ component of the energy flux vector evaluated at the flat plate is readily seen to be

$$q_2\bigg|_{z_2=0} = -k\frac{\partial T}{\partial z_2}\bigg|_{z_2=0} = \frac{-k(T_0 - T_\infty)\sqrt{N_{\rm Re}}}{L\sqrt{z_1^*}}[1 - \Theta_2(\eta = 0)]\Theta_1'\big|_{\eta=0} \qquad (2\text{-}38)$$

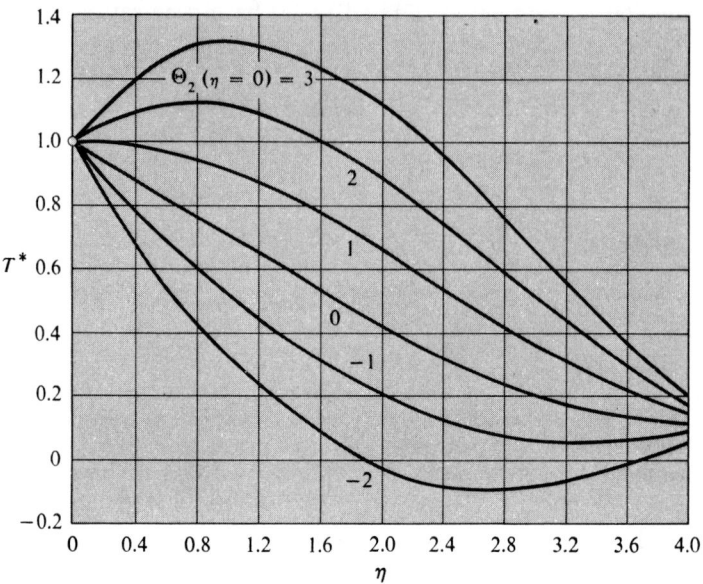

**Fig. 6.7.2-1** Temperature distribution in a laminar boundary layer, on a flat plate, at zero incidence to a parallel stream as predicted by Eq. (2-37) for $N_{Pr} = 0.7$ (air). The temperature of the wall is constant at $T_0$; the temperature of the stream is $T_\infty$. The curve $\Theta_2(\eta = 0) = 1$ corresponds to an adiabatic wall. If we assume $(T_0 - T_\infty) > 0$, the wall is cooled when $\Theta_2(\eta = 0) < 1$; the wall is heated when $\Theta_2(\eta = 0) > 1$. (*From Schlichting* [1, p. 284, fig. 12.13].)

Schlichting [1, p. 281] has shown that for a wide range of $N_{Pr}$ the quantity $-\Theta_1|_{\eta=0}$ is positive. We conclude that, if $(T_0 - T_\infty) > 0$, the plate is cooled when

$$\Theta_2(\eta = 0) < 1 \tag{2-39}$$

Schlichting [1, p. 282] has further shown that

$$\Theta_2(\eta = 0) \doteq \frac{N_{Br}}{2\sqrt{N_{Pr}}} \tag{2-40}$$

is an excellent approximation. Consequently, from Eq. (2-39) when

$$\frac{N_{Br}}{2\sqrt{N_{Pr}}} < 1 \tag{2-41}$$

the wall will be cooled. In order for the wall to be cooled, it is not sufficient that $(T_0 - T_\infty) > 0$. In addition, viscous dissipation must be small. For a stream of air flowing at $v_0 = 200$ m/s, $N_{Pr} = 0.7$, $\mu = 21 \times 10^{-6}$ kg/m s, $k = 28 \times 10^{-3}$ kg/m s$^3$°K, Eq. (2-41) says that the wall will be cooled for

[6.7] BOUNDARY-LAYER THEORY

$$T_0 - T_\infty > 18°C \qquad (2\text{-}42)$$

Schlichting (1, p. 284) gives specific numerical results for the rate of heat transfer from the plate which follow directly from the temperature distributions developed here.

## REFERENCE

1. Schlichting, Hermann: "Boundary-Layer Theory," 6th ed., McGraw-Hill, New York, 1968.

## EXERCISE

**6.7.2-1** *Flow past an isothermal flat plate* For flow past an isothermal flat plate, determine that the dimensionless velocity and temperature distributions have the same form when $N_{Pr} = 1$. We say that under these circumstances, the velocity and temperature distributions are "similar."

### 6.7.3 Thermal boundary-layer theory for plane flow past a curved wall
In Sec. 6.7.1 we discussed the temperature distribution in plane flow past a flat plate. In the limit, as the Reynolds number becomes unbounded, we found that the differential energy balance may be considerably simplified for the fluid in a thin boundary layer next to the plate. This simplification involved neglecting some (though not all) of the conduction terms with respect to the convection terms.

In this section, we wish to consider the thermal boundary layer formed by an incompressible newtonian fluid in plane flow past a curved wall. For simplicity, we take both viscosity and thermal conductivity to be constants, independent of temperature. We anticipate that the temperature distribution in a sufficiently thin boundary layer will be much the same whether the wall is curved or flat. Our object here is to show in what sense this intuitive feeling is correct.

Our approach here follows closely that in Sec. 3.5.3, where we determined the form of the Navier-Stokes equation and the equation of continuity appropriate to plane flow past a curved wall.

A portion of a typical curved wall is shown in Fig. 3.5.3-1. With respect to the rectangular cartesian coordinate system indicated, the equation of this surface is

$$z_2 = f(z_1) \qquad (3\text{-}1)$$

In order to better compare flow past a curved wall with flow past a flat plate, let us view this problem in terms of an orthogonal curvilinear coordinate system such that:

$x \equiv x^1$ is defined to be the arc length measured along the wall in a plane of constant $z$.
$y \equiv x^2$ is defined to be the arc length measured along straight lines that are normal to the wall.
$z \equiv x^3 \equiv z_3$ is the coordinate normal to the plane of flow.

By plane flow, we mean here that

$$v_x \equiv v_{\langle 1 \rangle} = v_x(x,y,t)$$
$$v_y \equiv v_{\langle 2 \rangle} = v_y(x,y,t)$$
$$v_z \equiv v_{\langle 3 \rangle} = 0 \quad (3\text{-}2)$$
$$T = T(x,y,t)$$

With these restrictions, the differential energy balance may be written as (see Exercise 6.7.3-1)

$$\rho \hat{c}\left(\frac{\partial T}{\partial t} + \frac{\partial T}{\partial x}\frac{v_x}{1+\varkappa y} + \frac{\partial T}{\partial y}v_y\right) = k\left\{\frac{1}{1+\varkappa y}\frac{\partial}{\partial x}\left(\frac{1}{1+\varkappa y}\frac{\partial T}{\partial x}\right)\right.$$
$$+ \frac{1}{1+\varkappa y}\frac{\partial}{\partial y}\left[(1+\varkappa y)\frac{\partial T}{\partial y}\right]\right\} + 2\mu\left[\frac{1}{(1+\varkappa y)^2}\left(\frac{\partial v_x}{\partial x} + \varkappa v_y\right)^2\right.$$
$$\left. + \left(\frac{\partial v_y}{\partial y}\right)^2 + \frac{1}{2}\left(\frac{\partial v_x}{\partial y} + \frac{1}{1+\varkappa y}\frac{\partial v_y}{\partial x} - \frac{\varkappa v_x}{1+\varkappa y}\right)^2\right] \quad (3\text{-}3)$$

where we define

$$\varkappa \equiv \frac{-f''}{[1+(f')^2]^{\frac{3}{2}}} = \varkappa(x) \quad (3\text{-}4)$$

The primes here indicate differentiation with respect to $z_1$. We may think of $-\varkappa$ as twice the mean curvature of the surface [1, p. 201], the normal curvature [1, p. 210] of the surface in the direction $x$, or the only nonzero principal curvature [1, p. 211] of the surface.

In addition to the dimensionless velocity, dimensionless temperature, and dimensionless time introduced in Sec. 6.4.1, let us define

$$x^* \equiv \frac{x}{L} \quad y^* \equiv \frac{y}{L} \quad \varkappa^* \equiv \varkappa L \quad (3\text{-}5)$$

If we extend the arguments of Sec. 6.7.1 to this geometry, we are motivated to express our results in terms of

$$y^{**} \equiv \sqrt{N_{\text{Re}}}\, y^* \qquad v_y^{**} \equiv \sqrt{N_{\text{Re}}}\, v_y^* \quad (3\text{-}6)$$

As the Reynolds number becomes unbounded, Eq. (3-3) reduces to

$$\frac{1}{N_{\text{St}}}\frac{\partial T^*}{\partial t^*} + \frac{\partial T^*}{\partial x^*}\frac{v_x^*}{1+\varkappa^{**}y^{**}} + \frac{\partial T^*}{\partial y^{**}}v_y^{**}$$
$$= \frac{1}{N_{\text{Pr}}}\frac{1}{1+\varkappa^{**}y^{**}}\frac{\partial}{\partial y^{**}}\left[(1+\varkappa^{**}y^{**})\frac{\partial T^*}{\partial y^{**}}\right]$$
$$+ \frac{N_{\text{Br}}}{N_{\text{Pr}}}\left(\frac{\partial v_x^*}{\partial y^{**}} - \frac{\varkappa^{**}v_x^*}{1+\varkappa^{***}y^{**}}\right)^2 \quad (3\text{-}7)$$

Here we define

$$\varkappa^{**} \equiv N_{\text{Re}}^{-\frac{1}{2}}\varkappa^* \tag{3-8}$$

For a fixed-wall configuration, $\varkappa^{**} \to 0$ in the limit as the Reynolds number becomes unbounded. Equation (3-7) further simplifies under these conditions to

$$\frac{1}{N_{\text{St}}}\frac{\partial T^*}{\partial t^*} + \frac{\partial T^*}{\partial x^*}v_x^* + \frac{\partial T^*}{\partial y^{**}}v_y^{**} = \frac{1}{N_{\text{Pr}}}\frac{\partial^2 T^*}{\partial y^{**2}} + \frac{N_{\text{Br}}}{N_{\text{Pr}}}\left(\frac{\partial v_x^*}{\partial y^{**}}\right)^2 \tag{3-9}$$

As we suggested in Sec. 6.7.1,

As $y^{**} \to \infty$: $\quad T^* \to \tilde{T}^* \tag{3-10}$

where $\tilde{T}^*$ is the dimensionless temperature distribution at the curved wall for the corresponding nonviscous, nonconducting flow:

$$\frac{1}{N_{\text{St}}}\frac{\partial \tilde{T}^*}{\partial t^*} + \frac{\partial \tilde{T}^*}{\partial x^*}\tilde{v}_x^* = 0 \tag{3-11}$$

By $\tilde{v}_x^*$ we mean the dimensionless $x$ component of velocity at the curved wall for the corresponding nonviscous flow.

Since we assume that $v_x^*$ and $v_y^{**}$ are known a priori, Eq. (3-9) can be looked upon as the differential equation to be solved for the dimensionless temperature distribution in the thermal boundary layer. As our intuition suggested, the differential energy balance appropriate to the thermal boundary layer developed in plane flow past a curved wall has the same form as that appropriate to plane flow past a flat plate found in Sec. 6.7.1.

### REFERENCE

1. McConnell, A. J.: "Applications of Tensor Analysis," Dover, New York, 1957.

### EXERCISES

**6.7.3-1** *Derivation of Eq. (3-3)* Noting the results of Exercise 3.5.3-1, derive Eq. (3-3) starting from Eq. (E) of Table 5.6.1-1.

**6.7.3-2** *Derivation of Eq. (3-7)* Introduce in Eq. (3-3) the dimensionless velocity, dimensionless temperature, and dimensionless time defined in Sec. 6.4.1 as well as those dimensionless variables defined in Eqs. (3-5) and (3-6). Construct the reasoning that leads to Eq. (3-7).

**6.7.4 The temperature distribution in flow past a wedge** As an illustration of the development given in the previous section, let us consider plane flow of an incompressible newtonian fluid past the wedge shown in Fig. 6.7.4-1. The temperature of the wedge's wall is maintained constant at $T_0$; the temperature of the gas stream at $x = 0$ is known to be $T_\infty$. Our object here is to determine the temperature distribution within the gas in the immediate neighborhood of the wedge, as well as

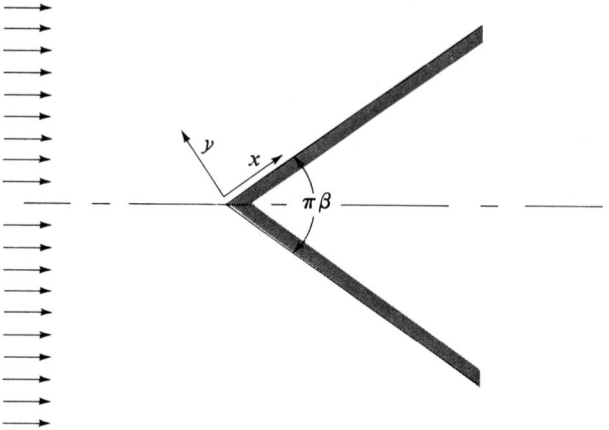

**Fig. 6.7.4-1** Flow past a wedge.

the local rate of the energy transfer from the wedge to the gas. For simplicity, we shall assume that both the viscosity and thermal conductivity of the gas are independent of temperature and we shall neglect viscous dissipation within the gas.

The nonviscous, potential flow appropriate to the fluid outside the boundary layer predicts that at the wall of the wedge and in the immediate neighborhood of the apex the $x$ component of velocity is [1, p. 141]

$$\tilde{v}_x = ux^m \tag{4-1}$$

where the included angle of the wedge is

$$\pi\beta = \frac{2\pi m}{1+m} \tag{4-2}$$

In Sec. 6.7.3 let $L$ be a characteristic length associated with the wedge, perhaps its length. Take the characteristic velocity to be $uL^m$ and the characteristic temperature to be $(T_0 - T_\infty)$. With this understanding, the dimensionless components of velocity, the dimensionless temperature, and the dimensionless coordinates become

$$v_x^* \equiv \frac{v_x}{uL^m} \qquad v_y^{**} \equiv \sqrt{N_{\text{Re}}}\,\frac{v_y}{uL^m}$$

$$T^* \equiv \frac{T - T_\infty}{T_0 - T_\infty} \qquad x^* \equiv \frac{x}{L} \qquad y^{**} \equiv \sqrt{N_{\text{Re}}}\,\frac{y}{L} \tag{4-3}$$

where

$$N_{\text{Re}} \equiv \frac{uL^{1+m}\rho}{\mu} \tag{4-4}$$

We have been purposely vague in introducing the characteristic length $L$, since we will find it drops out of the final results.

We found in Sec. 6.7.3 that the dimensionless form of the differential energy balance appropriate to the boundary layer can be written as

$$\frac{\partial T^*}{\partial x^*} v_x^* + \frac{\partial T^*}{\partial y^{**}} v_y^{**} = \frac{1}{N_{\text{Pr}}} \frac{\partial^2 T^*}{\partial y^{**2}} + \frac{N_{\text{Br}}}{N_{\text{Pr}}} \left(\frac{\partial v_x^*}{\partial y^{**}}\right)^2 \tag{4-5}$$

For this problem

$$N_{\text{Pr}} \equiv \frac{\hat{c}\mu}{k} \qquad N_{\text{Br}} \equiv \frac{\mu u^2 L^{2m}}{k(T_0 - T_\infty)} \tag{4-6}$$

In introducing this problem we said that we would neglect viscous effects. More precisely, we will restrict ourselves to the limit as $N_{\text{Br}}/N_{\text{Pr}} \to 0$, in which it appears reasonable to approximate Eq. (4-5) as

$$\frac{\partial T^*}{\partial x^*} v_x^* + \frac{\partial T^*}{\partial y^{**}} v_y^{**} = \frac{1}{N_{\text{Pr}}} \frac{\partial^2 T^*}{\partial y^{**2}} \tag{4-7}$$

The velocity distribution for the boundary layer was discussed in Sec. 3.5.4 as part of Case 1a. In view of our definitions for dimensionless variables in Eq. (4-3), Eq. (4-1) may be written as

$$\hat{v}_x^* = x^{*m} \tag{4-8}$$

This means that in Sec. 3.5.4 we are forced to define the as yet unspecified constant

$$K \equiv \left(\frac{1+m}{2}\right)^{m/1+m} \tag{4-9}$$

It follows immediately that

$$v_x^* = \frac{\partial \psi}{\partial y^{**}} = x^{*m} \frac{df}{d\eta} \tag{4-10}$$

and

$$v_y^{**} = -\frac{\partial \psi}{\partial x^*}$$

$$= -\left(\frac{1+m}{2}\right)^{\frac{1}{2}} x^{*(m-1)/2} f - \left(\frac{2}{1+m}\right)^{\frac{1}{2}} \frac{m-1}{2} x^{*(m-1)/2} \eta \frac{df}{d\eta} \tag{4-11}$$

Here $f$, a function of

$$\eta \equiv \frac{y^{**}}{g} = \frac{y^{**}}{[2/(1+m)]^{\frac{1}{2}} x^{*(1-m)/2}} \tag{4-12}$$

is a solution to

$$\frac{d^3f}{d\eta^3} + f\frac{d^2f}{d\eta^2} + \frac{2m}{1+m}\left[1 - \left(\frac{df}{d\eta}\right)^2\right] = 0 \qquad (4\text{-}13)$$

consistent with the boundary conditions

$$\text{At } \eta = 0: \quad f = \frac{df}{d\eta} = 0 \qquad (4\text{-}14)$$

and

$$\text{As } \eta \to \infty: \quad \frac{df}{d\eta} \to 1 \qquad (4\text{-}15)$$

We shall assume that the solution to this problem is already available to us.
If we anticipate a solution of the form

$$T^* = \Theta(\eta) \qquad (4\text{-}16)$$

Eq. (4-7) becomes

$$\frac{d^2\Theta}{d\eta^2} + N_{\text{Pr}} f \frac{d\Theta}{d\eta} = 0 \qquad (4\text{-}17)$$

The appropriate boundary conditions are

$$\text{At } \eta = 0: \quad \Theta = 1 \qquad (4\text{-}18)$$

and

$$\text{As } \eta \to \infty: \quad \Theta \to 0 \qquad (4\text{-}19)$$

Equation (4-17) is easily integrated consistent with Eqs. (4-18) and (4-19) to find

$$\Theta = \frac{\int_\eta^\infty \exp\left(-N_{\text{Pr}} \int_0^\xi f\, d\tau\right) d\xi}{\int_0^\infty \exp\left(-N_{\text{Pr}} \int_0^\xi f\, d\tau\right) d\xi} \qquad (4\text{-}20)$$

The $y$ component of the energy flux vector evaluated at the wall of the wedge is computed to be

$$q_y|_{y=0} = -k \frac{\partial T}{\partial y}\bigg|_{y=0}$$

$$= -\frac{k(T_0 - T_\infty)}{L} N_{\text{Re}}^{\frac{1}{2}} \left(\frac{1+m}{2}\right)^{\frac{1}{2}} x^{*(m-1)/2} \Theta'|_{\eta=0}$$

$$= \frac{k(T_0 - T_\infty)}{L} N_{\text{Re}}^{\frac{1}{2}} \left(\frac{1+m}{2}\right)^{\frac{1}{2}} x^{*(m-1)/2} \left[\int_0^\infty \exp\left(-N_{\text{Pr}} \int_0^\xi f\, d\tau\right) d\xi\right]^{-1}$$

$$(4\text{-}21)$$

or

$$\frac{N_{\text{Nu}} x^*}{(x^{*m+1} N_{\text{Re}})^{\frac{1}{2}}} = \left(\frac{1+m}{2}\right)^{\frac{1}{2}} \left[\int_0^\infty \exp\left(-N_{\text{Pr}} \int_0^\xi f\, d\tau\right) d\xi\right]^{-1} \qquad (4\text{-}22)$$

## [6.7] BOUNDARY-LAYER THEORY

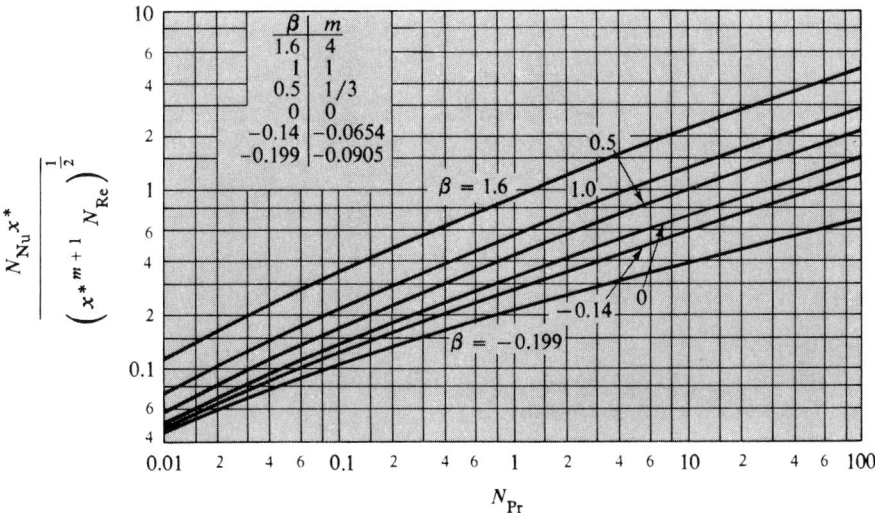

**Fig. 6.7.4-2** Nusselt number as a function of the Prandtl number for flow past an isothermal-walled wedge. (From Schlichting [1, p. 288, fig. 12.14]; Schlichting's variables $\bar{f}$ and $\bar{\eta}$ in his Eq. (12.87) are $\bar{f} = (2/(1 + m))^{\frac{1}{2}} f$ and $\bar{\eta} = (2/(1 + m))^{\frac{1}{2}} \eta$.)

where we define the Nusselt number as

$$N_{\text{Nu}} \equiv \frac{g_y|_{y=0} L}{k(T_0 - T_\infty)} \qquad (4\text{-}23)$$

Equation (4-22) is shown in Fig. 6.7.4-2.

Notice that the characteristic length $L$ has dropped out of Eqs. (4-20) and (4-22) ($\eta$ is independent of $L$). The results are not applicable to a semi-infinite wedge, however, since Eq. (4-1) is applicable only in the neighborhood of the apex.

### REFERENCE

1. Schlichting, Hermann: "Boundary-Layer Theory," 6th ed., McGraw-Hill, New York, 1968.

### 6.7.5 Thermal boundary-layer theory for flow past a body of revolution

In Secs. 6.7.1 and 6.7.3 we discussed the temperature distribution in plane flow past a flat plate and plane flow past a curved wall. We found that the thermal boundary layer could be described by the same set of equations in both cases.

In what follows, we investigate the thermal boundary layer formed by an incompressible newtonian fluid flowing past a body of revolution. We again take the viscosity and thermal conductivity to be constants, independent of temperature. In some sense we expect the temperature distribution in a sufficiently thin boundary layer on a body of revolution to be much the same as that on a flat plate.

Our approach here follows closely that in Sec. 3.5.7, where we determined the form of the Navier-Stokes equation and the equation of continuity appropriate to flow past a body of revolution.

A portion of a typical body of revolution is shown in Fig. 3.5.7-1. With respect to the cylindrical coordinate system indicated, the equation of the axially symmetric surface is

$$r = f(z) \tag{5-1}$$

In order to better compare flow past a body of revolution with flow past a flat plate, let us view this problem in terms of an orthogonal curvilinear coordinate system such that:

$x \equiv x^3$ is defined to be arc length measured along the wall (in the direction of flow) in a plane of constant $\theta$.
$y \equiv x^1$ is defined to be arc length measured along straight lines that are normal to the wall.
$\theta \equiv x^2$ is the aximuthal cylindrical coordinate (measured around the axis of the body).

The shape of the wall suggests

$$\begin{aligned} v_x &\equiv v_{\langle 3 \rangle} = v_x(x,y,t) \\ v_y &\equiv v_{\langle 1 \rangle} = v_y(x,y,t) \\ v_\theta &\equiv v_{\langle 2 \rangle} = 0 \\ T &= T(x,y,t) \end{aligned} \tag{5-2}$$

With these restrictions, the differential energy balance may be written as (see Exercise 6.7.5-1)

$$\rho \hat{c} \left( \frac{\partial T}{\partial t} + \frac{\partial T}{\partial x} \frac{v_x}{1 + \varkappa y} + \frac{\partial T}{\partial y} v_y \right)$$

$$= k \left\{ \frac{1}{1 + \varkappa y} \frac{\partial}{\partial x} \left( \frac{1}{1 + \varkappa y} \frac{\partial T}{\partial x} \right) + \frac{1}{1 + \varkappa y} \frac{\partial}{\partial y} \left[ (1 + \varkappa y) \frac{\partial T}{\partial y} \right] \right.$$

$$+ \frac{f'}{g(1 + \varkappa y)} \frac{\partial T}{\partial x} + \frac{1}{g} \frac{\partial T}{\partial y} \right\} + 2\mu \left\{ \frac{1}{(1 + \varkappa y)^2} \left( \frac{\partial v_x}{\partial x} + \varkappa v_y \right)^2 + \left( \frac{\partial v_y}{\partial y} \right)^2 \right.$$

$$+ \left( \frac{v_y}{g} \right)^2 + \frac{1}{2} \left[ (1 + \varkappa y) \frac{\partial}{\partial y} \left( \frac{v_x}{1 + \varkappa y} \right) + \frac{1}{1 + \varkappa y} \frac{\partial v_y}{\partial x} \right]^2 \right\} \tag{5-3}$$

where we define

$$\varkappa \equiv \frac{-f''}{[1 + (f')^2]^{\frac{3}{2}}} = \varkappa(x) \tag{5-4}$$

and

$$g \equiv f[1 + (f')^2]^{\frac{1}{2}} + y \tag{5-5}$$

## [6.7] BOUNDARY-LAYER THEORY

The primes here indicate differentiation with respect to the cylindrical coordinate $z$ measured along the axis of revolution. We may think of $-\varkappa$ as the normal curvature [1, p. 210] of the surface in the direction $x$; it is also one of the principal curvatures [1, p. 211] of the surface.

In addition to the dimensionless velocity, dimensionless temperature, and dimensionless time introduced in Sec. 6.4.1, let us define

$$x^* \equiv \frac{x}{L} \quad y^* \equiv \frac{y}{L} \quad f^* \equiv \frac{f}{L} \quad \varkappa^* \equiv \varkappa L \tag{5-6}$$

If we extend the arguments of Sec. 6.7.1 to this geometry, we are motivated to express our results in terms of

$$y^{**} \equiv \sqrt{N_{\text{Re}}}\, y^* \qquad v_y^{**} \equiv \sqrt{N_{\text{Re}}}\, v_y^* \tag{5-7}$$

As the Reynolds number becomes unbounded, Eq. (5-3) appears to reduce to

$$\frac{1}{N_{\text{St}}} \frac{\partial T^*}{\partial t^*} + \frac{\partial T^*}{\partial x^*} \frac{v_x^*}{1 + \varkappa^{**} y^{**}} + \frac{\partial T^*}{\partial y^{**}} v_y^{**}$$

$$= \frac{1}{N_{\text{Pr}}} \left\{ \frac{1}{1 + \varkappa^{**} y^{**}} \frac{\partial}{\partial y^{**}} \left[ (1 + \varkappa^{**} y^{**}) \frac{\partial T^*}{\partial y^{**}} \right] \right\}$$

$$+ \frac{N_{\text{Br}}}{N_{\text{Pr}}} \left[ (1 + \varkappa^{**} y^{**}) \frac{\partial}{\partial y^{**}} \left( \frac{v_x^*}{1 + \varkappa^{**} y^{**}} \right) \right]^2 \tag{5-8}$$

where

$$\varkappa^{**} \equiv N_{\text{Re}}^{-\frac{1}{2}} \varkappa^* \tag{5-9}$$

For a fixed-wall configuration, $\varkappa^{**} \to 0$ in the limit as the Reynolds number becomes unbounded and Eq. (5-8) further simplifies to

$$\frac{1}{N_{\text{St}}} \frac{\partial T^*}{\partial t^*} + \frac{\partial T^*}{\partial x^*} v_x^* + \frac{\partial T^*}{\partial y^{**}} v_y^{**} = \frac{1}{N_{\text{Pr}}} \frac{\partial^2 T^*}{\partial y^{**2}} + \frac{N_{\text{Br}}}{N_{\text{Pr}}} \left( \frac{\partial v_x^*}{\partial y^{**}} \right)^2 \tag{5-10}$$

Equation (5-10) indicates that the differential energy balance for the thermal boundary layer has the same form for plane flow past flat plates, plane flow past curved walls, and axisymmetric flow past bodies of revolution. But if the differential energy balance must be solved simultaneously with the equation of continuity and the form of the Navier-Stokes equation appropriate to the boundary layer, we see from Sec. 3.5.7 that the equation of continuity appropriate to the boundary layer of a body of revolution has a somewhat different form from that for a flat plate. In other words, the total boundary-value problem describing simultaneous momentum and energy transfer within a boundary layer on a body of revolution differs from the similar boundary-value problem appropriate to a flat plate.

In Sec. 3.5.7 we discussed the transformation suggested by Mangler [2, p. 235; see also Exercises 3.5.7-3 and 3.5.7-4] by means of which the boundary-layer equations appropriate to a body of revolution can be transformed into those for a flat plate.

Mangler suggests the introduction of the following variables:

$$\bar{x}^* \equiv \int_0^{x^*} f^{*2}\, dx^* \tag{5-11}$$

$$\bar{y}^{**} \equiv f^* y^{**} \tag{5-12}$$

$$\bar{t}^* \equiv f^{*2} t^* \tag{5-13}$$

$$\bar{v}_y^{**} \equiv \frac{1}{f^*} v_y^{**} + \frac{f^{*\prime}}{f^{*2}[1 + (f^{*\prime})^2]^{\frac{1}{2}}} y^{**} v_x^* \tag{5-14}$$

With this change of variables, Eq. (5-10) becomes

$$\frac{1}{N_{\mathrm{St}}} \frac{\partial T^*}{\partial \bar{t}^*} + \frac{\partial T^*}{\partial \bar{x}^*} v_x^* + \frac{\partial T^*}{\partial \bar{y}^{**}} \bar{v}_y^{**} = \frac{1}{N_{\mathrm{Pr}}} \frac{\partial^2 T^*}{\partial \bar{y}^{**2}} + \frac{N_{\mathrm{Br}}}{N_{\mathrm{Pr}}} \left( \frac{\partial v_x^*}{\partial \bar{y}^{**}} \right)^2 \tag{5-15}$$

It is in this sense that our original intuitive feelings are confirmed. The mathematical problems that describe boundary-layer flow past a body of revolution and plane flow past a flat plate have the same form.

## REFERENCES

1. McConnell, A. J.: "Applications of Tensor Analysis," Dover, New York, 1957.
2. Schlichting, Hermann: "Boundary-Layer Theory," 6th ed., McGraw-Hill, New York, 1968.

## EXERCISES

**6.7.5-1** *Derivation of Eq. (5-3)* Noting the results of Exercise 3.5.7-1, derive Eq. (5-3) starting from Eq. (E) of Table 5.6.1-1.

**6.7.5-2** *Derivation of Eq. (5-8)* Introduce in Eq. (5-3) the dimensionless velocity, dimensionless temperature, and dimensionless time defined in Sec. 6.4.1, as well as those dimensionless variables defined in Eqs. (5-6) and (5-7). Construct the reasoning that leads to Eq. (5-8).

**6.7.5-3** *Mangler's transformation* [2, p. 235] Starting with Eq. (5-10), make the change of variables indicated by Eqs. (5-11) to (5-14) to arrive at Eq. (5-15). The results of Exercise 3.5.7-3 are helpful.

### 6.7.6 Energy transfer in the entrance of a heated section of a tube

An incompressible newtonian fluid with constant viscosity and thermal conductivity flows through a tube of radius $R$. For $z < 0$, the wall of the tube is insulated:

$$\text{At } r = R, \text{ for } z < 0: \quad \frac{\partial T}{\partial r} = 0 \tag{6-1}$$

For $z > 0$, the temperature of the wall is maintained constant at $T_1$:

$$\text{At } r = R, \text{ for } z > 0: \quad T = T_1 \tag{6-2}$$

Very far upstream of the entrance to this heated section, the fluid is known to be at a uniform temperature $T_0$:

$$\text{As } z \to -\infty, \text{ for } r \leq R: \quad T \to T_0 \tag{6-3}$$

We wish to determine the rate of energy transfer to the fluid in the heated portion of the tube. For the moment we will focus our attention on the entrance to this heated section.

Since the viscosity and density of this fluid are taken to be constants independent of temperature, the velocity distribution is that found in Sec. 3.2.1:

$$v_r = v_\theta = 0 \qquad v_z = v_{z(\text{max})}\left[1 - \left(\frac{r}{R}\right)^2\right] \tag{6-4}$$

where $v_{z(\text{max})}$ is the $z$ component of velocity along the centerline.

The boundary conditions on temperature and the known velocity distribution suggest that the temperature distribution in the heated portion of the tube is axisymmetric:

$$T = T(r,z) \tag{6-5}$$

From Table 5.6.1-3, we see that the differential energy balance for this situation is

$$\rho \hat{c} v_z \frac{\partial T}{\partial z} = k\left[\frac{1}{r}\frac{\partial}{\partial r}\left(r\frac{\partial T}{\partial r}\right) + \frac{\partial^2 T}{\partial z^2}\right] + \mu\left(\frac{\partial v_z}{\partial r}\right)^2 \tag{6-6}$$

or

$$(1 - r^{*2})\frac{\partial T^*}{\partial z^*} = \frac{1}{N_{\text{Pe}}}\left[\frac{1}{r^*}\frac{\partial}{\partial r^*}\left(r^*\frac{\partial T^*}{\partial r^*}\right) + \frac{\partial^2 T^*}{\partial z^{*2}}\right] + \frac{4N_{\text{Br}}}{N_{\text{Pe}}} r^{*2} \tag{6-7}$$

Here the dimensionless temperature $T^*$, dimensionless radial coordinate $r^*$, and dimensionless axial coordinate $z^*$ are defined, respectively, as

$$T^* \equiv \frac{T - T_0}{T_1 - T_0} \qquad r^* \equiv \frac{r}{R} \qquad z^* \equiv \frac{z}{R} \tag{6-8}$$

The Peclet and Brinkman numbers are

$$N_{\text{Pe}} \equiv \frac{\hat{c} R v_{z(\text{max})} \rho}{k} \qquad N_{\text{Br}} \equiv \frac{\mu (v_{z(\text{max})})^2}{k(T_1 - T_0)} \tag{6-9}$$

In the initial statement of this problem, we limit ourselves to the entrance of the heated portion of this tube. As the Reynolds number tends to infinity for a specified value of the Prandtl number (as the Peclet number $N_{\text{Pe}}$ becomes unbounded), we anticipate a thermal boundary layer developing along the wall of the tube. It is this thermal boundary layer with which we are primarily concerned in this section.

Our approach in describing this thermal boundary layer is very similar to that taken in the preceding sections, where we were concerned with thermal boundary layers on submerged bodies. But there is one important change. In the preceding sections, the velocity distribution was obtained by making the boundary-layer approximations in Cauchy's first law. Here we know the velocity distribution throughout the fluid a priori. We will see that requires a different definition for the expanded coordinate to be used in describing the boundary layer.

Since we are primarily concerned with the thermal boundary layer along the wall of the tube, let us introduce

$$s^* \equiv 1 - r^* \tag{6-10}$$

a dimensionless distance measured from the wall. In terms of $s^*$, Eq. (6-7) becomes

$$(2s^* - s^{*2})\frac{\partial T^*}{\partial z^*}$$

$$= \frac{1}{N_{\text{Pe}}}\left(\frac{\partial^2 T^*}{\partial s^{*2}} - \frac{1}{1-s^*}\frac{\partial T^*}{\partial s^*} + \frac{\partial^2 T^*}{\partial z^{*2}}\right) + \frac{4N_{\text{Br}}}{N_{\text{Pe}}}(1-s^*)^2 \tag{6-11}$$

We are concerned with a very thin boundary layer that we intuitively feel must get thinner at a fixed value of $z^*$ as the Reynolds number becomes unbounded for a fixed value of the Prandtl number. The discussion in Secs. 3.5.1 and 6.7.1 motivates us to introduce as an expanded coordinate either

$$\tilde{s}^{**} \equiv (N_{\text{Re}})^a s^* \tag{6-12}$$

or

$$s^{**} \equiv (N_{\text{Pe}})^a s^* \tag{6-13}$$

There is a slight advantage in working in terms of $s^{**}$, in that the Prandtl number is eliminated from the final result. In terms of $s^{**}$, Eq. (6-11) becomes

$$(2s^{**} - N_{\text{Pe}}^{-a}s^{**2})\frac{\partial T^*}{\partial z^*}$$

$$= N_{\text{Pe}}^{3a-1}\frac{\partial^2 T^*}{\partial s^{**2}} - \frac{N_{\text{Pe}}^{2a-1}}{1 - N_{\text{Pe}}^{-a}s^{**}}\frac{\partial T^*}{\partial s^{**}} + N_{\text{Pe}}^{a-1}\frac{\partial^2 T^*}{\partial z^{*2}}$$

$$+ 4N_{\text{Br}}N_{\text{Pe}}^{a-1}(1 - N_{\text{Pe}}^{-a}s^{**})^2 \tag{6-14}$$

In order that some of the conduction terms survive as the Peclet number becomes unbounded for finite values of the Brinkman number, we conclude that

$$a = \tfrac{1}{3} \tag{6-15}$$

The form of the differential energy balance appropriate to the thermal boundary layer in this problem is

$$\text{As } N_{\text{Pe}} \to \infty: \quad 2s^{**}\frac{\partial T^*}{\partial z^*} = \frac{\partial^2 T^*}{\partial s^{**2}} \tag{6-16}$$

In view of Eq. (6-15), the dependence upon the Reynolds number in the expanded coordinate for the thermal boundary layer differs from that ($a = \tfrac{1}{2}$) introduced in Sec. 6.7.1. The primary difference is that in Secs. 6.7.1 to 6.7.5 the velocity distribution is to be determined by a boundary-layer analysis. In this problem we have an exact solution for the velocity distribution throughout the entire flow, both within and outside of the thermal boundary layer.

[6.7] BOUNDARY-LAYER THEORY

Since we are neglecting entirely both axial conduction and viscous dissipation in this analysis, we may rewrite boundary condition (6-3) as

At $z = 0$, for all $r < R$: $\quad T = T_0$  (6-17)

If $\tilde{T}$ represents the temperature distribution for the nonviscous, nonconducting external flow evaluated at the wall, the differential energy balance requires

$$\tilde{v}_z \frac{\partial \tilde{T}}{\partial z} = 0 \tag{6-18}$$

We conclude that

As $s^{**} \to \infty$, for all $z^* > 0$: $\quad T^* \to \tilde{T}^* = 0$  (6-19)

This, together with Eq. (6-2), in the form of

At $s^{**} = 0$, for all $z^* > 0$: $\quad T^* = 1$  (6-20)

gives the boundary conditions that must be satisfied in solving Eq. (6-16).

With the change of variable

$$\eta \equiv \frac{s^{**}}{\sqrt[3]{\frac{9}{2}z^*}} \tag{6-21}$$

Eq. (6-16) is reduced to an ordinary differential equation:

$$-3\eta^2 \frac{dT^*}{d\eta} = \frac{d^2 T^*}{d\eta^2} \tag{6-22}$$

Boundary conditions (6-19) and (6-20) are transformed, respectively, into

As $\eta \to \infty$: $\quad T^* \to 0$  (6-23)

and

At $\eta = 0$: $\quad T^* = 1$  (6-24)

A solution consistent with Eqs. (6-22) to (6-24) is readily found to be

$$T^* = \frac{\int_\eta^\infty \exp(-\eta^3)\, d\eta}{\int_0^\infty \exp(-\eta^3)\, d\eta} \tag{6-25}$$

In terms of the gamma function

$$\Gamma(n) \equiv \int_0^\infty x^{n-1} e^{-x}\, dx \tag{6-26}$$

Eq. (6-25) can be written somewhat more conveniently as

$$T^* = \frac{1}{\Gamma(\frac{4}{3})} \int_\eta^\infty \exp(-\eta^3)\, d\eta \tag{6-27}$$

Since

$$-q_r = k\frac{\partial T}{\partial r}$$

$$= -\frac{k(T_1 - T_0)}{R} N_{Pe}^{\frac{1}{3}} \frac{1}{\sqrt[3]{\frac{9}{2}z^*}} \frac{dT^*}{d\eta}$$

$$= \frac{k(T_1 - T_0)}{R} N_{Pe}^{\frac{1}{3}} \frac{1}{\sqrt[3]{\frac{9}{2}z^*}} \frac{1}{\Gamma(\frac{4}{3})} \exp(-\eta^3) \qquad (6\text{-}28)$$

we can readily calculate the average energy flux from the wall to the fluid in a heated portion of tube of length $L$ to be

$$(-q_r|_{r=R})_{\text{av}} \equiv \frac{R}{L}\int_0^{L/R} -q_r|_{r=R}\, dz^* = \left(\frac{9}{2}\right)^{\frac{2}{3}} \frac{k(T_1 - T_0)}{R} N_{Pe}^{\frac{1}{3}} \frac{1}{\Gamma(\frac{1}{3})} \left(\frac{R}{L}\right)^{\frac{1}{3}} \qquad (6\text{-}29)$$

In terms of the Nusselt number

$$N_{Nu} \equiv 2\frac{(-q_r|_{r=R})_{\text{av}} R}{(T_1 - T_0)k} \qquad (6\text{-}30)$$

this is somewhat more conveniently written as

$$N_{Nu} = \left(\frac{9}{2}\right)^{\frac{2}{3}} \frac{2}{\Gamma(\frac{1}{3})} N_{Pe}^{\frac{1}{3}} \left(\frac{R}{L}\right)^{\frac{1}{3}} \qquad (6\text{-}31)$$

With the abrupt change in boundary conditions described by Eqs. (6-1) and (6-2), we can anticipate that axial conduction neglected in the above analysis must in fact be significant in a small region near $z = 0$ and $r = R$. This is confirmed by Newman's [1] numerical solution for this region. Not unexpectedly, he finds that this region where axial conduction cannot be neglected becomes smaller as the Peclet number increases.

For more about this and similar problems, see the exercises that follow as well as Secs. 6.8.1 and 6.8.2.

## REFERENCES

1. Newman, John: Preprint 18646, Lawrence Radiation Laboratory, University of California, Berkeley, Calif., 1969.
2. Bird, R. B., W. E. Stewart, and E. N. Lightfoot: "Transport Phenomena," 7th printing, Wiley, New York, 1960.
3. Schlicting, Hermann: "Boundary-Layer Theory," 6th ed., McGraw-Hill, New York, 1968.

## EXERCISES

**6.7.6–1** *Derivation of Eq. (6-27)* Integrate Eq. (6-22) consistent with boundary conditions (6-23) and (6-24) to arrive at Eq. (6-27).

## [6.7] BOUNDARY-LAYER THEORY

**6.7.6-2** *Forced convection in a tube with constant energy flux at the wall*  Let us repeat the problem of the text, replacing boundary conditions (6-2) by

$$\text{At } r = R, \text{ for } z > 0: \quad k\frac{\partial T}{\partial r} = q = \text{a constant} \tag{1}$$

(a) Introduce the dimensionless temperature

$$T^* = \frac{T - T_0}{T_1} \tag{2}$$

where $T_1$ is a characteristic temperature that will be defined in such a way as to make the boundary value problem as simple as possible. Repeat the discussion of the text to conclude that, for the thermal boundary layer in the entrance of the heated portion of the tube, $T^*$ must satisfy

$$2s^{**}\frac{\partial T^*}{\partial z^*} = \frac{\partial^2 T^*}{\partial s^{**2}} \tag{3}$$

as the Peclet number

$$N_{\text{Pe}} \equiv \frac{\hat{c}Rv_{z(\max)}\rho}{k}$$

becomes unbounded, consistent with the boundary conditions

$$\text{As } s^{**} \to \infty, \text{ for all } z^* > 0: \quad T^* \to 0 \tag{4}$$

and

$$\text{At } s^{**} = 0, \text{ for all } z^* > 0: \quad \frac{\partial T^*}{\partial s^{**}} = -1 \tag{5}$$

In arriving at this form of the problem, you will find it necessary to define

$$T_1 \equiv \frac{Rq}{kN_{\text{Pe}}^{\frac{1}{3}}} \tag{6}$$

(b) Determine that

$$u \equiv \frac{\partial T^*}{\partial s^{**}} \tag{7}$$

must satisfy

$$2\frac{\partial u}{\partial z^*} = \frac{\partial}{\partial s^{**}}\left(\frac{1}{s^{**}}\frac{\partial u}{\partial s^{**}}\right) \tag{8}$$

consistent with the boundary conditions

$$\text{As } s^{**} \to \infty, \text{ for all } z^* > 0: \quad u \to 0 \tag{9}$$

and

$$\text{At } s^{**} = 0: \quad u = -1 \tag{10}$$

(c) Anticipate a solution to the problem posed in (b) of the form

$$u = u(\eta) \tag{11}$$

where

$$\eta \equiv \frac{s^{**}}{\sqrt[3]{\frac{9}{2}z^*}} \tag{12}$$

Conclude that

$$\frac{\partial T^*}{\partial s^{**}} = u = -\frac{\int_\eta^\infty \eta \exp(-\eta^3)\, d\eta}{\int_0^\infty \eta \exp(-\eta^3)\, d\eta} = \frac{-3}{\Gamma(\frac{2}{3})} \int_\eta^\infty \eta \exp(-\eta^3)\, d\eta \qquad (13)$$

(d) It is now clear that

$$T^* = T^*(\eta, s^{**}) \qquad (14)$$

or

$$T^* = T^*(\eta, z^*) \qquad (15)$$

Let us take this later point of view and write Eq. (13) in the form

$$\frac{\partial T^*}{\partial \eta} \frac{1}{\sqrt[3]{\frac{9}{2}z^*}} = \frac{-3}{\Gamma(\frac{2}{3})} \int_\eta^\infty \eta \exp(-\eta^3)\, d\eta \qquad (16)$$

The last boundary condition to be satisfied is that

As $\eta \to \infty$, for all $z^* > 0$: $\quad T^* \to 0 \qquad (17)$

We may, consequently, integrate Eq. (16) consistent with Eq. (17) to find[1]

$$\frac{T - T_0}{Rq/k} = \sqrt[3]{\frac{\frac{9}{2}z^*}{N_{\text{Pe}}}} \left\{ \eta \left[ \frac{\Gamma(\frac{2}{3}; \eta^3)}{\Gamma(\frac{2}{3})} - 1 \right] + \frac{\exp(-\eta^3)}{\Gamma(\frac{2}{3})} \right\} \qquad (18)$$

We have introduced here the incomplete gamma function

$$\Gamma(\tfrac{2}{3}; \eta^3) \equiv \int_0^{\eta^3} x^{\frac{2}{3}-1} \exp(-x)\, dx \qquad (19)$$

(e) If we introduce the Nusselt number

$$N_{\text{Nu}} \equiv \frac{2qR}{[(T|_{r=R})_{\text{av}} - T_0]k}$$

conclude that

$$N_{\text{Nu}} = \frac{8\Gamma(\frac{2}{3})}{3(\frac{9}{2})^{\frac{1}{3}}} N_{\text{Pe}}^{\frac{1}{3}} \left(\frac{R}{L}\right)^{\frac{1}{3}}$$

**6.7.6-3** *Heat transfer from an isothermal wall to a falling film* An incompressible newtonian fluid flows down an inclined plane as shown in Fig. 3.2.5-3 (see also Exercise 3.2.5-3). The wall is insulated for $z_1 < 0$:

At $z_2 = 0$, for $z_1 < 0$: $\quad \dfrac{\partial T}{\partial z_2} = 0 \qquad (1)$

For $z_1 > 0$, the wall is maintained at a constant temperature $T_1$:

At $z_2 = 0$, for $z_1 > 0$: $\quad T = T_1 \qquad (2)$

Very far upstream from the entrance to the heated portion of the wall, the fluid has a uniform temperature $T_0$:

As $z_1 \to -\infty$, for $0 \le z_2 \le \delta$: $\quad T \to T_0 \qquad (3)$

---

[1] In Bird, Stewart, and Lightfoot's [2, p. 309] solution to this problem,

$$v_0 = 2v_{z(\text{max})}$$

Following the general outline of the discussion of the text, determine that the temperature distribution in the fluid near the entrance to the heated section has the form

$$T^* = \frac{T - T_0}{T_1 - T_0} = \frac{1}{\Gamma(\frac{4}{3})} \int_\eta^\infty \exp(-\eta^3)\, d\eta \tag{4}$$

in the limit as the Peclet number

$$N_{\text{Pe}} \equiv \frac{\hat{c}\, \delta v_{1(\max)} \rho}{k}$$

becomes unbounded, where

$$\eta \equiv \frac{z_2^{**}}{\sqrt[3]{\tfrac{9}{2} z_1^*}} \tag{5}$$

and

$$z_1^* \equiv \frac{z_1}{\delta} \qquad z_2^{**} \equiv (N_{\text{Pe}})^{\frac{1}{3}} \frac{z_2}{\delta} \tag{6}$$

Here $v_{1(\max)}$ is the maximum velocity of the fluid in the film:

$$v_{1(\max)} \equiv \frac{\delta^2 \rho g \cos \alpha}{2\mu} \tag{7}$$

**6.7.6-4** *Heat transfer from a wall to a falling film with constant energy flux at the wall* Repeat Exercise 6.7.6-3, replacing the isothermal boundary condition with

$$\text{At } z_2 = 0, \text{ for } z_1 > 0: \qquad k\frac{\partial T}{\partial z_2} = q$$

Conclude that as the Peclet number grows very large the temperature distribution in the thermal boundary layer, near the entrance of the heated portion of the wall, is exactly the same as that found in Exercise 6.7.6-2 when $R$ is replaced by $\delta$ and $v_{z(\max)}$ by $v_{1(\max)}$ defined in Exercise 3.2.5-3.

**6.7.6-5** *Heat transfer from a gas stream to a falling film* An incompressible newtonian fluid flows down an inclined plane as shown in Fig. 3.2.5-3 (see Exercise 3.2.5-3). Let us assume that there is no energy transfer from the gas stream to the falling film for $z_1 < 0$:

$$\text{At } z_2 = \delta, \text{ for } z_1 < 0: \qquad \frac{\partial T}{\partial z_2} = 0$$

Outside the immediate neighborhood of the liquid film, the gas stream has a uniform temperature $T_1$. In order to simplify the problem somewhat, we will assume that for $z_1 > 0$ the temperature of the gas-liquid phase interface is $T_1$:

$$\text{At } z_2 = \delta, \text{ for } z_1 > 0: \qquad T = T_1$$

Very far upstream, the temperature of the fluid is uniform at $T_0$:

$$\text{As } z_1 \to -\infty, \text{ for } 0 \leq z_2 \leq \delta: \qquad T \to T_0$$

Determine the temperature distribution in the thermal boundary layer near the entrance to the heated portion of the film as the Peclet number becomes unbounded.

*Answer*

$$T^* \equiv \frac{T - T_0}{T_1 - T_0} = 1 - \operatorname{erf}\left(\frac{1 - z_2/\delta}{\sqrt{4kz_1/\hat{c}\,\delta^2 v_{1(\max)} \rho}}\right)$$

*Hint:* This problem is similar to the one discussed in the text, but the expanded $z_2$ coordinate is defined differently.

**6.7.6-6** *More on heat transfer from a gas stream to a falling film* Let us repeat Exercise 6.7.6-5 and attempt to describe the boundary condition at the gas-liquid phase interface more realistically. Rather than saying that the phase interface is in equilibrium with the gas very far away from it, let us describe the energy transfer in terms of Newton's law of cooling (see Sec. 6.2.3).

*Answer*

$$T^* \equiv \frac{T - T_1}{T_0 - T_1} = \mathrm{erf}\left(\frac{s^{**}}{\sqrt{4z_1^*}}\right) + \exp\left(\frac{s^{**}}{B} + \frac{z_1^*}{B^2}\right)\left[1 - \mathrm{erf}\left(\frac{s^{**}}{\sqrt{4z_1^*}} + \frac{\sqrt{z_1^*}}{B}\right)\right]$$

$$s^{**} \equiv \sqrt{N_{\mathrm{Pe}}}\left(1 - \frac{z_2}{\delta}\right) \qquad z_1^* \equiv \frac{z_1}{\delta}$$

$$B \equiv \frac{N_{\mathrm{Pe}}^{\frac{1}{2}}}{N_{\mathrm{Nu}}} \qquad N_{\mathrm{Pe}} \equiv \frac{\hat{c}\,\delta v_{1(\max)}\rho}{k} \qquad N_{\mathrm{Nu}} \equiv \frac{h\,\delta}{k}$$

*Hint:* See Sec. 6.2.3 as well as the hint for Exercise 6.7.6-5.

**6.7.6-7** *Energy transfer in the entrance of a heated section of a tube for a power-model fluid* Repeat the problem discussed in the text for a power-model fluid to conclude that the temperature distribution is again given by Eq. (6-27), where now

$$\eta \equiv \frac{s^{**}}{\sqrt[3]{9nz^*/(1+n)}}$$

This means that the Nusselt number as defined by Eq. (6-30) becomes

$$N_{\mathrm{Nu}} = 9\left(\frac{1+n}{9n}\right)^{\frac{1}{3}} \frac{1}{\Gamma(\frac{1}{3})} N_{\mathrm{Pe}}^{\frac{1}{3}}\left(\frac{R}{L}\right)^{\frac{1}{3}}$$

**6.7.6-8** *Natural convection at a vertical flat plate* [3, p. 300] In Exercise 6.7.1-1, we introduced the boundary-layer equations appropriate to natural convection in the limit as the Grashof number $N_{\mathrm{Gr}} \to \infty$. Let us apply these equations to analyze the heat transfer from a vertical hot plate.

A vertical flat plate of uniform temperature $T_1$ is immersed in a newtonian fluid that has a uniform temperature $T_2$ very far away from the plate. The coordinate $z_1$ is measured along the plate in the opposite direction from gravity starting at the leading edge of the plate; $z_2$ is measured into the fluid from the plate.

(a) What are the boundary conditions to be satisfied by the simultaneous solution of the boundary-layer equations?

(b) The equation of continuity can be identically satisfied by the introduction of a dimensionless stream function $\psi^*$. The resulting two partial differential equations can be reduced to ordinary differential equations by looking for a combination-of-variables solution of the form

$$\psi^* = (4z_1^*)^{\frac{3}{4}} Z(\eta) \qquad \theta = \theta(\eta)$$

where

$$\eta = \frac{z_2^{**}}{(4z_1^*)^{\frac{1}{4}}}$$

Determine that the equation of motion and the differential energy balance reduce respectively to

$$Z''' + 3ZZ'' - 2(Z')^2 + \theta = 0$$

and

$$\theta'' + 3N_{\mathrm{Pr}}Z\theta' = 0$$

## 6.8 DOWNSTREAM IN A TUBE

### 6.8.1 More about energy transfer in a heated section of a tube

Let us go back and take another look at the problem discussed in Sec. 6.7.6. An incompressible newtonian fluid with constant viscosity and thermal conductivity flows through a tube of radius $R$. For $z < 0$, the wall of the tube is insulated:

At $r = R$, for $z < 0$: $\quad \dfrac{\partial T}{\partial r} = 0 \quad$ (1-1)

For $z > 0$, the temperature of the wall is maintained constant at $T_1$:

At $r = R$, for $z > 0$: $\quad T = T_1 \quad$ (1-2)

Very far upstream from the entrance to this heated section, the fluid is known to be at a uniform temperature $T_0$:

As $z \to -\infty$, for $r \leq R$: $\quad T \to T_0 \quad$ (1-3)

In Sec. 6.7.6 we examined the temperature distribution and the rate of energy transfer to the fluid near the entrance to this heated portion of the tube for very large values of the Peclet number. In what follows, we give our attention to the temperature distribution in the fluid somewhat downstream from the entrance. We will still restrict ourselves to the limit as the Peclet number becomes unbounded.

We continue to assume that the temperature distribution is axisymmetric:

$$T = T(r,z) \quad (1\text{-}4)$$

Applying the velocity distribution found in Sec. 3.2.1, we conclude that the dimensionless differential energy balance has the same form as that found in Sec. 6.7.6:

$$(1 - r^{*2}) \frac{\partial T^*}{\partial z^*} = \frac{1}{N_{\text{Pe}}}\left[\frac{1}{r^*}\frac{\partial}{\partial r^*}\left(r^* \frac{\partial T^*}{\partial r^*}\right) + \frac{\partial^2 T^*}{\partial z^{*2}}\right] + \frac{4 N_{\text{Br}}}{N_{\text{Pe}}} r^{*2} \quad (1\text{-}5)$$

where

$$T^* \equiv \frac{T - T_0}{T_1 - T_0} \qquad r^* \equiv \frac{r}{R} \qquad z^* \equiv \frac{z}{R} \quad (1\text{-}6)$$

Intuitively, we expect that, sufficiently far downstream from the entrance, effects attributable to the curvature of the tube wall should be important, although axial conduction should continue to be a negligible effect. Since we are primarily concerned with what is happening at relatively large values of $z^*$, we are motivated to introduce a contracted dimensionless axial coordinate

$$z^{**} \equiv \frac{z^*}{(N_{\text{Pe}})^b} \quad (1\text{-}7)$$

In terms of $z^{**}$, Eq. (1-5) may be written

$$(1 - r^{*2})\frac{\partial T^*}{\partial z^{**}} = \frac{(N_{Pe})^{b-1}}{r^*}\frac{\partial}{\partial r^*}\left(r^*\frac{\partial T^*}{\partial r^*}\right) + (N_{Pe})^{-1-b}\frac{\partial^2 T^*}{\partial z^{**2}} + 4N_{Br}N_{Pe}^{b-1}r^{*2} \quad (1\text{-}8)$$

In order that the convection and radial conduction terms be of the same order of magnitude as the Peclet number becomes unbounded, we choose

$$b = 1 \quad (1\text{-}9)$$

With this understanding, Eq. (1-8) becomes

$$(1 - r^{*2})\frac{\partial T^*}{\partial z^{**}} = \frac{1}{r^*}\frac{\partial}{\partial r^*}\left(r^*\frac{\partial T^*}{\partial r^*}\right) + (N_{Pe})^{-2}\frac{\partial^2 T^*}{\partial z^{**2}} + 4N_{Br}r^{*2} \quad (1\text{-}10)$$

As the Peclet number becomes unbounded, it is intuitively appealing to neglect the axial conduction term with respect to the radial conduction term and simplify Eq. (1-10) to

$$(1 - r^{*2})\frac{\partial T^*}{\partial z^{**}} = \frac{1}{r^*}\frac{\partial}{\partial r^*}\left(r^*\frac{\partial T^*}{\partial r^*}\right) + 4N_{Br}r^{*2} \quad (1\text{-}11)$$

In the limit as the Peclet number goes to infinity, this is the differential equation to be solved for the temperature distribution in the fluid relatively far downstream from the entrance to the heating section. The appropriate boundary conditions are

At $r^* = 1$, for $z^{**} < 0$:  $\quad \dfrac{\partial T^*}{\partial r^*} = 0 \quad (1\text{-}12)$

At $r^* = 1$, for $z^{**} > 0$:  $\quad T^* = 1 \quad (1\text{-}13)$

and

As $z^{**} \to -\infty$, for $r^* \leq 1$:  $\quad T^* \to 0 \quad (1\text{-}14)$

For sufficiently small values of the Brinkman number, it would appear that we could neglect viscous dissipation and write Eq. (1-11) as

$$(1 - r^{*2})\frac{\partial T^*}{\partial z^{**}} = \frac{1}{r^*}\frac{\partial}{\partial r^*}\left(r^*\frac{\partial T^*}{\partial r^*}\right) \quad (1\text{-}15)$$

Since we are neglecting both axial conduction and viscous dissipation, boundary conditions (1-12) and (1-14) imply

At $z^{**} = 0$, for $r^* < 1$:  $\quad T^* = 0 \quad (1\text{-}16)$

Equations (1-13) and (1-16) are the boundary conditions to be satisfied by the solution to Eq. (1-15). This special case, known as the Graetz problem, has received considerable attention in the literature. The most accurate solution known at the present time is due to Brown [1]. An excellent survey of the literature has been presented by Newman [2].

## REFERENCES

1. Brown, G. M.: *A.I.Ch.E.J.*, **6**:179 (1960).
2. Newman, John: Preprint 18646, Lawrence Radiation Laboratory, University of California, Berkeley, Calif., 1969.

## EXERCISE

**6.8.1-1** Formalize the argument that takes boundary conditions (1-12) and (1-14) into (1-16).

### 6.8.2 Still more about energy transfer in a heated section of a tube
Instead of specifying the temperature of the wall as in Sec. 6.8.1, let us instead specify a constant energy flux.

To be specific, we again have an incompressible newtonian fluid with constant viscosity and thermal conductivity flowing through a tube of radius $R$. For $z < 0$, the wall of the tube is insulated:

$$\text{At } r = R, \text{ for } z < 0: \qquad \frac{\partial T}{\partial r} = 0 \tag{2-1}$$

But for $z > 0$, the radial component of the energy flux vector is a constant, $q$:

$$\text{At } r = R, \text{ for } z > 0: \qquad k\frac{\partial T}{\partial r} = q \tag{2-2}$$

Very far upstream of the entrance to this heated section, the fluid is known to be at a uniform temperature $T_0$:

$$\text{As } z \to -\infty, \text{ for } r \leq R: \qquad T \to T_0 \tag{2-3}$$

We will focus our attention upon the temperature distribution in the fluid very far downstream from the entrance to the heated section. As in Sec. 6.8.1, we will still restrict ourselves to the limit as the Peclet number becomes unbounded.

Repeating the argument in Sec. 6.8.1, we find that as the Peclet number becomes unbounded the dimensionless form of the differential energy balance appropriate somewhat downstream from the entrance of the tube is

$$(1 - r^{*2})\frac{\partial T^*}{\partial z^{**}} = \frac{1}{r^*}\frac{\partial}{\partial r^*}\left(r^*\frac{\partial T^*}{\partial r^*}\right) + 4N_{Br}r^{*2} \tag{2-4}$$

where

$$T^* \equiv \frac{T - T_0}{qR/k} \qquad r^* \equiv \frac{r}{R} \qquad z^{**} \equiv \frac{z}{RN_{Pe}} \tag{2-5}$$

and

$$N_{Pe} \equiv \frac{\hat{c}Rv_{z(\max)}\rho}{k} \qquad N_{Br} \equiv \frac{\mu(v_{z(\max)})^2}{qR} \tag{2-6}$$

Equation (2-4) must be solved consistent with the boundary conditions

$$\text{At } r^* = 1, \text{ for } z^{**} < 0: \quad \frac{\partial T^*}{\partial r^*} = 0 \tag{2-7}$$

$$\text{At } r^* = 1, \text{ for } z^{**} > 0: \quad \frac{\partial T^*}{\partial r^*} = 1 \tag{2-8}$$

and

$$\text{As } z^{**} \to -\infty, \text{ for } r^* \leq 1: \quad T^* \to 0 \tag{2-9}$$

For very small values of the Brinkman number, it seems reasonable to neglect the effects of viscous dissipation in Eq. (2-4):

$$(1 - r^{*2}) \frac{\partial T^*}{\partial z^{**}} = \frac{1}{r^*} \frac{\partial}{\partial r^*} \left( r^* \frac{\partial T^*}{\partial r^*} \right) \tag{2-10}$$

Since both axial conduction and viscous dissipation are neglected, boundary conditions (2-7) and (2-9) imply

$$\text{At } z^{**} = 0, \text{ for } r^* < 1: \quad T^* = 0 \tag{2-11}$$

Equation (2-10) is to be solved using Eqs. (2-8) and (2-11) as boundary conditions. Such a solution has been presented [1].

If our primary concern is the temperature distribution in the fluid very far downstream from the entrance to the heated section, a somewhat simpler approach is appropriate. We might expect that sufficiently far downstream from the entrance to the heated section the temperature of the fluid should be a linear function of axial position:

$$T^* = z^{**}\Theta_1(r^*) + \Theta_2(r^*) \tag{2-12}$$

With this assumption, Eq. (2-10) becomes

$$(1 - r^{*2})\Theta_1 = \frac{z^{**}}{r^*} \frac{d}{dr^*}\left(r^* \frac{d\Theta_1}{dr^*}\right) + \frac{1}{r^*} \frac{d}{dr^*}\left(r^* \frac{d\Theta_2}{dr^*}\right) \tag{2-13}$$

But if $\Theta_2$ is to be a function only of $r^*$, it appears that we must require

$$\frac{d}{dr^*}\left(r^* \frac{d\Theta_1}{dr^*}\right) = 0 \tag{2-14}$$

or

$$\Theta_1 = C_1 \ln r^* + C_2 \tag{2-15}$$

Since $T^*$ must be finite at $r^* = 0$, we must set $C_1 = 0$ and conclude that

$$\Theta_1 = C_2 = \text{a constant} \tag{2-16}$$

and Eq. (2-13) reduces to

$$\frac{1}{r^*} \frac{d}{dr^*}\left(r^* \frac{d\Theta_2}{dr^*}\right) = C_2(1 - r^{*2}) \tag{2-17}$$

## [6.8] DOWNSTREAM IN A TUBE

Unfortunately, a solution of the form required by Eqs. (2-12) and (2-16) is inconsistent with boundary conditions (2-8) and (2-11). One interpretation of this inconsistency is that we were incorrect in assuming that $T^*$ is of the form indicated by Eq. (2-12). Another possibility might be that boundary condition (2-11) is not relevant to the temperature distribution very far downstream from the entrance to the heated section. Let us explore this latter possibility.

As an alternative to boundary condition (2-11), we can observe that the energy entering with the fluid at cross section $z^{**} = 0$ must leave either through the cross section at $z^{**}$ or through the bounding walls of the tube:

$$\int_0^{2\pi} \int_0^R \rho \hat{c}(T - T_0) v_z r \, dr \, d\theta - 2\pi R z q = 0 \tag{2-18}$$

Applying the velocity distribution from Sec. 3.2.1, we can write Eq. (2-18) in terms of our dimensionless variables as

$$\int_0^1 T^*(1 - r^{*2}) r^* \, dr^* - z^{**} = 0 \tag{2-19}$$

In view of Eqs. (2-12) and (2-16), Eq. (2-19) requires

$$C_2 = 4 \tag{2-20}$$

and

$$\int_0^1 \Theta_2 (1 - r^{*2}) r^* \, dr^* = 0 \tag{2-21}$$

This and

At $r^* = 1$:  $\dfrac{d\Theta_2}{dr^*} = 1$ \hfill (2-22)

are the two boundary conditions to be satisfied by the required solution to Eq. (2-17).

We may now integrate Eq. (2-17) to find

$$\int_0^{r^*(d\Theta_2/dr^*)} d\left(r^* \frac{d\Theta_2}{dr^*}\right) = 4 \int_0^{r^*} (r^* - r^{*3}) \, dr^*$$

$$\frac{d\Theta_2}{dr^*} = 2r^* - r^{*3} \tag{2-23}$$

Notice that boundary condition (2-22) is automatically satisfied. Carrying out another integration, we find

$$\Theta_2 = r^{*2} - \frac{r^{*4}}{4} + C_3 \tag{2-24}$$

In view of Eq. (2-21), we must require

$$C_3 = -\tfrac{7}{24} \tag{2-25}$$

To summarize, Eqs. (2-12), (2-16), (2-24), and (2-25) tell us that the temperature distribution very far downstream from the entrance to the heated section should be approximately [2, p. 622; 3, p. 296]

$$T^* = 4z^{**} + r^{*2} - \tfrac{1}{4}r^{*4} - \tfrac{7}{24} \tag{2-26}$$

It is interesting to obtain an expression for the Nusselt number

$$N_{\text{Nu}} \equiv \frac{2qR}{(T|_{r=R} - T_b)k} = \frac{2}{T^*|_{r^*=1} - T_b^*} \tag{2-27}$$

where $T_b$ indicates the *bulk* temperature, the temperature that one would measure if the tube were chopped off at $z$ and the fluid flowing out were collected in a bucket and thoroughly mixed (this is often referred to as the *cup-mixing* temperature or *flow-average* temperature):

$$T_b^* \equiv \frac{T_b - T_0}{qR/k} = \frac{\int_0^1 T^*(1 - \xi^2)\xi\, d\xi}{\int_0^1 (1 - \xi^2)\xi\, d\xi} = 4z^{**} \tag{2-28}$$

Since

$$T^*|_{r^*=1} = 4z^{**} + \tfrac{11}{24} \tag{2-29}$$

we conclude that

$$N_{\text{Nu}} = \tfrac{48}{11} \tag{2-30}$$

Upon comparison with Siegel, Sparrow, and Hallman's solution [1], we see that Eq. (2-30) is exact in the limit as $z^{**} \to \infty$. Better still, their solution indicates that by $z^{**} = 0.085$ the Nusselt number has approached within 5 percent of the fully developed value $\tfrac{48}{11}$. This excellent agreement for small values of $z^{**}$ is somewhat surprising, in view of our replacement of boundary condition (2-11) by the somewhat less specific Eq. (2-18).

## REFERENCES

1. Siegel, R., E. M. Sparrow, and T. M. Hallman: *Appl. Sci. Res., Sect. A*, **7**:386 (1958).
2. Goldstein, S.: "Modern Developments in Fluid Mechanics," Oxford University Press, London, 1938.
3. Bird, R. B., W. E. Stewart, and E. N. Lightfoot: "Transport Phenomena," 7th printing, Wiley, New York, 1960.

## EXERCISE

**6.8.2-1** *Energy transfer in a heated section between two parallel walls* An incompressible newtonian fluid with constant viscosity and thermal conductivity flows between the two parallel walls shown in Fig. 6.8.2-1. For $z_1 < 0$, both walls are insulated. For $z_1 > 0$, the energy flux to the top wall is $q_{(1)}$ and the energy flux to the bottom wall is $q_{(2)}$. Very far upstream from this heated section, the fluid is known to be at a uniform temperature $T_0$.

## [6.8] DOWNSTREAM IN A TUBE

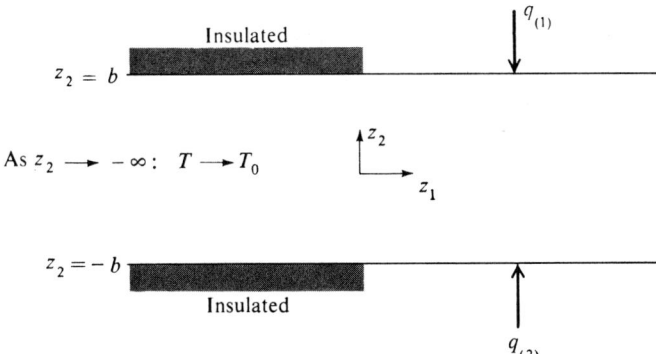

**Fig. 6.8.2-1** Energy transfer in a heated section between two parallel walls.

Determine that very far downstream from the entrance to this heated section

$$\frac{T - T_0}{q_{(1)}b/k} = \tfrac{3}{4}(1 + N_q)[z_1^{**} - \tfrac{1}{12}(z_2^*)^4 + \tfrac{1}{2}(z_2^*)^2 + \tfrac{2}{3}z_2^* - \tfrac{13}{140}] - N_q z_2^*$$

where

$$z_1^{**} \equiv \frac{z_1}{bN_{\text{Pe}}} \qquad z_2^* \equiv \frac{z_2}{b}$$

$$N_q \equiv \frac{q_{(2)}}{q_{(1)}} \qquad N_{\text{Pe}} \equiv \frac{\hat{c}b v_{1(\max)}\rho}{k}$$

and $v_{1(\max)}$ is the centerline $z_1$ component of velocity. Calculate also that

$$N_{\text{Nu}} \equiv \frac{q_{(1)}b}{(T|_{z_2=b} - T_B)k} + \frac{q_{(2)}b}{(T|_{z_2=-b} - T_B)k}$$

$$= \frac{1}{\tfrac{26}{53} - \tfrac{9}{35}N_q} + \frac{N_q}{-\tfrac{9}{35} + \tfrac{26}{35}N_q}$$

Here $T_B$ is the bulk temperature of the fluid [see discussion following Eq. (2-27)].

# 7
# Applications of Integral Averaging Techniques to Energy Transfer

The discussion of integral averaging techniques begun in Chap. 4 is continued here, but the emphasis is now upon energy transfer. If you have not already done so, I recommend that you read Chap. 4 before beginning this material. The derivations are more detailed there and a good deal more is said by way of motivation.

## 7.1. TIME AVERAGING

**7.1.1 Turbulent energy transfer** *Turbulent energy transfer* implies that at least one of the materials involved in the energy transfer process is in turbulent flow. Everything that we said in Secs. 4.1.1 to 4.1.5 is equally applicable here. The only additional complication is that, like the velocity and pressure distributions, the temperature distribution varies randomly with time in all possible frames of reference.

In the next few sections we shall be concerned with the time average of the differential energy balance and its solutions.

**7.1.2 The time-averaged differential energy balance** As noted in Sec. 4.1.2, for simplicity we shall limit ourselves to incompressible fluids. We shall also assume that there is no external or mutual energy transmission, that the ratio of the Brinkman number to the Peclet number is sufficiently small for viscous dissipation to be neglected

[7.1] TIME AVERAGING

(see Sec. 6.4.1), and that the heat capacity per unit mass may be treated as a constant. Under these circumstances, Eq. (E) of Table 5.6.1-1 reduces to

$$\rho \hat{c}\left[\frac{\partial T}{\partial t} + \text{div}\,(T\mathbf{v})\right] + \text{div}\,\mathbf{q} = 0 \tag{2-1}$$

Using the definition introduced in Sec. 4.1.2, let us take the time average of this equation:

$$\frac{1}{\Delta t}\int_{t}^{t+\Delta t}\left[\rho\hat{c}\frac{\partial T}{\partial t'} + \text{div}\,(\rho\hat{c}T\mathbf{v} + \mathbf{q})\right]dt' = 0 \tag{2-2}$$

The time-averaging operation commutes with partial differentiation with respect to time (see Sec. 4.1.2) and the divergence operation:

$$\rho\hat{c}\frac{\partial \bar{T}}{\partial t} + \text{div}\,(\overline{\rho\hat{c}T\mathbf{v}} + \bar{\mathbf{q}}) = 0 \tag{2-3}$$

It is more common to write this result as

$$\rho\hat{c}\left(\frac{\partial \bar{T}}{\partial t} + \nabla \bar{T}\cdot\mathbf{v}\right) = -\text{div}\,(\bar{\mathbf{q}} + \mathbf{q}^{(t)}) \tag{2-4}$$

where we have introduced the turbulent energy flux vector $\mathbf{q}^{(t)}$:

$$\mathbf{q}^{(t)} \equiv \rho\hat{c}(\overline{T\mathbf{v}} - \bar{T}\bar{\mathbf{v}}) \tag{2-5}$$

When we recognize Fourier's law as representing the energy flux vector, Eq. (2-4) takes the form

$$\rho\hat{c}\left(\frac{\partial \bar{T}}{\partial t} + \nabla \bar{T}\cdot\bar{\mathbf{v}}\right) = \text{div}\,(k\nabla \bar{T} - \mathbf{q}^{(t)}) \tag{2-6}$$

Our problem here is very similar to that encountered in Sec. 4.1.2. Just as there we had to stop and propose empirical data correlations for the Reynolds stress tensor $\mathbf{S}^{(t)}$, we must stop here and consider empirical representations for the turbulent energy flux vector $\mathbf{q}^{(t)}$.

**7.1.3 Empirical correlation for the turbulent energy flux vector $\mathbf{q}^{(t)}$** Three examples are used here to illustrate how empirical data correlations for the turbulent energy flux vector $\mathbf{q}^{(t)}$ can be prepared. Three ideas are used in this discussion.

1. For changes of frames such that

$$\bar{\mathbf{Q}} \doteq \mathbf{Q} \tag{3-1}$$

we may use the result of Sec. 4.1.3 to find that $\mathbf{q}^{(t)}$ is frame indifferent:

$$\mathbf{q}^{(t)*} \equiv \rho\hat{c}(\overline{T^*\mathbf{v}^*} - \bar{T}^*\bar{\mathbf{v}}^*) = \rho\hat{c}[\overline{T^*(\mathbf{v}^* - \bar{\mathbf{v}}^*)}] = \rho\hat{c}\mathbf{Q}\cdot[\overline{T(\mathbf{v}-\bar{\mathbf{v}})}] = \mathbf{Q}\cdot\mathbf{q}^{(t)} \tag{3-2}$$

Here **Q** is a (possibly) time-dependent orthogonal second-order tensor. In arriving at this result, we make use of the fact that a velocity difference is frame indifferent (see Exercise 1.2.2-1).

2. We shall assume that the principle of material frame indifference discussed in Sec. 2.3.1 applies to any empirical correlations developed for $\mathbf{q}^{(t)}$ so long as the changes of frame considered satisfy Eq. (3-1).
3. We shall use the Buckingham-Pi theorem to further limit the form of any expression for $\mathbf{q}^{(t)}$.

### EXAMPLE 1  PRANDTL'S MIXING LENGTH THEORY

Let us attempt to develop an empirical correlation for $\mathbf{q}^{(t)}$ appropriate to the fully developed flow regime in wall turbulence.

Influenced by the manner in which we approached empirical correlations for the Reynolds stress tensor in Sec. 4.1.4, we propose that the turbulent energy flux vector be regarded as a function of the density of the fluid, its heat capacity per unit mass $\hat{c}$, the distance $l$ from the wall, $\mathbf{\bar{D}}$, and $\nabla \bar{T}$:

$$\mathbf{q}^{(t)} = \mathbf{q}^{(t)}(\rho, \hat{c}, l, \mathbf{\bar{D}}, \nabla \bar{T}) \tag{3-3}$$

We specifically do not include thermal conductivity or viscosity as independent variables, because we are considering the fully developed turbulent flow regime. The most general expression of this form that satisfies our restricted form of the principle of material frame indifference is rather lengthy (see Exercise 7.1.3-1). It is perhaps sufficient to say that the literature has been primarily concerned with the special case

$$\mathbf{q}^{(t)} = -\varkappa \nabla \bar{T} \tag{3-4}$$

where

$$\varkappa = \varkappa(\rho, \hat{c}, l, \sqrt{2 \operatorname{tr} \mathbf{\bar{D}}^2}) \tag{3-5}$$

The Buckingham-Pi theorem [1] requires that Eq. (3-5) be of the form

$$\varkappa = \varkappa^* \rho \hat{c} l^2 \sqrt{2 \operatorname{tr} \mathbf{\bar{D}}^2} \tag{3-6}$$

and Eq. (3-4) becomes

$$\mathbf{q}^{(t)} = -\varkappa^* \rho \hat{c} l^2 \sqrt{2 \operatorname{tr} \mathbf{\bar{D}}^2} \nabla \bar{T} \tag{3-7}$$

Here $\varkappa^*$ is a dimensionless constant. Equation (3-7) should be viewed as the tensorial form of *Prandtl's mixing length theory* for energy transfer [2, p. 648].

I would like to emphasize that we do not expect the Prandtl mixing length theory to be appropriate to the laminar sublayer or buffer zone. We assumed at the beginning that we were constructing a representation for the turbulent energy flux vector in the fully developed turbulent flow regime.

### EXAMPLE 2  DEISSLER'S EXPRESSION FOR THE REGION NEAR THE WALL

Let us now consider that portion of the turbulent flow of an incompressible newtonian fluid in the immediate vicinity of a bounding wall: the laminar sublayer and the buffer zone. Looking back at our discussion of Example 2 in Sec. 4.1.4, it seems not unreasonable to propose

$$\mathbf{q}^{(t)} = \mathbf{q}^{(t)}(\rho, \hat{c}, \mu, l, \bar{\mathbf{v}} - \mathbf{v}_{(s)}, \nabla \bar{T}) \tag{3-8}$$

Remember that $\mathbf{v}_{(s)}$ indicates the velocity of the bounding wall.

Let us focus our attention in this relationship on the dependence of $\mathbf{q}^{(t)}$ upon the two vectors:

$$\mathbf{q}^{(t)} = \hat{\mathbf{q}}^{(t)}(\bar{\mathbf{v}} - \mathbf{v}_{(s)}, \nabla \bar{T}) \tag{3-9}$$

By our restricted form of the principle of material frame indifference, the functional relationship between these variables should be the same in every frame of reference. This means that

$$\mathbf{q}^{(t)*} = \mathbf{Q} \cdot \mathbf{q}^{(t)} = \mathbf{Q} \cdot \hat{\mathbf{q}}^{(t)}(\bar{\mathbf{v}} - \mathbf{v}_{(s)}, \nabla \bar{T}) = \hat{\mathbf{q}}^{(t)}(\mathbf{Q} \cdot [\bar{\mathbf{v}} - \mathbf{v}_{(s)}], \mathbf{Q} \cdot \nabla \bar{T}) \tag{3-10}$$

or $\hat{\mathbf{q}}^{(t)}$ is an isotropic function [3, p. 22]:

$$\hat{\mathbf{q}}^{(t)}(\bar{\mathbf{v}} - \mathbf{v}_{(s)}, \nabla \bar{T}) = \mathbf{Q}^T \cdot \hat{\mathbf{q}}^{(t)}(\mathbf{Q} \cdot [\bar{\mathbf{v}} - \mathbf{v}_{(s)}], \mathbf{Q} \cdot \nabla \bar{T}) \tag{3-11}$$

By representation theorems of Spencer and Rivlin [4, sec. 7] and of Smith [5], the most general polynomial isotropic vector function of two vectors has the form

$$\mathbf{q}^{(t)} = \varkappa_{(1)} \nabla \bar{T} + \varkappa_{(2)} (\bar{\mathbf{v}} - \mathbf{v}_{(s)}) \tag{3-12}$$

Here $\varkappa_{(1)}$ and $\varkappa_{(2)}$ are scalar-valued polynomials of the general form

$$\text{For } i = 1, 2: \quad \varkappa_{(i)} = \varkappa_{(i)}(\rho, \hat{c}, \mu, l, |\nabla \bar{T}|, |\bar{\mathbf{v}} - \mathbf{v}_{(s)}|, [\bar{\mathbf{v}} - \mathbf{v}_{(s)}] \cdot \nabla \bar{T}) \tag{3-13}$$

(In applying the theorem of Spencer and Rivlin, we identify the vector $\mathbf{b}$ that has covariant components $b_i$ with the skew-symmetric tensor which has contravariant components $\epsilon^{ijk} b_i$. Their theorem requires an additional term in Eq. (3-12) proportional to the vector product $(\bar{\mathbf{v}} - \mathbf{v}_{(s)}) \wedge \nabla \bar{T}$. This term is not consistent with the requirement that $\mathbf{q}^{(t)}$ be isotropic [3, p. 24] and consequently is dropped.)

An application of the Buckingham-Pi theorem [1] tells us that

$$\frac{\varkappa_{(1)}}{\rho \hat{c} l |\bar{\mathbf{v}} - \mathbf{v}_{(s)}|} = \varkappa_{(1)}^* \tag{3-14}$$

and

$$\frac{\varkappa_{(2)}}{\rho \hat{c} l |\nabla \bar{T}|} = \varkappa_{(2)}^* \tag{3-15}$$

Here

$$\text{For } i = 1, 2: \quad \varkappa_{(i)}^* = \varkappa_{(i)}^* \left( N, \frac{\hat{c} |\nabla \bar{T}| l}{|\bar{\mathbf{v}} - \mathbf{v}_{(s)}|^2}, \frac{[\bar{\mathbf{v}} - \mathbf{v}_{(s)}] \cdot \nabla \bar{T}}{|\bar{\mathbf{v}} - \mathbf{v}_{(s)}| |\nabla \bar{T}|} \right) \tag{3-16}$$

and

$$N \equiv \frac{\rho l |\bar{\mathbf{v}} - \mathbf{v}_{(s)}|}{\mu} \tag{3-17}$$

Deissler [6] has proposed on empirical grounds that

$$\mathbf{q}^{(t)} = -n^2 \rho \hat{c} l \, |\bar{\mathbf{v}} - \mathbf{v}_{(s)}| \, [1 - \exp(-n^2 N)] \nabla \bar{T} \tag{3-18}$$

The $n$ appearing here is meant to be the same as that used in Eq. (4-21) of Sec. 4.1.4 and evaluated in Sec. 4.1.5.

**EXAMPLE 3  EDDY CONDUCTIVITY IN FREE TURBULENCE**

If we move very far away from any walls into a region of free turbulence, it is tempting to look at Eq. (3-3) and assume instead

$$\mathbf{q}^{(t)} = \mathbf{q}^{(t)}(\rho, \hat{c}, l, \nabla \bar{T}) \tag{3-19}$$

Referring to Sec. 5.4.2, we can say by analogy that our restricted form of the principle of material frame indifference requires

$$\mathbf{q}^{(t)} = -\varkappa(\rho, \hat{c}, l, |\nabla \bar{T}|) \nabla \bar{T} \tag{3-20}$$

But the Buckingham-Pi theorem [1] tells us that $\varkappa$ must in fact be a constant scalar under the assumptions made:

$$\mathbf{q}^{(t)} = -\varkappa \nabla \bar{T} \tag{3-21}$$

The scalar $\varkappa$ is usually known as the *eddy conductivity*.

Notice that the type of theoretical objection raised against the use of an eddy viscosity in Sec. 4.1.4 does not apply here.

**REFERENCES**

1. Brand, L.: *Arch. Rational Mech. Analysis*, **1**:35 (1957).
2. Goldstein, S.: "Modern Developments in Fluid Dynamics," vol. 2, Oxford University Press, London, 1938.
3. Truesdell, C., and W. Noll: In S. Flügge (ed.), "Handbuch der Physik," vol. 3/3, Springer-Verlag, Berlin, 1965.
4. Spencer, A. J., and R. S. Rivlin: *Arch. Rational Mech. Analysis*, **4**:214 (1959/60).
5. Smith, G. F.: *Arch. Rational Mech. Analysis*, **18**:282 (1965).
6. Deissler, R. G.: *NACA Rept.* 1210, 1955.

**EXERCISE**

**7.1.3-1** Use a representation theorem due to Noll [3, p. 35] to conclude that the most general expression of the form of Eq. (3-3) which satisfies our restricted form of the principle of material frame indifference is

$$\mathbf{q}^{(t)} = (\varphi_{(0)} \mathbf{I} + \varphi_{(1)} \bar{\mathbf{D}} + \varphi_{(2)} \bar{\mathbf{D}}^2) \cdot \nabla \bar{T}$$

where

$$\varphi_{(k)} = \varphi_{(k)}(\rho, \hat{c}, l, \operatorname{div} \bar{\mathbf{v}}, \operatorname{tr} \bar{\mathbf{D}}^2, \det \bar{\mathbf{D}}, |\nabla \bar{T}|, \nabla \bar{T} \cdot [\bar{\mathbf{D}} \cdot \nabla \bar{T}], \nabla \bar{T} \cdot [\bar{\mathbf{D}}^2 \cdot \nabla \bar{T}])$$

## [7.1] TIME AVERAGING

**7.1.4 Turbulent energy transfer in a heated section of a tube** An incompressible newtonian fluid with constant viscosity and thermal conductivity is in turbulent flow through a tube of radius $R$. For $z < 0$, the wall of the tube is insulated:

$$\text{At } r = R, \text{ for } z < 0: \quad \frac{\partial \bar{T}}{\partial r} = 0 \tag{4-1}$$

But for $z > 0$, the radial component of the energy flux vector is constant:

$$\text{At } r = R, \text{ for } z > 0: \quad k\frac{\partial \bar{T}}{\partial r} = q \tag{4-2}$$

Very far upstream from the entrance to this heated section, the fluid has a uniform temperature $T_0$:

$$\text{As } z \to -\infty, \text{ for } r \leq R: \quad T \to T_0 \tag{4-3}$$

For the moment let us direct our attention to the temperature distribution in the laminar sublayer and buffer zone.

In Sec. 4.1.5, we examined the velocity distribution that is developed in turbulent flow through a tube. We found there that there is only one nonzero component of the time-averaged velocity vector:

$$\bar{v}_r = \bar{v}_\theta = 0 \qquad \bar{v}_z = \bar{v}_z(r) \tag{4-4}$$

In the laminar sublayer and buffer zone where

$$s^* \equiv \frac{s}{R} \ll 1 \tag{4-5}$$

the velocity distribution can be obtained by integrating

$$\{1 + n^2 v^* s^{**}[1 - \exp(-n^2 v^* s^{**})]\}\frac{dv^*}{ds^{**}} = 1 \tag{4-6}$$

consistent with the boundary condition

$$\text{At } r = R: \quad v^* = 0 \tag{4-7}$$

Here

$$v^* \equiv \frac{\bar{v}_z}{v_0} \equiv \sqrt{\frac{\rho}{S_0}}\,\bar{v}_z \tag{4-8}$$

$$s^{**} \equiv \frac{s}{R} N_{(t)} \tag{4-9}$$

and

$$N_{(t)} \equiv \frac{\rho v_0 R}{\mu} = \frac{\sqrt{\rho S_0}\, R}{\mu} \tag{4-10}$$

The result of this integration for $n = 0.124$ is shown in Fig. 4.1.5-1. Since we are concerned here with an incompressible fluid of constant viscosity, the velocity distribution is unaffected by the temperature distribution. We will consequently assume that Eqs. (4-4) and (4-6) again apply.

Let us assume that the time-averaged temperature distribution in the laminar sublayer and buffer zone has the form

$$\bar{T} = \bar{T}(r,z) \tag{4-11}$$

As indicated in Example 2 of Sec. 7.1.3, Deissler's expression for the turbulent energy flux is currently recommended for the region next to the wall. Consequently, the time-averaged differential energy balance for this region described by inequality (4-5) becomes

$$N_{(t)}^{-1} v^* \frac{\partial T^*}{\partial z^*} = \frac{\partial}{\partial s^{**}} \left( \left\{ \frac{1}{N_{\mathrm{Pr}}} + n^2 s^{**} v^* [1 - \exp(-n^2 s^{**} v^*)] \right\} \frac{\partial T^*}{\partial s^{**}} \right)$$
$$+ \frac{1}{N_{(t)}^2} \frac{\partial}{\partial z^*} \left( \left\{ \frac{1}{N_{\mathrm{Pr}}} + n^2 s^{**} v^* [1 - \exp(-n^2 s^{**} v^*)] \right\} \frac{\partial T^*}{\partial z^*} \right) \tag{4-12}$$

where

$$T^* \equiv \frac{\bar{T} - T_0}{qR/k} \qquad z^* \equiv \frac{z}{R} \tag{4-13}$$

Since we are interested in the turbulent flow, it is not unreasonable to confine our attention to the limit $N_{(t)} \to \infty$, in which case Eq. (4-12) reduces to

$$\frac{\partial}{\partial s^{**}} \left( \left\{ \frac{1}{N_{\mathrm{Pr}}} + n^2 s^{**} v^* [1 - \exp(-n^2 s^{**} v^*)] \right\} \frac{\partial T^*}{\partial s^{**}} \right) = 0 \tag{4-14}$$

Within the heated portion of the tube, this equation must be integrated consistent with Eq. (4-2) which now takes the form

$$\text{At } s^{**} = 0: \quad \frac{\partial T^*}{\partial s^{**}} = -\frac{1}{N_{(t)}} \tag{4-15}$$

The result is

$$\left\{ \frac{1}{N_{\mathrm{Pr}}} + n^2 s^{**} v^* [1 - \exp(-n^2 s^{**} v^*)] \right\} \frac{\partial T^*}{\partial s^{**}} = -\frac{1}{N_{\mathrm{Pr}} N_{(t)}} \tag{4-16}$$

Either the next integration may be carried out using a matching condition with the fully developed turbulent flow at the edge of the buffer zone, or the integration may be expressed in terms of the temperature at the wall of the tube as shown in Fig. 7.1.4-1.

This is perhaps made more obvious when we examine the special case

For $N_{Pr} = 1$:  $\{1 + n^2 s^{**} v^* [1 - \exp(-n^2 s^{**} v^*)]\} \dfrac{\partial T^*}{\partial s^{**}} = -\dfrac{1}{N_{(t)}}$  (4-17)

Comparing this with Eq. (4-6), we find

$$\frac{dT^*}{dv^*} = -\frac{1}{N_{(t)}} \tag{4-18}$$

This is easily integrated consistent with boundary condition (4-7) and

At $s^{**} = 0$:  $T^* = T_w^*$   (4-19)

The result is that, within the laminar sublayer and buffer zone,

$$T^* - T_w^* = -\frac{1}{N_{(t)}} v^* \tag{4-20}$$

This prompts us to say that, for $N_{Pr} = 1$, the velocity and temperature distributions in the region next to the wall are similar.

This analysis is carried somewhat further in Exercise 7.1.4-1, where we find the following expression for the temperature distribution in the fully developed portion of the flow that is near the wall (although outside the laminar sublayer and buffer zone) and somewhat downstream from the entrance to the heated section:

$$T^* - T_w^* = T_1^* - T_w^* - \frac{1}{N_{Pr} N_{(t)}} \frac{1}{\sqrt{\eta_1}} \ln\left(\frac{s^{**}}{s_1^{**}}\right) \tag{4-21}$$

This result is illustrated in Fig. 7.1.4-1.

There is at least one troublesome aspect to this discussion: the axial dependence of $T^*$ has not been spelled out, with the result that no attempt has been made to satisfy boundary conditions (4-1) and (4-3). Clearly $T_w^*$ is a function of axial position in view of boundary condition (4-2).


## REFERENCES

1. Deissler, R. G.: *NACA Rept.* 1210, 1955.
2. Colburn, A. P.: *Trans. A.I.Ch.E.*, **29**:174 (1933).


## EXERCISES

**7.1.4-1** *The temperature distribution in the fully developed turbulent flow*  Continue the discussion in the text, now focusing your attention upon the fully developed portion of the turbulent flow.

(a) Use the Prandtl mixing length theory discussed in Example 1 of Sec. 7.1.3 to determine that in the fully developed portion of the flow near the wall (although outside the buffer zone)

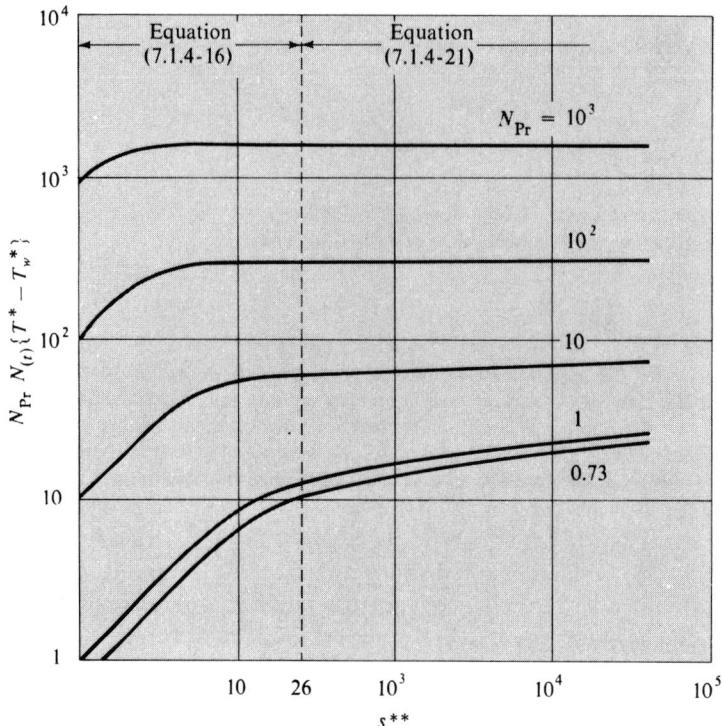

**Fig. 7.1.4-1** Theoretically calculated temperature distribution for energy transfer in turbulent flow through smooth circular tubes; $N_{Pr} = 0.73$ corresponds to air [1, p. 4]. These curves have a positive slope for all values of $s^{**}$. This is not surprising, since their development implies that they should be used only near the wall.

As $N_{(t)} \to \infty$:
$$\frac{\partial}{\partial s^{**}}\left[\left(\frac{1}{N_{Pr}} + \varkappa^* s^{**2}\frac{dv^*}{ds^{**}}\right)\frac{\partial T^*}{\partial s^{**}}\right] = 0 \tag{1}$$

(b) Formulate the appropriate boundary condition at the edge of the buffer zone $s^{**} = s_1^{**}$, and integrate Eq. (1) to find

$$\left(\frac{1}{N_{Pr}} + \varkappa^* s^{**2}\frac{dv^*}{ds^{**}}\right)\frac{\partial T^*}{\partial s^{**}} = -\frac{1}{N_{Pr}N_{(t)}} \tag{2}$$

(c) Prompted by a similar statement in Sec. 4.1.5 assume that

$$\varkappa^* s^{**2}\frac{dv^*}{ds^{**}} \gg \frac{1}{N_{Pr}} \tag{3}$$

Integrate Eq. (2) with the boundary condition

## [7.2] AREA AVERAGING

At $s^{**} = s_1^{**}$: $\quad T^* = T_1^*$ (4)

to find

$$T^* - T_1^* = -\frac{1}{N_{\mathrm{Pr}}N_{(t)}} \frac{\sqrt{\eta_1^*}}{\varkappa^*} \ln \frac{s^{**}}{s_1^{**}} \tag{5}$$

(d) We are again prompted by Sec. 4.1.5 to assume that

$$s_1^{**} = 26 \tag{6}$$

and

$$\varkappa^* = \eta_1^* = (0.36)^2 \tag{7}$$

Under these conditions, verify inequality (3).

Equation (7) also allows us to express Eq. (5) in the form of Eq. (4-21). Notice that this predicts a pointed profile at the center of the tube. This is not surprising, since the development of Eq. (1) (see the discussion in this section) implies that it should be used only near the wall.

(e) Assuming Eq. (7), calculate that

For $N_{\mathrm{Pr}} = 1$:
$$\frac{T^* - T_1^*}{v^* - v_1^*} = -\frac{1}{N_{(t)}} \tag{8}$$

Using the result of this section for the laminar sublayer and buffer zone, calculate that

For $N_{\mathrm{Pr}} = 1$: $\quad T^* - T_w^* = -\frac{v^*}{N_{(t)}}$

$$= -\frac{1}{N_{(t)}}\left(\frac{1}{0.36} \ln s^{**} + 3.8\right) \tag{9}$$

**7.1.4-2** *The Reynolds analogy for $N_{\mathrm{Pr}} = 1$* Multiply Eq. (9) of Exercise 7.1.4-1 by $v^*$ and integrate over the cross section of the tube to find

$$\langle v^* T^* \rangle - \langle v^* \rangle T_w^* = -\frac{\langle v^{*2} \rangle}{N_{(t)}} \tag{1}$$

where we define for any scalar $\varphi$

$$\langle \varphi \rangle \equiv \frac{1}{\pi R^2} \int_0^{2\pi} \int_0^R \varphi r \, dr \, d\theta \tag{2}$$

Rearrange Eq. (1) to conclude that

$$\frac{N_{\mathrm{Nu}}}{N_{\mathrm{Re}}} = \frac{c}{2} \tag{3}$$

Here

$$N_{\mathrm{Nu}} \equiv \frac{2hR}{k} \quad N_{\mathrm{Re}} \equiv \frac{2\rho \langle \bar{v}_z \rangle R}{\mu} \tag{4}$$

and the drag coefficient $c$ is defined as

$$c \equiv \frac{S_0}{\tfrac{1}{2}\rho\langle\bar{v}_z\rangle^2} \tag{5}$$

The film coefficient $h$ has been defined in terms of the bulk mixing temperature $\langle\bar{v}_z\bar{T}\rangle/\langle\bar{v}_z\rangle$:

$$h \equiv \frac{q}{\bar{T}_w - \langle\bar{v}_z\bar{T}\rangle/\langle\bar{v}_z\rangle} \tag{6}$$

Equation (3) represents a special case of an empirical analogy between the film coefficient for heat transfer and the drag coefficient first pointed out by Colburn [2].

## 7.2 AREA AVERAGING

### 7.2.1 Area averaging in energy transfer

The concept of area averaging introduced in Secs. 4.2.1 to 4.2.3 is generalized here to apply to energy transfer. The essence of this approach is to average the differential energy balance over a cross section normal to the gross energy transfer.

As before, you will find here that there are two ways in which empiricisms are used in conjunction with the area-averaged energy balance. When we are primarily concerned with evaluating the time or the position dependence of an area-averaged variable, we will normally be required to make some statement about the energy flux at a bonding phase interface, perhaps employing Newton's law of cooling. This class of problems is illustrated in Sec. 7.2.2. Approximate thermal boundary-layer theory, introduced in Sec. 7.2.3, is probably better known. In these problems an approximate temperature distribution in the boundary layer is postulated in terms of a function $\delta_T(x)$ of the arc length $x$. The area-averaged energy balance for the boundary layer provides an ordinary differential equation for $\delta_T$.

Because of the similarity of the material, I encourage you to review Secs. 4.2.1 to 4.2.3 before proceeding.

### 7.2.2 A straight cooling fin of rectangular profile [1, p. 221]

What follows is a problem in which we are primarily concerned with an area-averaged variable.

Let us determine the efficiency of the cooling fin pictured in Fig. 7.2.2-1. We will assume that we know the temperature of the wall upon which the fin is mounted:

At $z_1 = 0$:  $\quad T = T_w \tag{2-1}$

We will further neglect the energy lost through the end of the fin:

At $z_1 = L$:  $\quad \dfrac{\partial T}{\partial z_1} = 0 \tag{2-2}$

Our first problem is to determine the temperature distribution in the fin. It seems reasonable to recognize that the temperature will be a function of all three coordinates:

$$T = T(z_1, z_2, z_3) \tag{2-3}$$

# [7.2] AREA AVERAGING

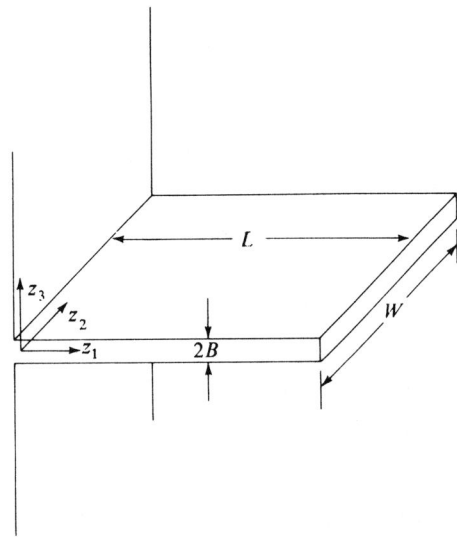

**Fig. 7.2.2-1** A straight cooling fin of rectangular profile.

The differential energy balance to be solved consequently takes the form

$$\frac{\partial^2 T}{\partial z_1^2} + \frac{\partial^2 T}{\partial z_2^2} + \frac{\partial^2 T}{\partial z_3^2} = 0 \qquad (2\text{-}4)$$

This seems to be a situation where we are not primarily concerned with the details of the temperature distribution in the fin. It will be sufficient if we determine the dependence of the average temperature

$$\bar{T} \equiv \frac{1}{2BW} \int_{-B}^{B} \int_{0}^{W} T \, dz_2 \, dz_3 \qquad (2\text{-}5)$$

upon $z_1$. With this definition for the area average, we find from Eq. (2-4)

$$\frac{\partial^2 \bar{T}}{\partial z_1^2} + \frac{1}{2BW} \int_{-B}^{B} \left( \frac{\partial T}{\partial z_2} \bigg|_{z_2=W} - \frac{\partial T}{\partial z_2} \bigg|_{z_2=0} \right) dz_3 \\ + \frac{1}{2BW} \int_{0}^{W} \left( \frac{\partial T}{\partial z_3} \bigg|_{z_3=B} - \frac{\partial T}{\partial z_3} \bigg|_{z_3=-B} \right) dz_2 = 0 \quad (2\text{-}6)$$

Let us apply Newton's law of cooling in the form

At a surface: $\quad \mathbf{q} \cdot \mathbf{n} = h(\bar{T} - T_a) \qquad (2\text{-}7)$

where $T_a$ is the ambient temperature of the surrounding air. This implies that

At $z_2 = 0$: $\quad k \dfrac{\partial T}{\partial z_2} = h(\bar{T} - T_a) \qquad (2\text{-}8)$

At $z_2 = W$: $\quad -k\dfrac{\partial T}{\partial z_2} = h(\bar{T} - T_a)$ \hfill (2-9)

At $z_3 = -B$: $\quad k\dfrac{\partial T}{\partial z_3} = h(\bar{T} - T_a)$ \hfill (2-10)

and

At $z_3 = B$: $\quad -k\dfrac{\partial T}{\partial z_3} = h(\bar{T} - T_a)$ \hfill (2-11)

In view of these relationships, Eq. (2-6) reduces to

$$\frac{\partial^2 \bar{T}}{\partial z_1^2} = \frac{h}{k}\left(\frac{2}{W} + \frac{1}{B}\right)(\bar{T} - T_a) \tag{2-12}$$

We find it convenient to introduce as dimensionless variables

$$\bar{T}^* \equiv \frac{\bar{T} - T_a}{T_w - T_a} \tag{2-13}$$

and

$$z_1^* \equiv \frac{z_1}{L} \tag{2-14}$$

This allows us to write Eq. (2-12) in the considerably simpler form

$$\frac{\partial^2 \bar{T}^*}{\partial z_1^2} = N^2 \bar{T}^* \tag{2-15}$$

where

$$N^2 \equiv \frac{hL^2}{k}\left(\frac{2}{W} + \frac{1}{B}\right) \tag{2-16}$$

From Eqs. (2-1) and (2-2), the appropriate boundary conditions are

At $z_1^* = 0$: $\quad \bar{T}^* = 1$ \hfill (2-17)

and

At $z_1^* = 1$: $\quad \dfrac{d\bar{T}^*}{dz_1^*} = 0$ \hfill (2-18)

The solution to Eq. (2-15) that is consistent with boundary conditions (2-17) and (2-18) can easily be determined to be

$$\bar{T}^* = \frac{\cosh(N[1 - z_1^*])}{\cosh N} \tag{2-19}$$

The effectiveness of a fin has been defined to be [1, p. 235]

## [7.2] AREA AVERAGING

$$\eta \equiv \frac{\text{heat which is actually dissipated by the fin surface}}{\text{heat which would be dissipated if (without change in } h\text{) the fin surface were held at } T_w} \quad (2\text{-}20)$$

From Eq. (2-19), we can readily calculate

$$\eta = \frac{\int_0^W \int_0^L h(\bar{T} - T_a)\, dz_1\, dz_2 + \int_{-B}^{B}\int_0^L h(\bar{T} - T_a)\, dz_1\, dz_3}{\int_0^W \int_0^L h(T_w - T_a)\, dz_1\, dz_2 + \int_{-B}^{B}\int_0^L h(T_w - T_a)\, dz_1\, dz_3}$$

$$= \int_0^L \bar{T}^*\, dz_1^* = \frac{\tanh(N)}{N} \quad (2\text{-}21)$$

For a further discussion of fins, see [1, 2].

## REFERENCES

1. Jakob, Max: "Heat Transfer," vol. 1, Wiley, New York, 1949.
2. Eckert, E. R., and R. M. Drake, Jr.: "Heat and Mass Transfer," 2d ed., McGraw-Hill, New York, 1959.

## EXERCISES

**7.2.2-1** Solve Eq. (2-15) consistent with the boundary conditions (2-17) and (2-18) to find as a result Eq. (2-19).

**7.2.2-2** *A circular cooling fin of rectangular profile* [1, p. 232] Let us consider a circular cooling fin of width $2B$ and radius $R$ mounted on a pipe, the outside radius of which is $\varkappa R$. The temperature at the base of the fin is known:

At $r = \varkappa R$: $\quad T = T_w$

As in the problem discussed in the text, we will neglect the energy lost through the outer edge:

At $r = R$: $\quad \dfrac{\partial T}{\partial r} = 0$

Follow the discussion in the text as a model in determining that the area-averaged temperature distribution in the fin is given by

$$\frac{\bar{T} - T_a}{T_w - T_a} = \frac{I_1(N)K_0(Nr/R) + I_0(Nr/R)K_1(N)}{I_1(N)K_0(N\varkappa) + I_0(N\varkappa)K_1(N)}$$

where

$$\bar{T} \equiv \frac{1}{4\pi B} \int_0^{2\pi}\int_{-B}^{B} T\, dz\, d\theta$$

and

$$N^2 \equiv \frac{hR^2}{kB}$$

By $I_n(x)$ and $K_n(x)$ we mean respectively the $n$th-order modified Bessel functions of the first and second kinds. Determine also that the efficiency of a circular fin of rectangular profile is

$$\eta = \frac{2\varkappa}{N^2(1-\varkappa^2)}\left[\frac{I_1(N)K_1(N\varkappa) - I_1(N\varkappa)K_1(N)}{I_1(N)K_0(N\varkappa) + I_0(N\varkappa)K_1(N)}\right]$$

### 7.2.3 Approximate thermal boundary-layer theory for plane flow past a curved wall

Thermal boundary-layer theory is developed in Secs. 6.7.1 to 6.7.6 for the limiting case as $N_{\text{Re}} \to \infty$. The boundary-layer energy balance is certainly easier to solve than the full differential energy balance, but sometimes the expense required in solving a partial differential equation is not warranted. Approximate boundary-layer theory is tailored for these circumstances.

We will restrict ourselves here to steady-state plane flow past a curved wall for an incompressible newtonian fluid. We will say that all material properties are independent of temperature. One effect of this assumption is that the velocity distribution from Secs. 4.2.1 to 4.2.3 is applicable.

We found in Sec. 6.7.3 that under these circumstances the differential energy balance reduces to

$$v_x^* \frac{\partial T^*}{\partial x^*} + v_y^{**} \frac{\partial T^*}{\partial y^{**}} = \frac{1}{N_{\text{Pr}}} \frac{\partial^2 T^*}{\partial y^{**2}} \tag{3-1}$$

where

$$y^{**} \equiv (N_{\text{Re}})^{\frac{1}{2}} y^* \qquad v_y^{**} \equiv (N_{\text{Re}})^{\frac{1}{2}} v_y^* \tag{3-2}$$

For plane boundary-layer flow past a curved wall, we seek a solution to Eq. (3-1) that is consistent with these conditions:

$$\text{At } y^{**} = 0: \qquad T^* = T_w^* \tag{3-3}$$

and

$$\text{As } y^{**} \to \infty: \qquad T^* \to \tilde{T}^* \tag{3-4}$$

By $\tilde{T}^*$ we mean the dimensionless temperature distribution at the curved wall for the corresponding nonviscous, nonconducting flow:

$$\frac{d\tilde{T}^*}{dx^*} = 0 \tag{3-5}$$

We shall begin by integrating Eq. (3-1) over the thermal boundary layer:

$$\int_0^\infty \left(v_x^* \frac{\partial T^*}{\partial x^*} + v_y^{**} \frac{\partial T^*}{\partial y^{**}} - \frac{1}{N_{\text{Pr}}} \frac{\partial^2 T^*}{\partial y^{**2}}\right) dy^{**} = 0 \tag{3-6}$$

Some rearrangement is in order.

We found in Sec. 4.2.3 that the differential equation of continuity implies

$$v_y^{**} = -\int_0^{y^{**}} \frac{\partial v_x^*}{\partial x^*} dy^{**} \tag{3-7}$$

## [7.2] AREA AVERAGING

This, together with an integration by parts, may be used to express the second term on the left of Eq. (3-6) as

$$\int_0^\infty v_y^* \frac{\partial T^*}{\partial y^{**}} dy^{**} = -\int_0^\infty \frac{\partial T^*}{\partial y^{**}} \left( \int_0^{y^{**}} \frac{\partial v_x^*}{\partial x^*} dy^{**} \right) dy^{**}$$

$$= -\tilde{T}^* \int_0^\infty \frac{\partial v_x^*}{\partial x^*} dy^{**} + \int_0^\infty T^* \frac{\partial v_x^*}{\partial x^*} dy^{**} \qquad (3\text{-}8)$$

The third term on the left of Eq. (3-6) may be integrated directly to find

$$\int_0^\infty \frac{\partial^2 T^*}{\partial y^{**2}} dy^{**} = -\left. \frac{\partial T^*}{\partial y^{**}} \right|_{y^{**}=0} \qquad (3\text{-}9)$$

In reaching this result, it has been necessary to require that

As $y^{**} \to \infty$: $\quad \dfrac{\partial T^*}{\partial y^{**}} \to 0 \qquad (3\text{-}10)$

Equations (3-8) and (3-9) can be used to rewrite Eq. (3-6) as

$$\frac{d}{dx^*} \int_0^\infty v_x^* (T^* - \tilde{T}^*) \, dy^{**} = -\frac{1}{N_{\text{Pr}}} \left. \frac{\partial T^*}{\partial y^{**}} \right|_{y^{**}=0} \qquad (3\text{-}11)$$

In Sec. 4.2.3 we introduced an approximate velocity distribution

$$\frac{v_x^*}{\tilde{v}_x^*} = h(x^*, \eta) \qquad (3\text{-}12)$$

where we defined

$$\eta \equiv \frac{y^{**}}{\delta^{**}(x^*)} = \frac{y}{\delta(x)} \qquad (3\text{-}13)$$

This suggests that we introduce here an approximate temperature distribution

$$\frac{T^* - \tilde{T}^*}{T_w^* - \tilde{T}^*} = g(\eta_T) \qquad (3\text{-}14)$$

By definition,

$$\eta_T \equiv \frac{y^{**}}{\delta_T^{**}(x^*)} = \frac{y}{\delta_T(x)} \qquad (3\text{-}15)$$

Like $\delta(x)$, $\delta_T(x)$ does not have a carefully defined physical meaning, though it is usually loosely thought of as the approximate thickness of the thermal boundary layer. This implies that we require the function $g(\eta_T)$ to be such that

For $\eta_T \geq 1$: $\quad g = 0 \qquad (3\text{-}16)$

These approximate velocity and temperature distributions allow us to evaluate in Eq. (3-11).

$$\int_0^\infty v_x^*(T^* - \tilde{T}^*)\, dy^{**} = \tilde{v}_x^*(T_w^* - \tilde{T}^*)\delta_T^{**}\gamma_1 \tag{3-17}$$

and

$$\left.\frac{\partial T^*}{\partial y^{**}}\right|_{y^{**}=0} = \frac{T_w^* - \tilde{T}^*}{\delta_T^{**}}\gamma_2 \tag{3-18}$$

If we define

$$\Delta \equiv \frac{\delta_T^{**}}{\delta^{**}} = \frac{\delta_T}{\delta} \tag{3-19}$$

then we should understand here that

$$\text{For } \Delta < 1: \quad \gamma_1 \equiv \int_0^1 h(x^*,\eta) g(\eta_T)\, d\eta_T \tag{3-20}$$

$$\text{For } \Delta > 1: \quad \gamma_1 \equiv \frac{1}{\Delta}\int_0^1 [h(x^*,\eta) - 1] g(\eta_T)\, d\eta + \int_0^1 g(\eta_T)\, d\eta_T \tag{3-21}$$

and

$$\gamma_2 \equiv \left.\frac{\partial T^*}{\partial \eta_T}\right|_{\eta_T=0} \tag{3-22}$$

Equations (3-17) and (3-18) enable us to express Eq. (3-11) as

$$\frac{d}{dx^*}[\tilde{v}_x^*(T_w^* - \tilde{T}^*)\delta_T^{**}\gamma_1] = -\frac{T_w^* - \tilde{T}^*}{N_{\text{Pr}}\delta_T^{**}}\gamma_2 \tag{3-23}$$

Influenced by the approach taken in Sec. 4.2.3, let us use a particular approximate temperature distribution

$$\frac{T^* - \tilde{T}^*}{T_w^* - \tilde{T}^*} = g(\eta_T) = a + b\eta_T + c\eta_T^2 + d\eta_T^3 + e\eta_T^4 \tag{3-24}$$

We evaluate the five parameters by requiring Eq. (3-24) to satisfy Eqs. (3-3), (3-4), and (3-10) as well as

$$\text{At } y^{**} = 0: \quad \frac{\partial^2 T^*}{\partial y^{**2}} = 0 \tag{3-25}$$

and

$$\text{As } y^{**} \to \infty: \quad \frac{\partial^2 T^*}{\partial y^{**2}} \to 0 \tag{3-26}$$

As a result we find

$$\frac{T^* - \tilde{T}^*}{T_w^* - \tilde{T}^*} = g(\eta_T) = 1 - 2\eta_T + 2\eta_T^3 - \eta_T^4 \tag{3-27}$$

This enables us to obtain explicit expressions for $\gamma_1$ and $\gamma_2$ in Eqs. (3-20) to (3-22):

## [7.2] AREA AVERAGING

For $\Delta < 1$:
$$\gamma_1 = \frac{2}{15}\Delta - \frac{3}{140}\Delta^3 + \frac{1}{180}\Delta^4$$
$$+ \frac{\Lambda}{6}\left(\frac{1}{15}\Delta - \frac{1}{14}\Delta^2 + \frac{9}{280}\Delta^3 - \frac{1}{180}\Delta^4\right) \quad (3\text{-}28)$$

For $\Delta > 1$:
$$\gamma_1 = \frac{3}{10} - \frac{3}{10\Delta} + \frac{2}{15\Delta^2} - \frac{3}{140\Delta^4} + \frac{1}{180\Delta^5}$$
$$+ \frac{\Lambda}{6}\left(\frac{1}{20\Delta} - \frac{1}{30\Delta^2} + \frac{1}{140\Delta^4} - \frac{1}{504\Delta^5}\right) \quad (3\text{-}29)$$

and
$$\gamma_2 = -2 \quad (3\text{-}30)$$

The quantity $\Lambda$ was defined in Sec. 4.2.3 as

$$\Lambda \equiv \delta^{**2}\frac{d\tilde{v}_x^*}{dx^*} \quad (3\text{-}31)$$

It may be more convenient to write Eq. (3-23) as

$$\frac{d}{dx^*}[\tilde{v}_x^*(T_w^* - \tilde{T}^*)\delta_T^{**}\gamma_1]^2 = \frac{4}{N_{\text{Pr}}}\tilde{v}_x^*(T_w^* - \tilde{T}^*)^2\gamma_1 \quad (3\text{-}32)$$

This should be regarded as a differential equation for $\delta_T^{**}$ to be solved consistent with

At $x^* = 0$:  $\delta_T^{**} = 0$ \quad (3-33)

since the temperature distribution at the wall is assumed to be known. In integrating Eq. (3-32), it is necessary to make a guess as to whether $\Delta$ is less than or greater than unity. We found in Exercise 6.7.2-1 that, for flow past an isothermal flat plate with no viscous dissipation, the velocity and temperature distributions have the same form when $N_{\text{Pr}} = 1$. This means that, for flow past an isothermal flat plate, $\Delta = 1$ for $N_{\text{Pr}} = 1$. Note that Eqs. (3-28) and (3-29) imply $\Delta \neq 1$ for flow past a curved wall ($\Lambda \neq 0$).

There are two cases that may be worthy of special note. For flow past an isothermal curved wall, Eq. (3-32) simplifies to

Isothermal wall: $\quad \dfrac{d}{dx^*}(\tilde{v}_x^*\delta_T^{**}\gamma_1)^2 = \dfrac{4}{N_{\text{Pr}}}\tilde{v}_x^*\gamma_1 \quad (3\text{-}34)$

For flow past an isothermal flat plate, this further reduces to

Isothermal flat plate: $\quad \dfrac{d}{dx^*}(\delta_T^{**}\gamma_1)^2 = \dfrac{4\gamma_1}{N_{\text{Pr}}} \quad (3\text{-}35)$

This last can be integrated consistent with (3-33) to find

$$\delta_T^{**} = 2\left(\frac{x^*}{\gamma_1 N_{\text{Pr}}}\right)^{\frac{1}{2}} \tag{3-36}$$

From Eq. (3-31) of Sec. 4.2.3, for flow past a flat flat at zero incidence,

$$\delta^{**} = 5.836 x^{*\frac{1}{2}} \tag{3-37}$$

In view of (3-19), Eqs. (3-36) and (3-37) provide us with a relationship for $\Delta$:

$$\Delta = 0.3427 (\gamma_1 N_{\text{Pr}})^{-\frac{1}{2}} \tag{3-38}$$

Schlichting (1, p. 293) gives

$$\Delta = N_{\text{Pr}}^{-\frac{1}{3}} \tag{3-39}$$

as an approximate solution to (3-38), in error by no more than 5%.

Recognizing (3-27), we can express the local Nusselt number for flow past an isothermal plate at zero incidence as

$$\begin{aligned} N_{\text{Nu}x} &= \frac{q_y x}{(T_w - \tilde{T})k} \\ &= \frac{2N_{\text{Re}}^{\frac{1}{2}} x^*}{\Delta \delta^{**}} \end{aligned} \tag{3-40}$$

Incorporating (3-37) and (3-39), we conclude

$$N_{\text{Nu}x} = 0.3427 N_{\text{Re}}^{\frac{1}{2}} N_{\text{Pr}}^{\frac{1}{3}} x^{*\frac{1}{2}} \tag{3-41}$$

This compares with a numerical coefficient 0.332 for $0.6 < N_{\text{Pr}} < 10$ (1, p. 285), which follows from the exact solution of Sec. 6.7.2 for heat transfer from an isothermal wall under conditions such that viscous dissipation can be neglected.

### REFERENCE

1. Schlichting, Hermann: "Boundary-Layer Theory," 6th ed., McGraw-Hill, New York, 1968.

### EXERCISES

**7.2.3-1** *Steady-state flow past a body of revolution*  Follow the example of the text as well as Exercise 4.2.3-2 in developing approximate boundary-layer theory for the steady-state flow of an incompressible newtonian fluid past a body of revolution.

(a) Starting with the boundary-layer equations derived in Sec. 6.7.5, determine that

$$\frac{d}{dx^*}[\tilde{v}_x^*(T_w^* - \tilde{T}^*)\delta_T^{**}\gamma_1] + \frac{1}{f^*}\frac{df^*}{dx^*}[\tilde{v}_x^*(T_w^* - \tilde{T}^*)\delta_T^{**}\gamma_1] = -\frac{(T_w^* - \tilde{T}^*)}{N_{\text{Pr}}\delta_T^{**}}$$

where $\gamma_1$ and $\gamma_2$ are again defined by Eqs. (3-20) to (3-22).

(b) Introduce the approximate temperature distribution of Eq. (3-24) and conclude that $\gamma_1$ and $\gamma_2$ are still represented by Eqs. (3-28) to (3-30).

(c) Conclude that

$$\frac{1}{f^{*2}}\frac{d}{dx^*}[f^*\tilde{v}_x^*(T_w^* - \tilde{T}^*)\delta_T^{**}\gamma_1]^2 = \frac{4}{N_{\text{Pr}}}\tilde{v}_x^*(T_w^* - \tilde{T}^*)^2\gamma_1$$

is to be solved as explained in the text consistent with the boundary condition (3-33).

**7.2.3-2** *Steady-state flow past a body of revolution of an incompressible power-model fluid* Repeat Exercise 7.2.3-1 for an incompressible power-model fluid. Determine that all of the above results apply so long as we replace $\Lambda$ by $\Lambda_n$, which was defined in Exercise 4.2.3-3 to be the solution to

$$\Lambda_n = \frac{1}{n}\delta^{**n+1}\tilde{v}_x^{*1-n}\frac{d\tilde{v}_x^*}{dx^*}\left(2 + \frac{\Lambda_n}{6}\right)^{1-n}$$

## 7.3 LOCAL VOLUME AVERAGING

### 7.3.1 Energy transfer in porous media

We are commonly concerned with exothermic chemical reactions in beds of porous pellets impregnated with a catalyst. We have witnessed underground nuclear explosions designed to increase the production of natural gas from the low permeability rock formations in which it sometimes occurs. Both in the west and in Canada there are very large deposits in sandstone formations of nearly solid asphalts; it has been suggested that controlled underground combustion might be used to thermally decompose these asphalts and distill the products in place. All of these processes have one thing in common. In order to analyze them one must understand energy transfer in porous media.

Perhaps the most common approach to energy transfer in porous media is to view the porous solid and whatever gases and liquids it contains as a continuum and then simply to employ the usual differential energy balance discussed in Sec. 5.3.2 [1, p. 328].[1] No attempt is made to distinguish between the energy transfer in the fluid and the energy transfer in the solid. A more serious objection is that the velocity and temperature used are not defined in terms of the actual velocity and temperature distributions in the solid and fluid phases.

Our successful discussion of momentum transfer in Secs. 4.3.1 to 4.3.11 suggests that we take the same point of view here in studying energy transfer. This means that we should begin by developing the local volume average of the differential energy balance.

For simplicity, we shall restrict this discussion to a single incompressible fluid flowing through a stationary, rigid porous medium.

---

[1] Jakob's [2, p. 399] discussion of transpiration cooling is a notable exception. He does distinguish between heat transfer in the solid and heat transfer in the fluid; however, he does not define his temperatures and velocities as averages of the actual velocity and temperature distributions in the solid and fluid phases.

**REFERENCES**

1. Bird, R. B., W. E. Stewart, and E. N. Lightfoot: "Transport Phenomena," 7th printing, Wiley, New York, 1960.
2. Jakob, Max: "Heat Transfer," vol. 2, Wiley, New York, 1957.

### 7.3.2 The local volume average of the differential energy balance

In Secs. 4.3.2, 4.3.4, and 4.3.5 we developed the local volume averages of the equation of continuity and of Cauchy's first law. To be more specific, these were local volume averages of the equation of continuity and of Cauchy's first law for the fluid flowing through the porous structure. For our purposes there, it was not important to examine the implications of conservation of mass and of Euler's first law for the solid material. But here it is essential that we account for energy transfer in both the solid and the fluid. Let us begin by taking the local volume average of the differential energy balance for each phase.

I said that we would limit ourselves to the flow of an incompressible fluid. Let us also neglect the effect of viscous dissipation in the differential energy balance (in principle we should argue that $N_{Br}/N_{Pr}N_{Re} \to 0$; see Sec. 6.4.1) and let us explicitly rule out the possibility of mutual or external energy transmission (radiation). With these restrictions, the local volume average of the differential energy balance in the form of Eq. (E) of Table 5.6.1-1 says for the fluid that

$$\frac{1}{V} \int_{V_{(f)}} \left( \rho_{(f)} \hat{c}_{(f)} \frac{\partial T}{\partial t} + \rho_{(f)} \hat{c}_{(f)} \text{ div } (T\mathbf{v}) + \text{div } \mathbf{q} \right) dV = 0 \tag{2-1}$$

The operations of volume integration and differentiation with respect to time may be interchanged in the first term on the left:

$$\frac{1}{V} \int_{V_{(f)}} \rho_{(f)} \hat{c}_{(f)} \frac{\partial T}{\partial t} dV = \rho_{(f)} \hat{c}_{(f)} \frac{\partial \bar{T}^{(f)}}{\partial t} \tag{2-2}$$

Here we take the heat capacity per unit mass $\hat{c}_{(f)}$ for the fluid to be a constant. If $B$ is any scalar, spatial vector, or second-order tensor field, we will define its local volume average for the fluid as

$$\bar{B}^{(f)} \equiv \frac{1}{V} \int_{V_{(f)}} B \, dV \tag{2-3}$$

The theorem of Sec. 4.3.3 can be used to express the second and third terms on the left of Eq. (2-1) as

$$\frac{1}{V} \int_{V_{(f)}} \rho_{(f)} \hat{c}_{(f)} \text{ div } (T\mathbf{v}) \, dV = \rho_{(f)} \hat{c}_{(f)} \text{ div } (\overline{T\mathbf{v}})^{(f)} \tag{2-4}$$

and

$$\frac{1}{V} \int_{V_{(f)}} \text{div } \mathbf{q} \, dV = \text{div } \bar{\mathbf{q}}^{(f)} + \frac{1}{V} \int_{S_w} \mathbf{q}_{(f)} \cdot \mathbf{n} \, dS \tag{2-5}$$

We define $\mathbf{n}$ to be the unit normal to $S_w$ directed from the fluid phase into the solid.

## [7.3] LOCAL VOLUME AVERAGING

In arriving at Eq. (2-4), we have observed that the velocity vector must be zero at the fluid-solid phase interface $S_w$. We shall presume that Fourier's law describes the thermal behavior of both the fluid and the solid. Further, we shall take the thermal conductivities for both phases to be constants. Applying the theorem from Sec. 4.3.3, we learn

$$\bar{\mathbf{q}}^{(f)} \equiv \frac{1}{V} \int_{V_{(f)}} \mathbf{q}_{(f)} \, dV = -\frac{k_{(f)}}{V} \int_{V_{(f)}} \nabla T_{(f)} \, dV = -k_{(f)} \nabla \bar{T}^{(f)} - \frac{k_{(f)}}{V} \int_{S_w} T\mathbf{n} \, dS \quad (2\text{-}6)$$

In view of Eqs. (2-2) and (2-4) to (2-6), Eq. (2-1) becomes

$$\rho_{(f)} \hat{c}_{(f)} \frac{\partial \bar{T}^{(f)}}{\partial t} + \rho_{(f)} \hat{c}_{(f)} \,\text{div}\, (\overline{T\mathbf{v}})^{(f)} - k_{(f)} \,\text{div}\, (\nabla \bar{T}^{(f)})$$

$$- k_{(f)} \,\text{div}\, \left( \frac{1}{V} \int_{S_w} T\mathbf{n} \, dS \right) + \frac{1}{V} \int_{S_w} \mathbf{q}_{(f)} \cdot \mathbf{n} \, dS = 0 \quad (2\text{-}7)$$

Turning our attention now to the stationary solid phase, essentially the same development yields as the local volume average of the differential energy balance

$$\rho_{(s)} \hat{c}_{(s)} \frac{\partial \bar{T}^{(s)}}{\partial t} - k_{(s)} \,\text{div}\, (\nabla \bar{T}^{(s)}) + k_{(s)} \,\text{div}\, \left( \frac{1}{V} \int_{S_w} T\mathbf{n} \, dS \right) - \frac{1}{V} \int_{S_w} \mathbf{q}_{(s)} \cdot \mathbf{n} \, dS = 0 \quad (2\text{-}8)$$

Comparing this with Eq. (2-7), notice that the last two terms have different signs. This is the result of our definition of $\mathbf{n}$ as the unit normal to $S_w$ directed from the fluid phase into the solid phase.

We shall not concern ourselves here with whatever small temperature differences there are between the solid and fluid phases and will consequently assume (Jakob [1, p. 399] made a somewhat similar assumption in his discussion of transpiration cooling)

$$\frac{1}{\Psi} \bar{T}^{(f)} = \frac{1}{1-\Psi} \bar{T}^{(s)} = \langle T \rangle^{(m)} \quad (2\text{-}9)$$

where we define the local volume average of temperature over both phases as

$$\langle T \rangle^{(m)} \equiv \frac{1}{V} \int_V T \, dV \quad (2\text{-}10)$$

By $\Psi$, I mean the porosity or void fraction

$$\Psi = \frac{V_{(f)}}{V} \quad (2\text{-}11)$$

The sum of Eqs. (2-7) and (2-8) requires

$$[\Psi' \rho_{(f)} \hat{c}_{(f)} + (1 - \Psi') \rho_{(s)} \hat{c}_{(s)}] \frac{\partial \langle T \rangle^{(m)}}{\partial t} + \rho_{(f)} \hat{c}_{(f)} \text{ div} (\langle T \rangle^{(m)} \bar{\mathbf{v}}^{(f)})$$

$$= k_{(f)} \text{ div} [\nabla(\Psi \langle T \rangle^{(m)})] + k_{(s)} \text{ div} \{\nabla[(1 - \Psi)\langle T \rangle^{(m)}]\} + \text{div } \mathbf{h} \quad (2\text{-}12)$$

Here we define the *thermal tortuosity vector*

$$\mathbf{h} \equiv \rho_{(f)} \hat{c}_{(f)} (\langle T \rangle^{(m)} \bar{\mathbf{v}}^{(f)} - \overline{T \mathbf{v}^{(f)}}) + (k_{(f)} - k_{(s)}) \frac{1}{V} \int_{S_w} T \mathbf{n} \, dS \quad (2\text{-}13)$$

In arriving at Eq. (2-12), we have satisfied the jump energy balance of Exercise 5.3.2-1 at the fluid-solid phase interface $S_w$.

The physical meaning of the thermal tortuosity vector is clarified by noting that, if both $T$ and $\Psi$ are independent of position,

$$\mathbf{h} = (k_{(f)} - k_{(s)}) \frac{T}{V} \int_{S_w} \mathbf{n} \, dS$$

$$= 0 \quad (2\text{-}14)$$

In reaching this conclusion, we have used the theorem of Sec. 4.3.3 applied to a constant.

It may also be helpful to express Eq. (2-12) as

$$[\Psi' \rho_{(f)} \hat{c}_{(f)} + (1 - \Psi') \rho_{(s)} \hat{c}_{(s)}] \frac{\partial \langle T \rangle^{(m)}}{\partial t}$$

$$+ \rho_{(f)} \hat{c}_{(f)} \text{ div} (\langle T \rangle^{(m)} \bar{\mathbf{v}}^{(f)}) = -\text{div } \mathbf{q}^{(e)} \quad (2\text{-}15)$$

where

$$\mathbf{q}^{(e)} \equiv -k_{(f)} \nabla(\Psi \langle T \rangle^{(m)}) - k_{(s)} \nabla[(1 - \Psi)\langle T \rangle^{(m)}] - \mathbf{h} \quad (2\text{-}16)$$

may be thought of as an "effective" energy flux vector. If $\Psi$ is independent of position,

$$\mathbf{q}^{(e)} = -[\Psi k_{(f)} + (1 - \Psi) k_{(s)}] \nabla \langle T \rangle^{(m)} - \mathbf{h} \quad (2\text{-}17)$$

Because of the simplifications shown in Eqs. (2-14) and (2-17), we shall direct our attention to structures of uniform porosity in the sections that immediately follow.

In the next section, I discuss the form that I might expect empirical correlations for **h** to take.

## REFERENCE

1. Jakob, Max: "Heat Transfer," vol. 2, Wiley, New York, 1957.

### 7.3.3 Empirical correlations for h
In this section, we give three examples of how experimental data can be used to prepare correlations for **h**, introduced in Sec. 7.3.2. Four points form the foundation for this discussion.

1. The thermal tortuosity vector **h** is frame indifferent:

$$\mathbf{h}^* \equiv \rho_{(f)}\hat{c}_{(f)}(\overline{\psi^{-1}\bar{T}^{*(f)}\overline{\mathbf{v}}^{*(f)}} - \overline{T^*\mathbf{v}^{*(f)}}) + (k_{(f)} - k_{(s)})\frac{1}{V}\int_{S_w} T^*\mathbf{n}^* \, dS$$

$$= \rho_{(f)}\hat{c}_{(f)}\overline{[T^*(\psi^{-1}\overline{\mathbf{v}}^{*(f)} - \mathbf{v}^*)]}^{(f)} + (k_{(f)} - k_{(s)})\frac{1}{V}\int_{S_w} T^*\mathbf{n}^* \, dS$$

$$= \rho_{(f)}\hat{c}_{(f)}\overline{[T\mathbf{Q}\cdot(\psi^{-1}\overline{\bar{\mathbf{v}}}^{(f)} - \mathbf{v})]}^{(f)} + (k_{(f)} - k_{(s)})\frac{1}{V}\int_{S_w} T\mathbf{Q}\cdot\mathbf{n} \, dS = \mathbf{Q}\cdot\mathbf{h} \quad (3\text{-}1)$$

In the second line we observe that the local volume average of a local volume average is simply the local volume average (see Exercise 4.3.8-1); in the third line we employ the frame indifference of temperature and of the velocity difference. Here **Q** is a (possibly) time-dependent, orthogonal, second-order tensor.

2. We assume that the principle of material frame indifference introduced in Sec. 2.3.1 applies to any empirical correlation developed for **h**.
3. The Buckingham-Pi theorem serves to further restrict the form of any expression for **h**.
4. The averaging surface $S$ is so large that **h** may be assumed *not* to be an *explicit* function of position in the porous structure, though it very well may be an *implicit* function of position as a result of its dependence upon other variables.

**EXAMPLE I  NONORIENTED POROUS SOLIDS FILLED WITH A STAGNANT FLUID**

By a stagnant fluid, we mean that there is no gross motion of the fluid:

$$\bar{\mathbf{v}}^{(f)} = 0 \quad (3\text{-}2)$$

This suggests that we might neglect the first term on the right of Eq. (2-13) and write

$$\mathbf{h} \doteq (k_{(f)} - k_{(s)})\frac{1}{V}\int_{S_w} T\mathbf{n} \, dS \quad (3\text{-}3)$$

For geometrically similar nonoriented porous media, **h** might be thought of as a function of the local "particle diameter" $l_0$, the thermal conductivities $k_{(f)}$ and $k_{(s)}$, the porosity $\Psi$, as well as some measure of the local temperature distribution such as $\nabla\langle T\rangle^{(m)}$:

$$\mathbf{h} = \mathbf{h}(l_0, k_{(f)}, k_{(s)}, \Psi, \nabla\langle T\rangle^{(m)}) \quad (3\text{-}4)$$

For the moment, let us fix our attention on the dependence of **h** upon $\nabla\langle T\rangle^{(m)}$:

$$\mathbf{h} = \hat{\mathbf{h}}(\nabla\langle T\rangle^{(m)}) \quad (3\text{-}5)$$

By the principle of material frame indifference, the functional relationship between these two variables should be the same in every frame of reference. This means that

$$\mathbf{h}^* = \mathbf{Q}\cdot\mathbf{h} = \mathbf{Q}\cdot\hat{\mathbf{h}}(\nabla\langle T\rangle^{(m)}) = \hat{\mathbf{h}}(\mathbf{Q}\cdot\nabla\langle T\rangle^{(m)}) \quad (3\text{-}6)$$

or $\hat{\mathbf{h}}$ is an isotropic function [1, p. 22]:

$$\hat{\mathbf{h}}(\nabla \langle T \rangle^{(m)}) = \mathbf{Q}^T \cdot \hat{\mathbf{h}}(\mathbf{Q} \cdot \nabla \langle T \rangle^{(m)}) \tag{3-7}$$

By a representation theorem for a vector-valued isotropic function of one vector [1, p. 35], we may write

$$\mathbf{h} = \hat{\mathbf{h}}(\nabla \langle T \rangle^{(m)}) = H \nabla \langle T \rangle^{(m)} \tag{3-8}$$

where

$$H = \hat{H}(|\nabla \langle T \rangle^{(m)}|) \tag{3-9}$$

Comparing Eqs. (3-8) and (3-9) with Eq. (3-4), we see that

$$H = H(l_0, k_{(f)}, k_{(s)}, \Psi, |\nabla \langle T \rangle^{(m)}|) \tag{3-10}$$

An application of the Buckingham-Pi theorem [2] allows us to conclude that

$$H = k_{(s)} K^* \tag{3-11}$$

Here

$$K^* = K^* \left( \frac{k_{(f)}}{k_{(s)}}, \Psi \right) \tag{3-12}$$

To summarize, Eqs. (3-8), (3-11), and (3-12) represent possibly the simplest form that empirical correlations for the thermal tortuosity vector $\mathbf{h}$ can take in a nonoriented porous medium.

### EXAMPLE 2 NONORIENTED POROUS SOLIDS FILLED WITH A FLOWING FLUID

For geometrically similar nonoriented porous media through which a fluid is flowing, $\mathbf{h}$ might be thought of as a function of the local particle diameter $l_0$, the thermal conductivities $k_{(f)}$ and $k_{(s)}$, the porosity $\Psi$, the density $\rho_{(f)}$ of the fluid, the heat capacity $\hat{c}_{(f)}$ per unit mass of the fluid, the local volume-averaged velocity of the fluid with respect to the local-averaged velocity of the solid $\bar{\mathbf{v}}^{(f)} - \bar{\mathbf{v}}^{(s)}$, as well as some measure of the local temperature distribution such as $\nabla \langle T \rangle^{(m)}$:

$$\mathbf{h} = \mathbf{h}(l_0, k_{(f)}, k_{(s)}, \Psi, \rho_{(f)} \hat{c}_{(f)}, \bar{\mathbf{v}}^{(f)} - \bar{\mathbf{v}}^{(s)}, \nabla \langle T \rangle^{(m)}) \tag{3-13}$$

Let us first examine the dependence of $\mathbf{h}$ upon the two vectors:

$$\mathbf{h} = \hat{\mathbf{h}}(\bar{\mathbf{v}}^{(f)} - \bar{\mathbf{v}}^{(s)}, \nabla \langle T \rangle^{(m)}) \tag{3-14}$$

By the principle of material frame indifference, the functional relationship between these variables should be the same in every frame of reference. This means that

$$\mathbf{h}^* = \mathbf{Q} \cdot \mathbf{h} = \mathbf{Q} \cdot \hat{\mathbf{h}}(\bar{\mathbf{v}}^{(f)} - \bar{\mathbf{v}}^{(s)}, \nabla \langle T \rangle^{(m)})$$
$$= \hat{\mathbf{h}}(\mathbf{Q} \cdot [\bar{\mathbf{v}}^{(f)} - \bar{\mathbf{v}}^{(s)}], \mathbf{Q} \cdot \nabla \langle T \rangle^{(m)}) \tag{3-15}$$

or $\hat{\mathbf{h}}$ is an isotropic function [1, p. 22]:

## [7.3] LOCAL VOLUME AVERAGING

$$\hat{\mathbf{h}}(\bar{\mathbf{v}}^{(f)} - \bar{\mathbf{v}}^{(s)}, \nabla\langle T\rangle^{(m)}) = \mathbf{Q}^T \cdot \hat{\mathbf{h}}(\mathbf{Q} \cdot [\bar{\mathbf{v}}^{(f)} - \bar{\mathbf{v}}^{(s)}], \mathbf{Q} \cdot \nabla\langle T\rangle^{(m)}) \qquad (3\text{-}16)$$

By representation theorems of Spencer and Rivlin [3, sec. 7] and of Smith [4], the most general polynomial isotropic vector function of two vectors has the form

$$\mathbf{h} = H_{(1)}\nabla\langle T\rangle^{(m)} + H_{(2)}(\bar{\mathbf{v}}^{(f)} - \bar{\mathbf{v}}^{(s)}) \qquad (3\text{-}17)$$

Here $H_{(1)}$ and $H_{(2)}$ are scalar-valued polynomials of $|\bar{\mathbf{v}}^{(f)} - \bar{\mathbf{v}}^{(s)}|$, $|\nabla\langle T\rangle^{(m)}|$, and $(\bar{\mathbf{v}}^{(f)} - \bar{\mathbf{v}}^{(s)}) \cdot \nabla\langle T\rangle^{(m)}$ as well as the other scalar variables indicated in Eq. (3-13):

For $i = 1, 2$: $\quad H_{(i)} = H_{(i)}(|\bar{\mathbf{v}}^{(f)} - \bar{\mathbf{v}}^{(s)}|, |\nabla\langle T\rangle^{(m)}|, [\bar{\mathbf{v}}^{(f)} - \bar{\mathbf{v}}^{(s)}] \cdot \nabla\langle T\rangle^{(m)},$

$$l_0, k_{(f)}, k_{(s)}, \Psi, \rho_{(f)}\hat{c}_{(f)}) \qquad (3\text{-}18)$$

(In applying the theorem of Spencer and Rivlin, we identify a vector $\mathbf{b}$ that has covariant components $b_i$ with the skew-symmetric tensor that has contravariant components $\epsilon^{ijk}b_i$. Their theorem requires an additional term in Eq. (3-17) proportional to the vector product $(\bar{\mathbf{v}}^{(f)} - \bar{\mathbf{v}}^{(s)}) \wedge \nabla\langle T\rangle^{(m)}$. This term is not consistent with the requirement that $\mathbf{h}$ be isotropic [1, p. 24] and consequently is dropped.)

An application of the Buckingham-Pi theorem [2] allows us to conclude that

$$H_{(1)} = k_{(s)} H^*_{(1)} \qquad (3\text{-}19)$$

and

$$H_{(2)} = \rho_{(f)}\hat{c}_{(f)} l_0 |\nabla\langle T\rangle^{(m)}| H^*_{(2)} \qquad (3\text{-}20)$$

Here

For $i = 1, 2$: $\quad H^*_{(i)} = H^*_{(i)}\left(\dfrac{k_{(f)}}{k_{(s)}}, N_{\text{Pe}}, \dfrac{[\bar{\mathbf{v}}^{(f)} - \bar{\mathbf{v}}^{(s)}] \cdot \nabla\langle T\rangle^{(m)}}{|\bar{\mathbf{v}}^{(f)} - \bar{\mathbf{v}}^{(s)}||\nabla\langle T\rangle^{(m)}|}, \Psi\right) \qquad (3\text{-}21)$

and

$$N_{\text{Pe}} \equiv \dfrac{\rho_{(f)}\hat{c}_{(f)} |\bar{\mathbf{v}}^{(f)} - \bar{\mathbf{v}}^{(s)}| l_0}{k_{(s)}} \qquad (3\text{-}22)$$

As we might expect, $H_{(2)} = 0$ for $|\nabla\langle T\rangle^{(m)}| = 0$, with the result that $\mathbf{h} = 0$ in this limit.

In summary, Eqs. (3-17) and (3-19) to (3-21) represent perhaps the simplest form that empirical correlations for the thermal tortuosity vector $\mathbf{h}$ can take when a fluid flows through a nonoriented porous medium.

**EXAMPLE 3 ORIENTED POROUS SOLIDS FILLED WITH A STAGNANT FLUID**

We should not expect Eqs. (3-8), (3-11), and (3-12) to represent the thermal tortuosity vector for a porous structure in which particle diameter $l$ is a function of position. For such a structure, Eq. (3-4) must be altered to include a dependence upon additional vector and possibly tensor quantities. For example, one might postulate a dependence of $\mathbf{h}$ upon a local gradient of particle diameter as well as $\nabla\langle T\rangle^{(m)}$:

$$\mathbf{h} = \mathbf{h}(l, k_{(f)}, k_{(s)}, \Psi, \nabla\langle T\rangle^{(m)}, \nabla l) \qquad (3\text{-}23)$$

The principle of material frame indifference and the Buckingham-Pi theorem may be employed as we employed them in Example 2 to conclude that

$$\mathbf{h} = K_{(1)}\nabla\langle T\rangle^{(m)} + K_{(2)}\nabla l \tag{3-24}$$

where

$$K_{(1)} = k_{(s)}K_{(1)}^* \tag{3-25}$$

$$K_{(2)} = k_{(s)}|\nabla\langle T\rangle^{(m)}|\, K_{(2)}^* \tag{3-26}$$

and

$$\text{For } i = 1, 2: \quad K_{(i)}^* = K_{(i)}^*\left(\frac{k_{(f)}}{k_{(s)}}, \frac{\nabla l \cdot \nabla\langle T\rangle^{(m)}}{|\nabla l|\,|\nabla\langle T\rangle^{(m)}|}, |\nabla l|, \Psi\right) \tag{3-27}$$

Equations (3-24) to (3-27) represent perhaps the simplest form that empirical correlations for thermal tortuosity vector **h** can take in a porous structure whose orientation can be described by the local gradient in particle diameter.

## REFERENCES

1. Truesdell, C., and W. Noll: In S. Flügge (ed.), "Handbuch der Physik," vol. 3/3, Springer-Verlag, Berlin, 1965.
2. Brand, Louis: *Arch. Rational Mech. Analysis*, **1**:35 (1957).
3. Spencer, A. J., and R. S. Rivlin: *Arch. Rational Mech. Analysis*, **4**:214 (1959/60).
4. Smith, G. F.: *Arch. Rational Mech. Analysis*, **18**:282 (1965).

### 7.3.4 Summary of results for a nonoriented, uniform-porosity structure

In Section 7.3.2 we found that, when the porosity $\Psi$ is independent of position in the porous medium, the local volume average of the differential energy balance requires

$$[\Psi \rho_{(f)}\hat{c}_{(f)} + (1 - \Psi)\rho_{(s)}\hat{c}_{(s)}]\frac{\partial \langle T\rangle^{(m)}}{\partial t}$$

$$+ \rho_{(f)}\hat{c}_{(f)} \,\text{div}\,(\langle T\rangle^{(m)}\bar{\mathbf{v}}^{(f)}) = -\text{div}\,\mathbf{q}^{(e)} \tag{4-1}$$

where

$$\mathbf{q}^{(e)} \equiv -[\Psi k_{(f)} + (1 - \Psi)k_{(s)}]\nabla\langle T\rangle^{(m)} - \mathbf{h} \tag{4-2}$$

should be thought of as the effective energy flux vector. In arriving at this result, we have neglected any effects attributable to viscous dissipation and radiation.

For a porous medium filled with a stagnant fluid, we suggested in Sec. 7.3.3 that the thermal tortuosity vector **h** can be described by Eqs. (3-8) and (3-11). This result suggests that we may write Eq. (4-2) in terms of an "effective" thermal conductivity $k^{(e)}$:

$$\mathbf{q}^{(e)} = -k^{(e)}\nabla\langle T\rangle^{(m)} \tag{4-3}$$

Here

$$k^{(e)} \equiv \Psi k_{(f)} + (1 - \Psi)k_{(s)} - k_{(s)}K^* \tag{4-4}$$

and

$$K^* = K^*\left(\frac{k_{(f)}}{k_{(s)}}, \Psi\right) \tag{4-5}$$

Experimental studies [1 to 7] for the stagnant fluid case confirm the general form of this expression for the effective thermal conductivity $k^{(e)}$.

When there is simultaneous flow through the porous structure, we suggested in Sec. 7.3.3 that the thermal tortuosity vector **h** might be represented by Eqs. (3-17), (3-19), and (3-20). In these terms, the effective energy flux vector can be expressed as

$$\mathbf{q}^{(e)} = -[\Psi k_{(f)} + (1 - \Psi)k_{(s)} - k_{(s)}K^*_{(1)}]\nabla\langle T\rangle^{(m)} + \rho_{(f)}\hat{c}_{(f)}l_0 |\nabla\langle T\rangle^{(m)}| K^*_{(2)}\bar{\mathbf{v}}^{(f)} \tag{4-6}$$

where

$$\text{For } i = 1, 2: \quad K^*_{(i)} = K^*_{(i)}\left(\frac{k_{(f)}}{k_{(s)}}, N_{\text{Pe}}, \frac{\bar{\mathbf{v}}^{(f)} \cdot \nabla\langle T\rangle^{(m)}}{|\bar{\mathbf{v}}^{(f)}| |\nabla\langle T\rangle^{(m)}|}, \Psi\right) \tag{4-7}$$

and

$$N_{\text{Pe}} \equiv \frac{\rho_{(f)}\hat{c}_{(f)} |\bar{\mathbf{v}}^{(f)}| l_0}{k_{(f)}} \tag{4-8}$$

In arriving at this expression we have assumed that the porous medium is stationary. Sometimes it may be more convenient to think of the effective energy flux vector $\mathbf{q}^{(e)}$ in terms of an effective thermal conductivity tensor $\mathbf{K}^{(e)}$:

$$\mathbf{q}^{(e)} = -\mathbf{K}^{(e)} \cdot \nabla\langle T\rangle^{(m)} \tag{4-9}$$

Here

$$\mathbf{K}^{(e)} \equiv [\Psi k_{(f)} + (1 - \Psi)k_{(s)} - k_{(s)}K^*_{(1)}]\mathbf{I} + \frac{\rho_{(f)}\hat{c}_{(f)}l_0 |\nabla\langle T\rangle^{(m)}| K^*_{(2)}}{\bar{\mathbf{v}}^{(f)} \cdot \nabla\langle T\rangle^{(m)}}\bar{\mathbf{v}}^{(f)}\bar{\mathbf{v}}^{(f)} \tag{4-10}$$

The results of experimental studies [8 to 11] may be thought of in terms of the components of this tensor.

For the sake of simplicity, in what follows we take

$$\text{For } i = 1, 2: \quad K^*_{(i)} = K^*_{(i)}\left(\frac{k_{(f)}}{k_{(s)}}, \Psi\right) \tag{4-11}$$

and write Eq. (4-6) as

$$\mathbf{q}^{(e)} = -\alpha\nabla\langle T\rangle^{(m)} + \beta |\nabla\langle T\rangle^{(m)}| \bar{\mathbf{v}}^{(f)} \tag{4-12}$$

where $\alpha$ and $\beta$ are independent of position in any particular situation.

## REFERENCES

1. Kunii, Daizo, and J. M. Smith: *A.I.Ch.E.J.*, **6**:71 (1960).

2. Kunii, Daizo, and J. M. Smith: *Soc. Petrol. Eng. J.*, **1961**:37.
3. Mischke, R. A., and J. M. Smith: *Ind. Eng. Chem., Fundamentals*, **1**:288 (1962).
4. Masamune, Shinobu, and J. M. Smith: *J. Chem. Eng. Data*, **8**:54 (1963).
5. Huang, Jinn-Huie, and J. M. Smith: *J. Chem. Eng. Data*, **8**:437 (1963).
6. Masamune, Shinobu, and J. M. Smith: *Ind. Eng. Chem., Fundamentals*, **2**:136 (1963).
7. Huang, Jinn-Huie, and J. M. Smith: *Ind. Eng. Chem., Fundamentals*, **2**:189 (1963).
8. Kunii, Daizo, and J. M. Smith: *A.I.Ch.E.J.*, **7**:29 (1961).
9. Willhite, G. P., Daizo Kunii, and J. M. Smith: *A.I.Ch.E.J.*, **8**:340 (1962).
10. Adivarahan, P., Daizo Kunii, and J. M. Smith: *Soc. Petrol. Eng. J.*, **1962**:290.
11. Willhite, G. P., J. S. Dranoff, and J. M. Smith: *Soc. Petrol. Eng. J.*, **1963**:185.

**7.3.5 Transpiration cooling** By *transpiration*, we mean that there is simultaneous flow and energy transfer in a porous structure. Transpiration sometimes can be used to reduce the rate of heat transfer or to decrease the amount of insulation needed for a particular application. The following problem illustrates this idea.

It has been proposed [1, p. 345] that the rate of evaporation of liquefied oxygen in small containers might be reduced by taking advantage of transpiration. The liquid could be stored in a spherical container surrounded by a spherical shell of porous insulating material like that shown in Fig. 7.3.5-1. A small gap is to be left between the container and the insulation, and the opening through the insulation is to be plugged. In operation we can visualize the evaporating oxygen leaving the spherical flask, moving through the gap between the flask and insulation, and then flowing uniformly out through the porous structure. Let us say that we have set as our design criterion that the rate of energy transfer to the oxygen flask should be no more than $\mathcal{Q}$. Oxygen enters the insulation at $r = \varkappa R$ at approximately the boiling point, $T_\varkappa$; the temperature of the oxygen leaving the insulation at $r = R$ is estimated to be $T_1$. The inner radius $\varkappa R$ of the insulation shell is fixed by the diameter of the

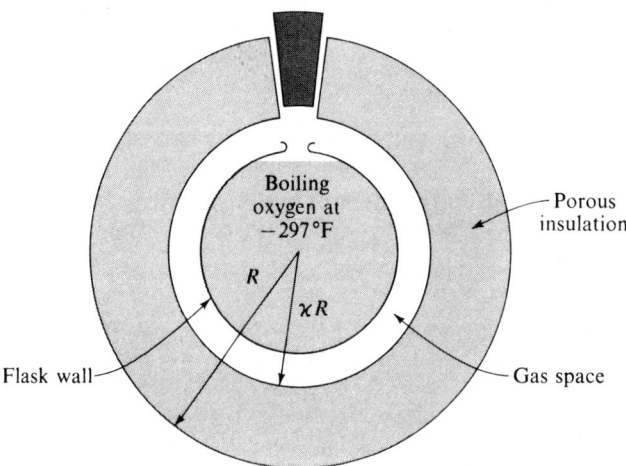

**Fig. 7.3.5-1** Reduction of evaporation rate by transpiration.

[7.3] LOCAL VOLUME AVERAGING     415

oxygen flask. We wish to determine the outer radius $R$ and in this way the thickness of insulation required.

It seems reasonable to begin by assuming in spherical coordinates

$$\bar{v}_r^{(f)} = \bar{v}_r^{(f)}(r) \qquad \bar{v}_\theta^{(f)} = \bar{v}_\varphi^{(f)} = 0 \tag{5-1}$$

and

$$\langle T \rangle^{(m)} = \langle T \rangle^{(m)}(r) \tag{5-2}$$

We estimate the pressure gradient through the porous insulation to be so small that the oxygen may be treated as an incompressible fluid. Under these circumstances, the local volume-averaged equation of continuity in the form of Eq. (4-3) of Sec. 4.3.4 requires

$$\frac{1}{r^2}\frac{d}{dr}(r^2 \bar{v}_r^{(f)}) = 0 \tag{5-3}$$

or

$$\bar{v}_r^{(f)} = \frac{\mathcal{Q}}{4\pi \lambda_{(x)} \rho_{(x)} r^2} \tag{5-4}$$

Here $\lambda_{(x)}$ and $\rho_{(x)}$ are, respectively, the heat of vaporization of oxygen and the density of oxygen evaluated at the temperature of the oxygen flask $T_x$.

We will further assume that everywhere

$$\left| \frac{d\langle T \rangle^{(m)}}{dr} \right| = \frac{d\langle T \rangle^{(m)}}{dr} \tag{5-5}$$

In view of Eqs. (5-1), (5-2), and (5-5), we see that Eq. (4-12) requires

$$q_r^{(e)} = -\alpha \frac{d\langle T \rangle^{(m)}}{dr} + \beta \frac{d\langle T \rangle^{(m)}}{dr} \bar{v}_r^{(f)} \tag{5-6}$$

and

$$q_\theta^{(e)} = q_\varphi^{(e)} = 0 \tag{5-7}$$

As suggested in Sec. 7.3.4, for the sake of simplicity, we will take both $\alpha$ and $\beta$ to be independent of position in the porous insulation.

The local volume-averaged differential energy balance in the form of Eq. (2-15) reduces under these circumstances to

$$\rho_{(f)} \hat{c}_{(f)} \bar{v}_r^{(f)} \frac{d\langle T \rangle^{(m)}}{dr} = -\frac{1}{r^2}\frac{d}{dr}(r^2 q_r^{(e)}) \tag{5-8}$$

Because of Eqs. (5-4) and (5-6), this becomes

$$\frac{d}{dr}\left[ \frac{\hat{c}_{(f)} \mathcal{Q}}{4\pi \lambda_{(x)}} \langle T \rangle^{(m)} - r^2 \left( \alpha \frac{d\langle T \rangle^{(m)}}{dr} - \beta \frac{d\langle T \rangle^{(m)}}{dr} \frac{\mathcal{Q}}{4\pi \lambda_{(x)} \rho_{(x)} r^2} \right) \right] = 0 \tag{5-9}$$

This is more simply expressed in terms of dimensionless variables as

$$\frac{d}{d\xi}\left[\Theta - (\xi^2 - B)\frac{d\Theta}{d\xi}\right] = 0 \qquad (5\text{-}10)$$

where

$$\Theta \equiv \frac{\langle T \rangle^{(m)} - T_1}{T_\varkappa - T_1} \qquad (5\text{-}11)$$

$$\xi \equiv \frac{r}{R_0} \qquad (5\text{-}12)$$

$$R_0 \equiv \frac{\hat{c}_{(f)} \mathscr{Q}}{4\pi\alpha\lambda_{(\varkappa)}} \qquad (5\text{-}13)$$

and

$$B \equiv \frac{\beta \mathscr{Q}}{4\pi\alpha\lambda_{(\varkappa)}\rho_{(\varkappa)}R_0^2} \qquad (5\text{-}14)$$

The corresponding boundary conditions describe the temperatures that exist in the gas phases surrounding the spherical shell of insulation:

$$\text{At } \xi = \xi_\varkappa \equiv \frac{\varkappa R - \epsilon}{R_0}: \quad \Theta = 1 \qquad (5\text{-}15)$$

and

$$\text{At } \xi = \xi_1 \equiv \frac{R + \epsilon}{R_0}: \quad \Theta = 0 \qquad (5\text{-}16)$$

By $\epsilon$, I mean the diameter of the averaging surface $S$.

Integrating once, we have

$$\Theta - (\xi^2 - B)\frac{d\Theta}{d\xi} = C_1 \qquad (5\text{-}17)$$

in which $C_1$ is a constant. Integrating again and applying boundary conditions (5-15) and (5-16), we learn

$$\Theta = \frac{f(\xi) - f(\xi_1)}{f(\xi_\varkappa) - f(\xi_1)} \qquad (5\text{-}18)$$

Here we define

$$\text{For } B \neq 0: \quad f(\xi) \equiv \left|\frac{\sqrt{B} - \xi}{\sqrt{B} + \xi}\right|^{1/(2\sqrt{B})} \qquad (5\text{-}19)$$

and

$$\text{For } B = 0: \quad f(\xi) \equiv \exp\left(-\frac{1}{\xi}\right) \qquad (5\text{-}20)$$

## [7.3] LOCAL VOLUME AVERAGING

The rate of heat transfer to the oxygen flask is specified as

$$\mathcal{Q} = -4\pi(\varkappa R - \epsilon)^2 q_r^{(e)}\Big|_{r=\varkappa R-\epsilon} = 4\pi\alpha R_0(T_\varkappa - T_1)(\xi_\varkappa^2 - B)\frac{d\Theta}{d\xi}\Big|_{\xi=\xi_\varkappa}$$

$$= 4\pi\alpha R_0(T_1 - T_\varkappa)\frac{f(\xi_\varkappa)}{f(\xi_1) - f(\xi_\varkappa)} \quad (5\text{-}21)$$

This is more conveniently expressed as

$$\mathcal{Q}^* = \frac{f(\xi_\varkappa)}{f(\xi_1) - f(\xi_\varkappa)} \quad (5\text{-}22)$$

where we introduce as a definition

$$\mathcal{Q}^* \equiv \frac{\mathcal{Q}}{4\pi\alpha R_0(T_1 - T_\varkappa)} \quad (5\text{-}23)$$

Equation (5-22) can in turn be solved for the unknown $\xi_1$ to find

For $B \neq 0$:   $\xi_1 = \sqrt{B}\,\dfrac{1-A}{1+A}$ (5-24)

and

For $B = 0$:   $\xi_1 = \left(\dfrac{1}{\xi_\varkappa} + \ln\dfrac{\mathcal{Q}^*}{1+\mathcal{Q}^*}\right)^{-1}$ (5-25)

where

$$A \equiv \left(\frac{\mathcal{Q}^*}{1+\mathcal{Q}^*}\right)^{-2\sqrt{B}} \frac{\sqrt{B}+\xi_\varkappa}{\sqrt{B}-\xi_\varkappa} \quad (5\text{-}26)$$

As a basis for comparison, let us ask what happens when there is no transpiration. We require the inner radius of the insulation $\varkappa R$ to be the same, but we allow the outer radius of the insulation $R_{(wt)}$ to take a different value in order to compensate for the lack of transpiration. Equation (5-9) reduces to

$$\frac{d}{dr}\left(r^2 \frac{d\langle T\rangle^{(m)}}{dr}\right) = 0 \quad (5\text{-}27)$$

This is solved consistent with the boundary conditions

At $r = \varkappa R - \epsilon$:   $\langle T\rangle^{(m)} = T_\varkappa$ (5-28)

and

At $r = R_{(wt)} + \epsilon$:   $\langle T\rangle^{(m)} = T_1$ (5-29)

to find

$$\Theta = \frac{1 - \xi_{1(wt)}/\xi}{1 - \xi_{1(wt)}/\xi_\varkappa} \quad (5\text{-}30)$$

This allows us to predict the dimensionless rate of energy transfer to the oxygen flask as

$$\mathscr{Q}^* = \frac{\xi_{1(wt)} \xi_\varkappa}{\xi_{1(wt)} - \xi_\varkappa} \qquad (5\text{-}31)$$

Here we define

$$\xi_{1(wt)} \equiv \frac{R_{(wt)} + \epsilon}{R_0} \qquad (5\text{-}32)$$

Equation (5-31) gives us the relation for

$$\xi_{1(wt)} = \frac{\xi_\varkappa \mathscr{Q}^*}{\mathscr{Q}^* - \xi_\varkappa} \qquad (5\text{-}33)$$

which can be compared with Eqs. (5-24) and (5-25).

Perhaps the effect of transpiration is best appreciated by considering a specific example:

$\mathscr{Q} = 60$ Btu/h
$\varkappa R = 0.5$ ft
$T_\varkappa = -297°$F
$T_1 = 30°$F
$\alpha = 0.02$ Btu/h ft °F
$\beta = 0$
$\hat{c}_{(f)} = 0.22$ Btu/lb$_m$ °F
$\lambda_{(\varkappa)} = 91.7$ Btu/lb$_m$
$\Psi = 0.7$
$\epsilon \doteq 0$ ft

From Eqs. (5-25) and (5-33), we find the thickness of insulation.

With transpiration: $\quad R - \varkappa R = 6$ in

Without transpiration: $\quad R_{(wt)} - \varkappa R = 13$ in

Clearly, transpiration can be of practical importance.

In order to help us further to evaluate this effect, let us ask what happens when the exterior of the insulation is sealed (perhaps with paint) in order to prevent transpiration. Without transpiration, we learn by analogy with Eq. (5-31) that the dimensionless rate of energy transfer to the oxygen flask is

$$\mathscr{Q}^*_{(wt)} = \frac{\xi_1 \xi_\varkappa}{\xi_1 - \xi_\varkappa} \qquad (5\text{-}34)$$

Here we have introduced

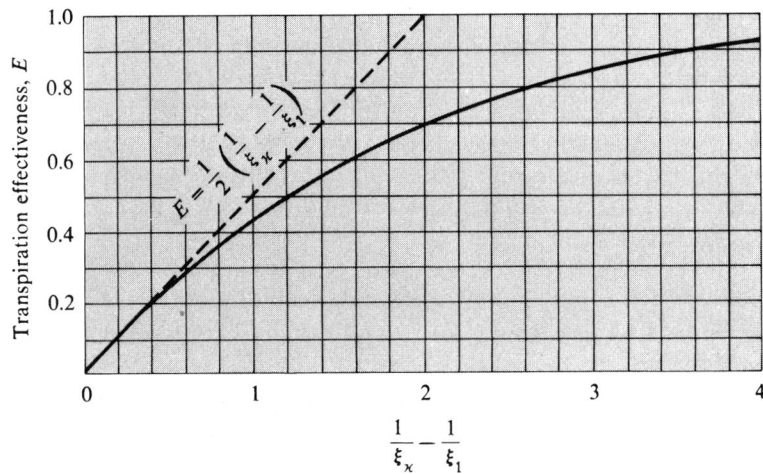

**Fig. 7.3.5-2** Effect of transpiration for $B = 0$. (*From Bird, Stewart, and Lightfoot* [1, *fig.* 10.5-2].)

$$\mathcal{Q}^*_{(wt)} \equiv \frac{\mathcal{Q}_{(wt)}}{4\pi\alpha R_0(T_1 - T_\varkappa)} \tag{5-35}$$

If we define the effectiveness $E$ of the transpiration as

$$E \equiv \frac{\mathcal{Q}_{(wt)} - \mathcal{Q}}{\mathcal{Q}_{(wt)}} \tag{5-36}$$

we can compute from Eqs. (5-22) and (5-34) that

$$E = 1 - \left(\frac{1}{\xi_\varkappa} - \frac{1}{\xi_1}\right) \frac{f(\xi_\varkappa)}{f(\xi_1) - f(\xi_\varkappa)} \tag{5-37}$$

This takes a particularly simple form:

For $B = 0$: $\quad E = 1 - \dfrac{1/\xi_\varkappa - 1/\xi_1}{\exp(1/\xi_\varkappa - 1/\xi_1) - 1} \tag{5-38}$

Equation (5-38) is shown in Fig. 7.3.5-2.

## REFERENCE

1. Bird, R. B., W. E. Stewart, and E. N. Lightfoot: "Transport Phenomena," 7th printing, Wiley, New York, 1960.

## 7.4 INTEGRAL BALANCES

**7.4.1 More on integral balances** In the usual undergraduate course treating thermodynamics for engineers, the problems do not center around determining velocity, temperature, and concentration distributions in materials. Rather, the student is asked to determine gross heat-transfer rates and power requirements or he may be asked to comment on whether a given process is feasible. The student is given as tools the integral mass and energy balances, the mechanical energy balance, and the integral entropy inequality. My purpose here is to indicate how the integral energy balance and the integral entropy inequality arise in the context of the preceding discussion. At the same time I shall indicate how one may obtain two additional limiting cases of the mechanical energy balance.

The discussion in the next few sections is closely related to that of Secs. 4.4.1 to 4.4.15. It might be worthwhile at this point for the reader to review these sections or at least to reread Sec. 4.4.1, which discusses the place of integral balances in engineering.

As I pointed out in Sec. 4.4.1, by an integral balance I mean an equation that describes an accumulation of any quantity in terms of influx and outflow. The implication is that the system over which the balance is made need not be a collection of material particles. The system might be the fluid on the shell side of a heat exchanger aboard a space capsule that is moving along some arbitrary trajectory through space. The system in this example is not a collection of material particles, since fluid is continuously entering and leaving the heat exchanger.

In the sections that follow, I will assume that the system has a volume $V_{(s)}$ and a closed bounding surface $S_{(s)}$. I will denote the velocity of the closed bounding surface as $\mathbf{v}_{(s)}$; in general, this velocity may be a function of position on the surface. I shall refer to that portion of $S_{(s)}$ across which mass moves as being entrances and exits of the system; I shall denote the collection of entrances and exits by $S_{(\text{ent ex})}$.

Many of the ideas associated with integral balances which are presented here originate with Bird [1].

**REFERENCE**

1. Bird, R. B.: *Chem. Eng. Sci.*, 6:123 (1957).

**7.4.2 The integral energy balance** We wish to develop here an energy balance that describes the time rate of change of the energy associated with an arbitrary system.

We will take the same approach that we used in arriving at the integral mass and momentum balances in Secs. 4.4.2 and 4.4.5. In the generalized transport theorem of Sec. 1.3.2, let us identify

$$\Psi = \rho(\hat{U} + \tfrac{1}{2}v^2) \tag{2-1}$$

internal and kinetic energy per unit volume, to obtain

$$\frac{d}{dt}\int_{V_{(s)}} \rho(\hat{U} + \tfrac{1}{2}v^2)\, dV$$

## [7.4] INTEGRAL BALANCES

$$= \int_{V_{(s)}} \frac{\partial}{\partial t}[\rho(\hat{U} + \tfrac{1}{2}v^2)]\, dV + \int_{S_{(s)}} \rho(\hat{U} + \tfrac{1}{2}v^2)(\mathbf{v}_{(s)} \cdot \mathbf{n})\, dS \quad (2\text{-}2)$$

The value of the first integral on the right will not be immediately obvious in any given problem. This difficulty is essentially the same as those we encountered in Secs. 4.4.2, 4.4.5, and 4.4.9.

By means of the differential equation of continuity, the differential energy balance in the form of Eq. (B) in Table 5.6-1 may be written as

$$\frac{\partial}{\partial t}[\rho(\hat{U} + \tfrac{1}{2}v^2)] + \operatorname{div}[\rho(\hat{U} + \tfrac{1}{2}v^2)\mathbf{v}]$$
$$= -\operatorname{div}\mathbf{q} + \operatorname{div}(\mathbf{T}\cdot\mathbf{v}) + \rho(\mathbf{v}\cdot\mathbf{f}) + \rho Q \quad (2\text{-}3)$$

Integrating this over the volume of our arbitrary system, we have

$$\int_{V_{(s)}}\left\{\frac{\partial}{\partial t}[\rho(\hat{U} + \tfrac{1}{2}v^2)] + \operatorname{div}[\rho(\hat{U} + \tfrac{1}{2}v^2)\mathbf{v}] + \operatorname{div}\mathbf{q} - \operatorname{div}(\mathbf{T}\cdot\mathbf{v}) \right.$$
$$\left. - \rho(\mathbf{v}\cdot\mathbf{f}) - \rho Q\right\} dV = 0 \quad (2\text{-}4)$$

The first term on the left is exactly the one needed in Eq. (2-2). The physical meaning of the second, third, and fourth terms is clearer after an application of Green's transformation:

$$\int_{V_{(s)}}\left\{\operatorname{div}(\rho[\hat{U} + \tfrac{1}{2}v^2]\mathbf{v}) + \operatorname{div}\mathbf{q} - \operatorname{div}(\mathbf{T}\cdot\mathbf{v})\right\} dV$$
$$= \int_{S_{(s)}}[\rho(\hat{U} + \tfrac{1}{2}v^2)\mathbf{v}\cdot\mathbf{n} + \mathbf{q}\cdot\mathbf{n} - \mathbf{v}\cdot(\mathbf{T}^T\cdot\mathbf{n})]\, dS \quad (2\text{-}5)$$

Equations (2-4) and (2-5) allow us to express Eq. (2-2) as

$$\frac{d}{dt}\int_{V_{(s)}}\rho(\hat{U} + \tfrac{1}{2}v^2)\, dV = \int_{S_{(s)}}\rho(\hat{U} + \tfrac{1}{2}v^2)(\mathbf{v} - \mathbf{v}_{(s)})\cdot(-\mathbf{n})\, dS$$
$$+ \mathcal{Q} - \int_{S_{(s)}}\mathbf{v}\cdot[\mathbf{T}\cdot(-\mathbf{n})]\, dS + \int_{V_{(s)}}\rho(\mathbf{v}\cdot\mathbf{f} + Q)\, dV \quad (2\text{-}6)$$

The first term on the right is the net rate at which internal and kinetic energy is brought into the system with whatever material is crossing the boundaries. By $\mathcal{Q}$ we mean the rate of (contact) energy transfer to the system across the bounding surfaces of the system:

$$\mathcal{Q} \equiv \int_{S_{(s)}} \mathbf{q}\cdot(-\mathbf{n})\, dS \quad (2\text{-}7)$$

The last term in Eq. (2-6) describes the rate at which work is done on the system by the external force, usually gravity, and the rate of external energy transmission to the system, usually in the form of radiation.

The third term on the right of Eq. (2-6) describes the rate at which work is done by the system upon its bounding surfaces or, in other words, the rate at which work is done by the system upon its surroundings. This is the total work done by the contact forces upon the boundary, whereas we usually speak in terms of the work done beyond that attributable to a constant ambient pressure $p_0$. In order to correct for the effect of $p_0$, we may begin by noting that

$$\int_{S_{(s)}} \mathbf{v} \cdot [\mathbf{T} \cdot (-\mathbf{n})] \, dS = \int_{S_{(s)}} \mathbf{v} \cdot [(\mathbf{T} + p_0 \mathbf{I}) \cdot (-\mathbf{n})] \, dS + \int_{S_{(s)}} p_0 (\mathbf{v} \cdot \mathbf{n}) \, dS \qquad (2\text{-}8)$$

Since $p_0$ is independent of time, we may use the generalized transport theorem to write the second term on the right of Eq. (2-8) as

$$\int_{S_{(s)}} p_0 (\mathbf{v} \cdot \mathbf{n}) \, dS = \int_{V_{(s)}} \frac{\partial p_0}{\partial t} \, dV + \int_{S_{(s)}} p_0 (\mathbf{v} \cdot \mathbf{n}) \, dS$$

$$= \frac{d}{dt} \int_{V_{(s)}} p_0 \, dV - \int_{S_{(s)}} p_0 (\mathbf{v} - \mathbf{v}_{(s)}) \cdot (-\mathbf{n}) \, dS \qquad (2\text{-}9)$$

Using Eqs. (2-8) and (2-9), we may replace Eq. (2-6) by

$$\frac{d}{dt} \int_{V_{(s)}} \rho \left( \hat{U} + \tfrac{1}{2} v^2 + \frac{p_0}{\rho} \right) dV = \int_{S_{(s)}} \rho \left( \hat{U} + \tfrac{1}{2} v^2 + \frac{p_0}{\rho} \right) (\mathbf{v} - \mathbf{v}_{(s)}) \cdot (-\mathbf{n}) \, dS + \mathcal{Q}$$
$$- \int_{S_{(s)}} \mathbf{v} \cdot [(\mathbf{T} + p_0 \mathbf{I}) \cdot (-\mathbf{n})] \, dS + \int_{V_{(s)}} \rho (\mathbf{v} \cdot \mathbf{f} + Q) \, dV \qquad (2\text{-}10)$$

or

$$\frac{d}{dt} \int_{V_{(s)}} \rho \left( \hat{U} + \tfrac{1}{2} v^2 + \frac{p_0}{\rho} \right) dV = \int_{S_{(\text{ent ex})}} \rho (\hat{H} + \tfrac{1}{2} v^2)(\mathbf{v} - \mathbf{v}_{(s)}) \cdot (-\mathbf{n}) \, dS$$
$$+ \mathcal{Q} - \mathcal{W} + \int_{V_{(s)}} \rho (\mathbf{v} \cdot \mathbf{f} + Q) \, dV + \int_{S_{(\text{ent ex})}} [-(P - p_0)(\mathbf{v}_{(s)} \cdot \mathbf{n}) + \mathbf{v} \cdot (\mathbf{S} \cdot \mathbf{n})] \, dS \qquad (2\text{-}11)$$

Introduced in Sec. 4.4.9,

$$\mathcal{W} \equiv \int_{S_{(s)} - S_{(\text{ent ex})}} \mathbf{v} \cdot [(\mathbf{T} + p_0 \mathbf{I}) \cdot (-\mathbf{n})] \, dS \qquad (2\text{-}12)$$

is the work done by the system on the surroundings at the impermeable surfaces of the system. Equation (2-11) will be referred to as a *general form of the integral energy balance*.

This is only one form of the integral energy balance for single-phase systems. If we had started with a different expression for the differential energy balance in Table 5.6.1-1, our final result would be somewhat different. Various forms of the

## [7.4] INTEGRAL BALANCES

integral energy balance appropriate to single-phase systems are present in Table 7.4.2-1.

As we pointed out in Sec. 4.4.2, we are more commonly concerned with multiphase systems. Using the approach and notation of Sec. 4.4.2 and making no additional assumptions, we find that the parallel of Eq. (2-11) for multiphase systems is

$$\frac{d}{dt} \int_{V_{(s)}} \rho \left( \hat{U} + \tfrac{1}{2}v^2 + \frac{p_0}{\rho} \right) dV = \int_{S_{(\text{ent ex})}} \rho(\hat{H} + \tfrac{1}{2}v^2)(\mathbf{v} - \mathbf{v}_{(s)}) \cdot (-\mathbf{n}) \, dS$$

$$+ \mathcal{Q} - \mathcal{W} + \int_{V_{(s)}} \rho(\mathbf{v} \cdot \mathbf{f} + Q) \, dV$$

$$+ \int_{S_{(\text{ent ex})}} [-(P - p_0)(\mathbf{v}_{(s)} \cdot \mathbf{n}) - \mathbf{v} \cdot (\mathbf{S} \cdot \mathbf{n})] \, dS$$

$$+ \int_{S_{(\text{sing})}} [\rho(\hat{U} + \tfrac{1}{2}v^2)(\mathbf{v} \cdot \boldsymbol{\xi} - u_{(\xi)}) + \mathbf{q} \cdot \boldsymbol{\xi} - \mathbf{v} \cdot (\mathbf{T} \cdot \boldsymbol{\xi})] \, dS \quad (2\text{-}13)$$

If we assume that the jump energy balance of Exercise 5.3.2-1 applies, this reduces to the equivalent result for single-phase systems, Eq. (2-11).

As before, if we start with a different form of the differential energy balance from Table 5.6.1-1, we will find a somewhat different form for the integral energy balance applicable to a multiphase system. The various possibilities are shown in Table 7.4.2-2.

Normally a number of assumptions are made in the course of analyzing a particular physical situation with the help of the integral energy balance. The most commonly invoked assumptions are these:

1. No mass transfer across internal phase interfaces.
2. The jump mass, momentum, and energy balances of Sec. 1.3.5, Exercise 2.2.3-1, and Exercise 5.3.2-1 apply.
3. No mutual or external energy transmission.
4. Entrances and exits are fixed in space.
5. Work done by viscous forces (as described by the extra-stress tensor) may be neglected at entrances and exits.

With assumptions 2 through 5, Eq. (2-13) simplifies to

$$\frac{d}{dt} \int_{V_{(s)}} \rho \left( \hat{U} + \tfrac{1}{2}v^2 + \frac{p_0}{\rho} \right) dV = \int_{S_{(\text{ent ex})}} \rho(\hat{H} + \tfrac{1}{2}v^2)(-\mathbf{v} \cdot \mathbf{n}) \, dS$$

$$+ \mathcal{Q} - \mathcal{W} + \int_{V_{(s)}} \rho \mathbf{v} \cdot \mathbf{f} \, dV \quad (2\text{-}14)$$

Table 7.4.2-3 indicates a number of other possibilities.

There are three common types of problems in which the integral energy balance is applied: the rate of energy transfer $\mathcal{Q}$ may be neglected, it may be the unknown

**Table 7.4.2-1** General forms of the integral energy balance applicable to a single-phase system

---

$$\frac{d}{dt} \int_{V_{(s)}} \rho \left( \hat{U} + \tfrac{1}{2} v^2 + \varphi + \frac{p_0}{\rho} \right) dV = \int_{S_{(\text{ent ex})}} \rho(\hat{H} + \tfrac{1}{2} v^2 + \varphi)(\mathbf{v} - \mathbf{v}_{(s)}) \cdot (-\mathbf{n}) \, dS + \mathcal{Q} - \mathcal{W}$$

$$+ \int_{V_{(s)}} \rho Q \, dV + \int_{S_{(\text{ent ex})}} [-(P - p_0)(\mathbf{v}_{(s)} \cdot \mathbf{n}) + \mathbf{v} \cdot (\mathbf{S} \cdot \mathbf{n})] \, dS \quad \text{(A)†}$$

$$\frac{d}{dt} \int_{V_{(s)}} \rho \left( \hat{U} + \tfrac{1}{2} v^2 + \frac{p_0}{\rho} \right) dV = \int_{S_{(\text{ent ex})}} \rho(\hat{H} + \tfrac{1}{2} v^2)(\mathbf{v} - \mathbf{v}_{(s)}) \cdot (-\mathbf{n}) \, dS + \mathcal{Q} - \mathcal{W}$$

$$+ \int_{V_{(s)}} \rho(\mathbf{v} \cdot \mathbf{f} + Q) \, dV + \int_{S_{(\text{ent ex})}} [-(P - p_0)(\mathbf{v}_{(s)} \cdot \mathbf{n}) + \mathbf{v} \cdot (\mathbf{S} \cdot \mathbf{n})] \, dS \quad \text{(B)}$$

$$\frac{d}{dt} \int_{V_{(s)}} \rho \hat{S} \, dV = \int_{S_{(\text{ent ex})}} \rho \hat{S} (\mathbf{v} - \mathbf{v}_{(s)}) \cdot (-\mathbf{n}) \, dS$$

$$+ \int_{S_{(s)}} \frac{1}{T} \mathbf{q} \cdot (-\mathbf{n}) \, dS + \int_{V_{(s)}} \left[ -\frac{1}{T^2} \mathbf{q} \cdot \nabla T + \frac{1}{T} \operatorname{tr}(\mathbf{S} \cdot \nabla \mathbf{v}) + \frac{1}{T} \rho Q \right] dV \quad \text{(C)}$$

$$\frac{d}{dt} \int_{V_{(s)}} \rho \left( \hat{U} + \frac{p_0}{\rho} \right) dV = \int_{S_{(\text{ent ex})}} \rho \left( \hat{U} + \frac{p_0}{\rho} \right) (\mathbf{v} - \mathbf{v}_{(s)}) \cdot (-\mathbf{n}) \, dS$$

$$+ \mathcal{Q} + \int_{V_{(s)}} -[(P - p_0) \operatorname{div} \mathbf{v} + \operatorname{tr}(\mathbf{S} \cdot \nabla \mathbf{v}) + \rho Q] \, dV \quad \text{(D)}$$

*For an incompressible fluid:*

$$\frac{d}{dt} \int_{V_{(s)}} \rho \hat{U} \, dV = \int_{S_{(\text{ent ex})}} \rho \hat{U} (\mathbf{v} - \mathbf{v}_{(s)}) \cdot (-\mathbf{n}) \, dS + \mathcal{Q} + \int_{V_{(s)}} [\operatorname{tr}(\mathbf{S} \cdot \nabla \mathbf{v}) + \rho Q] \, dV \quad \text{(E)}$$

*For an isothermal fluid:*

$$\frac{d}{dt} \int_{V_{(s)}} \rho \hat{S} \, dV = \int_{S_{(\text{ent ex})}} \rho \hat{S} (\mathbf{v} - \mathbf{v}_{(s)}) \cdot (-\mathbf{n}) \, dS + \frac{\mathcal{Q}}{T} + \frac{1}{T} \int_{V_{(s)}} [\operatorname{tr}(\mathbf{S} \cdot \nabla \mathbf{v}) + \rho Q] \, dV \quad \text{(F)}$$

*For an isentropic fluid:*

$$\mathcal{Q} + \int_{V_{(s)}} [\operatorname{tr}(\mathbf{S} \cdot \nabla \mathbf{v}) + \rho Q] \, dV = 0 \quad \text{(G)}$$

*For an isobaric fluid:*

$$\frac{d}{dt} \int_{V_{(s)}} \rho \hat{H} \, dV = \int_{S_{(\text{ent ex})}} \rho \hat{H} (\mathbf{v} - \mathbf{v}_{(s)}) \cdot (-\mathbf{n}) \, dS + \mathcal{Q} + \int_{V_{(s)}} [\operatorname{tr}(\mathbf{S} \cdot \nabla \mathbf{v}) + \rho Q] \, dV \quad \text{(H)}$$

---

† We assume that $\partial \varphi / \partial t = 0$.

## [7.4] INTEGRAL BALANCES

**Table 7.4.2-2** General forms of the integral energy balance applicable to a multiphase system where the jump energy balance of Exercise 5.3.2-1 applies†

$$\frac{d}{dt}\int_{V_{(s)}} \rho\left(\hat{U} + \tfrac{1}{2}v^2 + \varphi + \frac{p_0}{\rho}\right) dV = \int_{S_{(\text{ent ex})}} \rho(\hat{H} + \tfrac{1}{2}v^2 + \varphi)(\mathbf{v} - \mathbf{v}_{(s)}) \cdot (-\mathbf{n}) \, dS + \mathcal{Q} + \mathcal{W}$$

$$+ \int_{V_{(s)}} \rho Q \, dV + \int_{S_{(\text{ent ex})}} [-(P - p_0)(\mathbf{v}_{(s)} \cdot \mathbf{n}) + \mathbf{v} \cdot (\mathbf{S} \cdot \mathbf{n})] \, dS + \int_{S_{(\text{sing})}} [\rho\varphi(\mathbf{v} \cdot \boldsymbol{\xi} - u_{(\xi)})] \, dS \quad (\text{A})\ddagger$$

$$\frac{d}{dt}\int_{V_{(s)}} \rho\left(\hat{U} + \tfrac{1}{2}v^2 + \frac{p_0}{\rho}\right) dV = \int_{S_{(\text{ent ex})}} \rho(\hat{H} + \tfrac{1}{2}v^2)(\mathbf{v} - \mathbf{v}_{(s)}) \cdot (-\mathbf{n}) \, dS$$

$$+ \mathcal{Q} - \mathcal{W} + \int_{V_{(s)}} \rho(\mathbf{v} \cdot \mathbf{f} + Q) \, dV + \int_{S_{(\text{ent ex})}} [-(P - p_0)(\mathbf{v}_{(s)} \cdot \mathbf{n}) + \mathbf{v} \cdot (\mathbf{S} \cdot \mathbf{n})] \, dS \quad (\text{B})$$

$$\frac{d}{dt}\int_{V_{(s)}} \rho\hat{S} \, dV = \int_{S_{(\text{ent ex})}} \rho\hat{S}(\mathbf{v} - \mathbf{v}_{(s)}) \cdot (-\mathbf{n}) \, dS + \int_{S_{(s)}} \frac{1}{T}\mathbf{q} \cdot (-\mathbf{n}) \, dS + \int_{V_{(s)}} \left[-\frac{1}{T^2}\mathbf{q} \cdot \nabla T\right.$$

$$\left.+ \frac{1}{T} \text{tr}(\mathbf{S} \cdot \nabla \mathbf{v}) + \frac{1}{T}\rho Q\right] dV + \int_{S_{(\text{sing})}} \left[\rho\hat{S}(\mathbf{v} \cdot \boldsymbol{\xi} - u_{(\xi)}) + \frac{1}{T}\mathbf{q} \cdot \boldsymbol{\xi}\right] dS \quad (\text{C})$$

$$\frac{d}{dt}\int_{V_{(s)}} \rho\left(\hat{U} + \frac{p_0}{\rho}\right) dV = \int_{S_{(\text{ent ex})}} \rho\left(\hat{U} + \frac{p_0}{\rho}\right)(\mathbf{v} - \mathbf{v}_{(s)}) \cdot (-\mathbf{n}) \, dS + \mathcal{Q} + \int_{V_{(s)}} [-(P - p_0) \, \text{div} \, \mathbf{v}$$

$$+ \text{tr}(\mathbf{S} \cdot \nabla \mathbf{v}) + \rho Q] \, dV + \int_{S_{(\text{sing})}} \left[\rho\left(\hat{U} + \frac{p_0}{\rho}\right)(\mathbf{v} \cdot \boldsymbol{\xi} - u_{(\xi)}) + \mathbf{q} \cdot \boldsymbol{\xi}\right] dS \quad (\text{D})$$

*For incompressible fluids:*

$$\frac{d}{dt}\int_{V_{(s)}} \rho\hat{U} \, dV = \int_{S_{(\text{ent ex})}} \rho\hat{U}(\mathbf{v} - \mathbf{v}_{(s)}) \cdot (-\mathbf{n}) \, dS + \mathcal{Q} + \int_{V_{(s)}} [\text{tr}(\mathbf{S} \cdot \nabla \mathbf{v}) + \rho Q] \, dV$$

$$+ \int_{S_{(\text{sing})}} [\rho\hat{U}(\mathbf{v} \cdot \boldsymbol{\xi} - u_{(\xi)}) + \mathbf{q} \cdot \boldsymbol{\xi}] \, dS \quad (\text{E})$$

*For an isothermal system:*

$$\frac{d}{dt}\int_{V_{(s)}} \rho\hat{S} \, dV = \int_{S_{(\text{ent ex})}} \rho\hat{S}(\mathbf{v} - \mathbf{v}_{(s)}) \cdot (-\mathbf{n}) \, dS + \frac{1}{T}\mathcal{Q} + \frac{1}{T}\int_{V_{(s)}} [\text{tr}(\mathbf{S} \cdot \nabla \mathbf{v}) + \rho Q] \, dV$$

$$+ \int_{S_{(\text{sing})}} \left[\rho\hat{S}(\mathbf{v} \cdot \boldsymbol{\xi} - u_{(\xi)}) + \frac{1}{T}\mathbf{q} \cdot \boldsymbol{\xi}\right] dS \quad (\text{F})$$

*For an isentropic system:*

$$\mathcal{Q} + \int_{V_{(s)}} [\text{tr}(\mathbf{S} \cdot \nabla \mathbf{v}) + \rho Q] \, dV + \int_{S_{(\text{sing})}} [\mathbf{q} \cdot \boldsymbol{\xi}] \, dS = 0 \quad (\text{G})$$

*For an isobaric system:*

$$\frac{d}{dt}\int_{V_{(s)}} \rho\hat{H} \, dV = \int_{S_{(\text{ent ex})}} \rho\hat{H}(\mathbf{v} - \mathbf{v}_{(s)}) \cdot (-\mathbf{n}) \, dS + \mathcal{Q} + \int_{V_{(s)}} [\text{tr}(\mathbf{S} \cdot \nabla \mathbf{v}) + \rho Q] \, dV$$

$$+ \int_{S_{(\text{sing})}} [\rho\hat{H}(\mathbf{v} \cdot \boldsymbol{\xi} - u_{(\xi)}) + \mathbf{q} \cdot \boldsymbol{\xi}] \, dS \quad (\text{H})$$

† Here $u_{(\xi)}$ is the speed of displacement of the phase interface [1, p. 499]; the boldface bracket notation is defined by Eq. (4-7) of Sec. 1.3.4; $S_{(\text{sing})}$ refers to those phase interfaces in the region enclosed by $S_{(s)}$ which do not coincide with $S_{(s)}$.

‡ We assume that $\partial\varphi/\partial t = 0$.

**Table 7.4.2-3 Restricted forms of the integral energy balance applicable to a multiphase system†**

$$\frac{d}{dt}\int_{V_{(s)}} \rho\left(\hat{U} + \tfrac{1}{2}v^2 + \varphi + \frac{p_0}{\rho}\right) dV = \int_{S_{(\text{ent ex})}} \rho(\hat{H} + \tfrac{1}{2}v^2 + \varphi)(-\mathbf{v}\cdot\mathbf{n})\, dS + \mathcal{Q} + \mathcal{W} \quad \text{(A)‡}$$

$$\frac{d}{dt}\int_{V_{(s)}} \rho\left(\hat{U} + \tfrac{1}{2}v^2 + \frac{p_0}{\rho}\right) dV = \int_{S_{(\text{ent ex})}} \rho(\hat{H} + \tfrac{1}{2}v^2)(-\mathbf{v}\cdot\mathbf{n})\, dS + \mathcal{Q} - \mathcal{W} + \int_{V_{(s)}} \rho\mathbf{v}\cdot\mathbf{f}\, dV \quad \text{(B)}$$

$$\frac{d}{dt}\int_{V_{(s)}} \rho\hat{S}\, dV = \int_{S_{(\text{ent ex})}} \rho\hat{S}(-\mathbf{v}\cdot\mathbf{n})\, dS + \int_{S_{(s)}} \frac{1}{T}\mathbf{q}\cdot(-\mathbf{n})\, dS$$
$$+ \int_{V_{(s)}} \left[-\frac{1}{T^2}\mathbf{q}\cdot\nabla T + \frac{1}{T}\operatorname{tr}(\mathbf{S}\cdot\nabla\mathbf{v})\right] dV \quad \text{(C)}$$

$$\frac{d}{dt}\int_{V_{(s)}} \rho\left(\hat{U} + \frac{p_0}{\rho}\right) dV = \int_{S_{(\text{ent ex})}} \rho\left(\hat{U} + \frac{p_0}{\rho}\right)(-\mathbf{v}\cdot\mathbf{n})\, dS$$
$$+ \mathcal{Q} + \int_{V_{(s)}} [-(P - p_0)\operatorname{div}\mathbf{v} + \operatorname{tr}(\mathbf{S}\cdot\nabla\mathbf{v})]\, dV \quad \text{(D)}$$

*For incompressible fluids:*

$$\frac{d}{dt}\int_{V_{(s)}} \rho\hat{U}\, dV = \int_{S_{(\text{ent ex})}} \rho\hat{U}(-\mathbf{v}\cdot\mathbf{n})\, dS + \mathcal{Q} + \int_{V_{(s)}} \operatorname{tr}(\mathbf{S}\cdot\nabla\mathbf{v})\, dV \quad \text{(E)}$$

*For an isothermal system:*

$$\frac{d}{dt}\int_{V_{(s)}} \rho\hat{S}\, dV = \int_{S_{(\text{ent ex})}} \rho\hat{S}(-\mathbf{v}\cdot\mathbf{n})\, dS + \frac{1}{T}\mathcal{Q} + \frac{1}{T}\int_{V_{(s)}} \operatorname{tr}(\mathbf{S}\cdot\nabla\mathbf{v})\, dV \quad \text{(F)}$$

*For an isentropic system:*

$$\mathcal{Q} + \int_{V_{(s)}} \operatorname{tr}(\mathbf{S}\cdot\nabla\mathbf{v})\, dV = 0 \quad \text{(G)}$$

*For an isobaric system:*

$$\frac{d}{dt}\int_{V_{(s)}} \rho\hat{H}\, dV = \int_{S_{(\text{ent ex})}} \rho\hat{H}(-\mathbf{v}\cdot\mathbf{n})\, dS + \mathcal{Q} + \int_{V_{(s)}} \operatorname{tr}(\mathbf{S}\cdot\nabla\mathbf{v})\, dV \quad \text{(H)}$$

† These forms are applicable following assumptions 1 to 5 given in the text.
‡ We assume that $\partial\varphi/\partial t = 0$.

to be determined, or it may be known from previous experimental data. In this last case one employs an empirical correlation of data for $\mathcal{Q}$. In Sec. 7.4.4, we discuss the form that these empirical correlations should take.

### REFERENCE

1. Truesdell, C., and R. A. Toupin: In S. Flügge (ed.), "Handbuch der Physik," vol. 3/1, Springer-Verlag, Berlin, 1960.

## EXERCISES

**7.4.2-1** (a) Starting with Eq. (A) of Table 5.6.1-1, derive Eqs. (A) of Tables 7.4.2-1 to 7.4.2-3.
(b) Starting with Eq. (G) of Table 5.6.1-1, derive Eqs. (C) of Tables 7.4.2-1 to 7.4.2-3.
(c) Starting with Eq. (C) of Table 5.6.1-1, derive Eqs. (D) of Tables 7.4.2-1 to 7.4.2-3.
(d) Derive Eqs. (E), (F), and (H) of Tables 7.4.2-1 to 7.4.2-3.

**7.4.2-2** *An isentropic fluid* Let us define an isentropic fluid to be one in which specific entropy is independent of time and position. Prove that sufficient conditions for a fluid to be isentropic are that its specific internal energy and thermodynamic pressure are independent of time and position.

**7.4.2-3** *More about an isentropic fluid* If one is willing to assume that entropy is independent of time and position, Eqs. (D) of Tables 7.4.2-1, 7.4.2-2, and 7.4.2-3 simplify to Eqs. (G) of these tables
(a) If entropy is independent of time and position, prove that

$$\nabla P = \rho \nabla \hat{H}$$

and

$$\frac{\partial P}{\partial t} = \rho \frac{\partial \hat{H}}{\partial t}$$

(b) Use Green's transformation to prove that

$$\int_{V_{(s)}} -(P - p_0) \operatorname{div} \mathbf{v} \, dV = \int_{S_{(s)}} (P - p_0)(\mathbf{v} - \mathbf{v}_{(s)}) \cdot (-\mathbf{n}) \, dS$$
$$- \int_{S_{(s)}} (P - p_0) \mathbf{v}_{(s)} \cdot \mathbf{n} \, dS + \int_{V_{(s)}} \mathbf{v} \cdot \nabla P \, dV$$

(c) Use Green's transformation and the differential equation of continuity to prove that

$$\int_{V_{(s)}} \mathbf{v} \cdot \nabla P \, dV = \int_{S_{(s)}} \rho \hat{H} \mathbf{v} \cdot \mathbf{n} \, dS + \int_{V_{(s)}} \frac{\partial (\rho \hat{H})}{\partial t} \, dV - \int_{V_{(s)}} \rho \frac{\partial \hat{H}}{\partial t} \, dV$$

(d) Use the generalized transport theorem to find that

$$\int_{V_{(s)}} \mathbf{v} \cdot \nabla P \, dV = \frac{d}{dt} \int_{V_{(s)}} \rho \left( \hat{U} + \frac{p_0}{\rho} \right) dV$$
$$- \int_{S_{(\text{ent ex})}} \rho \hat{H}(\mathbf{v} - \mathbf{v}_{(s)}) \cdot (-\mathbf{n}) \, dS + \int_{S_{(s)}} (P - p_0) \mathbf{v}_{(s)} \cdot \mathbf{n} \, dS$$

(e) Deduce that

$$\int_{V_{(s)}} -(P - p_0) \operatorname{div} \mathbf{v} \, dV = \frac{d}{dt} \int_{V_{(s)}} \rho \left( \hat{U} + \frac{p_0}{\rho} \right) dV$$
$$- \int_{S_{(\text{ent ex})}} \rho \left( \hat{U} + \frac{p_0}{\rho} \right) (\mathbf{v} - \mathbf{v}_{(s)}) \cdot (-\mathbf{n}) \, dS$$

and that Eq. (G) of Table 7.4.2-1 follows from Eq. (D).
The arguments leading to Eqs. (G) of Tables 7.4.2-2 and 7.4.2-3 are very similar.

## 7.4.3 The integral energy balance for turbulent flows

In Sec. 4.4.3, I pointed out that some of the most important applications of the integral balances are to systems, portions of which are in turbulent flow. The remarks made there concerning the integral mass balance are equally applicable here.

We could take one of the approaches illustrated in Secs. 4.4.3, 4.4.6, and 4.4.10 and derive an integral energy balance starting with one of the time-averaged versions of the differential energy balances found in Sec. 7.1.1. See, for example, Exercise 7.4.3-1.

Because of the relative complexity of the time averages of some forms of differential energy balance shown in Table 5.6.1-1, it generally seems to be more practical to time average the various forms of the integral energy balance found in Tables 7.4.2-1 to 7.4.2-3. For either single-phase or multiphase systems, the time average of Eqs. (B) of Tables 7.4.2-1 and 7.4.2-2 yields

$$\frac{d}{dt}\int_{V_{(s)}} \overline{\rho\left(\hat{U} + \tfrac{1}{2}v^2 + \frac{p_0}{\rho}\right)} dV = \int_{S_{(ent\ ex)}} \overline{\rho(\hat{H} + \tfrac{1}{2}v^2)(\mathbf{v} - \mathbf{v}_{(s)}) \cdot (-\mathbf{n})} dS + \overline{\mathcal{Q}}$$

$$- \overline{\mathcal{W}} + \int_{V_{(s)}} \overline{\rho(\mathbf{v} \cdot \mathbf{f} + Q)} dV + \int_{S_{(ent\ ex)}} \overline{[-(P - p_0)(\mathbf{v}_{(s)} \cdot \mathbf{n}) + \mathbf{v} \cdot (\mathbf{S} \cdot \mathbf{n})]} dS \quad (3\text{-}1)$$

From Eq. (B) of Table 7.4.2-3, it is clear that Eq. (3-1) reduces to

$$\frac{d}{dt}\int_{V_{(s)}} \overline{\rho\left(\hat{U} + \tfrac{1}{2}v^2 + \frac{p_0}{\rho}\right)} dV = \int_{S_{(ent\ ex)}} \overline{\rho(\hat{H} + \tfrac{1}{2}v^2)(-\mathbf{v} \cdot \mathbf{n})} dS$$

$$+ \overline{\mathcal{Q}} - \overline{\mathcal{W}} + \int_{V_{(s)}} \overline{\rho(\mathbf{v} \cdot \mathbf{f})} dV \quad (3\text{-}2)$$

under the following restrictions:

1. No mutual or external energy transmission.
2. Entrances and exits are fixed in space.
3. Work done by viscous forces (as described by the extra-stress tensor) may be neglected at entrances and exits.

Somewhat simpler results can be obtained when we restrict ourselves to single-phase or multiphase systems that do not involve fluid-fluid phase interfaces. Under these circumstances, we can use Eqs. (3-7), (3-8), and (3-9) of Sec. 4.4.3 to show that Eqs. (3-1) and (3-2) reduce, respectively, to

$$\frac{d}{dt}\int_{V_{(s)}} \rho\left(\hat{U} + \tfrac{1}{2}v^2 + \frac{p_0}{\rho}\right) dV = \int_{S_{(ent\ ex)}} \overline{\rho(\hat{H} + \tfrac{1}{2}v^2)(\mathbf{v} - \mathbf{v}_{(s)}) \cdot (-\mathbf{n})} dS$$

$$+ \overline{\mathcal{Q}} - \overline{\mathcal{W}} + \int_{V_{(s)}} \overline{\rho(\mathbf{v} \cdot \mathbf{f} + Q)} dV$$

$$+ \int_{S_{(ent\ ex)}} \overline{[-(\bar{P} - p_0)(\mathbf{v}_{(s)} \cdot \mathbf{n}) + \mathbf{v} \cdot (\mathbf{S} \cdot \mathbf{n})]} dS \quad (3\text{-}3)$$

and

[7.4] INTEGRAL BALANCES

$$\frac{d}{dt}\int_{V_{(s)}} \overline{\rho\left(\hat{U} + \tfrac{1}{2}v^2 + \frac{p_0}{\rho}\right)}\, dV = \int_{S_{(ent\ ex)}} \overline{\rho(\hat{H} + \tfrac{1}{2}v^2)(-\mathbf{v}\cdot\mathbf{n})}\, dS$$

$$+ \overline{\mathscr{Q}} - \overline{\mathscr{W}} + \int_{V_{(s)}} \overline{\rho(\mathbf{v}\cdot\mathbf{f})}\, dV \qquad (3\text{-}4)$$

Other forms of the integral energy balance appropriate to systems that are wholly or partially in turbulent flow may be found in a similar fashion by time averaging other equations given in Tables 7.4.2-1 to 7.4.2-3.

### EXERCISES

**7.4.3-1** (a) Time average Eq. (B) of Table 5.6.1-1.

(b) Let us limit ourselves (for reasons explained in Sec. 4.4.3) to single-phase or multiphase systems that do not involve fluid-fluid phase interfaces. Starting with the result of (a), repeat the analysis presented in Sec. 7.4.2 to obtain a result that parallels Eq. (3-3).

**7.4.3-2** (a) Starting with another series of equations from Tables 7.4.2-1 to 7.4.2-3, derive further alternate forms for the time-averaged integral energy balance.

(b) Starting with the form of the differential energy balance appropriate to (a), repeat Exercise 7.4.3-1.

### 7.4.4 Empirical correlations for $\mathscr{Q}$

By means of two illustrations, we indicate here how empirical data correlations for $\mathscr{Q}$ ($\overline{\mathscr{Q}}$ when dealing with turbulent flows), introduced in Sec. 7.4.2, can be constructed. There are three principal ideas to be considered in this discussion.

1. The total rate of contact energy transmission to the system is frame indifferent:

$$\mathscr{Q}^* = \int_{S_{(s)}} \mathbf{q}^* \cdot (-\mathbf{n}^*)\, dS = \int_{S_{(s)}} \mathbf{q}\cdot(-\mathbf{n})\, dS = \mathscr{Q} \qquad (4\text{-}1)$$

2. We assume that the principle of material frame indifference, introduced in Sec. 2.3.1, applies to any empirical correlation developed for $\mathscr{Q}$.
3. The form of any expression for $\mathscr{Q}$ must satisfy the Buckingham-Pi theorem [1].

#### EXAMPLE I  FORCED CONVECTION IN PLANE FLOW PAST A CYLINDRICAL BODY

An infinitely long cylindrical body (the surface of which is traced by a straight line moving parallel to a fixed straight line and intersecting a fixed closed curve) is submerged in a large mass of an incompressible newtonian fluid. The surface temperature of the solid body is $T_0$; the fluid has a nearly uniform temperature $T_\infty$ outside the immediate neighborhood of the body. In a frame of reference that is fixed with respect to the earth, the cylindrical body translates without rotation at a constant velocity $\mathbf{v}_0$; the fluid at a very large distance from the body moves with a uniform velocity $\mathbf{v}_\infty$. The vectors $\mathbf{v}_0$ and $\mathbf{v}_\infty$ are normal to the axis of the cylinder so that we may expect that the fluid moves in a plane flow. One unit vector $\boldsymbol{\alpha}$ is sufficient to describe the orientation of the cylinder with respect to $\mathbf{v}_0$ and $\mathbf{v}_\infty$.

It seems reasonable to say that $\mathscr{Q}$ should be a function of the fluid density $\rho$, the fluid viscosity $\mu$, the fluid heat capacity per unit mass $\hat{c}$, the fluid's thermal conductivity $k$, a length $L$ that is characteristic of the cylinder's cross section, $\mathbf{v}_\infty - \mathbf{v}_0$, $\boldsymbol{\alpha}$, and $\Delta T \equiv T_\infty - T_0$:

$$\mathscr{Q} = f(\rho, \mu, \hat{c}, k, L, \mathbf{v}_\infty - \mathbf{v}_0, \boldsymbol{\alpha}, \Delta T) \tag{4-2}$$

Let us concentrate our attention upon the independent variables $\mathbf{v}_\infty - \mathbf{v}_0$ and $\boldsymbol{\alpha}$:

$$\mathscr{Q} = \tilde{f}(\mathbf{v}_\infty - \mathbf{v}_0, \boldsymbol{\alpha}) \tag{4-3}$$

By the principle of material frame indifference, we conclude that $\tilde{f}$ is a scalar-valued isotropic function of two vectors:

$$\tilde{f}(\mathbf{v}_\infty - \mathbf{v}_0, \boldsymbol{\alpha}) = \tilde{f}(\mathbf{Q} \cdot [\mathbf{v}_\infty - \mathbf{v}_0], \mathbf{Q} \cdot \boldsymbol{\alpha}) \tag{4-4}$$

Here $\mathbf{Q}$ is an orthogonal second-order tensor that describes in part a change of frame. A representation theorem due to Cauchy [2, p. 29] tells us that the most general isotropic scalar-valued function of two vectors has the form

$$\tilde{f}(\mathbf{v}_\infty - \mathbf{v}_0, \boldsymbol{\alpha}) = \tilde{F}(|\mathbf{v}_\infty - \mathbf{v}_0|, [\mathbf{v}_\infty - \mathbf{v}_0] \cdot \boldsymbol{\alpha}) \tag{4-5}$$

This allows us to express Eq. (4-2) as

$$\mathscr{Q} = F(\rho, \mu, \hat{c}, k, L, |\mathbf{v}_\infty - \mathbf{v}_0|, [\mathbf{v}_\infty - \mathbf{v}_0] \cdot \boldsymbol{\alpha}, \Delta T) \tag{4-6}$$

But the Buckingham-Pi theorem [1] requires that this last be of the form

$$N_{\text{Nu}} = N_{\text{Nu}}\left(N_{\text{Re}}, N_{\text{Pr}}, N_{\text{Br}}, \frac{\mathbf{v}_\infty - \mathbf{v}_0}{|\mathbf{v}_\infty - \mathbf{v}_0|} \cdot \boldsymbol{\alpha}\right), \tag{4-7}$$

where the Nusselt, Reynolds, Prandtl, and Brinkman numbers are defined as

$$N_{\text{Nu}} \equiv \frac{\mathscr{Q}}{kL\Delta T} \qquad N_{\text{Re}} \equiv \frac{L\rho|\mathbf{v}_\infty - \mathbf{v}_0|}{\mu}$$
$$N_{\text{Pr}} \equiv \frac{\hat{c}\mu}{k} \qquad N_{\text{Br}} \equiv \frac{\mu|\mathbf{v}_\infty - \mathbf{v}_0|^2}{k\Delta T} \tag{4-8}$$

In the literature, it is traditional to define a heat-transfer coefficient $h$ as

$$h \equiv \frac{\mathscr{Q}}{A\Delta T} \tag{4-9}$$

where $A$ is proportional to $L^2$ and denotes the area available for contact energy transfer. The Nusselt number is in turn expressed in terms of this heat-transfer coefficient:

$$N_{\text{Nu}} = \frac{hL}{k} \tag{4-10}$$

One then computes the rate of contact energy transfer to a system as

$$\mathcal{Q} = hA\Delta T \tag{4-11}$$

estimating the heat-transfer coefficient $h$ from an empirical data correlation of the form of Eq. (4-7).

Most empirical correlations for the Nusselt number are not as general as Eq. (4-7) indicates. Commonly, the Brinkman number $N_{Br}$ is quite small, suggesting that viscous dissipation may be neglected. Further, most studies are for a single orientation of a body (or a set of bodies such as a tube bundle) with respect to a fluid stream. Under these conditions, Eq. (4-7) assumes a simpler form [3; 4, p. 408]:

$$N_{Nu} = N_{Nu}(N_{Re}, N_{Pr}) \tag{4-12}$$

EXAMPLE 2 NATURAL CONVECTION FROM A SUBMERGED SPHERE

Consider a sphere of radius $a$ and surface temperature $T_0$ that is submerged in a large body of a newtonian fluid. Outside the immediate neighborhood of the sphere, the temperature of the fluid has a constant value $T_\infty$. No relative motion between the sphere and the fluid is imposed, although a circulation pattern is set up in the fluid as the result of natural convection.

In addition to saying that $\mathcal{Q}$ is a function of $a$, the fluid viscosity $\mu$, the fluid heat capacity per unit mass $\hat{c}$, and the local magnitude of the acceleration of gravity $g$, we must account for the temperature dependence of the fluid density, since this is the primary cause of natural convection. We can do this by saying that $\mathcal{Q}$ must also be a function of the fluid density $\rho$ and the coefficient of volume expansion of the fluid $\beta$:

$$\beta \equiv -\frac{1}{\rho}\left(\frac{\partial \rho}{\partial T}\right)_P \tag{4-13}$$

evaluated as some temperature characteristic of the experiment. This characteristic temperature is usually chosen to be the "film temperature" $T_f = (T_0 - T_\infty)/2$. To summarize, we postulate that

$$\mathcal{Q} = f(\mu, \hat{c}, a, \Delta T, \rho, \beta) \tag{4-14}$$

This relationship automatically satisfies the principle of material frame indifference.

The Buckingham-Pi theorem requires that Eq. (4-14) assume the general form

$$N_{Nu} = N_{Nu}(\beta \Delta T, N_{Fr}, N_{Pr}, N_{Br}) \tag{4-15}$$

where the Nusselt and Froude numbers are defined as

$$N_{Nu} \equiv \frac{\mathcal{Q}}{ka\Delta T} \qquad N_{Fr} \equiv \frac{\mu^2}{\rho^2 a^3 g} \tag{4-16}$$

The Prandtl and Brinkman numbers have the same forms as given in Eqs. (4-8). (This definition for the Froude number is consistent with the more common definition

$N_{\text{Fr}} = v^2/ga$, if we define the characteristic speed $v$ to be such that the Reynolds number $N_{\text{Re}} = \rho a v/\mu = 1$.)

As suggested in the discussion of Example 1, it is traditional in the literature to define a heat-transfer coefficient $h$ by Eq. (4-9) or in this case

$$h \equiv \frac{\mathcal{Q}}{4\pi a^2 \Delta T} \tag{4-17}$$

The rate of contact energy transfer to the sphere is consequently to be calculated by setting

$$\mathcal{Q} = h 4\pi a^2 \Delta T \tag{4-18}$$

where $h$ is to be determined from the empirical correlation of data of the form of Eq. (4-15).

As one might expect, the Brinkman number $N_{\text{Br}}$ is so small for most situations as to suggest that viscous dissipation may be neglected and that Eq. (4-15) may be approximated by

$$N_{\text{Nu}} = N_{\text{Nu}}(\beta \Delta T, N_{\text{Fr}}, N_{\text{Pr}}) \tag{4-19}$$

Ranz and Marshall [5] found that a data correlation of this form

$$N_{\text{Nu}} \equiv \frac{ha}{k} = 2 + 0.60 \left(\frac{\beta \Delta T}{N_{\text{Fr}}}\right)^{\frac{1}{4}} N_{\text{Pr}}^{\frac{1}{3}} \tag{4-20}$$

agrees well with available experimental data for $(\beta \Delta T/N_{\text{Fr}})^{\frac{1}{4}} N_{\text{Pr}}^{\frac{1}{3}} < 200$.

**REFERENCES**
1. Brand, Louis: *Arch. Rational Mech. Analysis*, **1**:35 (1957).
2. Truesdell, C., and W. Noll: In S. Flügge (ed.), "Handbuch der Physik," vol. 3/3, Springer-Verlag, Berlin, 1965.
3. Kays, W. M., and A. L. London: "Compact Heat Exchangers," 2d ed., McGraw-Hill, New York, 1964.
4. Bird, R. B., W. E. Stewart, and E. N. Lightfoot: "Transport Phenomena," 7th printing, Wiley, New York, 1960.
5. Ranz, W. E., and W. R. Marshall, Jr.: *Chem. Eng. Progr.*, **48**:141 and 173 (1952).

**7.4.5 The integral energy balance: An example**[1] An insulated, evacuated tank is connected through a valved pipe to a constant-pressure line containing an ideal diatomic gas maintained at a constant pressure $P_0$ and a constant temperature $T_0$. We may assume that the constant-pressure heat capacity per unit mass is

$$\hat{c}_P = \frac{7}{2}\frac{R}{M} \tag{5-1}$$

where $R$ is the gas law constant and $M$ is the molecular weight of the gas. The

---
[1] This example was suggested by Prof. G. M. Brown, Northwestern University.

## [7.4] INTEGRAL BALANCES

volume of the tank $\mathscr{V}$ is known.

The valve between the tank and the line is suddenly opened, admitting the gas to the tank. We wish to compute the amount and the temperature of the gas in the tank when the pressure in the tank is $P_{(\text{final})}$.

Let us choose our system to be the gas in the tank. This system has only one entrance, through the pipeline, and no exits. The boundary of the system is fixed in space.

For simplicity, we shall neglect the effects of turbulence.

From the integral mass balance,

$$\frac{d\mathscr{M}}{dt} = -\int_{S_{(\text{ent})}} \rho \mathbf{v} \cdot \mathbf{n}\, dS \tag{5-2}$$

Here $\mathscr{M}$ indicates the mass of the gas in the tank:

$$\mathscr{M} \equiv \int_{V_{(s)}} \rho\, dV \tag{5-3}$$

which is a function of time. By $S_{(\text{ent})}$ we refer to the tank's entrance.

If we neglect the changes in kinetic energy and potential energy and if we make the assumptions noted in Sec. 7.4.2, the integral energy balance of the form of Eq. (A) of Table 7.4.2-3 requires

$$\frac{d\mathscr{U}}{dt} = -\int_{S_{(\text{ent})}} \rho \hat{H} \mathbf{v} \cdot \mathbf{n}\, dS \tag{5-4}$$

By $\mathscr{U}$, we mean the internal energy associated with the system:

$$\mathscr{U} \equiv \int_{V_{(s)}} \rho \hat{U}\, dV \tag{5-5}$$

The specific enthalpy should be very nearly a constant with respect to position in the entrance, so that Eq. (5-4) may be combined with Eq. (5-2) to obtain

$$\frac{d\mathscr{U}}{dt} = -\hat{H}_{(\text{ent})} \int_{S_{(\text{ent})}} \rho \mathbf{v} \cdot \mathbf{n}\, dS = \hat{H}_{\text{ent}} \frac{d\mathscr{M}}{dt} \tag{5-6}$$

Furthermore, the specific enthalpy of the incoming gas should be nearly a constant as a function of time. With this assumption, Eq. (5-6) may be integrated to find that at any particular time

$$\mathscr{U}_{(\text{final})} = \hat{H}_{(\text{ent})} \mathscr{M}_{(\text{final})} \tag{5-7}$$

If the gas is well mixed in the tank, this last becomes

$$\hat{V}_{(\text{final})} = \hat{H}_{(\text{ent})} \tag{5-8}$$

It is easily shown that

$$\hat{c}_P \equiv T\left(\frac{\partial \hat{S}}{\partial T}\right)_P = \left(\frac{\partial \hat{H}}{\partial T}\right)_P = \left(\frac{\partial \hat{U}}{\partial T}\right)_P + P\left(\frac{\partial \hat{V}}{\partial T}\right)_P \quad (5\text{-}9)$$

For an ideal gas,
$$\hat{U} = \hat{U}(T) \quad (5\text{-}10)$$
and
$$p\hat{V} = \frac{RT}{M} \quad (5\text{-}11)$$

Consequently, Eq. (5-9) may be rearranged to read
$$\frac{d\hat{U}}{dT} = \hat{c}_P - \frac{R}{M} \quad (5\text{-}12)$$

In view of Eq. (5-1), this last may be easily integrated to find that
$$\hat{U}_{(\text{final})} - \hat{U}_{(\text{ent})} = \frac{5}{2}\frac{R}{M}(T_{(\text{final})} - T_{(\text{ent})}) \quad (5\text{-}13)$$

By means of Eqs. (5-8), (5-11), and (5-13), we get
$$\hat{U}_{(\text{final})} - \hat{U}_{(\text{ent})} = \hat{H}_{(\text{ent})} - \hat{U}_{(\text{ent})} = P_{(\text{ent})}\hat{V}_{(\text{ent})}$$
$$= \frac{R}{M}T_{(\text{ent})} = \frac{5}{2}\frac{R}{M}(T_{(\text{final})} - T_{(\text{ent})}) \quad (5\text{-}14)$$

or
$$T_{(\text{final})} = \tfrac{7}{5}T_{(\text{ent})} = \tfrac{7}{5}T_0 \quad (5\text{-}15)$$

It follows immediately that the mass of gas in the tank at the end of the process is
$$\mathcal{M}_{(\text{final})} = \frac{5}{7}\frac{M}{R}\frac{P_{(\text{final})}}{T_0}\mathcal{V} \quad (5\text{-}16)$$

## EXERCISES

**7.4.5-1** (a) In arriving at Eqs. (5-2) and (5-4), any effects attributable to turbulence were neglected. Discuss the solution to the example problem of this section, attempting to account for the effects of turbulence.

(b) Justify dropping certain terms in Eq. (A) of Table 7.4.2-3 to arrive at Eq. (5-4).

**7.4.5-2[1]** An insulated, evacuated vessel of 100 ft³ capacity is connected to a steam line that transports 200 lb$_f$/in² absolute of saturated steam. The valve between the steam line and the vessel is opened to admit steam to the vessel. Compute the amount of steam in the vessel and its temperature, if the valve is closed when the pressure in the vessel reaches 50 lb$_f$/in² absolute.

**7.4.5-3[1]** A well-insulated tank of 100 ft³ capacity is connected to a valved line containing an unlimited supply of saturated steam at 200 lb$_f$/in² absolute. Initially the tank is filled with saturated steam at 14.7 lb$_f$/in² absolute. At a given time, the valve is opened and steam is allowed to flow from the supply line into the tank. The valve is closed again when the tank pressure reaches 100 lb/in² absolute. How much steam flows into the tank? What is the final temperature of the steam in the tank?

**7.4.5-4**[1] A valve connects two identical insulated vessels each with a volume of 1 ft$^3$. The valve is initially closed. One vessel contains steam at 600°F and 1000 lb$_f$/in$^2$ absolute; the other vessel is evacuated.

(a) The valve is opened and flow occurs until the pressures are identical. The valve is then closed. Determine the final temperature in each vessel.

(b) If the valve is left open so that thermal equilibrium is eventually attained between the two vessels, calculate the final temperature.

**7.4.5-5**[1] An exit high-pressure line from a chemical reactor contains almost pure Freon-12 at 1000 lb$_f$/in$^2$ absolute and 280°F, according to the instruments. However, an operator opens a small valve in the side of the line and claims the recorded temperature must be wrong, since the gas issuing from the line feels cold. Resolve this question if possible by appropriate calculations.

**7.4.5-6**[1] Freon-12 at 10 lb$_f$/in$^2$ gauge and 0°F enters our plant at the rate of 1000 lb$_m$/h. An adiabatic compressor raises the pressure to 200 lb$_f$/in$^2$ at which point a thermometer in the line reads 240°F. A heat exchanger cools the stream to 140°F, the pressure remaining constant. Calculate the power input to the compressor.

**7.4.5-7**[1] Steam flows in a large uninsulated pipeline at the rate of 10 lb$_m$/s. At the first station, temperature and pressure gauges indicate 600°F and 1000 lb$_f$/in$^2$ absolute, respectively. Downstream at the second station the pressure is 750 lb$_f$/in$^2$ absolute and the quality is 0.85 (fraction vapor). What is the temperature at the second station (in degrees Fahrenheit) and what is the rate at which heat is transferred to the pipeline?

**7.4.5-8**[1] Oxygen passes through an adiabatic steady-flow compressor at the rate of 1000 lb$_m$/h, entering as a saturated vapor at 36.6 lb$_f$/in$^2$ absolute and emerging at 17.5 atm and 175°K. Determine the shaft power per unit of mass of $O_2$ and the required motor horsepower.

**7.4.5-9**[1] Carbon dioxide passes through an adiabatic steady-flow turbine at the rate of 500 lb$_m$/min. It enters at 400 lb$_f$/in$^2$ absolute and 100°F and emerges as a saturated vapor at 100 lb$_f$/in$^2$ absolute. What is the shaft-work output per pound-mass of carbon dioxide, and what is the power delivered by the turbine?

**7.4.5-10** One pound-mole of a monatomic ideal gas is enclosed in a rigid, insulated container at 500°R and 10 atm. Another identical container is evacuated and connected to the first by a short-valved pipe. The valve is opened to allow the gas to expand slowly into the second container. As soon as the pressures in the two containers are equal, the valve is closed. Compute the final pressure and temperature of the gas in each container.

**7.4.5-11** *Stagnation temperature* A "total temperature" probe illustrated in Fig. 7.4.5-1 can be used to measure the temperature $T_1$ of an ideal gas that moves with a speed $v_1$. A portion of the gas

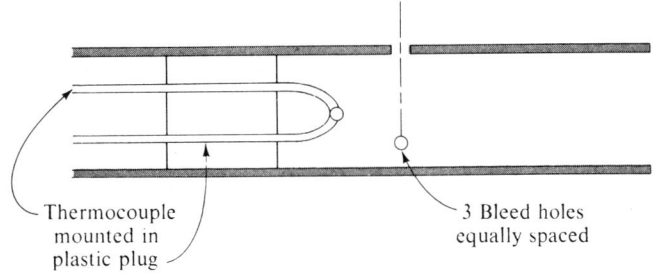

**Fig. 7.4.5-1** A total temperature probe.

---

[1] Suggested by Prof. W. W. Graessley, Northwestern University.

enters the open end of the probe and decelerates to nearly zero velocity before slowly leaking out of the bleed holes. You may assume that the tubing surrounding the thermocouple has a small thermal conductivity.

Determine that the temperature $T_2$ measured by the thermocouple is

$$T_2 = T_1 + \frac{(v_1)^2}{2\hat{c}_P}$$

*Hint:* Choose your system so as to include a portion of the gas stream that is undisturbed by the presence of the probe. What other assumptions must be made?

**7.4.6 More about the mechanical energy balance** In Sec. 4.4.9 we derived one of the general forms of the mechanical energy balance for single-phase systems:

$$\frac{d}{dt}\int_{V_{(s)}} \rho(\tfrac{1}{2}v^2 + \varphi)\, dV = \int_{S_{(\text{ent ex})}} \rho\left(\tfrac{1}{2}v^2 + \varphi + \frac{P - p_0}{\rho}\right)(\mathbf{v} - \mathbf{v}_{(s)})\cdot(-\mathbf{n})\, dS$$

$$+ \int_{V_{(s)}} (P - p_0)\,\text{div}\,\mathbf{v}\, dV - \mathscr{W} - \mathscr{E} + \int_{S_{(\text{ent ex})}} [-(P - p_0)\mathbf{v}_{(s)} \cdot \mathbf{n}$$

$$+ \mathbf{v}\cdot(\mathbf{S}\cdot\mathbf{n})]\, dS \quad (6\text{-}1)$$

This is not one of the more useful forms of the mechanical energy balance in the sense that the value of the second integral on the right will not be immediately obvious for most situations. We get around this difficulty in Sec. 4.4.9 by restricting ourselves to incompressible fluids, in which case

$$\int_{V_{(s)}} (P - p_0)\,\text{div}\,\mathbf{v}\, dV = 0 \quad (6\text{-}2)$$

Our object here is to indicate that there are other useful forms of the mechanical energy balance that are not restricted to incompressible fluids. We illustrate this point by devoting the bulk of this section to isothermal fluids.

For the moment, let us concentrate on rearranging the second integral on the right of Eq. (6-1) for the case of an isothermal fluid, that is, for a fluid in which temperature is independent of time and position. We can begin by using Green's transformation to find that

$$\int_{V_{(s)}} (P - p_0)\,\text{div}\,\mathbf{v}\, dV = \int_{V_{(s)}} \text{div}\,([P - p_0]\mathbf{v})\, dV - \int_{V_{(s)}} \mathbf{v}\cdot\nabla P\, dV$$

$$= -\int_{S_{(s)}} (P - p_0)(\mathbf{v} - \mathbf{v}_{(s)})\cdot(-\mathbf{n})\, dS + \int_{S_{(s)}} (P - p_0)\mathbf{v}_{(s)}\cdot\mathbf{n}\, dS - \int_{V_{(s)}} \mathbf{v}\cdot\nabla P\, dV$$

$$(6\text{-}3)$$

Since

$$\nabla P = \rho\nabla\hat{G} \quad (6\text{-}4)$$

we can use another application of Green's transformation and the differential equation of continuity to arrive at

## [7.4] INTEGRAL BALANCES

$$\int_{V_{(s)}} \mathbf{v} \cdot \nabla P \, dV = \int_{V_{(s)}} \rho \mathbf{v} \cdot \nabla \hat{G} \, dV = \int_{V_{(s)}} [\text{div}\,(\rho \hat{G} \mathbf{v}) - \hat{G}\,\text{div}\,(\rho \mathbf{v})]\, dV$$

$$= \int_{S_{(s)}} \rho \hat{G} \mathbf{v} \cdot \mathbf{n}\, dS + \int_{V_{(s)}} \hat{G}\frac{\partial \rho}{\partial t}\, dV = \int_{S_{(s)}} \rho \hat{G} \mathbf{v} \cdot \mathbf{n}\, dS + \int_{V_{(s)}} \frac{\partial(\rho \hat{G})}{\partial t}\, dV$$

$$- \int_{V_{(s)}} \rho\frac{\partial \hat{G}}{\partial t}\, dV \quad (6\text{-}5)$$

Noting that

$$\rho\frac{\partial \hat{G}}{\partial t} = \frac{\partial P}{\partial t} = \frac{\partial (P - P_0)}{\partial t} \quad (6\text{-}6)$$

we may use the generalized transport theorem to express Eq. (6-5)$_4$ as

$$\int_{V_{(s)}} \mathbf{v} \cdot \nabla P\, dV = \frac{d}{dt}\int_{V_{(s)}} \rho\!\left(\hat{A} + \frac{P_0}{\rho}\right) dV - \int_{S_{(\text{ent ex})}} \rho \hat{G}(\mathbf{v} - \mathbf{v}_{(s)}) \cdot (-\mathbf{n})\, dS$$

$$+ \int_{S_{(s)}} (P - P_0)\mathbf{v}_{(s)} \cdot \mathbf{n}\, dS \quad (6\text{-}7)$$

Substituting this into Eq. (6-3), we have

$$\int_{V_{(s)}} (P - P_0)\,\text{div}\,\mathbf{v}\, dV = -\frac{d}{dt}\int_{V_{(s)}} \rho\!\left(\hat{A} + \frac{P_0}{\rho}\right) dV$$

$$+ \int_{S_{(\text{ent ex})}} \rho\!\left(\hat{A} + \frac{P_0}{\rho}\right)(\mathbf{v} - \mathbf{v}_{(s)}) \cdot (-\mathbf{n})\, dS \quad (6\text{-}8)$$

This last allows us to express Eq. (6-1) as

$$\frac{d}{dt}\int_{V_{(s)}} \rho\!\left(\hat{A} + \tfrac{1}{2}v^2 + \varphi + \frac{P_0}{\rho}\right) dV = \int_{S_{(\text{ent ex})}} \rho(\hat{G} + \tfrac{1}{2}v^2 + \varphi)(\mathbf{v} - \mathbf{v}_{(s)}) \cdot (-\mathbf{n})\, dS$$

$$-\mathscr{W} - \mathscr{E} + \int_{S_{(\text{ent ex})}} [-(P - P_0)\mathbf{v}_{(s)} \cdot \mathbf{n} + \mathbf{v} \cdot (\mathbf{S} \cdot \mathbf{n})]\, dS \quad (6\text{-}9)$$

a general form of the mechanical energy balance appropriate to single-phase systems composed of isothermal fluids.

Still other general forms of the integral mechanical energy balance can be derived. They are presented in Table 7.4.6-1.

The generalization of these relations to multiphase systems closely follows the discussion of Sec. 4.4.9. The results are given in Table 7.4.6-2.

The most common applications of the integral mechanical energy balances are to systems such that the following restrictions are reasonable:

1. There is no mass transfer across internal phase interfaces.
2. The jump mass and momentum balances of Sec. 1.3.5 and Exercise 2.2.3-1 apply.
3. Entrances and exits are fixed in space.

**Table 7.4.6-1** General forms of the integral mechanical energy balance applicable to a single-phase system

$$\frac{d}{dt}\int_{V_{(s)}} \rho(\tfrac{1}{2}v^2 + \varphi)\,dV = \int_{S_{(\text{ent ex})}} \rho\left(\tfrac{1}{2}v^2 + \varphi + \frac{P - p_0}{\rho}\right)(\mathbf{v} - \mathbf{v}_{(s)}) \cdot (-\mathbf{n})\,dS$$

$$+ \int_{V_{(s)}} (P - p_0)\,\text{div}\,\mathbf{v}\,dV - \mathscr{W} - \mathscr{E} + \int_{S_{(\text{ent ex})}} [-(P - p_0)\mathbf{v}_{(s)} \cdot \mathbf{n} + \mathbf{v} \cdot (\mathbf{S} \cdot \mathbf{n})]\,dS \quad \text{(A)†}$$

$$\frac{d}{dt}\int_{V_{(s)}} \tfrac{1}{2}\rho v^2\,dV = \int_{S_{(\text{ent ex})}} \rho\left(\tfrac{1}{2}v^2 + \frac{P - p_0}{\rho}\right)(\mathbf{v} - \mathbf{v}_{(s)}) \cdot (-\mathbf{n})\,dS$$

$$+ \int_{V_{(s)}} (P - p_0)\,\text{div}\,\mathbf{v}\,dV - \mathscr{W} - \mathscr{E} + \int_{V_{(s)}} \mathbf{v} \cdot \rho \mathbf{f}\,dV$$

$$+ \int_{S_{(\text{ent ex})}} [-(P - p_0)\mathbf{v}_{(s)} \cdot \mathbf{n} + \mathbf{v} \cdot (\mathbf{S} \cdot \mathbf{n})]\,dS \quad \text{(B)}$$

*For an incompressible fluid:*

$$\frac{d}{dt}\int_{V_{(s)}} \rho(\tfrac{1}{2}v^2 + \varphi)\,dV = \int_{S_{(\text{ent ex})}} \rho\left(\tfrac{1}{2}v^2 + \varphi + \frac{P - p_0}{\rho}\right)(\mathbf{v} - \mathbf{v}_{(s)}) \cdot (-\mathbf{n})\,dS$$

$$- \mathscr{W} - \mathscr{E} + \int_{S_{(\text{ent ex})}} [-(p - p_0)\mathbf{v}_{(s)} \cdot \mathbf{n} + \mathbf{v} \cdot (\mathbf{S} \cdot \mathbf{n})]\,dS \quad \text{(C)†}$$

*For an isothermal fluid:*

$$\frac{d}{dt}\int_{V_{(s)}} \rho\left(\hat{A} + \tfrac{1}{2}v^2 + \varphi + \frac{p_0}{\rho}\right)dV = \int_{S_{(\text{ent ex})}} \rho(\hat{G} + \tfrac{1}{2}v^2 + \varphi)(\mathbf{v} - \mathbf{v}_{(s)}) \cdot (-\mathbf{n})\,dS$$

$$- \mathscr{W} - \mathscr{E} + \int_{S_{(\text{ent ex})}} [-(P - p_0)\mathbf{v}_{(s)} \cdot \mathbf{n} + \mathbf{v} \cdot (\mathbf{S} \cdot \mathbf{n})]\,dS \quad \text{(D)†‡}$$

*For an isentropic fluid:*

$$\frac{d}{dt}\int_{V_{(s)}} \rho\left(\hat{U} + \tfrac{1}{2}v^2 + \varphi + \frac{p_0}{\rho}\right)dV = \int_{S_{(\text{ent ex})}} \rho(\hat{H} + \tfrac{1}{2}v^2 + \varphi)(\mathbf{v} - \mathbf{v}_{(s)}) \cdot (-\mathbf{n})\,dS$$

$$- \mathscr{W} - \mathscr{E} + \int_{S_{(\text{ent ex})}} [-(P - p_0)\mathbf{v}_{(s)} \cdot \mathbf{n} + \mathbf{v} \cdot (\mathbf{S} \cdot \mathbf{n})]\,dS \quad \text{(E)†‡}$$

*For an isobaric fluid:*

$$\frac{d}{dt}\int_{V_{(s)}} \rho(\tfrac{1}{2}v^2 + \varphi)\,dV = \int_{S_{(\text{ent ex})}} \rho(\tfrac{1}{2}v^2 + \varphi)(\mathbf{v} - \mathbf{v}_{(s)}) \cdot (-\mathbf{n})\,dS$$

$$- \mathscr{W} - \mathscr{E} + \int_{S_{(s)} - S_{(\text{ent ex})}} (P - p_0)\mathbf{v} \cdot \mathbf{n}\,dS + \int_{S_{(\text{ent ex})}} \mathbf{v} \cdot (\mathbf{S} \cdot \mathbf{n})\,dS \quad \text{(F)†}$$

† We assume that $\partial \varphi / \partial t = 0$.
‡ Applicable to systems where composition is independent of time and position.

## [7.4] INTEGRAL BALANCES

**Table 7.4.6-2** General forms of the integral mechanical energy balance applicable to a multiphase system†

---

$$\frac{d}{dt}\int_{V_{(s)}} \rho(\tfrac{1}{2}v^2 + \varphi)\, dV = \int_{S_{(\text{ent ex})}} \rho\left(\tfrac{1}{2}v^2 + \varphi + \frac{P - p_0}{\rho}\right)(\mathbf{v} - \mathbf{v}_{(s)}) \cdot (-\mathbf{n})\, dS$$

$$+ \int_{V_{(s)}} (P - p_0)\,\text{div}\,\mathbf{v}\, dV - \mathscr{W} - \mathscr{E} + \int_{S_{(\text{ent ex})}} [-(P - p_0)\mathbf{v}_{(s)} \cdot \mathbf{n} + \mathbf{v} \cdot (\mathbf{S} \cdot \mathbf{n})]\, dS$$

$$+ \int_{S_{(\text{sing})}} [\rho(\tfrac{1}{2}v^2 + \varphi)(\mathbf{v} \cdot \boldsymbol{\xi} - u_{(\xi)}) - \mathbf{v} \cdot (\mathbf{T} + p_0\mathbf{I}) \cdot \boldsymbol{\xi}]\, dS \quad (A)\ddagger$$

$$\frac{d}{dt}\int_{V_{(s)}} \tfrac{1}{2}\rho v^2\, dV = \int_{S_{(\text{ent ex})}} \rho\left(\tfrac{1}{2}v^2 + \frac{P - p_0}{\rho}\right)(\mathbf{v} - \mathbf{v}_{(s)}) \cdot (-\mathbf{n})\, dS$$

$$+ \int_{V_{(s)}} (P - p_0)\,\text{div}\,\mathbf{v}\, dV - \mathscr{W} - \mathscr{E} + \int_{V_{(s)}} \mathbf{v} \cdot \rho \mathbf{f}\, dV$$

$$+ \int_{S_{(\text{ent ex})}} [-(P - p_0)\mathbf{v}_{(s)} \cdot \mathbf{n} + \mathbf{v} \cdot (\mathbf{S} \cdot \mathbf{n})]\, dS$$

$$+ \int_{S_{(\text{sing})}} [\tfrac{1}{2}\rho v^2(\mathbf{v} \cdot \boldsymbol{\xi} - u_{(\xi)}) - \mathbf{v} \cdot (\mathbf{T} + p_0\mathbf{I}) \cdot \boldsymbol{\xi}]\, dS \quad (B)$$

*For incompressible fluids:*

$$\frac{d}{dt}\int_{V_{(s)}} \rho(\tfrac{1}{2}v^2 + \varphi)\, dV = \int_{S_{(\text{ent ex})}} \rho\left(\tfrac{1}{2}v^2 + \varphi + \frac{P - p_0}{\rho}\right)(\mathbf{v} - \mathbf{v}_{(s)}) \cdot (-\mathbf{n})\, dS$$

$$- \mathscr{W} - \mathscr{E} + \int_{S_{(\text{ent ex})}} [-(p - p_0)\mathbf{v}_{(s)} \cdot \mathbf{n} + \mathbf{v} \cdot (\mathbf{S} \cdot \mathbf{n})]\, dS$$

$$+ \int_{S_{(\text{sing})}} [\rho(\tfrac{1}{2}v^2 + \varphi)(\mathbf{v} \cdot \boldsymbol{\xi} - u_{(\xi)}) - \mathbf{v} \cdot (\mathbf{T} + p_0\mathbf{I}) \cdot \boldsymbol{\xi}]\, dS \quad (C)\ddagger$$

*For an isothermal system:*

$$\frac{d}{dt}\int_{V_{(s)}} \rho\left(\hat{A} + \tfrac{1}{2}v^2 + \varphi + \frac{p_0}{\rho}\right) dV = \int_{S_{(\text{ent ex})}} \rho(\hat{G} + \tfrac{1}{2}v^2 + \varphi)(\mathbf{v} - \mathbf{v}_{(s)}) \cdot (-\mathbf{n})\, dS$$

$$- \mathscr{W} - \mathscr{E} + \int_{S_{(\text{ent ex})}} [-(P - p_0)\mathbf{v}_{(s)} \cdot \mathbf{n} + \mathbf{v} \cdot (\mathbf{S} \cdot \mathbf{n})]\, dS$$

$$+ \int_{S_{(\text{sing})}} [\rho(\hat{A} + \tfrac{1}{2}v^2 + \varphi)(\mathbf{v} \cdot \boldsymbol{\xi} - u_{(\xi)}) - \mathbf{v} \cdot (\mathbf{T} \cdot \boldsymbol{\xi})]\, dS \quad (D)\ddagger\S$$

---

(*Continued on next page*)

**Table 7.4.6-2** (*Continued*)

*For an isentropic system:*

$$\frac{d}{dt}\int_{V_{(s)}} \rho\left(\hat{U} + \tfrac{1}{2}v^2 + \varphi + \frac{p_0}{\rho}\right) dV = \int_{S_{\text{(ent ex)}}} \rho(\hat{H} + \tfrac{1}{2}v^2 + \varphi)(\mathbf{v} - \mathbf{v}_{(s)}) \cdot (-\mathbf{n})\, dS$$

$$- \mathscr{W} - \mathscr{E} + \int_{S_{\text{(ent ex)}}} [-(P - p_0)\mathbf{v}_{(s)} \cdot \mathbf{n} + \mathbf{v} \cdot (\mathbf{S} \cdot \mathbf{n})]\, dS$$

$$+ \int_{S_{\text{(sing)}}} [\![\rho(\hat{U} + \tfrac{1}{2}v^2 + \varphi)(\mathbf{v} \cdot \boldsymbol{\xi} - u_{(\xi)}) - \mathbf{v} \cdot (\mathbf{T} \cdot \boldsymbol{\xi})]\!]\, dS \quad \text{(E)}\ddagger\S$$

*For an isobaric system:*

$$\frac{d}{dt}\int_{V_{(s)}} \rho(\tfrac{1}{2}v^2 + \varphi)\, dV = \int_{S_{\text{(ent ex)}}} \rho(\tfrac{1}{2}v^2 + \varphi)(\mathbf{v} - \mathbf{v}_{(s)}) \cdot (-\mathbf{n})\, dS$$

$$- \mathscr{W} - \mathscr{E} + \int_{S_{(s)} - S_{\text{(ent ex)}}} (P - p_0)\mathbf{v} \cdot \mathbf{n}\, dS + \int_{S_{\text{(ent ex)}}} \mathbf{v} \cdot (\mathbf{S} \cdot \mathbf{n})\, dS$$

$$+ \int_{S_{\text{(sing)}}} [\![\rho(\tfrac{1}{2}v^2 + \varphi)(\mathbf{v} \cdot \boldsymbol{\xi} - u_{(\xi)}) - \mathbf{v} \cdot (\mathbf{S} \cdot \boldsymbol{\xi})]\!]\, dS \quad \text{(F)}\ddagger$$

† Here $u_{(\xi)}$ is the speed of displacement of the phase interface [1, p. 499]; the boldface bracket notation is defined by Eq. (4-7) of Sec. 1.3.4; $S_{\text{(sing)}}$ refers to those phase interfaces in the region enclosed by $S_{(s)}$ that do not coincide with $S_{(s)}$.
‡ We assume that $\partial \varphi / \partial t = 0$.
§ Applicable only to systems where composition is independent of position and time.

4. Work done by viscous forces (as described by the extra-stress tensor) may be neglected at entrances and exits.

The forms of the integral mechanical energy balance applicable under these restrictions are shown in Table 7.4.6-3.

The extension of this discussion to turbulent flows follows along the lines of Sec. 4.4.10. No further remarks appear necessary.

The remarks made concerning $\mathscr{E}$ at the conclusion of Sec. 4.4.9 are still applicable. Empirical data correlations for $\mathscr{E}$ are often useful. The approach recommended in preparing these correlations is outlined in Sec. 4.4.11.

Finally, I wish to call particular attention to R. B. Bird's discussion of the mechanical energy balance [2]. With only minor extensions, I have adopted his viewpoint.

### REFERENCES

1. Truesdell, C., and R. A. Toupin: In S. Flügge (ed), "Handbuch der Physik," vol. 3/1, Springer-Verlag, Berlin, 1960.
2. Bird, R. B.: *Chem. Eng. Sci.*, 6:123 (1957).

## [7.4] INTEGRAL BALANCES

**Table 7.4.6-3** Restricted forms of the integral mechanical energy balance applicable to a multiphase system†

$$\frac{d}{dt}\int_{V_{(s)}} \rho(\tfrac{1}{2}v^2 + \varphi)\,dV = \int_{S_{(\text{ent ex})}} \rho\left(\tfrac{1}{2}v^2 + \varphi + \frac{P - p_0}{\rho}\right)(-\mathbf{v}\cdot\mathbf{n})\,dS + \int_{V_{(s)}} (P - p_0)\,\text{div}\,\mathbf{v}\,dV$$
$$- \mathscr{W} - \mathscr{E} \quad \text{(A)}\ddagger$$

$$\frac{d}{dt}\int_{V_{(s)}} \tfrac{1}{2}\rho v^2\,dV = \int_{S_{(\text{ent ex})}} \rho\left(\tfrac{1}{2}v^2 + \frac{P - p_0}{\rho}\right)(-\mathbf{v}\cdot\mathbf{n})\,dS$$
$$+ \int_{V_{(s)}} (P - p_0)\,\text{div}\,\mathbf{v}\,dV - \mathscr{W} - \mathscr{E} + \int_{V_{(s)}} \mathbf{v}\cdot\rho\mathbf{f}\,dV \quad \text{(B)}$$

*For incompressible fluids:*

$$\frac{d}{dt}\int_{V_{(s)}} \rho(\tfrac{1}{2}v^2 + \varphi)\,dV = \int_{S_{(\text{ent ex})}} \rho\left(\tfrac{1}{2}v^2 + \varphi + \frac{P - p_0}{\rho}\right)(-\mathbf{v}\cdot\mathbf{n})\,dS - \mathscr{W} - \mathscr{E} \quad \text{(C)}\ddagger$$

*For an isothermal system:*

$$\frac{d}{dt}\int_{V_{(s)}} \rho\left(\hat{A} + \tfrac{1}{2}v^2 + \varphi + \frac{p_0}{\rho}\right)dV = \int_{S_{(\text{ent ex})}} \rho(\hat{G} + \tfrac{1}{2}v^2 + \varphi)(-\mathbf{v}\cdot\mathbf{n})\,dS - \mathscr{W} - \mathscr{E} \quad \text{(D)}\ddagger\S$$

*For an isentropic system:*

$$\frac{d}{dt}\int_{V_{(s)}} \rho\left(\hat{U} + \tfrac{1}{2}v^2 + \varphi + \frac{p_0}{\rho}\right)dV = \int_{S_{(\text{ent ex})}} \rho(\hat{H} + \tfrac{1}{2}v^2 + \varphi)(-\mathbf{v}\cdot\mathbf{n})\,dS - \mathscr{W} - \mathscr{E} \quad \text{(E)}\ddagger\S$$

*For an isobaric system:*

$$\frac{d}{dt}\int_{V_{(s)}} \rho(\tfrac{1}{2}v^2 + \varphi)\,dV = \int_{S_{(\text{ent ex})}} \rho(\tfrac{1}{2}v^2 + \varphi)(-\mathbf{v}\cdot\mathbf{n})\,dS - \mathscr{W} - \mathscr{E}$$
$$+ \int_{S_{(s)} - S_{(\text{ent ex})}} (P - p_0)\mathbf{v}\cdot\mathbf{n}\,dS \quad \text{(F)}\ddagger$$

† These forms are applicable following assumptions 1 to 4 given in the text.
‡ We assume that $\partial \varphi / \partial t = 0$.
§ Applicable only to systems where composition is independent of time and position.

### EXERCISE

**7.4.6-1** (a) Use the results of Exercise 7.4.2-3 to derive Eqs. (E) of Tables 7.4.6-1 to 7.4.6-3.
(b) Derive Eqs. (F) of Tables 7.4.6-1 to 7.4.6-3.

**7.4.7 The integral entropy inequality** The entropy inequality is the only postulate for which we have not as yet derived the corresponding integral relationship.

In Sec. 5.5.2 we derive two forms of the differential entropy inequality:

$$\frac{\partial(\rho\hat{S})}{\partial t} + \text{div}\,(\rho\hat{S}\mathbf{v}) + \text{div}\left(\frac{1}{T}\mathbf{q}\right) - \rho\frac{Q}{T} \geq 0 \quad (7\text{-}1)$$

and

$$-\frac{1}{T^2}\mathbf{q}\cdot\nabla T + \frac{1}{T}\operatorname{tr}(\mathbf{S}\cdot\nabla\mathbf{v}) \geq 0 \tag{7-2}$$

We can integrate this last over the region occupied by an arbitrary system to immediately obtain

$$\int_{V_{(s)}}\left[-\frac{1}{T^2}\mathbf{q}\cdot\nabla T + \frac{1}{T}\operatorname{tr}(\mathbf{S}\cdot\nabla\mathbf{v})\right]dV \geq 0 \tag{7-3}$$

This is one form of the *integral entropy inequality* applicable to both single-phase and multiphase systems.

A somewhat more useful form of the integral entropy inequality can be arrived at by integrating Eq. (7-1) over an arbitrary system:

$$\int_{V_{(s)}}\left[\frac{\partial(\rho\hat{S})}{\partial t} + \operatorname{div}(\rho\hat{S}\cdot\mathbf{v}) + \operatorname{div}\left(\frac{1}{T}\mathbf{q}\right) - \rho\frac{Q}{T}\right]dV \geq 0 \tag{7-4}$$

The generalized transport theorem can be used to express the first term on the left as

$$\int_{V_{(s)}}\frac{\partial(\rho\hat{S})}{\partial t}dV = \frac{d}{dt}\int_{V_{(s)}}\rho\hat{S}\,dV - \int_{S_{(s)}}\rho\hat{S}\mathbf{v}_{(s)}\cdot\mathbf{n}\,dS \tag{7-5}$$

After an application of Green's transformation, the second and third terms on the left of Eq. (7-4) become

$$\int_{V_{(s)}}\left[\operatorname{div}(\rho\hat{S}\mathbf{v}) + \operatorname{div}\left(\frac{1}{T}\mathbf{q}\right)\right]dV = \int_{S_{(s)}}\left[\rho\hat{S}\mathbf{v} + \frac{1}{T}\mathbf{q}\right]\cdot\mathbf{n}\,dS \tag{7-6}$$

By means of Eqs. (7-5) and (7-6), we are able to write Eq. (7-4) as

$$\frac{d}{dt}\int_{V_{(s)}}\rho\hat{S}\,dV \geq \int_{S_{(\text{ent ex})}}\rho\hat{S}(\mathbf{v}-\mathbf{v}_{(s)})\cdot(-\mathbf{n})\,dS$$

$$+ \int_{S_{(s)}}\frac{1}{T}\mathbf{q}\cdot(-\mathbf{n})\,dS + \int_{V_{(s)}}\frac{\rho Q}{T}\,dV \tag{7-7}$$

This says that the time rate change of the entropy associated with an arbitrary system is greater than or equal to the net rate at which entropy is brought into the system with whatever material flows across the boundaries of this system, the net rate at which entropy is transferred to the system as a result of contact energy transfer to the system, and the net rate at which entropy is produced in the system as a result of mutual and external energy transmission to the system. We will refer to Eq. (7-7) as a form of the *integral entropy inequality* appropriate to single-phase systems.

For multiphase systems we can take the approach of Sec. 4.4.2 and immediately arrive at

## [7.4] INTEGRAL BALANCES

$$\frac{d}{dt}\int_{V_{(s)}} \rho \hat{S}\, dV \geq \int_{S_{(\text{ent ex})}} \rho \hat{S}(\mathbf{v} - \mathbf{v}_{(s)}) \cdot (-\mathbf{n})\, dS + \int_{S_{(s)}} \frac{1}{T}\mathbf{q} \cdot (-\mathbf{n})\, dS$$

$$+ \int_{V_{(s)}} \frac{\rho Q}{T}\, dV + \int_{S_{(\text{sing})}} \left[\rho \hat{S}(\mathbf{v} \cdot \boldsymbol{\xi} - u_{(\xi)}) + \frac{1}{T}\mathbf{q} \cdot \boldsymbol{\xi}\right] dS \quad (7\text{-}8)$$

This is another form of the *general integral entropy inequality* appropriate to multiphase systems.

We usually will be willing to say that the jump entropy inequality of Exercise 5.5.2-2 applies, in which case Eq. (7-8) reduces to (7-7). In other words, Eq. (7-7) applies equally to single-phase and multiphase systems so long as the jump entropy inequality of Exercise 5.5.2-2 is valid for all phase interfaces involved.

**7.4.8 The integral entropy inequality for turbulent flows** Our approach here is basically the same as that we have taken in arriving at forms of the other integral balances appropriate to turbulent flows. See, for example, Sec. 4.4.3.

We could repeat the analysis of Sec. 7.4.7 using time averages of the differential entropy inequalities derived in Sec. 5.5.2.

It seems much more straightforward to time average the integral entropy inequality of Sec. 7.4.7 to find for any single-phase or multiphase system

$$\frac{d}{dt}\int_{V_{(s)}} \overline{\rho \hat{S}}\, dV \geq \int_{(\text{ent ex})} \overline{\rho \hat{S}(\mathbf{v} - \mathbf{v}_{(s)}) \cdot (-\mathbf{n})}\, dS + \int_{S_{(s)}} \overline{\frac{1}{T}\mathbf{q} \cdot (-\mathbf{n})}\, dS$$

$$+ \int_{V_{(s)}} \overline{\rho \frac{Q}{T}}\, dV \quad (8\text{-}1)$$

This is the *integral entropy inequality* for turbulent flows. The only assumption that we have made in arriving at this result is that the jump entropy inequality of Exercise 5.5.2-2 is applicable at all phase interfaces involved.

A somewhat simpler result can be obtained for single-phase or multiphase systems that do not involve fluid-fluid phase interfaces. Under these circumstances, we can use Eqs. (3-7) to (3-9) of Sec. 4.4.3 to show that Eq. (8-1) reduces to

$$\frac{d}{dt}\int_{V_{(s)}} \overline{\rho \hat{S}}\, dV \geq \int_{S_{(\text{ent ex})}} (\overline{\rho \hat{S} \mathbf{v}} - \overline{\rho \hat{S}} \overline{\mathbf{v}_{(s)}}) \cdot (-\mathbf{n})\, dS + \int_{S_{(s)}} \overline{\frac{1}{T}\mathbf{q}} \cdot (-\mathbf{n})\, dS$$

$$+ \int_{V_{(s)}} \overline{\frac{\rho Q}{T}}\, dV \quad (8\text{-}2)$$

**444**   APPLICATIONS OF INTEGRAL AVERAGING TECHNIQUES TO ENERGY TRANSFER

### EXERCISE

**7.4.8-1**   Derive Eq. (8-2) by repeating the analysis of Sec. 7.4.7 using a time-averaged form of the differential entropy inequality.

### 7.4.9   The integral entropy inequality: An example
J. H. Keenan [1, p. 73] states:

> It is impossible to construct an engine to work between two heat reservoirs, each having a fixed and uniform temperature, which will exceed in efficiency a reversible engine working between the same reservoirs.

Let us prove this statement as a corollary of the integral entropy inequality.

Before beginning, let us stop and note that a *reversible* system (device, engine, etc.) is one for which an entropy inequality reduces to an equality.

Let us assume that the engine operates at steady state and that no turbulent flows are involved. (In Exercise 7.4.9-1 you are asked to extend these results to an engine that operates on a periodic cycle and to an engine that involves turbulent flows.) We interpret Keenan's statement to refer to an engine that has no entrances or exits for flow. Referring to Fig. 7.4.9-1, we shall say that

$$T_1 > T_2 \tag{9-1}$$

and that, consequently, the efficiency of the engine may be defined to be $\mathcal{W}/\mathcal{Q}_1$.

For the irreversible engine, the integral energy balance in the form of Eq. (B) of Table 7.4.2-3 requires that

$$\mathcal{Q}_1 - \mathcal{Q}_2 - \mathcal{W} = 0 \tag{9-2}$$

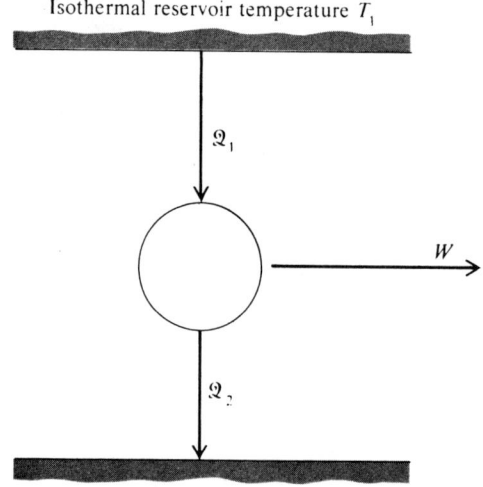

Fig. 7.4.9-1   Heat engine.

## [7.4] INTEGRAL BALANCES

The integral entropy inequality of Sec. 7.4.7 further says that

$$\frac{\mathcal{Q}_1}{T_1} - \frac{\mathcal{Q}_2}{T_2} \leq 0 \tag{9-3}$$

Together they tell us that

$$\frac{\mathcal{W}}{\mathcal{Q}_1} \leq 1 - \frac{T_2}{T_1} \tag{9-4}$$

The same analysis applied to the reversible engine yields

$$\frac{\mathcal{W}_{(\text{rev})}}{\mathcal{Q}_{1(\text{rev})}} = 1 - \frac{T_2}{T_1} \tag{9-5}$$

Equations (9-4) and (9-5) imply the required result:

$$\frac{\mathcal{W}}{\mathcal{Q}_1} \leq \frac{\mathcal{W}_{(\text{rev})}}{\mathcal{Q}_{1(\text{rev})}} \tag{9-6}$$

### REFERENCES

1. Keenan, J. H.: "Thermodynamics," Wiley, New York, 1941.
2. Smith, J. M., and H. C. Van Ness: "Introduction to Chemical Engineering Thermodynamics," 2d ed., McGraw-Hill, New York, 1959.

### EXERCISES

**7.4.9-1** (a) Repeat the analysis of this section assuming that the engine operates according to a periodic cycle.
(b) Repeat the analysis of this section assuming that the engine involves a turbulent flow.

**7.4.9-2** J. M. Smith and H. C. Van Ness [2, p. 176] state that:

No apparatus can operate in such a way that its only effect (in system and surroundings) is to convert the heat taken in completely into work.

Prove this statement, starting with the integral energy balance and the integral entropy inequality.
*Hint:* In carrying out this proof one must interpret what is meant by the authors' statement.

**7.4.9-3** [2, p. 338] An inventor has devised a process that makes energy continuously available at an elevated temperature. Saturated steam at 220°F is the only source of energy. There is an abundance of cooling water at 75°F. What is the maximum amount of energy that could be made available at 400°F/Btu of energy given up by the steam?

**7.4.9-4** J. M. Smith and H. C. Van Ness [2, p. 176] state that:

Any process which consists solely in the transfer of heat from one temperature to a higher one is impossible.

Use the integral entropy inequality to prove this statement.

# 8
# Foundations for Mass Transfer

This is an excellent time for review. In fact, a review is practically forced upon you in this chapter.

I think that it is reasonable to think of a body which is composed of a single species as being a particular type of multicomponent material. In what follows, we reformulate the fundamental postulates made in Chaps. 1, 2, and 5, in order that they now apply to multicomponent bodies. We are fortunate that, with a few relatively minor modifications, all we have said about single-component systems can be applied to multicomponent ones.

After outlining for you the structure to be used in discussing multicomponent materials, I go on to develop the concepts of *equilibrium* and *stable equilibrium*. These ideas are discussed in the context of nonequilibrium thermodynamics (see Sec. 5.1.2). The presentation here is a departure from the standard one given in the context of Gibbs thermostatics. I have done this knowingly, hoping to relate these ideas more closely to experimental reality. All of the results derived, including the criteria for a stable equilibrium, appear to be consistent with our knowledge of experimental reality as well as with the criteria for stability derived in the context of Gibbs thermostatics.

## 8.1 VIEWPOINT

**8.1.1 Viewpoint in considering multicomponent materials** Up to this point we have been primarily concerned with single-component materials or materials of uniform composition. Hereafter, we shall be treating a material consisting of $N$ species or constituents, which is undergoing an arbitrary number of homogeneous and heterogeneous chemical reactions. Not only are we interested in the velocity and temperature distributions in such a material, but we wish to follow its composition as a function of time and position. We may ask, for example, how rapidly a particular species formed by a catalytic reaction at an adjacent surface will distribute itself throughout a material; or we may wish to determine the rate at which a liquid droplet will evaporate into a surrounding gas stream.

Our first task is to choose a continuum model for an $N$-component material. I have indicated that we will wish to follow each species individually as the $N$-component mixture goes through some operation, possibly involving deformation, flow, and chemical reactions. This suggests that we view each species as a continuous medium with a variable mass-density field. The model for the $N$-component mixture is, consequently, a superposition of these $N$ continuous media.

One feature of this model initially may seem confusing. At any point in space occupied by the mixture, $N$ material particles (one from each of the continuous media representing individual species) coexist. The confusion seems to arise from a dangerous and incorrect identification of a material particle in a continuous medium of species $A$ with a molecule of species $A$. On the other hand, this superposition of $N$ constituent media is consistent with our usual practice of identifying compositions with each point in a multicomponent mixture.

**8.1.2 Body, motion, and material coordinates of a species** The ideas we introduced in Sec. 1.1.1 for a single-component material may be extended easily to a particular species $A$ in an $N$-constituent mixture.

A *body of species* $A$ is a set, the elements $\zeta_{(A)}$ of which are called particles of species $A$ or material particles of species $A$. A one-to-one continuous mapping of this set onto a region of three-dimensional euclidean point space exists and is called a *configuration* of the body of constituent $A$:

$$\mathbf{z} = \chi_{(A)}(\zeta_{(A)}) \tag{2-1}$$

or

$$\zeta_{(A)} = \chi_{(A)}^{-1}(\mathbf{z}) \tag{2-2}$$

Here $\chi_{(A)}^{-1}$ indicates the inverse mapping of $\chi_{(A)}$. The point $\mathbf{z} = \chi_{(A)}(\zeta_{(A)})$ is called the place occupied by the particle $\zeta_{(A)}$, and $\zeta_{(A)} = \chi_{(A)}^{-1}(\mathbf{z})$ is called the particle of species $A$ whose place is $\mathbf{z}$.

For the moment, we will not concern ourselves directly with a body of an $N$-component mixture, but rather with the $N$ constituent bodies. It is the *superposition* of these $N$ constituent bodies that forms the model for the $N$-component mixture.

A *motion* of a body of species $A$ is a one-parameter family of configurations; the real parameter $t$ is time. We write

$$\mathbf{z} = \boldsymbol{\chi}_{(A)}(\zeta_{(A)}, t) \tag{2-3}$$

and

$$\zeta_{(A)} = \boldsymbol{\chi}_{(A)}^{-1}(\mathbf{z}, t) \tag{2-4}$$

Although we do not wish to confuse a body of species $A$ with any of its spatial configurations, we must recognize that it is available to us for observation and study only in those configurations. It is often convenient to take advantage of this by using positions in some particular configuration as a means of specifying the particles of species $A$ which form the body. This reference configuration may be, but need not be, one actually occupied by the body in the course of its motion. The place of a particle of species $A$ in the reference configuration $\boldsymbol{\varkappa}_{(A)}$ will be denoted by

$$\mathbf{z}_{\varkappa(A)} = \boldsymbol{\varkappa}_{(A)}(\zeta_{(A)}) \tag{2-5}$$

The particle at the place $\mathbf{z}_{\varkappa(A)}$ in the configuration $\boldsymbol{\varkappa}_{(A)}$ may be expressed as

$$\zeta_{(A)} = \boldsymbol{\varkappa}_{(A)}^{-1}(\mathbf{z}_{\varkappa(A)}) \tag{2-6}$$

If $\boldsymbol{\chi}_{(A)}$ is a motion of the body of species $A$, then

$$\mathbf{z} = \boldsymbol{\chi}_{(A)}(\zeta_{(A)}, t) = \boldsymbol{\chi}_{(A)}(\boldsymbol{\varkappa}_{(A)}^{-1}(\mathbf{z}_{\varkappa(A)}), t) = \boldsymbol{\chi}_{\varkappa(A)}(\mathbf{z}_{\varkappa(A)}, t) \tag{2-7}$$

This defines a family of *deformations* from the reference configuration of constituent $A$. The subscript $\varkappa$ is to remind one that the form of $\boldsymbol{\chi}_{\varkappa(A)}(\mathbf{z}_{\varkappa(A)}, t)$ depends upon the choice of reference configuration $\boldsymbol{\varkappa}_{(A)}$.

The coordinates $z_{\varkappa(A)i}$ identify the place $\mathbf{z}_{\varkappa(A)}$ in an arbitrary coordinate system which is stationary as a function of time with respect to the reference configuration $\boldsymbol{\varkappa}_{(A)}$; these coordinates are referred to as the *material coordinates* of species $A$.

Let $B$ be any quantity, scalar or tensor. We shall wish to talk about the time derivative of $B$ following the motion of a particle of species $A$. We define

$$\frac{d_{(A)}B}{dt} \equiv \left(\frac{\partial B}{\partial t}\right)_{z_{\varkappa(A)}} = \left(\frac{\partial B}{\partial t}\right)_{z_{\varkappa(A)1}, z_{\varkappa(A)2}, z_{\varkappa(A)3}} \tag{2-8}$$

As an example, the velocity vector of species $A$, $\mathbf{v}_{(A)}$, represents the time rate of change of position of a material particle of species $A$:

$$\mathbf{v}_{(A)} \equiv \frac{d_{(A)}\mathbf{z}}{dt} = \left[\frac{\partial \boldsymbol{\chi}_{\varkappa(A)}(\mathbf{z}_{\varkappa(A)}, t)}{\partial t}\right]_{z_{\varkappa(A)}} \tag{2-9}$$

## EXERCISES

**8.1.2-1** Let $B$ be any real scalar field, spatial vector field, or second-order tensor field. Show that

$$\frac{d_{(A)}B}{dt} = \frac{\partial B}{\partial t} + (\nabla B) \cdot \mathbf{v}_{(A)}$$

*8.1.2-2* Show that the contravariant components of the velocity vector of species $A$ are

$$v^i_{(A)} = \frac{d_{(A)}x^i}{dt}$$

*8.1.2-3* Let $\mathbf{a} = \mathbf{a}(\mathbf{z},t)$ be some vector field which is a function both of position and time.
(a) Show that

$$\frac{d_{(A)}\mathbf{a}}{dt} = \frac{\delta_{(A)}a^n}{\delta t}\,\mathbf{g}_n$$

where

$$\frac{\delta_{(A)}a^n}{\delta t} \equiv \frac{\partial a^n}{\partial t} + a^n{}_{,i}v^i_{(A)}$$

(b) Show that

$$\frac{d_{(A)}\mathbf{a}}{dt} = \frac{\delta_{(A)}a_n}{\delta t}\,\mathbf{g}^n$$

where

$$\frac{\delta_{(A)}a_n}{\delta t} \equiv \frac{\partial a_n}{\partial t} + a_{n,i}v^i_{(A)}$$

The quantity $\delta_{(A)}a^n/\delta t$ may be referred to as the *species (A) intrinsic derivative* of the contravariant vector-field component $a^n$; it should be viewed as the contravariant component of the vector field $d_{(A)}\mathbf{a}/dt$. The quantity $\delta_{(A)}a_n/\delta t$ may be designated as the species $(A)$ intrinsic derivative of the covariant component $a_n$; it is the covariant component of the vector field $d_{(A)}\mathbf{a}/dt$.

*8.1.2-4* The species intrinsic derivative introduced in Exercise 8.1.2-3 may be extended readily to higher-order tensor fields. Consider the second-order tensor field $\mathbf{T} = \mathbf{T}(\mathbf{z},t)$.
(a) Show that

$$\frac{d_{(A)}\mathbf{T}}{dt} = \frac{\delta_{(A)}T_{ij}}{\delta t}\,\mathbf{g}^i\mathbf{g}^j$$

where

$$\frac{\delta_{(A)}T_{ij}}{\delta t} \equiv \frac{\partial T_{ij}}{\partial t} + T_{ij,k}v^k_{(A)}$$

(b) Show that

$$\frac{d_{(A)}\mathbf{T}}{dt} = \frac{\delta_{(A)}T_i{}^j}{\delta t} \cdot \mathbf{g}^i\mathbf{g}_j$$

where

$$\frac{\delta_{(A)}T_i{}^j}{\delta t} \equiv \frac{\partial T_i{}^j}{\partial t} + T_i{}^j{}_{,k}v^k_{(A)}$$

The quantity $\delta_{(A)}T_{ij}/\delta t$ may be referred to as the *species (A) intrinsic derivative* of the doubly covariant tensor-field component $T_{ij}$; it should be thought of as the doubly covariant component of the tensor field $d_{(A)}\mathbf{T}/dt$. The quantity $\delta_{(A)}T_i{}^j/\delta t$ may be referred to as the *species (A) intrinsic derivative* of the mixed tensor-field component $T_i{}^j$; it is a mixed component of the tensor field $d_{(A)}\mathbf{T}/dt$.

## 8.2 MASS BALANCE

**8.2.1 The species mass balance** In order to discuss the movements of the various constituents in a multicomponent mixture, we require a sixth postulate, the *mass balance for an individual species*.

The time rate of the change of mass of each species $A$ ($A = 1, 2, \ldots, N$) in a multicomponent mixture is equal to the rate at which the mass of species $A$ is produced by homogeneous chemical reactions.

Let $V_{(A)}$ denote the volume associated with a body of species $A$ (a set of particles of species $A$), $\rho_{(A)}$ denote the mass density of species $A$, and $r_{(A)}$ denote the rate of production of species $A$ per unit volume by homogeneous chemical reactions. This sixth postulate states that

$$\frac{d}{dt}\int_{V_{(A)}} \rho_{(A)}\, dV = \int_{V_{(A)}} r_{(A)}\, dV \tag{1-1}$$

It is understood here that the limits on these volume integrations in spatial coordinates are functions of time.

For the moment, we assume that all quantities are continuous and differentiable as many times as desired.

The transport theorem of Sec. 1.3.2 may be modified to apply to a collection of particles of species $A$ (see Exercise 8.2.1-1). When applied to the left side of Eq. (1-1), we obtain

$$\int_{V_{(A)}} \left(\frac{d_{(A)}\rho_{(A)}}{dt} + \rho_{(A)} \operatorname{div} \mathbf{v}_{(A)}\right) dV = \int_{V_{(A)}} r_{(A)}\, dV \tag{1-2}$$

or

$$\int_{V_{(A)}} \left(\frac{d_{(A)}\rho_{(A)}}{dt} + \rho_{(A)} \operatorname{div} \mathbf{v}_{(A)} - r_{(A)}\right) dV = 0 \tag{1-3}$$

Since this applies to an arbitrary body of species $A$, we conclude that at every point in the material

$$\frac{d_{(A)}\rho_{(A)}}{dt} + \rho_{(A)} \operatorname{div} \mathbf{v}_{(A)} = r_{(A)} \tag{1-4}$$

This is usually referred to as the *equation of continuity for species A*.

A commonly used alternate form of this equation is

$$\frac{\partial \rho_{(A)}}{\partial t} + \operatorname{div}(\rho_{(A)}\mathbf{v}_{(A)}) = r_{(A)} \tag{1-5}$$

Sometimes it is more convenient to work in terms of molar density $c_{(A)} \equiv \rho_{(A)}/M_{(A)}$, where $M_{(A)}$ is the molecular weight of species $A$. In this case, Eq. (1-5)

## [8.2] MASS BALANCE

may be rewritten as

$$\frac{\partial c_{(A)}}{\partial t} + \text{div}\,(c_{(A)}\mathbf{v}_{(A)}) = \frac{r_{(A)}}{M_{(A)}} \tag{1-6}$$

It is perhaps worth emphasizing that $r_{(A)}$ is the rate of production of species $A$ per unit volume by *homogeneous* chemical reactions. It says nothing at all about *heterogeneous* chemical reactions. Since heterogeneous chemical reactions take place at phase interfaces, they must be described in terms of boundary conditions.[1] This will be discussed further in Sec. 9.1.1.

In what follows, we shall assume that a constitutive equation or empirical data correlation for $r_{(A)}$ as a function of composition, temperature, pressure, etc., is known. We will not discuss the preparation of these constitutive equations, since so much attention is devoted to this point in the variety of kinetics texts available.

### EXERCISES

**8.2.1-1** *Transport theorem for species A*   Let $\Phi$ be any scalar, vector, or tensor field. Show that

$$\frac{d}{dt}\int_{V_{(A)}} \Phi\, dV = \int_{V_{(A)}} \left( \frac{d_{(A)}\Phi}{dt} + \Phi\,\text{div}\,\mathbf{v}_{(A)} \right) dV$$

This is a form of the *transport theorem* applicable to a body of species $A$.

*Hint:* Review Sec. 1.3.2.

**8.2.1-2** *Transport theorem for a discontinuous body of species A*   Let $\Phi$ be any scalar, vector, or tensor field. Derive the following *transport theorem* for a body of species $A$ which contains a singular surface $S_{(\text{sing})}$:

$$\frac{d}{dt}\int_{V_{(A)}} \Phi\, dV = \int_{V_{(A)}} \frac{\partial \Phi}{\partial t}\, dV + \int_{S_{(A)}} \Phi(\mathbf{v}_{(A)}\cdot\mathbf{n})\, dS - \int_{S_{(\text{sing})}} [\![\Phi u_{(\xi)}]\!]\, dS$$

The notation has the same meaning as in Sec. 1.3.4 with the understanding that we are now restricting ourselves to a singular surface which occurs in a body of species $A$.

**8.2.1-3** *Jump mass balance for species A*   We identify a *phase interface* in a multicomponent mixture with a singular surface in a body of species $A$.

Use the approach suggested in Sec. 1.3.5 to derive the following *jump mass balance* for species $A$ at a phase interface:

$$[\![\rho_{(A)}(\mathbf{v}_{(A)}\cdot\boldsymbol{\xi} - u_{(\xi)})]\!] = r_{(A)}^{(\sigma)}$$

where $r_{(A)}^{(\sigma)}$ is the rate at which species $A$ is produced by heterogeneous or catalytic chemical reactions per unit area.

**8.2.1-4**   Use the approach suggested in Exercise 1.3.5-2 to derive the jump mass of Exercise 8.2.1-3.

---

[1] One of the most common practical discussions of heterogeneous chemical reactions is in the context of porous solids, where the reaction takes place at the gas-solid or liquid-solid phase interfaces distributed throughout the porous structure. When the local volume average of the equation of continuity of species $A$ is used to describe mass transfer in the porous solid, the rate of production of species $A$ by heterogeneous chemical reactions is described as a source term. See Sec. 10.3.1 for further details.

**8.2.1-5** Let us define the density $\rho$ of a multicomponent mixture in the usual manner:

$$\rho \equiv \sum_{A=1}^{N} \rho_{(A)}$$

Let us also define the mass-averaged velocity $\mathbf{v}$ of a multicomponent mixture:

$$\mathbf{v} \equiv \frac{1}{\rho} \sum_{A=1}^{N} \rho_{(A)} \mathbf{v}_{(A)}$$

(a) Show that the sum of Eq. (1-5) over all $N$ species may be written as

$$\frac{\partial \rho}{\partial t} + \text{div}(\rho \mathbf{v}) = \sum_{A=1}^{N} r_{(A)}$$

Our conception of conservation of mass, formalized in Sec. 8.3.2, tells us that

$$\sum_{A=1}^{N} r_{(A)} = 0$$

Consequently, we have that

$$\frac{\partial \rho}{\partial t} + \text{div}(\rho \mathbf{v}) = 0$$

This equation, known as the overall equation of continuity, is formally identical to the equation of continuity derived in Sec. 1.3.3 for a single-component material.

(b) Show that the above result may be expressed as

$$\frac{d_{(\mathbf{v})} \rho}{dt} + \rho \, \text{div} \, \mathbf{v} = \sum_{A=1}^{N} r_{(A)} = 0$$

where

$$\frac{d_{(\mathbf{v})} \rho}{dt} \equiv \frac{\partial \rho}{\partial t} + (\nabla \rho) \cdot \mathbf{v}$$

**8.2.1-6** Derive Eq. (1-5).

## 8.2.2 Concentrations, velocities, and mass fluxes

One of the most confusing aspects of mass-transfer problems is that there are several sets of terminology in common use.

In discussing the concentration of species $A$ in a multicomponent mixture, one is free to refer to the mass density $\rho_{(A)}$, the molar density $c_{(A)}$, the mass fraction $\omega_{(A)}$, or the mole fraction $x_{(A)}$. The relations between these quantities are explored in Tables 8.2.2-1 and 8.2.2-2.

Workers are not content to use the velocity of species $A$, $\mathbf{v}_{(A)}$. Frequently, one refers to the velocity of species $A$ with respect to the mass-averaged velocity:

$$\mathbf{u}_{(A)} \equiv \mathbf{v}_{(A)} - \mathbf{v} \qquad (2\text{-}1)$$

where the mass-averaged velocity $\mathbf{v}$ is defined as

$$\mathbf{v} \equiv \sum_{A=1}^{N} \omega_{(A)} \mathbf{v}_{(A)} \qquad (2\text{-}2)$$

[8.2] MASS BALANCE

**Table 8.2.2-1  Notation for concentrations†**

$$\rho = \sum_{A=1}^{N} \rho_{(A)} = \text{mass density of solution}$$

$$\omega_{(A)} = \frac{\rho_{(A)}}{\rho} = \text{mass fraction of component } A$$

$$c_{(A)} = \frac{\rho_{(A)}}{M_{(A)}} = \text{molar density of component } A$$

$$c = \sum_{A=1}^{N} c_{(A)} = \text{molar density of solution}$$

$$x_{(A)} = \frac{c_{(A)}}{c} = \text{mole fraction of component } A$$

$$M = \frac{\rho}{c} = \sum_{A=1}^{N} x_{(A)} M_{(A)} = \text{molar-averaged molecular weight of mixture}$$

$$\frac{1}{M} = \frac{c}{\rho} = \sum_{A=1}^{N} \frac{\omega_{(A)}}{M_{(A)}}$$

† Suggested by Table 16.1-1 of [1].

In Tables 8.2.2-3 and 8.2.2-4, we see that we may also work with the velocity of species $A$ with respect to the molar-averaged velocity:

$$\mathbf{u}_{(A)}^{\star} \equiv \mathbf{v}_{(A)} - \mathbf{v}^{\star} \tag{2-3}$$

We define the molar-averaged velocity $\mathbf{v}^{\star}$ as

$$\mathbf{v}^{\star} \equiv \sum_{A=1}^{N} x_{(A)} \mathbf{v}_{(A)} \tag{2-4}$$

**Table 8.2.2-2  Relations between mass and mole fractions**

| | |
|---|---|
| $x_{(A)} = \dfrac{\omega_{(A)}/M_{(A)}}{\sum_{B=1}^{N} \omega_{(B)}/M_{(B)}}$ | $\omega_{(A)} = \dfrac{x_{(A)} M_{(A)}}{\sum_{B=1}^{N} x_{(B)} M_{(B)}}$ |
| *For a binary system:* | *For a binary system:* |
| $dx_{(A)} = \dfrac{d\omega_{(A)}}{M_{(A)} M_{(B)} (\omega_{(A)}/M_{(A)} + \omega_{(B)}/M_{(B)})^2}$ | $d\omega_{(A)} = \dfrac{M_{(A)} M_{(B)} \, dx_{(A)}}{(x_{(A)} M_{(A)} + x_{(B)} M_{(B)})^2}$ |
| $= \left(\dfrac{\rho}{c}\right)^2 \dfrac{d\omega_{(A)}}{M_{(A)} M_{(B)}}$ | $= \left(\dfrac{c}{\rho}\right)^2 M_{(A)} M_{(B)} \, dx_{(A)}$ |

**Table 8.2.2-3 Various velocities and relations between them†**

$\mathbf{v}_{(A)}$ = velocity of species $A$ with respect to the fixed frame of reference

$\mathbf{v} \equiv \sum_{A=1}^{N} \omega_{(A)} \mathbf{v}_{(A)}$ = mass-averaged velocity

$\mathbf{v}^{\star} \equiv \sum_{A=1}^{N} x_{(A)} \mathbf{v}_{(A)}$ = molar-averaged velocity

$\mathbf{u}_{(A)} \equiv \mathbf{v}_{(A)} - \mathbf{v}$ = velocity of species $A$ relative to $\mathbf{v}$

$\mathbf{u}_A^{\star} \equiv \mathbf{v}_{(A)} - \mathbf{v}^{\star}$ = velocity of species $A$ relative to $\mathbf{v}^{\star}$

$\mathbf{v} - \mathbf{v}^{\star} = \sum_{A=1}^{N} \omega_{(A)} (\mathbf{v}_{(A)} - \mathbf{v}^{\star})$

$\mathbf{v}^{\star} - \mathbf{v} = \sum_{A=1}^{N} x_{(A)} (\mathbf{v}_{(A)} - \mathbf{v})$

† Suggested by Table 16.1-2 of [1].

In Sec. 8.2.1 we derived the differential mass balance for species $A$ in the form

$$\frac{\partial \rho_{(A)}}{\partial t} + \text{div}\,(\rho_{(A)} \mathbf{v}_{(A)}) = r_{(A)} \tag{2-5}$$

The quantity

$$\mathbf{n}_{(A)} \equiv \rho_{(A)} \mathbf{v}_{(A)} \tag{2-6}$$

is usually called the *mass flux* of species $A$ *with respect to the fixed frame of reference* (the fixed stars or the laboratory walls). The alternate ways we have of looking at concentrations and velocities suggest the introduction of the molar flux of species $A$ with respect to the fixed frame of reference, as well as mass and molar fluxes with respect to the mass- and molar-averaged velocities. A tabulation of the relations between these quantities is given in Table 8.2.2-4.

Just as we may write the differential mass balance in terms of $\mathbf{n}_{(A)}$,

$$\frac{\partial \rho_{(A)}}{\partial t} + \text{div}\,\mathbf{n}_{(A)} = r_{(A)} \tag{2-7}$$

we may rearrange it in terms of all the other mass and molar flux vectors. For example, the mass flux of species $A$ *with respect to the mass-averaged velocity* is defined as

$$\mathbf{j}_{(A)} \equiv \rho_{(A)} (\mathbf{v}_{(A)} - \mathbf{v}) \tag{2-8}$$

Equation (2-5) may be rewritten as

$$\frac{\partial \rho_{(A)}}{\partial t} + \text{div}\,(\rho_{(A)} \mathbf{v}) + \text{div}\,\mathbf{j}_{(A)} = r_{(A)} \tag{2-9}$$

## [8.2] MASS BALANCE

**Table 8.2.2-4 Mass and molar fluxes†**

| Quantity | With respect to fixed frame of reference | | With respect to $\mathbf{v}$ | | With respect to $\mathbf{v}^\star$ | |
|---|---|---|---|---|---|---|
| Velocity of species $A$ | $\mathbf{v}_{(A)}$ | (A) | $\mathbf{u}_{(A)} = \mathbf{v}_{(A)} - \mathbf{v}$ | (B) | $\mathbf{u}_{(A)}^\star = \mathbf{v}_{(A)} - \mathbf{v}^\star$ | (C) |
| Mass flux of species $A$ | $\mathbf{n}_{(A)} = \rho_{(A)}\mathbf{v}_{(A)}$ | (D) | $\mathbf{j}_{(A)} = \rho_{(A)}(\mathbf{v}_{(A)} - \mathbf{v})$ | (E) | $\mathbf{j}_{(A)}^\star = \rho_{(A)}(\mathbf{v}_{(A)} - \mathbf{v}^\star)$ | (F) |
| Molar flux of species $A$ | $\mathbf{N}_{(A)} = c_{(A)}\mathbf{v}_{(A)}$ | (G) | $\mathbf{J}_{(A)} = c_{(A)}(\mathbf{v}_{(A)} - \mathbf{v})$ | (H) | $\mathbf{J}_{(A)}^\star = c_{(A)}(\mathbf{v}_{(A)} - \mathbf{v}^\star)$ | (I) |
| Sum of mass fluxes | $\sum_{A=1}^{N} \mathbf{n}_{(A)} = \rho\mathbf{v}$ | (J) | $\sum_{A=1}^{N} \mathbf{j}_{(A)} = 0$ | (K) | $\sum_{A=1}^{N} \mathbf{j}_{(A)}^\star = \rho(\mathbf{v} - \mathbf{v}^\star)$ | (L) |
| Sum of molar fluxes | $\sum_{A=1}^{N} \mathbf{N}_{(A)} = c\mathbf{v}^\star$ | (M) | $\sum_{A=1}^{N} \mathbf{J}_{(A)} = c(\mathbf{v}^\star - \mathbf{v})$ | (N) | $\sum_{A=1}^{N} \mathbf{J}_{(A)}^\star = 0$ | (O) |
| Fluxes in terms of $\mathbf{n}_{(A)}$ ($A=1,2,\ldots,N$) | $\mathbf{N}_{(A)} = \dfrac{\mathbf{n}_{(A)}}{M_{(A)}}$ | (P) | $\mathbf{j}_{(A)} = \mathbf{n}_{(A)} - \omega_{(A)}\sum_{B=1}^{N}\mathbf{n}_{(B)}$ | (Q) | $\mathbf{j}_{(A)}^\star = \mathbf{n}_{(A)} - M_{(A)}x_{(A)}\sum_{B=1}^{N}\dfrac{\mathbf{n}_{(B)}}{M_{(B)}}$ | (R) |
| Fluxes in terms of $\mathbf{N}_{(A)}$ ($A=1,2,\ldots,N$) | $\mathbf{n}_{(A)} = M_{(A)}\mathbf{N}_{(A)}$ | (S) | $\mathbf{J}_{(A)} = \mathbf{N}_{(A)} - \dfrac{\omega_{(A)}}{M_{(A)}}\sum_{B=1}^{N}M_{(B)}\mathbf{N}_{(B)}$ | (T) | $\mathbf{J}_{(A)}^\star = \mathbf{N}_{(A)} - x_{(A)}\sum_{B=1}^{N}\mathbf{N}_{(B)}$ | (U) |
| Fluxes in terms of $\mathbf{j}_{(A)}$ and $\mathbf{v}$ | $\mathbf{n}_{(A)} = \mathbf{j}_{(A)} + \rho_{(A)}\mathbf{v}$ | (V) | $\mathbf{J}_{(A)} = \dfrac{\mathbf{j}_{(A)}}{M_{(A)}}$ | (W) | $\mathbf{j}_{(A)}^\star = \mathbf{j}_{(A)} - \omega_{(A)}M\sum_{B=1}^{N}\dfrac{\mathbf{j}_{(B)}}{M_{(B)}}$ | (X) |
| | | | *For a binary system:* $\mathbf{J}_{(A)} = \dfrac{M_{(B)}}{M}\mathbf{J}_{(A)}^\star$ | | *For a binary system:* $\mathbf{j}_{(A)} = \dfrac{M}{M_{(B)}}\mathbf{j}_{(A)}^\star$ | |
| Fluxes in terms of $\mathbf{J}_{(A)}^\star$ and $\mathbf{v}^\star$ | $\mathbf{N}_{(A)} = \mathbf{J}_{(A)}^\star + c_{(A)}\mathbf{v}^\star$ | (Y) | $\mathbf{J}_{(A)} = \mathbf{J}_{(A)}^\star - \dfrac{x_{(A)}}{M}\sum_{B=1}^{N}M_{(B)}\mathbf{J}_{(B)}^\star$ | (Z) | $\mathbf{j}_{(A)}^\star = M_{(A)}\mathbf{J}_{(A)}^\star$ | (AA) |

† Suggested by Table 16.1-3 of [1].

From the overall equation of continuity (Exercise 8.2.1-5 and Sec. 8.3.2):

$$\frac{\partial \rho_{(A)}}{\partial t} + \text{div}\,(\rho_{(A)}\mathbf{v}) = \frac{\partial \rho_{(A)}}{\partial t} + \nabla \rho_{(A)} \cdot \mathbf{v} + \rho_{(A)}(\text{div}\,\mathbf{v})$$

$$= \frac{d_{(v)}\rho_{(A)}}{dt} - \omega_{(A)}\frac{d_{(v)}\rho}{dt} = \rho\frac{d_{(v)}\omega_{(A)}}{dt} \quad (2\text{-}10)$$

If $B$ is any quantity, scalar or tensor, we use the notation

$$\frac{d_{(v)}B}{dt} \equiv \frac{\partial B}{\partial t} + \nabla B \cdot \mathbf{v} \quad (2\text{-}11)$$

to denote the derivative with respect to time following a fictitious particle which moves with the local mass-averaged velocity of the multicomponent mixture. Equations (2-9) and (2-10) give

$$\rho\frac{d_{(v)}\omega_{(A)}}{dt} + \text{div}\,\mathbf{j}_{(A)} = r_{(A)} \quad (2\text{-}12)$$

Several other alternative forms of the differential equation of continuity for species $A$ are presented in Table 8.2.2-5.

To assist the reader in formulating problems, Eq. (2-7) is presented for three specific coordinate systems in Table 8.2.2-6.

**Table 8.2.2-5 Forms of the equation of continuity for Species A**

$$\frac{\partial \rho_{(A)}}{\partial t} + \text{div}\,\mathbf{n}_{(A)} = r_{(A)} \quad (A)$$

$$\frac{\partial c_{(A)}}{\partial t} + \text{div}\,\mathbf{N}_{(A)} = \frac{r_{(A)}}{M_{(A)}} \quad (B)$$

$$\rho\frac{d_{(v)}\omega_{(A)}}{dt} + \text{div}\,\mathbf{j}_{(A)} = r_{(A)} \quad (C)\dagger$$

$$\rho\frac{d_{(v)}}{dt}\left(\frac{c_{(A)}}{\rho}\right) + \text{div}\,\mathbf{J}_{(A)} = \frac{r_{(A)}}{M_{(A)}} \quad (D)\dagger$$

$$c\frac{d_{(v\star)}}{dt}\left(\frac{\rho_{(A)}}{c}\right) + \text{div}\,\mathbf{j}^{\star}_{(A)} = r_{(A)} - M_{(A)}x_{(A)}\sum_{B=1}^{N}\frac{r_{(B)}}{M_{(B)}} \quad (E)\ddagger$$

$$c\frac{d_{(v\star)}x_{(A)}}{dt} + \text{div}\,\mathbf{J}^{\star}_{(A)} = \frac{r_{(A)}}{M_{(A)}} - x_{(A)}\sum_{B=1}^{N}\frac{r_{(B)}}{M_{(B)}} \quad (F)\ddagger$$

$$\dagger \quad \frac{d_{(v)}}{dt}\Psi \equiv \frac{\partial \Psi}{\partial t} + \nabla\Psi \cdot \mathbf{v}$$

$$\ddagger \quad \frac{d_{(v\star)}}{dt}\Psi \equiv \frac{\partial \Psi}{\partial t} + \nabla\Psi \cdot \mathbf{v}^{\star}$$

**Table 8.2.2-6** The equation of continuity for species A in several coordinate systems [Eq. (A) of Table 8.2.2-5]

*Rectangular cartesian coordinates:*

$$\frac{\partial \rho_{(A)}}{\partial t} + \frac{\partial n_{(A)1}}{\partial z_1} + \frac{\partial n_{(A)2}}{\partial z_2} + \frac{\partial n_{(A)3}}{\partial z_3} = r_{(A)} \quad (A)$$

*Cylindrical coordinates:*

$$\frac{\partial \rho_{(A)}}{\partial t} + \frac{1}{r}\frac{\partial}{\partial r}(rn_{(A)r}) + \frac{1}{r}\frac{\partial n_{(A)\theta}}{\partial \theta} + \frac{\partial n_{(A)z}}{\partial z} = r_{(A)} \quad (B)$$

*Spherical coordinates:*

$$\frac{\partial \rho_{(A)}}{\partial t} + \frac{1}{r^2}\frac{\partial}{\partial r}(r^2 n_{(A)r}) + \frac{1}{r\sin\theta}\frac{\partial}{\partial \theta}(n_{(A)\theta}\sin\theta) + \frac{1}{r\sin\theta}\frac{\partial n_{(A)\varphi}}{\partial \varphi} = r_{(A)} \quad (C)$$

## REFERENCE

1. Bird, R. B., W. E. Stewart, and E. N. Lightfoot: "Transport Phenomena," 7th printing, Wiley, New York, 1960.

## EXERCISE

**8.2.2-1** (a) Given Table 8.2.2-1, derive the relations given in Table 8.2.2-2.
(b) Given Tables 8.2.2-1 and 8.2.2-3, derive the relations given in Table 8.2.2-4.
(c) Starting with Eq. (A) of Table 8.2.2-5, derive Eqs. (B) to (F).

## 8.3 REVISED POSTULATES

**8.3.1 Previous postulates** Our discussions of momentum and energy transfer in single-component systems (Chaps. 1 to 7) are based upon five postulates: conservation of mass (Sec. 1.3.1), Euler's first and second laws (Sec. 2.2.1), the energy balance (Sec. 5.3.1), and the entropy inequality (Sec. 5.5.1).

We wish to preserve these postulates for our work with multicomponent systems, but it is clear that some changes must be made. We have not defined what we might mean by a material particle of a multicomponent mixture or by a multicomponent body. Rather than talk about a multicomponent body, we have visualized a superposition of material bodies, one corresponding to each of the $N$ species present. We recognize that it is not physically appealing to define a multicomponent body to be the collection of $N$ constituent bodies, since each of the $N$ species present may move and deform with respect to the other $N-1$ species.

Exercise 8.2.1-5 suggests how we may proceed. We define a *mass-averaged material particle* of a multicomponent mixture to be a (artificial) particle which moves with the mass-averaged velocity of the mixture (see Table 8.2.2-3). A *multicomponent body* is defined to be a set, the elements of which are mass-averaged material particles.

As mentioned in Exercise 8.2.1-3, we view a *phase interface* in a multicomponent mixture as a singular surface in a body of species $A$. Since a surface of discontinuity with respect to $\rho_{(A)}$ or $\mathbf{v}_{(A)}$ will mean a surface of discontinuity with respect to $\rho$ or $\mathbf{v}$, we shall find it convenient in what follows to identify a phase interface with a singular surface in a multicomponent body.

In the following sections, the notation $d_{(v)}/dt$ indicates a derivative with respect to time following a mass-averaged material particle. By $V_{(v)}$ we mean the volume or region of space occupied by a set of mass-averaged material particles; $S_{(v)}$ denotes the closed bounding surface of this region.

## EXERCISES

**8.3.1-1** *Transport theorem* Let $\Phi$ be any scalar, vector, or tensor field. Show that

$$\frac{d}{dt}\int_{V_{(v)}} \Phi \, dV = \int_{V_{(v)}} \left(\frac{d_{(v)}\Phi}{dt} + \Phi \, \text{div } \mathbf{v}\right) dV$$

This is the *transport theorem* applicable to a multicomponent body.

**8.3.1-2** *Transport theorem for a discontinuous body* Let $\Phi$ be any scalar, vector, or tensor field. Derive the following *transport theorem* for a multicomponent body which contains a singular surface $S_{(\text{sing})}$:

$$\frac{d}{dt}\int_{V_{(v)}} \Phi \, dV = \int_{V_{(v)}} \frac{\partial \Phi}{\partial t} dV + \int_{S_{(v)}} \Phi \mathbf{v}\cdot\mathbf{n}\, dS - \int_{S_{(\text{sing})}} [\Phi u_{(\xi)}] dS$$

The notation has the same meaning it had in Sec. 1.3.4, with the understanding that we are concerned with a singular surface which occurs in a multicomponent body.

### 8.3.2 Conservation of mass
A restatement of our first postulate, *conservation of mass*, is:

The mass of a multicomponent body is independent of time.

We define the *mass* $M$ of the multicomponent body as

$$M \equiv \int_{V_{(v)}} \rho \, dV \tag{2-1}$$

where

$$\rho \equiv \sum_{A=1}^{N} \rho_{(A)} \tag{2-2}$$

Our statement that mass is conserved says that

$$\frac{d}{dt}\int_{V_{(v)}} \rho \, dV = 0 \tag{2-3}$$

Let us restrict ourselves to a multicomponent body in which all quantities are continuous and differentiable as many times as desired. This rules out the appearance

## Table 8.3.2-1 Forms of the overall equation of continuity

$$\frac{\partial \rho}{\partial t} + \operatorname{div}(\rho \mathbf{v}) = 0 \qquad \text{(A)}$$

$$\frac{\partial c}{\partial t} + \operatorname{div}(c \mathbf{v}^\star) = \sum_{A=1}^{N} \frac{r_{(A)}}{M_{(A)}} \qquad \text{(B)}$$

---

of shock waves and phase interfaces (see Sec. 8.3.1). The transport theorem of Exercise 8.3.1-1 allows us to rewrite Eq. (2-3) as

$$\int_{V_{(\mathbf{v})}} \left( \frac{d_{(\mathbf{v})}\rho}{dt} + \rho \operatorname{div} \mathbf{v} \right) dV = 0 \qquad (2\text{-}4)$$

Since this must be true for an arbitrary multicomponent body, we conclude that

$$\frac{d_{(\mathbf{v})}\rho}{dt} + \rho \operatorname{div} \mathbf{v} = 0 \qquad (2\text{-}5)$$

This equation may be referred to as the *overall equation of continuity* for a multicomponent mixture. Not surprisingly, it is of the same form as the equation of continuity for a single-component material developed in Sec. 1.3.3.

Equation (2-5) confirms the intuitive feeling expressed in Exercise 8.2.1-5 that

$$\sum_{A=1}^{N} r_{(A)} = 0 \qquad (2\text{-}6)$$

Two commonly used forms of the overall equation of continuity are presented in Table 8.3.2-1.

## EXERCISES

**8.3.2-1** *Overall jump mass balance* Use Exercise 8.3.1-2 and the approach suggested in Sec. 1.3.5 to derive the following *overall jump mass balance* at a phase interface:

$$[\rho(\mathbf{v} \cdot \boldsymbol{\xi} - u_{(\xi)})] = 0$$

As might have been expected, this jump mass balance has the same form as that for a single component material.

**8.3.2-2** *More on the overall jump mass balance* Starting with the jump mass balance for an individual species in Exercise 8.2.1-3, derive as an alternate form of the *overall jump mass balance* at a phase interface:

$$[c(\mathbf{v}^\star \cdot \boldsymbol{\xi} - u_{(\xi)})] = \sum_{A=1}^{N} \frac{r_{(A)}^{(\sigma)}}{M_{(A)}}$$

**8.3.2-3** Use Exercise 8.3.1-2 and the approach suggested in Exercise 1.3.5-2 to derive the overall jump mass balance of Exercise 8.3.2-1.

**8.3.2-4** *Alternate form of the transport theorem* Show that an alternate form of the transport theorem of Exercise 8.3.1-1 is

$$\frac{d}{dt} \int_{V_{(\mathbf{v})}} \rho \Phi \, dV = \int_{V_{(\mathbf{v})}} \rho \frac{d_{(\mathbf{v})}\Phi}{dt} dV$$

It is to be again understood that $\Phi$ is any scalar, vector, or tensor field.

**8.3.2-5** Derive Eq. (B) of Table 8.3.2-1 starting from the equation of continuity for an individual species.

### 8.3.3 Euler's first law

We may restate our second postulate, *Euler's first law*, as:

> The time rate of change of the momentum of a multicomponent body relative to the fixed stars is equal to the sum of the forces acting on the body.

Our discussion of forces in Sec. 2.1.1 continues to apply, but we must recognize that each constituent may be subject to different external forces. Consider for example a dilute solution of sodium chloride in water subjected to an electric field. The sodium chloride will be nearly fully ionized, which means that we must consider three separate species: the sodium ions, the chloride ions, and water. The force of the electric field upon the sodium ions will be equal in magnitude but opposite in direction to the force which it exerts on the chloride ions. The electric field does not act directly upon the water. Yet all three species are under the influence of gravity.

With this thought in mind, our statement of Euler's first law may be written as

$$\frac{d}{dt}\int_{V_{(v)}} \rho \mathbf{v}\, dV = \int_{S_{(v)}} \mathbf{t}\, dS + \int_{V_{(v)}} \sum_{A=1}^{N} \rho_{(A)}\mathbf{f}_{(A)}\, dV \tag{3-1}$$

Here $\mathbf{f}_{(A)}$ denotes the external force per unit mass acting upon species $A$.

Let us again restrict ourselves to a multicomponent body in which all quantities are continuous and differentiable as many times as desired. If we introduce the stress tensor $\mathbf{T}$ in the usual manner, we may use the transport theorem of Exercise 8.3.2-4 to conclude that

$$\rho \frac{d_{(v)}\mathbf{v}}{dt} = \text{div } \mathbf{T} + \sum_{A=1}^{N} \rho_{(A)}\mathbf{f}_{(A)} \tag{3-2}$$

This equation is known as *Cauchy's first law* for a multicomponent mixture.

If we define the *mass-averaged external force per unit mass* as

$$\mathbf{f} = \frac{1}{\rho} \sum_{A=1}^{N} \rho_{(A)}\mathbf{f}_{(A)} \tag{3-3}$$

Eq. (3-2) has the same form as Cauchy's first law for a single-component material:

$$\rho \frac{d_{(v)}\mathbf{v}}{dt} = \text{div } \mathbf{T} + \rho\mathbf{f} \tag{3-4}$$

### EXERCISES

**8.3.3-1** *Overall jump momentum balance* Use Exercise 8.3.1-2 and the approach suggested in Sec. 1.3.5 to derive the following *overall jump momentum balance* at a phase interface:

$$[\rho\mathbf{v}(\mathbf{v}\cdot\boldsymbol{\xi} - u_{(\xi)}) - \mathbf{T}\cdot\boldsymbol{\xi}] = 0$$

[8.3] REVISED POSTULATES 461

This has the same form as the jump momentum balance for a single-component material (see Exercise 2.2.3-1).

**8.3.3-2** Use Exercise 8.3.1-2 and the approach suggested in Exercise 1.3.5-2 to derive the overall jump momentum balance of Exercise 8.3.3-1.

### 8.3.4 Euler's second law
We shall restate our third postulate, *Euler's second law*, as:

> The time rate of change of the moment of momentum of a multicomponent body relative to the fixed stars is equal to the sum of the moments of all forces acting on the body.

This thought may be expressed as

$$\frac{d}{dt}\int_{V_{(v)}} \rho(\mathbf{p}\wedge\mathbf{v})\,dV = \int_{S_{(v)}} \mathbf{p}\wedge\mathbf{t}\,dS + \int_{V_{(v)}} \rho(\mathbf{p}\wedge\mathbf{f})\,dV \tag{4-1}$$

Note that we have introduced here the mass-averaged external force per unit mass **f** defined in Sec. 8.3.3.

Let us again restrict ourselves to a multicomponent body in which all quantities are continuous and differentiable as many times as desired. If we introduce the stress tensor **T**, we may use the transport theorem of Exercise 8.3.2-4 and Cauchy's first law for a multicomponent mixture to conclude that

$$\mathbf{T} = \mathbf{T}^T \tag{4-2}$$

This expresses *Cauchy's second law* for a multicomponent mixture:

> A necessary and sufficient condition for the balance of moment of momentum in a multicomponent body where momentum is balanced is that the stress tensor be symmetric.

### 8.3.5 The energy balance
Our fourth postulate, the *energy balance*, may be revised to read:

> The time rate of change of the internal and kinetic energy of a multicomponent body relative to the fixed stars is equal to the rate at which work is done on the body by the forces which act upon the body plus the rate of energy transmission to the body.

This idea may be stated symbolically as

$$\frac{d}{dt}\int_{V_{(v)}} \rho\left(\hat{U} + \sum_{A=1}^{N} \tfrac{1}{2}\omega_{(A)}v_{(A)}^2\right) dV = \int_{S_{(v)}} \mathbf{v}\cdot(\mathbf{T}\cdot\mathbf{n})\,dS$$
$$+ \int_{V_{(v)}} \sum_{A=1}^{N} \rho_{(A)}(\mathbf{v}_{(A)}\cdot\mathbf{f}_{(A)})\,dV + \int_{S_{(v)}} h\,dS + \int_{V_{(v)}} \rho Q\,dV \tag{5-1}$$

It is entirely possible that the external and mutual energy transmission rates may vary among the species; in this event we understand that $Q$ is the mass average of the external and mutual energy transmission rates per unit mass which correspond to each of the $N$ species.

Let us limit ourselves to a multicomponent body in which all quantities are continuous and differentiable as many times as desired. If we introduce the energy flux vector (see Exercise 5.3.1-2), the transport theorem of Exercise 8.3.2-4 and Cauchy's first law allow us to obtain two important forms of the *overall differential energy balance* for a multicomponent mixture:

$$\rho \frac{d_{(v)}}{dt}\left(\hat{U} + \sum_{A=1}^{N} \tfrac{1}{2}\omega_{(A)}u_{(A)}^2 + \tfrac{1}{2}v^2\right) = -\text{div } \mathbf{q}$$

$$+ \text{div } (\mathbf{T} \cdot \mathbf{v}) + \sum_{A=1}^{N} \rho_{(A)}(\mathbf{v}_{(A)} \cdot \mathbf{f}_{(A)}) + \rho Q \quad (5\text{-}2)$$

and

$$\rho \frac{d_{(v)}\hat{U}}{dt} = -\text{div } \mathbf{q} + \text{tr } (\mathbf{T} \cdot \nabla \mathbf{v}) + \sum_{A=1}^{N} \rho_{(A)}(\mathbf{u}_{(A)} \cdot \mathbf{f}_{(A)})$$

$$+ \rho Q - \rho \frac{d_{(v)}}{dt} \sum_{A=1}^{N} \tfrac{1}{2}\omega_{(A)}u_{(A)}^2 \quad (5\text{-}3)$$

Here we introduce $\mathbf{u}_{(A)}$, the velocity of species $A$ with respect to the mass-averaged velocity $\mathbf{v}$ (see Table 8.2.2-3).

For most situations, we expect the last term on the right of Eq. (5-3) to be small compared with the term on the left and we are prepared to neglect it.

One can define a new internal energy $\hat{U}^*$ [1, p. 613; 2, p. 28]

$$\hat{U}^* \equiv \hat{U} + \sum_{A=1}^{N} \tfrac{1}{2}\omega_{(A)}u_{(A)}^2 \quad (5\text{-}4)$$

and write the overall energy balance as

$$\rho \frac{d_{(v)}\hat{U}^*}{dt} = -\text{div } \mathbf{q} + \text{tr } (\mathbf{T} \cdot \nabla \mathbf{v}) + \sum_{A=1}^{N} \rho_{(A)}(\mathbf{u}_{(A)} \cdot \mathbf{f}_{(A)}) + \rho Q \quad (5\text{-}5)$$

Unfortunately, this simplification disappears when we employ the fundamental constitutive equation for a thermodynamically homogeneous material, Eq. (2-2) or (2-3) of Sec. 5.1.2. [Remember that our fundamental constitutive equation for $\hat{U}$, Eq. (2-1) of Sec. 5.1.2, assumes that it is not an explicit function of the motion of the material. Equation (5-4) requires $\hat{U}^*$ to depend explicitly upon the material's motion through the last term on the right.]

We are usually more interested in determining the temperature distribution within a material than the internal energy distribution. From Eq. (2-15) of Sec.

[8.3] REVISED POSTULATES

5.1.2, we have that

$$d\hat{U} = T\,d\hat{S} - P\,d\hat{V} + \sum_{A=1}^{N-1}(\mu_{(A)} - \mu_{(N)})\,d\omega_{(A)}$$

$$= T\left(\frac{\partial \hat{S}}{\partial T}\right)_{\hat{V},\omega_{(B)}} dT + \left[T\left(\frac{\partial \hat{S}}{\partial \hat{V}}\right)_{T,\omega_{(B)}} - P\right]d\hat{V}$$

$$+ \sum_{A=1}^{N-1}\left[T\left(\frac{\partial \hat{S}}{\partial \omega_{(A)}}\right)_{T,\hat{V},\omega_{(B)}(B \neq A,N)} + \mu_{(A)} - \mu_{(N)}\right]d\omega_{(A)}$$

$$= \hat{c}_V\,dT + \left[T\left(\frac{\partial P}{\partial T}\right)_{\hat{V},\omega_{(B)}} - P\right]d\hat{V}$$

$$+ \sum_{A=1}^{N-1}\left\{-T\left[\frac{\partial(\mu_{(A)} - \mu_{(N)})}{\partial T}\right]_{\hat{V},\omega_{(B)}} + \mu_{(A)} - \mu_{(N)}\right\}d\omega_{(A)}$$

$$= \hat{c}_V\,dT + \left[T\left(\frac{\partial P}{\partial T}\right)_{\hat{V},\omega_{(B)}} - P\right]d\hat{V} + \sum_{A=1}^{N}\left[\mu_{(A)} - T\left(\frac{\partial \mu_{(A)}}{\partial T}\right)_{\hat{V},\omega_{(B)}}\right]d\omega_{(A)}$$

(5-6)

Here we introduce the heat capacity per unit mass at constant specific volume (see Exercise 5.1.2-7),

$$\hat{c}_V \equiv T\left(\frac{\partial \hat{S}}{\partial T}\right)_{\hat{V},\omega_{(B)}} \tag{5-7}$$

and we employ one of the Maxwell relations (see Exercise 5.1.2-2):

$$\left(\frac{\partial \hat{S}}{\partial \omega_{(A)}}\right)_{T,\hat{V},\omega_{(B)}(B \neq A,N)} = -\left[\frac{\partial(\mu_{(A)} - \mu_{(N)})}{\partial T}\right]_{\hat{V},\omega_{(B)}} \tag{5-8}$$

The overall equation of continuity tells us that

$$\rho \frac{d_{(v)}\hat{V}}{dt} = -\frac{1}{\rho}\frac{d_{(v)}\rho}{dt} = \text{div }\mathbf{v} \tag{5-9}$$

and the equation of continuity for species $A$ gives (see Table 8.2.2-5)

$$\rho \frac{d_{(v)}\omega_{(A)}}{dt} = -\text{div }\mathbf{j}_{(A)} + r_{(A)} \tag{5-10}$$

**Table 8.3.5-1  Various forms of the overall differential energy balance†**

$$\rho \frac{d_{(v)}}{dt}(\hat{U} + \tfrac{1}{2}v^2 + \varphi) = -\text{div } \mathbf{q} + \text{div }(\mathbf{T}\cdot\mathbf{v}) + \rho Q + \sum_{A=1}^{N} \mathbf{j}_{(A)}\cdot\mathbf{f}_{(A)} - \rho \frac{d_{(v)}}{dt}\sum_{A=1}^{N} \tfrac{1}{2}\omega_{(A)}u_{(A)}^2 \quad \text{(A)}\ddagger$$

$$\rho \frac{d_{(v)}}{dt}(\hat{U} + \tfrac{1}{2}v^2) = -\text{div } \mathbf{q} + \text{div }(\mathbf{T}\cdot\mathbf{v}) + \sum_{A=1}^{N} \mathbf{n}_{(A)}\cdot\mathbf{f}_{(A)} + \rho Q - \rho \frac{d_{(v)}}{dt}\sum_{A=1}^{N} \tfrac{1}{2}\omega_{(A)}u_{(A)}^2 \quad \text{(B)}$$

$$\rho \frac{d_{(v)}\hat{U}}{dt} = -\text{div } \mathbf{q} - P\,\text{div } \mathbf{v} + \text{tr}(\mathbf{S}\cdot\nabla\mathbf{v}) + \rho Q + \sum_{A=1}^{N} \mathbf{j}_{(A)}\cdot\mathbf{f}_{(A)} - \rho \frac{d_{(v)}}{dt}\sum_{A=1}^{N} \tfrac{1}{2}\omega_{(A)}u_{(A)}^2 \quad \text{(C)}$$

$$\rho \frac{d_{(v)}\hat{H}}{dt} = -\text{div } \mathbf{q} + \frac{d_{(v)}P}{dt} + \text{tr}(\mathbf{S}\cdot\nabla\mathbf{v}) + \rho Q + \sum_{A=1}^{N} \mathbf{j}_{(A)}\cdot\mathbf{f}_{(A)} - \rho \frac{d_{(v)}}{dt}\sum_{A=1}^{N} \tfrac{1}{2}\omega_{(A)}u_{(A)}^2 \quad \text{(D)}$$

$$\rho \hat{c}_V \frac{d_{(v)}T}{dt} = -\text{div } \mathbf{q} - T\left(\frac{\partial P}{\partial T}\right)_{\hat{V},\omega_{(B)}}\text{div } \mathbf{v} + \text{tr}(\mathbf{S}\cdot\nabla\mathbf{v}) + \rho Q + \sum_{A=1}^{N} \mathbf{j}_{(A)}\cdot\mathbf{f}_{(A)}$$
$$+ \sum_{A=1}^{N}\left[\mu_{(A)} - T\left(\frac{\partial \mu_{(A)}}{\partial T}\right)_{\hat{V},\omega_{(B)}}\right](\text{div }\mathbf{j}_{(A)} - r_{(A)}) - \rho \frac{d_{(v)}}{dt}\sum_{A=1}^{N} \tfrac{1}{2}\omega_{(A)}u_{(A)}^2 \quad \text{(E)}$$

$$\rho \hat{c}_V \frac{d_{(v)}T}{dt} = -\text{div } \mathbf{q} - T\left(\frac{\partial P}{\partial T}\right)_{\hat{V},\omega_{(B)}}\text{div } \mathbf{v} + \text{tr}(\mathbf{S}\cdot\nabla\mathbf{v}) + \rho Q + \sum_{A=1}^{N} \mathbf{j}_{(A)}\cdot\mathbf{f}_{(A)}$$
$$+ \sum_{A=1}^{N}\left\{\overline{U}_{(A)} + \left[P - T\left(\frac{\partial P}{\partial T}\right)_{\hat{V},\omega_{(B)}}\right]\overline{V}_{(A)}\right\}(\text{div }\mathbf{j}_{(A)} - r_{(A)}) - \rho \frac{d_{(v)}}{dt}\sum_{A=1}^{N} \tfrac{1}{2}\omega_{(A)}u_{(A)}^2 \quad \text{(F)}$$

$$\rho \hat{c}_P \frac{d_{(v)}T}{dt} = -\text{div } \mathbf{q} + \left(\frac{\partial \ln \hat{V}}{\partial \ln T}\right)_{P,\omega_{(B)}}\frac{d_{(v)}P}{dt} + \text{tr}(\mathbf{S}\cdot\nabla\mathbf{v}) + \rho Q + \sum_{A=1}^{N} \mathbf{j}_{(A)}\cdot\mathbf{f}_{(A)}$$
$$+ \sum_{A=1}^{N}\left[\mu_{(A)} - T\left(\frac{\partial \mu_{(A)}}{\partial T}\right)_{P,\omega_{(B)}}\right](\text{div }\mathbf{j}_{(A)} - r_{(A)}) - \rho \frac{d_{(v)}}{dt}\sum_{A=1}^{N} \tfrac{1}{2}\omega_{(A)}u_{(A)}^2 \quad \text{(G)}$$

$$\rho \hat{c}_P \frac{d_{(v)}T}{dt} = -\text{div } \mathbf{q} + \left(\frac{\partial \ln \hat{V}}{\partial \ln T}\right)_{P,\omega_{(B)}}\frac{d_{(v)}P}{dt} + \text{tr}(\mathbf{S}\cdot\nabla\mathbf{v}) + \rho Q + \sum_{A=1}^{N} \mathbf{j}_{(A)}\cdot\mathbf{f}_{(A)}$$
$$+ \sum_{A=1}^{N}\overline{H}_{(A)}(\text{div }\mathbf{j}_{(A)} - r_{(A)}) - \rho \frac{d_{(v)}}{dt}\sum_{A=1}^{N} \tfrac{1}{2}\omega_{(A)}u_{(A)}^2 \quad \text{(H)}$$

$$\rho \frac{d_{(v)}\hat{S}}{dt} = -\text{div}\left(\frac{1}{T}\left[\mathbf{q} - \sum_{A=1}^{N}(\mu_{(A)} + \tfrac{1}{2}u_{(A)}^2)\mathbf{j}_{(A)}\right]\right) - \frac{1}{T^2}\mathbf{q}\cdot\nabla T + \frac{1}{T}\text{tr}(\mathbf{S}\cdot\nabla\mathbf{v}) + \rho \frac{Q}{T}$$
$$- \frac{1}{T}\sum_{A=1}^{N}\mathbf{j}_{(A)}\cdot\left[T\nabla\left(\frac{\mu_{(A)} + \tfrac{1}{2}u_{(A)}^2}{T}\right) - \mathbf{f}_{(A)} + \frac{d_{(v)}\mathbf{u}_{(A)}}{dt}\right] - \sum_{A=1}^{N}\frac{1}{T}(\mu_{(A)} + \tfrac{1}{2}u_{(A)}^2)r_{(A)} \quad \text{(I)}$$

$$\frac{\partial}{\partial t}\left(\sum_{A=1}^{N}\rho_{(A)}\overline{H}_{(A)}\right) + \text{div}\left(\sum_{A=1}^{N}\overline{H}_{(A)}\mathbf{n}_{(A)}\right) = -\text{div}\left(\mathbf{q} - \sum_{A=1}^{N}\overline{H}_{(A)}\mathbf{j}_{(A)}\right) + \frac{d_{(v)}P}{dt} + \text{tr}(\mathbf{S}\cdot\nabla\mathbf{v})$$
$$+ \rho Q + \sum_{A=1}^{N}\mathbf{j}_{(A)}\cdot\mathbf{f}_{(A)} - \rho \frac{d_{(v)}}{dt}\sum_{A=1}^{N}\tfrac{1}{2}\omega_{(A)}u_{(A)}^2 \quad \text{(J)}$$

† Most authors (for example [3, p. 562]) neglect $\rho \frac{d_{(v)}}{dt}\sum_{A=1}^{N}\tfrac{1}{2}\omega_{(A)}u_{(A)}^2$ appearing on the right side of these equations. This point is discussed following Eq. (5-3) in the text.

‡ We assume here that $\mathbf{f} \equiv \sum_{A=1}^{N}\omega_{(A)}\mathbf{f}_{(A)} = -\nabla\varphi$, where $\varphi$ is not an explicit function of time.

## [8.3] REVISED POSTULATES

Equations (5-6), (5-9), and (5-10) allow us to rewrite Eq. (5-3) as

$$\rho \hat{c}_V \frac{d_{(v)} T}{dt} = -\text{div } \mathbf{q} + \text{tr } (\mathbf{S} \cdot \nabla \mathbf{v}) + \sum_{A=1}^{N} \rho_{(A)} \mathbf{u}_{(A)} \cdot \mathbf{f}_{(A)}$$

$$+ \rho Q - T \left( \frac{\partial P}{\partial T} \right)_{\hat{V}, \omega_{(B)}} \text{div } \mathbf{v}$$

$$+ \sum_{A=1}^{N} \left[ \mu_{(A)} - T \left( \frac{\partial \mu_{(A)}}{\partial T} \right)_{\hat{V}, \omega_{(B)}} \right] (\text{div } \mathbf{j}_{(A)} - r_{(A)})$$

$$- \rho \frac{d_{(v)}}{dt} \sum_{A=1}^{N} \tfrac{1}{2} \omega_{(A)} u_{(A)}^2 \tag{5-11}$$

Various forms of the overall differential energy balance are presented in Table 8.3.5-1.

## REFERENCES

1. Truesdell, C., and R. A. Toupin: In S. Flügge (ed.), "Handbuch der Physik," vol. 3/1, Springer-Verlag, Berlin, 1960.
2. Groot, S. R. de, and P. Mazur: "Non-equilibrium Thermodynamics," Interscience, New York, 1962.
3. Bird, R. B., W. E. Stewart, and E. N. Lightfoot: "Transport Phenomena," 7th printing, Wiley, New York, 1960.

## EXERCISES

**8.3.5-1** *Overall jump energy balance*  Use Exercise 8.3.1-2 and the approach suggested in Sec. 1.3.5 to derive the following *overall jump energy balance* at a phase interface:

$$\left[ \rho \left( \hat{U} + \tfrac{1}{2} v^2 + \sum_{A=1}^{N} \tfrac{1}{2} \omega_{(A)} u_{(A)}^2 \right) (\mathbf{v} \cdot \boldsymbol{\xi} - u_{(\xi)}) + (\mathbf{q} - \mathbf{T} \cdot \mathbf{v}) \cdot \boldsymbol{\xi} \right] = 0$$

Note that this result is somewhat different from that obtained for a single-component material (see Exercise 5.3.2-1).

**8.3.5-2**  Use Exercise 8.3.1-2 and the approach suggested in Exercise 1.3.5-4 to derive the overall jump energy balance of Exercise 8.3.5-1.

**8.3.5-3**  (*a*) Starting with the definition of a partial mass variable given in Exercise 5.1.2-4, derive

$$\bar{U}_{(A)} - \bar{U}_{(N)} + \left[ P - T \left( \frac{\partial P}{\partial T} \right)_{\hat{V}, \omega_{(B)}} \right] (\bar{V}_{(A)} - \bar{V}_{(N)}) = \mu_{(A)} - \mu_{(N)} - T \left[ \frac{\partial (\mu_{(A)} - \mu_{(N)})}{\partial T} \right]_{\hat{V}, \omega_{(B)}}$$

(b) In the same manner find that

$$\bar{H}_{(A)} - \bar{H}_{(N)} = \mu_{(A)} - \mu_{(N)} - T\left(\frac{\partial(\mu_{(A)} - \mu_{(N)})}{\partial T}\right)_{P,\omega_{(B)}}$$

**8.3.5-4** Derive the additional forms of the overall differential energy balance given in Table 8.3.5-1.

**8.3.6 The entropy inequality** Our fifth postulate, the *entropy inequality*, may be revised to read:

> The minimum rate of production of entropy in a multicomponent body is locally proportional to the rates of energy transmission and of mass transfer for all species to the body.

In order to obtain a more explicit interpretation of this postulate, let us follow the procedure of Sec. 5.5.1 and express the energy balance in the form of an entropy balance for a continuous body.

From the Gibbs equation, Eq. (2-15) of Sec. 5.1.2, from the equation of continuity for species $A$, Eq. (C) of Table 8.2.2-5, and from the overall equation of continuity of Sec. 8.3.2, we have that

$$\frac{\rho}{T}\frac{d_{(v)}\hat{U}}{dt} = \rho\frac{d_{(v)}\hat{S}}{dt} + \frac{P}{T\rho}\frac{d_{(v)}\rho}{dt} + \sum_{A=1}^{N}\frac{\mu_{(A)}\rho}{T}\frac{d_{(v)}\omega_{(A)}}{dt}$$

$$= \rho\frac{d_{(v)}\hat{S}}{dt} - \frac{P}{T}\operatorname{div}\mathbf{v} + \sum_{A=1}^{N}\frac{\mu_{(A)}}{T}(r_{(A)} - \operatorname{div}\mathbf{j}_{(A)}) \qquad (6\text{-}1)$$

The differential energy balance, Eq. (C) of Table 8.3.5-1, may be written as

$$\frac{\rho}{T}\frac{d_{(v)}\hat{U}}{dt} = -\frac{1}{T}\operatorname{div}\mathbf{q} + \frac{1}{T}\operatorname{tr}(\mathbf{T}\cdot\nabla\mathbf{v}) + \frac{1}{T}\sum_{A=1}^{N}\mathbf{j}_{(A)}\cdot\mathbf{f}_{(A)}$$

$$+ \rho\frac{Q}{T} - \sum_{A=1}^{N}\frac{\mathbf{j}_{(A)}}{T}\cdot\frac{d_{(v)}\mathbf{u}_{(A)}}{dt} - \sum_{A=1}^{N}\frac{u_{(A)}^{2}\rho}{2T}\frac{d_{(v)}\omega_{(A)}}{dt}$$

$$= -\operatorname{div}\left(\frac{\mathbf{q}}{T}\right) - \frac{1}{T^{2}}\mathbf{q}\cdot\nabla T + \frac{1}{T}\operatorname{tr}(\mathbf{T}\cdot\nabla\mathbf{v})$$

$$+ \frac{1}{T}\sum_{A=1}^{N}\mathbf{j}_{(A)}\cdot\mathbf{f}_{(A)} + \frac{\rho Q}{T} - \sum_{A=1}^{N}\frac{\mathbf{j}_{(A)}}{T}\cdot\frac{d_{(v)}\mathbf{u}_{(A)}}{dt}$$

$$- \sum_{A=1}^{N}\frac{u_{(A)}^{2}}{2T}(r_{(A)} - \operatorname{div}\mathbf{j}_{(A)}) \qquad (6\text{-}2)$$

In this rearrangement, Eq. (C) of Table 8.2.2-5 has again been employed. We may

## [8.3] REVISED POSTULATES

now eliminate $d_{(v)}\hat{U}/dt$ between Eqs. (6-1) and (6-2) to obtain

$$\rho \frac{d_{(v)}\hat{S}}{dt} = -\text{div}\left(\frac{\mathbf{q}}{T}\right) - \frac{1}{T^2}\mathbf{q}\cdot\nabla T + \frac{1}{T}\text{tr}(\mathbf{S}\cdot\nabla\mathbf{v})$$

$$+ \frac{1}{T}\sum_{A=1}^{N}\mathbf{j}_{(A)}\cdot\mathbf{f}_{(A)} + \frac{\rho Q}{T} - \sum_{A=1}^{N}\frac{\mathbf{j}_{(A)}}{T}\cdot\frac{d_{(v)}\mathbf{u}_{(A)}}{dt}$$

$$- \sum_{A=1}^{N}\frac{1}{T}(\mu_{(A)} + \tfrac{1}{2}u_{(A)}^2)(r_{(A)} - \text{div}\,\mathbf{j}_{(A)})$$

$$= -\text{div}\left(\frac{1}{T}\left[\mathbf{q} - \sum_{A=1}^{N}(\mu_{(A)} + \tfrac{1}{2}u_{(A)}^2)\mathbf{j}_{(A)}\right]\right)$$

$$- \frac{1}{T^2}\mathbf{q}\cdot\nabla T + \frac{1}{T}\text{tr}(\mathbf{S}\cdot\nabla\mathbf{v}) + \frac{\rho Q}{T}$$

$$- \frac{1}{T}\sum_{A=1}^{N}\mathbf{j}_{(A)}\cdot\left[T\nabla\left(\frac{\mu_{(A)} + \tfrac{1}{2}u_{(A)}^2}{T}\right) - \mathbf{f}_{(A)} + \frac{d_{(v)}\mathbf{u}_{(A)}}{dt}\right]$$

$$- \sum_{A=1}^{N}\frac{1}{T}(\mu_{(A)} + \tfrac{1}{2}u_{(A)}^2)r_{(A)} \tag{6-3}$$

We may finally integrate this result over a continuous body and employ the transport theorem of Exercise 8.3.1-1 to obtain as an entropy balance for any continuous body:

$$\frac{d}{dt}\int_{V_{(v)}}\rho\hat{S}\,dV = -\int_{S_{(v)}}\frac{1}{T}\left[\mathbf{q} - \sum_{A=1}^{N}(\mu_{(A)} + \tfrac{1}{2}u_{(A)}^2)\mathbf{j}_{(A)}\right]\cdot\mathbf{n}\,dS$$

$$+ \int_{V_{(v)}}\left\{-\frac{1}{T^2}\mathbf{q}\cdot\nabla T + \frac{1}{T}\text{tr}(\mathbf{S}\cdot\nabla\mathbf{v}) + \frac{\rho Q}{T}\right.$$

$$- \frac{1}{T}\sum_{A=1}^{N}\mathbf{j}_{(A)}\cdot\left[T\nabla\left(\frac{\mu_{(A)} + \tfrac{1}{2}u_{(A)}^2}{T}\right) - \mathbf{f}_{(A)} + \frac{d_{(v)}\mathbf{u}_{(A)}}{dt}\right]$$

$$\left. - \sum_{A=1}^{N}\frac{1}{T}(\mu_{(A)} + \tfrac{1}{2}u_{(A)}^2)r_{(A)}\right\}dV \tag{6-4}$$

This result suggests that the *entropy inequality* may be expressed symbolically for *any* body as

$$\frac{d}{dt}\int_{V_{(v)}}\rho\hat{S}\,dV + \int_{S_{(v)}}\frac{1}{T}\left[\mathbf{q} - \sum_{A=1}^{N}(\mu_{(A)} + \tfrac{1}{2}u_{(A)}^2)\mathbf{j}_{(A)}\right]\cdot\mathbf{n}\,dS$$

$$- \int_{V_{(v)}}\frac{\rho Q}{T}\,dV \geq 0 \tag{6-5}$$

The second term on the left of this inequality expresses the net rate at which entropy is transferred from the body to the surroundings through the bounding surface of the multicomponent body. The third term on the left indicates the supply of entropy to the body as the result of external energy transmission.

A particular constitutive equation for specific internal energy as a function of entropy, Eq. (2-3) of Sec. 5.1.2, has been employed in *motivating* inequality (6-5). But the entropy inequality in the form of inequality (6-5) is stated as a fundamental postulate, like mass conservation. It should be satisfied by all multicomponent bodies, no matter what constitutive equations are used to describe their behavior.

Inequality (6-5) and Eq. (6-3) together imply the *overall differential entropy inequality* valid at each point in a multicomponent material,

$$\rho \frac{d_{(v)}\hat{S}}{dt} + \text{div}\left(\frac{1}{T}\left[\mathbf{q} - \sum_{A=1}^{N}(\mu_{(A)} + \tfrac{1}{2}u_{(A)}^2)\mathbf{j}_{(A)}\right]\right) - \rho \frac{Q}{T} = -\frac{1}{T^2}\mathbf{q}\cdot\nabla T$$

$$+ \frac{1}{T}\text{tr}(\mathbf{S}\cdot\nabla\mathbf{v}) - \frac{1}{T}\sum_{A=1}^{N}\mathbf{j}_{(A)}\cdot\left[T\nabla\left(\frac{\mu_{(A)} + \tfrac{1}{2}u_{(A)}^2}{T}\right) - \mathbf{f}_{(A)} + \frac{d_{(v)}\mathbf{u}_{(A)}}{dt}\right]$$

$$-\sum_{A=1}^{N}\frac{1}{T}(\mu_{(A)} + \tfrac{1}{2}u_{(A)}^2)r_{(A)} \geq 0 \quad (6\text{-}6)$$

Inequality (6-5) is a special case of the entropy inequality proposed by Bowen [1]. Inequality (6-6) is in substantial agreement with the form of the overall differential entropy inequality for multicomponent bodies proposed by de Groot and Mazur [2, pp. 22 and 29]. If we neglect terms involving the kinetic energy of diffusion ($\tfrac{1}{2}u_{(A)}^2$), inequality (6-6) reduces to the suggestion of Meixner and Reik [3]. However, other forms of the entropy inequality have been postulated [4; 5, p. 646; 6] and this remains an active area of current research.

**REFERENCES**

1. Bowen, R. M.: *Arch. Rational Mech. Analysis*, **24**:370 (1967).
2. de Groot, S. R., and P. Mazur: "Non-equilibrium Thermodynamics," Interscience, New York, 1962.
3. Meixner, J., and H. G. Reik: In S. Flügge (ed.), "Handbuch der Physik," vol. 3/2, Springer-Verlag, Berlin, 1959.
4. Eckart, Carl: *Phys. Rev.*, **58**:269, 924 (1940).
5. Truesdell, C., and R. A. Toupin: In S. Flügge (ed.), "Handbuch der Physik," vol. 3/1, Springer-Verlag, Berlin, 1960.
6. Müller, Ingo: *Arch. Rational Mech. Analysis*, **28**:1 (1968).
7. Hirschfelder, J. O., C. F. Curtiss, and R. B. Bird: "Molecular Theory of Gases and Liquids," Wiley, New York, 1954: corrected with notes added 1964.

**EXERCISES**

**8.3.6-1** *Overall jump entropy inequality* Use Exercise 8.3.1-2 and the approach suggested in Exercise 1.3.5-2 to derive the *overall jump entropy inequality*:

$$\left[\rho\hat{S}(\mathbf{v}\cdot\boldsymbol{\xi} - u_{(\xi)}) + \frac{1}{T}\left\{\mathbf{q} - \sum_{A=1}^{N}(\mu_{(A)} + \tfrac{1}{2}u_{(A)}^2)\mathbf{j}_{(A)}\right\}\cdot\boldsymbol{\xi}\right] \geq 0$$

Note that this result is somewhat different from that obtained for a single-component material (see Exercise 5.5.2-2).

## [8.3] REVISED POSTULATES

**8.3.6-2** Discuss the difficulty encountered in deriving the overall jump entropy inequality of Exercise 8.3.6-1 when the approach of Sec. 1.3.5 is used.

**8.3.6-3** Let us neglect interphase mass transfer, assume temperature is continuous across a phase interface, and assume that the tangential components of velocity of all the species present are continuous across a phase interface. Show that under these circumstances the overall jump entropy inequality reduces to

$$[\mathbf{q} \cdot \boldsymbol{\xi}] \geq 0$$

or

$$[\mathbf{T} \cdot \boldsymbol{\xi}] \geq 0$$

These same forms were obtained under similar restrictions for a single-component material in Exercise 5.5.2-4.

We show later (Sec. 8.5.3) that temperature and the velocities of all species are continuous across a singular surface or phase interface in stable equilibrium. (The velocities of all species are identically zero everywhere at stable equilibrium.) This may explain the apparent contradiction between the results obtained here and those obtained previously for the overall jump momentum balance (Exercise 8.3.3-1) and for the overall jump energy balance (Exercise 8.3.5-1).

**8.3.6-4** Let us define the *flux of thermal energy* $\boldsymbol{\epsilon}$ as [7, p. 702 and note for p. 701 on p. 1204]

$$\boldsymbol{\epsilon} \equiv \mathbf{q} - \sum_{A=1}^{N} (\bar{H}_{(A)} + \tfrac{1}{2}u_{(A)}^2)\mathbf{j}_{(A)}$$

(a) Determine that

$$\mathbf{q} = \boldsymbol{\epsilon} + \sum_{A=1}^{N} (\mu_{(A)} + T\bar{S}_{(A)} + \tfrac{1}{2}u_{(A)}^2)\mathbf{j}_{(A)}$$

(b) Determine that

$$\rho \frac{d_{(\mathbf{v})}\hat{S}}{dt} + \text{div}\left(\frac{1}{T}\boldsymbol{\epsilon} + \sum_{A=1}^{N}\bar{S}_{(A)}\mathbf{j}_{(A)}\right) - \frac{\rho Q}{T}$$

$$= -\frac{1}{T^2}\boldsymbol{\epsilon} \cdot \nabla T + \frac{1}{T}\text{tr}\,(\mathbf{S} \cdot \mathbf{D}) - \frac{1}{T}\sum_{A=1}^{N-1}\mathbf{j}_{(A)} \cdot (\boldsymbol{\Lambda}_{(A)} - \boldsymbol{\Lambda}_{(N)})$$

$$- \sum_{A=1}^{N}\frac{1}{T}(\mu_{(A)} + \tfrac{1}{2}u_{(A)}^2)r_{(A)} \geq 0$$

where

$$\boldsymbol{\Lambda}_{(A)} \equiv \bar{S}_{(A)}\nabla T + \nabla(\mu_{(A)} + \tfrac{1}{2}u_{(A)}^2) - \mathbf{f}_{(A)} + \frac{d_{(\mathbf{v})}\mathbf{u}_{(A)}}{dt}$$

(c) If we define

$$\mathbf{d}_{(A)} \equiv \frac{\rho_{(A)}}{cRT}\left(\boldsymbol{\Lambda}_{(A)} - \frac{1}{\rho}\nabla P + \frac{1}{\rho}\sum_{B=1}^{N}\rho_{(B)}\mathbf{f}_{(B)}\right)$$

determine that

$$-\frac{1}{T^2}\boldsymbol{\epsilon} \cdot \nabla T + \frac{1}{T}\text{tr}\,(\mathbf{S} \cdot \mathbf{D}) - \sum_{A=1}^{N}\frac{cR}{\rho_{(A)}}\mathbf{j}_{(A)} \cdot \mathbf{d}_{(A)} - \sum_{A=1}^{N}\frac{1}{T}(\mu_{(A)} + \tfrac{1}{2}u_{(A)}^2)\,r_{(A)} \geq 0$$

(d) Use the Gibbs-Duhem equation to show that

$$\sum_{A=1}^{N}\mathbf{d}_{(A)} = \frac{1}{cRT}\left[\sum_{A=1}^{N}\rho_{(A)}\nabla(\tfrac{1}{2}u_{(A)}^2) + \sum_{A=1}^{N}\rho_{(A)}\frac{d_{(\mathbf{v})}\mathbf{u}_{(A)}}{dt}\right]$$

Hirschfelder, Curtiss, and Bird neglect terms which account for changes in the kinetic energy of diffusion [for example, $\nabla(\frac{1}{2}u_{(A)}^2)$ and $d_{(v)}\mathbf{u}_{(A)}/dt$ in the definition of $\mathbf{\Lambda}_{(A)}$]. It is for this reason that they find that the $\mathbf{d}_{(A)}$ ($A = 1, \ldots, N$) are linearly dependent [7, p. 714].

(e) Show that

$$\mathbf{\Lambda}_{(A)} = \sum_{\substack{B=1 \\ B \neq A}}^{N} \left(\frac{\partial \mu_{(A)}}{\partial \omega_{(B)}}\right)_{T, P, \omega_{(C)}(C \neq A, B)} \nabla\omega_{(B)} + \overline{V}_{(A)}\nabla P - \mathbf{f}_{(A)} + \nabla(\tfrac{1}{2}u_{(A)}^2) + \frac{d_{(v)}\mathbf{u}_{(A)}}{dt}$$

This means that we can write

$$\mathbf{d}_{(A)} = \frac{\rho_{(A)}}{cRT}\left[\sum_{\substack{B=1 \\ B \neq A}}^{N}\left(\frac{\partial \mu_{(A)}}{\partial \omega_{(B)}}\right)_{T, P, \omega_{(C)}(C \neq A, B)}\nabla\omega_{(B)} + \left(\overline{V}_{(A)} - \frac{1}{\rho}\right)\nabla P\right.$$
$$\left. - \left(\mathbf{f}_{(A)} - \sum_{B=1}^{N}\omega_{(B)}\mathbf{f}_{(B)}\right) + \nabla(\tfrac{1}{2}u_{(A)}^2) + \frac{d_{(v)}\mathbf{u}_{(A)}}{dt}\right]$$

(f) Show that we may also write

$$\mathbf{d}_{(A)} = \frac{x_{(A)}}{RT}\left[\sum_{\substack{B=1 \\ B \neq A}}^{N}\left(\frac{\partial \mu_{(A)}^{(m)}}{\partial x_{(B)}}\right)_{T, P, \omega_{(C)}(C \neq A, B)}\nabla x_{(B)} + M_{(A)}\left(\frac{\overline{V}_{(A)}^{(m)}}{M_{(A)}} - \frac{1}{\rho}\right)\nabla P\right.$$
$$\left. - M_{(A)}\left(\mathbf{f}_{(A)} - \sum_{B=1}^{N}\omega_{(B)}\mathbf{f}_{(B)}\right) + M_{(A)}\nabla(\tfrac{1}{2}u_{(A)}^2) + M_{(A)}\frac{d_{(v)}\mathbf{u}_{(A)}}{dt}\right]$$

**8.3.6-5** *Alternate forms of the overall differential entropy inequality* Derive the following alternative forms of the overall differential entropy inequality:

$$\rho\frac{d_{(v)}}{dt}\left((\hat{U} + \tfrac{1}{2}v^2 + \sum_{B=1}^{N}\tfrac{1}{2}\omega_{(B)}u_{(B)}^2 + \varphi\right) + \text{div }\mathbf{q} - \text{div }(\mathbf{T}\cdot\mathbf{v}) - \sum_{B=1}^{N}\mathbf{j}_{(B)}\cdot\mathbf{f}_{(B)}$$
$$- T\,\text{div}\left(\frac{1}{T}\left[\mathbf{q} - \sum_{B=1}^{N}(\mu_{(B)} + \tfrac{1}{2}u_{(B)}^2)\mathbf{j}_{(B)}\right]\right) \leq \rho T\frac{d_{(v)}\hat{S}}{dt} \quad \text{(A)}$$

$$\rho\frac{d_{(v)}\hat{H}}{dt} + T\,\text{div}\left(\frac{1}{T}\mathbf{q}\right) - \frac{d_{(v)}P}{dT} - \rho Q - \sum_{B=1}^{N}T\mathbf{j}_{(B)}\cdot\nabla\left(\frac{\mu_{(B)} + \tfrac{1}{2}u_{(B)}^2}{T}\right) - \sum_{B=1}^{N}\mu_{(B)}r_{(B)}$$
$$- \sum_{B=1}^{N}\tfrac{1}{2}u_{(B)}^2\,\text{div }\mathbf{j}_{(B)} \geq 0 \quad \text{(B)}$$

$$\rho\frac{d_{(v)}}{dt}\left(\hat{A} + \tfrac{1}{2}v^2 + \sum_{B=1}^{N}\tfrac{1}{2}\omega_{(B)}u_{(B)}^2 + \varphi\right) + \text{div }\mathbf{q} - \text{div }(\mathbf{T}\cdot\mathbf{v}) + \rho\hat{S}\frac{d_{(v)}T}{dt} - \sum_{B=1}^{N}\mathbf{j}_{(B)}\cdot\mathbf{f}_{(B)}$$
$$- T\,\text{div}\left(\frac{1}{T}\left[\mathbf{q} - \sum_{B=1}^{N}(\mu_{(B)} + \tfrac{1}{2}u_{(B)}^2)\mathbf{j}_{(B)}\right]\right) \leq 0 \quad \text{(C)}$$

$$\rho\frac{d_{(v)}}{dt}\left(\hat{G} + \tfrac{1}{2}v^2 + \sum_{B=1}^{N}\tfrac{1}{2}\omega_{(B)}u_{(B)}^2 + \varphi\right) - P\,\text{div }\mathbf{v} - \frac{d_{(v)}P}{dt} + \text{div }\mathbf{q} - \text{div }(\mathbf{T}\cdot\mathbf{v})$$
$$+ \rho\hat{S}\frac{d_{(v)}T}{dt} - \sum_{B=1}^{N}\mathbf{j}_{(B)}\cdot\mathbf{f}_{(B)} - T\,\text{div}\left(\frac{1}{T}\left[\mathbf{q} - \sum_{B=1}^{N}(\mu_{(B)} + \tfrac{1}{2}u_{(B)}^2)\mathbf{j}_{(B)}\right]\right) \leq 0 \quad \text{(D)}$$

## 8.4 BEHAVIOR

**8.4.1 Behavior of multicomponent materials** Let us consider a multicomponent body consisting of $N$ species. For simplicity, we shall assume that all quantities are continuous and differentiable as many times as required. We shall also assume that constitutive equations for $\mathbf{T}$ and $\mathbf{q}$ similar to those discussed in Sec. 2.3.2 and 5.4.2 exist for this material.

We have $(N + 14)$ scalar equations which describe the response of this body to various stimuli:

The $N$ mass balances for individual species (Sec. 8.2.1).
Cauchy's first law (Sec. 8.3.3).
The overall differential energy balance (Sec. 8.3.5).
The constitutive equation for $\mathbf{T}$.
The constitutive equation for $\mathbf{q}$.
A caloric equation of state which describes $\hat{U}$ in terms of $T$ and $\rho_{(A)}$ ($A = 1, \ldots, N$) (Sec. 5.1.2).

Unfortunately, these equations involve $(4N + 11)$ unknowns: $\rho_{(A)}$ ($A = 1, \ldots, N$), $\mathbf{j}_{(A)}$ ($A = 1, \ldots, N - 1$), $\mathbf{v}$, $\mathbf{T}$, $\hat{U}$, $T$, and $\mathbf{q}$. One premise of our discussion here is that constitutive equations describing local reaction rates, external forces, and external energy transmission are given a priori. We are forced to conclude that further information is required.

The usual approach is to write constitutive equations for $\mathbf{j}_{(A)}$ ($A = 1, \ldots, N - 1$) with the explanation that these equations describe contact mass transmission in much the same way as constitutive equations for $\mathbf{q}$ and $\mathbf{T}$ describe contact energy transmission and contact forces. This is the approach which we take here and which we currently recommend to the reader for the solution of mass-transfer problems which have some practical interest. Another approach to this problem is explained in Sec. 8.6.1.

### EXERCISE

**8.4.1-1** Show that the $\mathbf{j}_{(A)}$ ($A = 1, \ldots, N - 1$) are frame-indifferent spatial vector fields.

**8.4.2 More on the behavior of multicomponent materials** In Secs. 2.3.1 and 5.4.1 we saw that our common experience suggests three principles which we subsequently found to be useful in constructing constitutive equations for the stress tensor and for the energy flux vector.

These principles may easily be extended to apply to the stress tensor and the energy flux vector for multicomponent materials. The essential elements of these principles should also apply to representations for $\mathbf{j}_{(A)}$ ($A = 1, \ldots, N - 1$) in the sense that such constitutive equations describe one aspect of the behavior of multicomponent materials.

Whatever is to happen in the future should have no influence on present values of **T**, **q**, and $\mathbf{j}_{(A)}$ ($A = 1, \ldots, N-1$). This implies a *principle of determinism:*

> The stress tensor, the energy flux vector, and the mass-flux vector with respect to the mass-averaged velocity for every species $A$ in a multicomponent body are determined by the history of the motions of all species $B$ ($B = 1, \ldots, N$) which the multicomponent body has undergone, as well as by the thermal history of the multicomponent body.

It seems reasonable to say that the temperature distribution and motions of individual species in one portion of a multicomponent body should not necessarily affect **T**, **q**, and $\mathbf{j}_{(A)}$ ($A = 1, \ldots, N-1$) in another portion of the body. We may state this as a *principle of local action:*

> The temperature distribution and the motions of individual species outside an arbitrarily small neighborhood of a mass-averaged material particle may be ignored in determining the stress tensor and the energy flux vector at this mass-averaged material particle. The temperature distribution and the motions of individual species outside an arbitrarily small neighborhood of a particle of species $A$ may be ignored in determining the mass flux of species $A$ with respect to the mass-averaged velocity at this material particle.

In view of Exercise 8.4.1-1, we may extend the *principle of material frame indifference:*

> Constitutive equations must be invariant under changes of frame of reference. If a constitutive equation is satisfied for a process in which the stress tensor, the energy flux vector, the mass flux with respect to the mass-averaged velocity for every species $A$, and the motion for every species $A$ are given by
>
> $$\begin{aligned} \mathbf{T} &= \mathbf{T}(\mathbf{z}_\varkappa, t) \qquad \mathbf{q} = \mathbf{q}(\mathbf{z}_\varkappa, t) \\ \mathbf{j}_{(A)} &= \mathbf{j}_{(A)}(\mathbf{z}_\varkappa, t) \qquad \mathbf{z} = \boldsymbol{\chi}_{\varkappa(A)}(\mathbf{z}_{\varkappa(A)}, t) \end{aligned} \qquad (2\text{-}1)$$
>
> then it must also be satisfied for any equivalent process described with respect to another frame of reference. In particular, the constitutive equation must be satisfied for a process in which the stress tensor, the energy flux vector, the mass flux with respect to the mass-averaged velocity for every species $A$, and the motion for every species $A$ are given by
>
> $$\begin{aligned} \mathbf{T}^* &= \mathbf{T}^*(\mathbf{z}_\varkappa^*, t^*) = \mathbf{Q}(t) \cdot \mathbf{T}(\mathbf{z}_\varkappa, t) \cdot \mathbf{Q}(t)^T \\ \mathbf{q}^* &= \mathbf{q}^*(\mathbf{z}_\varkappa^*, t^*) = \mathbf{Q}(t) \cdot \mathbf{q}(\mathbf{z}_\varkappa, t) \\ \mathbf{j}_{(A)}^* &= \mathbf{j}_{(A)}^*(\mathbf{z}_\varkappa^*, t^*) = \mathbf{Q}(t) \cdot \mathbf{j}_{(A)}(\mathbf{z}_\varkappa, t) \\ \mathbf{z}^* &= \boldsymbol{\chi}_{\varkappa(A)}^*(\mathbf{z}_{\varkappa(A)}^*, t^*) = \mathbf{c}(t) + \mathbf{Q}(t) \cdot \boldsymbol{\chi}_{\varkappa(A)}(\mathbf{z}_{\varkappa(A)}, t) \\ t^* &= t - a \end{aligned} \qquad (2\text{-}2)$$

Here $\mathbf{z}_\varkappa$ denotes the position occupied by the mass-averaged material particle $\zeta$ in the reference configuration $\varkappa$.

In the next sections we use these principles when constructing constitutive equations for $\mathbf{T}$, $\mathbf{q}$, and $\mathbf{j}_{(A)}$.

**8.4.3 Constitutive equations for the stress tensor** Let us go back and review for a moment what we found to be true for a single-component material. The constitutive equations discussed in Secs. 2.3.2 and 2.3.3 expressed $\mathbf{S}$ as a function of $\mathbf{D}$. Fourier's law, introduced in Sec. 5.4.2, gives $\mathbf{q}$ as a function of $\nabla T$. These same terms arise as products when describing the rate of production of entropy beyond that supplied by the outside world in inequality $(2-1)_2$ of Sec. 5.5.2.

This suggests that our intuition may be helped in formulating constitutive equations for multicomponent materials by rearranging the rate of production of entropy per unit volume in a multicomponent body beyond that supplied by the external world. Exercise 8.3.6-4, for example, suggests the following set of fluxes and corresponding affinities:

| Flux | Corresponding affinity |
|---|---|
| $\mathbf{S}$ | $\mathbf{D}$ |
| $\boldsymbol{\epsilon} = \mathbf{q} - \sum_{A=1}^{N} (\overset{\blacksquare}{H}_{(A)} + \tfrac{1}{2}u_{(A)}^2)\mathbf{j}_{(A)}$ | $\nabla \ln T$ |
| $\mathbf{j}_{(A)}$ | $\dfrac{cRT}{\rho_{(A)}} \mathbf{d}_{(A)} = \sum_{\substack{B=1 \\ B\ne A}}^{N} \left(\dfrac{\partial \mu_{(A)}}{\partial \omega_{(A)}}\right)_{T,P,\omega_{(c)}(C\ne A,B)} \nabla \omega_{(B)}$ |
| | $+ \left(\overset{\blacksquare}{V}_{(A)} - \dfrac{1}{\rho}\right)\nabla P$ |
| | $- \left(\mathbf{f}_{(A)} - \sum_{B=1}^{N} \omega_{(B)}\mathbf{f}_{(B)}\right)$ |
| | $+ \nabla(\tfrac{1}{2}u_{(A)}^2) + \dfrac{d_{(v)}\mathbf{u}_{(A)}}{dt}$ |
| $r_{(A)}$ | $\mu_{(A)} + \tfrac{1}{2}u_{(A)}^2$ |

The linear theory of irreversible processes [1, p. 34], which has received considerable attention during the last 20 years, states that each flux is a linear homogeneous function of all affinities with the restriction that quantities whose tensorial order differ by an odd integer cannot interact in nonoriented media. {A nonoriented or isotropic material is one whose behavior is unaffected by orthogonal transformations (rotations or reflections) of some reference configuration of the material [2, p. 60]. It may also be thought of as a substance which has no natural direction when it assumes its reference configuration.} There are also symmetries attributed to the coefficients appearing in these linear constitutive equations. This theory is without firm foundations [3; 4, pp. 643 and 646; 5, p. 49] and will not be used here.

While in general we should expect the stress tensor in a multicomponent material to be a function of the motions of all the species present in the material as well as the temperature distribution, this is a matter of current research [6, 7]. Since there is little or no experimental evidence to guide us, at this writing we recommend current practice in engineering which is to use the constitutive equations for **T** discussed in Secs. 2.3.2 to 2.3.4, recognizing that all parameters should be functions of the local thermodynamic state variables: $T$, $\rho_{(1)}, \ldots, \rho_{(N)}$.

## REFERENCES

1. Groot, S. R. de: "Thermodynamics of Irreversible Processes," Interscience, New York, 1951.
2. Truesdell, C.: "The Elements of Continuum Mechanics," Springer-Verlag, New York, 1966.
3. Coleman, B., and C. Truesdell: *J. Chem. Phys.*, **33**:28 (1960).
4. Truesdell, C., and R. A. Toupin: In S. Flügge (ed.), "Handbuch der Physik," vol. 3/1, Springer-Verlag, Berlin, 1960.
5. Truesdell, C.: "Six Lectures on Modern Natural Philosophy," Springer-Verlag, New York, 1966.
6. Bowen, R. M.: *Arch. Rational Mech. Analysis*, **24**:370 (1967).
7. Müller, Ingo: *Arch. Rational Mech. Analysis*, **28**:1 (1968).

### 8.4.4 Constitutive equations for the energy flux vector

Kinetic theory [1, pp. 483 and 715] and our discussion in Sec. 8.4.3 suggest that the *flux of thermal energy* $\boldsymbol{\epsilon}$,

$$\boldsymbol{\epsilon} \equiv \mathbf{q} - \sum_{A=1}^{N} (\bar{H}_{(A)} + \tfrac{1}{2} u_{(A)}^2) \mathbf{j}_{(A)} \tag{4-1}$$

should be a function of the local thermodynamic state variables, $\nabla \ln T$, and $\dfrac{cRT}{\rho_{(B)}} \mathbf{d}_{(B)}$ ($B = 1, \ldots, N$):

$$\boldsymbol{\epsilon} = \boldsymbol{\epsilon}\left(T, P, \omega_{(1)}, \ldots, \omega_{(N-1)}, \nabla \ln T, \frac{cRT}{\rho_{(1)}} \mathbf{d}_{(1)}, \ldots, \frac{cRT}{\rho_{(N)}} \mathbf{d}_{(N)}\right) \tag{4-2}$$

where (see Exercise 8.3.6-4)

$$\mathbf{d}_{(A)} = \frac{\rho_{(A)}}{cRT} \Bigg[ \sum_{\substack{B=1 \\ B \neq A}}^{N} \left(\frac{\partial \mu_{(A)}}{\partial \omega_{(B)}}\right)_{T, P, \omega_{(C)}(C \neq A, B)} \nabla \omega_{(B)}$$

$$+ \left(\bar{V}_{(A)} - \frac{1}{\rho}\right) \nabla P - \left(\mathbf{f}_{(A)} - \sum_{B=1}^{N} \omega_{(B)} \mathbf{f}_{(B)}\right) + \nabla(\tfrac{1}{2} u_{(A)}^2) + \frac{d_{(v)} \mathbf{u}_{(A)}}{dt} \Bigg] \tag{4-3}$$

There is no experimental evidence to suggest that $\boldsymbol{\epsilon}$ should be a function of the rate of deformation tensor **D** or the rates of production of individual species by chemical reaction $\mathbf{r}_{(A)}$ ($A = 1, \ldots, N-1$).

Equation (4-2) automatically satisfies the principles of determinism and local action discussed in Sec. 8.4.2. If we ask for the most general polynomial vector function of this form that also satisfies the principle of material frame indifference, we find (using the representation theorems of Spencer and Rivlin [2, sec. 7] and of Smith [3] in the manner suggested by Exercise 5.4.2-1)

$$\boldsymbol{\epsilon} = \alpha \nabla \ln T + cRT \sum_{A=1}^{N} \frac{\alpha_{(A)}}{\rho_{(A)}} \mathbf{d}_{(A)} \tag{4-4}$$

It should be understood here that the coefficients $\alpha$ and $\alpha_{(A)}$ ($A = 1, \ldots, N$) are all functions of the local thermodynamic state variables as well as of all the scalar products involving $\nabla \ln T$ and $\mathbf{d}_{(B)}$ ($B = 1, \ldots, N$) (including the magnitudes of these vectors).

The kinetic theory result is a special case of Eq. (4-4):

$$\boldsymbol{\epsilon} = -\lambda \nabla T - cRT \sum_{A=1}^{N} \frac{D_{(A)}^T}{\rho_{(A)}} \mathbf{d}_{(A)} \tag{4-5}$$

Here the coefficients $\lambda$ and $D_{(A)}^T$ ($A = 1, \ldots, N$) are functions only of the thermodynamic state variables, and Eq. (4-3) is approximated by

$$\mathbf{d}_{(A)} \doteq \frac{\rho_{(A)}}{cRT} \left[ \sum_{\substack{B=1 \\ B \neq A}}^{N} \left( \frac{\partial \mu_{(A)}}{\partial \omega_{(B)}} \right)_{T, P, \omega_{(C)}(C \neq A, B)} \nabla \omega_{(B)} \right.$$

$$\left. + \left( \hat{V}_{(A)} - \frac{1}{\rho} \right) \nabla P - \mathbf{f}_{(A)} + \sum_{B=1}^{N} \omega_{(B)} \mathbf{f}_{(B)} \right] \tag{4-6}$$

The direct dependence of $\boldsymbol{\epsilon}$ upon concentration and pressure gradients through the $\mathbf{d}_{(A)}$ is usually referred to as the *Dufour effect*. It is generally believed to be small [1, p. 717] and will often be neglected in what follows. Under these conditions we will write Eq. (4-5) as

$$\boldsymbol{\epsilon} = -k \nabla T \tag{4-7}$$

The scalar $k = k(T, \rho_{(1)}, \ldots, \rho_{(N)})$ is referred to as the *thermal conductivity*.

## REFERENCES

1. Hirschfelder, J. O., C. F. Curtiss, and R. B. Bird: "Molecular Theory of Gases and Liquids," Wiley, New York, 1954; corrected with notes added 1964.
2. Spencer, A. J., and R. S. Rivlin: *Arch. Rational Mech. Analysis*, 4:214 (1959/60).
3. Smith, G. F.: *Arch. Rational Mech. Analysis*, 18:282 (1965).

**8.4.5 Constitutive equations for the mass flux vector** Kinetic theory [1, p. 479] and our discussion in Sec. 8.4.3 suggest that the mass flux of species $A$ with respect to the mass-averaged velocity should be a function of the local thermodynamic state variables, $\nabla \ln T$, and $\frac{cRT}{\rho_{(B)}} \mathbf{d}_{(B)}$ ($B = 1, \ldots, N$):

$$\mathbf{j}_{(A)} = \mathbf{j}_{(A)}(T, P, \omega_{(1)}, \ldots, \omega_{(N-1)}, \nabla \ln T, \frac{cRT}{\rho_{(1)}} \mathbf{d}_{(1)}, \ldots, \frac{cRT}{\rho_{(N)}} \mathbf{d}_{(N)}) \tag{5-1}$$

This relationship automatically satisfies the principles of determinism and local action discussed in Sec. 8.4.2. If we ask for the most general polynomial vector function of this form that also satisfies the principle of material frame indifference, we find (using the representation theorem of Spencer and Rivlin [2, sec. 7] in the manner suggested by Exercise 5.4.2-1)

$$\mathbf{j}_{(A)} = \alpha_{(A0)} \nabla \ln T + cRT \sum_{B=1}^{N} \frac{\alpha_{(AB)}}{\rho_{(B)}} \mathbf{d}_{(B)} \tag{5-2}$$

It should be understood here that the coefficients $\alpha_{(AC)}$ $(C = 0, 1, \ldots, N)$ are functions of the local thermodynamic state variables as well as of all the scalar products which can be formed among $\Delta \ln T$ and $\frac{cRT}{\rho_{(B)}} \mathbf{d}_{(B)}$ $(B = 1, \ldots, N)$ (including hte magnitudes of these vectors).

The kinetic theory result [1, p. 479]

$$\mathbf{j}_{(A)} = -D_{(A)}^T \nabla \ln T + \frac{c^2}{\rho} \sum_{B=1}^{N} M_{(A)} M_{(B)} D_{(AB)} \mathbf{d}_{(B)} \tag{5-3}$$

is a special case of Eq. (5-2) where the multicomponent *thermal diffusion coefficients* $D_{(A)}^T$ and the *multicomponent diffusion coefficients* $D_{(AB)}$ are functions only of the local thermodynamic state variables, and where $\mathbf{d}_{(A)}$ is again approximated by Eq. (4-6). This last constraint means that

$$\sum_{A=1}^{N} \mathbf{d}_{(A)} \doteq 0 \tag{5-4}$$

The resulting indeterminateness in the coefficients $D_{(AB)}$ is removed by requiring

$$D_{(AA)} = 0 \quad (A = 1, \ldots, N) \tag{5-5}$$

Because

$$\sum_{A=1}^{N} \mathbf{j}_{(A)} = 0 \tag{5-6}$$

we must require

$$\sum_{A=1}^{N} D_{(A)}^T = 0 \tag{5-7}$$

and

$$\sum_{A=1}^{N} (M_{(A)} M_{(B)} D_{(AB)} - M_{(A)} M_{(C)} D_{(AC)}) = 0 \tag{5-8}$$

For $N > 2$, the quantities $D_{(AB)}$ and $D_{(BA)}$ are not in general equal.

Since as yet no other constitutive equation for $\mathbf{j}_{(A)}$ has received as much attention, the discussion in this text will center upon Eq. (5-3).

It is common to attribute Eq. (5-3) to four separate effects. We write

$$\mathbf{j}_{(A)} = \mathbf{j}_{(A)}^{(\omega)} + \mathbf{j}_{(A)}^{(P)} + \mathbf{j}_{(A)}^{(f)} + \mathbf{j}_{(A)}^{(T)} \tag{5-9}$$

where

$$\mathbf{j}_{(A)}^{(\omega)} \equiv \frac{c}{RT} \sum_{B=1}^{N} M_{(A)} M_{(B)} D_{(AB)} \left[ \omega_{(B)} \sum_{\substack{C=1 \\ C \neq B}}^{N} \left( \frac{\partial \mu_{(B)}}{\partial \omega_{(C)}} \right)_{T,P,\omega_{(D)}(D \neq B,C)} \nabla \omega_{(C)} \right] \tag{5-10}$$

$$\mathbf{j}_{(A)}^{(P)} \equiv \frac{c}{RT} \sum_{B=1}^{N} M_{(A)} M_{(B)} D_{(AB)} \left[ \omega_{(B)} \left( \bar{V}_{(B)} - \frac{1}{\rho} \right) \nabla P \right] \tag{5-11}$$

$$\mathbf{j}_{(A)}^{(f)} \equiv -\frac{c}{RT} \sum_{B=1}^{N} M_{(A)} M_{(B)} D_{(AB)} \left[ \omega_{(B)} \left( \mathbf{f}_{(B)} - \sum_{C=1}^{N} \omega_{(C)} \mathbf{f}_{(C)} \right) \right] \tag{5-12}$$

and

$$\mathbf{j}_{(A)}^{(T)} \equiv -D_{(A)}^T \nabla \ln T \tag{5-13}$$

The term $\mathbf{j}_{(A)}^{(\omega)}$ is referred to as *ordinary diffusion*. This is the mass flux which is attributable to concentration gradients of the various species present in the mixture. We denote by $\mathbf{j}_{(A)}^{(P)}$ the effect of *pressure diffusion*. This term indicates that there may be movement of species $A$ relative to the mixture as the result of a pressure gradient which exists in the mixture. While for most situations this is a negligibly small effect, steep pressure gradients can be developed in a centrifuge and used to separate a multicomponent mixture [3, p. 356; for correction see 4, p. 576]. We speak of $\mathbf{j}_{(A)}^{(f)}$ as being *forced diffusion*. It is of primary importance in ionic systems where the external force on an ion is given by the product of the ionic charge and the local electric field strength. This means that a different force may be acting upon each of the ionic species. When gravity is the only external force, the $\mathbf{f}_{(A)}$ ($A = 1, \ldots, N$) are all the same and the $\mathbf{j}_{(A)}^{(f)}$ are identically zero. We refer to $\mathbf{j}_{(A)}^{(T)}$ as the Soret effect or *thermal diffusion*. For most situations this effect is quite small, but devices can be arranged to produce large temperature gradients so as to take advantage of it for separations [5, 6, 7, 8].

Hirschfelder, Curtiss, and Bird [1, p. 487] have attempted to invert Eq. (5-3). They succeeded when the coefficients $D_{(AB)}$ are replaced by their first approximations (which result when a trial function consisting of a single Sonine polynomial is used in a variational statement of the integral equations which determine the first Chapman-Enskog iterate [1, p. 474]) to obtain

$$\mathbf{d}_{(A)} = \sum_{B=1}^{N} \frac{x_{(A)} x_{(B)}}{\mathscr{D}_{(AB)}} (\mathbf{v}_{(B)} - \mathbf{v}_{(A)}) + \sum_{B=1}^{N} \frac{x_{(A)} x_{(B)}}{\mathscr{D}_{(AB)}} \left( \frac{D_{(B)}^T}{\rho_{(B)}} - \frac{D_{(A)}^T}{\rho_{(A)}} \right) \nabla \ln T$$

$$= \sum_{B=1}^{N} \frac{1}{c \mathscr{D}_{(AB)}} (x_{(A)} \mathbf{N}_{(B)} - x_{(B)} \mathbf{N}_{(A)}) + \sum_{B=1}^{N} \frac{x_{(A)} x_{(B)}}{\mathscr{D}_{(AB)}} \left( \frac{D_{(B)}^D}{\rho_{(B)}} - \frac{D_{(A)}^T}{\rho_{(A)}} \right) \nabla \ln T \tag{5-14}$$

The $\mathscr{D}_{(AB)}$ which appear here are the *binary diffusion coefficients*; $\mathscr{D}_{(AB)} \equiv D_{(AB)}$

when Eq. (5-3) is written for a mixture of two species. [Actually, it is the first approximation to $\mathscr{D}_{(AB)}$ in the sense described above which appears in Eq. (5-14).] These equations may be referred to as the *generalized Stefan-Maxwell* equations. They are often used as the starting point when considering ordinary diffusion in multicomponent gas mixtures. One advantage of these equations is that the $\mathscr{D}_{(AB)}$ are nearly independent of composition for low-density gas mixtures [4, p. 511].

## REFERENCES

1. Hirschfelder, J. O., C. F. Curtiss, and R. B. Bird: "Molecular Theory of Gases and Liquids," Wiley, New York, 1954: corrected with notes added 1964.
2. Spencer, A. J., and R. S. Rivlin: *Arch. Rational Mech. Analysis*, **4**:214 (1959/60).
3. Guggenheim, E. A.: "Thermodynamics," North-Holland Publishing, Amsterdam, 1950.
4. Bird, R. B., W. E. Stewart, and E. N. Lightfoot: "Transport Phenomena," 7th printing, Wiley, New York, 1960.
5. Jones, R. C., and W. H. Furry: *Rev. Mod. Phys.*, **18**:151 (1946).
6. Grew, K. E., and T. L. Ibbs: "Thermal Diffusion in Gases," Cambridge University Press, New York, 1952.
7. Jones, A. L., and R. W. Foreman: *Ind. Eng. Chem.*, **44**:2249 (1952).
8. Jones, A. L., and E. C. Milberger: *Ind. Eng. Chem.*, **45**:2689 (1953).

## EXERCISES

**8.4.5-1** (a) Show that Eq. (5-4) does imply an indeterminateness in the coefficients $D_{(AB)}$ of Eq. (5-3) which can be removed by requiring Eq. (5-5).

(b) Show that Eq. (5-6) does require Eq. (5-8).

(c) Show that for $N = 2$,

$$D_{(AB)} = D_{(BA)}$$

**8.4.5-2** Show that Eqs. (5-10) to (5-12) may also be written [4 p. 567] as

$$\mathbf{j}_{(A)}^{(\omega)} = \frac{c^2}{\rho RT} \sum_{B=1}^{N} M_{(A)} M_{(B)} D_{(AB)} \left[ x_{(B)} \sum_{\substack{C=1 \\ C \neq B}}^{N} \left( \frac{\partial \mu_{(B)}^{(m)}}{\partial x_{(C)}} \right)_{T,P,x_{(D)}(D \neq B,C)} \nabla x_{(C)} \right]$$

$$\mathbf{j}_{(A)}^{(P)} = \frac{c^2}{\rho RT} \sum_{B=1}^{N} M_{(A)} M_{(B)} D_{(AB)} \left[ x_{(B)} M_{(B)} \left( \frac{\bar{V}_{(B)}^{(m)}}{M_{(B)}} - \frac{1}{\rho} \right) \nabla P \right]$$

$$\mathbf{j}_{(A)}^{(f)} = -\frac{c^2}{\rho RT} \sum_{B=1}^{N} M_{(A)} M_{(B)} D_{(AB)} \left[ x_{(B)} M_{(B)} \left( \mathbf{f}_{(B)} - \sum_{C=1}^{N} \omega_{(C)} \mathbf{f}_{(C)} \right) \right]$$

**8.4.5-3** Rearrange Eq. (4-5) to read

$$\boldsymbol{\epsilon} = -\left[ \lambda - \frac{R}{2c} \sum_{A=1}^{N} \sum_{B=1}^{N} \frac{c_{(A)} c_{(B)}}{\mathscr{D}_{(AB)}} \left( \frac{D_{(A)}^T}{\rho_{(A)}} - \frac{D_{(B)}^T}{\rho_{(B)}} \right)^2 \right] \nabla T$$

$$- \frac{RT}{c} \sum_{A=1}^{N} \frac{D_{(A)}^T}{\rho_{(A)}} \sum_{B=1}^{N} \frac{c_{(A)} c_{(B)}}{\mathscr{D}_{(AB)}} (\mathbf{u}_{(B)} - \mathbf{u}_{(A)})$$

**8.4.5-4** *A more convenient form for the generalized Stefan-Maxwell equation*  Rearrange Eq. (4-6) as

$$\mathbf{d}_{(A)} = x_{(A)} \sum_{\substack{B=1 \\ B \neq A}}^{N} \left( \frac{\partial \ln a_{(A)}^{(m)}}{\partial x_{(B)}} \right)_{T,P,x_{(C)}(C \neq A,B)} \nabla x_{(B)}$$

$$+ \frac{x_{(A)} M_{(A)}}{RT} \left[ \left( \overline{V}_{(A)} - \frac{1}{\rho} \right) \nabla P - \mathbf{f}_{(A)} + \sum_{B=1}^{N} \omega_{(B)} \mathbf{f}_{(B)} \right]$$

$$= -x_{(A)} \sum_{\substack{B=1 \\ B \neq A}}^{N} \left( \frac{\partial \ln a_{(A)}^{(m)}}{\partial x_{(A)}} \right)_{T,P,x_{(C)}(C \neq A,B)} \nabla x_{(B)}$$

$$+ \frac{x_{(A)} M_{(A)}}{RT} \left[ \left( \overline{V}_{(A)} - \frac{1}{\rho} \right) \nabla P - \mathbf{f}_{(A)} + \sum_{B=1}^{N} \omega_{(B)} \mathbf{f}_{(B)} \right]$$

$$= - \sum_{\substack{B=1 \\ B \neq A}}^{N} \left( \frac{\partial \ln a_{(A)}^{(m)}}{\partial \ln x_{(A)}} \right)_{T,P,x_{(C)}(C \neq A,B)} \nabla x_{(B)} + \frac{x_{(A)} M_{(A)}}{RT} \left[ \left( \overline{V}_{(A)} - \frac{1}{\rho} \right) \nabla P - \mathbf{f}_{(A)} + \sum_{B=1}^{N} \omega_{(B)} \mathbf{f}_{(B)} \right]$$

where we have defined the *relative activity* (on a molar basis) as

$$a_{(A)}^{(m)} \equiv \exp \frac{\mu_{(A)}^{(m)} - \mu_{(A)}^{(m)\circ}}{RT}$$

By $\mu_{(A)}^{(m)\circ}$, we mean the chemical potential (on a molar basis) for pure species $A$ at the same temperature and pressure.

For ideal solutions,

$$\left( \frac{\partial \ln a_{(A)}^{(m)}}{\partial \ln x_{(A)}} \right)_{T,P,x_{(C)}(C \neq A,B)} = 1$$

and the result above reduces to

$$\mathbf{d}_{(A)} = \nabla x_{(A)} + \frac{x_{(A)} M_{(A)}}{RT} \left[ \left( \overline{V}_{(A)} - \frac{1}{\rho} \right) \nabla P - \mathbf{f}_{(A)} + \sum_{B=1}^{N} \omega_{(B)} \mathbf{f}_{(B)} \right]$$

This means that, for ideal solutions, the generalized Stefan-Maxwell equation becomes

$$\nabla x_{(A)} + \frac{x_{(A)} M_{(A)}}{RT} \left[ \left( \overline{V}_{(A)} - \frac{1}{\rho} \right) \nabla P - \mathbf{f}_{(A)} + \sum_{B=1}^{N} \omega_{(B)} \mathbf{f}_{(B)} \right]$$

$$= \sum_{B=1}^{N} \frac{1}{c \mathscr{D}_{(AB)}} (x_{(A)} \mathbf{N}_{(B)} - x_{(B)} \mathbf{N}_{(A)}) + \sum_{B=1}^{N} \frac{x_{(A)} x_{(B)}}{\mathscr{D}_{(AB)}} \left( \frac{D_{(B)}^T}{\rho_{(B)}} - \frac{D_{(A)}^T}{\rho_{(A)}} \right) \nabla \ln T$$

When we further neglect any effects attributable to thermal, pressure, and forced diffusion, we have what is commonly referred to as the *Stefan-Maxwell* equation:

$$\nabla x_{(A)} = \sum_{B=1}^{N} \frac{1}{c \mathscr{D}_{(AB)}} (x_{(A)} \mathbf{N}_{(B)} - x_{(B)} \mathbf{N}_{(A)})$$

## 8.4.6 Constitutive equations for the mass flux vector in binary solutions

For a mixture consisting of two components, Eq. (5-3) reduces to

$$\mathbf{j}_{(A)} = \frac{c^2}{\rho} M_{(A)} M_{(B)} \mathscr{D}_{(AB)} \mathbf{d}_{(B)} - D^T_{(A)} \nabla \ln T \tag{6-1}$$

Here we introduce a special notation $\mathscr{D}_{(AB)}$ for the *binary diffusion coefficient*. In arriving at this result we make use of Eq. (5-4) and the symmetry of the binary diffusion coefficient (Exercise 8.4.5-1). Equation (5-4) and Eq. (K) of Table 8.2.2-4 may also be used to express this result as

$$\mathbf{j}_{(A)} = -\mathbf{j}_{(B)} = -\frac{c^2}{\rho} M_{(A)} M_{(B)} \mathscr{D}_{(AB)} \mathbf{d}_{(A)} - D^T_{(A)} \nabla \ln T \tag{6-2}$$

From Eq. (4-6), we find that this may be put in the somewhat more useful form

$$\mathbf{j}_{(A)} = -\left(\frac{c}{RT}\right) M_{(A)} M_{(B)} \mathscr{D}_{(AB)} \omega_{(A)} \left[ \left(\frac{\partial \mu_{(A)}}{\partial \omega_{(B)}}\right)_{T,P} \nabla \omega_{(B)} \right.$$
$$\left. + \left(\bar{V}_{(A)} - \frac{1}{\rho}\right) \nabla P - \omega_{(B)} (\mathbf{f}_{(A)} - \mathbf{f}_{(B)}) \right] - D^T_{(A)} \nabla \ln T \tag{6-3}$$

This may also be written as

$$\mathbf{j}_{(A)} = -c M_{(A)} M_{(B)} \mathscr{D}_{(AB)} \left[ \left(\frac{\partial \ln a_{(A)}}{\partial \ln \omega_{(A)}}\right)_{T,P} \nabla \omega_{(A)} \right.$$
$$\left. + \frac{\omega_{(A)}}{RT} \left(\bar{V}_{(A)} - \frac{1}{\rho}\right) \nabla P - \frac{\omega_{(A)} \omega_{(B)}}{RT} (\mathbf{f}_{(A)} - \mathbf{f}_{(B)}) \right] - D^T_{(A)} \nabla \ln T \tag{6-4}$$

where we introduce the *relative activity (on a mass basis)* defined as

$$a_{(A)} \equiv \exp\left(\frac{\mu_{(A)} - \mu^\circ_{(A)}}{RT}\right) \tag{6-5}$$

By $\mu^\circ_{(A)}$, we mean the chemical potential for pure species $A$ at the same temperature and pressure.

Instead of Eq. (6-4) it is more common to express Eq. (6-3) as

$$\mathbf{j}_{(A)} = -\left(\frac{c^2}{\rho}\right) M_{(A)} M_{(B)} \mathscr{D}_{(AB)} \left[ \left(\frac{\partial \ln a^{(m)}_{(A)}}{\partial \ln x_{(A)}}\right)_{T,P} \nabla x_{(A)} \right.$$
$$\left. + \frac{M_{(A)} x_{(A)}}{RT} \left(\frac{\bar{V}^{(m)}_{(A)}}{M_{(A)}} - \frac{1}{\rho}\right) \nabla P - \frac{M_{(A)} x_{(A)} \omega_{(B)}}{RT} (\mathbf{f}_{(A)} - \mathbf{f}_{(B)}) \right] - D^T_{(A)} \nabla \ln T \tag{6-6}$$

where we define the *relative activity (on a molar basis)* as

$$a^{(m)}_{(A)} \equiv \exp \frac{\mu^{(m)}_{(A)} - \mu^{(m)\circ}_{(A)}}{RT} \tag{6-7}$$

[8.4] BEHAVIOR

By $\mu_{(A)}^{(m)\circ}$, we mean the chemical potential (on a molar basis) for pure species $A$ at the same temperature and pressure. The principal advantage of working in terms of $a_{(A)}^{(m)}$ is that

For ideal solutions: $\quad \left( \dfrac{\partial \ln a_{(A)}^{(m)}}{\partial \ln x_{(A)}} \right)_{T,P} = 1 \qquad (6\text{-}8)$

*Ordinary diffusion* is one of the most commonly discussed limiting cases of Eq. (6-6):

$$\mathbf{j}_{(A)} = -\left(\frac{c^2}{\rho}\right) M_{(A)} M_{(B)} \mathscr{D}_{(AB)} \left(\frac{\partial \ln a_{(A)}^{(m)}}{\partial \ln x_{(A)}}\right)_{T,P} \nabla x_{(A)} \qquad (6\text{-}9)$$

For an *ideal solution*, Eq. (6-8) indicates that this reduces to

$$\mathbf{j}_{(A)} = -\left(\frac{c^2}{\rho}\right) M_{(A)} M_{(B)} \mathscr{D}_{(AB)} \nabla x_{(A)} \qquad (6\text{-}10)$$

For nonideal mixtures, it is common practice in the literature to write Eq. (6-9) as

$$\mathbf{j}_{(A)} = -\left(\frac{c^2}{\rho}\right) M_{(A)} M_{(B)} \mathscr{D}_{(AB)}^0 \nabla x_{(A)} \qquad (6\text{-}11)$$

where we introduce

$$\mathscr{D}_{(AB)}^0 \equiv \left(\frac{\partial \ln a_{(A)}^{(m)}}{\partial \ln x_{(A)}}\right)_{T,P} \mathscr{D}_{(AB)} \qquad (6\text{-}12)$$

Equation (6-11) is commonly referred to as *Fick's first law of binary diffusion*. Various equivalent forms of Fick's first law are presented in Table 8.4.6-1.

Since much of the work which follows assumes Fick's first law, we give in Table 8.4.6-2 some important forms of the equation of continuity for species $A$ (see Table 8.2.2-5) which are consistent with it. For the limiting case of constant $\rho$ and constant $\mathscr{D}_{(AB)}^0$, Table 8.4.6-3 shows the equation of continuity for species $A$ in the three principal coordinate systems.

**Table 8.4.6-1  Equivalent forms of Fick's first law of binary diffusion†**

| | |
|---|---|
| $\mathbf{n}_{(A)} = \omega_{(A)}(\mathbf{n}_{(A)} + \mathbf{n}_{(B)}) - \rho \mathscr{D}_{(AB)}^0 \nabla \omega_{(A)}$ | (A) |
| $\mathbf{j}_{(A)} = -\rho \mathscr{D}_{(AB)}^0 \nabla \omega_{(A)}$ | (B) |
| $\mathbf{j}_{(A)} = -\left(\dfrac{c^2}{\rho}\right) M_{(A)} M_{(B)} \mathscr{D}_{(AB)}^0 \nabla x_{(A)}$ | (C) |
| $\mathbf{N}_{(A)} = x_{(A)}(\mathbf{N}_{(A)} + \mathbf{N}_{(B)}) - c \mathscr{D}_{(AB)}^0 \nabla x_{(A)}$ | (D) |
| $\mathbf{J}_{(A)}^\star = -c \mathscr{D}_{(AB)}^0 \nabla x_{(A)}$ | (E) |
| $\mathbf{J}_{(A)}^\star = -\left(\dfrac{\rho^2}{c M_{(A)} M_{(B)}}\right) \mathscr{D}_{(AB)}^0 \nabla \omega_{(A)}$ | (F) |

† Here $\mathscr{D}_{(AB)}^0 \equiv (\partial \ln a_{(A)}^{(m)} / \partial \ln x_{(A)})_{T,P} \mathscr{D}_{(AB)}$.

**Table 8.4.6-2 Some important forms of the equation of continuity for species $A$ consistent with Fick's first law of binary diffusion†**

$$\rho\left(\frac{\partial \omega_{(A)}}{\partial t} + \nabla \omega_{(A)} \cdot \mathbf{v}\right) = \text{div}\,(\rho \mathscr{D}^0_{(AB)} \nabla \omega_{(A)}) + r_{(A)} \quad \text{(A)†}$$

$$c\left(\frac{\partial x_{(A)}}{\partial t} + \nabla x_{(A)} \cdot \mathbf{v}^\star\right) = \text{div}\,(c\mathscr{D}^0_{(AB)} \nabla x_{(A)}) + x_{(B)}\frac{r_{(A)}}{M_{(A)}} - x_{(A)}\frac{r_{(B)}}{M_{(B)}} \quad \text{(B)}$$

For constant $\rho$ and $\mathscr{D}^0_{(AB)}$:

$$\frac{\partial \rho_{(A)}}{\partial t} + \nabla \rho_{(A)} \cdot \mathbf{v} = \mathscr{D}^0_{(AB)}\,\text{div}\,\nabla \rho_{(A)} + r_{(A)} \quad \text{(C)}$$

For constant $c$ and $\mathscr{D}^0_{(AB)}$:

$$\frac{\partial c_{(A)}}{\partial t} + \nabla c_{(A)} \cdot \mathbf{v}^\star = \mathscr{D}^0_{(AB)}\,\text{div}\,\nabla c_{(A)} + x_{(B)}\frac{r_{(A)}}{M_{(A)}} - x_{(A)}\frac{r_{(B)}}{M_{(B)}} \quad \text{(D)}$$

Either for constant $\rho$, constant $\mathscr{D}^0_{(AB)}$, $\mathbf{v} = 0$, and no chemical reactions or for constant $c$, constant $\mathscr{D}^0_{(AB)}$, $\mathbf{v}^\star = 0$, and no chemical reactions (Fick's second law of diffusion):

$$\frac{\partial \rho_{(A)}}{\partial t} = \mathscr{D}^0_{(AB)}\,\text{div}\,\nabla \rho_{(A)} \quad \text{(E)}$$

† Here $\mathscr{D}^0_{(AB)} \equiv (\partial \ln a^{(m)}_{(A)}/\partial \ln x_{(A)})_{T,P}\mathscr{D}_{(AB)}$.

‡ See also Exercise 8.4.6-4.

**Table 8.4.6-3 The equation of continuity for species $A$ for constant $\rho$ and $\mathscr{D}^0_{(AB)}$ [Eq. (C) of Table 8.4.6-2]**

*Rectangular cartesian coordinates:*

$$\frac{\partial \rho_{(A)}}{\partial t} + \frac{\partial \rho_{(A)}}{\partial z_1}v_1 + \frac{\partial \rho_{(A)}}{\partial z_2}v_2 + \frac{\partial \rho_{(A)}}{\partial z_3}v_3 = \mathscr{D}^0_{(AB)}\left(\frac{\partial^2 \rho_{(A)}}{\partial z_1^2} + \frac{\partial^2 \rho_{(A)}}{\partial z_2^2} + \frac{\partial^2 \rho_{(A)}}{\partial z_3^2}\right) + r_{(A)} \quad \text{(A)}$$

*Cylindrical coordinates:*

$$\frac{\partial \rho_{(A)}}{\partial t} + \frac{\partial \rho_{(A)}}{\partial r}v_r + \frac{\partial \rho_{(A)}}{\partial \theta}\frac{v_\theta}{r} + \frac{\partial \rho_{(A)}}{\partial z}v_z$$

$$= \mathscr{D}^0_{(AB)}\left[\frac{1}{r}\frac{\partial}{\partial r}\left(r\frac{\partial \rho_{(A)}}{\partial r}\right) + \frac{1}{r^2}\frac{\partial^2 \rho_{(A)}}{\partial \theta^2} + \frac{\partial^2 \rho_{(A)}}{\partial z^2}\right] + r_{(A)} \quad \text{(B)}$$

*Spherical coordinates:*

$$\frac{\partial \rho_{(A)}}{\partial t} + \frac{\partial \rho_{(A)}}{\partial r}v_r + \frac{\partial \rho_{(A)}}{\partial \theta}\frac{v_\theta}{r} + \frac{\partial \rho_{(A)}}{\partial \varphi}\frac{v_\varphi}{r \sin \theta}$$

$$= \mathscr{D}^0_{(AB)}\left[\frac{1}{r^2}\frac{\partial}{\partial r}\left(r^2\frac{\partial \rho_{(A)}}{\partial r}\right) + \frac{1}{r^2 \sin \theta}\frac{\partial}{\partial \theta}\left(\sin\theta\frac{\partial \rho_{(A)}}{\partial \theta}\right) + \frac{1}{r^2 \sin^2 \theta}\frac{\partial^2 \rho_{(A)}}{\partial \varphi^2}\right] + r_{(A)} \quad \text{(C)}$$

## REFERENCE

1. Bedingfield, C. H., Jr., and T. B. Drew: *Ind. Eng. Chem.*, **42**:1164 (1950).

## EXERCISES

**8.4.6-1** Starting with Eq. (6-3), derive Eq. (6-4).

**8.4.6-2** Starting with Eq. (6-3), derive Eq. (6-6).

**8.4.6-3** Show that the various forms of Fick's first law presented in Table 8.4.6-1 are equivalent.

**8.4.6-4** *An alternative form of the equation of continuity* Starting with Eq. (A) of Table 8.4.6-2, determine that

$$c\left(\frac{\partial \ln M}{\partial t} + \nabla \ln M \cdot \mathbf{v}\right) = \text{div}\,(c\mathscr{D}^0_{(AB)}\,\nabla \ln M) + \frac{r_{(A)}}{M_{(A)}} + \frac{r_{(B)}}{M_{(B)}}$$

A special case of this result has been suggested by Bedingfield and Drew [1].

### 8.4.7 Constitutive equations for the mass flux vector: Limiting cases in ideal solutions

The simplicity of the relationships developed in Sec. 8.4.6 explains the attention given in the literature to mass transfer in binary solutions. Our intuition suggests that there may be some limiting cases for which Eq. (6-6) may be extended for use in multicomponent ideal solutions in the form

$$\mathbf{N}_{(A)} = -c\mathscr{D}_{(Am)}\left\{\nabla x_{(A)} + \frac{M_{(A)}x_{(A)}}{RT}\left[\left(\bar{V}_{(A)} - \frac{1}{\rho}\right)\nabla P - \mathbf{f}_{(A)} + \sum_{B=1}^{N}\omega_{(B)}\mathbf{f}_{(B)}\right]\right\}$$

$$+ x_{(A)}\sum_{B=1}^{N}\mathbf{N}_{(B)} - c\sum_{B=1}^{N}x_{(A)}x_{(B)}\left(\frac{D^T_{(A)}}{\rho_{(A)}} - \frac{D^T_{(B)}}{\rho_{(B)}}\right)\nabla \ln T \quad (7\text{-}1)$$

Here $\mathscr{D}_{(Am)}$ should be thought of as the diffusion coefficient for species $A$ in the multicomponent mixture. Our purpose in this section is to establish some conditions under which we are justified in writing this constitutive equation as well as expressions for $\mathscr{D}_{(Am)}$ in terms of the appropriate binary diffusion coefficients.

From Exercise 8.4.5-4, the generalized Stefan-Maxwell equation for ideal solutions is

$$\nabla x_{(A)} + \frac{x_{(A)}M_{(A)}}{RT}\left[\left(\bar{V}_{(A)} - \frac{1}{\rho}\right)\nabla P - \mathbf{f}_{(A)} + \sum_{B=1}^{N}\omega_{(B)}\mathbf{f}_{(B)}\right]$$

$$= \sum_{B=1}^{N}\frac{1}{c\mathscr{D}_{(AB)}}(x_{(A)}\mathbf{N}_{(B)} - x_{(B)}\mathbf{N}_{(A)}) + \sum_{B=1}^{N}\frac{x_{(A)}x_{(B)}}{\mathscr{D}_{(AB)}}\left(\frac{D^T_{(B)}}{\rho_{(B)}} - \frac{D^T_{(A)}}{\rho_{(A)}}\right)\nabla \ln T$$

$$(7\text{-}2)$$

This suggests that we arrange Eq. (7-1) in the form

$$\nabla x_{(A)} + \frac{M_{(A)}x_{(A)}}{RT}\left[\left(\bar{V}_{(A)} - \frac{1}{\rho}\right)\nabla P - \mathbf{f}_{(A)} + \sum_{B=1}^{N}\omega_{(B)}\mathbf{f}_{(B)}\right]$$

$$= \frac{1}{c\mathscr{D}_{(Am)}}\left(-\mathbf{N}_{(A)} + x_{(A)}\sum_{B=1}^{N}\mathbf{N}_{(B)}\right) - \frac{1}{\mathscr{D}_{(Am)}}\sum_{B=1}^{N}x_{(A)}x_{(B)}\left(\frac{D^T_{(A)}}{\rho_{(A)}} - \frac{D^T_{(B)}}{\rho_{(B)}}\right)\nabla \ln T$$

$$(7\text{-}3)$$

From Eqs. (7-2) and (7-3), it follows that

$$\frac{1}{c\mathscr{D}_{(Am)}}\left(\mathbf{N}_{(A)} - x_{(A)} \sum_{B=1}^{N} \mathbf{N}_{(B)}\right) + \frac{1}{\mathscr{D}_{(Am)}} \sum_{B=1}^{N} x_{(A)}x_{(B)}\left(\frac{D_{(A)}^{T}}{\rho_{(A)}} - \frac{D_{(B)}^{T}}{\rho_{(B)}}\right) \nabla \ln T$$
$$= \sum_{B=1}^{N} \frac{1}{c\mathscr{D}_{(AB)}}(x_{(B)}\mathbf{N}_{(A)} - x_{(A)}\mathbf{N}_{(B)}) + \sum_{B=1}^{N} \frac{x_{(A)}x_{(B)}}{\mathscr{D}_{(AB)}}\left(\frac{D_{(A)}^{T}}{\rho_{(A)}} - \frac{D_{(B)}^{T}}{\rho_{(B)}}\right) \nabla \ln T$$

(7-4)

Now let us examine some special cases.

For trace components 2, 3, ..., $N$ in nearly pure species 1, Eq. (7-4) simplifies to

$$\frac{\mathbf{N}_{(A)} + (1/M_{(A)})D_{(A)}^{T}\nabla \ln T}{c\mathscr{D}_{(Am)}} \doteq \frac{\mathbf{N}_{(A)} + (1/M_{(A)})D_{(A)}^{T}\nabla \ln T}{c\mathscr{D}_{(A1)}} \qquad (7\text{-}5)$$

In other words, Eq. (7-1) describes the mass flux vector for species $A$ in nearly pure component 1 when we interpret

$$\mathscr{D}_{(Am)} \doteq \mathscr{D}_{(A1)} \qquad (7\text{-}6)$$

This is a well-known result for ordinary diffusion [1, p. 571].

For ideal solutions in which all the binary diffusion coefficients are the same, Eq. (7-4) requires

$$\mathscr{D}_{(Am)} = \mathscr{D}_{(AB)} \qquad (7\text{-}7)$$

This is a well-known result for ordinary diffusion [1, p. 571].

For ideal solutions in which species 2, 3, ..., $N$ all move with the same velocity (or are stationary) and in which thermal diffusion may be neglected, Eq. (7-4) may be arranged as follows:

$$\frac{x_{(1)}\mathbf{v}_{(1)} - x_{(1)}\sum_{B=1}^{N}x_{(B)}\mathbf{v}_{(B)}}{\mathscr{D}_{(1m)}} = \sum_{B=1}^{N} \frac{x_{(1)}x_{(B)}}{\mathscr{D}_{(1B)}}(\mathbf{v}_{(1)} - \mathbf{v}_{(B)})$$

$$\frac{x_{(1)}(1 - x_{(1)})\mathbf{v}_{(1)} - x_{(1)}\mathbf{v}_{(2)}\sum_{B=2}^{N}x_{(B)}}{\mathscr{D}_{(1m)}} = (\mathbf{v}_{(1)} - \mathbf{v}_{(2)})\sum_{B=2}^{N} \frac{x_{(1)}x_{(B)}}{\mathscr{D}_{(1B)}}$$

$$\frac{(1 - x_{(1)})\mathbf{v}_{(1)} - \mathbf{v}_{(2)}\sum_{B=2}^{N}x_{(B)}}{\mathscr{D}_{(1m)}} = (\mathbf{v}_{(1)} - \mathbf{v}_{(2)})\sum_{B=2}^{N} \frac{x_{(B)}}{\mathscr{D}_{(1B)}}$$

$$\frac{(1 - x_{(1)})(\mathbf{v}_{(1)} - \mathbf{v}_{(2)})}{\mathscr{D}_{(1m)}} = (\mathbf{v}_{(1)} - \mathbf{v}_{(2)})\sum_{B=2}^{N} \frac{x_{(B)}}{\mathscr{D}_{(1B)}}$$

$$\frac{1 - x_{(1)}}{\mathscr{D}_{(1m)}} = \sum_{B=2}^{N} \frac{x_{(B)}}{\mathscr{D}_{(1B)}} \qquad (7\text{-}8)$$

It is clear that Eq. (7-1) again applies with the interpretation

$$\mathscr{D}_{(1m)} = \frac{1 - x_{(1)}}{\sum_{B=2}^{N} x_{(B)}/\mathscr{D}_{(1B)}} \qquad (7\text{-}9)$$

This represents an extension of Wilke's [2] result for ordinary diffusion.

Of these three limiting cases, the first involving the use of Eqs. (7-1) and (7-6) to describe the diffusion of a trace contaminant in nearly pure species 1 is without question the most important. It is the basis for most of the work in the literature involving multicomponent solutions and will be used several times in this text.

For sufficiently dilute solutions, $c$ and $\rho$ are very nearly constants and

$$\mathbf{v} \doteq \mathbf{v}^\star \qquad (7\text{-}10)$$

In this limit, we can express Eq. (7-1) also as

$$\mathbf{n}_{(A)} = -\rho \mathscr{D}_{(Am)} \bigg\{ \nabla \omega_{(A)} + \frac{M_{(A)}\omega_{(A)}}{RT} \bigg[ \bigg( \bar{V}_{(A)} - \frac{1}{\rho} \bigg) \nabla P - \mathbf{f}_{(A)} + \sum_{B=1}^{N} \omega_{(B)} \mathbf{f}_{(B)} \bigg] \bigg\}$$
$$+ \omega_{(A)} \sum_{B=1}^{N} \mathbf{n}_{(B)} - \rho \sum_{B=1}^{N} \omega_{(A)} x_{(B)} \bigg( \frac{D_{(A)}^T}{\rho_{(A)}} - \frac{D_{(B)}^T}{\rho_{(B)}} \bigg) \nabla \ln T \qquad (7\text{-}11)$$

**REFERENCES**

1. Bird, R. B., W. E. Stewart, and E. N. Lightfoot: "Transport Phenomena," 7th printing, Wiley, New York, 1960.
2. Wilke, C. R.: *Chem. Eng. Progr.*, **46**:95 (1950).

## 8.5 INTRINSICALLY STABLE EQUILIBRIUM

**8.5.1 Stable equilibrium**[1] Consider a multiphase, multicomponent mixture that we have achieved experimentally as the result of one or more processing steps. Perhaps it sits on our laboratory bench in a small beaker. We find that, after a few taps on the beaker wall with our stirring rod or after a brief thermal shock with our bunsen burner, the system returns to its original condition. Our motivation for what follows is this picture of a relatively isolated body that always returns to the same condition after a minor disturbance of an arbitrary form.

We define *equilibrium* to be achieved by a body when the entropy inequality (sometimes referred to as the second law) becomes an equality. We define an equilibrium to be *intrinsically stable* if, after any disturbance characterized by a parameter $\epsilon$, the entropy inequality remains an equality to the approximation of the lowest-

---

[1] Sections 8.5.1 to 8.5.6 were coauthored with Prof. G. M. Brown, Department of Chemical Engineering, Northwestern University, Evanston, Illinois 60201.

order terms in $\epsilon$. In the example mentioned above, $\epsilon$ might be chosen to reflect the magnitude of the initial velocity disturbance at the boundary. A body in a locally stable equilibrium will return to it after a small disturbance.

A body may achieve a *metastable* equilibrium. Such an equilibrium is stable with respect to small disturbances and satisfies the definition for an intrinsically stable equilibrium given above. However, it is unstable with respect to somewhat larger disturbances; the entropy inequality does not remain an equality when higher-order terms in $\epsilon$ are included.

Some multicomponent systems are at equilibrium in the sense that we have defined, but are unstable with respect to a class of disturbances to the concentration distributions. Mixtures of hydrogen, oxygen, and water vapor at room temperature and atmospheric pressure appear to fall in this category. So long as a mechanism for any potential chemical reaction is absent (in the case mentioned, no spark or catalyst), the system may appear stable to all but very large disturbances in the other variables. Such a system is said to have achieved a *frozen* equilibrium.

A body may reach an equilibrium that is stable with respect to very large disturbances, in which case it is sometimes said to have obtained *global stability*.

In the following sections, we wish to develop necessary and sufficient criteria for the achievement of an intrinsically stable equilibrium in a multiphase, multicomponent body capable of undergoing a number of simultaneous chemical reactions and totally enclosed by fixed, impermeable, adiabatic walls. Each component of the body is acted upon by what may be a different external force; we restrict the analysis somewhat by requiring each of these external forces to be representable by a potential.

**8.5.2 An isolated system approaching equilibrium** We visualize that an isolated system, initially at equilibrium, is disturbed and that all intensive variables (temperature, mass fractions, thermodynamic pressure, velocity of each species, etc.) may be arbitrary functions of time and position in this body (subject only to the laws of mechanics). The magnitude of the disturbance is characterized by a perturbation parameter $\epsilon$. We ask what conditions must be satisfied at equilibrium in order that the equilibrium be considered intrinsically stable or, in other words, in order that the entropy inequality remain an equality to the approximation of the lowest-order terms in $\epsilon$.

Let us prepare to determine these criteria by first examining individually the constraints imposed on this body by conservation of mass, by Euler's first law, by the energy balance, and by the entropy inequality.

**CONSERVATION OF MASS**

From our postulated mass balance for an individual species in Sec. 8.2.1, we know that the time rate of change of the mass of species $A$ in the system is equal to the rate at which component $A$ is produced by each of the $K$ chemical reactions:

$$\frac{d}{dt}\int_V \rho_{(A)}\, dV = \int_V \sum_{j=1}^{K} r_{(A,j)}\, dV \tag{2-1}$$

## [8.5] INTRINSICALLY STABLE EQUILIBRIUM

Here $r_{(A,j)}$ is the rate per unit volume at which mass of species $A$ is produced by chemical reaction $j$. Let us introduce the $j$th reaction coordinate $\psi_{(j)}$ by defining for all $A, B = 1, 2, \ldots, N$:

$$\frac{\partial \psi_{(j)}}{\partial t} \equiv \frac{r_{(A,j)}}{M_{(A)}\nu_{(A,j)}} = \frac{r_{(B,j)}}{M_{(B)}\nu_{(B,j)}} \tag{2-2}$$

The right side of this equation represents a normalized rate of production of moles of species $A$ by chemical reaction $j$; $M_{(A)}$ is the molecular weight of species $A$ and $\nu_{(A,j)}$ is the stoichiometric coefficient for species $A$ in chemical reaction $j$. The stoichiometric coefficient is taken to be a positive number for a species consumed in the chemical reaction.

The transport theorem for a region containing a singular surface or phase interface (Exercise 8.3.1-2) may be applied to the left side of Eq. (2-1) to find

$$\int_V \left( \frac{\partial \rho_{(A)}}{\partial t} - \sum_{j=1}^K r_{(A,j)} \right) dV - \int_{S_{(sing)}} [\rho_{(A)}] u_{(\xi)} \, dS = 0 \tag{2-3}$$

or, in view of Eq. (2-2),

$$\int_V \frac{\partial}{\partial t} \left( \rho_{(A)} - \sum_{j=1}^K M_{(A)}\nu_{(A,j)}\psi_{(j)} \right) dV - \int_{S_{(sing)}} [\rho_{(A)}] u_{(\xi)} \, dS = 0 \tag{2-4}$$

In arriving at Eqs. (2-3) and (2-4), we have observed that the velocity of all species must be zero at the fixed impermeable wall that completely encloses the body.

Overall conservation of mass is satisfied by requiring no mass be produced by chemical reaction. This means that for every $j = 1, 2, \ldots, K$,

$$\sum_{A=1}^N r_{(A,j)} = \frac{\partial \psi_{(j)}}{\partial t} \sum_{A=1}^N M_{(A)}\nu_{(A,j)} = 0 \tag{2-5}$$

If

$$\frac{\partial \psi_{(j)}}{\partial t} \neq 0 \tag{2-6}$$

this indicates that

$$\sum_{A=1}^N M_{(A)}\nu_{(A,j)} = 0 \tag{2-7}$$

### EULER'S FIRST LAW

By an isolated body, we mean that the sum of the forces exerted upon the body is zero. From Sec. 8.3.3, Euler's first law consequently requires that the time rate of change momentum of the body must be zero:

$$\frac{d}{dt} \int_V \left( \sum_{A=1}^N \rho_{(A)} \mathbf{v}_{(A)} \right) dV = 0 \tag{2-8}$$

An application of the transport theorem (Exercise 8.3.1-2) yields

$$\int_V \frac{\partial}{\partial t}\left(\sum_{A=1}^N \rho_{(A)}\mathbf{v}_{(A)}\right)dV - \int_{S_{(sing)}}\left[\sum_{A=1}^N \rho_{(A)}\mathbf{v}_{(A)}\right]u_{(\xi)}\,dS = 0 \tag{2-9}$$

In obtaining this result we realize that the mass-averaged velocity **v** must be zero at the fixed impermeable wall enclosing the body.

### ENERGY BALANCE

For this isolated body, the energy balance of Sec. 8.3.5 states that

$$\frac{d}{dt}\int_V \rho\left(\hat{U} + \sum_{A=1}^N \tfrac{1}{2}\omega_{(A)}v_{(A)}^2\right)dV = \int_V \sum_{A=1}^N \rho_{(A)}(\mathbf{v}_{(A)} \cdot \mathbf{f}_{(A)})\,dV \tag{2-10}$$

The contact forces do no work, since the boundary of the body is fixed in space. The boundary is adiabatic, which we interpret here as meaning that there is neither contact energy transfer with the surroundings nor external energy transmission in the form of radiation. We neglect the possibility of mutual energy transmission. The time rate of change of the internal and kinetic energy of the body is the result only of work done by the external forces upon the body.

If we assume that the external force for each species $A$ is representable by a potential:

$$\mathbf{f}_{(A)} = -\nabla \varphi_{(A)} \tag{2-11}$$

and if we assume that this potential is independent of time for a fixed position in space, we may rewrite the integrand on the right of Eq. (2-10) as

$$\rho_{(A)}(\mathbf{v}_{(A)} \cdot \mathbf{f}_{(A)}) = -\mathrm{div}\,(\rho_{(A)}\mathbf{v}_{(A)}\varphi_{(A)}) + \varphi_{(A)}\,\mathrm{div}\,(\rho_{(A)}\mathbf{v}_{(A)})$$
$$= -\mathrm{div}\,(\rho_{(A)}\mathbf{v}_{(A)}\varphi_{(A)})$$
$$+ \varphi_{(A)}\left[\frac{\partial \rho_{(A)}}{\partial t} + \mathrm{div}\,(\rho_{(A)}\mathbf{v}_{(A)})\right] - \frac{\partial(\rho_{(A)}\varphi_{(A)})}{\partial t} \tag{2-12}$$

The differential mass balance for species $A$ together with Eq. (2-2) says that

$$\frac{\partial \rho_{(A)}}{\partial t} + \mathrm{div}\,(\rho_{(A)}\mathbf{v}_{(A)}) = \sum_{j=1}^K r_{(A,j)} = \sum_{j=1}^K \frac{\partial}{\partial t}(M_{(A)}\nu_{(A,j)}\psi_{(j)}) \tag{2-13}$$

This allows us to express Eq. (2-12) as

$$\rho_{(A)}(\mathbf{v}_{(A)} \cdot \mathbf{f}_{(A)}) = -\mathrm{div}\,(\rho_{(A)}\mathbf{v}_{(A)}\varphi_{(A)}) + \frac{\partial}{\partial t}\left(\sum_{j=1}^K \varphi_{(A)}M_{(A)}\nu_{(A,j)}\psi_{(j)} - \rho_{(A)}\varphi_{(A)}\right) \tag{2-14}$$

Green's transformation and the jump mass balance for species $A$, developed in Exercise 8.2.1-3, allow us to rewrite the volume integral of the first term on the right

## [8.5] INTRINSICALLY STABLE EQUILIBRIUM

of this last equation as

$$-\int_V \operatorname{div}(\rho_{(A)} \mathbf{v}_{(A)} \varphi_{(A)}) \, dV = \int_{S_{(\text{sing})}} \varphi_{(A)} [\rho_{(A)} \mathbf{v}_{(A)} \cdot \boldsymbol{\xi}] \, dS$$

$$= \int_{S_{(\text{sing})}} \varphi_{(A)} [\rho_{(A)}] u_{(\xi)} \, dS \quad (2\text{-}15)$$

In arriving at this expression, we have noted that the velocity of species $A$ on the bounding surface of the system must be identically zero.

The left side of Eq. (2-10) may be rewritten, using the transport theorem for a region containing a singular surface (Exercise 8.3.1-2); the right side of this equation may be transformed, using Eqs. (2-14) and (2-15). As a result of these operations, we have

$$\int_V \frac{\partial}{\partial t} \left[ \rho \left( \hat{U} + \sum_{A=1}^N \tfrac{1}{2} \omega_{(A)} v_{(A)}^2 \right) \right] dV - \int_{S_{(\text{sing})}} \left[ \rho \left( \hat{U} + \sum_{A=1}^N \tfrac{1}{2} \omega_{(A)} v_{(A)}^2 \right) \right] u_{(\xi)} \, dS$$

$$= \int_{S_{(\text{sing})}} \sum_{A=1}^N \varphi_{(A)} [\rho_{(A)}] u_{(\xi)} \, dS$$

$$+ \int_V \frac{\partial}{\partial t} \left[ \sum_{A=1}^N \sum_{j=1}^N \left( \varphi_{(A)} M_{(A)} \nu_{(A,j)} \psi_{(j)} - \rho_{(A)} \varphi_{(A)} \right) \right] dV \quad (2\text{-}16)$$

In applying the transport theorem, we again have made use of the fact that the boundaries of the body are impermeable and fixed in space.

If we define the total energy per unit volume $E$ as

$$\check{E} \equiv \rho \left[ \hat{U} + \sum_{A=1}^N \left( \tfrac{1}{2} \omega_{(A)} v_{(A)}^2 + \omega_{(A)} \varphi_{(A)} \right) \right] \quad (2\text{-}17)$$

Eq. (2-16) may be expressed as

$$\int_V \frac{\partial}{\partial t} \left( \check{E} - \sum_{A=1}^N \sum_{j=1}^K \varphi_{(A)} M_{(A)} \nu_{(A,j)} \psi_{(j)} \right) dV - \int_{S_{(\text{sing})}} [\check{E}] u_{(\xi)} \, dS \quad (2\text{-}18)$$

### ENTROPY INEQUALITY

For the isolated body under consideration here, the entropy inequality of Sec. 8.3.6 says that the time rate of change of the body's entropy must be greater than or equal to zero:

$$\frac{d}{dt} \int_V \check{S} \, dV \geq 0 \quad (2\text{-}19)$$

The transport theorem for a region containing a singular surface (Exercise 8.3.1-2) allows us to write this as

$$\int_V \frac{\partial \check{S}}{\partial t} \, dV - \int_{S_{(\text{sing})}} [\check{S}] u_{(\xi)} \, dS \geq 0 \quad (2\text{-}20)$$

In this application of the transport theorem, we have again noted that velocity must be zero on the bounding surfaces of the body.

As indicated previously, we define *equilibrium* to be attained at time $t = t_e$, if the entropy inequality becomes an equality. For the isolated body, we have from Eq. (2-19)

$$\text{At } t = t_e: \quad \frac{d}{dt}\int_V \check{S}\, dV = 0 \tag{2-21}$$

This together with Eq. (2-19) implies an additional condition to be satisfied at equilibrium by our isolated body:

$$\text{At } t = t_e: \quad \frac{d^2}{dt^2}\int_V \check{S}\, dV \equiv \lim_{t \to t_e} \frac{1}{t_e - t}\left(\frac{d}{dt}\int_V \check{S}\, dV\bigg|_{t_e} - \frac{d}{dt}\int_V \check{S}\, dV\bigg|_t\right) \leq 0 \tag{2-22}$$

**8.5.3 Implications of Eq. (2-21) in Sec. 8.5.2** If equilibrium is to be achieved at $t = t_e$ by the isolated multicomponent body considered here, Eq. (2-21) must be satisfied within the constraints imposed by conservation of mass, Euler's first law, and the energy balance: Eqs. (2-4), (2-9), and (2-18). We may recognize these constraints by means of lagrangian multipliers:

$$\frac{d}{dt}\int_V \check{S}\, dV = \int_V \frac{\partial \check{S}}{\partial t}\, dV - \int_{S_{(\text{sing})}} [\check{S}]u_{(\xi)}\, dS$$

$$+ \sum_{A=1}^{N} \lambda_{(A)} \int_V \frac{\partial}{\partial t}\left(\rho_{(A)} - \sum_{j=1}^{K} M_{(A)}\nu_{(A,j)}\psi_{(j)}\right) dV$$

$$- \sum_{A=1}^{N} \lambda_{(A)} \int_{S_{(\text{sing})}} [\rho_{(A)}]u_{(\xi)}\, dS + \boldsymbol{\lambda}_m \cdot \int_V \frac{\partial(\rho\mathbf{v})}{\partial t}\, dV$$

$$- \boldsymbol{\lambda}_m \cdot \int_{S_{(\text{sing})}} [\rho\mathbf{v}]u_{(\xi)}\, dS$$

$$+ \lambda_e \int_V \frac{\partial}{\partial t}\left(\check{E} - \sum_{A=1}^{N}\sum_{j=1}^{K}\varphi_{(A)}M_{(A)}\nu_{(A,j)}\psi_{(j)}\right) dV - \lambda_e \int_{S_{(\text{sing})}} [\check{E}]u_{(\xi)}\, dS$$

$$= 0 \tag{3-1}$$

Here $\lambda_{(A)}$ ($A = 1, 2, \ldots, N$) and $\lambda_e$ are constants or lagrangian multipliers; $\boldsymbol{\lambda}_m$ is a constant spatial vector, the components of which are lagrangian multipliers. From Eq. (2-9) of Sec. 5.1.2, we have

$$\frac{\partial \check{S}}{\partial t} = \frac{1}{T}\frac{\partial \check{U}}{\partial t} - \sum_{A=1}^{N} \frac{\mu_{(A)}}{T}\frac{\partial \rho_{(A)}}{\partial t} \tag{3-2}$$

Equation (2-17) yields

$$\frac{\partial \check{E}}{\partial t} = \frac{\partial \check{U}}{\partial t} + \sum_{A=1}^{N} \rho_{(A)}\mathbf{v}_{(A)} \cdot \frac{\partial \mathbf{v}_{(A)}}{\partial t} + \sum_{A=1}^{N} \tfrac{1}{2}v_{(A)}^2 \frac{\partial \rho_{(A)}}{\partial t} + \sum_{A=1}^{N} \varphi_{(A)}\frac{\partial \rho_{(A)}}{\partial t} \tag{3-3}$$

## [8.5] INTRINSICALLY STABLE EQUILIBRIUM

Here we have assumed that the potentials $\varphi_{(A)}$ ($A = 1, 2, \ldots, N$) are independent of time. After rearranging Eq. (3-1) by means of Eqs. (3-2) and (3-3), we have

$$\int_V \left[ \left(\frac{1}{T} + \lambda_e\right)\frac{\partial \breve{U}}{\partial t} + \sum_{A=1}^{N}\left(-\frac{\mu_{(A)}}{T} + \lambda_{(A)} + \boldsymbol{\lambda}_m \cdot \mathbf{v}_{(A)} + \lambda_e \tfrac{1}{2}v^2_{(A)} + \lambda_e \varphi_{(A)}\right)\frac{\partial \rho_{(A)}}{\partial t} \right.$$

$$+ \sum_{A=1}^{N} \rho_{(A)}(\lambda_e \mathbf{v}_{(A)} + \boldsymbol{\lambda}_m) \cdot \frac{\partial \mathbf{v}_{(A)}}{\partial t}$$

$$\left. - \sum_{j=1}^{K}\sum_{A=1}^{N}(\lambda_e \varphi_{(A)} M_{(A)} \nu_{(A,j)} + \lambda_{(A)} M_{(A)} \nu_{(A,j)}) \frac{\partial \psi_{(j)}}{\partial t} \right] dV$$

$$- \int_{S_{(\text{sing})}} \left[\left(\frac{1}{T} + \lambda_e\right)\breve{U} + \frac{P}{T} + \sum_{A=1}^{N}\left(-\frac{\mu_{(A)}}{T} + \lambda_{(A)} + \lambda_e \tfrac{1}{2}v^2_{(A)}\right.\right.$$

$$\left.\left. + \lambda_e \varphi_{(A)} + \boldsymbol{\lambda}_m \cdot \mathbf{v}_{(A)}\right)\rho_{(A)}\right] u_{(\xi)} \, dS = 0 \quad (3\text{-}4)$$

Sufficient conditions that this last be satisfied for *equilibrium* at $t = t_e$ are that $\breve{U}$, $\rho_{(A)}$, $\mathbf{v}_{(A)}$, and $\psi_{(j)}$ be independent of time and that $u_{(\xi)} = 0$ everywhere on $S_{(\text{sing})}$.

Now let us ask what must be true at equilibrium in order that the equilibrium be *intrinsically stable*. Let $\epsilon$ be a parameter that represents the disturbance from equilibrium (perhaps the magnitude of an initial disturbance to a boundary condition). Expand all dependent variables as power series in $\epsilon$. For example,

$$T = T^{(0)} + \epsilon T^{(1)} + \epsilon^2 T^{(2)} + \cdots$$

$$\mu_{(A)} = \mu^{(0)}_{(A)} + \epsilon \mu^{(1)}_{(A)} + \epsilon^2 \mu^{(2)}_{(A)} + \cdots$$

The functions $T^{(0)}$, $\mu^{(0)}_{(A)}$, etc., indicate the temperature, chemical potential, etc., at $t = t_e$ (equilibrium). For an intrinsically stable equilibrium, the zero- and first-order terms (coefficients of $\epsilon^0$ and $\epsilon^1$, respectively) on the left side of Equation (3-4) must be zero. This ensures that Eq. (3-4) is satisfied in the limit of very small disturbances. Necessary and sufficient conditions for these terms to be zero are that throughout the system, at $t = t_e$, $\breve{U}$, $\rho_{(A)}$, $\mathbf{v}_{(A)}$, and $\psi_{(j)}$ be independent of time (for simplicity we drop the superscript 0, indicating the zeroth perturbation, and specify the time to be $t_e$, at which equilibrium is achieved), $u_{(\xi)} = 0$ everywhere on $S_{(\text{sing})}$

$$T = -\frac{1}{\lambda_e} = \text{a constant} \quad (3\text{-}5)$$

$$\mu_{(A)} + \varphi_{(A)} + \tfrac{1}{2}v^2_{(A)} - T\boldsymbol{\lambda}_m \cdot \mathbf{v}_{(A)} = T\lambda_{(A)} = \text{a constant} \quad (3\text{-}6)$$

$$\mathbf{v}_{(A)} = T\boldsymbol{\lambda}_m = \text{a constant vector} \quad (A = 1, 2, \ldots, N) \quad (3\text{-}7)$$

$$\sum_{A=1}^{N}\left(-\frac{1}{T}\varphi_{(A)} M_{(A)} \nu_{(A,j)} + \lambda_{(A)} M_{(A)} \nu_{(A,j)}\right) = 0 \quad (j = 1, 2, \ldots, K) \quad (3\text{-}8)$$

and at every point on the surfaces of discontinuity (which may be associated with phase interfaces) we have

$$\text{On } S_{(\text{sing})}: \quad \left[\frac{P}{T} + \sum_{A=1}^{N}\left(-\frac{\mu_{(A)}}{T} + \lambda_{(A)} - \frac{1}{2T}v_{(A)}^2\right.\right.$$

$$\left.\left. - \frac{1}{T}\varphi_{(A)} + \boldsymbol{\lambda}_m \cdot \mathbf{v}_{(A)}\right)\rho_{(A)}\right] = 0 \quad (3\text{-}9)$$

Since the boundaries of the body must be stationary, Eq. (3-7) requires that at every point throughout the body

$$\mathbf{v}_{(A)} = T\boldsymbol{\lambda}_m = 0 \quad (A = 1, 2, \ldots, N) \quad (3\text{-}10)$$

Equations (3-6), (3-8), and (3-10) imply that at every point in the body

$$\mu_{(A)} + \varphi_{(A)} = T\lambda_{(A)} = \text{a constant} \quad (A = 1, 2, \ldots, N) \quad (3\text{-}11)$$

and

$$\sum_{A=1}^{N}[-\varphi_{(A)}M_{(A)}v_{(A,j)} + (\mu_{(A)} + \varphi_{(A)})M_{(A)}v_{(A,j)}]$$

$$= \sum_{A=1}^{N}\mu_{(A)}M_{(A)}v_{(A,j)} = 0 \quad (j = 1, 2, \ldots, K) \quad (3\text{-}12)$$

Equations (3-5) and (3-9) to (3-11) imply that pressure is continuous across any surfaces of discontinuity present:

$$\text{On } S_{(\text{sing})}: \quad [P] = 0 \quad (3\text{-}13)$$

To summarize, for the isolated body under consideration, the necessary and sufficient conditions that Eq. (2-21) be satisfied for *an intrinsically stable equilibrium* at time $t = t_e$ are Eqs. (3-5) and (3-10) to (3-13), as well as the more obvious requirement that, at $t = t_e$, $\check{U}$, $\rho_{(A)}$, $\mathbf{v}_{(A)}$, and $\psi_{(j)}$ be independent of time throughout the system and $u_{(\xi)} = 0$ everywhere on $S_{(\text{sing})}$.

Equations (3-5) and (3-10) to (3-13) represent constraints upon the various intensive variables in a system at a locally stable equilibrium. In the next section, we determine constraints that must be satisfied by any caloric equation of state [of the form of Eqs. (2-2) or (2-3) of Sec. 5.1.2] at those conditions capable of supporting an intrinsically stable equilibrium.

## REFERENCE

1. Lewis, G. N., Merle Randall, K. S. Pitzer, and Leo Brewer: "Thermodynamics," 2d ed., McGraw-Hill, New York, 1961.

## EXERCISE

**8.5.3-1** *The composition and pressure in a natural gas well* A natural gas well has been closed for some time. The mole fraction $x_{(A)0}$ of species $A$ and the pressure $P_0$ are known at the surface. We

## [8.5] INTRINSICALLY STABLE EQUILIBRIUM

wish to determine the pressure and composition in the gas as a function of depth below the surface.

In order to simplify the computations, let us assume that we are dealing with a binary gas mixture of species $A$ and $B$, that temperature is uniform throughout the gas, and that the mixture obeys the ideal-gas law and, consequently, may be assumed to behave as an ideal solution.

(a) Let us assume that gravity acts in the $z_3$ direction. If at the surface

At $z_3 = 0$: $\quad \mu_{(A)} = \mu_{(A)0}$

then determine that

$$\mu_{(A)} - \mu_{(A)0} - gz_3 = 0$$

(b) Express the chemical potential $\mu_{(A)}^{(m)}$ in terms of the fugacity $f_{(A)}$ [1, p. 204]:

$$\mu_{(A)}^{(m)} = RT \ln f_{(A)} + B_{(A)}$$

Here $B_{(A)}$ is a function only of temperature. Conclude that

$$\frac{x_{(A)} P}{x_{(A)0} P_0} = \exp \frac{gz_3 M_{(A)}}{RT}$$

(c) Reason from Eq. (3-11) that

$$\frac{dP}{P} + \frac{dx_{(A)}}{x_{(A)}} = \frac{g M_{(A)}}{RT} dz_3$$

Use Cauchy's first law to say that

$$\frac{1}{P} \frac{dP}{dz_3} = \frac{g}{RT} [M_{(A)} x_{(A)} + M_{(B)}(1 - x_{(A)})]$$

Conclude that

$$\frac{1}{x_{(A)}} \frac{dx_{(A)}}{dz_3} = \frac{g}{RT} (M_{(A)} - M_{(B)})(1 - x_{(A)})$$

which may be readily integrated to find

$$\frac{1 - x_{(A)0}}{x_{(A)0}} \frac{x_{(A)}}{1 - x_{(A)}} = \exp \frac{gz_3 (M_{(A)} - M_{(B)})}{RT}$$

This result together with the result of part (b) may be used to determine pressure as a function of depth.

A much different approach to this problem is discussed in Sec. 9.2.5, using the concept of pressure diffusion. It is reassuring that the expressions obtained for composition and pressure as a function of depth are exactly the same.

### 8.5.4 Implications of Inequality (2-22) in Sec. 8.5.2

For equilibrium at $t = t_e$, inequality (2-22) must be satisfied within the constants imposed by conservation of mass, Euler's first law, and the energy balance: Eqs. (2-4), (2-9), and (2-18). A convenient way of recognizing these constraints is to evaluate the second derivative on the left of inequality (2-22) by differentiating Eq. (3-1)$_1$ with respect to time. The details of this differentiation are given in Appendix C.

From inequality (1-3) in Appendix C, sufficient conditions that inequality (2-22) be satisfied for *equilibrium* at $t = t_e$ are that $\check{U}$, $\rho_{(A)}$, $\mathbf{v}_{(A)}$, and $\psi'_{(j)}$ be independent of time and that $u_{(\xi)} = 0$ everywhere on $S_{(\text{sing})}$.

Now let us ask what else must be true at equilibrium in order that the equilibrium be *intrinsically stable*. As in discussing the implications of Eq. (2-21) in Sec. 8.5.2, we let $\epsilon$ be a parameter that represents the disturbance from equilibrium and we expand all dependent variables as power series in $\epsilon$. For an intrinsically stable equilibrium, inequality (2-22) must be satisifed to the approximation of its lowest-order terms in $\epsilon$. This ensures that inequality (2-22) is satisfied in the limit of very small disturbances.

Using Eq. (3-1), we show that inequality (2-22) can be written in a somewhat more convenient form, inequality (1-3) in Appendix C. Equations (2-9) and (2-13) of Sec. 5.1.2, and Eqs. (2-17), (3-5), and (3-10) to (3-13), as well as the requirement that, at $t = t_e$, $u_{(\xi)} = 0$ on $S_{(\text{sing})}$, indicate that the zero- and first-order contributions to the surface integrals on the left of inequality (1-3) in Appendix C are identically zero. We can concentrate our attention on the lowest-order terms in $\epsilon$ arising from the volume integral:

$$\int_V \left[ \frac{\partial^2 \check{S}}{\partial t^2} + \sum_{A=1}^{N} \left( \frac{\mu_{(A)} + \varphi_{(A)}}{T} \right) \frac{\partial^2}{\partial t^2} \left( \rho_{(A)} - \sum_{j=1}^{K} M_{(A)} \nu_{(A,j)} \Psi_{(j)} \right) \right. $$
$$\left. - \frac{1}{T} \frac{\partial^2}{\partial t^2} \left( \check{E} - \sum_{A=1}^{N} \sum_{j=1}^{K} \varphi_{(A)} M_{(A)} \nu_{(A,j)} \Psi_{(j)} \right) \right] dV \leq 0 \quad (4\text{-}1)$$

Differentiating Eqs. (3-2) and (3-3), we have

$$\frac{\partial^2 \check{S}}{\partial t^2} = \frac{1}{T} \frac{\partial^2 \check{U}}{\partial t^2} + \left( \frac{\partial T^{-1}}{\partial \check{U}} \right)_{\rho_{(B)}} \left( \frac{\partial \check{U}}{\partial t} \right)^2 + \sum_{A=1}^{N} \left( \frac{\partial T^{-1}}{\partial \rho_{(A)}} \right)_{\check{U}, \rho_{(B)}(B \neq A)} \frac{\partial \rho_{(A)}}{\partial t} \frac{\partial \check{U}}{\partial t}$$
$$- \sum_{A=1}^{N} \left( \frac{\partial \mu_{(A)} T^{-1}}{\partial \check{U}} \right)_{\rho_{(B)}} \frac{\partial \check{U}}{\partial t} \frac{\partial \rho_{(A)}}{\partial t}$$
$$- \sum_{B=1}^{N} \sum_{A=1}^{N} \left( \frac{\partial \mu_{(A)} T^{-1}}{\partial \rho_{(B)}} \right)_{\check{U}, \rho_{(C)}(C \neq B)} \frac{\partial \rho_{(B)}}{\partial t} \frac{\partial \rho_{(A)}}{\partial t} - \sum_{A=1}^{N} \frac{\mu_{(A)}}{T} \frac{\partial^2 \rho_{(A)}}{\partial t^2} \quad (4\text{-}2)$$

and

$$\frac{\partial^2 \check{E}}{\partial t^2} = \frac{\partial^2 \check{U}}{\partial t^2} + 2\mathbf{v} \cdot \sum_{A=1}^{N} \frac{\partial \rho_{(A)}}{\partial t} \frac{\partial \mathbf{v}}{\partial t} + \rho \left( \frac{\partial v}{\partial t} \right)^2$$
$$+ \rho \mathbf{v} \cdot \frac{\partial^2 \mathbf{v}}{\partial t^2} + \tfrac{1}{2} v^2 \sum_{A=1}^{N} \frac{\partial^2 \rho_{(A)}}{\partial t^2} + \sum_{A=1}^{N} \varphi_{(A)} \frac{\partial^2 \rho_{(A)}}{\partial t^2} \quad (4\text{-}3)$$

Substituting these in inequality (4-1) and making use of Eqs. (3-10) and (3-12) we see that, at $t = t_e$,

$$\int_V \left\{ \sum_{B=1}^{N} \sum_{A=1}^{N} \left( \frac{\partial \mu_{(A)} T^{-1}}{\partial \rho_{(B)}} \right)_{\check{U}, \rho_{(C)}(C \neq B)} \frac{\partial \rho_{(B)}}{\partial t} \frac{\partial \rho_{(A)}}{\partial t} + \sum_{A=1}^{N} \left[ -\left( \frac{\partial T^{-1}}{\partial \rho_{(A)}} \right)_{\check{U}, \rho_{(B)}(B \neq A)} \right. \right.$$
$$\left. \left. + \left( \frac{\partial \mu_{(A)} T^{-1}}{\partial \check{U}} \right)_{\rho_{(B)}} \right] \frac{\partial \rho_{(A)}}{\partial t} \frac{\partial \check{U}}{\partial t} - \left( \frac{\partial T^{-1}}{\partial \check{U}} \right)_{\rho_{(B)}} \left( \frac{\partial \check{U}}{\partial t} \right)^2 + \frac{\rho}{T} \left( \frac{\partial v}{\partial t} \right)^2 \right\} dV \geq 0 \quad (4\text{-}4)$$

Consider the vector space whose elements are ordered sets of $N + 2$, real-valued functions of time and position (as the field of scalars we take all real-valued functions of time and position with the usual rules for addition and multiplication of functions); this is a generalization of the vector space of $N + 2$-tuples of real numbers

[8.5] INTRINSICALLY STABLE EQUILIBRIUM

[1, pp. 1 and 5]. Let **x** be an element of this vector space:

$$\mathbf{x} \equiv \left( \frac{\partial \rho_{(1)}}{\partial t}, \frac{\partial \rho_{(2)}}{\partial t}, \ldots, \frac{\partial \rho_{(N)}}{\partial t}, \frac{\partial \check{U}}{\partial t}, \frac{\partial v}{\partial t} \right) \tag{4-5}$$

In these terms, inequality (4-4) may be written as

$$\text{At } t = t_e: \quad \int_V (\mathbf{x}, \mathbf{Sx}) \, dV \geq 0 \tag{4-6}$$

where $(\mathbf{x}, \mathbf{Sx})$ represents an inner product of the vectors $\mathbf{x}$ and $\mathbf{Sx}$ of this vector space. Here $\mathbf{S}$ is a transformation of the vector space into itself. If we take $\mathbf{\gamma}_1, \mathbf{\gamma}_2, \ldots, \mathbf{\gamma}_{N+2}$ as a basis for this vector space,

$$\mathbf{\gamma}_1 \equiv (1, 0, \ldots, 0) \quad \mathbf{\gamma}_2 \equiv (0, 1, 0, \ldots, 0) \quad \cdots \quad \mathbf{\gamma}_{N+2} \equiv (0, \ldots, 1) \tag{4-7}$$

then the elements of the matrix of the transformation $\mathbf{S}$ with respect to this basis are [1, p. 65]

$$S_{AB} = S_{BA} = \left( \frac{\partial \mu_{(A)} T^{-1}}{\partial \rho_{(B)}} \right)_{\check{U}, \rho_{(C)}(C \neq B)} \quad \text{for} \quad A, B = 1, 2, \ldots, N \tag{4-8}$$

$$S_{N+1,A} = S_{A,N+1} = -\left( \frac{\partial T^{-1}}{\partial \rho_{(A)}} \right)_{\check{U}, \rho_{(B)}(B \neq A)} \quad \text{for} \quad A = 1, 2, \ldots, N \tag{4-9}$$

$$S_{N+1,N+1} = -\left( \frac{\partial T^{-1}}{\partial \check{U}} \right)_{\rho_{(B)}} \tag{4-10}$$

$$S_{N+2,A} = S_{A,N+2} = 0 \quad \text{for} \quad A = 1, 2, \ldots, N+1 \tag{4-11}$$

and

$$S_{N+2,N+2} = \frac{\rho}{T} \tag{4-12}$$

In arriving at these expressions, we make use of two expressions easily obtainable from Eq. (2-9) of Sec. 5.1.2:

$$\left( \frac{\partial \mu_{(A)} T^{-1}}{\partial \rho_{(B)}} \right)_{\check{U}, \rho_{(C)}(C \neq B)} = -\frac{\partial^2 \check{S}}{\partial \rho_{(B)} \partial \rho_{(A)}}$$

$$= \left( \frac{\partial \mu_{(B)} T^{-1}}{\partial \rho_{(A)}} \right)_{\check{U}, \rho_{(C)}(C \neq A)} \tag{4-13}$$

and

$$-\left( \frac{\partial T^{-1}}{\partial \rho_{(A)}} \right)_{\check{U}, \rho_{(B)}(B \neq A)} = -\frac{\partial^2 \check{S}}{\partial \rho_{(A)} \partial \check{U}}$$

$$= \left( \frac{\partial \mu_{(A)} T^{-1}}{\partial \check{U}} \right)_{\rho_{(B)}} \tag{4-14}$$

For an intrinsically stable equilibrium, the zero-order contribution to the term on the left of inequality (4-6) is identically zero. There is no first-order term. The integrand of the second-order term should be thought of as $(\mathbf{x}^{(1)}, \mathbf{S}^{(0)}\mathbf{x}^{(1)})$, where $\mathbf{x}^{(1)}$ is the first perturbation of Eq. (4-5). Since $\mathbf{x}^{(1)}$ is arbitrary, the second-order contribution to inequality (4-6), consequently, implies that at each point in the multicomponent, multiphase body (suppressing the superscripts),

$$\text{At } t = t_e: \quad (\mathbf{x}, \mathbf{Sx}) \geq 0 \tag{4-15}$$

or that $\mathbf{S}$ is a positive transformation [1, p. 140]. [Since the matrix of $\mathbf{S}$ with respect to the basis defined by Eq. (4-7) is symmetric, $\mathbf{S}$ is self-adjoint.] A necessary and sufficient condition for the transformation $\mathbf{S}$ to be positive is that the descending principal minors of the matrix $\|S_{AB}\|$ of $\mathbf{S}$ should all be positive [1, p. 167].

$$S_{11} \geq 0 \qquad \begin{vmatrix} S_{11} & S_{12} \\ S_{21} & S_{22} \end{vmatrix} \geq 0$$

$$\begin{vmatrix} S_{11} & S_{12} & S_{13} \\ S_{21} & S_{22} & S_{23} \\ S_{31} & S_{32} & S_{33} \end{vmatrix} \geq 0 \quad \cdots \quad \det \|S_{AB}\| \geq 0 \tag{4-16}$$

As a helpful additional piece of information, we know that the diagonal elements of the matrix $\|S_{AB}\|$ ($S_{11}, S_{22}, \ldots, S_{N+2, N+2}$) are also positive (since $\mathbf{S} = \mathbf{B}^2$ for some self-adjoint transformation $\mathbf{B}$; see [1, pp. 140 and 166]).

We can think of inequalities (4-16) as constraints that must be satisfied by any caloric equation of state [of the form of Eqs. (2-2) or (2-3) of Sec. 5.1.2] at those conditions capable of supporting an intrinsically stable equilibrium.

The limiting criteria for an intrinsically stable equilibrium are discussed by Beegle, Modell, and Reid [2] and by Slattery [3].

**REFERENCE**

1. Halmos, P. R.: "Finite-dimensional Vector Spaces," Van Nostrand, New York, 1958.
2. Beegle, R. L., M. Modell, and R. C. Reid: *A:I.Ch.E.J.*, **20**: 1194 and 1200 (1974).
3. Slattery, J. C.: *A:I.Ch.E.J.*, **23**: 275 (1977).

**8.5.5 Specific results for a single-component system** As an example, let us examine the necessary and sufficient conditions that inequality (2-22) be satisfied for an *intrinsically stable equilibrium* of a single-component (possibly multiphase) isolated body.

Inequalities (4-16) require for this case

$$S_{11} = \left( \frac{\partial \mu T^{-1}}{\partial \rho} \right)_{\check{U}}$$

$$= -\left( \frac{\partial^2 \check{S}}{\partial \rho^2} \right)_{\check{U}} \geq 0 \tag{5-1}$$

## [8.5] INTRINSICALLY STABLE EQUILIBRIUM

$$S_{11}S_{22} - (S_{12})^2 = \frac{1}{T^2}\left(\frac{\partial \mu T^{-1}}{\partial \rho}\right)_{\check{U}}\left(\frac{\partial T}{\partial \check{U}}\right)_\rho - \left[\frac{1}{T^2}\left(\frac{\partial T}{\partial \rho}\right)_{\check{U}}\right]^2$$

$$= \left(\frac{\partial^2 \check{S}}{\partial \rho^2}\right)_{\check{U}}\left(\frac{\partial^2 \check{S}}{\partial \check{U}^2}\right)_\rho - \left(\frac{\partial^2 \check{S}}{\partial \check{U} \partial \rho}\right)^2 \geq 0 \tag{5-2}$$

and

$$S_{33} = \frac{\rho}{T} \geq 0 \tag{5-3}$$

Since the diagonal elements of the matrix $\|S_{ij}\|$ must also be positive, it is necessary that

$$S_{22} = -\left(\frac{\partial T^{-1}}{\partial \check{U}}\right)_\rho = \frac{1}{T^2}\left(\frac{\partial T}{\partial \check{U}}\right)_\rho$$

$$= \frac{1}{\rho T^2}\left(\frac{\partial T}{\partial \hat{U}}\right)_{\hat{V}} = \frac{1}{\rho T^2 \hat{c}_V} \geq 0 \tag{5-4}$$

or, in view of inequality (5-3),

$$\hat{c}_V \equiv \left(\frac{\partial \hat{U}}{\partial T}\right)_{\hat{V}} \geq 0 \tag{5-5}$$

By interchanging $\gamma_1$ and $\gamma_2$ in the basis defined by Eqs. (4-7), it is easy to show that inequalities (5-2), (5-3), and (5-5) are necessary and sufficient conditions that inequality (5-2) be satisfied for *an intrinsically stable equilibrium* of a single-component, multiphase, isolated body.

Inequality (5-2) may be put into a more useful form. Using Eq. (2-9) of Sec. 5.1.2, we have

$$\left(\frac{\partial^2 \check{S}}{\partial \rho^2}\right)_{\check{U}} - \frac{(\partial^2 \check{S}/\partial \check{U} \partial \rho)^2}{(\partial^2 \check{S}/\partial \check{U}^2)_\rho} = -\left(\frac{\partial \mu T^{-1}}{\partial \rho}\right)_{\check{U}} + \frac{(\partial T^{-1}/\partial \rho)_{\check{U}} \cdot (\partial \mu T^{-1}/\partial \check{U})_\rho}{(\partial T^{-1}/\partial \check{U})_\rho}$$

$$= -\left(\frac{\partial \mu T^{-1}}{\partial \rho}\right)_{\check{U}} - \left(\frac{\partial \mu T^{-1}}{\partial \check{U}}\right)_\rho\left(\frac{\partial \check{U}}{\partial \rho}\right)_T$$

$$= -\left(\frac{\partial \mu T^{-1}}{\partial \rho}\right)_T \tag{5-6}$$

This last and the Gibbs-Duhem equation, Eq. (2-16) of Sec. 5.1.2, allow us to express inequality (5-2) as

$$S_{11}S_{22} - (S_{12})^2 = \frac{1}{T^3}\left(\frac{\partial T}{\partial \check{U}}\right)_\rho\left(\frac{\partial \mu}{\partial \rho}\right)_T$$

$$= \frac{1}{\rho^2 T^3}\left(\frac{\partial T}{\partial \hat{U}}\right)_{\hat{V}}\left(\frac{\partial P}{\partial \rho}\right)_T$$

$$= -\frac{1}{T^3 \rho^4 \hat{c}_V}\left(\frac{\partial P}{\partial \hat{V}}\right)_T \geq 0 \tag{5-7}$$

But in view of inequalities (5-3) and (5-5), inequality (5-7) becomes

$$\left(\frac{\partial P}{\partial \hat{V}}\right)_T \leq 0 \tag{5-8}$$

Inequalities (5-1), (5-3), and (5-8), or (5-3), (5-5), and (5-8) are constraints that must be satisfied by any caloric equation of state of the form

$$\hat{U} = \hat{U}(\hat{S}, \rho) \tag{5-9}$$

at those conditions capable of supporting an intrinsically stable equilibrium.

### 8.5.6 Relation to classic questions of thermostatics

Every body with which Gibbs [1] dealt in his treatment of thermostatics was automatically at *equilibrium*, in the sense in which we use the term, since all quantities including the configuration of the body were independent of time. We might say that all bodies considered in thermostatics are in a *state of rest* (the term *equilibrium state* is often used).

In his definition of a *stable state of rest* (stable equilibrium state) of an isolated body, Gibbs [1, p. 100] restricted his comparisons to all states of rest that the body might achieve with a fixed volume and internal energy. This is made particularly clear in the clarification given by Coleman and Noll [2]. The stable state of rest is defined to have a corresponding entropy that is greater than the entropy associated with any other possible state of rest.

An *intrinsically stable equilibrium,* discussed in the preceding sections, is conceptually different. In asking whether or not an equilibrium is stable, we compare it with a sequence of nonequilibrium states obtained by imposing a small disturbance on the system at equilibrium. *Intrinsically stable equilibrium* is obviously outside the scope of thermostatics and in this sense has not been investigated previously.

It is interesting, though perhaps not surprising, that the criteria developed here for an *intrinsically stable equilibrium* in an isolated body are identical to the criteria developed by Gibbs and others for an *intrinsically stable state of rest* (stable equilibrium state). Gibbs [1, pp. 65 and 145] showed that temperature should be uniform throughout an isolated body in an intrinsically stable state of rest [compare Eq. (3-5)]. He also derived an equation that is similar to Eq. (3-11) [1, p. 146]. Relationships similar to Eq. (3-12) have been presented previously by Prigogine and Defay [3, p. 76], Denbigh [4, p. 138], and Haase [5, p. 79]. Inequalities (5-2) and (5-5) are presented by Prigogine and Defay [3, p. 211]. Gibbs [1, pp. 100 to 108] and Callen [6, pp. 134 and 361] obtain inequalities (5-5) and (5-8).

Equation (3-13) and the fact that Eqs. (3-5), (3-10) to (3-12), (4-16), (5-1) to (5-3), (5-5), and (5-8) hold throughout a multiphase system in an intrinsically stable equilibrium are not unexpected since we neglect all interfacial effects here. Interfacial effects are taken into account by Slattery [7].

### REFERENCES

1. Gibbs, J. W.: *Trans. Conn. Acad.*, **2**:309 (1873); **2**:382 (1873); **3**:108 (1876); **3**:343 (1878); page numbers refer to "Collected Works," vol. 1, Longmans, New York, 1928.

2. Coleman, B. D., and Walter Noll: *Arch. Rational Mech. Analysis*, **4**:97 (1960).
3. Prigogine, I., and R. Defay: "Chemical Thermodynamics," Wiley, New York, 1954.
4. Denbigh, K. G.: "The Principles of Chemical Equilibrium," Cambridge, London, 1963.
5. Haase, Rolf: "Thermodynamik der Mischphasen," Springer-Verlag, Berlin, 1956.
6. Callen, H. B.: "Thermodynamics," Wiley, New York, 1963.
7. Slattery, J. C.: *A.I.Ch.E.J.*, **23**: 275 (1977).

## 8.6 ALTERNATIVE VIEWPOINT

### 8.6.1 An alternative approach to the mechanics of multicomponent systems

In this chapter I have laid the foundations for what is currently the most popular approach to the mechanics of multicomponent systems. I would be less than honest with you if I did not point out that there is another set of basic postulates for multicomponent systems that has received serious attention in the literature.

In Sec. 8.4.5 we discussed constitutive equations for the mass-flux vector $\mathbf{j}_{(A)}$ or $\mathbf{n}_{(A)}$. Since $\rho_{(A)}$ is one of our acknowledged dependent variables, this is equivalent to writing a constitutive equation for $\mathbf{v}_{(A)}$, the velocity of species $A$. Viewed in the context of Chaps. 1 and 2, this is a strange situation. In discussing the isothermal flow of a single-component incompressible material, we found that pressure and the three components of the velocity are to be determined by the simultaneous solution of the equation of continuity and Cauchy's first law. We write a constitutive equation for the extra-stress tensor rather than for the velocity vector. For nonisothermal flow of a compressible single-component material, this situation is a little more complex: density, the three components of velocity, thermodynamic pressure, and temperature must be determined as a result of simultaneous solution of the equation of continuity, Cauchy's first law, the differential energy balance, and a caloric equation of state (see Sec. 5.1.2). This situation is basically the same in that velocity is viewed as a dependent variable to be determined as a result of satisfying Euler's first law and the postulate that mass is conserved.

Why have we taken a different approach in laying the foundations for multicomponent systems? The answer is supplied in Sec. 8.4.1. The postulates we have introduced are apparently not sufficient, for we find ourselves in the position of having too few equations to be solved for too many dependent variables. In Sec. 8.4.1 our answer was to write constitutive equations for $(N-1)$ of the $\mathbf{j}_{(A)}$. An alternative approach is to begin with more postulates.

In Sec. 8.2.1 we began by postulating a mass balance for each individual species. It would have been quite natural to have postulated a momentum balance, moment of momentum balance, energy balance, and entropy inequality for each species as well [1, pp. 469, 567, 612, and 645; 2; 3]. This approach has been successfully used to justify the constitutive equation for the mass flux vector advocated in Sec. 8.4.5 for ordinary diffusion [4].

The principal problem with this approach is that many more constitutive equations must be formulated. Constitutive equations are needed for each of the "partial" stress tensors that are introduced in writing the momentum balances for the individual species and for the rate of production per unit volume of momentum associated with each species as the result of interactions with the other components present in the mixture, etc. Up to the present time, no one has been able to propose a method for experimentally measuring each of these quantities. This means that, if one did propose constitutive equations for each of these quantities, they could not be confirmed individually with experimental measurements. One would have to rely upon gross measurements that would depend upon several of the constitutive assumptions made for the mixture.

To summarize, it does not appear to be practical at the present time to write individual momentum balances, energy balances, and entropy inequalities, unless one is attempting to use this approach to investigate constitutive equations for $\mathbf{j}_{(A)}$, as did Truesdell [4]. For the analysis of mass-transfer problems involving even a small degree of complexity, the best approach at the present time appears to be the one described in the prior sections of this chapter. Its usefulness has been demonstrated over a long period of time in physics and engineering.

**REFERENCES**

1. Truesdell, C., and R. A. Toupin: In S. Flügge (ed.), "Handbuch der Physik," vol. 3/1, Springer-Verlag, Berlin, 1960.
2. Bowen, R. M.: *Arch. Rational Mech. Analysis*, **24**:370 (1967).
3. Müller, Ingo: *Arch. Rational Mech. Analysis*, **28**:1 (1968).
4. Truesdell, C.: *J. Chem. Phys.*, **37**:2336 (1962).

# 9
# Applications of the Differential Balances to Mass Transfer

This chapter parallels Chaps. 3 and 6. While the applications are different, the approach is really the same. For this reason, I feel that one is better prepared to begin here after studying these earlier chapters.

In the absence of forced convection or of natural convection resulting from a density gradient in a gravitational field, the convective terms are often neglected in the differential mass balance for species $A$. Usually no explanation is given. Sometimes the qualitative argument is made that, since diffusion is a very slow process, the resulting diffusion induced convection must certainly be negligible with respect to it. I hope that you will understand as a result of reading this chapter that this argument is too simplistic. Under certain circumstances, diffusion induced convection is a major feature of the problem. Sometimes we are able to show that diffusion induced convection is identically zero. More generally, I suggest that diffusion induced convection can be neglected with respect to diffusion as equilibrium is approached in the limit of dilute solutions.

## 9.1 PHILOSOPHY

**9.1.1 The philosophy of solving mass-transfer problems** In principle, the problems we are now about to take up are considerably more complex than those we dealt with in either Chap. 3 or Chap. 6. I say this because we should now begin to think in terms of simultaneous momentum, energy, and mass transfer. In its full generality, a problem of this type would require simultaneous solution of the differential mass balance for each species present, Cauchy's first law, and the differential energy balance, with particular constitutive equations for the mass flux vectors, the stress tensor, and the energy flux vector.

If I have frightened you, I apologize, because the specific examples we are about to take up are not all that complicated. A great deal can be learned about problems of engineering importance by making simplifying assumptions. For example, in the problems that follow we shall usually neglect thermal effects and treat the fluids involved as though they were isothermal. This has the advantage of allowing us to ignore the energy balance, but it should be clearly understood that we lose a certain degree of realism in the process. Although in every problem we should be quite concerned with the velocity distribution, we often will not say anything about the pressure distribution. In this way we shall often avoid satisfying Cauchy's first law.

As we said in Sec. 3.1.1, the first step in analyzing a physical situation is to decide just exactly what the problem is. In part, this means that constitutive equations for the mass flux vectors, the stress tensor, and the energy flux vector must be chosen. With respect to the mass and energy flux vectors, the literature to date gives us very little choice beyond those constitutive equations described in Secs. 8.4.4 and 8.4.5 (or the special cases taken up in Secs. 8.4.6 and 8.4.7). Though Sec. 8.4.3 suggests a variety of constitutive equations for the stress tensor, they are basically all of the same form as those introduced in Secs. 2.3.2 to 2.3.4. Unfortunately, mass transfer in viscoelastic fluids has been largely neglected until now.

To complete the specification of a particular problem, we must describe the geometry of the material or the geometry through which the material moves, the homogeneous and heterogeneous chemical reactions (see Sec. 8.2.1), the forces that cause the material to move, and any energy transfer to the material. Just as in Chaps. 3 and 6, every problem requires a statement of boundary conditions in its formulation. Beyond those indicated in Secs. 3.1.1 and 6.1.1, there are several common types of boundary conditions for which one should look in an unfamiliar physical situation.

1. We shall assume that at an interface the phases are in equilibrium. It might be somewhat more natural to say that the chemical potentials of all species present are continuous across the phase boundary. This is suggested by anticipating that in a sense local equilibrium is established at the phase boundary. In the limit of a stable equilibrium, Sec. 8.5.3 tells us that the chemical potentials are continuous across an interface. The use of chemical potentials in describing a physical problem is not generally recommended because of the scarcity of experimental data for chemical potential as a function of solution concentration.
2. The jump mass balance discussed in Exercise 8.2.1-3 must be satisfied for every species at every phase interface.
3. We assume that concentrations and mass fluxes remain finite at all points in the material.

## [9.2] COMPLETE SOLUTIONS

The advice we gave in Sec. 3.1.1 is still applicable. Sometimes it will be relatively simple to formulate a problem, but either impossible to come up with an analytic solution or very expensive to run out a numerical solution. It is often worthwhile to approximate a realistic, difficult problem by one that is somewhat easier to handle. This may be all that is needed in some cases, or perhaps it can serve as a useful check on whatever numerical work is being done. We confine our attention in this chapter to some problems that can be solved exactly. We do this with the firm conviction that in this way a variety of concepts can be grasped in a minimum length of time. We also believe that problems of this type must be mastered before venturing into those requiring considerable numerical work.

As in our discussions of solutions for Cauchy's first law and for the energy balance, we do not say here that the results we determine are unique. We are most interested in finding *a* solution. Sometimes experiments will suggest that the solutions obtained are unique, but often this evidence is not available.

We begin by examining in Secs. 9.2.1 to 9.3.4 diffusion induced convection problems (those in which the mass-average or molar-average velocity distribution is solely attributable to concentration gradients in the fluid). On the one hand, the variety of mass-transfer problems is possibly best emphasized in the absence of forced convection. On the other hand, one should realize that diffusion induced convection in mass transfer is distinctly different from natural convection. Natural convection is impossible without a density gradient. By contrast, diffusion induced convection can play a major role in mass transfer even though the material is assumed to have a uniform density.

We conclude in Sec. 9.4.1 by examining analogies between forced-convection mass transfer and forced-convection energy transfer.

## 9.2 COMPLETE SOLUTIONS

### 9.2.1 Unsteady-state evaporation

A very long vertical tube is partially filled with a pure liquid $A$. This liquid is isolated from the remainder of the tube, which is filled with a maxture of $A$ and $B$, by a closed diaphragm. The entire apparatus is maintained at a constant temperature and pressure (neglecting the very small hydrostatic effect). At time $t = 0$, the diaphragm is carefully opened and the evaporation of $A$ commences. The apparatus is arranged in such a manner that the liquid-gas phase interface remains fixed in space as the evaporation takes place. We wish to determine the concentration distribution of $A$ in the gas phase as a function of time.

We will assume that $A$ and $B$ form an ideal-gas mixture. This allows us to say that the molar density $c$ is a constant throughout the gas phase. We will further assume that $B$ is insoluble in $A$.

In order that this problem may be as simple as possible, let us replace the finite gas phase with a semi-infinite gas that occupies all space corresponding to $z_2 > 0$. The initial and boundary conditions become

At $t = 0$, for all $z_2 > 0$: $\quad x_{(A)} = x_{(A)0}$ \hfill (1-1)

At $z_2 = 0$, for all $t > 0$: $\quad N_{(B)2} = 0$ \hfill (1-2)

and

At $z_2 = 0$, for all $t > 0$: $\quad x_{(A)} = x_{(A)\mathrm{eq}}$ (1-3)

By $x_{(A)\mathrm{eq}}$ we mean the mole fraction of species $A$ in the $AB$ gas mixture that is in equilibrium with pure liquid $A$ at the existing temperature and pressure.

It seems reasonable to postulate

$$v_1^\star = v_3^\star = 0 \quad v_2^\star = v_2^\star(t, z_2) \quad x_{(A)} = x_{(A)}(t, z_2) \tag{1-4}$$

Since $c$ can be taken to be a constant and since there is no homogeneous chemical reaction, the overall equation of continuity in the form of Eq. (B) of Table 8.3.2-1 requires

$$\frac{\partial v_2^\star}{\partial z_2} = 0 \tag{1-5}$$

This implies

$$v_2^\star = v_2^\star(t) \tag{1-6}$$

By means of Eq. (1-2), the definition for the molar-averaged velocity $\mathbf{v}^\star$, and Fick's first law in the form of Eq. (D) of Table 8.4.6-1, we can reason that

At $z_2 = 0$, for $t > 0$: $\quad v_2^\star = \dfrac{1}{c} N_{(A)2}$

$$= \frac{x_{(A)}}{c} N_{(A)2} - \mathscr{D}_{(AB)} \frac{\partial x_{(A)}}{\partial z_2}$$

$$= -\frac{\mathscr{D}_{(AB)}}{1 - x_{(A)}} \frac{\partial x_{(A)}}{\partial z_2} \tag{1-7}$$

Equations (1-3), (1-6), and (1-7) allow us to say that everywhere in the gas phase

Everywhere for $t > 0$: $\quad v_2^\star = -\dfrac{\mathscr{D}_{(AB)}}{1 - x_{(A)\mathrm{eq}}} \dfrac{\partial x_{(A)}}{\partial z_2}\bigg|_{z_2=0}$ (1-8)

Note that $(\partial x_{(A)}/\partial z_2)_{z_2=0}$, although independent of position, is a function of time.

Let us assume that the diffusion coefficient may be taken to be a constant. In view of Eqs. (1-4) and (1-8), we may write the equation of continuity for species $A$ consistent with Fick's first law, Eq. (D) of Table 8.4.6-2, as

$$\frac{\partial x_{(A)}}{\partial t} - \frac{\mathscr{D}_{(AB)}}{1 - x_{(A)\mathrm{eq}}} \frac{\partial x_{(A)}}{\partial z_2}\bigg|_{z_2=0} \frac{\partial x_{(A)}}{\partial z_2} - \mathscr{D}_{(AB)} \frac{\partial^2 x_{(A)}}{\partial z_2^2} = 0 \tag{1-9}$$

We seek a solution to this equation consistent with boundary conditions (1-1) and (1-3).

Let us look for a solution by first transforming Eq. (1-9) into an ordinary differential equation. An earlier experience (see Sec. 3.2.4) suggests defining a new independent variable

$$\eta \equiv \frac{z_2}{\sqrt{4\mathscr{D}_{(AB)} t}} \tag{1-10}$$

## [9.2] COMPLETE SOLUTIONS

In terms of this variable, Eq. (1-9) may be expressed as

$$\frac{d^2 x_{(A)}}{d\eta^2} + \left(2\eta + \varphi\right)\frac{dx_{(A)}}{d\eta} = 0 \qquad (1\text{-}11)$$

where

$$\varphi \equiv \frac{1}{1 - x_{(A)\text{eq}}} \frac{dx_{(A)}}{d\eta}\bigg|_{\eta=0} \qquad (1\text{-}12)$$

The appropriate boundary conditions for Eq. (1-11) are

$$\text{At } \eta = 0: \qquad x_{(A)} = x_{(A)\text{eq}} \qquad (1\text{-}13)$$

and

$$\text{As } \eta \to \infty: \qquad x_{(A)} \to x_{(A)0} \qquad (1\text{-}14)$$

Integrating Eq. (1-11) once, we find

$$\frac{dx_{(A)}}{d\eta} = C_1 \exp\left(-\left[\eta + \frac{\varphi}{2}\right]^2\right) \qquad (1\text{-}15)$$

Here $C_1$ is a constant to be determined. Carrying out a second integration, consistent with Eq. (1-13), we learn

$$x_{(A)} - x_{(A)\text{eq}} = C_1 \int_{\varphi/2}^{\eta + \varphi/2} e^{-x^2}\, dx \qquad (1\text{-}16)$$

Boundary condition (1-14) requires

$$x_{(A)0} - x_{(A)\text{eq}} = C_1 \int_{\varphi/2}^{\infty} e^{-x^2}\, dx$$

$$= C_1 \left(\int_0^\infty e^{-x^2} dx - \int_0^{\varphi/2} e^{-x^2} dx\right)$$

$$= C_1 \frac{\sqrt{\pi}}{2}\left[1 - \text{erf}\left(\frac{\varphi}{2}\right)\right] \qquad (1\text{-}17)$$

We have as a final result that [1]

$$\frac{x_{(A)} - x_{(A)\text{eq}}}{x_{(A)0} - x_{(A)\text{eq}}} = \frac{2/\sqrt{\pi}}{1 - \text{erf}(\varphi/2)} \int_{\varphi/2}^{\eta + \varphi/2} e^{-x^2} dx$$

$$= \frac{\text{erf}(\eta + \varphi/2) - \text{erf}(\varphi/2)}{1 - \text{erf}(\varphi/2)} \qquad (1\text{-}18)$$

The quantity $\varphi$ is determined from Eqs. (1-12) and (1-18) as

$$\varphi = \frac{-2(x_{(A)\text{eq}} - x_{(A)0})}{\sqrt{\pi}(1 - x_{(A)\text{eq}})} \frac{\exp[-(\varphi/2)^2]}{[1 - \text{erf}(\varphi/2)]} \qquad (1\text{-}19)$$

This equation has been used to prepare Table 9.2.1-1, which presents $\varphi$ as a function of

$$\frac{x_{(A)0} - x_{(A)eq}}{1 - x_{(A)eq}}.$$

We originally set out to determine the evaporation rate from the surface. Starting with Fick's first law in the form of Eq. (D) of Table 8.4.6-1, as well as Eqs. (1-2) and (1-3), we calculate the evaporation rate as

$$N_{(A)2}|_{z_2=0} = \frac{-c\mathscr{D}_{(AB)}}{1 - x_{(A)eq}} \frac{\partial x_{(A)}}{\partial z_2}\bigg|_{z_2=0}$$

$$= \frac{-c}{1 - x_{(A)eq}} \sqrt{\frac{\mathscr{D}_{(AB)}}{4t}} \frac{dx_{(A)}}{d\eta}\bigg|_{\eta=0}$$

$$= -c\varphi \sqrt{\frac{\mathscr{D}_{(AB)}}{4t}}$$

$$= c(x_{(A)eq} - x_{(A)0}) \sqrt{\frac{\mathscr{D}_{(AB)}}{\pi t}} \left[ -\frac{\sqrt{\pi}}{2} \frac{\varphi}{(x_{(A)eq} - x_{(A)0})} \right] \quad (1\text{-}20)$$

It is rather natural to ask about the effect of convection in the differential mass balance for species $A$. We have the mental picture that diffusion takes place rather slowly and that, consequently, $v_2^\star$ may be small (it is always dangerous to refer to a dimensional quantity as being small, since its magnitude depends upon the system of units chosen). If we arbitrarily set $v_2^\star = 0$, we see from Eqs. (1-8) and (1-12) that this has the effect of setting $\varphi = 0$ in the solution obtained above:

No convection:

$$\frac{x_{(A)} - x_{(A)eq}}{x_{(A)0} - x_{(A)eq}} = \text{erf}(\eta) \quad (1\text{-}21)$$

No convection:

$$N_{(A)2}|_{z_2=0} = -c\mathscr{D}_{(AB)} \frac{\partial x_{(A)}}{\partial z_2}\bigg|_{z_2=0}$$

$$= -c \sqrt{\frac{\mathscr{D}_{(AB)}}{4t}} \frac{dx_{(A)}}{d\eta}\bigg|_{\eta=0}$$

$$= c(x_{(A)eq} - x_{(A)0}) \sqrt{\frac{\mathscr{D}_{(AB)}}{\pi t}} \quad (1\text{-}22)$$

Upon comparison with this last, we see that the term in brackets in Eq. (1-20) may be

**Table 9.2.1-1** [1]

| $\dfrac{x_{(A)eq} - x_{(A)0}}{1 - x_{(A)eq}}$ | $-\dfrac{\varphi}{2}$ | $-\dfrac{\sqrt{\pi}\varphi}{2(x_{(A)eq} - x_{(A)0})}(1 - x_{(A)eq})$ |
|---|---|---|
| 0 | 0.0000 | 1.000 |
| $\frac{1}{3}$ | 0.1562 | 0.8306 |
| 1 | 0.3578 | 0.6342 |
| 3 | 0.6618 | 0.3910 |
| $\infty$ | $\infty$ | 0 |

regarded as a correction to the evaporation rate accounting for convection. Table 9.2.1-1 shows us that this correction for convection can be neglected only when the equilibrium composition $x_{(A)\text{eq}}$ is very nearly the initial composition $x_{(A)0}$ in the limit of dilute solutions (as $x_{(A)\text{eq}} \to 0$).

We shall have more to say about the appropriate conditions for neglecting convection in Secs. 9.3.1 and 9.3.2.

## REFERENCES

1. Arnold, J. H.: *Trans. Am. Inst. Chem. Engrs.*, **40**:361 (1944).
2. Carslaw, H. S., and J. C. Jaeger: "Conduction of Heat in Solids," 2d ed., Oxford University Press, London, 1959.
3. Stevenson, W. H.: *A.I.Ch.E.J.*, **14**:350 (1968).
4. Whitaker, Stephen: *Ind. Eng. Chem., Fundamentals*, **6**:476 (1967).
5. Bird, R. B., W. E. Stewart, and E. N. Lightfoot: "Transport Phenomena," 7th printing, Wiley, New York, 1960.
6. Wright, W. A.: *J. Phys. Chem.*, **37**:233 (1933).
7. Robinson, Ernest, W. A. Wright, and G. W. Bennett: *J. Phys. Chem.*, **36**:658 (1932).
8. Slattery, J. C., and C. Y. Lin: *Chem. Eng. Commun.*, **2**:245 (1978).

## EXERCISES

**9.2.1-1** *Steady-state evaporation through a stagnant gas film* Consider the system shown in Fig. 9.2.1-1. Pure liquid $A$ continuously evaporates into an ideal-gas mixture of $A$ and $B$. The apparatus is arranged in such a manner that the liquid-gas phase interface remains fixed in space as the evaporation takes place. Species $B$ is assumed to be insoluble in liquid $A$. For the existing conditions, the equilibrium composition of the gas phase is $x_{(A)\text{eq}}$. The composition of the gas phase at the top of the column is maintained constant at $x_{(A)L}$. Determine the concentration distribution of $A$ in the gas phase as well as the rate of evaporation of $A$ from the surface.

This is one of the classic experiments for determining the diffusivity of a vapor $A$ through an insoluble gas $B$.

*Answer*

$$\frac{1 - x_{(A)}}{1 - x_{(A)\text{eq}}} = \left(\frac{1 - x_{(A)L}}{1 - x_{(A)\text{eq}}}\right)^{z_2/L}$$

$$N_{(A)2}\big|_{z_2=0} = \frac{c\mathscr{D}_{(AB)}}{L} \ln \frac{1 - x_{(A)L}}{1 - x_{(A)\text{eq}}}$$

**9.2.1-2** *Mass transfer within a solid sphere (constant surface composition)* A solid sphere of species $B$ contains a uniformly distributed trace of species $A$; the mass fraction of $A$ is $\omega_{(A)0}$. The radius of the sphere is $R$. At time $t = 0$, this sphere is placed in a large, well-stirred container of species $A$ (either vapor or liquid) containing a trace of species $B$. If such a solid were at equilibrium with this fluid, its composition would be $\omega_{(A)\text{eq}}$. Determine the composition distribution in the sphere as a function of time.

In analyzing this problem, assume that the density $\rho$ of the sphere and the diffusion coefficient $\mathscr{D}^0_{(AB)}$ are constants independent of composition. This should be nearly true in the limit as $\omega_{(A)0} \to \omega_{(A)\text{eq}}$. Do *not* assume $\mathbf{v} = 0$ merely because we are concerned with diffusion in a solid (see Sec. 9.3.1 for a discussion of this point).

For a complete solution of this problem, we would have to solve for the concentration distributions in the fluid and the solid simultaneously. In carrying out such a solution, we would assume that the chemical potentials of both species are continuous across the phase interface and that the jump mass balances for both species must be satisfied at the phase interface. This would be a very

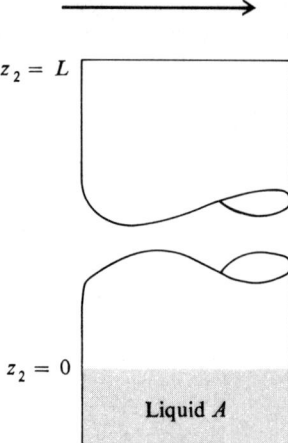

**Fig. 9.2.1-1** Steady-state evaporation through a stagnant gas film.

difficult problem, similar to the one that we encountered in Sec. 6.2.2 where we studied the temperature distribution in a semi-infinite slab. I suggest that you make the same kind of simplifying approximation that we made there. Assume that for the solid phase

At $r = R$, for $t > 0$: $\quad \omega_{(A)} = \omega_{(A)eq}$

*Answer*

$$\frac{\omega_{(A)} - \omega_{(A)eq}}{\omega_{(A)0} - \omega_{(A)eq}} = \frac{2R}{\pi} \sum_{n=1}^{\infty} \frac{(-1)^{n+1}}{nr} \sin\left(n\pi \frac{r}{R}\right) \exp\left(\frac{-n^2\pi^2 \mathscr{D}^0_{(AB)} t}{R^2}\right)$$

*Hint:* See Exercise 6.2.4-4.

**9.2.1-3** *Mass transfer within a solid sphere (Newton's law of mass transfer)* In Exercise 9.2.1-2, we assumed that the surface of the solid sphere was in equilibrium with the fluid very far away from it. As suggested by our treatment of a somewhat similar heat-transfer problem in Sec. 6.2.3, there is a preferred approach.

A somewhat better approximation would be to employ *Newton's law of mass transfer*,[1] an empirical observation that:

> The mass flux across a fluid-solid phase interface is roughly proportional to the difference between the composition of the surface and the composition that the surface would attain if it were in equilibrium with the surrounding medium.

In equation form we write

$$\mathbf{j}_{(A)} \cdot \mathbf{n} = k_{(A)\omega}(\omega_{(A)} - \omega_{(A)eq}) \tag{1}$$

[1] I have adopted this name to stress the analogy with Newton's law of cooling introduced in Sec. 6.2.3. The relationship was originally suggested by A. N. Shchukarev and W. Nernst (see V. G. Levich, "Physicochemical Hydrodynamics," Prentice-Hall, Englewood Cliffs, N.J., 1962, page 41).

or

$$\mathbf{J}_{(A)}^{\star} \cdot \mathbf{n} = k_{(A)x}(x_{(A)} - x_{(A)\text{eq}}) \qquad (2)$$

The understanding here is that $\mathbf{n}$ is the unit normal to the phase interface that is directed into these surroundings. The coefficients $k_{(A)\omega}$ and $k_{(A)x}$ are usually referred to as *mass-transfer coefficients*. Experimentally, neither of these mass-transfer coefficients is found to be a constant, although they are often assumed to be constants in constructing a mathematical model for a real situation. For a somewhat different use of the concept of a mass-transfer coefficient see Sec. 10.4.4.

Repeat Exercise 9.2.1-2 using Newton's law of mass transfer in the form of Eq. (1) to constrain the concentration distribution at the phase interface.

(a) Having made the change of variables suggested in Exercise 6.2.4-4, look for a solution by the method of separation of variables. Satisfy all but the initial condition and determine that the concentration distribution has the form

$$u \equiv \xi W = \sum_{n=1}^{\infty} E_n \sin(\lambda_n \xi) \exp(-\lambda_n^2 \tau) \qquad (3)$$

where

$$W = \frac{\omega_{(A)} - \omega_{(A)\text{eq}}}{\omega_{(A)0} - \omega_{(A)\text{eq}}} \qquad \xi \equiv \frac{r}{R} \qquad \tau \equiv \frac{t \mathscr{D}_{(AB)}^0}{R^2} \qquad (4)$$

and the $\lambda_n$ ($n = 1, 2, \ldots$) are the roots of

$$\lambda_n \cot(\lambda_n) = 1 - A \qquad (5)$$

Here

$$A \equiv \frac{R k_{(A)\omega}}{\rho \mathscr{D}_{(AB)}^0} \qquad (6)$$

The roots of Eq. (5) have been tabulated by Carslaw and Jaeger [2, p. 492].

(b) Take essentially the same approach as we did in parts (c) and (d) of Exercise 6.2.4-1 to show that

When $n \neq m$:
$$\int_0^1 \sin(\lambda_m \xi) \sin(\lambda_n \xi) \, d\xi = 0 \qquad (7)$$

and

$$\int_0^1 \sin^2(\lambda_n \xi) \, d\xi = \frac{\lambda_n^2 + A(A-1)}{2[\lambda_n^2 + (1-A)^2]} \qquad (8)$$

(c) Use the results of (b) in determining the coefficients $E_n$. Determine that the final expression for the concentration distribution is [2, p. 238]

$$u \equiv \xi W = \sum_{n=1}^{\infty} \frac{2A[\lambda_n^2 + (1-A)^2] \sin \lambda_n}{\lambda_n^2 [\lambda_n^2 + A(A-1)]} \sin(\lambda_n \xi) \exp(-\lambda_n^2 \tau) \qquad (9)$$

**9.2.1-4** *Binary diffusion in a stagnant gas* [3]  (a) Let us assume that species $B$ is stagnant:

$$\mathbf{N}_{(B)} = 0$$

Determine that

$$\mathbf{v}^{\star} = -\frac{\mathscr{D}_{(AB)}^0}{1 - x_{(A)}} \nabla x_{(A)}$$

(b) Let us limit ourselves to steady-state diffusion with no chemical reactions under conditions such that the total molar density $c$ and diffusion coefficient $\mathscr{D}_{(AB)}^0$ are constants. Conclude that the equation of continuity for species $A$ reduces to

$$\nabla x_{(B)} \cdot \nabla x_{(B)} = x_{(B)} \, \text{div} \, \nabla x_{(B)}$$

(c) Introduce as a new dependent variable

$$\alpha \equiv -\ln x_{(B)}$$

Prove that in terms of $\alpha$, the equation of continuity for species $A$ becomes

$$\text{div } \nabla \alpha = 0$$

**9.2.1-5** *More on steady-state diffusion through a stagnant gas film* Reexamine the problem described in Exercise 9.2.1-1, assuming now that the diffusion takes place in a cylindrical tube of radius $R$. Use the approach suggested in Exercise 9.2.1-4.

Conclude that we must relax the requirement that the tangential components of velocity must be zero at $r = R$ if species $B$ is stagnant. When the tangential components of velocity are not required to be zero at $r = R$, the solution found in Exercise 9.2.1-1 is correct for the finite geometry [4].

**9.2.1-6** *Constant evaporating mixtures* [5, p. 587] Let us examine a situation that is similar to that discussed in Exercise 9.2.1-1. A mixture of ethanol and toluene evaporates into an ideal-gas mixture of ethanol, toluene, and nitrogen. The apparatus is arranged in such a manner that the liquid-gas phase interface remains fixed in space as the evaporation takes place. Nitrogen is taken to be insoluble in the evaporating liquid. At the top of the column, the gas is maintained as essentially pure nitrogen. The entire system is maintained at 60°C and constant pressure. The ratio of the binary diffusivities of toluene and ethanol in nitrogen ($\mathscr{D}_{(TN_2)}/\mathscr{D}_{(EN_2)}$) is about 0.695. The vapor-liquid equilibrium data for the ethanol-toluene system at 60°C are given in Table 9.2.1-2.

(a) Use mass balances on the liquid phase to determine that for a constant-evaporating mixture (the composition of the liquid phase does not change as evaporation takes place)

$$\frac{\mathscr{D}_{(Em)}}{\mathscr{D}_{(Tm)}} = \frac{\ln|1 - y^*_{(T)}/x_{(T)}|}{\ln|1 - y^*_{(E)}/x_{(E)}|}$$

Here $\mathscr{D}_{(Em)}$ and $\mathscr{D}_{(Tm)}$ are the diffusion coefficients for ethanol and toluene in the gas mixture as discussed in Sec. 8.4.7; $x_{(E)}$ and $x_{(T)}$ are the mole fractions of ethanol and toluene in the liquid phase; $y^*_{(E)}$ and $y^*_{(T)}$ are the mole fractions of ethanol and toluene in the gas at the phase interface.

(b) Determine that at high pressures

$$\frac{\mathscr{D}_{(EN_2)}}{\mathscr{D}_{(TN_2)}} = \frac{x_{(E)} y^*_{(T)}}{x_{(T)} y^*_{(E)}}$$

Use this relationship to estimate that at high pressures the constant-evaporating mixture is $x_{(T)} = 0.13$.

(c) Use the relationship developed in (a) to *estimate* that at 760 mmHg the constant-evaporating mixture is $2_{(T)} = 0.17$. Robinson, Wright, and Bennett [7] obtained experimentally $x_{(T)} = 0.20$.
For a more complete discussion of this problem, see Slattery and Lin [8].

**Table 9.2.1-2 Vapor-liquid equilibrium data for the ethanol-toluene system at 60°C [6]**

| Mole fraction toluene in liquid | 0.096 | 0.155 | 0.233 | 0.274 | 0.375 |
|---|---|---|---|---|---|
| Mole fraction toluene in vapor | 0.147 | 0.198 | 0.242 | 0.256 | 0.277 |
| Total pressure, mmHg | 388 | 397 | 397 | 395 | 390 |

## 9.2.2 Unsteady diffusion with a first-order homogeneous reaction

At time $t = 0$, a gas of pure species $A$ is brought into contact with a liquid $B$. Component $A$ diffuses into the liquid phase, where it undergoes an irreversible first-order reaction $(A + B \rightarrow 2C)$. Let us determine the rate at which species $A$ is absorbed by the liquid phase. For the period of time of particular concern to us, it may be assumed that species $A$ and $C$ are never present in the liquid solution in more than trace amounts.

To somewhat simplify the analysis, let us take the liquid-gas phase interface to be the plane $z_2 = 0$ and let us say that the liquid phase occupies the half-space $z_2 > 0$. The initial condition is that

At $t = 0$, for all $z_2 > 0$: $\quad x_{(A)} = 0$ \hfill (2-1)

Since the liquid and gas phases are assumed to be in equilibrium at the phase interface, we require

At $z_2 = 0$, for all $t > 0$: $\quad x_{(A)} = x_{(A)\text{eq}}$ \hfill (2-2)

where $x_{(A)\text{eq}}$ is presumed to be known a priori. In order to recognize that the liquid must be supported by an impermeable container, we specify that

As $z_2 \rightarrow \infty$, for all $t$: $\quad \mathbf{v}^\star \rightarrow 0$ \hfill (2-3)

Because we are dealing with a dilute liquid solution, it seems reasonable to assume both that the solution is ideal and that the density $\rho$ is a constant. But for this dilute solution

$$c \approx \frac{\rho}{M_{(B)}} \tag{2-4}$$

which suggests that we may assume that the molar density $c$ is nearly a constant as well.

It seems reasonable to postulate

$$v_1^\star = v_3^\star = 0 \qquad v_2^\star = v_2^\star(t, z_2) \qquad x_{(A)} = x_{(A)}(t, z_2) \tag{2-5}$$

From the overall equation of continuity in the form of Eq. (B) of Table 8.3.2-1, we find

$$\frac{\partial v_2^\star}{\partial z_2} = 0 \tag{2-6}$$

This implies

$$v_2^\star = v_2^\star(t) \tag{2-7}$$

In order to be consistent with boundary condition (2-3), we must require

Everywhere:

$$v_2^\star = 0 \tag{2-8}$$

It should now become clear to you that we have specified a very specialized problem in that Eq. (2-8) requires that the number of moles of components $B$ and $C$ leaving the liquid through the phase interface must be exactly equal to the number of moles of $A$ entering the liquid. We limited ourselves to this physical situation when we said both that the phase interface must be fixed in space at the plane $z_2 = 0$

and that the liquid must be bounded by an impermeable wall as $z_2 \to \infty$, Eq. (2-3). We will examine a somewhat more realistic situation in Sec. 9.3.3.

Since we are concerned with the concentration distribution of the trace quantity $A$ in an ideal ternary solution, we may use Eq. (7-1) of Sec. 8.4.7 to describe the mass flux vector:

$$\mathbf{N}_{(A)} = c_{(A)}\mathbf{v}^{\star} - c\mathscr{D}_{(Am)} \nabla x_{(A)} \tag{2-9}$$

We will further simplify the problem by taking $\mathscr{D}_{(Am)}$ to be a constant. In view of Eqs. (2-5) and (2-7) to (2-9), the equation of continuity for species $A$ in the form of Eq. (B) of Table 8.2.2-5 specifies

$$\frac{\partial x_{(A)}}{\partial t} = \mathscr{D}_{(Am)} \frac{\partial^2 x_{(A)}}{\partial^2 z_2} + \frac{r_{(A)}}{cM_{(A)}} \tag{2-10}$$

Since this is a first-order, irreversible, homogeneous reaction in a dilute solution, we assume

$$\frac{r_{(A)}}{M_{(A)}} = -k_1''' c_{(A)} \tag{2-11}$$

The required concentration distribution for species $A$ is, consequently, a solution to

$$\frac{\partial x_{(A)}}{\partial t} = \mathscr{D}_{(Am)} \frac{\partial^2 x_{(A)}}{\partial^2 z_2} - k_1''' x_{(A)} \tag{2-12}$$

that satisfies both Eqs. (2-1) and (2-2).

Let us begin by taking the Laplace transform of Eq. (2-12):

$$sg = \mathscr{D}_{(Am)} \frac{\partial^2 g}{\partial^2 z_2} - k_1''' g \tag{2-13}$$

Here we define

$$g = g(s,z_2) \equiv \mathscr{L}[x_{(A)}(t,z_2)] \tag{2-14}$$

It is readily seen that one solution to Eq. (2-13) is of the form

$$g = A \exp(\sqrt{K} z_2) + B \exp(-\sqrt{K} z_2) \tag{2-15}$$

where

$$K \equiv \frac{s + k_1'''}{\mathscr{D}_{(Am)}} \tag{2-16}$$

and the constants $A$ and $B$ are as yet unspecified. Since we must require that

As $z_2 \to \infty$: $g$ be finite $\tag{2-17}$

we have quickly

$$A = 0 \tag{2-18}$$

In terms of the transformed variable $g$, boundary condition (2-2) says

## [9.2] COMPLETE SOLUTIONS

At $z_2 = 0$: $\quad g = \dfrac{1}{s} x_{(A)eq}$ \hfill (2-19)

Consequently,

$$B = \frac{1}{s} x_{(A)eq} \tag{2-20}$$

To summarize,

$$g = \frac{1}{s} x_{(A)eq} \exp(-\sqrt{K}\, z_2) \tag{2-21}$$

Taking the inverse Laplace transform of this, we have

$$x_{(A)} = \mathcal{L}^{-1}[g]$$

$$= x_{(A)eq} \int_0^t \frac{z_2}{2\sqrt{\pi \mathscr{D}_{(Am)} u^3}} \exp\left(-k_1''' u - \frac{z_2^2}{4\mathscr{D}_{(Am)} u}\right) du \tag{2-22}$$

or

$$\frac{x_{(A)}}{x_{(A)eq}} = \frac{2}{\sqrt{\pi}} \int_{z_2/\sqrt{4\mathscr{D}_{(Am)} t}}^{\infty} \exp\left(-\lambda^2 - \frac{k_1''' z_2^2}{4\mathscr{D}_{(Am)} \lambda^2}\right) d\lambda \tag{2-23}$$

Noting that [1, p. 140]

$$\frac{4}{\sqrt{\pi}} \int_r^{\infty} \exp\left(-\lambda^2 - \frac{a^2}{\lambda^2}\right) d\lambda = e^{2a} \operatorname{erfc}\left(r + \frac{a}{r}\right) + e^{-2a} \operatorname{erfc}\left(r - \frac{a}{r}\right) \tag{2-24}$$

we may write Eq. (2-23) in the somewhat more useful form [2]

$$\frac{2 x_{(A)}}{x_{(A)eq}} = \exp\left(z_2 \sqrt{\frac{k_1'''}{\mathscr{D}_{(Am)}}}\right) \operatorname{erfc}(\zeta + \sqrt{k_1''' t})$$

$$+ \exp\left(-z_2 \sqrt{\frac{k_1'''}{\mathscr{D}_{(Am)}}}\right) \operatorname{erfc}(\zeta - \sqrt{k_1''' t}) \tag{2-25}$$

Here we have introduced the complementary error function

$$\operatorname{erfc}(x) \equiv 1 - \operatorname{erf} x = \frac{2}{\sqrt{\pi}} \int_x^{\infty} e^{-\lambda^2} d\lambda \tag{2-26}$$

and we have defined

$$\zeta \equiv \frac{z_2}{\sqrt{4\mathscr{D}_{(Am)} t}} \tag{2-27}$$

We set out to determine the rate at which species $A$ is absorbed by the liquid phase. This is the same as asking for the flux of species $A$ through the liquid-gas phase interface:

$$N_{(A)2}\bigg|_{z_2=0} = -c\mathcal{D}_{(Am)}\frac{\partial x_{(A)}}{\partial z_2}\bigg|_{z_2=0}$$

$$= cx_{(A)\text{eq}}\sqrt{k_1'''\mathcal{D}_{(Am)}}\left[\text{erf}\sqrt{k_1'''t} + \frac{\exp(-k_1'''t)}{\sqrt{\pi k_1'''t}}\right] \tag{2-28}$$

The total amount of $A$ absorbed between time $t = 0$ and time $t = t_0$ is, consequently,

$$\int_0^{t_0} N_{(A)}\bigg|_{z_2=0} dt = cx_{(A)\text{eq}}\sqrt{k_1'''\mathcal{D}_{(Am)}}\left[\left(t_0 + \frac{1}{2k_1'''}\right)\text{erf}\sqrt{k_1'''t_0}\right.$$

$$\left. + \sqrt{\frac{t_0}{\pi k_1'''}}\exp(-k_1'''t_0)\right] \tag{2-29}$$

We are often particularly interested in the limit

$$\text{As } k_1'''t_0 \to \infty: \quad \int_0^{t_0} N_{(A)}\bigg|_{z_2=0} dt \to cx_{(A)\text{eq}}\sqrt{k_1'''\mathcal{D}_{(Am)}}\left(t_0 + \frac{1}{2k_1'''}\right) \tag{2-30}$$

## REFERENCES

1. Churchill, R. V.: "Operational Mathematics," 2d ed., McGraw-Hill, New York, 1958.
2. Danckwerts, P. V.: *Trans. Faraday Soc.*, **46**:300 (1950).
3. Danckwerts, P. V.: *Trans. Faraday Soc.*, **47**:1014 (1951).
4. Crank, J.: "The Mathematics of Diffusion," Oxford University Press, London, 1956.
5. Lightfoot, E. N.: *A.I.Ch.E.J.*, **10**:278 (1964).
6. Bird, R. B., W. E. Stewart, and E. N. Lightfoot: "Transport Phenomena," 7th printing, Wiley, New York, 1960.
7. Irving, J., and N. Mullineux: "Mathematics in Physics and Engineering," Academic, New York, 1959.

## EXERCISES

**9.2.2-1** Fill in the details in going from Eq. (2-21) to Eq. (2-25).
*Hint:* Use the convolution theorem.

**9.2.2-2** Derive Eq. (2-29), starting with Eq. (2-25).

**9.2.2-3** Repeat the problem discussed in this section assuming that the liquid has a finite depth $L$. The plane $z_2 = 0$ represents the gas-liquid phase interface; the plane $z_2 = L$ is a wall that is impermeable to all three species.

(a) Begin by introducing as dimensionless variables

$$t^* \equiv k_1'''t \quad x_{(A)}^* \equiv \frac{x_{(A)}}{x_{(A)\text{eq}}} \quad y \equiv \alpha - z_2\sqrt{\frac{k_1'''}{\mathcal{D}_{(Am)}}}$$

where

$$\alpha \equiv L\sqrt{\frac{k_1'''}{\mathcal{D}_{(Am)}}}$$

(b) Take the Laplace transform with respect to $t^*$ to find

$$g(s) = \frac{\cosh(y\sqrt{s+1})}{s\cosh(\alpha\sqrt{s+1})}$$

(c) Take the inverse transform to learn that

$$x^*_{(A)} = 4\pi \sum_{n=1}^{\infty} \left[ \frac{(-1)^n(2n-1)}{(2n-1)^2\pi^2 + 4\alpha^2} \right] \left[ \exp\left(-\left[\frac{(2n-1)^2\pi^2 + 4\alpha^2}{4\alpha^2}\right]t^*\right) - 1 \right] \cos\frac{(2n-1)\pi y}{2\alpha}$$

**9.2.2-4** *A general solution for unsteady diffusion with a first-order homogeneous reaction* Let us assume that the equation of continuity for species $A$ in a system may be shown to take the form

$$\frac{\partial \omega_{(A)}}{\partial t} + \nabla\omega_{(A)} \cdot \mathbf{v} = \mathscr{D}_{(Am)} \text{ div } \nabla\omega_{(A)} + k_1''' \omega_{(A)} \tag{1}$$

where $\mathbf{v}$ is known to be independent of time. This equation is to be solved for $\omega_{(A)}$ subject to the conditions that

At $t = 0$: $\quad \omega_{(A)} = 0$ \hfill (2)

and

At some surfaces: $\quad \omega_{(A)} = \omega_{(A)s}$ \hfill (3)

We wish to show that [3; 4, p. 124; 5]

$$\omega_{(A)} = f \exp(k_1''' t) - k_1''' \int_0^t f \exp(k_1''' \tau) \, d\tau$$

$$= \int_0^t \frac{\partial f}{\partial \tau} \exp(k_1''' \tau) \, d\tau \tag{4}$$

Here $f$ is a solution to the same problem with $k_1''' = 0$.

(a) Begin by introducing as dimensionless variables

$$t^* \equiv k_1''' t \qquad z_i^* \equiv z_i \sqrt{\frac{k_1'''}{\mathscr{D}_{(Am)}}} \qquad \mathbf{v}^* \equiv \frac{\mathbf{v}}{\sqrt{k_1''' \mathscr{D}_{(Am)}}}$$

(b) Take the Laplace transform of both problems (with and without reaction).
(c) Assume a solution to the original problem of the form

$$\mathscr{L}[\omega_{(A)}] = a\mathscr{L}[f]$$

where

$$a = a(s)$$

(d) Invert this expression for $\mathscr{L}[\omega_{(A)}]$ to obtain the desired result.

**9.2.2-5** *More on a general solution for unsteady diffusion with a first-order homogeneous reaction* [3; 4, p. 124] Repeat Exercise 9.2.2-4 assuming that boundary condition (3) is replaced by

At some surfaces: $\quad \nabla\omega_{(A)} \cdot \mathbf{n} = K(\omega_{(A)\infty} - \omega_{(A)})$ \hfill (3a)

Determine that the solution has the same form as Eq. (4) of Exercise 9.2.2-4.

**9.2.2-6** *Still more on a general solution for unsteady diffusion with a first-order homogeneous reaction* Let us assume that a solution to Eq. (1) of Exercise 9.2.2-4 is to be found consistent with the conditions that

At $t = 0$: $\quad \omega_{(A)} = \omega_{(A)0}$

At some surfaces: $\quad \omega_{(A)} = 0$

and

At some surfaces: $\quad \nabla\omega_{(A)} \cdot \mathbf{n} = -K\omega_{(A)}$

Use the approach suggested in Exercise 9.2.2-4 to determine that the solution has the form [6, p. 621]

$$\omega_{(A)} = g \exp(k_1''' t)$$

where $g$ is a solution to the same problem with $k_1''' = 0$.

**9.2.2-7 And still more on a general solution for unsteady diffusion with a first-order homogeneous reaction** Let us assume that a solution to Eq. (1) of Exercise 9.2.2-4 is to be found consistent with the conditions that

At $t = 0$: $\quad \omega_{(A)} = \omega_{(A)0}$ \hfill (1)

At surfaces I: $\quad \omega_{(A)} = \omega_{(A)s}$ \hfill (2)

and

At surfaces II: $\quad \nabla \omega_{(A)} \cdot \mathbf{n} = K(\omega_{(A)\infty} - \omega_{(A)})$ \hfill (3)

Begin by assuming

$$\omega_{(A)} = \omega_{(A)1} + \omega_{(A)2}$$

The function $\omega_{(A)1}$ is understood to satisfy the equation of continuity consistent with Eqs. (2) and (3), and

At $t = 0$: $\quad \omega_{(A)1} = 0$

The function $\omega_{(A)2}$ is a solution to the equation of continuity consistent with Eq. (1) and

At surfaces I: $\quad \omega_{(A)2} = 0$

and

At surfaces II: $\quad \nabla \omega_{(A)2} \cdot \mathbf{n} = -K\omega_{(A)2}$

Conclude that a solution to the equation of continuity that satisfies Eqs. (1) to (3) is [5; corrected by C. Y. Lin and J. D. Chen 1977].

$$\omega_{(A)} = f \exp(k_1''' t) - k_1''' \int_0^t f \exp(k_1''' t)\, dt + g \exp(k_1''' t)$$

Here $f$ is a solution to the system of equations describing $\omega_{(A)1}$ with $k_1''' = 0$; $g$ is a solution to the system of equations describing $\omega_{(A)2}$ with $k_1''' = 0$.

**9.2.2-8 Critical size of an autocatalytic system [6, p. 623]** Acetylene gas is thermodynamically unstable. It tends to decompose:

$$H_2C_2 \text{ (gas)} \rightarrow H_2 \text{ (gas)} + 2C \text{ (solid)}$$

One of the steps in this reaction appears to involve a free radical. Since free radicals are effectively neutralized by contact with an iron surface, their concentration is essentially zero at such a surface. This suggests that acetylene gas can be safely stored in steel cylinders of sufficiently small diameter. If the cylinder is too large, the formation of even a small concentration of free radicals is likely to cause a rapidly increasing rate of decomposition according to the overall reaction described above. Since this reaction is exothermic, an explosion may result.

Let us determine this critical diameter for a storage cylinder, assuming that the decomposition may be described as a first-order homogeneous reaction.

(a) Let us assume that we have an ideal-gas mixture at constant temperature and pressure, so that we may say the molar density $c$ is a constant. Supply an argument to conclude

$$\mathbf{v}^\star = 0$$

(b) Let $x_{(A)}$ denote the mole fraction of the free radical in the gas mixture. Reason that

$$\frac{\partial x_{(A)}}{\partial t} = \mathscr{D}_{(Am)} \frac{1}{r} \frac{\partial}{\partial r}\left(r \frac{\partial x_{(A)}}{\partial r}\right) + k_1''' x_{(A)}$$

[9.2] COMPLETE SOLUTIONS

We wish to solve this equation for the free-radical concentration, using as initial and boundary conditions

At $t = 0$, for $r < R$: $\quad x_{(A)} = x_{(A)0}(r)$

and

At $r = R$, for $t > 0$: $\quad x_{(A)} = 0$

Here $x_{(A)0}$ represents any reasonable function of $r$; $R$ is the radius of the storage cylinder.
(c) Introduce as dimensionless variables

$$t^* \equiv k_1''' t$$

and

$$r^* \equiv \frac{r}{R}$$

Use Exercise 9.2.2-6 to determine

$$x_{(A)} = \sum_{n=1}^{\infty} A_n \exp\left[(1 - \lambda_n^2 K)t^*\right] J_0(\lambda_n r^*)$$

where

$$J_0(\lambda_n) = 0 \quad n = 1, 2, \ldots$$

$$A_n = \frac{\int_0^1 r^* J_0(\lambda_n r^*) x_{(A)0}(r^*)\, dr^*}{\tfrac{1}{2}[J_1(\lambda_n)]^2} \quad n = 1, 2, \ldots$$

and

$$K \equiv \frac{\mathscr{D}_{(Am)}}{R^2 k_1'''}$$

(d) Argue that the acetylene gas can be safely stored so long as

$$R < \lambda_1 \sqrt{\frac{\mathscr{D}_{(Am)}}{k_1'''}} = 2.4048 \sqrt{\frac{\mathscr{D}_{(Am)}}{k_1'''}}$$

Here $\lambda_1$ is the first and smallest zero of $J_0(x)$ [7, p. 130].

**9.2.2-9** Repeat Exercise 9.2.2-3 using Exercise 9.2.2-7.

## 9.2.3 Unsteady diffusion with a slow catalytic reaction

A gas consisting of a mixture of species of $A$ and $B$ is brought into contact with a solid surface that acts as a catalyst for the isomerization reaction $A \to B$. Let us determine the rate at which species $A$ is consumed.

To make the problem as simple as possible, we will take the gas-solid phase interface to be the plane $z_2 = 0$, and we will say that the gas phase occupies the half-space $z_2 > 0$. The initial condition is that

At $t = 0$, for all $z_2 > 0$: $\quad x_{(A)} = x_{(A)0}$ \hfill (3-1)

At the gas-solid phase interface, the isomerization reaction takes place. We will say that this catalytic reaction can be adequately described as first order. From the jump mass balance for species $A$ in Exercise 8.2.1-3.

At $z_2 = 0$, for all $t > 0$:
$$N_{(A)2} = \frac{r_{(A)}^{(\sigma)}}{M_{(A)}} = -k_1'' c_{(A)} \tag{3-2}$$

We will further assume that we are dealing with an ideal-gas mixture at a constant temperature and pressure, so that we may take the molar density $c$ to be a constant.

It seems reasonable to postulate
$$v_1^\star = v_3^\star = 0 \qquad v_2^\star = v_2^\star(t, z_2) \qquad x_{(A)} = x_{(A)}(t, z_2) \tag{3-3}$$

From the overall equation of continuity in the form of Eq. (B) of Table 8.3.2-1, we see
$$\frac{\partial v_2^\star}{\partial z_2} = 0 \tag{3-4}$$

This means
$$v_2^\star = v_2^\star(t) \tag{3-5}$$

By means of Eq. (3-2)$_1$ and the definition for the molar-averaged velocity $\mathbf{v}^\star$, we can say that

At $z_2 = 0$, for all $t > 0$:
$$v_2^\star = \frac{1}{c}(N_{(A)2} + N_{(B)2}) = 0 \tag{3-6}$$

Equations (3-5) and (3-6) allow us to conclude that everywhere in the gas phase

Everywhere for $t > 0$:
$$v_2^\star = 0 \tag{3-7}$$

Let us take the diffusion coefficient to be a constant. In view of Eqs. (3-3) and (3-7), we may write the equation of continuity for species $A$ consistent with Fick's first law, Eq. (D) of Table 8.4.6-2, as

$$\frac{\partial x_{(A)}}{\partial t} = \mathscr{D}_{(AB)} \frac{\partial^2 x_{(A)}}{\partial z_2^2} \tag{3-8}$$

We seek a solution to this equation that satisfies boundary conditions (3-1) and (3-2)$_2$. We will find it convenient to use Eq. (3-7) in rewriting this last boundary condition as

At $z_2 = 0$, for all $t > 0$:
$$\frac{\mathscr{D}_{(AB)}}{k_{(A)}''} \frac{\partial x_{(A)}}{\partial z_2} = x_{(A)} \tag{3-9}$$

Let us introduce the following dimensionless variables:
$$t^\star \equiv \frac{(k_{(A)}'')^2}{\mathscr{D}_{(AB)}} t \qquad z_2^\star \equiv \frac{k_{(A)}'' z_2}{\mathscr{D}_{(AB)}} \qquad x_{(A)}^{\star\star} \equiv \frac{x_{(A)}}{x_{(A)0}} \tag{3-10}$$

In these terms, Eqs. (3-8), (3-1), and (3-9) become, respectively,

$$\frac{\partial x_{(A)}^{\star\star}}{\partial t^\star} = \frac{\partial^2 x_{(A)}^{\star\star}}{\partial z_2^{\star 2}} \tag{3-11}$$

At $t^\star = 0$, for all $z_2^\star > 0$: $\quad x_{(A)}^{\star\star} = 1 \tag{3-12}$

## [9.2] COMPLETE SOLUTIONS

and

$$\text{At } z_2^* = 0, \text{ for all } t^* > 0: \quad \frac{\partial x_{(A)}^{**}}{\partial z_2^*} = x_{(A)}^{**} \tag{3-13}$$

It is clear now that this problem is essentially the same as that considered in Sec. 6.2.3. Consequently, we may use the result there to immediately express the concentration distribution as

$$x_{(A)}^{**} = \text{erf}\left(\frac{z_2^*}{\sqrt{4t^*}}\right) + \exp\left(z_2^* + t^*\right)\left[1 - \text{erf}\left(\frac{z_2^*}{\sqrt{4t^*}} + \sqrt{t^*}\right)\right] \tag{3-14}$$

The rate at which species $A$ is consumed follows easily from Eqs. (3-2) and (3-14) as

$$-N_{(A)2}\big|_{z_2=0} = k''_{(A)} c_{(A)}\big|_{z_2=0}$$

$$= k''_{(A)} c x_{(A)0} \exp\left(\frac{(k''_{(A)})^2 t}{\mathscr{D}_{(AB)}}\right)\left[1 - \text{erf}\left(k''_{(A)}\sqrt{\frac{t}{\mathscr{D}_{(AB)}}}\right)\right] \tag{3-15}$$

### EXERCISES

**9.2.3-1** *Unsteady diffusion with an instantaneous catalytic reaction* At time $t = 0$, a gas mixture of species $A$ and $A_2$ is brought into contact with a solid surface that acts as a catalyst for the dimerization reaction $2A \to A_2$. The reaction occurs instantaneously. Let us determine the rate at which species $A$ is consumed.

To simplify the analysis, let us take the gas-solid phase interface to be the plane $z_2 = 0$ and let us say that the gas phase occupies the half-space $z_2 > 0$. The initial condition is that

$$\text{At } t = 0, \text{ for all } z_2 > 0: \quad x_{(A)} = x_{(A)0} \tag{1}$$

At the gas-solid phase interface, the dimerization reaction takes place instantaneously:

$$\text{At } z_2 = 0, \text{ for all } t > 0: \quad N_{(A)2} = -2N_{(A_2)2} \quad x_{(A)} = 0 \tag{2}$$

We will assume that we are dealing with an ideal-gas mixture at a constant temperature and pressure, so that we may take the molar density $c$ to be a constant.

(a) Determine that the equation for continuity for species $A$ simplifies under these circumstances to

$$\frac{\partial x_{(A)}}{\partial t} = \frac{\mathscr{D}_{(AA_2)}}{2} \frac{\partial x_{(A)}}{\partial z_2}\bigg|_{z_2=0} \frac{\partial x_{(A)}}{\partial z_2} + \mathscr{D}_{(AA_2)} \frac{\partial^2 x_{(A)}}{\partial z_2^{\,2}} \tag{3}$$

(b) Introduce as new independent and dependent variables

$$\eta \equiv \frac{z_2}{\sqrt{4\mathscr{D}_{(AA_2)} t}} \tag{4}$$

and

$$x_{(A)}^{**} \equiv \frac{x_{(A)}}{x_{(A)0}} \tag{5}$$

Determine that

$$x_{(A)}^{**} = \frac{\text{erf}(\eta + \varphi/2) - \text{erf}(\varphi/2)}{1 - \text{erf}(\varphi/2)} \tag{6}$$

**Table 9.2.3-1**

| $\varphi$ | $x_{(A)0}$ | $\dfrac{\exp[-(\varphi/2)^2]}{1-\mathrm{erf}(\varphi/2)}$ |
|---|---|---|
| 0.0 | 0.0 | 1.00 |
| 0.1 | 0.167 | 1.06 |
| 0.3 | 0.453 | 1.18 |
| 0.5 | 0.695 | 1.28 |
| (1.0) | (1.09) | (1.62) |

where $\varphi$ is determined by the relation

$$\varphi = \frac{x_{(A)0}}{\sqrt{\pi}} \frac{\exp[-(\varphi/2)^2]}{1-\mathrm{erf}(\varphi/2)} \tag{7}$$

(c) Conclude that the rate at which species $A$ is consumed is

$$-N_{(A)2}\big|_{z_2=0} = cx_{(A)0} \sqrt{\frac{\mathscr{D}_{(AA_2)}}{\pi t}} \frac{\exp[-(\varphi/2)^2]}{1-\mathrm{erf}(\varphi/2)} \tag{8}$$

(d) Anticipating Sec. 9.3.1, we can reason that convection should be negligibly small when the system is nearly at equilibrium for all time $t > 0$. This corresponds to examining the limit as $x_{(A)0} \to 0$. Conclude that in this limit

$$-N_{(A)2}\big|_{z_2=0} = cx_{(A)0} \sqrt{\frac{\mathscr{D}_{(AA_2)}}{\pi t}} \tag{9}$$

Comparison with Eq. (9) indicates that the last term in Eq. (8) may be thought of as a correction for convection. Table 9.2.3-1 indicates that diffusion induced convection has a surprisingly small effect upon the rate at which species $A$ disappears, even when its initial concentration is relatively large.

**9.2.3-2** *More on diffusion with an instantaneous catalytic reaction* Repeat Exercise 9.2.3-1 for the instantaneous catalytic polymerization $nA \to A_n$. Determine that the rate at which species $A$ disappears is

$$-N_{(A)2}\big|_{z_2=0} = cx_{(A)0} \sqrt{\frac{\mathscr{D}_{(AA_n)}}{\pi t}} \frac{\exp(-[\varphi/2]^2)}{1-\mathrm{erf}[\varphi/2]}$$

where $\varphi$ is determined by the relation

$$\varphi = \frac{2(n-1)}{n} \frac{x_{(A)0}}{\sqrt{\pi}} \frac{\exp(-[\varphi/2]^2)}{1-\mathrm{erf}[\varphi/2]}$$

Table 9.2.3-1 is again applicable here if we replace $x_{(A)0}$ by $[2(n-1)/n]x_{(A)0}$.

## 9.2.4 Thermal diffusion in the two-bulb experiment

Figure 9.2.4-1 depicts two bulbs joined by an insulated tube. The bulb on the left is maintained at a constant temperature $T_1$; the bulb on the right is maintained at a constant temperature $T_2$. The diameter of the tube is sufficiently small that thermal convection may be

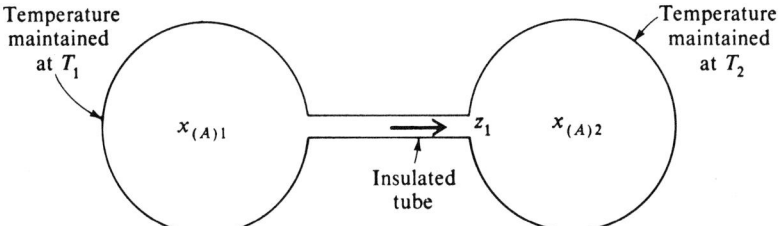

**Fig. 9.2.4-1** Gradient in the mole fraction of species $A$ developed at steady state as the result of a balance between thermal diffusion and ordinary diffusion.

neglected. We wish to develop a relation between the steady-state compositions in these bulbs.

Since the system is closed and nothing is changing as a function of time, it seems reasonable to postulate

$$\mathbf{v} = 0 \quad x_{(A)} = x_{(A)}(z_1) \quad T = T(z_1) \tag{4-1}$$

From the equation of continuity for species $A$ in the form of Eq. (C) of Table 8.2.2-5,

$$\frac{\partial j_{(A)1}}{\partial z_1} = 0 \tag{4-2}$$

With the realization that the system is closed, we conclude

$$j_{(A)1} = 0 \tag{4-3}$$

In view of Eq. (6-8) of Sec. 8.4.6, Eq. (6-6) of Sec. 8.4.6 requires for an ideal solution under these circumstances

$$j_{(A)1} = -\left(\frac{c^2}{\rho}\right) M_{(A)} M_{(B)} \mathscr{D}_{(AB)} \frac{\partial x_{(A)}}{\partial z_1} - D_{(A)}^T \frac{\partial \ln T}{\partial z_1} = 0 \tag{4-4}$$

For gases [1, p. 520],

$$\frac{\rho D_{(A)}^T}{c^2 M_{(A)} M_{(B)} \mathscr{D}_{(AB)}} = \alpha x_{(A)} x_{(B)} \tag{4-5}$$

where the thermal diffusion factor $\alpha$ is very nearly independent of concentration. Equations (4-4) and (4-5) imply

$$\frac{\partial x_{(A)}}{\partial z_1} + \alpha x_{(A)} x_{(B)} \frac{\partial \ln T}{\partial z_1} = 0 \tag{4-6}$$

This is readily integrated to find

$$\int_{x_{(A)1}}^{x_{(A)2}} \frac{dx_{(A)}}{x_{(A)}(1 - x_{(A)})} = -\alpha \int_{T_1}^{T_2} d \ln T$$

$$\frac{x_{(A)1}}{x_{(A)2}} \frac{1 - x_{(A)2}}{1 - x_{(A)1}} = \left(\frac{T_2}{T_1}\right)^\alpha \tag{4-7}$$

In carrying out this integration, we have treated $\alpha$ as a constant. While it is very nearly independent of concentration, its temperature dependence may be complex. It has been recommended [2] that $\alpha$ in Eq. (4-7) be evaluated at a mean temperature $T_m$:

$$T_m \equiv \frac{T_1 T_2}{T_1 - T_2} \ln \frac{T_2}{T_1} \tag{4-8}$$

If the two bulbs in Fig. 9.2.4-1 are of equal volume, if we are able to neglect the volume of gas in the connecting tube, and if the initial uniform composition of the gas is $x_{(A)0}$, we may say that

$$x_{(A)1} + x_{(A)2} = 2x_{(A)0} \tag{4-9}$$

Equations (4-7) and (4-9) give us two equations to be solved for the two unknowns $x_{(A)1}$ and $x_{(A)2}$.

## REFERENCES

1. Hirschfelder, J. O., C. F. Curtiss, and R. B. Bird: "Molecular Theory of Gases and Liquids," corrected printing, Wiley, New York, 1964.
2. Brown, Harrison: *Phys. Rev.*, **58**:661 (1940).
3. Grew, K. E., and T. L. Ibbs: "Thermal Diffusion in Gases," Cambridge, London, 1952.

## EXERCISES

**9.2.4-1** *Natural convection between vertical heated plates*  Let us assume that we have a binary solution of ideal gases undergoing natural convection between the vertical heated plates of Sec. 6.4.2. Determine that Eq. (4-7) applies also in this situation at steady state.

**9.2.4-2** *Natural convection between concentric vertical heated cylinders*  Assume that we have a binary solution of ideal gases undergoing natural convection between concentric vertical heated cylinders as described in Exercise 6.4.2-1. Conclude that Eq. (4-7) applies here also.

The geometry described here is very similar to the Clusius-Dickel thermal diffusion column [3, p.91], which has been used successfully for the separation of isotopes. But there is one important difference. The Clusius-Dickel column has a finite length, whereas in the problem described in Exercise 6.4.2-1 end effects are neglected. The reversal of flow at the top and bottom of the Clusius-Dickel column reinforces the separation, with the result that the primary concentration difference is not radial but rather axial. This particular aspect of the problem has been nicely explained by Grew and Ibbs [3, p. 92].

### 9.2.5 Pressure diffusion in a natural gas well[1]

A natural gas well of depth $L$ has been closed for some time. The mole fraction $x_{(A)0}$ of species $A$ and the pressure $P_0$ at the top of the well are known. We wish to determine the composition and pressure at the bottom of the well.

In order to simplify the computations we will assume that we are dealing with a binary gas mixture of species $A$ and $B$, that temperature is uniform throughout the well, and that the mixture obeys the ideal-gas law and, consequently, may be

---

[1] This problem was suggested by Prof. G. M. Brown, Department of Chemical Engineering, Northwestern University, Evanston, Illinois.

[9.2] COMPLETE SOLUTIONS

regarded as an ideal solution.

Since the system is closed and nothing is changing as a function of time, it seems reasonable to postulate

$$\mathbf{v} = 0 \qquad x_{(A)} = x_{(A)}(z_3) \qquad P = P(z_3) \tag{5-1}$$

We assume here that gravity acts in the $z_3$ direction. From the equation of continuity for species $A$ in the form of Eq. (C) in Table 8.2.2-5,

$$\frac{\partial j_{(A)3}}{\partial z_3} = 0 \tag{5-2}$$

Because the system is closed, we may conclude

$$j_{(A)3} = 0 \tag{5-3}$$

In view of Eq. (6-8) of Sec. 8.4.6, Eq. (6-6) of Sec. 8.4.6 requires for an ideal solution under these circumstances

$$j_{(A)3} = -\left(\frac{c^2}{\rho}\right) M_{(A)} M_{(B)} \mathscr{D}_{(AB)} \left[\frac{dx_{(A)}}{dz_3} + \frac{M_{(A)} x_{(A)}}{RT}\left(\frac{\bar{V}_{(A)}^{(m)}}{M_{(A)}} - \frac{1}{\rho}\right)\frac{dP}{dz_3}\right] = 0 \tag{5-4}$$

or

$$\frac{dx_{(A)}}{dz_3} = -\frac{M_{(A)} x_{(A)}}{RT}\left(\frac{\bar{V}_{(A)}^{(m)}}{M_{(A)}} - \frac{1}{\rho}\right)\frac{dP}{dz_3} \tag{5-5}$$

From the ideal-gas law and the definitions introduced in Exercises 5.1.2-4 and 5.1.2-5,

$$\bar{V}_{(A)}^{(m)} = \bar{V}_{(B)}^{(m)} = \tilde{V} = \frac{RT}{P} \tag{5-6}$$

Cauchy's first law requires

$$\frac{dP}{dz_3} = \rho g \tag{5-7}$$

Equations (5-6) and (5-7) allow us to express Eq. (5-5) as

$$\frac{dx_{(A)}}{dz_3} = -\frac{g x_{(A)}}{RT}\left(\frac{RT\rho}{P} - M_{(A)}\right) \tag{5-8}$$

But the ideal-gas law further requires

$$\frac{RT\rho}{P} = x_{(A)} M_{(A)} + x_{(B)} M_{(B)} \tag{5-9}$$

so that we may eliminate $\rho$ and $P$ from Eq. (5-8) and say

$$\frac{dx_{(A)}}{dz_3} = \frac{g x_{(A)}}{RT}(M_{(A)} - M_{(B)})(1 - x_{(A)}) \tag{5-10}$$

This is readily integrated:

$$\int_{x_{(A)0}}^{x_{(A)}} \frac{dx_{(A)}}{x_{(A)}(1 - x_{(A)})} = \frac{g}{RT}(M_{(A)} - M_{(B)}) \int_0^{z_3} dz_3 \qquad (5\text{-}11)$$

to find

$$\frac{1 - x_{(A)0}}{x_{(A)0}} \frac{x_{(A)}}{1 - x_{(A)}} = \exp\left(\frac{gz_3}{RT}[M_{(A)} - M_{(B)}]\right) \qquad (5\text{-}12)$$

The composition at the bottom of the well is easily seen to be

At $z = L$: 
$$x_{(A)} = \frac{x_{(A)0} \exp\left([gL/RT][M_{(A)} - M_{(B)}]\right)}{1 - x_{(A)0} + x_{(A)0} \exp\left([gL/RT][M_{(A)} - M_{(B)}]\right)} \qquad (5\text{-}13)$$

From Eqs. (5-7) and (5-9), we see

$$\frac{dP}{dx_{(A)}} \frac{dx_{(A)}}{dz_3} = \frac{dP}{dz_3} = \rho g = \frac{gP}{RT}(M_{(A)}x_{(A)} + M_{(B)}x_{(B)}) \qquad (5\text{-}14)$$

In view of Eq. (5-10), this last may be expressed as

$$\frac{dP}{dx_{(A)}} = \frac{[x_{(A)}(M_{(A)} - M_{(B)}) + M_{(B)}]P}{x_{(A)}(1 - x_{(A)})(M_{(A)} - M_{(B)})} \qquad (5\text{-}15)$$

Upon integration, we learn

$$\frac{P}{P_0} = \left(\frac{x_{(A)}}{x_{(A)0}}\right)^{M_{(B)}/(M_{(A)} - M_{(B)})} \left(\frac{1 - x_{(A)0}}{1 - x_{(A)}}\right)^{M_{(A)}/(M_{(A)} - M_{(B)})} \qquad (5\text{-}16)$$

In view of Eq. (5-12), we can say, alternatively,

$$\frac{x_{(A)}P}{x_{(A)0}P_0} = \exp\frac{gz_3 M_{(A)}}{RT} \qquad (5\text{-}17)$$

At the bottom of the well

At $z = L$: 
$$\frac{P}{P_0} = \left\{1 - x_{(A)0} + x_{(A)0}\exp\left(\frac{gL}{RT}[M_{(A)} - M_{(B)}]\right)\right\}$$
$$\times \frac{\exp(gLM_{(A)}/RT)}{\exp([gL/RT][M_{(A)} - M_{(B)}])} \qquad (5\text{-}18)$$

In Exercise 8.5.3-1 this same problem is analyzed assuming that the gas has achieved a thermodynamically stable equilibrium. It is reassuring that the same results are obtained using two apparently radically different approaches.

## EXERCISE

**9.2.5-1** *The ultracentrifuge* Figure 9.2.5-1 shows a binary liquid solution mounted in a cylindrical cell on a high-speed centrifuge. We wish to determine the concentration distribution of the two

# [9.2] COMPLETE SOLUTIONS

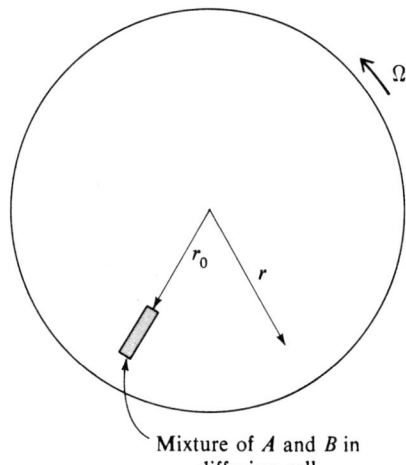

**Fig. 9.2.5-1** Steady-state pressure diffusion in a centrifuge.

components $A$ and $B$ at steady state.

In order to somewhat simplify the analysis, we will assume that the two species form an ideal solution and that the partial molal volumes may be taken to be independent of composition.

(a) Determine that

$$\bar{V}_{(B)}^{(m)} \frac{\partial \ln x_{(A)}}{\partial r} = \frac{pr\Omega^2}{RT}\left(\frac{M_{(A)}\bar{V}_{(B)}^{(m)}}{\rho} - \bar{V}_{(A)}^{(m)}\bar{V}_{(B)}^{(m)}\right)$$

and

$$\bar{V}_{(A)}^{(m)} \frac{\partial \ln x_{(B)}}{\partial r} = \frac{pr\Omega^2}{RT}\left(\frac{M_{(B)}V_{(A)}^{(m)}}{\rho} - \bar{V}_{(A)}^{(m)}\bar{V}_{(B)}^{(m)}\right)$$

to conclude

$$\bar{V}_{(B)}^{(m)} \frac{\partial \ln x_{(A)}}{\partial r} - \bar{V}_{(A)}^{(m)} \frac{\partial \ln x_{(B)}}{\partial r} = \frac{r\Omega^2}{RT}(M_{(A)}\bar{V}_{(B)}^{(m)} - M_{(B)}\bar{V}_{(A)}^{(m)})$$

(b) If

At $r = r_0$: $x_{(A)} = x_{(A)0}$ and $x_{(B)} = x_{(B)0}$

then we may integrate the result of part (a) to learn

$$\left(\frac{x_{(A)}}{x_{(A)0}}\right)^{\bar{V}_{(B)}^{(m)}} \left(\frac{x_{(B)0}}{x_{(B)}}\right)^{\bar{V}_{(A)}^{(m)}} = \exp\left(\frac{\Omega^2}{2RT}[M_{(A)}\bar{V}_{(B)}^{(m)} - M_{(B)}\bar{V}_{(A)}^{(m)}][r^2 - r_0^2]\right)$$

(c) How does this simplify for the case in which the mole fraction of species $A$ is negligible?

## 9.2.6 Forced diffusion in electrochemical systems

By forced diffusion, we refer to a situation in which the individual species in a solution are subjected to unequal external forces. As a result of these force differences, the various species are accelerated with respect to one another and a separation occurs.

Perhaps the most common example of forced diffusion occurs when a salt solution is subjected to an electric field. When a salt such as $AgNO_3$ is dissolved in water, it ionizes. From our present viewpoint we should almost certainly consider this to be a ternary rather than a binary solution; we should regard $Ag^+$ and $NO_3^-$ as individual species. The necessity for regarding these ions as individual species becomes more obvious when a solution is placed in an electric field. In this case, the force beyond gravity acting on $Ag^+$ is in the opposite direction to that acting on $NO_3^-$.

For simplicity, we shall neglect any effect attributable to pressure diffusion or thermal diffusion. We shall furthermore confine our attention to dilute solutions for which Eq. (7-1) of Sec. 8.4.7 is applicable and simplifies to

$$\mathbf{N}_{(A)} = -c\mathscr{D}_{(Am)}\left[\nabla x_{(A)} + \frac{M_{(A)}x_{(A)}}{RT}\left(-\mathbf{f}_{(A)} + \sum_{B=1}^{N}\omega_{(B)}\mathbf{f}_{(B)}\right)\right] + c_{(A)}\mathbf{v}^\star \quad (6\text{-}1)$$

We shall assume that this solution is subjected to an electric field for which the electrostatic potential is $\Phi$. Under these conditions, an ionic species $A$ is subjected to two external forces, gravity and that attributable to the electric field:

$$\mathbf{f}_{(A)} = \mathbf{g} - \frac{\epsilon_{(A)}}{m_{(A)}}\nabla\Phi \quad (6\text{-}2)$$

Here $\mathbf{g}$ is the acceleration of gravity, $\epsilon_{(A)}$ is the ionic charge, and $m_{(A)}$ is the ionic mass. In principle, we should satisfy the equations of continuity for all the species present in the solution consistent with Eq. (6-1), the appropriate boundary conditions, and Poisson's equation for the electrostatic potential $\Phi$. In practice, we avoid the use of Poisson's equation by requiring that any electrolytic solution be electrically neutral:

$$\sum_{A=1}^{N} a_{(A)}c_{(A)} = 0 \quad (6\text{-}3)$$

where $a_{(A)}$ is the valence or charge number of species $A$. Newman [1, p. 101] gives an excellent discussion of the justification for electroneutrality. Consequently, our approach to the analysis of electrolytic solutions in the presence of an electric field will be to satisfy the equations of continuity consistent with Eqs. (6-1) and (6-3).

A binary electrolyte is the simplest case to handle. By a binary electrolyte, I mean the solution of a single salt composed of one kind of cation (+) and one kind of anion (−). Let $\nu_{(+)}$ and $\nu_{(-)}$ be the numbers of cations and anions produced by dissociation of one molecule of electrolyte. This suggests that, if we define

$$\varkappa \equiv \frac{c_{(+)}}{\nu_{(+)}} = \frac{c_{(-)}}{\nu_{(-)}} \quad (6\text{-}4)$$

we can automatically satisfy the electroneutrality requirement, Eq. (6-3). Assuming there are no homogeneous chemical reactions, we can write the equations of continuity for the cation and anion as

## [9.2] COMPLETE SOLUTIONS

$$\frac{\partial \varkappa}{\partial t} + \text{div}\, \frac{\mathbf{N}_{(+)}}{\nu_{(+)}} = 0 \tag{6-5}$$

and

$$\frac{\partial \varkappa}{\partial t} + \text{div}\, \frac{\mathbf{N}_{(-)}}{\nu_{(-)}} = 0 \tag{6-6}$$

Since we are dealing with a dilute solution, Eq. (6-1) requires for these two ions

$$\frac{\mathbf{N}_{(+)}}{\nu_{(+)}} = -\mathscr{D}_{(+m)}\left(\nabla \varkappa + \frac{\varkappa \epsilon_{(+)}}{kT}\nabla \Phi\right) + \varkappa \mathbf{v}^\star \tag{6-7}$$

and

$$\frac{\mathbf{N}_{(-)}}{\nu_{(-)}} = -\mathscr{D}_{(-m)}\left(\nabla \varkappa + \frac{\varkappa \epsilon_{(-)}}{kT}\nabla \Phi\right) + \varkappa \mathbf{v}^\star \tag{6-8}$$

Here we have introduced the Boltzmann constant $k = R/\tilde{N}$; $\tilde{N}$ is Avogadro's number. By taking the difference between these last two equations and rearranging, we can find

$$\frac{-\mathscr{D}_{(+m)}\varkappa \epsilon_{(+)}}{kT}\nabla \Phi = \frac{\epsilon_{(+)}\mathscr{D}_{(+m)}}{\epsilon_{(+)}\mathscr{D}_{(+m)} - \epsilon_{(-)}\mathscr{D}_{(-m)}}\left(\frac{\mathbf{N}_{(+)}}{\nu_{(+)}} - \frac{\mathbf{N}_{(-)}}{\nu_{(-)}}\right)$$

$$+ \frac{\epsilon_{(+)}\mathscr{D}_{(+m)}(\mathscr{D}_{(+m)} - \mathscr{D}_{(-m)})}{\epsilon_{(+)}\mathscr{D}_{(+m)} - \epsilon_{(-)}\mathscr{D}_{(-m)}}\nabla \varkappa \tag{6-9}$$

This allows us to eliminate the electrostatic potential $\Phi$ between Eqs. (6-7) and (6-9) to find after some rearrangement

$$\frac{1}{a_{(+)}\mathscr{D}_{(+m)} - a_{(-)}\mathscr{D}_{(-m)}}\left(\frac{a_{(+)}\mathscr{D}_{(+m)}}{\nu_{(-)}}\mathbf{N}_{(-)} - \frac{a_{(-)}\mathscr{D}_{(-m)}}{\nu_{(+)}}\mathbf{N}_{(+)}\right) = -D\nabla \varkappa + \varkappa \mathbf{v}^\star \tag{6-10}$$

where

$$D \equiv \frac{\mathscr{D}_{(+m)}\mathscr{D}_{(-m)}(a_{(+)} - a_{(-)})}{a_{(+)}\mathscr{D}_{(+m)} - a_{(-)}\mathscr{D}_{(-m)}} \tag{6-11}$$

Taking the divergence of Eq. (6-10) and employing Eqs. (6-5) and (6-6), we can finally say that

$$\frac{\partial \varkappa}{\partial t} + \text{div}\,(\varkappa \mathbf{v}^\star) = D\,\text{div}\,\nabla \varkappa \tag{6-12}$$

Equation (7-1) of Sec. 8.4.7 and the results here are applicable to dilute solutions as well as to two other cases explained in Sec. 8.4.7. For dilute solutions, it would appear that $c$ is nearly a constant and Eq. (6-12) can be further simplified to

$$\frac{\partial \varkappa}{\partial t} + \nabla \varkappa \cdot \mathbf{v}^\star = D\,\text{div}\,\nabla \varkappa \tag{6-13}$$

or

$$\frac{\partial \varkappa}{\partial t} + \nabla \varkappa \cdot \mathbf{v} = D \text{ div } \nabla \varkappa \tag{6-14}$$

For a dilute binary electrolyte, we need solve only Eq. (6-13) or Eq. (6-14) consistent with the overall equation of continuity and the appropriate boundary conditions. The resulting solution for $\varkappa$ can be related to the desired concentration distributions through Eqs. (6-4).

For a more complete introduction to the transport processes in electrolytic solutions, with particular attention to the variety of possible boundary conditions, see Newman [1] and Levich [2].

## REFERENCES

1. Newman, John: In C. W. Tobias (ed.), "Advances in Electrochemistry and Electrochemical Engineering," vol. 5, Interscience, New York, 1967.
2. Levich, V. G.: "Physicochemical Hydrodynamics," Prentice-Hall, Englewood Cliffs, N.J., 1962.
3. Bird, R. B., W. E. Stewart, and E. N. Lightfoot: "Transport Phenomena," 7th printing, Wiley, New York, 1960.

## EXERCISES

**9.2.6-1** *The maximum current density in a simple cell*  The simple cell shown in Fig. 9.2.6-1 is filled with a dilute binary electrolyte formed by dissolving a small amount of a salt $MX$ in water. We wish to relate the current density $\mathbf{i}$,

$$\mathbf{i} \equiv F \sum_{A=1}^{N} a_{(A)} \mathbf{N}_{(A)}$$

to the concentration distributions for the ions $M^+$ and $X^-$. Here $F$ is Faraday's constant. The cell may be assumed to be operating at steady state.

(a) Using the approach described in the text, determine that

$$\frac{i_2 z_2}{cDF} = \ln \left( \frac{\mathscr{D}_{(-m)} - [\mathscr{D}_{(+m)} + \mathscr{D}_{(-m)}] x_{(+)}}{\mathscr{D}_{(-m)} - [\mathscr{D}_{(+m)} + \mathscr{D}_{(-m)}] x_{(+)\text{cath}}} \right)$$

(b) Conclude that for a very dilute solution we should be justified in approximating this expression by

$$\frac{-i_2 z_2}{2c \mathscr{D}_{(+m)} F} = x_{(+)} - x_{(+)\text{cath}}$$

(c) Determine that the maximum current density is

$$-i_{2(\text{max})} = 4c \mathscr{D}_{(+m)} \frac{F x_{(MX)\text{avg}}}{L}$$

**9.2.6-2** *A simple cell filled with a ternary electrolyte*  Let us now visualize that the cell shown in Fig. 9.2.6-1 contains both $Ag^+NO_3^-$ at an average concentration $10^{-6}$ $N$ and $K^+NO_3^-$ at an average concentration of 0.1 $N$. A voltage is imposed upon the cell that is just sufficient to cause the silver ion concentration at the cathode to drop essentially to zero.

[9.2] COMPLETE SOLUTIONS 529

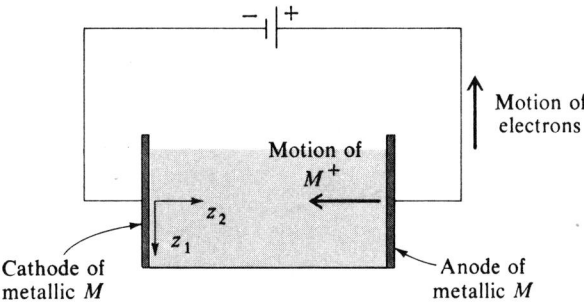

**Fig. 9.2.6-1** A simple cell filled with a binary electrolyte.

(a) Determine that the concentration distributions of the three ionic species present are

$$x_{(NO_3^-)} - x_{(NO_3^-)\text{cath}} = \frac{\alpha z_2}{L} \tag{1}$$

$$x_{(K^+)} = \frac{(x_{(NO_3^-)\text{cath}})^2}{x_{(NO_3^-)\text{cath}} + \alpha z_2/L} \exp\left(-2\alpha \left[\frac{\mathscr{D}_{(Ag+m)}}{\mathscr{D}_{(NO_3^-m)}} + \frac{\mathscr{D}_{(Ag+m)}}{\mathscr{D}_{(K+m)}}\right] \frac{z_2}{L}\right) \tag{2}$$

$$x_{(Ag^+)} = x_{(NO_3^-)\text{cath}} + \frac{\alpha z_2}{L}$$

$$- \frac{(x_{(NO_3^-)\text{cath}})^2}{x_{(NO_3^-)\text{cath}} + \alpha z_2/L} \exp\left(-2\alpha \left[\frac{\mathscr{D}_{(Ag+m)}}{\mathscr{D}_{(NO_3^-m)}} + \frac{\mathscr{D}_{(Ag+m)}}{\mathscr{D}_{(K+m)}}\right] \frac{z_2}{L}\right) \tag{3}$$

where

$$\alpha \equiv \frac{-N_{(Ag^+)2}L}{2c\mathscr{D}_{(Ag+m)}} \tag{4}$$

(b) Reason that for very small values of $\alpha$, Eqs. (2) and (3) above reduce to

$$x_{(K^+)} - x_{(K^+)\text{cath}} = \frac{-\alpha z_2}{L} \tag{5}$$

and

$$x_{(Ag^+)} = \frac{2\alpha z_2}{L} \tag{6}$$

Determine that, for the conditions described, the concentration distributions are those represented in Fig. 9.2.6-2 [3, p. 588].

The limit as $\alpha \to 0$ corresponds to neglecting convection with respect to diffusion.

## 9.2.7 Steady-state evaporation through a multicomponent stagnant film

Let us consider a system that is similar to that shown in Fig. 9.2.1-1. Pure liquid $A$ continuously evaporates into an ideal-gas mixture of $A$, $E$, and $F$. The apparatus is arranged in such a manner that the liquid-gas phase interface remains fixed in space as the evaporation takes place. Species $E$ and $F$ are assumed to be insoluble in liquid $A$:

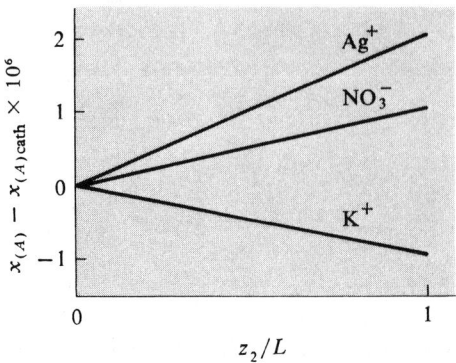

**Fig. 9.2.6-2** Ionic concentrations for very small values of $\alpha$.

At $z_2 = 0$: $\quad N_{(E)2} = N_{(F)2} = 0$ \hfill (7-1)

For the existing conditions, the equilibrium composition of the gas phase is $x_{(A)\text{eq}}$:

At $z_2 = 0$: $\quad x_{(A)} = x_{(A)\text{eq}}$ \hfill (7-2)

The composition of the gas phase at the top of the column is maintained constant:

At $z_2 = L$: $\quad x_{(E)} = x_{(E)L} \quad x_{(F)} = x_{(F)L}$ \hfill (7-3)

We wish to determine the rate of evaporation of $A$ from the surface. Let us begin by asking for the mole fraction distribution of each species in the gas phase.

If we assume that our ideal-gas mixture is at a constant temperature and pressure (forgetting about any hydrostatic effect), then the total molar density $c$ is a constant. This suggests that it may be reasonable to look for a solution to this problem of the form

$$v_2^\star = v_2^\star(z_2) \quad v_1^\star = v_3^\star = 0 \quad (7\text{-}4)$$

$$x_{(E)} = x_{(E)}(z_2) \quad (7\text{-}5)$$

and

$$x_{(F)} = x_{(F)}(z_2) \quad (7\text{-}6)$$

In view of Eqs. (7-4) to (7-6), it follows immediately from Exercise 8.4.5-2 that only the $z_2$ components of $\mathbf{N}_{(A)}$, $\mathbf{N}_{(E)}$, and $\mathbf{N}_{(F)}$ are nonzero. From the equations of continuity for these three species, as well as from Eq. (7-1), we conclude

$$N_{(A)2} = \text{a constant} \quad N_{(E)2} = N_{(F)2} = 0 \quad (7\text{-}7)$$

When we neglect any effects attributable to thermal, pressure, and forced diffusion, Exercise 8.4.5-4 indicates that for an ideal gas the generalized Stefan-Maxwell Eq. (5-14) of Sec. 8.4.5 reduces to

$$\nabla x_{(C)} = \sum_{B=1}^{N} \frac{1}{c\mathscr{D}_{(CB)}} (x_{(C)}\mathbf{N}_{(B)} - x_{(B)}\mathbf{N}_{(C)}) \quad (7\text{-}8)$$

Because of Eqs. (7-7), Eq. (7-8) says that for species $E$ and $F$,

$$\frac{dx_{(E)}}{dz_2} = \frac{N_{(A)2}}{c\mathscr{D}_{(AE)}} x_{(E)} \tag{7-9}$$

and

$$\frac{dx_{(F)}}{dz_2} = \frac{N_{(A)2}}{c\mathscr{D}_{(AF)}} x_{(F)} \tag{7-10}$$

Equations (7-9) and (7-10) may be integrated consistent with boundary conditions (7-3) to find

$$\frac{x_{(E)}}{x_{(E)L}} = \exp\left(-\alpha\left[1 - \frac{z_2}{L}\right]\right) \tag{7-11}$$

and

$$\frac{x_{(F)}}{x_{(F)L}} = \exp\left(-\alpha\beta\left[1 - \frac{z_2}{L}\right]\right) \tag{7-12}$$

Here

$$\alpha \equiv \frac{N_{(A)2} L}{c\mathscr{D}_{(AE)}} \tag{7-13}$$

and

$$\beta \equiv \frac{\mathscr{D}_{(AE)}}{\mathscr{D}_{(AF)}} \tag{7-14}$$

If we assume that $\alpha$ and $\beta$ are known, then Eqs. (7-11) and (7-12) determine the mole fraction distributions in the gas phase.

Our principal object here is to determine $\alpha$ or $N_{(A)2}$. This may be accomplished by requiring Eqs. (7-11) and (7-12) to satisfy Eq. (7-2):

$$x_{(A)\text{eq}} = 1 - x_{(E)L} \exp(-\alpha) - x_{(F)L} \exp(-\alpha\beta) \tag{7-15}$$

Given $\beta$, we may finally solve Eq. (7-15) for $\alpha$.

For further information on multicomponent ordinary diffusion, I suggest reading [1 to 6].

## REFERENCES

1. Cussler, E. L., Jr., and E. N. Lightfoot, Jr.: *A.I.Ch.E.J.*, **9**:702 (1963).
2. Cussler, E. L., Jr., and E. N. Lightfoot, Jr.: *A.I.Ch.E.J.*, **9**:783 (1963).
3. Toor, H. L.: *A.I.Ch.E.J.*, **10**:448 (1964).
4. Toor, H. L.: *A.I.Ch.E.J.*, **10**:460 (1964).
5. Toor, H. L., C. V. Seshadri, and K. R. Arnold: *A.I.Ch.E.J.*, **11**:746 (1965).
6. Arnold, K. R., and H. L. Toor: *A.I.Ch.E.J.*, **13**:909 (1967).
7. Hsu, H. W., and R. B. Bird: *A.I.Ch.E.J.*, **6**:516 (1960).

## EXERCISE

**9.2.7-1** *Steady-state diffusion with a slow catalytic reaction* [7] A gas consisting of a mixture of species $A$, $E$, and $F$ is brought into contact with a solid surface that acts as a catalyst for the isomerization reaction $A \rightarrow E$. We wish to determine the rate at which species $A$ is consumed.

In order to make the problem as simple as possible, let us visualize an idealized steady-state situation in which species $A$ diffuses through a stagnant gas film of thickness $L$ to a catalytic surface as shown in Fig. 9.2.7-1. We will say that this catalytic reaction can be adequately described as first order:

At $z_2 = 0$: $\quad N_{(A)2} = \dfrac{r_{(A)}^{(\sigma)}}{M_{(A)}} = -k_1'' c_{(A)}$

Find that

$$x_{(F)} = x_{(F)L} \exp(\alpha[\beta - 1][z_2^* - 1])$$

and

$$x_{(A)} - x_{(A)L} = \frac{1-\gamma}{\beta - 1} x_{(F)L} \{\exp(\alpha[\beta - 1][z_2^* - 1]) - 1\} + \alpha\gamma(z_2^* - 1)$$

where

$$\alpha = \frac{-N_{(A)2} L}{c\mathscr{D}_{(AF)}} \quad \beta = \frac{\mathscr{D}_{(AF)}}{\mathscr{D}_{(EF)}} \quad \gamma = \frac{\mathscr{D}_{(AF)}}{\mathscr{D}_{(AE)}} \quad z_2^* = \frac{z_2}{L}$$

Conclude that the rate at which species $A$ is consumed by the chemical reaction can be determined by solving the implicit equation

$$\alpha(\nu + \gamma) = \frac{1-\gamma}{\beta - 1} x_{(F)L} \{\exp(\alpha[1 - \beta]) - 1\} + x_{(A)L}$$

Here

$$\nu \equiv \frac{\mathscr{D}_{(AF)}}{L k_1''}$$

Fig. 9.2.7-1  Stagnant gas film on catalytic surface.

### 9.2.8 Condensation of mixed vapors [1, p. 586]
Chloroform and benzene condense continuously as shown in Fig. 9.2.8-1 from an ideal-gas mixture of known composition at 1 atm. We wish to relate the composition of the condensate to the total molar rate of condensation and to the energy flux to the condensate film.

Let us begin by assuming

$$x_{(B)} = x_{(B)}(z_2) \qquad x_{(C)} = x_{(C)}(z_2) \tag{8-1}$$

and

$$v_1^\star = v_3^\star = 0 \qquad v_2^\star = v_2^\star(z_2) \tag{8-2}$$

From the overall equation of continuity, we conclude that

## [9.2] COMPLETE SOLUTIONS

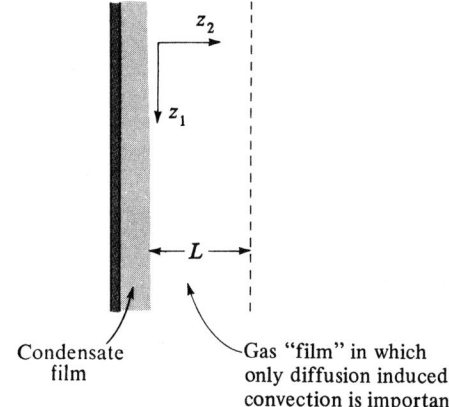

**Fig. 9.2.8-1** Condensation of mixed vapors.

$$v_2^\star = \text{a constant} \tag{8-3}$$

In view of Eqs. (8-1) and (8-2), Fick's first law requires that

$$N_{(B)1} = N_{(B)3} = 0 \qquad N_{(B)2} = N_{(B)2}(z_2) \tag{8-4}$$

From the equation of continuity for species $B$, we learn that

$$N_{(B)2} = \text{a constant} \tag{8-5}$$

It follows in a similar manner that

$$N_{(C)1} = N_{(C)3} = 0 \qquad N_{(C)2} = \text{a constant} \tag{8-6}$$

It seems reasonable to assume that the ratio of benzene to chloroform in the condensate may be considered to be equal to the ratio of the rates of condensation:

$$\frac{N_{(B)2}}{N_{(C)2}} = \frac{x_{(B)\text{cond}}}{x_{(C)\text{cond}}} \tag{8-7}$$

This allows us to rewrite the $z_2$ component of Fick's first law as

$$\frac{N_{(B)2}}{c\mathscr{D}_{(BC)}} = -\frac{1}{1 - (x_{(B)}/x_{(B)\text{cond}})} \frac{dx_{(B)}}{dz_2} \tag{8-8}$$

This in turn can be integrated consistent with boundary conditions

At $z_2 = 0$: $\qquad x_{(B)} = x_{(B)}^* \qquad x_{(C)} = x_{(C)}^* \tag{8-9}$

and

At $z_2 = L$: $\qquad x_{(B)} = x_{(B)L} \qquad x_{(C)} = x_{(C)L} \tag{8-10}$

to find

$$\frac{N_{(B)2}L}{c\mathscr{D}_{(BC)}x_{(B)\text{cond}}} = \ln\left|\frac{x_{(B)\text{cond}} - x_{(B)L}}{x_{(B)\text{cond}} - x_{(B)}^*}\right| \qquad (8\text{-}11)$$

In much the same manner, we can calculate that

$$\frac{N_{(C)2}L}{c\mathscr{D}_{(BC)}x_{(C)\text{cond}}} = \ln\left|\frac{x_{(C)\text{cond}} - x_{(C)L}}{x_{(C)\text{cond}} - x_{(C)}^*}\right| \qquad (8\text{-}12)$$

If we are primarily interested in the total molar rate of condensation, from Eq. (8-7) it follows that

$$v_2^* = \frac{N_{(B)2}}{cx_{(B)\text{cond}}} = \frac{N_{(C)2}}{cx_{(C)\text{cond}}} \qquad (8\text{-}13)$$

$$= \frac{\mathscr{D}_{(BC)}}{L}\ln\left|\frac{x_{(B)\text{cond}} - x_{(B)L}}{x_{(B)\text{cond}} - x_{(B)}^*}\right| = \frac{\mathscr{D}_{(BC)}}{L}\ln\left|\frac{x_{(C)\text{cond}} - x_{(C)L}}{x_{(C)\text{cond}} - x_{(C)}^*}\right| \qquad (8\text{-}14)$$

The differential energy balance in the form of Eq. (J) of Table 8.3.5-1 simplifies considerably for this case, when we neglect viscous dissipation, pressure gradients, and the time rate of change of the kinetic energy of diffusion:

$$\text{div}\left(\sum_{A=1}^{N}\overline{H}_{(A)}^{(m)}\mathbf{N}_{(A)} + \mathbf{q} - \sum_{A=1}^{N}\overline{H}_{(A)}^{(m)}\mathbf{J}_{(A)}\right) = 0 \qquad (8\text{-}15)$$

From Sec. 8.4.4, we have that the thermal energy flux may be described by

$$\boldsymbol{\epsilon} \equiv \mathbf{q} - \sum_{A=1}^{N}(\overline{H}_{(A)} + \tfrac{1}{2}u_{(A)}^2)\mathbf{j}_{(A)}$$

$$\doteq \mathbf{q} - \sum_{A=1}^{N}\overline{H}_{(A)}^{(m)}\mathbf{J}_{(A)}$$

$$= -k\,\nabla T \qquad (8\text{-}16)$$

where we again neglect the kinetic energy of diffusion. In view of Eqs. (8-1) and (8-2), it seems reasonable to assume that

$$T = T(z_2) \qquad (8\text{-}17)$$

This enables us to conclude from Eqs. (8-15) and (8-16) that

$$\sum_{A=1}^{N}\overline{H}_{(A)}^{(m)}N_{(A)2} - k\frac{dT}{dz_2} = C_1 = \text{a constant} \qquad (8\text{-}18)$$

If we measure enthalpy with respect to pure liquids at the saturation temperature $T_0$, then the partial molal enthalpies of benzene and chloroform in the ideal-gas mixture are

[9.2] COMPLETE SOLUTIONS

$$\overline{H}^{(m)}_{(B)} = \tilde{c}_{P(B)}(T - T_0) + \tilde{\lambda}_{(B)} \qquad (8\text{-}19)$$

and

$$\overline{H}^{(m)}_{(C)} = \tilde{c}_{P(C)}(T - T_0) + \tilde{\lambda}_{(C)} \qquad (8\text{-}20)$$

Here $c_{P(A)}$ and $\lambda_{(A)}$ are the molar heat capacity at constant pressure and molar heat of vaporization for species $A$. Using these expressions, we can rearrange Eq. (8-18) in the form

$$k \frac{dT}{dz_2} = cv_2^\star \tilde{c}_{P,\text{mix}}(T - T_0) + cv_2^\star \tilde{\lambda}_{\text{mix}} - C_1 \qquad (8\text{-}21)$$

where for convenience we introduce

$$\tilde{c}_{P,\text{mix}} \equiv x_{(B)\text{cond}} \tilde{c}_{P(B)} + x_{(C)\text{cond}} \tilde{c}_{P(C)} \qquad (8\text{-}22)$$

and

$$\tilde{\lambda}_{\text{mix}} \equiv x_{(B)\text{cond}} \tilde{\lambda}_{(B)} + x_{(C)\text{cond}} \tilde{\lambda}_{(C)} \qquad (8\text{-}23)$$

We can finally integrate Eq. (8-21) consistent with the boundary conditions

At $z_2 = 0$: $\quad T = T_0 \qquad (8\text{-}24)$

and

At $z_2 = L$: $\quad T = T_L \qquad (8\text{-}25)$

to learn

$$\frac{T - T_0}{T_L - T_0} = \frac{1 - \exp(-z_2 A/L)}{1 - \exp(-A)} \qquad (8\text{-}26)$$

Here

$$A \equiv \frac{-cv_2^\star \tilde{c}_{P,\text{mix}} L}{k} \qquad (8\text{-}27)$$

Equation (J) of Table 8.3.5-1, indicates that the thermal energy flux $\mathbf{e}$ with respect to a fixed frame of reference is

$$\mathbf{e} \equiv \boldsymbol{\epsilon} + \sum_{A=1}^{N} \overline{H}_{(A)} \mathbf{n}_{(A)} = \mathbf{q} - \sum_{A=1}^{N} \overline{H}_{(A)} \mathbf{j}_{(A)} + \sum_{A=1}^{N} \overline{H}_{(A)} \mathbf{n}_{(A)}$$

$$= \mathbf{q} + \sum_{A=1}^{N} \rho_{(A)} \overline{H}_{(A)} \mathbf{v}$$

$$= \mathbf{q} + \rho \hat{H} \mathbf{v} \qquad (8\text{-}28)$$

In terms of the constitutive equation suggested in Sec. 8.4.4, this becomes

$$\mathbf{e} = -k\,\nabla T + \sum_{A=1}^{N} \overline{H}_{(A)}^{(m)} \mathbf{N}_{(A)} \tag{8-29}$$

Equations (8-19), (8-20), and (8-26) allow us to compute the thermal energy flux into the condensate film as

$$\text{At } z_2 = 0: \quad -e_2 = k\frac{dT}{dz_2} - cv_2^\star \tilde{\lambda}_{\text{mix}}$$

$$= \frac{Ak(T_L - T_0)}{L[1 - \exp(-A)]} - cv_2^\star \tilde{\lambda}_{\text{mix}} \tag{8-30}$$

We suggest in Ex. 9.2.8-1 how this expression for the thermal energy flux into the condensate film may be used to estimate the temperature of the condenser wall.

For an early discussion of the condensation of mixed vapors, see Colburn and Drew [2].

## REFERENCES

1. Bird, R. B., W. E. Stewart, and E. N. Lightfoot: "Transport Phenomena," 7th printing, Wiley, New York, 1960.
2. Colburn, A. P., and T. B. Drew: *Trans. Am. Inst. Chem. Engrs.*, 33:197 (1937).
3. Chu, Ju Chin, R. L. Getty, L. F. Brennecke, and Rajendra Paul: "Distillation Equilibrium Data," Reinhold, New York, 1950.

## EXERCISES

**9.2.8-1** *More on the condensation of mixed vapors* [1, p. 586] Let us represent the rate of heat transfer to the condenser surface by Newton's law of cooling, so that we may say

$$\text{At } z_2 = 0: \quad -e_2 = h(T_0 - T_c)$$

If we can estimate a value for the heat-transfer coefficient $h$ for the condensate film, then we may use this expression, together with Eq. (8-30), to compute the temperature $T_c$ of the condenser wall.

Let us consider a particular example:

$$x_{(C)L} = 0.5$$
$$T_L = 76.4°C$$
$$L = 0.01 \text{ cm}$$
$$h = 200 \text{ Btu/h ft}^2 \text{ °F}$$
$$\mathscr{D}_{(BC)} = 0.050 \text{ cm}^2/\text{s}$$
$$\tilde{c}_{P(B)} = 22.8 \text{ cal/g mol °C}$$
$$\tilde{c}_{P(C)} = 16.5 \text{ cal/g mol °C}$$
$$k = 0.007 \text{ Btu/h ft °F}$$
$$\tilde{\lambda}_{\text{mix}} = 12{,}800 \text{ Btu/lb mol}$$

Using the vapor-liquid equilibrium data in Table 9.2.8-1, compute that if $x_{(C)\text{cond}} = 0.44$, then $T_c = 87°F$.

[9.2] COMPLETE SOLUTIONS

**Table 9.2.8-1 Vapor-liquid equilibrium data for chloroform-benzene system at 1 atm [3, p. 61]**

| $x_{(C)}$, vapor | $x_{(C)}$, liquid | Saturation temperature, °C |
|---|---|---|
| 0 | 0 | 80.6 |
| 0.10 | 0.08 | 79.8 |
| 0.20 | 0.15 | 79.0 |
| 0.30 | 0.22 | 78.2 |
| 0.40 | 0.29 | 77.3 |
| 0.50 | 0.36 | 76.4 |
| 0.60 | 0.44 | 75.3 |
| 0.70 | 0.54 | 74.0 |
| 0.80 | 0.66 | 71.9 |
| 0.90 | 0.79 | 68.9 |
| 1.00 | 1.00 | 61.4 |

**9.2.8-2** *Condensation of a single species from a gas mixture* [1, p. 572] Repeat the analysis of the text for an ideal-gas mixture of two species $A$ and $B$, assuming now that species $B$ is noncondensable. Determine that

$$-\alpha \equiv \frac{N_{(A)2}L}{c\mathcal{D}_{(AB)}} = \ln\left|\frac{1-x_{(A)L}}{1-x_{(A)0}}\right| \tag{1}$$

$$\frac{x_{(A)} - x_{(A)0}}{x_{(A)L} - x_{(A)0}} = \frac{1 - \exp(-\alpha z_2/L)}{1 - \exp(-\alpha)} \tag{2}$$

and

$$\frac{T - T_0}{T_L - T_0} = \frac{1 - \exp(N_{(A)2}\tilde{c}_{P(A)}z_2/k)}{1 - \exp(N_{(A)2}\tilde{c}_{P(A)}L/k)} \tag{3}$$

It is interesting to compare the expression for the energy flux to that which would exist in the absence of mass transfer. Using a superscript zero to indicate conditions in the absence of mass transfer, compute that

$$\frac{\epsilon|_{z_2=0}}{\epsilon^0|_{z_2=0}} = \frac{-k(dT/dz_2)|_{z_2=0}}{-k(dT/dz_2)^0_{z_2=0}} = \frac{-N_{(A)2}\tilde{c}_{P(A)}L/k}{1 - \exp(N_{(A)2}\tilde{c}_{P(A)}L/k)} \tag{4}$$

Comparing Eqs. (1) and (4), notice that, while the rate of heat transfer is directly affected by simultaneous mass transfer, the mass flux is not directly affected by heat transfer.

For many applications, the effect of mass transfer is small and the right-hand side of Eq. (4) is nearly unity. Consider a particular example such as you might encounter in air conditioning:

At $z_2 = L$: $T = 80°F$ $x_{(H_2O)} = 0.018$

and

At $z_2 = 0$: $T = 50°F$

Estimate the right-hand side of Eq. (4), noting that $N_{Pr} = 0.73$ for nitrogen [1, p. 256] and that $N_{Sc} = 0.2$ to 5 for most gas pairs [1, p. 512].

## 9.3 NO CONVECTION

**9.3.1 When diffusion induced convection is neglected** In our discussion of momentum and energy transfer in Chaps. 3 and 6, we found that analytic solutions for a wider variety of problems are available when we consider limiting cases. Sometimes, the analysis for a limiting case is all that a situation demands. At worst, it is the starting point for further numerical work.

We have already looked at the dimensionless forms of Cauchy's first law and the differential energy balance in Secs. 3.3.1 and 6.4.1 (there are a few additional terms in the overall differential energy balance to be considered; see Sec. 8.3.5), so we will confine our attention here to the differential mass balance for an individual species $A$.

Let us define the following dimensionless variables:

$$\rho^* \equiv \frac{\rho}{\rho_0} \qquad t^* \equiv \frac{t}{t_0} \qquad z_i^* \equiv \frac{z_i}{L_0}$$
$$\mathbf{v}^* \equiv \frac{\mathbf{v}}{v_0} \qquad \mathbf{j}_{(A)}^* \equiv \frac{L_0 \mathbf{j}_{(A)}}{\rho_0 \mathscr{D}_0} \qquad r_{(A)}^* \equiv \frac{r_{(A)}}{r_{(A)0}} \qquad (1\text{-}1)$$

Here the quantities distinguished by a subscript 0 are reference (or characteristic) quantities having the same dimension as the variables being made dimensionless. For example, $v_0$ is a characteristic magnitude of velocity and $r_{(A)0}$ is a characteristic rate at which the mass of species $A$ is produced per unit volume by homogeneous chemical reactions. By $\mathscr{D}_0$ we mean a characteristic diffusion coefficient. As we have seen previously, the choice of reference quantities in any particular problem is arbitrary. The best choices are those that simplify the equations and boundary conditions for a problem as much as possible. In terms of these dimensionless variables, the equation of continuity for species $A$ in the form of Eq. (C) of Table 8.2.2-5 becomes

$$\frac{1}{N_{\text{St}}} \rho^* \frac{\partial \omega_{(A)}}{\partial t^*} + \rho^* \nabla \omega_{(A)} \cdot \mathbf{v}^* + \frac{1}{N_{\text{Sc}} N_{\text{Re}}} \text{div } \mathbf{j}_{(A)}^* = \frac{N_{\text{Da}}}{N_{\text{Sc}} N_{\text{Re}}} r_{(A)}^* \qquad (1\text{-}2)$$

where we define the Strouhal, Schmidt, Reynolds, and Damköhler numbers as

$$N_{\text{St}} \equiv \frac{t_0 v_0}{L_0} \qquad N_{\text{Sc}} \equiv \frac{\mu_0}{\rho_0 \mathscr{D}_0}$$
$$N_{\text{Re}} \equiv \frac{\rho_0 v_0 L_0}{\mu_0} \qquad N_{\text{Da}} \equiv \frac{r_{(A)0} L_0^2}{\rho_0 \mathscr{D}_0} \qquad (1\text{-}3)$$

From our experience in Secs. 3.3.1 and 6.5.1, we would be inclined to say that, in the limit as $N_{\text{Re}} \to 0$ for a fixed value of the Schmidt number, the second term on the left of Eq. (1-2), representing convection, can be neglected with respect to the third term on the left, representing diffusion.

The trouble with this argument is that eliminating forced convection does not necessarily send the second term on the left of Eq. (1-2) to zero. In general, **v**

## [9.3] NO CONVECTION

(and $\mathbf{v}^*$) will be different from zero as the result either of natural convection or of diffusion-induced convection. In natural convection, the motion is driven by density gradients in an external force field. The density gradients were attributable to temperature gradients in Sec. 6.4.2; here they would be due to concentration gradients. In diffusion-induced convection, the motion results from the very fact that the various species are moving relative to one another. I would like to focus our attention here on diffusion-induced convection. For that reason, we will eliminate the possibility of natural convection by assuming that $\rho^*$ is a constant.

Our experience in Sec. 9.2.1 suggests that we should be able to neglect the effects of diffusion induced convection in sufficiently dilute solutions. Let $\omega_{(A)m}$ characterize the maximum mass fraction of species $A$ in the system under study. It is helpful in examining a sequence of progressively more dilute solutions to introduce the expanded variable

$$\omega_{(A)}^{**} \equiv \frac{\omega_{(A)}}{\omega_{(A)m}} \tag{1-3}$$

From Sec. 8.4.7, we see that for dilute solutions

$$\mathbf{j}_{(A)}^{**} \equiv \frac{1}{\omega_{(A)m}} \mathbf{j}_{(A)}^{*} = -\nabla \omega_{(A)}^{**} \tag{1-4}$$

In view of (1-3) and (1-4), we can rewrite the equation of continuity for species $A$ (1-2) as

$$\frac{1}{N_{\text{St}}} \frac{\partial (\rho^* \omega_{(A)})}{\partial t^*} + \text{div}\left( \rho^* \omega_{(A)} \mathbf{v}_{(A)}^* + \frac{1}{N_{\text{Sc}} N_{\text{Re}}} \mathbf{j}_{(A)}^* \right) = \frac{N_{\text{Da}}}{N_{\text{Sc}} N_{\text{Re}}} r_{(A)}^* \tag{1-5}$$

$$\frac{\omega_{(A)m}}{N_{\text{St}}} \frac{\partial (\rho^* \omega_{(A)}^{**})}{\partial t^*} + \text{div}\left( \omega_{(A)m} \rho^* \omega_{(A)}^{**} \mathbf{v}_{(A)}^* + \frac{\omega_{(A)m}}{N_{\text{Sc}} N_{\text{Re}}} \mathbf{j}_{(A)}^{**} \right) = \frac{N_{\text{Da}}}{N_{\text{Sc}} N_{\text{Re}}} r_{(A)}^* \tag{1-6}$$

For the limit of dilute solutions $\omega_{(A)m} \ll 1$, this reduces to

$$\frac{\omega_{(A)m}}{N_{\text{St}}} \frac{\partial (\rho^* \omega_{(A)}^{**})}{\partial t^*} + \text{div}\left( \frac{\omega_{(A)m}}{N_{\text{Sc}} N_{\text{Re}}} \mathbf{j}_{(A)}^{**} \right) = \frac{N_{\text{Da}}}{N_{\text{Sc}} N_{\text{Re}}} r_{(A)}^* \tag{1-7}$$

or

$$\frac{\partial \rho_{(A)}}{\partial t} + \text{div}\, \mathbf{j}_{(A)} = r_{(A)} \tag{1-8}$$

Here we have recognized that the definitions of the characteristic time $t_0$ and characteristic speed $v_0$ are arbitrary and that consequently the values of $N_{\text{St}}$ and $N_{\text{Re}}$ are arbitrary.

As suggested in Exercise 9.3.1-1, the same argument can be repeated in molar terms to conclude that for dilute solutions the equation of continuity for species $A$ takes the form

$$\frac{\partial c_{(A)}}{\partial t} + \text{div}\, \mathbf{J}^* = \frac{r_{(A)}}{M_{(A)}} - x_{(A)} \sum_{B=1}^{N} \frac{r_{(B)}}{M_{(B)}} \tag{1-9}$$

Let us consider a binary system in which there are no chemical reactions, for which the diffusion coefficient is a constant, and for which either $\rho$ or $c$ is a constant. Under these circumstances, both Eqs. (1-8) and (1-9) reduce to

$$\frac{\partial c_{(A)}}{\partial t} = \mathscr{D}_{(AB)}\,\mathrm{div}\,(\nabla c_{(A)}) \tag{1-10}$$

This is usually referred to as Fick's *second law of diffusion* or sometimes simply as the *diffusion equation*.

This approach to neglecting convection is perhaps best understood in terms of specific examples such as those provided in the sections that follow.

**EXERCISE**

**9.3.1-1** Let $x_{(A)m}$ characterize the maximum mole fraction of species $A$ in the system under study. Construct an argument similar to that given in the text to suggest that in the dilute solution limit as $x_{(A)m} \to 0$ the equation of continuity for species $A$ reduces to Eq. (1-9).

**9.3.2 More on unsteady-state evaporation** Let us reexamine our discussion of unsteady-state evaporation in Sec. 9.2.1 and let us focus our attention upon the limit as $x_{(A)\mathrm{eq}} \to 0$.

Our discussion in Sec. 9.3.1 suggests that we introduce the expanded dependent variable

$$x_{(A)}^{**} \equiv \frac{x_{(A)}}{x_{(A)\mathrm{eq}}} \tag{2-1}$$

in Eq. (1-11) of Sec. 9.2.1 to find

$$\frac{d^2 x_{(A)}^{**}}{d\eta^2} + (2\eta + x_{(A)\mathrm{eq}} \varphi^{**})\frac{d x_{(A)}^{**}}{d\eta} = 0 \tag{2-2}$$

where

$$\varphi^{**} \equiv \frac{1}{1 - x_{(A)\mathrm{eq}}}\left.\frac{d x_{(A)}^{**}}{d\eta}\right|_{\eta=0} \tag{2-3}$$

In the limit as $x_{(A)\mathrm{eq}} \to 0$, Eq. (2-2) appears to reduce to

$$\frac{d^2 x_{(A)}^{**}}{d\eta^2} + 2\eta\,\frac{d x_{(A)}^{**}}{d\eta} = 0 \tag{2-4}$$

From Sec. 9.2.1, we see that the appropriate boundary conditions are

At $\eta = 0$: $\quad x_{(A)}^{**} = 1$ \hfill (2-5)

and

As $\eta \to \infty$: $\quad x_{(A)}^{**} \to x_{(A)0}/x_{(A)\mathrm{eq}}$ \hfill (2-6)

## [9.3] NO CONVECTION

The solution to this problem is given by Eq. (1-21) of Sec. 9.2.1 and some typical results are presented in Table 9.2.1-1. As we would expect, the term in brackets in Eq. (1-20) of Sec. 9.2.1, which may be regarded as a correction to the evaporation rate accounting for convection, approaches unity in the limit of dilute solutions.

### 9.3.3 More on unsteady diffusion with a first-order homogenous reaction

Let us reconsider the problem discussed in Sec. 9.2.2. We will again take the liquid-gas phase interface to be the plane $z_2 = 0$. But rather than requiring

As $z_2 \to 0$, for all $t$: $\quad \mathbf{v}^\star \to 0$ $\qquad$ (3-1)

we will say instead that species $B$ and $C$ are nonvolatile:

At $z_2 = 0$, for all $t$: $\quad N_{(B)2} = N_{(C)2} = 0$ $\qquad$ (3-2)

In this way, we will avoid saying as we did in Sec. 9.2.2 that the number of moles of components $B$ and $C$ leaving the liquid through the phase interface must be exactly equal to the number of moles of $A$ entering the liquid. Of course, we will find that we must require $\mathbf{v}^\star$ to be nonzero as $z_2 \to \infty$, but this seems reasonable in view of the fact that we have required the phase interface to be fixed in space for all time.

Continuing as we did in Sec. 9.2.2, we can argue that the only nonzero component of the molar-averaged velocity vector is

$$v_2^\star = v_2^\star(t) \qquad (3\text{-}3)$$

By means of Eq. (3-2), the definition for the molar-averaged velocity, and Eq. (7-1) of Sec. 8.4.7, we can reason that

At $z_2 = 0$, for all $t$:
$$v_2^\star = \frac{1}{c} N_{(A)2} = \frac{x_{(A)}}{c} N_{(A)2} - \mathscr{D}_{(Am)} \frac{\partial x_{(A)}}{\partial z_2}$$

$$= -\frac{\mathscr{D}_{(Am)}}{1 - x_{(A)}} \frac{\partial x_{(A)}}{\partial z_2} \qquad (3\text{-}4)$$

Equation (2-2) of Sec. 9.2.2 and Eqs. (3-3) and (3-4) allow us to say that everywhere in the gas phase

Everywhere for all $t$: $\quad v_2^\star = \dfrac{-\mathscr{D}_{(Am)}}{1 - x_{(A)\text{eq}}} \dfrac{\partial x_{(A)}}{\partial z_2}\bigg|_{z_2=0}$ $\qquad$ (3-5)

Note that $(\partial x_{(A)}/\partial z_2)|_{z_2=0}$, although independent of position, is a function of time.

Equation (F) of Table 8.2.2-5 now requires

$$\frac{\partial x_{(A)}}{\partial t} - \frac{\mathscr{D}_{(Am)}}{1 - x_{(A)\text{eq}}} \frac{\partial x_{(A)}}{\partial z_2}\bigg|_{z_2=0} \frac{\partial x_{(A)}}{\partial z_2} - \mathscr{D}_{(Am)} \frac{\partial^2 x_{(A)}}{\partial z_2^2} = -k'''_{(A)} x_{(A)} \qquad (3\text{-}6)$$

Comparing this with the comparable Eq. (2-10) of Sec. 9.2.2, we see that the second term on the left is new. We would like to find a solution to this equation consistent with Eqs. (2-1) and (2-2) of Sec. 9.2.2. In order to simplify our search for a solution, let us consider the limit as $x_{(A)eq} \to 0$.

Our argument is facilitated by the introduction of the following dimensionless variables:

$$t^* \equiv k'''_{(A)} t \qquad z_2^* \equiv \frac{z_2}{L_0} \qquad x^{**}_{(A)} \equiv \frac{x_{(A)}}{x_{(A)eq}} \tag{3-7}$$

In these terms Eq. (3-6) becomes

$$\frac{\partial x^{**}_{(A)}}{\partial t^*} - \frac{x_{(A)eq}}{1 - x_{(A)eq}} \left.\frac{\partial x^{**}_{(A)}}{\partial z_2^*}\right|_{z_2=0} \frac{\partial x^{**}_{(A)}}{\partial z_2^*} - \frac{\partial^2 x^{**}_{(A)}}{\partial z_2^{*2}} = -x^{**}_{(A)} \tag{3-8}$$

where for the sake of simplicity we have defined our characteristic length as

$$L_0 \equiv \sqrt{\frac{\mathscr{D}_{(Am)}}{k'''_{(A)}}} \tag{3-9}$$

It seems reasonable to say that as $x_{(A)eq} \to 0$, Eq. (3-8) reduces to

$$\frac{\partial x^{**}_{(A)}}{\partial t^*} = \frac{\partial^2 x^{**}_{(A)}}{\partial z_2^{*2}} - x^{**}_{(A)} \tag{3-10}$$

The appropriate initial and boundary conditions, Eqs. (2-1) and (2-2) of Sec. 9.2.2, now become

At $t^* = 0$, for all $z_2^* > 0$: $\quad x^{**}_{(A)} = 0 \tag{3-11}$

and

At $z_2^* = 0$, for all $t^* > 0$: $\quad x^{**}_{(A)} = 1 \tag{3-12}$

Except for the introduction of dimensionless variables, Eqs. (3-10) to (3-12) represent exactly the same problem that was solved in Sec. 9.2.2. But remember that the solution found there logically applies here only in the limit as $x_{(A)eq} \to 0$.

### 9.3.4 More on unsteady diffusion with a slow catalytic reaction

At time $t = 0$, a gas consisting of a mixture of species $A$ and $A_2$ is brought into contact with a solid surface that acts as a catalyst for the dimerization reaction $2A \to A_2$. Let us determine the rate at which species $A$ is consumed.

To simplify the analysis, let us take the gas-solid phase interface to be the plane $z_2 = 0$ and let us say that the gas phase occupies the half-space $z_2 > 0$. The initial condition is that

At $t = 0$, for all $z_2 > 0$: $\quad x_{(A)} = x_{(A)0} \tag{4-1}$

At the gas-solid phase interface, the dimerization reaction takes place. We assume that this catalytic reaction can be adequately described as first order:

## [9.3] NO CONVECTION

At $z_2 = 0$, for all $t > 0$:  $\quad N_{(A)2} = \dfrac{r_{(A)}^{(o)}}{M_{(A)}} = -k''_{(A)} c_{(A)}$  (4-2)

Let us assume that we are dealing with an ideal-gas mixture at a constant temperature and pressure, so that we may take the molar density $c$ to be a constant.

It seems reasonable to postulate

$$v_1^\star = v_3^\star = 0 \qquad v_2^\star = v_2^\star(t, z_2) \qquad x_{(A)} = x_{(A)}(t, z_2) \qquad (4\text{-}3)$$

From the overall equation of continuity in the form of Eq. (B) of Table 8.3.2-1, we find

$$\frac{\partial v_2^\star}{\partial z_2} = 0 \qquad (4\text{-}4)$$

This implies

$$v_2^\star = v_2^\star(t) \qquad (4\text{-}5)$$

By means of Eq. (4-2)$_1$, the definition for the molar-averaged velocity $\mathbf{v}^\star$, and Fick's first law in the form of Eq. (D) of Table 8.4.6-1, we can reason that

At $z_2 = 0$, for $t > 0$:  $\quad v_2^\star = \dfrac{1}{2c} N_{(A)2}$

$$= \frac{x_{(A)}}{4c} N_{(A)2} - \frac{1}{2} \mathscr{D}_{(AA_2)} \frac{\partial x_{(A)}}{\partial z_2}$$

$$= \frac{-\mathscr{D}_{(AA_2)}}{2 - x_{(A)}} \frac{\partial x_{(A)}}{\partial z_2} \qquad (4\text{-}6)$$

Equations (4-5) and (4-6) allow us to say that everywhere in the gas phase

Everywhere for $t > 0$:  $\quad v_2^\star = \dfrac{-\mathscr{D}_{(AA_2)}}{2 - x_{(A)}\big|_{z_2=0}} \dfrac{\partial x_{(A)}}{\partial z_2}\bigg|_{z_2=0}$  (4-7)

Note that $(\partial x_{(A)}/\partial z_2)|_{z_2=0}$ and $x_{(A)}|_{z_2=0}$, although independent of position, are functions of time.

Let us assume that the diffusion coefficient may be taken to be a constant. In view of Eqs. (4-3) and (4-7), we may write the equation of continuity for species $A$ consistent with Fick's first law, Eq. (D) of Table 8.4.6-2, as

$$\frac{\partial x_{(A)}}{\partial t} = \frac{\mathscr{D}_{(AA_2)}}{2 - x_{(A)}\big|_{z_2=0}} \frac{\partial x_{(A)}}{\partial z_2}\bigg|_{z_2=0} \frac{\partial x_{(A)}}{\partial z_2} + \mathscr{D}_{(AA_2)} \frac{\partial^2 x_{(A)}}{\partial z_2^2} \qquad (4\text{-}8)$$

We seek a solution to this equation consistent with boundary conditions (4-1) and (4-2)$_2$. We will find it convenient to use Eq. (4-6) in rewriting this last boundary condition as

At $z_2 = 0$, for all $t > 0$:
$$\frac{2\mathscr{D}_{(AA_2)}}{k''_{(A)}}\frac{\partial x_{(A)}}{\partial z_2} = x_{(A)}(2 - x_{(A)}) \tag{4-9}$$

Since Eq. (4-8) is nonlinear, we cannot take its Laplace transform. The transformation

$$\eta \equiv \frac{z_2}{\sqrt{4\mathscr{D}_{(AA_2)}t}} \tag{4-10}$$

does not work either; boundary condition (4-9) would be a function of $t$ as well as of $\eta$.

Since we are having difficulty solving the complete problem, let us consider the limiting case of dilute solutions as $x_{(A)0} \to 0$. As suggested in Sec. 9.3.1, the argument is made more obvious by the introduction of the following dimensionless variables:

$$t^* \equiv \frac{(k''_{(A)})^2}{\mathscr{D}_{(AA_2)}}t \qquad z_2^* \equiv \frac{k''_{(A)}z_2}{\mathscr{D}_{(AA_2)}} \qquad x_{(A)}^{**} \equiv \frac{x_{(A)}}{x_{(A)0}} \tag{4-11}$$

In these terms, Eqs. (4-8), (4-1), and (4-9) become, respectively,

$$\frac{\partial x_{(A)}^{**}}{\partial t^*} = \frac{x_{(A)0}}{2 - x_{(A)0}x_{(A)}^{**}|_{z_2^*=0}}\frac{\partial x_{(A)}^{**}}{\partial z_2^*}\bigg|_{z_2=0}\frac{\partial x_{(A)}^{**}}{\partial z_2^*} + \frac{\partial^2 x_{(A)}^{**}}{\partial z_2^{*2}} \tag{4-12}$$

At $t^* = 0$, for all $z_2^* > 0$: $\qquad x_{(A)}^{**} = 1 \tag{4-13}$

and

At $z_2^* = 0$, for all $t^* > 0$: $\qquad 2\dfrac{\partial x_{(A)}^{**}}{\partial z_2^*} = x_{(A)}^{**}(2 - x_{(A)0}x_{(A)}^{**}) \tag{4-14}$

It seems reasonable to say that as $x_{(A)0} \to 0$, Eq. (4-12) reduces to

$$\frac{\partial x_{(A)}^{**}}{\partial t^*} = \frac{\partial^2 x_{(A)}^{**}}{\partial z_2^{*2}} \tag{4-15}$$

The appropriate initial condition is still Eq. (4-13), but the boundary condition (4-14) simplifies to

At $z_2^* = 0$, for all $t^* > 0$: $\qquad \dfrac{\partial x_{(A)}^{**}}{\partial z_2^*} = x_{(A)}^{**} \tag{4-16}$

This problem is essentially the same as that considered in Sec. 6.2.3. Consequently, we may use the result there to immediately express the concentration distribution as

$$x_{(A)}^{**} = \text{erf}\left(\frac{z_2^*}{\sqrt{4t^*}}\right) + \exp(z_2^* + t^*)\left[1 - \text{erf}\left(\frac{z_2^*}{\sqrt{4t^*}} + \sqrt{t^*}\right)\right] \tag{4-17}$$

The rate at which species $A$ is consumed follows easily from Eqs. (4-2) and (4-17) as

$$-N_{(A)2}|_{z_2=0} = k''_{(A)}cx_{(A)0}\exp\left(\frac{(k''_{(A)})^2 t}{\mathscr{D}_{(AA_2)}}\right)\left[1 - \text{erf}\left(k''_{(A)}\sqrt{\frac{t}{\mathscr{D}_{(AA_2)}}}\right)\right] \tag{4-18}$$

## 9.4 FORCED CONVECTION

**9.4.1 When diffusion induced convection can be neglected with respect to forced convection** Up to this point we have been concerned with problems in which the mass-averaged or molar-averaged velocity distribution is solely attributable to concentration gradients in the fluid. We notice that convection in mass transfer differs fundamentally from convection in energy transfer. Natural convection in energy transfer is possible only if there are density gradients in the fluid. In mass transfer, diffusion induced convection is possible even though the total mass density is a constant. The motions of the individual species are sufficient in general to require the mass-averaged or molar-averaged velocity distribution to differ from zero.

We have the same differences in considering forced convection. In energy transfer, we could say that the velocity distribution for a fluid with constant density and viscosity is independent of the temperature distribution. In mass transfer, it is not sufficient to say that the fluid has a constant density and a constant viscosity, in order for the velocity distribution to be independent of the concentration distribution. We must require in addition that diffusion induced convection be negligible at all surfaces on which velocity is not explicitly specified as part of the forced convection problem. Put a little differently, we must ask that diffusion induced convection be neglicated on all surfaces at which mass transfer occurs.

Our discussion in Secs. 9.2.1 and 9.3.1 suggest that diffusion induced convection can be neglected with respect to forced convection in sufficiently dilute solutions. Let us denote as species 1 the solvent, the component whose concentration is largest. From the definition of the mass-averaged velocity and from Sec. 8.4.7, we see that for sufficiently dilute solutions

$$\mathbf{v} = \sum_{A=1}^{N} \omega_{(A)} \mathbf{v}_{(A)}$$

$$= \frac{1}{\rho} \sum_{A=1}^{N} \mathbf{n}_{(A)}$$

$$= \frac{1}{\rho} \mathbf{n}_{(1)} + \sum_{A=2}^{N} \omega_{(A)} \mathbf{v} - \sum_{A=2}^{N} \mathscr{D}_{(A1)} \nabla \omega_{(A)} \tag{1-1}$$

In terms of the dimensionless variables introduced in Sec. 9.3.1, this becomes

$$\mathbf{v}^* = \frac{1}{\rho^*} \mathbf{n}_{(1)}^* + \sum_{A=2}^{N} \omega_{(A)m} \omega_{(A)}^{**} \mathbf{v}^* - \sum_{A=2}^{N} \frac{\omega_{(A)m}}{N_{\text{Sc}(A)} N_{\text{Re}}} \nabla \omega_{(A)}^{**} \tag{1-2}$$

where

$$\mathbf{n}_{(1)}^* \equiv \frac{\mathbf{n}_{(1)}^*}{\rho_0 v_0} \tag{1-3}$$

and

$$N_{\text{Sc}(a)} \equiv \frac{\mu_0}{\rho_0 \mathscr{D}_{(A1)}} \tag{1-4}$$

In the limit of dilute solutions for which $\omega_{(A)m} \ll 1$ ($A = 2, \ldots, N$), this reduces to

$$\mathbf{v}^* = \frac{1}{\rho^*} \mathbf{n}_{(1)}^*$$

$$= \mathbf{v}_{(1)}^* \tag{1-5}$$

Typically the solvent would not be subject to mass transfer, in which case on the boundaries of the system $\mathbf{v}_{(1)}$ reduces to the velocity of the boundary. Under these circumstances, the mass-averaged velocity distribution $\mathbf{v}$ no longer depends upon the concentration distribution.

Unfortunately, because of their complexity, very little attention has been devoted in the literature to solving problems in which both diffusion induced and forced convection are taken into account. In what follows we will assume that we are justified in neglecting diffusion induced convection with respect to forced convection.

**9.4.2 Similarities between energy and mass transfer** In Sec. 9.4.1, I pointed out that there is a fundamental difference between forced convection in energy transfer and forced convection in mass transfer. However, when the effects of diffusion induced convection are neglected with respect to forced convection, many mass-transfer problems have the same mathematical form as energy-transfer problems. If we take advantage of these similarities, we can save ourselves a great deal of work.

Very little theoretical attention has been devoted to the solution of mass-transfer problems involving compressible fluids. The principal difficulty is that little can be said about the solution to Cauchy's first law for compressible fluids. A few problems can be handled for compressible fluids undergoing isochoric motions (see Secs. 6.3.1 to 6.4.2), but numerical solutions of Cauchy's first law appear to be our only recourse at the present time for compressible fluids in nonisochoric motions. For this reason, we shall confine ourselves to incompressible fluids in talking about forced convection.

For an incompressible binary fluid with constant properties, the equation of continuity for species $A$ that is consistent with Fick's first law becomes in terms of dimensionless variables (see Sec. 9.3.1)

$$\frac{1}{N_{\text{St}}} \frac{\partial \omega_{(A)}}{\partial t^*} + \nabla \omega_{(A)} \cdot \mathbf{v}^* = \frac{1}{N_{\text{Sc}} N_{\text{Re}}} \text{div } \nabla \omega_{(A)} + \frac{N_{\text{Da}}}{N_{\text{Sc}} N_{\text{Re}}} r_{(A)}^* \tag{2-1}$$

The differential energy balance may be written in a dimensionless form for an incompressible fluid with constant properties as (see Sec. 6.4.1)

$$\frac{1}{N_{\text{St}}} \frac{\partial T^*}{\partial t^*} + \nabla T^* \cdot \mathbf{v}^* = \frac{1}{N_{\text{Pr}} N_{\text{Re}}} \text{div } \nabla T^* + \frac{N_{\text{Br}}}{N_{\text{Pr}} N_{\text{Re}}} \text{tr } (\mathbf{S}^* \cdot \nabla \mathbf{v}^*) \tag{2-2}$$

Comparing Eqs. (2-1) and (2-2), we see that they have identical forms when homogeneous chemical reactions are neglected in the first and viscous dissipation is neglected

[9.4] FORCED CONVECTION

in the second. Furthermore, for many common cases, there is no difficulty in finding an energy-transfer problem that has the same form of boundary conditions as the mass-transfer problem in which you are interested. Temperature is specified instead of concentration; Newton's law of cooling replaces either Newton's law of mass transfer or the description of the mass flux for a first-order catalytic reaction. There are two major exceptions: catalytic reactions other than first order have no parallel in energy transfer and, whereas temperature is continuous across a phase boundary, concentration is not. Beyond these exceptions, for incompressible fluids having constant properties, binary mass transfer in the absence of homogeneous reactions is mathematically the same as energy transfer without viscous dissipation. For this reason, we give little further attention to these problems except to refer the reader to Secs. 6.5.1 to 6.8.2.

Forced convection in fluids that are undergoing homogeneous reactions are another matter. In the next section, we consider a simple class of problems involving first-order homogeneous reactions.

Unfortunately, there is little in the way of precedent that we can recommend to the reader who is interested in forced convection of concentrated multicomponent solutions (see Sec. 8.4.7 for the approach to dilute solutions).

### 9.4.3 Gas absorption in a falling film with chemical reaction

An incompressible newtonian fluid of nearly pure species $B$ flows down an inclined plane as shown in Fig. 3.2.5-3 (see Exercise 3.2.5-3). Species $A$ is transferred from the surrounding gas stream to the liquid where it undergoes an irreversible first-order homogeneous reaction. Let us assume that there is no mass transfer from the gas stream to the falling film for $z_1 < 0$:

At $z_2 = \delta$, for $z_1 < 0$: $\quad \mathbf{n}_{(A)} \cdot \boldsymbol{\xi} = 0$ \hfill (3-1)

Here $\boldsymbol{\xi}$ is the unit normal to the phase interface pointed from the liquid to the gas. Outside the immediate neighborhood of the liquid film, the gas stream has a uniform concentration. If the liquid were in equilibrium with this gas stream, its concentration would be $\rho_{(A)\text{eq}}$. In order to simplify the problem somewhat, we will assume that for $z_1 > 0$, the concentration of the liquid at the phase interface is $\rho_{(A)\text{eq}}$:

At $z_2 = \delta$, for $z_1 > 0$: $\quad \rho_{(A)} = \rho_{(A)\text{eq}}$ \hfill (3-2)

Very far upstream, the liquid is pure species $B$:

As $z_1 \to -\infty$, for $0 \leq z_2 \leq \delta$: $\quad \rho_{(A)} \to 0$ \hfill (3-3)

We wish to determine the mass-density distribution in the concentration boundary layer near the entrance to that portion of the film absorbing species $A$ as the Peclet number (for mass transfer) becomes unbounded.

In the context of Sec. 9.4.1, let us confine our attention to the limit as

$$\Delta \omega_{(A)} \equiv \frac{\rho_{(A)\text{eq}}}{\rho} \to 0 \tag{3-4}$$

In this limit, I suggest that natural convection may be neglected with respect to forced convection and the velocity distribution in the fluid is the same as that found in Exercise 3.2.5-3:

$$v_1 = v_{1,\max}\left[2\frac{z_2}{\delta} - \left(\frac{z_2}{\delta}\right)^2\right] \tag{3-5}$$

Let us assume that

$$\rho_{(A)} = \rho_{(A)}(z_1, z_2) \tag{3-6}$$

The equation of continuity for species $A$ in the form of Eq. (C) of Table 8.4.6-2, consequently, requires

$$v_1 \frac{\partial \rho_{(A)}}{\partial z_1} = \mathscr{D}^0_{(AB)}\left(\frac{\partial^2 \rho_{(A)}}{\partial z_1^2} + \frac{\partial^2 \rho_{(A)}}{\partial z_2^2}\right) - k_1''' \rho_{(A)} \tag{3-7}$$

We anticipate that determining a solution to this equation which is consistent with boundary conditions (3-1) to (3-3) will be very difficult.

This suggests that we restrict our attention to the entrance of the adsorption region as we did in Ex. 6.7.6-5.

If we introduce as dimensionless variables

$$\omega_{(A)}^{**} \equiv \frac{\rho_{(A)}}{\rho_{(A)\text{eq}}} \qquad z_1^* \equiv \frac{z_1}{\delta} \qquad s^* \equiv 1 - \frac{z_2}{\delta} \tag{3-8}$$

Eq. (3-7) becomes

$$(1 - s^{*2})\frac{\partial \omega_{(A)}^{**}}{\partial z_1^*} = \frac{1}{N_{\text{Sc}} N_{\text{Re}}}\left(\frac{\partial^2 \omega_{(A)}^{**}}{\partial z_1^{*2}} + \frac{\partial^2 \omega_{(A)}^{**}}{\partial z_2^{*2}}\right) - \frac{N_{\text{Da}}}{N_{\text{Sc}} N_{\text{Re}}} \omega_{(A)}^{**} \tag{3-9}$$

where

$$N_{\text{Sc}} \equiv \frac{\mu}{\rho \mathscr{D}^0_{(AB)}} \qquad N_{\text{Re}} \equiv \frac{\delta v_{1,\max} \rho}{\mu} \qquad N_{\text{Da}} \equiv \frac{k_1''' \delta^2}{\mathscr{D}^0_{(AB)}} \tag{3-10}$$

If we are primarily interested in the entrance region to the absorption section, then our discussion in Ex. 6.7.6-5 suggests that we introduce as an "expanded" variable

$$s^{**} \equiv (N_{\text{Pe } M})^{\frac{1}{2}} s^* \tag{3-11}$$

where $N_{\text{Pe } M}$ is the Peclet number for mass transfer:

$$N_{\text{Pe } M} \equiv N_{\text{Sc}} N_{\text{Re}} \tag{3-12}$$

In terms of this expanded variable, Eq. (3-9) becomes

$$\left(1 - \frac{s^{**2}}{N_{\text{Pe } M}}\right)\frac{\partial \omega_{(A)}^{**}}{\partial z_1^*} = \frac{1}{N_{\text{Pe } M}}\frac{\partial^2 \omega_{(A)}^{**}}{\partial z_1^{*2}} + \frac{\partial^2 \omega_{(A)}^{**}}{\partial s^{**2}} - \frac{N_{\text{Da}}}{N_{\text{Pe } M}} \omega_{(A)}^{**} \tag{3-13}$$

In the limit as $N_{\text{Pe } M} \to \infty$, it appears that this last simplifies to

## [9.4] FORCED CONVECTION

$$\frac{\partial \omega_{(A)}^{**}}{\partial z_1^*} = \frac{\partial^2 \omega_{(A)}^{**}}{\partial s^{**2}} - \frac{N_{\text{Da}}}{N_{\text{Pe }M}} \omega_{(A)}^{**} \quad (3\text{-}14)$$

Since we are neglecting axial diffusion in Eq. (3-14), it seems reasonable to replace boundary conditions (3-1) and (3-3) by

$$\text{At } z_1^* = 0: \quad \omega_{(A)}^{**} = 0 \quad (3\text{-}15)$$

In terms of our dimensionless variables, Eq. (3-2) may be expressed as

$$\text{At } s^{**} = 0: \quad \omega_{(A)}^{**} = 1 \quad (3\text{-}16)$$

Our problem now is reduced to finding a solution to Eq. (3-14) that is consistent with boundary conditions (3-15) and (3-16).

Equations (3-14) to (3-16) belong to the class of problems discussed in Exercise 9.4.3-1. Consequently, the solution of interest here can be determined from Sec. 3.2.4 as

$$\omega_{(A)}^{**} = \exp\left(-\frac{N_{\text{Da}} z_1^*}{N_{\text{Pe }M}}\right)\left[1 - \text{erf}\left(\frac{s^{**}}{\sqrt{4z_1^*}}\right)\right] + \frac{N_{\text{Da}}}{N_{\text{Pe }M}} \int_0^{z_1^*} \exp\left(-\frac{N_{\text{Da}} z_1^*}{N_{\text{Pe }M}}\right)$$

$$\times \left[1 - \text{erf}\left(\frac{s^{**}}{\sqrt{4z_1^*}}\right)\right] dz_1^* \quad (3\text{-}17)$$

Finally, it is interesting to compute the rate at which mass of species $A$ is absorbed per unit width of a film of length $L$:

$$\mathcal{W}_{(A)} \equiv \int_0^L -n_{(A)2}|_{z_2 = \delta} \, dz_1 \quad (3\text{-}18)$$

Using Eq. (3-17), it follows that [1, p. 553]

$$\frac{\mathcal{W}_{(A)}}{\rho_{(A)\text{eq}} v_{1,\text{max}}} \sqrt{\frac{k_1'''}{\mathcal{D}_{(AB)}^0}} = (\tfrac{1}{2} + u)\,\text{erf}\sqrt{u} + \sqrt{\frac{u}{\pi}} \exp(-u) \quad (3\text{-}19)$$

where for the sake of convenience we have introduced

$$u \equiv \frac{N_{\text{Da}} L}{N_{\text{Pe }M}\, \delta} \quad (3\text{-}20)$$

In the limit where there is no chemical reaction ($u \to 0$), Eq. (3-19) requires

$$\frac{\mathcal{W}_{(A)}}{\rho_{(A)\text{eq}}\sqrt{\mathcal{D}_{(AB)}^0 v_{1,\text{max}} L}} = \sqrt{\frac{4}{\pi}} \quad (3\text{-}21)$$

## REFERENCE

1. Bird, R. B., W. E. Stewart, and E. N. Lightfoot: "Transport Phenomena," 7th printing, Wiley, New York, 1960.

## EXERCISES

**9.4.3-1** *Some further remarks on general solutions* Exercises 9.2.2-4 to 9.2.2-7 inspire us to make a few further remarks here on general solutions to problems involving first-order homogeneous reactions.

(a) Let us assume that the equation of continuity for species $A$ and the required boundary conditions may be expressed in terms of dimensionless variables as

$$\frac{\partial \omega_{(A)}}{\partial z_1^*} = \frac{\partial^2 \omega_{(A)}}{\partial^2 z_2^*} + N\omega_{(A)} \tag{1}$$

At $z_1^* = 0$: $\quad \omega_{(A)} = \omega_{(A)0}$ \hfill (2)

At $z_2^* = b^*, c^*, \ldots$: $\quad \omega_{(A)} = 0$ \hfill (3)

At $z_2^* = d^*, e^*, \ldots$: $\quad \dfrac{\partial \omega_{(A)}}{\partial z_2^*} = -K^* \omega_{(A)}$ \hfill (4)

Let $f$ be a solution for the same problem with no chemical reaction:

$$\frac{\partial f}{\partial z_1^*} = \frac{\partial^2 f}{\partial^2 z_2^*}$$

At $z_1^* = 0$: $\quad f = \omega_{(A)0}$

At $z_2^* = b^*, c^*, \ldots$: $\quad f = 0$

At $z_2^* = d^*, e^*, \ldots$: $\quad \dfrac{\partial f}{\partial z_2^*} = -K^* f$

Reason that

$$\omega_{(A)} = f \exp(Nz_1^*)$$

(b) Let us now assume that we wish to determine a solution to Eq. (1) consistent with the following boundary conditions:

At $z_1^* = 0$: $\quad \omega_{(A)} = 0$ \hfill (5)

At $z_2^* = b^*, c^*, \ldots$: $\quad \omega_{(A)} = \omega_{(A)s}$ \hfill (6)

At $z_2^* = d^*, e^*, \ldots$: $\quad \dfrac{\partial \omega_{(A)}}{\partial z_2^*} = K^*(\omega_{(A)\infty} - \omega_{(A)})$ \hfill (7)

Let $F$ be a solution to the same problem with no chemical reaction. Determine that

$$\omega_{(A)} = F \exp(Nz_1^*) - N \int_0^{z_1^*} F \exp(Nz_1^*) \, dz_1^*$$

(c) We now wish to determine a solution to Eq. (1) that is consistent with boundary conditions (2), (6), and (7). Conclude that

$$\omega_{(A)} = (f + F) \exp(Nz_1^*) - N \int_0^{z_1^*} F \exp(Nz_1^*) \, dz_1^*$$

where $f$ and $F$ are defined in (a) and (b) above.

(d) Let us now assume that our equation of continuity for species $A$ and required boundary conditions become in terms of dimensionless variables

$$\frac{\partial \omega_{(A)}}{\partial z^*} = \frac{1}{r^*} \frac{\partial}{\partial r^*} \left( r^* \frac{\partial \omega_{(A)}}{\partial r^*} \right) + N\omega_{(A)}$$

[9.4] FORCED CONVECTION

At $z^* = 0$: $\quad \omega_{(A)} = \omega_{(A)0}$

At $r^* = b^*, c^*, \ldots$: $\quad \omega_{(A)} = \omega_{(A)s}$

At $r^* = d^*, e^*, \ldots$: $\quad \dfrac{\partial \omega_{(A)}}{\partial r^*} = K^*(\omega_{(A)\infty} - \omega_{(A)})$

Construct an argument to conclude that

$$\omega_{(A)} = (f + F)\exp(Nz^*) - N\int_0^{z^*} F \exp(Nz^*)\, dz^*$$

Here $f$ is a solution to the same problem for no chemical reaction and $\omega_{(A)s} = \omega_{(A)\infty} = 0$. The function $F$ is a solution to the same problem for no chemical reaction and $\omega_{(A)0} = 0$.

**9.4.3-2** *More on gas absorption in a falling film with chemical reaction* Let us repeat the problem discussed in this section, attempting to describe the boundary condition at the gas-liquid phase interface more realistically. Rather than saying that the phase interface is in equilibrium with the gas very far away from it, let us describe the mass transfer by means of Newton's law of mass transfer (see Exercise 9.2.1-4):

At $z_2 = \delta$, for $z_1 > 0$: $\quad j_{(A)2} = k_{(A)\omega}(\omega_{(A)} - \omega_{(A)\mathrm{eq}})$

*Answer*

$$\omega^{**}_{(A)} \equiv \frac{\omega_{(A)}}{\omega_{(A)\mathrm{eq}}} = F \exp\left(-\frac{N_{\mathrm{Da}}}{N_{\mathrm{Pe}\,M}} z_1^*\right) + \frac{N_{\mathrm{Da}}}{N_{\mathrm{Pe}\,M}} \int_0^{z_1^*} F \exp\left(-\frac{N_{\mathrm{Da}}}{N_{\mathrm{Pe}\,M}} z_1^*\right) dz_1^*$$

$$F = 1 - \mathrm{erf}\left(\frac{s^{**}}{\sqrt{4z_1^*}}\right) - \exp\left(\frac{s^{**}}{B} + \frac{z_1^*}{B^2}\right)\left[1 - \mathrm{erf}\left(\frac{s^{**}}{\sqrt{4z_1^*}} + \frac{\sqrt{z_1^*}}{B}\right)\right]$$

*Hint:* See Exercise 6.7.6-6.

# 10
# Applications of Integral Averaging Techniques to Mass Transfer

This is our conclusion to integral averaging techniques begun in Chap. 4 and continued in Chap. 7. As I mentioned in introducing Chap. 7, the ideas presented here are best understood in the context of Chap. 4. It is in Chap. 4 that I try to spend a little extra time in discussing the motivation for some of the developments. It is also there that some of the key steps common to all the derivations are explained in detail.

## 10.1 TIME AVERAGING

**10.1.1 Turbulent mass transfer** By *turbulent mass transfer* I mean that at least one of the phases involved in the mass-transfer process is in turbulent flow. For a discussion of the basic concepts and terminology, please refer to Secs. 4.1.1 to 4.1.5.

In the next few sections we shall be concerned with the time average of the differential equation of continuity for an individual species $A$. Our approach here will be very similar to that taken in Secs. 7.1.1 to 7.1.4, where we discussed turbulent energy transfer.

## 10.1.2 The time-averaged differential equation of continuity for species A

As in our previous discussions of turbulence (Secs. 4.1.1 to 4.1.5 and 7.1.1 to 7.1.4), we will for simplicity limit ourselves to incompressible fluids. For this limiting case, Eq. (C) of Table 8.2.2-5 becomes

$$\rho\left[\frac{\partial \omega_{(A)}}{\partial t} + \text{div}\,(\omega_{(A)}\mathbf{v})\right] + \text{div}\,\mathbf{j}_{(A)} - r_{(A)} = 0 \tag{2-1}$$

Using the definition introduced in Sec. 4.1.2, let us take the time average of this equation:

$$\frac{1}{\Delta t}\int_{t}^{t+\Delta t}\left[\rho\frac{\partial \omega_{(A)}}{\partial t'} + \text{div}\,(\rho\omega_{(A)}\mathbf{v} + \mathbf{j}_{(A)}) - r_{(A)}\right] dt' = 0 \tag{2-2}$$

The time-averaging operation commutes with partial differentiation with respect to time (see Sec. 4.1.2) and with the divergence operation:

$$\rho\frac{\partial \bar{\omega}_{(A)}}{\partial t} + \text{div}\,(\overline{\rho\omega_{(A)}\mathbf{v}} + \overline{\mathbf{j}_{(A)}}) - \bar{r}_{(A)} = 0 \tag{2-3}$$

It is more common to write this result as

$$\rho\left(\frac{\partial \bar{\omega}_{(A)}}{\partial t} + \nabla\bar{\omega}_{(A)}\cdot\bar{\mathbf{v}}\right) + \text{div}\,(\overline{\mathbf{j}_{(A)}} + \mathbf{j}_{(A)}^{(t)}) = \bar{r}_{(A)} \tag{2-4}$$

where we have introduced the turbulent mass flux vector $\mathbf{j}_{(A)}^{(t)}$:

$$\mathbf{j}_{(A)}^{(t)} \equiv \rho(\overline{\omega_{(A)}\mathbf{v}} - \bar{\omega}_{(A)}\bar{\mathbf{v}}) \tag{2-5}$$

When we limit ourselves to binary diffusion and when we recognize that Fick's first law is an appropriate expression for the mass flux vector, Eq. (2-4) takes the form

$$\rho\left(\frac{\partial \bar{\omega}_{(A)}}{\partial t} + \nabla\bar{\omega}_{(A)}\cdot\bar{\mathbf{v}}\right) = \text{div}\,(\rho\mathscr{D}_{(AB)}^{0}\nabla\bar{\omega}_{(A)} - \mathbf{j}_{(A)}^{(t)}) + \bar{r}_{(A)} \tag{2-6}$$

Let us in particular assume that we have an $n$th-order homogeneous reaction:

$$r_{(A)} = k_n''' \rho_{(A)}^n \tag{2-7}$$

so that

$$\bar{r}_{(A)} = k_n''' \bar{\rho}_{(A)}^n + k_n'''(\overline{\rho_{(A)}^n} - \bar{\rho}_{(A)}^n) \tag{2-8}$$

Notice that for a first-order reaction,

$$\bar{r}_{(A)} = k_1''' \bar{\rho}_{(A)} \tag{2-9}$$

The rate of production of the mass of species $A$ per unit volume is not an *explicit* function of the concentration fluctuations. In contrast, $\bar{r}_{(A)}$ is explicitly dependent upon the concentration fluctuations for higher-order reactions.

The problem posed here by $\mathbf{j}_{(A)}^{(t)}$ is very similar to those encountered in Secs. 4.1.2 and 7.1.2. Just as we had there to stop and propose empirical data correlations for the Reynolds stress tensor $\mathbf{S}^{(t)}$ and the turbulent energy flux vector $\mathbf{q}^{(t)}$, we must here stop and formulate empirical representations for the turbulent mass flux vector $\mathbf{j}_{(A)}^{(t)}$.

**EXERCISE**

**10.1.2-1** *Turbulent diffusion in dilute electrolytes* Quite often it is convenient to arrange the computations for a mass-transfer problem a little differently from the arrangement suggested in the text.

(a) Determine that an alternative expression for the time-averaged equation of continuity for species $A$ is

$$\frac{\partial \bar{c}_{(A)}}{\partial t} + \text{div } \overline{\mathbf{N}}_{(A)} = \frac{\bar{r}_{(A)}}{M_{(A)}}$$

(b) For sufficiently dilute solutions of an electrolyte, find that

$$\overline{\mathbf{N}}_{(A)} = -c\mathscr{D}_{(Am)}\left(\nabla \bar{x}_{(A)} + \frac{\bar{x}_{(A)}\epsilon_{(A)}}{kT}\nabla \Phi\right) + \bar{c}_{(A)}\bar{\mathbf{v}} + \frac{1}{M_{(A)}}\mathbf{j}_{(A)}^{(t)}$$

$$= -c\mathscr{D}_{(Am)}\left(\nabla \bar{x}_{(A)} + \frac{\bar{x}_{(A)}\epsilon_{(A)}}{kT}\nabla \Phi\right) + \bar{x}_{(A)}\sum_{B=1}^{N}\overline{\mathbf{N}}_{(B)} + \frac{1}{M_{(A)}}\mathbf{j}_{(A)}^{(t)}$$

Here we have neglected pressure- and thermal-diffusion effects.

**10.1.3 Empirical correlations for the turbulent mass flux vector $\mathbf{j}_{(A)}^{(t)}$** Our discussion of empirical data correlations for the turbulent mass flux vector $\mathbf{j}_{(A)}^{(t)}$ will be relatively brief, inasmuch as it is essentially a duplication of Sec. 7.1.3.

Our approach is based upon three principles.

1. For changes of frame such that

$$\overline{\mathbf{Q}} \doteq \mathbf{Q} \tag{3-1}$$

we may use the result of Sec. 4.1.3 to find that $\mathbf{j}_{(A)}^{(t)}$ is frame indifferent:

$$\mathbf{j}_{(A)}^{(t)*} \equiv \rho(\overline{\omega_{(A)}^* \mathbf{v}^*} - \overline{\omega_{(A)}^*}\,\overline{\mathbf{v}^*}) = \rho[\overline{\omega_{(A)}^*(\mathbf{v}^* - \overline{\mathbf{v}^*})}]$$

$$= \rho \mathbf{Q} \cdot [\overline{\omega_{(A)}(\mathbf{v} - \overline{\mathbf{v}})}] = \mathbf{Q} \cdot \mathbf{j}_{(A)}^{(t)} \tag{3-2}$$

Here $\mathbf{Q}$ is a (possibly) time-dependent orthogonal second-order tensor. In order to obtain Eq. (3-2), we have made use of the fact that a velocity difference is frame indifferent (see Exercise 1.2.2-1).

2. We shall assume that the principle of material frame indifference discussed in Sec. 2.3.1 applies to any empirical correlations developed for $\mathbf{j}_{(A)}^{(t)}$ so long as the changes of frame considered satisfy Eq. (3-1).
3. The Buckingham-Pi theorem will be used to further limit the form of any expression for $\mathbf{j}_{(A)}^{(t)}$.

## EXAMPLE 1  PRANDTL'S MIXING LENGTH THEORY

Example 1 in Sec. 7.1.3 suggests that, for the fully developed flow regime in wall turbulence, we assume that the turbulent mass flux vector be regarded as a function of the density of the fluid, the distance $l$ from the wall, $\bar{\mathbf{D}}$, and $\nabla \bar{\omega}_{(A)}$:

$$\mathbf{j}_{(A)}^{(t)} = \mathbf{j}_{(A)}^{(t)}(\rho, l, \bar{\mathbf{D}}, \nabla \bar{\omega}_{(A)}) \qquad (3\text{-}3)$$

Because we are limiting ourselves to the fully developed flow regime, the diffusivity and viscosity are not included as independent variables. The implications of the principle of material frame indifference and of the Buckingham-Pi theorem are spelled out in Sec. 7.1.3.

A special case of Eq. (3-3) that is consistent with the principle of material frame indifference and the Buckingham-Pi theorem is

$$\mathbf{j}_{(A)}^{(t)} = -\mathscr{D}^* \rho l^2 \sqrt{2 \operatorname{tr} \bar{\mathbf{D}}^2} \nabla \bar{\omega}_{(A)} \qquad (3\text{-}4)$$

where $\mathscr{D}^*$ is a dimensionless constant. Equation (3-4) should be viewed as the tensorial form of *Prandtl's mixing length theory* for mass transfer. It is probably worth emphasizing that we should not expect the Prandtl mixing length theory to be appropriate to the laminar sublayer or buffer zone.

## EXAMPLE 2  DEISSLER'S EXPRESSION FOR THE REGION NEAR THE WALL

In view of our discussion of Example 2 in Sec. 7.1.3, we are motivated to propose for the laminar sublayer and the buffer zone

$$\mathbf{j}_{(A)}^{(t)} = \mathbf{j}_{(A)}^{(t)}(\rho, \mu, l, \bar{\mathbf{v}} - \mathbf{v}_{(s)}, \nabla \bar{\omega}_{(A)}) \qquad (3\text{-}5)$$

Remember that $\mathbf{v}_{(s)}$ indicates the velocity of the bounding wall. Deissler [1] has proposed on empirical grounds that

$$\mathbf{j}_{(A)}^{(t)} = -n^2 \rho l \, |\bar{\mathbf{v}} - \mathbf{v}_{(s)}| \, [1 - \exp(-n^2 N)] \nabla \bar{\omega}_{(A)} \qquad (3\text{-}6)$$

with the definition

$$N \equiv \frac{\rho l \, |\bar{\mathbf{v}} - \mathbf{v}_{(s)}|}{\mu} \qquad (3\text{-}7)$$

The $n$ appearing here is meant to be the same as that used in Eq. (4-21) of Sec. 4.1.4 and evaluated in Sec. 4.1.5. Of course, Eq. (3-6) satisfies the principle of material frame indifference and is consistent with the Buckingham-Pi theorem.

## EXAMPLE 3  EDDY DIFFUSIVITY IN FREE TURBULENCE

Very far away from any wall in a region of free turbulence, it is common to say that

$$\mathbf{j}_{(A)}^{(t)} = -\rho \mathscr{D}_{(AB)}^{(t)} \nabla \bar{\omega}_{(A)} \qquad (3\text{-}8)$$

The scalar $\mathscr{D}_{(AB)}^{(t)}$ is normally assumed to be independent of position. It is known as the *eddy diffusivity*.

In the next section, we will look at a technique that has been used to measure $\mathscr{D}_{(AB)}^{(t)}$.

**REFERENCE**

1. Deissler, R. G.: *NACA Rept.* 1210, 1955.

### 10.1.4 Turbulent diffusion from a point source in a moving stream [1: 2, p. 552]

In a region far removed from any bounding walls or surfaces, a fluid of pure species $B$ moves in a steady-state, turbulent flow with a uniform and constant speed $\bar{v}_0$. With respect to the cylindrical coordinate system $(r,\theta,z)$ shown in Fig. 10.1.4-1, the fluid moves in the $z$ direction. Species $A$ is continuously injected into the stream at the origin of this coordinate system. The rate of injection is $W_{(A)}$ (mass per unit time), which can be considered to be so small that the mass-averaged speed of the stream does not deviate appreciably from $\bar{v}_0$. As species $A$ moves downstream from the point of injection, it diffuses in both the axial and radial directions. We wish to determine the concentration distribution of $A$ in the stream.

Since the region of flow under consideration is very far away from any bounding walls, it seems reasonable to assume that the flow is in free turbulence and that the turbulent mass flux vector may be expressed in terms of a constant eddy diffusivity as described in Example 3 of Sec. 10.1.3. According to the assumptions above, we are justified in assuming that there is only one nonzero component of the time-averaged velocity vector in the cylindrical coordinate system indicated:

$$\bar{v}_r = \bar{v}_\theta = 0 \qquad \bar{v}_z = \bar{v}_0 \tag{4-1}$$

For the problem described, the time-averaged differential equation of continuity for species $A$, derived in Sec. 10.1.2, reduces to

$$\bar{v}_0 \left( \frac{\partial \bar{\omega}_{(A)}}{\partial z} \right)_r = (\mathscr{D}^0_{(AB)} + \mathscr{D}^{(t)}_{(AB)}) \operatorname{div} \nabla \bar{\omega}_{(A)} \tag{4-2}$$

Our intuition suggests and, as we shall see later, experimental evidence confirms that $\mathscr{D}^0_{(AB)} \ll \mathscr{D}^{(t)}_{(AB)}$.

Since the fluid very far downstream from the point of injection is pure species $B$, it seems reasonable to employ as one boundary condition that in the spherical

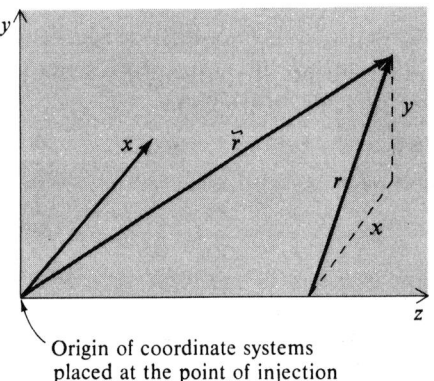

Origin of coordinate systems placed at the point of injection

Fig. 10.1.4-1 Coordinate systems used to describe turbulent diffusion from a point source in a constant velocity stream.

coordinate system $(r,\theta)$ suggested in Fig. 10.1.4-1

$$\text{As } \tilde{r} \to \infty: \quad \bar{\omega}_{(A)} \to 0 \tag{4-3}$$

We must also make a statement about the mass flow rate of species $A$ at the point of injection. For any constant value of $r$, we can say that

$$W_{(A)} = \int_0^{2\pi} \int_0^\pi \bar{n}_{(A)\tilde{r}} \tilde{r}^2 \sin\theta \, d\theta \, d\varphi$$

$$= 2\pi \int_0^\pi \left[ \bar{\rho}_{(A)} \bar{v}_{\tilde{r}} - \rho(\mathscr{D}_{(AB)}^0 + \mathscr{D}_{(AB)}^{(t)}) \frac{\partial \bar{\omega}_{(A)}}{\partial \tilde{r}} \right] \tilde{r}^2 \sin\theta \, d\theta \tag{4-4}$$

If for the moment we assume that

$$\text{As } \tilde{r} \to 0: \quad \bar{\rho}_{(A)} \tilde{r} \to C_1 = \text{a constant} \tag{4-5}$$

it follows that

$$\text{As } \tilde{r} \to 0: \quad \int_0^\pi \bar{\rho}_{(A)} \bar{v}_{\tilde{r}} \tilde{r}^2 \sin\theta \, d\theta = \int_0^\pi \bar{\rho}_{(A)} \bar{v}_0 \tilde{r}^2 \sin\theta \cos\theta \, d\theta$$

$$= \bar{\rho}_{(A)} \bar{v}_0 \tilde{r}^2 \int_0^\pi \sin\theta \cos\theta \, d\theta = 0 \tag{4-6}$$

and

$$\text{As } \tilde{r} \to 0: \quad W_{(A)} = -2\pi \int_0^\pi \rho(\mathscr{D}_{(AB)}^0 + \mathscr{D}_{(AB)}^{(t)}) \frac{\partial \bar{\omega}_{(A)}}{\partial \tilde{r}} \tilde{r}^2 \sin\theta \, d\theta$$

$$= -2\pi \rho(\mathscr{D}_{(AB)}^0 + \mathscr{D}_{(AB)}^{(t)}) \frac{\partial \bar{\omega}_{(A)}}{\partial \tilde{r}} \tilde{r}^2 \int_0^\pi \sin\theta \, d\theta$$

$$= -4\pi \rho(\mathscr{D}_{(AB)}^0 + \mathscr{D}_{(AB)}^{(t)}) \frac{\partial \bar{\omega}_{(A)}}{\partial \tilde{r}} \tilde{r}^2 \tag{4-7}$$

In arriving at Eq. (4-7), we have made use of Eq. (4-5) to reason that

$$\text{As } \tilde{r} \to 0: \quad \tilde{r}^2 \frac{\partial \bar{\omega}_{(A)}}{\partial \tilde{r}} = -\frac{C_1}{\rho} \tag{4-8}$$

In a moment we shall return to check Eq. (4-5).

Our next step is to find a solution to Eq. (4-2) that is consistent with boundary conditions (4-3) and (4-7). This is a little awkward, since Eq. (4-2) is stated in terms of cylindrical coordinates, whereas boundary conditions (4-3) and (4-7) are more naturally given in terms of spherical coordinates. Wilson [1] at this point made the clever suggestion that we look for a solution in the form of

$$\bar{\omega}_{(A)} = e^{-\alpha z} \varphi(\tilde{r}) \tag{4-9}$$

Employing this assumed form for the solution, we can calculate that

$$\text{div}\,(\nabla\bar{\omega}_{(A)}) = 2\nabla\varphi\cdot\nabla(e^{-\alpha z}) + \varphi\,\text{div}\,\nabla(e^{-\alpha z}) + e^{-\alpha z}\,\text{div}\,\nabla\varphi$$

$$= -2\alpha e^{-\alpha z}\left(\frac{\partial\varphi}{\partial z}\right)_r + \alpha^2\varphi e^{-\alpha z} + e^{-\alpha z}\,\text{div}\,\nabla\varphi \tag{4-10}$$

and

$$\left(\frac{\partial\bar{\omega}_{(A)}}{\partial z}\right)_r = -\alpha e^{-\alpha z}\varphi + e^{-\alpha z}\left(\frac{\partial\varphi}{\partial z}\right)_r \tag{4-11}$$

As a result, Eq. (4-2) becomes

$$\varphi\left(-\frac{\alpha\bar{v}_0}{\mathscr{D}^0_{(AB)} + \mathscr{D}^{(t)}_{(AB)}} - \alpha^2\right) + \left(\frac{\partial\varphi}{\partial z}\right)_r\left(\frac{\bar{v}_0}{\mathscr{D}^0_{(AB)} + \mathscr{D}^{(t)}_{(AB)}} + 2\alpha\right) = \text{div}\,\nabla\varphi \tag{4-12}$$

With the definition

$$\alpha \equiv -\frac{\bar{v}_0}{2(\mathscr{D}^0_{(AB)} + \mathscr{D}^{(t)}_{(AB)})} \tag{4-13}$$

Eq. (4-12) further reduces to

$$\alpha^2\varphi = \frac{2}{\tilde{r}}\frac{\partial\varphi}{\partial\tilde{r}} + \frac{\partial^2\varphi}{\partial\tilde{r}^2} \tag{4-14}$$

The standard change of variable

$$Y \equiv \tilde{r}\varphi \tag{4-15}$$

can be used to express Eq. (4-14) as

$$\alpha^2 Y = \frac{\partial^2 Y}{\partial\tilde{r}^2} \tag{4-16}$$

solutions to which have the form

$$Y \equiv \tilde{r}\varphi = A\exp(\alpha\tilde{r}) + B\exp(-\alpha\tilde{r}) \tag{4-17}$$

In order that boundary condition (4-3) be satisfied, we must require

$$B = 0 \tag{4-18}$$

(remember that $\alpha$ is negative).
    From Eqs. (4-9), (4-15), (4-17), and (4-18), we find

$$\bar{\omega}_{(A)} = \frac{A}{\tilde{r}}\exp[\alpha(\tilde{r} - z)] \tag{4-19}$$

Finally, boundary condition (4-7) demands

$$\text{As } \tilde{r} \to 0: \quad W_{(A)} \to 4\pi\rho(\mathscr{D}^0_{(AB)} + \mathscr{D}^{(t)}_{(AB)})A \tag{4-20}$$

or

$$A \equiv \frac{W_{(A)}}{4\pi\rho(\mathscr{D}^0_{(AB)} + \mathscr{D}^{(t)}_{(AB)})} \qquad (4\text{-}21)$$

To summarize, the mass-fraction distribution for species $A$ in the free-turbulence flow described is represented by Eqs. (4-13), (4-19), and (4-21). We see further that we were justified in assuming Eq. (4-5).

From an experimental point of view, the useful result here is

$$\frac{d \ln (\tilde{r}\bar{\omega}_{(A)})}{d(\tilde{r} - z)} = \alpha \equiv -\frac{\bar{v}_0}{2(\mathscr{D}^0_{(AB)} + \mathscr{D}^{(t)}_{(AB)})} \qquad (4\text{-}22)$$

If the slope on the left can be evaluated from experimental data, this expression may be used to calculate $\mathscr{D}^0_{(AB)} + \mathscr{D}^{(t)}_{(AB)}$. Towle and Sherwood [3] have done this for $CO_2$ injected into a stream of air to conclude that $\mathscr{D}^{(t)}_{(CO_2, \text{air})} \approx 23 \text{ cm}^2/\text{s}$, which is several orders of magnitude larger than $\mathscr{D}^0_{(CO_2, \text{air})}$. For a further discussion of this experimental technique, see Sherwood and Pigford [4, p. 42].

**REFERENCES**

1. Wilson, H. A.: *Cambridge Phil. Soc.*, **12**:406 (1904).
2. Bird, R. B., W. E. Stewart, and E. N. Lightfoot: "Transport Phenomena," 7th printing, Wiley, New York, 1960.
3. Towle, W. L., and T. K. Sherwood: *Ind. Eng. Chem.*, **31**:457 (1939).
4. Sherwood, T. K., and R. L. Pigford: "Absorption and Extraction," 2d ed., McGraw-Hill, New York, 1952.

## 10.2 AREA AVERAGING

**10.2.1 Area averaging in mass transfer** In what follows, we extend to mass transfer the concept of area averaging introduced in Secs. 4.2.1 to 4.2.3 and 7.2.1 to 7.2.3. The essential point is that sometimes it is advantageous to average the equation of continuity over a cross section normal to the gross mass transfer.

Keep in mind that whenever one of the integral averaging techniques is used some information is lost. We are always called upon to compensate for this loss of information by making an approximation or by applying an empirical data correlation. You will notice that the approximation employed in Sec. 10.2.2 is a little different from those used in Secs. 4.2.2 and 7.2.2. Because of the somewhat ad hoc nature of area averaging, I cannot give specific recommendations for the types of approximations to be employed that will be applicable in each and every problem you may encounter. My feeling is that, having been warned an approximation will be necessary, you will find the example problems in Secs. 4.2.2, 7.2.2, and 10.2.2 sufficient stimuli for your imagination.

I will not take up here approximate concentration boundary-layer theory. As I mentioned in Sec. 9.4.2, there are many similarities between energy and mass transfer. In the absence of homogeneous chemical reactions and natural convection, the equation of continuity for species $A$ in an incompressible binary fluid takes the same form as the differential energy balance in the absence of viscous dissipation. With these restrictions and with a few obvious changes in notation, the discussion in Sec. 7.2.3 can be interpreted as approximate *concentration* boundary-layer theory for plane flow past a curved wall.

I think you will gain the maximum benefit from the next section by reading it in the context of Secs. 4.2.2 and 7.2.2.

### 10.2.2 Longitudinal dispersion

At time $t = 0$, we find that for $z > 0$ a very long tube is filled with a pure solvent $\rho_{(A)} = 0$; for $z < 0$, the solvent has a uniform concentration of dissolved material $\rho_{(A)} = \rho_{(A)0}$. For $t > 0$, the fluid is forced to move in the $z$ direction through the tube with a constant volume flow rate. We wish to determine the concentration in the tube as a function of time and position.

In this analysis we will assume that the physical properties of the liquid are constants. It follows that the velocity distribution is independent of composition; its specific form is dictated by the constitutive equation chosen for the extra-stress tensor, as discussed in Sec. 3.2.1. For the analysis I have in mind here, it will not be necessary to choose a particular constitutive equation or to be any more explicit about the velocity distribution.

The implication is that there is no homogeneous chemical reaction. The equation of continuity for species $A$, consequently, requires

$$\frac{\partial \rho_{(A)}}{\partial t} + \frac{1}{r}\frac{\partial (r n_{(A)r})}{\partial r} + \frac{1}{r}\frac{\partial n_{(A)\theta}}{\partial \theta} + \frac{\partial n_{(A)z}}{\partial z} = 0 \tag{2-1}$$

It is clear that, if we use this as a basis for our analysis, we will be faced with solving a partial differential equation.

Let us assume that we are primarily interested in the area-averaged composition

$$\bar{\rho}_{(A)} \equiv \frac{1}{\pi R^2} \int_0^{2\pi} \int_0^R \rho_{(A)} r \, dr \, d\theta \tag{2-2}$$

as a function of time and axial position. This suggests that we take the area average of Eq. (2-1):

$$\frac{\partial \bar{\rho}_{(A)}}{\partial t} + \frac{1}{\pi R^2}\int_0^{2\pi}\int_0^R \left[\frac{\partial(rn_{(A)r})}{\partial r} + \frac{\partial n_{(A)\theta}}{\partial \theta}\right] dr\,d\theta + \frac{\partial \bar{n}_{(A)z}}{\partial z} = 0 \tag{2-3}$$

The second and third terms on the left can be integrated to find

$$\frac{1}{\pi R^2}\int_0^{2\pi}\int_0^R \frac{\partial(rn_{(A)r})}{\partial r}\, dr\, d\theta = 0 \tag{2-4}$$

and

$$\frac{1}{\pi R^2}\int_0^R \int_0^{2\pi} \frac{\partial n_{(A)\theta}}{\partial \theta}\, d\theta\, dr = 0 \tag{2-5}$$

## [10.2] AREA AVERAGING

As a result, Eq. (2-3) assumes the simpler form

$$\frac{\partial \bar{\rho}_{(A)}}{\partial t} + \frac{\partial \bar{n}_{(A)z}}{\partial z} = 0 \qquad (2\text{-}6)$$

We can express the second term on the left of this equation in terms of composition by using the area average of the $z$ component of Fick's first law:

$$\bar{n}_{(A)z} = \overline{\rho_{(A)} v_z} - \mathscr{D}^0_{(AB)} \frac{\partial \bar{\rho}_{(A)}}{\partial z} \qquad (2\text{-}7)$$

Unfortunately, we do not achieve in this way a differential equation for $\bar{\rho}_{(A)}$.

An approximation appears to be in order. Equation (2-7) suggests that the simplest approach is to say

$$\bar{n}_{(A)z} \doteq \bar{\rho}_{(A)} \bar{v}_z - \mathscr{K} \frac{\partial \bar{\rho}_{(A)}}{\partial z} \qquad (2\text{-}8)$$

In order to compensate for the fact that the area average of a product is generally not equal to the product of the area averages, we replace the diffusion coefficient by an empirical dispersion coefficient $\mathscr{K}$, which we will assume here to be a constant. Recognizing that

$$\bar{v}_z = \text{a constant} \qquad (2\text{-}9)$$

we see that Eq. (2-8) enables us to say from Eq. (2-6)

$$\frac{\partial \bar{\rho}_{(A)}}{\partial t} + \bar{v}_z \frac{\partial \bar{\rho}_{(A)}}{\partial z} - \mathscr{K} \frac{\partial^2 \bar{\rho}_{(A)}}{\partial z^2} = 0 \qquad (2\text{-}10)$$

Our problem now is to obtain a solution to this equation consistent with the initial conditions

$$\text{At } t = 0, \text{ for } z > 0: \qquad \bar{\rho}_{(A)} = 0 \qquad (2\text{-}11)$$

and

$$\text{At } t = 0, \text{ for } z < 0: \qquad \bar{\rho}_{(A)} = \rho_{(A)0} \qquad (2\text{-}12)$$

If we think of $\bar{\rho}_{(A)}$ as a function of $t$ and a new independent variable $\zeta$,

$$\zeta \equiv z - \bar{v}_z t \qquad (2\text{-}13)$$

Eq. (2-10) reduces to

$$\left(\frac{\partial \bar{\rho}_{(A)}}{\partial t}\right)_\zeta - \mathscr{K} \frac{\partial^2 \bar{\rho}_{(A)}}{\partial \zeta^2} = 0 \qquad (2\text{-}14)$$

Our experience in Sec. 9.2.1 suggests that we think of $\overline{\rho^*_{(A)}}$,

$$\overline{\rho^*_{(A)}} \equiv \frac{\bar{\rho}_{(A)}}{\rho_{(A)0}} \qquad (2\text{-}15)$$

as a function of a single independent variable $\eta$,

$$\eta \equiv \frac{\zeta}{\sqrt{4\mathcal{K}t}} \tag{2-16}$$

since this allows us to express Eq. (2-14) as an ordinary differential equation:

$$\frac{d^2\overline{\rho_{(A)}^*}}{d\eta^2} + 2\eta \frac{d\overline{\rho_{(A)}^*}}{d\eta} = 0 \tag{2-17}$$

From Eqs. (2-11) and (2-12), the corresponding boundary conditions are

$$\text{As } \eta \to \infty: \quad \overline{\rho_{(A)}^*} \to 0 \tag{2-18}$$

and

$$\text{As } \eta \to -\infty: \quad \overline{\rho_{(A)}^*} \to 1 \tag{2-19}$$

Integrating Eq. (2-17) once we find,

$$\frac{d\overline{\rho_{(A)}^*}}{d\eta} = C_1 \exp(-\eta^2) \tag{2-20}$$

A second integration consistent with boundary condition (2-18) yields

$$\overline{\rho_{(A)}^*} = -C_1 \int_\eta^\infty \exp(-\eta^2)\, d\eta = -\frac{C_1 \sqrt{\pi}}{2}(1 - \operatorname{erf}\eta) \tag{2-21}$$

In order that boundary condition (2-19) be satisfied, we must set

$$C_1 = -\frac{1}{\sqrt{\pi}} \tag{2-22}$$

Our final result for the area-averaged composition in the tube is, consequently,

$$\overline{\rho_{(A)}^*} = \tfrac{1}{2}(1 - \operatorname{erf}\eta) \tag{2-23}$$

One aspect of this solution for which we may have some intuitive feeling is the length $L$ of the transition zone in which $\bar{\rho}_{(A)}$ changes from $0.9\rho_{(A)0}$ to $0.1\rho_{(A)0}$. From Eq. (2-23) we can calculate

$$0.8 = \tfrac{1}{2}(\operatorname{erf}\eta_{0.1} - \operatorname{erf}\eta_{0.9}) = \operatorname{erf}\eta_{0.1} = \operatorname{erf}\left(\frac{L}{4\sqrt{\mathcal{K}t}}\right) \tag{2-24}$$

A table for the error function [1] tells us

$$L = 3.62\sqrt{\mathcal{K}t} \tag{2-25}$$

Taylor [2] has also analyzed longitudinal dispersion resulting from the introduction of a concentrated mass of solute in the cross section $z = 0$ at time $t = 0$.

A very interesting theoretical analysis of the dependence of the dispersion coefficient $\mathcal{K}$ upon the diffusion coefficient $\mathcal{D}_{(AB)}^0$ has been given by Taylor [2–4] and later more carefully by Aris [5], who concluded that

$$\mathscr{K} = \overline{\mathscr{D}^0_{(AB)}} + \chi \frac{\bar{v}_z^2 R^2}{\overline{\mathscr{D}^0_{(AB)}}} \qquad (2\text{-}26)$$

Here $\overline{\mathscr{D}^0_{(AB)}}$ is the area-averaged diffusion coefficient and $\chi$ is a factor that depends upon the shape of the cross section of the tube as well as the variation in the velocity and diffusion coefficient profiles. If the diffusion coefficient is taken to be a constant, the velocity profile parabolic, and the tube cross section circular, then they have found

$$\chi = \frac{1}{48} \qquad (2\text{-}27)$$

A further refinement has been offered by Gill and Sankarasubramanian [6].

## REFERENCES

1. Dwight, H. B.: "Mathematical Tables," 3d ed., Dover, New York, 1961.
2. Taylor, Geoffrey: *Proc. Roy. Soc. (London), Ser. A*, **219**:186 (1953).
3. Taylor, Geoffrey: *Proc. Roy. Soc. (London), Ser. A*, **223**:446 (1954).
4. Taylor, Geoffrey: *Proc. Roy. Soc. (London), Ser. A*, **225**:473 (1954).
5. Aris, Rutherford: *Proc. Roy. Soc. (London), Ser. A*, **235**:67 (1956).
6. Gill, W. N., and R. Sankarasubramanian: *Proc. Roy. Soc. (London), Ser. A*, **316**:341 (1970).
7. Wehner, J. F., and R. H. Wilhelm: *Chem. Eng. Sci.*, **6**:89 (1956).

## EXERCISE

**10.2.2.1** *A catalytic tubular reactor* The open tube shown in Fig. 10.2.2-1 is a very simple reactor. For $0 < z < L$, the wall of the tube is a catalyst for the reaction $A \to B$. You may assume that the physical properties of the liquid are constants and that the catalytic reaction can be described as first order:

At $r = R$: $\quad n_{(A)r} = r_{(A)}^{(\sigma)} = k_1'' \rho_{(A)}$

If the liquid very far upstream has a uniform mass density $\rho_{(A)0}$, what is the composition of $A$ in the product downstream?

Use the same approach taken in the text to determine the average composition of the liquid downstream from the reactor.

(*a*) The tubular reactor together with its connecting upstream and downstream sections is

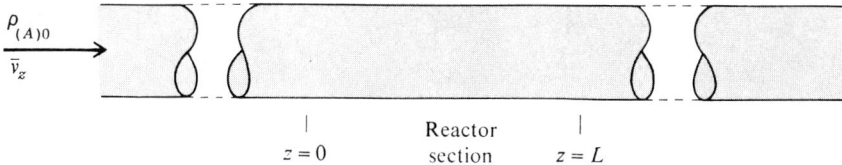

**Fig. 10.2.2-1** For $0 < z < L$, the wall of a very long open tube is a catalyst for the reaction $A \to B$.

illustrated in Fig. 10.2.2-1. Conclude that for the region upstream from the reactor:

$$\frac{\partial \overline{\rho^*_{(A)}}}{\partial z^*} = \frac{1}{N_{\text{Pe}}} \frac{\partial^2 \overline{\rho^*_{(A)}}}{\partial z^{*2}} \tag{1}$$

for the reactor itself:

$$\frac{\partial \overline{\rho^*_{(A)}}}{\partial z^*} = \frac{1}{N_{\text{Pe}}} \frac{\partial^2 \overline{\rho^*_{(A)}}}{\partial z^{*2}} - \frac{N_{\text{Da}}}{N_{\text{Pe}}} \overline{\rho^*_{(A)}} \tag{2}$$

and for the region downstream of the reactor:

$$\frac{\partial \overline{\rho^*_{(A)}}}{\partial z^*} = \frac{1}{N_{\text{Pe}}} \frac{\partial^2 \overline{\rho^*_{(A)}}}{\partial z^{*2}} \tag{3}$$

Here

$$N_{\text{Pe}} \equiv \frac{L \bar{v}_z}{\mathcal{K}} \quad N_{\text{Da}} \equiv \frac{2k''_1 L^2}{R\mathcal{K}}$$

$$\overline{\rho^*_{(A)}} \equiv \frac{\bar{\rho}_{(A)}}{\rho_{(A)0}} \quad z^* \equiv \frac{z}{L}$$

What assumptions have been made?

(b) What are the boundary conditions that must be satisfied by this system of equations at the entrance to the reactor, at the exit from the reactor, very far upstream, and very far downstream?

(c) Solve Eqs. (1) to (3) above individually and evaluate the six constants of integration to find the following concentration distribution.

For $z^* \leq 0$:

$$\frac{\overline{\rho^*_{(A)}} - 1}{\overline{\rho^*_{(A)}}(0) - 1} = \exp(N_{\text{Pe}} z^*)$$

where

$$\overline{\rho^*_{(A)}}(0) = g_0 \left[ (1 + a) \exp\left(\frac{aN_{\text{Pe}}}{2}\right) - (1 - a) \exp\left(\frac{-aN_{\text{Pe}}}{2}\right) \right]$$

$$g_0 \equiv 2 \left[ (1 + a)^2 \exp\left(\frac{aN_{\text{Pe}}}{2}\right) - (1 - a)^2 \exp\left(\frac{-aN_{\text{Pe}}}{2}\right) \right]^{-1}$$

and

$$a \equiv \left[ 1 + \frac{4N_{\text{Da}}}{(N_{\text{Pe}})^2} \right]^{\frac{1}{2}}$$

For $0 \leq z^* \leq 1$:

$$\overline{\rho^*_{(A)}} = g_0 \exp\left(\frac{N_{\text{Pe}} z^*}{2}\right) \left\{ (1 + a) \exp\left(\frac{aN_{\text{Pe}}}{2}[1 - z^*]\right) - (1 - a) \exp\left(-\frac{aN_{\text{Pe}}}{2}[1 - z^*]\right) \right\}$$

For $z^* \geq 1$:

$$\overline{\rho^*_{(A)}} = 2g_0 a \exp\frac{N_{\text{Pe}}}{2}$$

The analysis suggested here is essentially that given by Wehner and Wilhelm [7] for the fixed-bed catalytic reactor (see Exercise 10.3.9-4).

## 10.3 LOCAL VOLUME AVERAGING

**10.3.1 Mass transfer in porous media** We are commonly concerned with chemical reactions in beds of porous pellets impregnated with a catalyst. It is common practice in some localities to pump waste streams down a disposal well and into a layer of porous rock, but serious consideration must be given to the possibility that freshwater supplies for surrounding communities may be contaminated. When significant portions of a river's flow are diverted to distant localities (perhaps by a system of aqueducts), salt water may begin to encroach upon the river's delta region, destroying its previous fertility. Can fresh water be pumped down selected wells in the delta in order to limit the concentration of salt in the soil?

With one significant exception [1], the most common approach to mass transfer in porous media has been to view the porous solid and whatever gases and liquids it contains as a continuum and then simply to employ the usual differential equation of continuity discussed in Sec. 8.2.1. In other words, one treats mass transfer in a bed of porous catalyst pellets in the same manner as the diffusion of helium through Pyrex glass [2] or the diffusion of trichloromethane in a solid copolymer of ethylene and propylene [3]. But there is a fundamental difference. In the case of the bed of porous catalyst pellets, we are concerned with two distinct phases: diffusion takes place in the gas phase while a chemical reaction proceeds at the gas-solid phase boundary. In contrast, intermolecular forces control the rate at which helium moves through Pyrex glass and the rate at which trichloromethane diffuses through the copolymer. In the usual approach to mass transfer in porous media, the velocities and concentrations used are not defined in terms of the actual velocity and concentration distributions in the fluid phases.

Our successful discussions of momentum transfer in Secs. 4.3.1 to 4.3.11 and of energy transfer in Secs. 7.3.1 to 7.3.5 suggest that we take the same point of view here in studying mass transfer. This means that we should begin by developing the local volume average of the differential equation of continuity for species $A$.

For simplicity, we shall restrict this discussion to a single fluid flowing through a stationary, rigid porous medium.

### REFERENCES

1. Whitaker, Stephen: *A.I.Ch.E.J.*, **13**:420 (1967).
2. McAfee, K. B.: *Sci. Am.*, **199**(1):52 (1958).
3. Laurence, R. L., and J. C. Slattery: *J. Polymer Sci.*, (A-1) **5**:1327 (1967).

**10.3.2 The local volume average of the equation of continuity for species $A$**
We can begin as we did in Sec. 4.3.2, where we started a development for the local volume average of the equation of continuity. Let us think of a particular point $z$ in the porous medium and let us integrate the equation of continuity for species $A$ over $V_{(f)}$, the region of space occupied by the fluid within $S$ associated with $z$:

$$\frac{1}{V}\int_{V_{(f)}} \left( \frac{\partial \rho_{(A)}}{\partial t} + \text{div } \mathbf{n}_{(A)} - r_{(A)} \right) dV = 0 \tag{2-1}$$

We use here the equation of continuity for species $A$ in the form of Eq. (A) of Table 8.2.2-5. The operations of volume integration and differentiation with respect to time may be interchanged in the first term of the left of Eq. (2-1):

$$\frac{1}{V}\int_{V_{(f)}} \frac{\partial \rho_{(A)}}{\partial t} dV = \frac{\partial \bar{\rho}_{(A)}}{\partial t} \tag{2-2}$$

The theorem of Sec. 4.3.3 can be used to express the second term on the left of Eq. (2-1) as

$$\frac{1}{V}\int_{V_{(f)}} \operatorname{div} \mathbf{n}_{(A)} dV = \operatorname{div} \bar{\mathbf{n}}_{(A)} - r''_{(A)} \tag{2-3}$$

where we have introduced $r''_{(A)}$ as the rate at which species $A$ is produced by a catalytic chemical reaction at the fluid-solid phase interface (see Exercise 8.2.1-3):

$$r''_{(A)} \equiv -\frac{1}{V}\int_{S_w} \mathbf{n}_{(A)} \cdot \mathbf{n}\, dS = \frac{1}{V}\int_{S_w} r_{(A)}{}^{(\sigma)} dS \tag{2-4}$$

Equations (2-2) and (2-3) allow us to express Eq. (2-1) as

$$\frac{\partial \bar{\rho}_{(A)}}{\partial t} + \operatorname{div} \bar{\mathbf{n}}_{(A)} = r''_{(A)} + \bar{r}_{(A)} \tag{2-5}$$

This is one convenient form for the local volume-averaged equation of continuity for species $A$.

If we had started instead with Eq. (B) of Table 8.2.2-5, we would have found by an entirely analogous train of thought still another convenient form for the local volume-averaged equation of continuity for species $A$:

$$\frac{\partial \bar{c}_{(A)}}{\partial t} + \operatorname{div} \bar{\mathbf{N}}_{(A)} = \frac{r''_{(A)}}{M_{(A)}} + \frac{\bar{r}_{(A)}}{M_{(A)}} \tag{2-6}$$

This result can, of course, also be obtained by dividing Eq. (2-5) by the molecular weight for species $A$.

In the next sections, I discuss constitutive equations for the local volume-averaged mass flux vector $\bar{\mathbf{n}}_{(A)}$.

**10.3.3 When Fick's first law applies**  Let $l_0$ represent a characteristic pore diameter of the structure, and let $\lambda$ be the molecular mean free path [1, p. 10]. When the Knudsen number

$$N_{\text{Kn}} \equiv \frac{l_0}{\lambda} > 10 \tag{3-1}$$

Fick's first law can be used to describe binary diffusion within a gas in a porous medium [2] (it can also be used to describe diffusion in multicomponent solutions for the three limiting cases discussed in Sec. 8.4.7). Current practice is to always use

## [10.3] LOCAL VOLUME AVERAGING

Fick's first law when talking about binary diffusion in liquids. In this section we will assume conditions are such that it is appropriate to use Fick's first law in analyzing diffusion within the fluid contained by the permeable structure.

We assume that the diffusion coefficient $\mathscr{D}_{(AB)}$ is a constant. We can take the local volume average of Fick's first law in the form of Eq. (A) of Table 8.4.6-1 to find

$$\overline{\mathbf{n}}_{(A)} = \overline{\rho_{(A)}\mathbf{v}} - \mathscr{D}_{(AB)}\overline{\rho\nabla\omega_{(A)}} \tag{3-2}$$

The theorem of Sec. 4.3.3 allows us to say

$$\overline{\nabla\omega_{(A)}} = \nabla\overline{\omega}_{(A)} + \frac{1}{V}\int_{S_w}\omega_{(A)}\mathbf{n}\,dS \tag{3-3}$$

This allows us to write Eq. (3-2) as

$$\overline{\mathbf{n}}_{(A)} = \langle\rho_{(A)}\rangle\overline{\mathbf{v}} - \langle\rho\rangle\mathscr{D}_{(AB)}\nabla\overline{\omega}_{(A)} - \boldsymbol{\delta}_{(A)} \tag{3-4}$$

where we define the *mass density tortuosity* vector for species A

$$\boldsymbol{\delta}_{(A)} \equiv \langle\rho_{(A)}\rangle\overline{\mathbf{v}} - \overline{\rho_{(A)}\mathbf{v}} - \mathscr{D}_{(AB)}\overline{\rho\nabla\omega_{(A)}}$$

$$- \langle\rho\rangle\mathscr{D}_{(AB)}\overline{\nabla\omega}_{(A)} + \frac{\langle\rho\rangle\mathscr{D}_{(AB)}}{V}\int_{S_w}\omega_{(A)}\mathbf{n}\,dS \tag{3-5}$$

The local volume average of the equation of continuity for species A in the form of Eq. (2-5) may be expressed as

$$\frac{\partial\overline{\rho}_{(A)}}{\partial t} + \mathrm{div}\,(\langle\rho_{(A)}\rangle\overline{\mathbf{v}}) = -\mathrm{div}\,\mathbf{j}^{(e)}_{(A)} + \overline{r}''_{(A)} + \overline{r}_{(A)} \tag{3-6}$$

where the vector $\mathbf{j}^{(e)}_{(A)}$,

$$\mathbf{j}^{(e)}_{(A)} \equiv -\langle\rho\rangle\mathscr{D}_{(AB)}\nabla\overline{\omega}_{(A)} - \boldsymbol{\delta}_{(A)} \tag{3-7}$$

should be thought of as an "effective" mass flux vector with respect to $\overline{\mathbf{v}}$.

Very similar results can be obtained if we assume the diffusion coefficient $\mathscr{D}_{(AB)}$ is a constant and take the local volume average of Fick's first law in the form of Eq. (D) of Table 8.4.6-1:

$$\overline{\mathbf{N}}_{(A)} = \langle c_{(A)}\rangle\overline{\mathbf{v}^*} - \langle c\rangle\mathscr{D}_{(AB)}\nabla\overline{x}_{(A)} - \boldsymbol{\Delta}_{(A)} \tag{3-8}$$

$$\boldsymbol{\Delta}_{(A)} \equiv \langle c_{(A)}\rangle\overline{\mathbf{v}^*} - \overline{c_{(A)}\mathbf{v}^*} - \mathscr{D}_{(AB)}\overline{c\nabla x_{(A)}} - \langle c\rangle\mathscr{D}_{(AB)}\overline{\nabla x}_{(A)}$$

$$- \frac{\langle c\rangle\mathscr{D}_{(AB)}}{V}\int_{S_w}x_{(A)}\mathbf{n}\,dS \tag{3-9}$$

We will refer to $\boldsymbol{\Delta}_{(A)}$ as the molar-density tortuosity vector for species A. In these terms, the local volume average of the equation of continuity in the form of Eq. (2-6) becomes

**568** APPLICATIONS OF INTEGRAL AVERAGING TECHNIQUES TO MASS TRANSFER

$$\frac{\partial \overline{c}_{(A)}}{\partial t} + \text{div}(\langle c_{(A)}\rangle \overline{\mathbf{v}^*}) = -\text{div}\,\mathbf{J}_{(A)}^{\star(s)} + \frac{r''_{(A)}}{M_{(A)}} - \frac{\overline{r}_{(A)}}{M_{(A)}} \tag{3-10}$$

where the vector $\mathbf{J}_{(A)}^{\star(e)}$,

$$\mathbf{J}_{(A)}^{\star(s)} \equiv -\langle c \rangle \mathscr{D}_{(AB)} \overline{\nabla x}_{(A)} - \mathbf{\Delta}_{(A)} \tag{3-11}$$

should be thought of as an effective molar flux vector with respect to $\overline{\mathbf{v}^*}$.

One point worth emphasizing is that

$$\mathbf{j}_{(A)}^{(e)} \neq \frac{M_{(A)} M_{(B)}}{M} \mathbf{J}_{(A)}^{\star(e)} \tag{3-12}$$

The physical meaning of the mass-density tortuosity vector $\boldsymbol{\delta}_{(A)}$ is clarified by noting that, if $\rho_{(A)}$, $\rho$ and $\Psi$ are independent of position,

$$\boldsymbol{\delta}_{(A)} = \frac{\rho_{(A)} \mathscr{D}_{(AB)}}{V} \int_{S_w} \mathbf{n}\, dS$$

$$= 0 \tag{3-13}$$

We have used here the theorem of Sec. 4.3.3 applied to a constant. In the same way, if $c_{(A)}$, $c$ and $\Psi$ are independent of position,

$$\mathbf{\Delta}_{(A)} = 0 \tag{3-14}$$

Because of these simplifications, we shall direct our attention to structures of uniform porosity in the sections that immediately follow.

**REFERENCES**

1. Hirschfelder, J. O., C. F. Curtiss, and R. B. Bird: "Molecular Theory of Gases and Liquids," 2d printing, Wiley, New York, 1954.
2. Scott, D. S.: *Can. J. Chem. Eng.*, **40**:173 (1962).

**10.3.4 Empirical correlations for $\boldsymbol{\delta}_{(A)}$ and $\boldsymbol{\Delta}_{(A)}$** In this section, we give three examples of how experimental data can be used to prepare correlations for $\boldsymbol{\delta}_{(A)}$ and $\boldsymbol{\Delta}_{(A)}$, introduced in Sec. 10.3.3. Four points form the foundation for this discussion.

1. The tortuosity vectors $\boldsymbol{\delta}_{(A)}$ and $\boldsymbol{\Delta}_{(A)}$ are frame indifferent. For example,

$$\boldsymbol{\delta}_{(A)}^* \equiv \langle \rho_{(A)}^* \rangle \overline{\mathbf{v}^*} - \overline{\rho_{(A)}^* \mathbf{v}^*} + \mathscr{D}_{(AB)} \overline{\rho^* \nabla \omega_{(A)}^*}$$

$$- \langle \rho^* \rangle \mathscr{D}_{(AB)} \overline{\nabla \omega_{(A)}^*} + \frac{\langle \rho^* \rangle \mathscr{D}_{(AB)}}{V} \int_{S_w} \omega_{(A)}^* \mathbf{n}^* dS$$

$$= \overline{\rho_{(A)}^* (\Psi^{-1} \mathbf{v}^* - \mathbf{v}^*)} + \mathscr{D}_{(AB)} \overline{\rho^* \nabla \omega_{(A)}^*} - \langle \rho^* \rangle \mathscr{D}_{(AB)} \overline{\nabla \omega_{(A)}^*}$$

$$+ \frac{\langle \rho^* \rangle \mathscr{D}_{(AB)}}{V} \int_{S_w} \omega_{(A)}^* \mathbf{n}^* dS$$

## [10.3] LOCAL VOLUME AVERAGING

$$= \overline{\rho_{(A)}\mathbf{Q} \cdot (\Psi^{-1}\overline{\mathbf{v}} - \mathbf{v})} + \mathscr{D}_{(AB)}\overline{\rho\mathbf{Q} \cdot \nabla \omega_{(A)}}$$

$$- \langle\rho\rangle\mathscr{D}_{(AB)}\mathbf{Q} \cdot \overline{\nabla \omega_{(A)}} + \frac{\langle\rho\rangle\mathscr{D}_{(AB)}}{V}\int_{S_w} \omega_{(A)}\mathbf{Q} \cdot \mathbf{n}\, dS = \mathbf{Q} \cdot \boldsymbol{\delta}_{(A)} \quad (4\text{-}1)$$

In the second line we observe that the local volume average of a local volume average is simply the local volume average (see Exercise 4.3.8-1); in the third line, we employ the frame indifference of the mass density for species $A$ and the frame indifference of a velocity difference. Here $\mathbf{Q}$ is a (possibly) time-dependent, orthogonal, second-order tensor. The molar-density tortuosity vector $\boldsymbol{\Delta}_{(A)}$ can be proved to be frame indifferent in exactly the same manner.

2. We assume that the principle of material frame indifference introduced in Sec. 2.3.1 applies to any empirical correlation developed for $\boldsymbol{\delta}_{(A)}$ and $\boldsymbol{\Delta}_{(A)}$.
3. The Buckingham-Pi theorem serves to further restrict the form of any expression for $\boldsymbol{\delta}_{(A)}$ and $\boldsymbol{\Delta}_{(A)}$.
4. The averaging surface $S$ is so large that $\boldsymbol{\delta}_{(A)}$ and $\boldsymbol{\Delta}_{(A)}$ may be assumed not to be *explicit* functions of position in the porous structure, though they very well may be implicit functions of position as a result of their dependence upon other variables.

**EXAMPLE I  NONORIENTED POROUS SOLIDS WHEN CONVECTION CAN BE NEGLECTED**

In Sec. 9.3.1 we argued that, when there is no forced convection, natural convection may be neglected in the limit as equilibrium is approached. In this limit we may certainly neglect the first four terms on the right of Eq. (3-5):

$$\boldsymbol{\delta}_{(A)} = \frac{\langle\rho\rangle\mathscr{D}_{(AB)}}{V}\int_{S_w} \omega_{(A)}\mathbf{n}\, dS \quad (4\text{-}2)$$

For geometrically similar nonoriented porous media, $\boldsymbol{\delta}_{(A)}$ might be thought of as a function of the "particle diameter" $l_0$, the diffusion coefficient $\mathscr{D}_{(AB)}$, the porosity $\Psi$, as well as some measures of the local concentration distribution such as $\overline{\rho}_{(A)}$ and $\langle\rho\rangle\nabla\overline{\omega}_{(A)}$:

$$\boldsymbol{\delta}_{(A)} = \boldsymbol{\delta}_{(A)}(l_0\mathscr{D}_{(AB)}, \Psi, \overline{\rho}_{(A)}, \langle\rho\rangle\nabla\overline{\omega}_{(A)}) \quad (4\text{-}3)$$

For the moment, let us fix our attention on the dependence of $\boldsymbol{\delta}_{(A)}$ upon $\langle\rho\rangle\nabla\overline{\omega}_{(A)}$:

$$\boldsymbol{\delta}_{(A)} = \hat{\boldsymbol{\delta}}_{(A)}(\langle\rho\rangle\nabla\overline{\omega}_{(A)}) \quad (4\text{-}4)$$

By the principle of material frame indifference, the functional relationship between these two variables should be the same in every frame of reference. This means that

$$\boldsymbol{\delta}^*_{(A)} \equiv \mathbf{Q} \cdot \boldsymbol{\delta}_{(A)} = \mathbf{Q} \cdot \hat{\boldsymbol{\delta}}(\langle\rho\rangle\nabla\overline{\omega}_{(A)}) = \hat{\boldsymbol{\delta}}_{(A)}(\mathbf{Q} \cdot \langle\rho\rangle\nabla\overline{\omega}_{(A)}) \quad (4\text{-}5)$$

or $\hat{\boldsymbol{\delta}}_{(A)}$ is an isotropic function [1, p. 22]:

$$\hat{\boldsymbol{\delta}}_{(A)}(\langle\rho\rangle\nabla\overline{\omega}_{(A)}) = \mathbf{Q}^T \cdot \hat{\boldsymbol{\delta}}_{(A)}(\mathbf{Q} \cdot \langle\rho\rangle\nabla\overline{\omega}_{(A)}) \quad (4\text{-}6)$$

By a representation theorem for a vector-valued isotropic function of one vector [1, p. 35], we may write

$$\boldsymbol{\delta}_{(A)} = \hat{\boldsymbol{\delta}}_{(A)}(\langle\rho\rangle\nabla\bar{\omega}_{(A)}) = D_{(A)}\langle\rho\rangle\nabla\bar{\omega}_{(A)} \tag{4-7}$$

where

$$D_{(A)} = \hat{D}_{(A)}(\langle\rho\rangle|\nabla\bar{\omega}_{(A)}|) \tag{4-8}$$

Comparing Eqs. (4-7) and (4-8) with Eq. (4-3) we see

$$D_{(A)} = D_{(A)}(l_0, \mathscr{D}_{(AB)}, \Psi, \bar{\rho}_{(A)}, \langle\rho\rangle|\nabla\bar{\omega}_{(A)}|) \tag{4-9}$$

An application of the Buckingham-Pi theorem [2] allows us to conclude that

$$D_{(A)} = \mathscr{D}_{(AB)} D^*_{(A)} \tag{4-10}$$

Here

$$D^*_{(A)} = D^*_{(A)}\left(\Psi, \frac{l_0 \langle\rho\rangle |\nabla\bar{\omega}_{(A)}|}{\bar{\rho}_{(A)}}\right) \tag{4-11}$$

To summarize, Eqs. (4-7), (4-10), and (4-11) represent probably the simplest form that empirical correlations for the mass-density tortuosity vector $\boldsymbol{\delta}_{(A)}$ can take in a nonoriented porous medium.

### EXAMPLE 2 NONORIENTED POROUS SOLID FILLED WITH A FLOWING FLUID

For geometrically similar nonoriented porous media under conditions such that convection is not negligible, $\boldsymbol{\delta}_{(A)}$ may be thought of as a function of the local particle diameter $l_0$, the diffusion coefficient $\mathscr{D}_{(AB)}$, the porosity $\Psi$, the local volume-averaged velocity of the fluid with respect to the local volume-averaged velocity of the solid $\bar{\mathbf{v}} - \bar{\mathbf{v}}^{(s)}$, as well as some measures of the local mass-density distribution such as $\bar{\rho}_{(A)}$ and $\langle\rho\rangle\nabla\bar{\omega}_{(A)}$:

$$\boldsymbol{\delta}_{(A)} = \boldsymbol{\delta}_{(A)}(l_0, \mathscr{D}_{(AB)}, \Psi, \bar{\mathbf{v}} - \bar{\mathbf{v}}^{(s)}, \bar{\rho}_{(A)}, \langle\rho\rangle\nabla\bar{\omega}_{(A)}) \tag{4-12}$$

Let us begin by examining the dependence of $\boldsymbol{\delta}_{(A)}$ upon the two vectors:

$$\boldsymbol{\delta}_{(A)} = \hat{\boldsymbol{\delta}}_{(A)}(\bar{\mathbf{v}} - \bar{\mathbf{v}}^{(s)}, \langle\rho\rangle\nabla\bar{\omega}_{(A)}) \tag{4-13}$$

By the principle of material frame indifference, the functional relationship between these two variables should be the same in every frame of reference. This means that

$$\begin{aligned}\boldsymbol{\delta}^*_{(A)} &= \mathbf{Q} \cdot \boldsymbol{\delta}_{(A)} = \mathbf{Q} \cdot \hat{\boldsymbol{\delta}}_{(A)}(\bar{\mathbf{v}} - \bar{\mathbf{v}}^{(s)}, \langle\rho\rangle\nabla\bar{\omega}_{(A)}) \\ &= \hat{\boldsymbol{\delta}}_{(A)}(\mathbf{Q} \cdot (\bar{\mathbf{v}} - \bar{\mathbf{v}}^{(s)}), \mathbf{Q} \cdot \langle\rho\rangle\nabla\bar{\omega}_{(A)})\end{aligned} \tag{4-14}$$

or $\boldsymbol{\delta}_{(A)}$ is an isotropic function [1, p. 22]:

$$\hat{\boldsymbol{\delta}}_{(A)}(\bar{\mathbf{v}} - \bar{\mathbf{v}}^{(s)}, \langle\rho\rangle\nabla\bar{\omega}_{(A)}) = \mathbf{Q}^T \cdot \hat{\boldsymbol{\delta}}_{(A)}(\mathbf{Q} \cdot [\bar{\mathbf{v}} - \bar{\mathbf{v}}^{(s)}], \mathbf{Q} \cdot \langle\rho\rangle\nabla\bar{\omega}_{(A)}) \tag{4-15}$$

By an argument similar to that given in Sec. 7.3.3 (Example 2), we conclude

$$\boldsymbol{\delta}_{(A)} = D_{(A1)}\langle\rho\rangle\nabla\bar{\omega}_{(A)} - D_{(A2)}(\bar{\mathbf{v}} - \bar{\mathbf{v}}^{(s)}) \tag{4-16}$$

## [10.3] LOCAL VOLUME AVERAGING

where, for $i = 1, 2$,

$$D_{(Ai)} = D_{(Ai)}(|\bar{\mathbf{v}} - \bar{\mathbf{v}}^{(s)}|, \langle\rho\rangle|\nabla\bar{\omega}_{(A)}|, \langle\rho\rangle(\bar{\mathbf{v}} - \bar{\mathbf{v}}^{(s)}) \cdot \nabla\bar{\omega}_{(A)}, \bar{\rho}_{(A)}, l_0, \mathscr{D}_{(AB)}, \Psi) \quad (4\text{-}17)$$

An application of the Buckingham-Pi theorem [2] shows that

$$D_{(A1)} = \mathscr{D}_{(AB)} D^*_{(A1)} \quad (4\text{-}18)$$

and

$$D_{(A2)} = l_0 \langle\rho\rangle |\nabla\bar{\omega}_{(A)}| D^*_{(A2)} \quad (4\text{-}19)$$

Here, for $i = 1, 2$,

$$D^*_{(Ai)} = D^*_{(Ai)}\left(N_{\text{Pe}}, \frac{[\bar{\mathbf{v}} - \bar{\mathbf{v}}^{(s)}] \cdot \nabla\bar{\omega}_{(A)}}{|\bar{\mathbf{v}} - \bar{\mathbf{v}}^{(s)}||\nabla\bar{\omega}_{(A)}|}, \frac{l_0 \langle\rho\rangle |\nabla\bar{\omega}_{(A)}|}{\bar{\rho}_{(A)}}, \Psi\right) \quad (4\text{-}20)$$

and

$$N_{\text{Pe}} \equiv \frac{l_0 |\bar{\mathbf{v}} - \bar{\mathbf{v}}^{(s)}|}{\mathscr{D}_{(AB)}} \quad (4\text{-}21)$$

We expect that $D^*_{(A2)} = 0$ for $|\nabla\bar{\omega}_{(A)}| = 0$, with result that $\boldsymbol{\delta}_{(A)} = 0$ in this limit.

In summary, Eqs. (4-16) and (4-18) to (4-20) represent possibly the simplest form that empirical correlations for the mass-density tortuosity vector $\boldsymbol{\delta}_{(A)}$ can take when a fluid flows through a nonoriented porous medium.

**EXAMPLE 3: ORIENTED POROUS SOLIDS WHEN CONVECTION CAN BE NEGLECTED**

When convection can be neglected, we saw in Example 1 that Eq. (3-4) reduces to Eq. (4-2). But one should not expect Eqs. (4-7), (4-10), and (4-11) to describe the mass-density tortuosity vector for a porous structure in which particle diameter $l$ is a function of position. For such a structure, Eq. (4-3) must be altered to include a dependence upon additional vector and possibly tensor quantities. For example, one might postulate a dependence of $\boldsymbol{\delta}_{(A)}$ upon the local gradient in particle diameter as well as upon $\langle\rho\rangle\nabla\bar{\omega}_{(A)}$:

$$\boldsymbol{\delta}_{(A)} = \boldsymbol{\delta}_{(A)}(l, \mathscr{D}_{(AB)}, \Psi, \bar{\rho}_{(A)}, \langle\rho\rangle\nabla\bar{\omega}_{(A)}, \nabla l) \quad (4\text{-}22)$$

Following essentially the same argument given in Example 2 above, the principle of material frame indifference and the Buckingham-Pi theorem require

$$\boldsymbol{\delta}_{(A)} = E_{(A1)}\langle\rho\rangle\nabla\bar{\omega}_{(A)} + E_{(A2)}\nabla l \quad (4\text{-}23)$$

where

$$E_{(A1)} = \mathscr{D}_{(AB)} E^*_{(A1)} \quad (4\text{-}24)$$

$$E_{(A2)} = \mathscr{D}_{(AB)} \langle\rho\rangle |\nabla\bar{\omega}_{(A)}| E^*_{(A2)} \quad (4\text{-}25)$$

and, for $i = 1, 2$,

$$E^*_{(Ai)} = E^*_{(Ai)}\left(\frac{\nabla l \cdot \nabla \bar{\omega}_{(A)}}{|\nabla l||\nabla \bar{\omega}_{(A)}|}, |\nabla l|, \frac{l\langle\rho\rangle|\nabla\bar{\omega}_{(A)}|}{\bar{\rho}_{(A)}}, \Psi\right) \tag{4-26}$$

We expect that $E^*_{(A2)} = 0$ for $|\nabla \bar{\omega}_{(A)}| = 0$, with the result that $\boldsymbol{\delta}_{(A)} = 0$ in this limit.

Equations (4-23) to (4-26) represent possibly the simplest form that empirical correlations for the mass-density tortuosity vector $\boldsymbol{\delta}_{(A)}$ can take in an oriented porous medium, assuming that the orientation of the structure can be attributed to the local gradient of particle diameter.

### REFERENCES

1. Truesdell, C., and W. Noll: In S. Flügge (ed.), "Handbuch der Physik," vol. 3/3, Springer-Verlag, Berlin, 1965.
2. Brand, Louis: *Arch. Rational Mech. Analysis*, **1**:35 (1957).

### 10.3.5 Summary of results for a liquid or dense gas in a nonoriented, uniform-porosity structure

I would like to summarize here the results for the case with which the literature has been primarily concerned until now: a liquid or dense gas in a nonoriented, uniform-porosity structure.

In Sec. 10.3.3 we found that the local volume average of the differential equation of continuity for species $A$ requires that

$$\frac{\partial \bar{\rho}_{(A)}}{\partial t} + \text{div}(\langle\rho_{(A)}\rangle\bar{\mathbf{v}}) = -\text{div}\,\mathbf{j}^{(e)}_{(A)} + r''_{(A)} + \bar{r}_{(A)} \tag{5-1}$$

where

$$\mathbf{j}^{(e)}_{(A)} \equiv -\langle\rho\rangle\mathscr{D}_{(AB)}\nabla\bar{\omega}_{(A)} - \boldsymbol{\delta}_{(A)} \tag{5-2}$$

should be thought of as the effective mass flux vector with respect to the volume-averaged mass-averaged velocity $\bar{\mathbf{v}}$. In arriving at this result, we have assumed only that Fick's first law is applicable. In this way, we have limited the discussion to liquids and gases which are so dense that the molecular mean free path is small compared with the average pore diameter of the structure.

In Sec. 10.3.4 (Example 2), we suggest that for a nonoriented porous solid filled with a flowing fluid, the mass density tortuosity vector $\boldsymbol{\delta}_{(A)}$ might be represented by Eqs. (4-16) and (4-18) to (4-21). In these terms, the effective mass flux vector can be expressed as

$$\mathbf{j}^{(e)}_{(A)} = -\langle\rho\rangle\mathscr{D}_{(AB)}(1 + D^*_{(A1)})\nabla\bar{\omega}_{(A)} + l_0\langle\rho\rangle|\nabla\bar{\omega}_{(A)}|\,D^*_{(A2)}\bar{\mathbf{v}} \tag{5-3}$$

where, for $i = 1, 2$,

$$D^*_{(Ai)} = D^*_{(Ai)}\left(N_{\text{Pe}}, \frac{\bar{\mathbf{v}} \cdot \nabla\bar{\omega}_{(A)}}{|\bar{\mathbf{v}}||\nabla\bar{\omega}_{(A)}|}, \frac{l_0\langle\rho\rangle|\nabla\bar{\omega}_{(A)}|}{\bar{\rho}_{(A)}}, \Psi\right) \tag{5-4}$$

and

$$N_{\text{Pe}} \equiv \frac{l_0|\bar{\mathbf{v}}|}{\mathscr{D}_{(AB)}} \tag{5-5}$$

## [10.3] LOCAL VOLUME AVERAGING

In arriving at this expression, we have assumed that the porous medium is stationary.

Sometimes it is more convenient to think of the effective mass flux vector in terms of an effective diffusivity tensor $\mathbf{D}_{(AB)}^{(e)}$:

$$\mathbf{j}_{(A)}^{(e)} = -\langle \rho \rangle \mathbf{D}_{(AB)}^{(e)} \cdot \nabla \bar{\omega}_{(A)} \tag{5-6}$$

Here

$$\mathbf{D}_{(AB)}^{(e)} \equiv \mathscr{D}_{(AB)}(1 + D_{(A1)}^{*})\mathbf{I} - \frac{l_0 |\nabla \bar{\omega}_{(A)}| D_{(A2)}^{*}}{\bar{\mathbf{v}} \cdot \nabla \bar{\omega}_{(A)}} \bar{\mathbf{v}} \bar{\mathbf{v}} \tag{5-7}$$

We shall often find it more convenient to work in molar terms. Returning to Sec. 10.3.3, we found there that the local volume average of the differential equation of continuity could also be expressed as

$$\frac{\partial \bar{c}_{(A)}}{\partial t} + \text{div}\,(\langle c_{(A)} \rangle \overline{\mathbf{v}^*}) = -\text{div}\,\mathbf{J}_{(A)}^{\star(e)} + \frac{r''_{(A)}}{M_{(A)}} + \frac{\bar{r}_{(A)}}{M_{(A)}} \tag{5-8}$$

where

$$\mathbf{J}_{(A)}^{\star(e)} \equiv -\langle c \rangle \mathscr{D}_{(AB)} \nabla \bar{x}_{(A)} - \mathbf{\Delta}_{(A)} \tag{5-9}$$

should be thought of as the effective molar flux of species $A$ with respect to the volume-averaged molar-averaged velocity $\overline{\mathbf{v}^*}$. If we visualize repeating for $\mathbf{\Delta}_{(A)}$ the type of analysis given in Sec. 10.3.4 (Example 2) for a nonoriented porous solid filled with a flowing fluid, we would find by analogy with Eqs. (5-3) to (5-5) that

$$\mathbf{J}_{(A)}^{\star(e)} = -\langle c \rangle \mathscr{D}_{(AB)}(1 + D_{(A1)}^{\star\star})\nabla \bar{x}_{(A)} + l_0 \langle c \rangle |\nabla \bar{x}_{(A)}| D_{(A)}^{\star\star} \overline{\mathbf{v}^*} \tag{5-10}$$

For $i = 1, 2$:
$$D_{(Ai)}^{\star\star} = D_{(Ai)}^{\star\star}\left(N_{\text{Pe}}^{\star},\, \frac{\overline{\mathbf{v}^*} \cdot \nabla \bar{x}_{(A)}}{|\overline{\mathbf{v}^*}| |\nabla \bar{x}_{(A)}|},\, \frac{l_0 \langle c \rangle |\nabla \bar{x}_{(A)}|}{\bar{c}_{(A)}},\, \Psi\right) \tag{5-11}$$

$$N_{\text{Pe}}^{\star} \equiv \frac{l_0 |\overline{\mathbf{v}^*}|}{\mathscr{D}_{(AB)}} \tag{5-12}$$

These results can, of course, also be written in terms of an effective diffusivity tensor $\mathbf{D}_{(AB)}^{\star(e)}$:

$$\mathbf{J}_{(A)}^{\star(e)} = -\langle c \rangle \mathbf{D}_{(AB)}^{\star(e)} \cdot \nabla \bar{x}_{(A)} \tag{5-13}$$

$$\mathbf{D}_{(AB)}^{\star(e)} \equiv \mathscr{D}_{(AB)}(1 + D_{(A1)}^{\star\star})\mathbf{I} - \frac{l_0 |\nabla \bar{x}_{(A)}| D_{(A2)}^{\star\star}}{\overline{\mathbf{v}^*} \cdot \nabla \bar{x}_{(A)}} \overline{\mathbf{v}^*}\overline{\mathbf{v}^*} \tag{5-14}$$

For the sake of simplicity, in what follows we take

For $i = 1, 2$:
$$D_{(Ai)}^{*} = D_{(Ai)}^{*}(\Psi) \tag{5-15}$$

and

For $i = 1, 2$:
$$D_{(Ai)}^{\star\star} = D_{(Ai)}^{\star\star}(\Psi) \tag{5-16}$$

This allows us to express Eqs. (5-3) and (5-10) as

$$\mathbf{j}_{(A)}^{(e)} = -\langle \rho \rangle A_{(A)} \nabla \bar{\omega}_{(A)} + B_{(A)} \langle \rho \rangle |\nabla \bar{\omega}_{(A)}| \bar{\mathbf{v}} \tag{5-17}$$

and
$$\mathbf{J}_{(A)}^{\star(e)} = -\langle c \rangle A_{(A)}^{\star} \nabla \bar{x}_{(A)} + B_{(A)}^{\star} \langle c \rangle |\nabla \bar{x}_{(A)}| \overline{\mathbf{v}^{\star}} \tag{5-18}$$

where $A_{(A)}$, $B_{(A)}$, $A_{(A)}^{\star}$, and $B_{(A)}^{\star}$ are independent of position for any particular porous medium.

**10.3.6 When Fick's first law does not apply** Let $l_0$ represent a characteristic pore diameter of the structure, $\lambda$ be the molecular mean free path [1, p. 10], and $N_{\text{Kn}} \equiv l_0/\lambda$ be the Knudsen number. When

$$0.1 < N_{\text{Kn}} < 10 \tag{6-1}$$

Fick's first law cannot be used to describe binary diffusion in a porous medium [2]. This means that we must go back to Sec. 10.3.2 and prepare empirical data correlations for $\bar{\mathbf{n}}_{(A)}$ in Eq. (2-5). The only difficulty is that $\bar{\mathbf{n}}_{(A)}$ is not a frame-indifferent vector.

This suggests that we rewrite Eq. (2-5) in terms of the effective mass flux vector with respect to $\bar{\mathbf{v}}$:

$$\frac{\partial \bar{\rho}_{(A)}}{\partial t} + \text{div}\,(\langle \rho_{(A)} \rangle \bar{\mathbf{v}}) = -\text{div}\,\mathbf{j}_{(A)}^{(e)} + r_{(A)}'' + \bar{r}_{(A)} \tag{6-2}$$

$$\mathbf{j}_{(A)}^{(e)} \equiv \overline{\mathbf{j}_{(A)}} + \overline{\rho_{(A)}\mathbf{v}} - \langle \rho_{(A)} \rangle \bar{\mathbf{v}} \tag{6-3}$$

The same reasoning we used in Sec. 10.3.4 to prepare empirical correlations for $\boldsymbol{\delta}_{(A)}$ may be used here to formulate empirical correlations for $\mathbf{j}_{(A)}^{(e)}$. For example, for a nonoriented porous solid filled with a flowing fluid (see Sec. 10.3.4, Example 2) our initial guess might be that

$$\mathbf{j}_{(A)}^{(e)} = -\langle \rho \rangle \mathscr{D}_{(AB)} D_{(A1)}^{\star} \nabla \bar{\omega}_{(A)} + D_{(A2)}^{\star} \bar{\rho}_{(A)} \bar{\mathbf{v}} \tag{6-4}$$

where, for $i = 1, 2$,

$$D_{(Ai)}^{\star} = D_{(Ai)}^{\star}\left(N_{\text{Pe}}, \frac{\bar{\mathbf{v}} \cdot \nabla \bar{\omega}_{(A)}}{|\bar{\mathbf{v}}| |\nabla \bar{\omega}_{(A)}|}, \frac{l_0 \langle \rho \rangle |\nabla \bar{\omega}_{(A)}|}{\bar{\rho}_{(A)}}, \Psi\right) \tag{6-5}$$

and

$$N_{\text{Pe}} \equiv \frac{l_0 |\bar{\mathbf{v}}|}{\mathscr{D}_{(AB)}} \tag{6-6}$$

This also could be thought of in terms of an effective diffusivity tensor $\mathbf{D}_{(AB)}^{(e)}$:

$$\mathbf{j}_{(A)}^{(e)} = -\langle \rho \rangle \mathbf{D}_{(AB)}^{(e)} \cdot \nabla \bar{\omega}_{(A)} \tag{6-7}$$

$$\mathbf{D}_{(AB)}^{(e)} \equiv \mathscr{D}_{(AB)} D_{(A1)}^{\star} \mathbf{I} - \frac{D_{(A2)}^{\star} \bar{\rho}_{(A)}}{\langle \rho \rangle (\bar{\mathbf{v}} \cdot \nabla \bar{\omega}_{(A)})} \bar{\mathbf{v}}\bar{\mathbf{v}} \tag{6-8}$$

If we prefer to think in molar terms, we can introduce the effective molar flux vector with respect to $\overline{\mathbf{v}^{\star}}$ in Eq. (2-6):

$$\frac{\partial \bar{c}_{(A)}}{\partial t} + \text{div}\,(\overline{\langle c_{(A)}\rangle \mathbf{v}^\star}) = -\text{div}\,\mathbf{J}^{\star(e)}_{(A)} + \frac{\bar{r}''_{(A)}}{M_{(A)}} + \frac{\bar{r}_{(A)}}{M_{(A)}} \qquad (6\text{-}9)$$

$$\mathbf{J}^{\star(e)}_{(A)} \equiv \overline{\mathbf{J}^\star_{(A)}} + \overline{c_{(A)}\mathbf{v}^\star} - \overline{\langle c_{(A)}\rangle \mathbf{v}^\star} \qquad (6\text{-}10)$$

Again by analogy with Example 2 in Sec. 10.3.4, we would hypothesize that for a nonoriented porous solid filled with a flowing fluid

$$\mathbf{J}^{\star(e)}_{(A)} = -\langle c \rangle \mathscr{D}_{(AB)} D^{\star\star}_{(A1)} \nabla \bar{x}_{(A)} + D^{\star\star}_{(A2)} \bar{c}_{(A)} \overline{\mathbf{v}^\star} \qquad (6\text{-}11)$$

$$\text{For } i = 1, 2: \quad D^{\star\star}_{(Ai)} = D^{\star\star}_{(Ai)}\left(N^\star_{\text{Pe}}, \frac{\overline{\mathbf{v}^\star} \cdot \nabla \bar{x}_{(A)}}{|\overline{\mathbf{v}^\star}|\,|\nabla \bar{x}_{(A)}|}, \frac{l_0 \langle c \rangle |\nabla \bar{x}_{(A)}|}{\bar{c}_{(A)}}, \Psi\right) \qquad (6\text{-}12)$$

and

$$N^\star_{\text{Pe}} \equiv \frac{l_0\,|\overline{\mathbf{v}^\star}|}{\mathscr{D}_{(AB)}} \qquad (6\text{-}13)$$

In terms of an effective diffusivity tensor $\mathbf{D}^{\star(e)}_{(AB)}$, we can say

$$\mathbf{J}^{\star(e)}_{(A)} = -\langle c \rangle \mathbf{D}^{\star(e)}_{(AB)} \cdot \nabla \bar{x}_{(A)} \qquad (6\text{-}14)$$

where

$$\mathbf{D}^{\star(e)}_{(AB)} \equiv \mathscr{D}_{(AB)} D^{\star\star}_{(A1)} \mathbf{I} - \frac{D^{\star\star}_{(A2)} \bar{c}_{(A)}}{\langle c \rangle (\overline{\mathbf{v}^\star} \cdot \nabla \bar{x}_{(A)})} \overline{\mathbf{v}^\star}\,\overline{\mathbf{v}^\star} \qquad (6\text{-}15)$$

Evans, Watson, and Mason [3] visualize binary diffusion in a porous medium as being described by a ternary diffusion problem, the third species being the stationary porous structure. They refer to this as their "dusty" gas model. If we interpret their variables as being local volume averages, then their result can be viewed as a special case of Eq. (6-11) with

$$D^{\star\star}_{(A1)} \equiv \frac{\Psi}{q}\left[1 + \frac{\bar{c}_{(s)}}{\bar{c}}\left(\frac{M_{(A)} + M_{(B)}}{M_{(B)}}\right)^{\frac{1}{2}} k_1\right]^{-1} \qquad (6\text{-}16)$$

and

$$D^{\star\star}_{(A2)} + 1 \equiv \left[1 + \frac{\bar{c}_{(s)}}{\bar{c}}\left(\frac{M_{(A)} + M_{(B)}}{M_{(B)}}\right)^{\frac{1}{2}} k_2\right]^{-1} \qquad (6\text{-}17)$$

Here $\bar{c}_{(s)}$ is the local volume-averaged molar density of solids; $q$, $k_1$, and $k_2$ are constants characteristic of the porous structure; a slight dependence of $k_1$ and $k_2$ upon the properties of the gas mixture is possible.

One final word of caution. When $N_{\text{Kn}} < 10$, our local volume-averaged equation of motion is no longer applicable. The final form (Darcy's law or its equivalent; see Secs. 4.3.5 to 4.3.7) depends upon a constitutive equation for the stress tensor that is not applicable when molecular collisions with the walls of the porous structure become as important as intermolecular collisions. Arguments based upon values of the pressure gradient deduced from Darcy's law are almost certainly not valid.

## REFERENCES

1. Hirschfelder, J. O., C. F. Curtiss, and R. B. Bird: "Molecular Theory of Gases and Liquids," 2d printing, Wiley, New York, 1954.
2. Scott, D. S.: *Can. J. Chem. Eng.*, **40**:173 (1962).
3. Evans, R. B., G. M. Watson, and E. A. Mason: *J. Chem. Phys.*, **35**:2076 (1961).

**10.3.7 Knudsen diffusion** When the Knudsen number $N_{\text{Kn}} < 0.1$, we refer to mass transfer in a porous structure as being Knudsen diffusion [1]. If we think for the moment in terms of a molecular model, then in Knudsen diffusion collisions between the gas molecules and the walls of the porous structure are more important than collisions between two or more molecules. This suggests that, in a continuum description of Knudsen diffusion, the movement of each species should be independent of all other species present in the gas.

This goal of independence of movement of the various species present will be furthered if $\bar{\rho}$ and $\bar{\mathbf{v}}$ do not appear in the final form of the equation of continuity for any species $A$. Reasoning as we did in Secs. 10.3.4 and 10.3.6, we can propose an empirical data correlation for $\mathbf{j}_{(A)}^{(e)}$ that satisfies these conditions.

Let us begin by postulating that

$$\mathbf{j}_{(A)}^{(e)} = \mathbf{j}_{A}^{(e)}(\bar{\rho}_{(A)}, \nabla \bar{\rho}_{(A)}, \bar{\mathbf{v}} - \bar{\mathbf{v}}^{(s)}, l_0, \Psi, R, T, M_{(A)}) \tag{7-1}$$

where $R$ is the gas law constant, $T$ is the temperature, and $M_{(A)}$ is the molecular weight for species $A$. The principle of material frame indifference and the Buckingham-Pi theorem require for a stationary porous structure

$$\mathbf{j}_{(A)}^{(e)} = -\left(\frac{RT}{M_{(A)}}\right)^{\frac{1}{2}} l_0 D_{(A1)}^* \nabla \bar{\rho}_{(A)} + D_{(A2)}^* \bar{\rho}_{(A)} \bar{\mathbf{v}} \tag{7-2}$$

Here, for $i = 1, 2$,

$$D_{(Ai)}^* = D_{(Ai)}^* \left(\frac{l_0 |\nabla \bar{\rho}_{(A)}|}{\bar{\rho}_{(A)}}, \left(\frac{RT}{M_{(A)}}\right)^{\frac{1}{2}} \frac{1}{|\mathbf{v}|}, \Psi\right) \tag{7-3}$$

In order that $\bar{\mathbf{v}}$ drop out of the final form of the equation of continuity, we take

$$D_{(A2)}^* = -1 \tag{7-4}$$

and

$$D_{(A1)}^* = D_{(A1)}^* \left(\frac{l_0 |\nabla \bar{\rho}_{(A)}|}{\bar{\rho}_{(A)}}, \Psi\right) \tag{7-5}$$

In terms of the equation of continuity for species $A$ in the form of Eq. (2-5),

$$\frac{\partial \bar{\rho}_{(A)}}{\partial t} + \text{div } \bar{\mathbf{n}}_{(A)} = r_{(A)}'' + \bar{r}_{(A)} \tag{7-6}$$

Equations (7-2) and (7-4) imply

[10.3] LOCAL VOLUME AVERAGING

$$\bar{\mathbf{n}}_{(A)} = -D_{(A)\mathrm{Kn}} \nabla \bar{\rho}_{(A)} \tag{7-7}$$

where

$$D_{(A)\mathrm{Kn}} \equiv \left(\frac{RT}{M_{(A)}}\right)^{\frac{1}{2}} l_0 D^*_{(A1)} \tag{7-8}$$

is known as the *Knudsen diffusion coefficient*. By comparison, the Knudsen diffusion coefficient is usually said to have the form [2; 3, p. 78; 4; 5, p. 17]

$$D_{(A)\mathrm{Kn}} = \frac{4}{3}\left(\frac{8RT}{\pi M_{(A)}}\right)^{\frac{1}{2}} l_0 K^* \tag{7-9}$$

The dimensionless coefficient $K^*$ is characteristic of the porous medium.

Equations (7-6) and (7-7) are easily interpreted in molar terms as

$$\frac{\partial \bar{c}_{(A)}}{\partial t} + \mathrm{div}\,\bar{\mathbf{N}}_{(A)} = \frac{r''_{(A)}}{M_{(A)}} + \frac{\bar{r}_{(A)}}{M_{(A)}} \tag{7-10}$$

and

$$\bar{\mathbf{N}}_{(A)} = -D_{(A)\mathrm{Kn}} \nabla \bar{c}_{(A)} \tag{7-11}$$

As I mentioned in concluding the preceding section, for $N_{\mathrm{Kn}} < 10$, Darcy's law or its equivalent is no longer applicable. This means that Darcy's law cannot be used to make a statement about the pressure gradient in Knudsen diffusion.

**REFERENCES**

1. Scott, D. S.: *Can. J. Chem. Eng.*, **40**:173 (1962).
2. Pollard, W. G., and R. D. Present: *Phys. Rev.*, **73**:762 (1948).
3. Carman, P. C.: "Flow of Gases through Porous Media," Academic, New York, 1956.
4. Evans, R. B., G. M. Watson, and E. A. Mason: *J. Chem. Phys.*, **35**:2076 (1961).
5. Satterfield, C. N., and T. K. Sherwood: "The Role of Diffusion in Catalysis," Addison-Wesley, Reading, Mass., 1963.

**10.3.8 The local volume average of the equation of continuity** We have already seen in Chap. 9 that the overall equation of continuity is often very useful in solving mass-transfer problems. This motivates us now to look at the local volume average of the overall equation of continuity.

We could directly take the local volume average of the overall equation of continuity in the two forms shown in Table 8.3.2-1. It is easier and completely equivalent to sum Eqs. (2-5) and (2-6) over all species to conclude

$$\frac{\partial \bar{\rho}}{\partial t} + \mathrm{div}\,(\overline{\rho \mathbf{v}}) = 0 \tag{8-1}$$

$$\frac{\partial \bar{c}}{\partial t} + \mathrm{div}\,(\overline{c\mathbf{v}^\star}) = \sum_{A=1}^{N} \left(\frac{r''_{(A)}}{M_{(A)}} + \frac{\bar{r}_{(A)}}{M_{(A)}}\right) \tag{8-2}$$

# APPLICATIONS OF INTEGRAL AVERAGING TECHNIQUES TO MASS TRANSFER

We will hereafter refer to these equations as the local volume averages of the overall equation of continuity.

For an incompressible fluid, Eq. (8-1) simplifies considerably to

$$\text{div } \bar{\mathbf{v}} = 0 \qquad (8\text{-}3)$$

Incompressible fluids form one of the simplest classes of mass-transfer problems in porous media.

If we can assume the molar density $c$ is a constant (an ideal gas at constant temperature and pressure), Eq. (8-2) reduces to

$$c \text{ div } \overline{\mathbf{v}^*} = \sum_{A=1}^{N} \left( \frac{r''_{(A)}}{M_{(A)}} + \frac{\bar{r}_{(A)}}{M_{(A)}} \right) \qquad (8\text{-}4)$$

If $c$ is a constant and if the number of moles produced by chemical reactions is exactly equal to the number of moles consumed in these reactions,

$$\sum_{A=1}^{N} \left( \frac{r''_{(A)}}{M_{(A)}} + \frac{\bar{r}_{(A)}}{M_{(A)}} \right) = 0 \qquad (8\text{-}5)$$

Eq. (8-2) becomes

$$\text{div } \overline{\mathbf{v}^*} = 0 \qquad (8\text{-}6)$$

From a mathematical point of view, this class of mass-transfer problems is just as simple as those for incompressible fluids.

### 10.3.9 The effectiveness factor for spherical catalyst particles

A catalytic reaction ($A \to B$) takes place in the gas phase in either a fixed-bed or fluidized reactor. We shall assume that the catalyst is uniformly distributed throughout each of the porous spherical particles of radius $R$ with which the reactor is filled. We wish to focus our attention here upon one of these porous spherical catalyst particles.

We can anticipate that more of the chemical reaction takes place on the catalyst surface in the immediate vicinity of the surface of the sphere than on the catalyst surface distributed around the center of the sphere. This seems obvious when we look at the comparable diffusion paths. What I would like to do here is to examine the overall effectiveness of the catalyst surface in a porous spherical particle. Let us begin by asking about the rate at which species $A$ is consumed by a first-order chemical reaction in the particle:

$$\frac{r''_{(A)}}{M_{(A)}} = -k''_1 a \langle c_{(A)} \rangle \qquad (9\text{-}1)$$

assuming that the concentration of species $A$ at the surface of the particle has a uniform value $c_{(A)0}$. Here $a$ denotes the available catalytic surface area per unit volume.

Since we are dealing with a catalytic reaction,

## [10.3] LOCAL VOLUME AVERAGING

$$\sum_{C=1}^{N} \frac{\bar{r}_{(C)}}{M_{(C)}} = 0 \tag{9-2}$$

We can further say that

$$\sum_{C=1}^{N} \frac{r_{(C)}}{M_{(C)}} = 0 \tag{9-3}$$

since one mole of $A$ is consumed for every mole of $B$ produced. Because we are dealing with a gas, we will idealize the problem to the extent of assuming that the overall molar density $c$ is a constant. Consequently, the local volume average of the overall equation of continuity reduces to

$$\overline{\operatorname{div} \mathbf{v}^{\star}} = 0 \tag{9-4}$$

It seems reasonable to begin this problem by assuming in spherical coordinates that

$$\overline{v_r^{\star}} = \overline{v_r^{\star}}(r) \qquad \overline{v_\theta^{\star}} = \overline{v_\varphi^{\star}} = 0 \tag{9-5}$$

and

$$\bar{c}_{(A)} = \bar{c}_{(A)}(r) \tag{9-6}$$

In view of Eq. (9-5), Eq. (9-4) requires

$$\frac{d}{dr}(r^2 \overline{v_r^{\star}}) = 0 \tag{9-7}$$

or

$$\overline{v_r^{\star}} = 0 \tag{9-8}$$

since we must require $\overline{v_r^{\star}}$ to be finite at the center of the sphere.

Let us assume that the gas in this porous catalyst particle is so dense that Fick's first law applies. For simplicity, we shall assume that $\mathbf{J}_{(A)}^{\star(e)}$ can be represented by Eq. (5-18). Because of Eqs. (9-5), (9-6), and (9-8), there is only one nonzero component of this vector:

$$J_{(A)r}^{\star(e)} = -A_{(A)}^{\star} \frac{\partial \bar{c}_{(A)}}{\partial r} \qquad J_{(A)\theta}^{\star(e)} = J_{(A)\varphi}^{\star(e)} = 0 \tag{9-9}$$

Recognizing Eqs. (9-1) and (9-9), we can now express the local volume average of the equation of continuity for species $A$ in the form of Eq. (3-10) as

$$\frac{1}{r^2}\frac{d}{dr}\left(r^2 \frac{d\langle c_{(A)}\rangle}{dr}\right) = \frac{k_1'' a}{A_{(A)}^{\star}\Psi}\langle c_{(A)}\rangle \tag{9-10}$$

This differential equation is to be solved consistent with the boundary condition

$$\text{At } r = R + \epsilon: \qquad \langle c_{(A)} \rangle = c_{(A)0} \tag{9-11}$$

Here $\epsilon$ is the diameter of the averaging surface $S$. We shall generally be willing to say that $\epsilon \ll R$.

It is convenient to introduce as dimensionless variables

$$c^* \equiv \frac{\langle c_{(A)} \rangle}{c_{(A)0}} \qquad r^* \equiv \frac{r}{R+\epsilon} \tag{9-12}$$

This allows us to write Eqs. (9-10) and (9-11), respectively, as

$$\frac{1}{r^{*2}} \frac{d}{dr^*}\left(r^{*2} \frac{dc^*}{dr^*}\right) = 9\Lambda^2 c^* \tag{9-13}$$

and

$$\text{At } r^* = 1: \quad c^* = 1 \tag{9-14}$$

For convenience in comparing the results to be obtained here with those for other particle shapes, we have defined [1]

$$\Lambda \equiv \left(\frac{k_1'' a}{A_{(A)}^\star \Psi}\right)^{\frac{1}{2}} \frac{V_p}{S_p} \tag{9-15}$$

where $V_p$ and $S_p$ are the volume and bounding surface of the catalyst particle. For a spherical catalyst particle such as we have here,

$$\Lambda \equiv \left(\frac{k_1'' a}{A_{(A)}^\star \Psi}\right)^{\frac{1}{2}} \frac{R+\epsilon}{3} \tag{9-16}$$

If we introduce as a change variable

$$u \equiv r^* c^* \tag{9-17}$$

Eq. (9-13) becomes

$$\frac{d^2 u}{dr^{*2}} = 9\Lambda^2 u \tag{9-18}$$

This can now be solved consistent with the conditions that

$$\text{At } r^* = 1: \quad u = 1 \tag{9-19}$$

and

$$\text{At } r^* = 0: \quad u = 0 \tag{9-20}$$

to find

$$c^* \equiv \frac{\langle c_{(A)} \rangle}{c_{(A)0}} = \frac{1}{r^*} \frac{\sinh(3\Lambda r^*)}{\sinh(3\Lambda)} \tag{9-21}$$

Given this concentration distribution within the catalyst particle, we can now calculate the rate at which moles of species $A$ are consumed by chemical reaction:

$$\begin{aligned}
\mathcal{W}_{(A)} &\equiv -\int_0^{2\pi}\int_0^{\pi} N_{(A)r}\big|_{r^*=1}(R+\epsilon)^2 \sin\theta \, d\theta \, d\varphi \\
&= -4\pi(R+\epsilon)^2 \bar{N}_{(A)r}\big|_{r^*=1} \\
&= -4\pi(R+\epsilon)^2 J_{(A)r}^{\star(e)}\big|_{r^*=1} \\
&= 4\pi(R+\epsilon) A_{(A)}^\star \Psi c_{(A)0} \frac{dc^*}{dr^*}\bigg|_{r^*=1}
\end{aligned}$$

## [10.3] LOCAL VOLUME AVERAGING

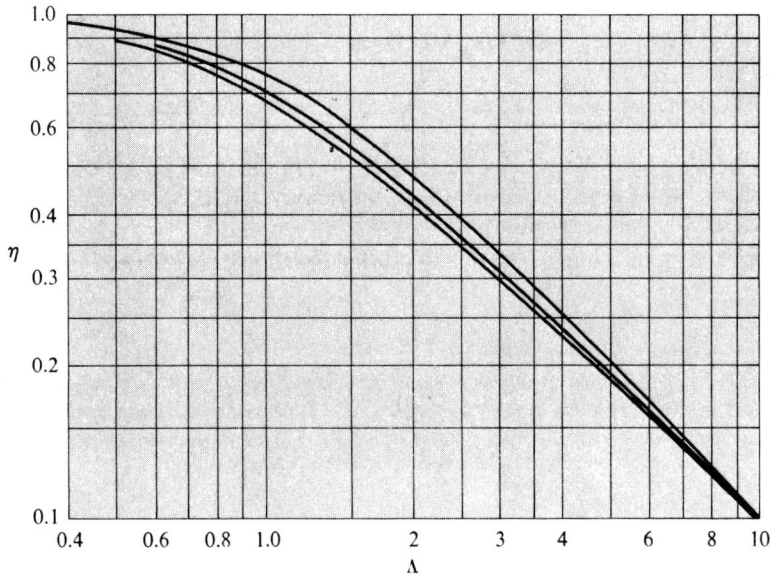

**Fig. 10.3.9-1** Effectiveness factors for porous solid catalysts: top curve, flat plates (sealed edges); middle curve, cylinders (sealed ends); bottom curve, spheres. (From Aris [1] and Bird, Stewart, and Lightfoot [2, fig. 17.6-3].)

$$= 4\pi(R + \epsilon)A_{(A)}^{\star}\Psi c_{(A)0}[3\Lambda \coth(3\Lambda) - 1] \tag{9-22}$$

In the first line, we have taken advantage of our discussion of integrals of volume-averaged variables in Sec. 4.3.8. If all the catalytic surface were exposed to fresh fluid, then the molar rate of consumption of species $A$ would be

$$\mathscr{W}_{(A)0} = \tfrac{4}{3}\pi R^3 a k_1'' c_{(A)0} \tag{9-23}$$

The *effectiveness factor* $\eta$ is defined as

$$\eta \equiv \frac{\mathscr{W}_{(A)}}{\mathscr{W}_{(A)0}} \tag{9-24}$$

From Eqs. (9-22) and (9-23), it is apparent that the effectiveness factor for spherical catalyst particles is

$$\eta = \left(\frac{R+\epsilon}{R}\right)^3 \frac{1}{3\Lambda^2}[3\Lambda \coth(3\Lambda) - 1] \tag{9-25}$$

or

$$\eta = \frac{1}{3\Lambda^2}[3\Lambda \coth(3\Lambda) - 1] \tag{9-26}$$

since we are generally willing to assume

$$\frac{R+\epsilon}{R} \doteq 1 \tag{9-27}$$

Figure 10.3.9-1 compares Eq. (9-26) for spheres with the analogous expressions for flat plates (Exercise 10.3.9-1) and cylinders (Exercise 10.3.9-2). From a practical point of view, we are fortunate that the effectiveness factor is nearly independent of the gross particle shape.

## REFERENCES

1. Aris, R.: *Chem. Eng. Sci.*, **6**:262 (1957).
2. Bird, R. B., W. E. Stewart, and E. N. Lightfoot: "Transport Phenomena," 7th printing, Wiley, New York, 1960.
3. Irving, J., and N. Mullineux: "Mathematics in Physics and Engineering," Academic, New York, 1959.
4. Wehner, J. F., and R. H. Wilhelm: *Chem. Eng. Sci.*, **6**:89 (1956).

## EXERCISES

**10.3.9-1** *The effectiveness factor for a flat plate* Repeat the discussion in the text for a first-order catalytic reaction $A \to B$ taking place in a flat plate (with sealed edges) of thickness $2b$. Conclude that the effectiveness factor is

$$\eta = \frac{1}{\Lambda} \tanh \Lambda$$

where

$$\Lambda^2 = \frac{k_1'' a (b + \epsilon)^2}{A_{(A)}^{\star} \Psi}$$

**10.3.9-2** *The effectiveness factor for cylinders* Repeat the analysis in the text for a first-order catalytic reaction $A \to B$ that takes place in a cylindrical catalyst particle (with sealed ends). Determine that the effectiveness factor is given by

$$\eta = \frac{1}{\Lambda} \frac{I_1(2\Lambda)}{I_0(2\Lambda)}$$

where

$$\Lambda^2 = \frac{k_1'' a (R + \epsilon)^2}{4 A_{(A)}^{\star} \Psi}$$

By $I_n(x)$, we mean the modified Bessel function of the first kind [3, p. 143].

**10.3.9-3** *More on the effectiveness factor for spheres* Again consider the problem described in the text, but this time assume the reaction is zero order:

$$\text{For } c_{(A)} > 0: \quad \frac{r_{(A)}''}{M_{(A)}} = -k_0'' a$$

What is the effectiveness factor?

**10.3.9-4** *A first-order catalytic reactor* A catalytic reaction $A \to B$ is carried out by passing a liquid through a tubular reactor of length $L$ that is packed with catalyst pellets. We wish to determine the volume-averaged mass density of species $A$ as a function of position in the reactor assuming that species $A$ is consumed by a first-order chemical reaction

$$r_{(A)}'' = -k_1'' a \langle \rho_{(A)} \rangle$$

and assuming that the mass density of species $A$ has a uniform value $\rho_{(A)0}$ very far upstream from the entrance to the reactor. Neglect any effects attributable to the development of the velocity profile at the entrance to the reactor.

(a) Wehner and Wilhelm [4] suggest that a tubular reactor should be analyzed, together with its connecting upstream and downstream sections, as illustrated in Fig. 10.3.9-2. Conclude that for the open tube upstream from the reactor:

$$\frac{\partial \rho_{(A)}^*}{\partial z^*} = \frac{1}{N_{\text{Pe}\,U}} \frac{\partial^2 \rho_{(A)}^*}{\partial z^{*2}} \quad (1)$$

for the reactor itself:

$$\frac{\partial \langle \rho_{(A)}^* \rangle}{\partial z^*} = \frac{1}{\bar{N}_{\text{Pe}}} \frac{\partial^2 \langle \rho_{(A)}^* \rangle}{\partial z^{*2}} - \frac{\bar{N}_{\text{Da}}}{\bar{N}_{\text{Pe}}} \langle \rho_{(A)}^* \rangle \quad (2)$$

and for the open tube downstream from the reactor:

$$\frac{\partial \rho_{(A)}^*}{\partial z^*} = \frac{1}{N_{\text{Pe}\,D}} \frac{\partial^2 \rho_{(A)}^*}{\partial z^{*2}} \quad (3)$$

Here

$$N_{\text{Pe}\,U} \equiv \frac{L\bar{v}_z}{\mathscr{D}_{(AB)U}} \qquad \bar{N}_{\text{Pe}} \equiv \frac{\Psi^{-1} L \bar{v}_z}{A_{(A)} + B_{(A)} \bar{v}_z}$$

$$N_{\text{Pe}\,D} \equiv \frac{L\bar{v}_z}{\mathscr{D}_{(AB)D}} \qquad \bar{N}_{\text{Da}} \equiv \frac{\Psi^{-1} k_1'' a L^2}{A_{(A)} + B_{(A)} \bar{v}_z}$$

$$\rho_{(A)}^* \equiv \frac{\rho_{(A)}}{\rho_{(A)0}} \qquad z^* \equiv \frac{z}{L}$$

What assumptions have been made in the upstream and downstream sections?

(b) What are the boundary conditions that must be satisfied by this system of equations at the entrance to the reactor, at the exit from the reactor, very far upstream, and very far downstream?

(c) Solve Eqs. (1) to (3) above individually and evaluate the six constants of integration to find the following concentration distribution [4; see also Exercise 10.2.2-1].

For $z^* \leq 0$: $\quad \dfrac{1 - \rho_{(A)}^*}{1 - \rho_{(A)}^*(0)} = \exp(N_{\text{Pe}\,U} z^*)$

where

$$\rho_{(A)}^*(0) \equiv g_0 \left[ (1 + b) \exp\left(\frac{b\bar{N}_{\text{Pe}}}{2}\right) - (1 - b) \exp\left(-\frac{b\bar{N}_{\text{Pe}}}{2}\right) \right]$$

$$g_0 \equiv 2 \left[ (1 + b)^2 \exp\left(\frac{b\bar{N}_{\text{Pe}}}{2}\right) - (1 - b)^2 \exp\left(-\frac{b\bar{N}_{\text{Pe}}}{2}\right) \right]^{-1}$$

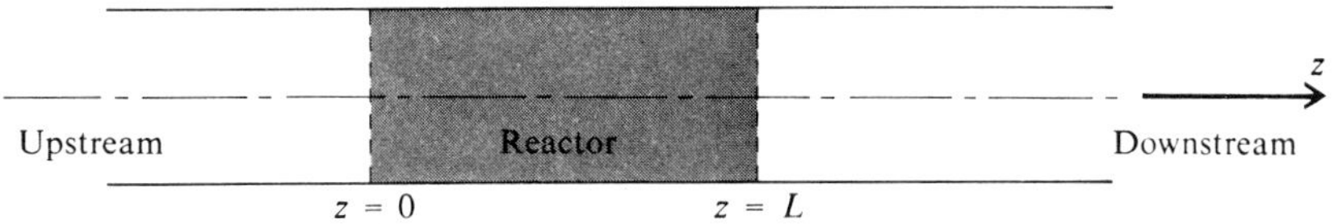

Fig. 10.3.9-2 Tubular reactor with connecting upstream and downstream sections.

and

$$b = \left[1 + \frac{4\bar{N}_{Da}}{(\bar{N}_{Pe})^2}\right]^{\frac{1}{2}}$$

For $0 \leq z^* \leq 1$: $\langle p^*_{(A)} \rangle = g_0 \exp\left(\frac{\bar{N}_{Pe}z^*}{2}\right)\left\{(1+b)\exp\left(\frac{b\bar{N}_{Pe}}{2}[1-z^*]\right)\right.$

$$\left. - (1-b)\exp\left(\frac{b\bar{N}_{Pe}}{2}[z^*-1]\right)\right\}$$

For $z^* \geq 1$: $p^*_{(A)} = 2bg_0 \exp\frac{\bar{N}_{Pe}}{2}$

## 10.4 INTEGRAL BALANCES

**10.4.1 Still more on integral balances** In the sections that follow we have two purposes. First, we have one integral balance left to discuss: the integral mass balance for an individual species in a multicomponent mixture. Second, and just as important, we must extend our previous discussions of integral balances to multicomponent systems.

By *multicomponent systems* I mean systems in which concentration is a function of time or position. If a system consists of more than one species, but concentration is independent of both time and position, the previously developed integral balances apply without change.

The sections that follow are closely related to Secs. 4.4.1 to 4.4.15 and 7.4.1 to 7.4.9. It might be helpful for the reader to review these sections or at least to reread Sec. 4.4.1, which discusses the place of integral balances in engineering, and Sec. 7.4.1, which in part reviews the notation to be used.

There is one point concerning the notation about which the reader should exercise a degree of caution. The entrance and exit surfaces $S_{(ent\ ex)}$ are to be interpreted in the broadest possible sense to include both:

1. Surfaces that are unobstructed for flow and across which the individual species are carried primarily by convection.
2. Phase interfaces (liquid-liquid, liquid-solid, etc.) across which the individual species are carried primarily by diffusion.

**10.4.2 The integral mass balance for species** $A$ Just as in Sec. 4.4.2 we developed a mass balance for a system consisting of a single species, we are now in the position to develop a mass balance for each individual species present in a multicomponent system.

Let us take the same approach that we have used in developing integral balances for single-component systems. In the generalized transport theorem of Sec. 1.3.2, let us identify

$$\Psi = \rho_{(A)} \qquad (2\text{-}1)$$

the mass of species $A$ per unit volume, to obtain

$$\frac{d}{dt}\int_{V_{(s)}} \rho_{(A)}\, dV = \int_{V_{(s)}} \frac{\partial \rho_{(A)}}{\partial t}\, dV + \int_{S_{(s)}} \rho_{(A)}(\mathbf{v}_{(s)} \cdot \mathbf{n})\, dS \qquad (2\text{-}2)$$

In general, the value of the first integral on the right will not be immediately clear in any given problem. This difficulty is essentially the same as the ones we found in Secs. 4.4.2, 4.4.5, and 4.4.9.

The differential equation of continuity for species $A$ in the form of Eq. (A) of Table 8.2.2-5 may be integrated over the system to obtain

$$\int_{V_{(s)}} \left(\frac{\partial \rho_{(A)}}{\partial t} + \operatorname{div} \mathbf{n}_{(A)} - r_{(A)}\right) dV = 0 \qquad (2\text{-}3)$$

The first integral on the left is exactly the one whose significance was not clear in Eq. (2-2). Green's transformation may be used to express the second term as

$$\int_{V_{(s)}} \operatorname{div} \mathbf{n}_{(A)}\, dV = \int_{S_{(s)}} \rho_{(A)} \mathbf{v}_{(A)} \cdot \mathbf{n}\, dS \qquad (2\text{-}4)$$

Equations (2-3) and (2-4) allow us to express Eq. (2-2) as

$$\frac{d}{dt}\int_{V_{(s)}} \rho_{(A)}\, dV = \int_{S_{(s)}} \rho_{(A)}(\mathbf{v}_{(A)} - \mathbf{v}_{(s)}) \cdot (-\mathbf{n})\, dS + \int_{V_{(s)}} r_{(A)}\, dV \qquad (2\text{-}5)$$

or

$$\frac{d}{dt}\int_{V_{(s)}} \rho_{(A)}\, dV = \int_{S_{(\text{ent ex})}} \rho_{(A)}(\mathbf{v}_{(A)} - \mathbf{v}_{(s)}) \cdot (-\mathbf{n})\, dS + \int_{V_{(s)}} r_{(A)}\, dV \qquad (2\text{-}6)$$

Here and in all the sections to follow we interpret $S_{(\text{ent ex})}$ in the broadest possible sense to include both:

1. Surfaces that are unobstructed for flow and across which $A$ is carried primarily by convection.
2. Phase interfaces (liquid-liquid, liquid-solid, etc.) across which species $A$ is carried primarily by diffusion. We shall refer to these last as the *diffusion surfaces* of the system, $S_{(\text{diff})}$.

Equation (2-6) is a *general* form of the *integral mass balance for species A* appropriate to single-phase systems.

We will generally find it convenient to account for the effects of diffusion explicitly and write Eq. (2-6) as

$$\frac{d}{dt}\int_{V_{(s)}} \rho_{(A)}\, dV = \int_{S_{(\text{ent ex})}} \rho_{(A)}(\mathbf{v} - \mathbf{v}_{(s)}) \cdot (-\mathbf{n})\, dS$$

$$+ \int_{S_{(\text{ent ex})}} \mathbf{j}_{(A)} \cdot (-\mathbf{n})\, dS + \int_{(s)} r_{(A)}\, dV \qquad (2\text{-}7)$$

Since we will almost always be willing to neglect diffusion with respect to convection on those portions of $S_{(ent\ ex)}$ unobstructed for flow:

$$\int_{S_{(ent\ ex)}} \mathbf{j}_{(A)} \cdot (-\mathbf{n})\, dS \doteq \int_{S_{(diff)}} \mathbf{j}_{(A)} \cdot (-\mathbf{n})\, dS \equiv \mathcal{J}_{(A)} \qquad (2\text{-}8)$$

we can express Eq. (2-7) as

$$\frac{d}{dt}\int_{V_{(s)}} \rho_{(A)}\, dV = \int_{S_{(ent\ ex)}} \rho_{(A)}(\mathbf{v} - \mathbf{v}_{(s)}) \cdot (-\mathbf{n})\, dS + \mathcal{J}_{(A)} + \int_{V_{(A)}} r_{(A)}\, dV \qquad (2\text{-}9)$$

In words, Eq. (2-9) says that the time rate of change of the mass of species $A$ in the system is equal to the net rate at which the mass of species $A$ is brought into the system by convection, the net rate at which the mass of species $A$ diffuses into the system (relative to the mass-averaged velocity), and the rate at which the mass of species $A$ is produced in the system by chemical reactions. Notice that $\mathcal{J}_{(A)}$ also includes the rate of production of species $A$ at the catalytic surfaces within or bounding the system. We will refer to Eq. (2-9) as the *integral mass balance for species $A$* appropriate to a single-phase system.

As we pointed out in Sec. 4.4.2, we are more commonly concerned with multiphase systems. Using the approach and notation of Sec. 4.4.2 and assuming only that we may neglect diffusion with respect to convection on those portions of $S_{(ent\ ex)}$ unobstructed for flow, we find that the *integral mass balance for species $A$* appropriate to a multiphase system is

$$\frac{d}{dt}\int_{V_{(s)}} \rho_{(A)}\, dV = \int_{(ent\ ex)} \rho_{(A)}(\mathbf{v} - \mathbf{v}_{(s)}) \cdot (-\mathbf{n})\, dS$$
$$+ \mathcal{J}_{(A)} + \int_{V_{(s)}} r_{(A)}\, dV + \int_{S_{(sing)}} [\rho_{(A)}(\mathbf{v}_{(A)} \cdot \boldsymbol{\xi} - u_{(\xi)})]\, dS \qquad (2\text{-}10)$$

We generally will be willing to say that the jump mass balance of Exercise 8.2.1-3 applies, in which case Eq. (2-10) reduces to Eq. (2-9). The integral mass balance (2-9) consequently applies equally well to single-phase and multiphase systems so long as the jump mass balance of Exercise 8.2.1-3 is valid for all phase interfaces involved.

There are three common types of problems in which the integral mass balance for species $A$ is applied: the rate of diffusion $\mathcal{J}_{(A)}$ may be neglected, it may be the unknown to be determined, or it may be known from previous experimental data. In this last case, one employs an empirical correlation of data for $\mathcal{J}_{(A)}$. In Sec. 10.4.4 we discuss the form that these empirical correlations should take.

### 10.4.3 The integral mass balance for species $A$ appropriate to turbulent flows

We pointed out in Sec. 4.4.3 that some of the most important and interesting applications of the integral balances are to systems, portions of which are in turbulent

## [10.4] INTEGRAL BALANCES

flow. Yet the integral mass balance just arrived at in Sec. 10.4.2 is not suitable for a problem in which turbulence is important. When turbulence is important, we should be talking in terms of time-averaged quantities.

There are two approaches we can take in developing a mass balance which is more appropriate to systems in turbulent flow. The first approach is limited to single-phase or multiphase systems that do not involve fluid-fluid phase interfaces. Starting with the time-averaged differential equation of continuity of Sec. 10.1.2,

$$\frac{\partial \bar{\rho}_{(A)}}{\partial t} + \text{div } \bar{\mathbf{n}}_{(A)} = \bar{r}_{(A)} \tag{3-1}$$

we can repeat the derivation of Sec. 10.4.2 to find

$$\frac{d}{dt} \int_{V_{(s)}} \bar{\rho}_{(A)}\, dV = \int_{S_{(\text{ent ex})}} (\overline{\rho_{(A)}\mathbf{v}} - \bar{\rho}_{(A)}\mathbf{v}_{(s)}) \cdot (-\mathbf{n})\, dS$$

$$+ \mathscr{J}_{(A)} + \int_{V_{(s)}} \bar{r}_{(A)}\, dV + \int_{S_{(\text{sing})}} [\overline{\rho_{(A)}\mathbf{v}_{(A)}} \cdot \boldsymbol{\xi} - \bar{\rho}_{(A)} u_{(\xi)}]\, dS \tag{3-2}$$

where

$$\mathscr{J}_{(A)} \equiv \int_{S_{(\text{diff})}} \bar{\mathbf{j}}_{(A)} \cdot (-\mathbf{n})\, dS \tag{3-3}$$

The time-averaged jump mass balance of Exercise 10.4.3-1 simplifies this to

$$\frac{d}{dt} \int_{V_{(s)}} \bar{\rho}_{(A)}\, dV = \int_{S_{(\text{ent ex})}} (\overline{\rho_{(A)}\mathbf{v}} - \bar{\rho}_{(A)}\mathbf{v}_{(s)}) \cdot (-\mathbf{n})\, dS + \mathscr{J}_{(A)} + \int_{V_{(s)}} \bar{r}_{(A)}\, dV \tag{3-4}$$

Note that in arriving at Eqs. (3-2) and (3-4), we have again neglected diffusion of the species $A$ with respect to convection on those portions of $S_{(\text{ent ex})}$ unobstructed for flow:

$$\int_{S_{(\text{ent ex})}} \bar{\mathbf{j}}_{(A)} \cdot (-\mathbf{n})\, dS \doteq \mathscr{J}_{(A)} \tag{3-5}$$

For single-phase or multiphase systems that include one or more fluid-fluid interfaces, I recommend time averaging the integral mass balance of Sec. 10.4.2:

$$\frac{d}{dt} \int_{V_{(s)}} \rho_{(A)}\, dV = \int_{S_{(\text{ent ex})}} \rho_{(A)}(\mathbf{v} - \mathbf{v}_{(s)}) \cdot (-\mathbf{n})\, dS + \mathscr{J}_{(A)} + \int_{V_{(s)}} r_{(A)}\, dV \tag{3-6}$$

We then find

$$\frac{d}{dt} \overline{\int_{V_{(s)}} \rho_{(A)}\, dV} = \overline{\int_{S_{(\text{ent ex})}} \rho_{(A)}(\mathbf{v} - \mathbf{v}_{(s)}) \cdot (-\mathbf{n})\, dS} + \overline{\mathscr{J}}_{(A)} + \overline{\int_{V_{(s)}} r_{(A)}\, dV} \tag{3-7}$$

where

$$\overline{\mathscr{J}}_{(A)} \equiv \overline{\int_{S_{(\text{diff})}} \mathbf{j}_{(A)} \cdot (-\mathbf{n})\, dS} \tag{3-8}$$

If we restrict ourselves to single-phase or multiphase systems that do not involve fluid-fluid phase interfaces, Eqs. (3-7) to (3-9) of Sec. 4.4.3 can be used to show that Eqs. (3-7) and (3-8) reduce to Eqs. (3-4) and (3-3), respectively.

## EXERCISE

**10.4.3-1** *Time-averaged jump mass balance for species A*  Determine that the time-averaged jump mass balance for species $A$ applicable to solid-fluid phase interfaces that bound turbulent flows is identical to the balance found in Exercise 8.2.1-3:

$$[\overline{\rho_{(A)}\mathbf{v}_{(A)}} \cdot \boldsymbol{\xi} - \bar{\rho}_{(A)}u_{(\xi)}] = [\rho_{(A)}(\mathbf{v}_{(A)} \cdot \boldsymbol{\xi} - u_{(\xi)})] = 0$$

## 10.4.4 Empirical correlations for $\mathscr{J}_{(A)}$

Empirical data correlations for $\mathscr{J}_{(A)}$ ($\overline{\mathscr{J}}_{(A)}$ when dealing with turbulent flows) are prepared in much the same way as our empirical correlations for $\mathscr{Q}$, discussed in Sec. 7.4.4. There are three principal thoughts to be kept in mind.

1. The rate of diffusion of species $A$ from the permeable or catalytic surfaces of the system is frame indifferent:

$$\begin{aligned}\mathscr{J}^*_{(A)} &\equiv \int_{S_{(\text{diff})}} \mathbf{j}^*_{(A)} \cdot (-\mathbf{n}^*)\, dS \\ &= \int_{S_{(\text{diff})}} \mathbf{j}_{(A)} \cdot (-\mathbf{n})\, dS \\ &= \mathscr{J}_{(A)} \end{aligned} \qquad (4\text{-}1)$$

2. We assume that the principle of material frame indifference, introduced in Sec. 2.3.1, applies to any empirical correlation developed for $\mathscr{J}_{(A)}$.
3. The form of any expression for $\mathscr{J}_{(A)}$ must satisfy the Buckingham-Pi theorem [1].

We illustrate the approach in terms of a specific situation.

### EXAMPLE: FORCED CONVECTION IN PLANE FLOW OF A BINARY FLUID PAST A CYLINDRICAL BODY

An infinitely long cylindrical body is submerged in a large mass of a binary newtonian fluid. We assume that the surface of the body is in equilibrium with the fluid at the surface and that the mass fraction of species $A$ at the surface is a constant $\omega_{(A)0}$. Outside the immediate neighborhood of the body, the mass fraction of species $A$ has a nearly uniform value $\omega_{(A)\infty}$. In a frame of reference that is fixed with respect to the earth, the cylindrical body translates without rotation at a constant velocity $\mathbf{v}_0$; the fluid at a very large distance from the body moves with a uniform velocity $\mathbf{v}_\infty$. The vectors $\mathbf{v}_0$ and $\mathbf{v}_\infty$ are normal to the axis of the cylinder, so that we may expect that the fluid moves in a plane flow. One unit vector $\boldsymbol{\alpha}$ is sufficient to describe the orientation of the cylinder with respect to $\mathbf{v}_0$ and $\mathbf{v}_\infty$.

[10.4] INTEGRAL BALANCES  589

It seems reasonable to assume that $\mathscr{J}_{(A)}$ should be a function of

$$\Delta\omega_{(A)} \equiv \omega_{(A)0} - \omega_{(A)\infty} \tag{4-2}$$

a characteristic fluid density $\rho$, a characteristic fluid viscosity $\mu$, a characteristic diffusion coefficient $\mathscr{D}_{(AB)}$, a length $L$ that is characteristic of the cylinder's cross section, $\mathbf{v}_\infty - \mathbf{v}_0$, and $\boldsymbol{\alpha}$:

$$\mathscr{J}_{(A)} = f(\rho, \mu, \mathscr{D}_{(AB)}, L, \mathbf{v}_\infty - \mathbf{v}_0, \boldsymbol{\alpha}, \Delta\omega_{(A)}) \tag{4-3}$$

We recognize that density, viscosity, and the diffusion coefficient may be dependent upon position as the result of their functional dependence upon composition. In referring to $\rho$, $\mu$, and $\mathscr{D}_{(AB)}$ as "characteristic" of the fluid, we mean that they are to be evaluated at some average or representative composition. Dependence upon

$$\Delta\omega_{(B)} \equiv \omega_{(B)0} - \omega_{(B)\infty} \tag{4-4}$$

is not included, since

$$\Delta\omega_{(B)} = -\Delta\omega_{(A)} \tag{4-5}$$

The same argument that we used in discussing Example 1 of Sec. 7.4.4 may be repeated here to show that the principle of material frame indifference and the Buckingham-Pi theorem require that this be of a form[1]

$$N_{\text{Nu}(A)} = N_{\text{Nu}(A)}\left(N_{\text{Re}}, N_{\text{Sc}}, \Delta\omega_{(A)}, \frac{\mathbf{v}_\infty - \mathbf{v}_0}{|\mathbf{v}_\infty - \mathbf{v}_0|} \cdot \boldsymbol{\alpha}\right) \tag{4-6}$$

where the Nusselt, Reynolds, and Schmidt numbers are defined as

$$N_{\text{Nu}(A)} \equiv \frac{\mathscr{J}_{(A)}}{\rho\mathscr{D}_{(AB)}L\Delta\omega_{(A)}} \qquad N_{\text{Re}} \equiv \frac{L\rho|\mathbf{v}_\infty - \mathbf{v}_0|}{\mu} \qquad N_{\text{Sc}} \equiv \frac{\mu}{\rho\mathscr{D}_{(AB)}} \tag{4-7}$$

We follow Bird, Stewart, and Lightfoot [2, p. 640] in defining a mass-transfer coefficient $k_{(A)\omega}$ as[2]

$$k_{(A)\omega} \equiv \frac{\mathscr{J}_{(A)}}{A\Delta\omega_{(A)}} \tag{4-8}$$

---

[1] We have anticipated our definition of the mass-transfer coefficient in Eq. (4-8) by our definition of the Nusselt number. The Buckingham-Pi theorem suggests $\mathscr{J}_{(A)}/\rho\mathscr{D}_{(AB)}L$ as a dimensionless group.

[2] Equation (4-8) can easily be rewritten as

$$k_{(A)\omega} = \frac{\mathscr{N}_{(A)} - \omega_{(A)0}(\mathscr{N}_{(A)} + \mathscr{N}_{(B)})}{A\,\Delta\omega_{(A)}} \tag{4-8a}$$

where

$$\mathscr{N}_{(A)} \equiv \int_{S(\text{diff})} \mathbf{n}_{(A)} \cdot (-\mathbf{n})\, dS \tag{4-8b}$$

and $\omega_{(A)0}$ is the mass fraction of species $A$ at $S_{(\text{diff})}$, assumed to be a constant.

The mass-transfer coefficient defined by Eq. (4-8) differs from that widely used in the literature

*footnote continued overleaf*

where $A$ is proportional to $L^2$ and denotes the area available for mass transfer. The Nusselt number for species $A$ is usually expressed in terms of this mass-transfer coefficient:

$$N_{\text{Nu}(A)} = \frac{k_{(A)\omega} L}{\rho \mathscr{D}_{(AB)}} \tag{4-9}$$

One then computes the rate of diffusion of species $A$ across the permeable surfaces of the system as

$$\mathscr{J}_{(A)} = k_{(A)\omega} A \Delta \omega_{(A)} \tag{4-10}$$

estimating the mass-transfer coefficient $k_{(A)\omega}$ from an empirical data correlation of the form of Eq. (4-6).

Most empirical correlations for the Nusselt number are not as general as Eq. (4-6) suggests. Most studies are for a single orientation of a body (or a set of bodies such as an array of particles) with respect to the fluid stream. Further, for sufficiently small rates of mass transfer, natural convection is not important and $\Delta \omega_{(A)}$ is so small that its influence can be neglected. Under these conditions, Eq. (4-6) assumes a simpler form [2, p. 647]:

$$N_{\text{Nu}(A)} = N_{\text{Nu}(A)}(N_{\text{Re}}, N_{\text{Sc}}) \tag{4-11}$$

When natural convection can be neglected, there is a strict analogy between energy and mass transfer (see Sec. 9.4.2). Since there are more data for energy transfer available in the literature, it is often convenient to identify Eq. (4-11) with the analogous relation in energy transfer.

When natural convection is important (larger rates of mass transfer), the dependence of $N_{\text{Nu}(A)}$ upon $\Delta \omega_{(A)}$ in Eq. (4-6) cannot be neglected. Because of a

---

prior to 1960. The traditional definition is suggested by writing the integral mass balance of Sec. 10.4.2 as

$$\frac{d}{dt} \int_{V_{(A)}} \rho_{(A)} \, dV = \int_{S_{(\text{ent ex})} - S_{(\text{diff})}} \rho_{(A)} (\mathbf{v} - \mathbf{v}_{(s)}) \cdot (-\mathbf{n}) \, dS + \mathscr{W}_{(A)} + \int_{V_{(s)}} r_{(A)} \, dV \qquad 4\text{-}8(\text{c})$$

where

$$\mathscr{W}_{(A)} \equiv \int_{S_{(\text{diff})}} \rho_{(A)} (\mathbf{v}_{(A)} - \mathbf{v}_{(s)}) \cdot (-\mathbf{n}) \, dS \tag{4-8d}$$

Equation (4-8c) again incorporates the assumption that diffusion can be neglected with respect to convection on those portions of $S_{(\text{ent ex})}$ unobstructed for flow: $S_{(\text{ent ex})} - S_{(\text{diff})}$. The traditional mass-transfer coefficient $K_{(A)\rho}$ is defined as

$$K_{(A)\rho} \equiv \frac{\mathscr{W}_{(A)}}{A \, \Delta \rho_{(A)}}$$

The advantage of working in terms of $\mathscr{W}_{(A)}$ and the traditional mass-transfer coefficient $K_{(A)\rho}$ is that the contribution of convection on $S_{(\text{diff})}$ is automatically accounted for. The disadvantage is that $K_{(A)\rho}$ shows a more complicated dependence upon concentration level and mass-transfer rates than does $k_{(A)\omega}$ [2, p. 640]. In our opinion, this loss outweighs the possible gain in computational ease.

shortage of experimental data, the recommended approach at the present time is to derive a simple correction to be applied to empirical correlations of the form of Eq. (4-11) that are restricted to small rates of mass transfer. Bird, Stewart, and Lightfoot [2, p. 658] give an excellent summary of the three most popular corrections: film theory, penetration theory, and an application of two-dimensional boundary-layer flow along a flat plate.

## REFERENCES

1. Brand, Louis: *Arch. Rational Mech. Analysis*, **1**:35 (1957).
2. Bird, R. B., W. E. Stewart, and E. N. Lightfoot: "Transport Phenomena," 7th printing, Wiley, New York, 1960.

**10.4.5 The integral mass balance for species $A$: An example** [1, p. 707] A fluid stream containing a waste material $A$ at concentration $\rho_{(A)0}$ is to be discharged into a river at a constant volume rate of flow $Q$. Material $A$ is unstable and decomposes at a rate proportional to its mass density:

$$-r_{(A)} = k_1 \rho_{(A)} \tag{5-1}$$

In order to reduce pollution, the stream is to pass through a holding tank of volume $V$ before it is discharged into the river. At time $t = 0$, the fluid begins to flow into the empty tank, which may be considered to have a perfect stirrer. No liquid leaves the tank until it is filled. We wish to develop an expression for the mass density of $A$ in the tank and in the effluent from the tank as a function of time.

This problem should be considered in two parts. First, we must determine the mass density of $A$ in the tank as a function of time during the filling process. The mass density of $A$ in the tank at the moment the tank becomes filled forms the boundary condition for the second portion of the problem: the mass density of $A$ in the tank and in the effluent stream as a function of time.

Figure 10.4.5-1 schematically describes the situation during the filling process.

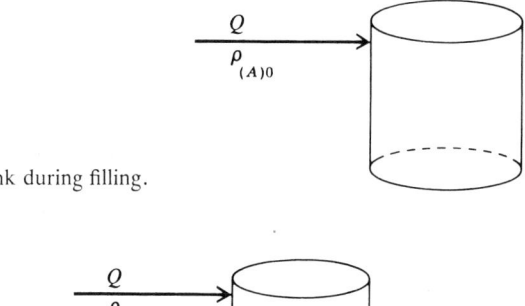

**Fig. 10.4.5-1** Waste tank during filling.

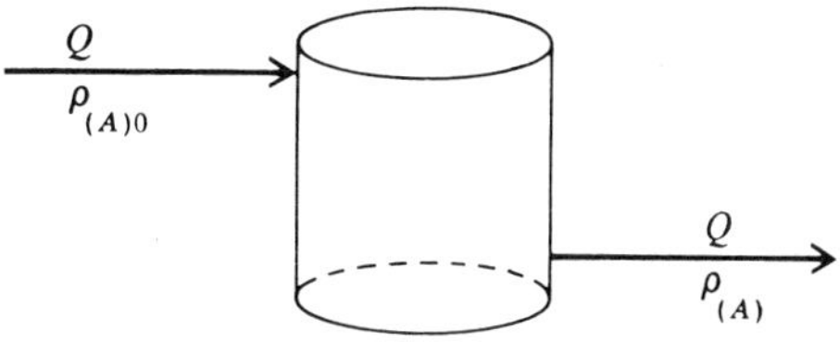

**Fig. 10.4.5-2** Waste tank after filling.

Let us choose our system to be the fluid in the tank. For this system, which has one entrance and no exit, the integral mass balance of Sec. 10.4.2 requires

$$\frac{d}{dt}\int_{V_{(s)}} \rho_{(A)} \, dV = \int_{S_{(\text{ent})}} \rho_{(A)0}\mathbf{v} \cdot (-\mathbf{n}) \, dS - k_1 \int_{V_{(s)}} \rho_{(A)} \, dV \tag{5-2}$$

or

$$\frac{d\mathscr{M}_{(A)}}{dt} = \rho_{(A)0} Q - k_1 \mathscr{M}_{(A)} \tag{5-3}$$

where we denote the mass of species $A$ in the system by $\mathscr{M}_{(A)}$:

$$\mathscr{M}_{(A)} \equiv \int_{V_{(s)}} \rho_{(A)} \, dV \tag{5-4}$$

Equation (5-3) is easily integrated to find $\mathscr{M}_{(A)}$ as a function of time:

$$\mathscr{M}_{(A)} = \frac{\rho_{(A)0} Q}{k_1} [1 - \exp(-k_1 t)] \tag{5-5}$$

This means that, when the tank is filled,

$$\text{At } t = \frac{V}{Q}: \quad \rho_{(A)} = \rho_{(A)f} \equiv \frac{\rho_{(A)0} Q}{V k_1}\left[1 - \exp\left(-\frac{k_1 V}{Q}\right)\right] \tag{5-6}$$

Once the waste tank is filled, the discharge line is opened as shown in Fig. 10.4.5-2. Our system is still the fluid in the tank, but now we have both an entrance and an exit. The integral mass balance for species $A$ requires

$$\frac{d\mathscr{M}_{(A)}}{dt} = \rho_{(A)0} Q - \frac{\mathscr{M}_{(A)}}{V} Q - k_1 \mathscr{M}_{(A)} \tag{5-7}$$

This can be integrated using Eq. (5-6) as the boundary condition to find

$$\frac{\rho_{(A)} - \rho_{(A)\infty}}{\rho_{(A)f} - \rho_{(A)\infty}} = \exp\left(-\left[\frac{Q}{V} + k_1\right]\left[t - \frac{V}{Q}\right]\right) \tag{5-8}$$

Here $\rho_{(A)\infty}$ is the steady-state mass density of species $A$ in the waste tank:

$$\text{As } t \to \infty: \quad \rho_{(A)} \to \rho_{(A)\infty} \equiv \frac{\rho_{(A)0} Q}{Q + k_1 V} \tag{5-9}$$

## REFERENCE

1. Bird, R. B., W. E. Stewart, and E. N. Lightfoot: "Transport Phenomena," 7th printing, Wiley, New York, 1960.

## [10.4] INTEGRAL BALANCES

### EXERCISES

**10.4.5-1** Carry out the required integrations and rearrangements to obtain Eqs. (5-5) and (5-8).

**10.4.5-2** *Irreversible first-order reaction in a continuous reactor* [1, p. 707] A solution of species $A$ at mass density $\rho_{(A)0}$ initially fills a well-stirred reactor of volume $V$. At time $t = 0$, an identical solution of $A$ is introduced at a constant volume rate of flow $Q$. At the same time, a constant stream of dissolved catalyst is introduced, causing $A$ to disappear according to the expression

$$-r_{(A)} = k_1 \rho_{(A)}$$

where the constant $k_1$ may be assumed to be independent of composition and time. Determine that the mass density of species $A$ in the reactor at any time is given by

$$\frac{\rho_{(A)} - \rho_{(A)\infty}}{\rho_{(A)\infty}} = \frac{k_1 V}{Q} \exp\left[-\left(\frac{Q}{V} + k_1\right)t\right]$$

where

As $t \to \infty$: $\quad \rho_{(A)} \to \rho_{(A)\infty} \equiv \dfrac{\rho_{(A)0} Q}{Q + k_1 V}$

**10.4.5-3** *Irreversible second-order reaction in a continuous reactor* [1, p. 708] Repeat Exercise 10.4.5-2 assuming now that species $A$ disappears according to the expression

$$-r_{(A)} = k_2 \rho_{(A)}^2$$

*Answer*

$$\rho_{(A)} = -\frac{B}{2k_2} + \left\{\frac{k_2 V}{BV - Q} + \left(\frac{2k_2}{B + 2k_2 \rho_{(A)0}} - \frac{k_2 V}{BV - Q}\right) \exp\left(-\left[B - \frac{Q}{V}\right]t\right)\right\}^{-1}$$

where

$$B \equiv \frac{Q}{V}\left(1 + \sqrt{1 + \frac{4k_2 V \rho_{(A)0}}{Q}}\right)$$

*Hint:* The differential equation to be solved can be put into the form of a Bernoulli differential equation with the change of variable

$$u \equiv \rho_{(A)} + \frac{Q}{2k_2 V}\left(1 + \sqrt{1 + \frac{4k_2 V \rho_{(A)0}}{Q}}\right)$$

The resulting Bernoulli equation can in turn be integrated by making another change of variable

$$v \equiv u^{-1}$$

**10.4.5-4** *Start-up of a chemical reactor* [1, pp. 701 and 708] Species $B$ is to be formed by a reversible reaction from a raw material $A$ in a chemical reactor of volume $V$ equipped with a perfect stirrer. Unfortunately, $B$ undergoes an irreversible first-order decomposition to a third species $C$. All reactions may be considered to be first order. We use the notation

$$A \underset{k_{1A}}{\overset{k_{1B}}{\rightleftarrows}} B \xrightarrow{k_{1C}} C$$

At time $t = 0$, a solution of $A$ at mass density $\rho_{(A)0}$ that is free of species $B$ is introduced into the initially empty reactor at a constant volume rate of flow $Q$.

(a) Determine that during the filling period the mass of species B in the reactor is the following function of time:

$$\mathcal{M}_{(B)} = \frac{\rho_{(A)0}Q}{k_{1C}}\left[1 + \frac{s_-}{(s_+ - s_-)}\exp(s_+t) - \frac{s_+}{(s_+ - s_-)}\exp(s_-t)\right]$$

where

$$2s_\pm = -(k_{1A} + k_{1B} + k_{1C}) \pm \sqrt{(k_{1A} + k_{1B} + k_{1C})^2 - 4k_{1B}k_{1C}}$$

*Hint:* Take Laplace transforms of the integral mass balances for species $A$ and $B$.
(b) Prove that $s_+$ and $s_-$ are always real and negative.
*Hint:* Start by showing that

$$(k_{1A} + k_{1B} + k_{1C})^2 - 4k_{1B}k_{1C} = (k_{1A} - k_{1B} + k_{1C})^2 + 4k_{1A}k_{1B}$$

**10.4.5-5 Continuous-flow stirred-tank reactors** Two successive first-order irreversible reactions $(A \rightarrow B \rightarrow C)$ are to be carried out in a series of continuous-flow stirred-tank reactors. Derive an expression from which we may find the number of reactors required to give a maximum concentration of $B$ in the product. All reactors are at the same temperature and have the same holding time.

Assuming that $k_1 = k_2$ and that the initial concentrations of $B$ and $C$ in the feed are zero, what is this number when $k_1 = k_2 = 0.1$ h$^{-1}$ and the holding time per tank is 1 h?

## 10.4.6 The integral overall mass balance

The derivation of the integral mass balance for a single-component system given in Sec. 4.4.2 may be repeated almost line for line for a multicomponent system. The only change necessary is that the equation of continuity of Sec. 1.3.3 must be replaced by the overall equation of continuity of Sec. 8.3.2.

Two forms of the *integral overall mass balance* are found, corresponding, respectively, to Eqs. (A) and (B) of Table 8.3.2-1:

$$\frac{d}{dt}\int_{V_{(s)}} \rho \, dV = \int_{S_{(\text{ent ex})}} \rho(\mathbf{v} - \mathbf{v}_{(s)}) \cdot (-\mathbf{n}) \, dS \tag{6-1}$$

and

$$\frac{d}{dt}\int_{V_{(s)}} c \, dV = \int_{S_{(\text{ent ex})}} c(\mathbf{v}^\star - \mathbf{v}_{(s)}) \cdot (-\mathbf{n}) \, dS + \int_{V_{(s)}} \sum_{A=1}^{N} \frac{r_{(A)}}{M_{(A)}} \, dV \tag{6-2}$$

The only assumption made in deriving these results is that the jump overall mass balance of Exercises 8.3.2-1 and 8.3.2-2 is assumed to apply at the phase interface.

Equation (6-1) looks deceptively like the integral mass balance found in Sec. 4.4.2. Don't forget that $S_{(\text{ent ex})}$ should be interpreted in the broadest possible sense to include both:

1. Surfaces that are unobstructed for flow across which the individual species are carried primarily by convection.

## [10.4] INTEGRAL BALANCES

2. Phase interfaces (liquid-liquid, liquid-solid, etc.) across which the individual species are carried primarily by diffusion.

The extension of this discussion to turbulent flows follows immediately along the lines of Sec. 4.4.3.

### 10.4.7 The integral overall momentum, mechanical energy, and moment of momentum balances

The integral overall momentum, mechanical energy, and moment of momentum balances are of exactly the same form as those derived for single-component systems in Secs. 4.4.5 to 4.4.15. In the derivations, it is necessary only to replace Cauchy's first law for single-component materials derived in Sec. 2.2.3 with Cauchy's first law for a multicomponent mixture discussed in Sec. 8.3.3. This means that we must interpret **v** as the mass-averaged velocity vector and **f** as the mass-averaged external force vector.

However, it is necessary to modify the further discussion of the mechanical energy balance in Sec. 7.4.6. All the results of Tables 7.4.6-1 through 7.4.6-3 are equally applicable to single-component and multicomponent systems, with the exceptions of Eq. (D) for isothermal systems and Eq. (E) for isentropic systems. Results comparable to Eqs. (D) and (E) can be prepared, but they are not presented because of their complexity.

At the risk of sounding like a worried mother, I again caution the reader that $S_{(ent\ ex)}$ is to be interpreted in the broadest possible sense to include those surfaces across which the individual species are carried primarily by diffusion.

### 10.4.8 The integral overall energy balance

The derivation of the integral energy balance given in Sec. 7.4.2 for single-component systems can be repeated here for multicomponent systems, replacing only the differential energy balance of Sec. 5.3.2 by the overall differential energy balance for a multicomponent mixture of Sec. 8.3.5. However, I do suggest that one neglect the terms attributable to the kinetic energies of the individual species measured with respect to the mass-averaged velocity of the material. I am not currently aware of any situation in which these terms are of practical importance.

Because of the form of the caloric equation of state for a multicomponent material, Eq. (2-1) of Sec. 5.1.2, not all the results of Tables 7.4.2-1 to 7.4.2-3 carry over immediately to multicomponent systems. In fact, some of the comparable results for multicomponent systems are sufficiently complex to be of marginal usefulness and are not given. For this reason, I thought it might be helpful to restate in Tables 10.4.8-1 to 10.4.8-3 those forms of the integral overall energy balance that are more likely to be useful.

Again the reader is cautioned to interpret $S_{(ent\ ex)}$ in the broadest possible sense as explained in Secs. 10.4.1 and 10.4.2.

### REFERENCE

1. Truesdell, C., and R. A. Toupin: In S. Flügge (ed.), "Handbuch der Physik," Springer-Verlag, Berlin, 1960.

**Table 10.4.8-1** General forms of the integral overall energy balance applicable to a single-phase system†

$$\frac{d}{dt}\int_{V_{(s)}} \rho\left(\hat{U} + \tfrac{1}{2}v^2 + \varphi + \frac{P_0}{\rho}\right) dV$$

$$= \int_{S_{(\text{ent ex})}} \rho(\hat{H} + \tfrac{1}{2}v^2 + \varphi)(\mathbf{v} - \mathbf{v}_{(s)}) \cdot (-\mathbf{n}) \, dS + \mathcal{Q} - \mathcal{W} + \int_{V_{(s)}} \left(\sum_{A=1}^{N} \mathbf{j}_{(A)} \cdot \mathbf{f}_{(A)} + \rho Q\right) dV$$

$$+ \int_{S_{(\text{ent ex})}} [-(P - P_0)(\mathbf{v}_{(s)} \cdot \mathbf{n}) + \mathbf{v} \cdot (\mathbf{S} \cdot \mathbf{n})] \, dS \quad \text{(A)}‡$$

$$\frac{d}{dt}\int_{V_{(s)}} \rho\left(\hat{U} + \tfrac{1}{2}v^2 + \frac{P_0}{\rho}\right) dV = \int_{S_{(\text{ent ex})}} \rho(\hat{H} + \tfrac{1}{2}v^2)(\mathbf{v} - \mathbf{v}_{(s)}) \cdot (-\mathbf{n}) \, dS + \mathcal{Q} - \mathcal{W}$$

$$+ \int_{V_{(s)}} \left(\sum_{A=1}^{N} \mathbf{n}_{(A)} \cdot \mathbf{f}_{(A)} + \rho Q\right) dV + \int_{S_{(\text{ent ex})}} [-(P - P_0)(\mathbf{v}_{(s)} \cdot \mathbf{n}) + \mathbf{v} \cdot (\mathbf{S} \cdot \mathbf{n})] \, dS \quad \text{(B)}$$

$$\frac{d}{dt}\int_{V_{(s)}} \rho\left(\hat{U} + \frac{P_0}{\rho}\right) dV = \int_{S_{(\text{ent ex})}} \rho\left(\hat{U} + \frac{P_0}{\rho}\right)(\mathbf{v} - \mathbf{v}_{(s)}) \cdot (-\mathbf{n}) \, dS$$

$$+ \mathcal{Q} + \int_{V_{(s)}} \left[-(P - P_0) \operatorname{div} \mathbf{v} + \operatorname{tr}(\mathbf{S} \cdot \nabla \mathbf{v}) + \sum_{A=1}^{N} \mathbf{j}_{(A)} \cdot \mathbf{f}_{(A)} + \rho Q\right] dV \quad \text{(C)}$$

*For an incompressible fluid:*

$$\frac{d}{dt}\int_{V_{(s)}} \rho \hat{U} \, dV = \int_{S_{(\text{ent ex})}} \rho \hat{U}(\mathbf{v} - \mathbf{v}_{(s)}) \cdot (-\mathbf{n}) \, dS$$

$$+ \mathcal{Q} + \int_{V_{(s)}} \left[\operatorname{tr}(\mathbf{S} \cdot \nabla \mathbf{v}) + \sum_{A=1}^{N} \mathbf{j}_{(A)} \cdot \mathbf{f}_{(A)} + \rho Q\right] dV \quad \text{(D)}$$

*For an isobaric fluid:*

$$\frac{d}{dt}\int_{V_{(s)}} \rho \hat{H} \, dV = \int_{S_{(\text{ent ex})}} \rho \hat{H}(\mathbf{v} - \mathbf{v}_{(s)}) \cdot (-\mathbf{n}) \, dS$$

$$+ \mathcal{Q} + \int_{V_{(s)}} \left[\operatorname{tr}(\mathbf{S} \cdot \nabla \mathbf{v}) + \sum_{A=1}^{N} \mathbf{j}_{(A)} \cdot \mathbf{f}_{(A)} + \rho Q\right] dV \quad \text{(E)}$$

† We neglect contributions attributable to the kinetic energies of the individual species measured with respect to the mass-averaged velocity of the material.

‡ We assume that $\mathbf{f} = \sum_{A=1}^{N} \omega_{(A)} \mathbf{f}_{(A)} = -\nabla \varphi$, where $\varphi$ is not an explicit function of time.

### 10.4.9 The integral overall energy balance: An example

Perhaps the principal problem associated with using the integral overall energy balance is the estimation of thermodynamic properties in the absence of directly applicable experimental data. Hougen, Watson, and Ragatz [1, 2] have done an excellent job in covering this point.

Stoichiometry texts [1] and thermodynamic texts [2, 3] give many examples employing the integral overall energy balance. The purpose of the problem solved here is not to illustrate how complex these problems can become but rather to indicate the point of view I advocate in attacking them.

## [10.4] INTEGRAL BALANCES

**Table 10.4.8-2** General forms of the integral overall energy balance applicable to a multi-phase system where the overall jump energy balance of Exercise 8.3.5-1 applies†‡

$$\frac{d}{dt}\int_{V_{(s)}} \rho\left(\hat{U} + \tfrac{1}{2}v^2 + \varphi + \frac{p_0}{\rho}\right) dV = \int_{S_{(\text{ent ex})}} \rho(\hat{H} + \tfrac{1}{2}v^2 + \varphi)(\mathbf{v} - \mathbf{v}_{(s)}) \cdot (-\mathbf{n})\, dS + \mathcal{Q} - \mathcal{W}$$

$$+ \int_{V_{(s)}} \left(\sum_{A=1}^{N} \mathbf{j}_{(A)} \cdot \mathbf{f}_{(A)} + \rho Q\right) dV + \int_{S_{(\text{ent ex})}} [-(P - p_0)(\mathbf{v}_{(s)} \cdot \mathbf{n}) + \mathbf{v} \cdot (\mathbf{S} \cdot \mathbf{n})]\, dS$$

$$+ \int_{S_{(\text{sing})}} [\rho\varphi(\mathbf{v} \cdot \boldsymbol{\xi} - u_{(\xi)})]\, dS \quad \text{(A)§}$$

$$\frac{d}{dt}\int_{V_{(s)}} \rho\left(\hat{U} + \tfrac{1}{2}v^2 + \frac{p_0}{\rho}\right) dV = \int_{S_{(\text{ent ex})}} \rho(\hat{H} + \tfrac{1}{2}v^2)(\mathbf{v} - \mathbf{v}_{(s)}) \cdot (-\mathbf{n})\, dS + \mathcal{Q} - \mathcal{W}$$

$$+ \int_{V_{(s)}} \left(\sum_{A=1}^{N} \mathbf{n}_{(A)} \cdot \mathbf{f}_{(A)} + \rho Q\right) dV + \int_{S_{(\text{ent ex})}} [-(P - p_0)(\mathbf{v}_{(s)} \cdot \mathbf{n}) + \mathbf{v} \cdot (\mathbf{S} \cdot \mathbf{n})]\, dS \quad \text{(B)}$$

$$\frac{d}{dt}\int_{V_{(s)}} \rho\left(\hat{U} + \frac{p_0}{\rho}\right) dV = \int_{S_{(\text{ent ex})}} \rho\left(\hat{U} + \frac{p_0}{\rho}\right)(\mathbf{v} - \mathbf{v}_{(s)}) \cdot (-\mathbf{n})\, dS$$

$$+ \mathcal{Q} + \int_{V_{(s)}} \left[-(P - p_0)\,\text{div}\,\mathbf{v} + \text{tr}\,(\mathbf{S} \cdot \nabla\mathbf{v}) + \sum_{A=1}^{N} \mathbf{j}_{(A)} \cdot \mathbf{f}_{(A)} + \rho Q\right] dV$$

$$+ \int_{S_{(\text{sing})}} \left[\rho\left(\hat{U} + \frac{p_0}{\rho}\right)(\mathbf{v} \cdot \boldsymbol{\xi} - u_{(\xi)}) + \mathbf{q} \cdot \boldsymbol{\xi}\right] dS \quad \text{(C)}$$

*For an incompressible fluid:*

$$\frac{d}{dt}\int_{V_{(s)}} \rho\hat{U}\, dV = \int_{S_{(\text{ent ex})}} \rho\hat{U}(\mathbf{v} - \mathbf{v}_{(s)}) \cdot (-\mathbf{n})\, dS$$

$$+ \mathcal{Q} + \int_{V_{(s)}} \left[\text{tr}\,(\mathbf{S} \cdot \nabla\mathbf{v}) + \sum_{A=1}^{N} \mathbf{j}_{(A)} \cdot \mathbf{f}_{(A)} + \rho Q\right] dV$$

$$+ \int_{S_{(\text{sing})}} [\rho\hat{U}(\mathbf{v} \cdot \boldsymbol{\xi} - u_{(\xi)}) + \mathbf{q} \cdot \boldsymbol{\xi}]\, dS \quad \text{(D)}$$

*For an isobaric fluid:*

$$\frac{d}{dt}\int_{V_{(s)}} \rho\hat{H}\, dV = \int_{S_{(\text{ent ex})}} \rho\hat{H}(\mathbf{v} - \mathbf{v}_{(s)}) \cdot (-\mathbf{n})\, dS$$

$$+ \mathcal{Q} + \int_{V_{(s)}} \left[\text{tr}\,(\mathbf{S} \cdot \nabla\mathbf{v}) + \sum_{A=1}^{N} \mathbf{j}_{(A)} \cdot \mathbf{f}_{(A)} + \rho Q\right] dV$$

$$+ \int_{S_{(\text{sing})}} [\rho\hat{H}(\mathbf{v} \cdot \boldsymbol{\xi} - u_{(\xi)}) + \mathbf{q} \cdot \boldsymbol{\xi}]\, dS \quad \text{(E)}$$

† We neglect contributions attributable to the kinetic energies of the individual species measured with respect to the mass-averaged velocity of the material.
‡ Here $u_{(\xi)}$ is the speed of displacement of the phase interface [1, p. 499]; the boldface bracket notation is defined by Eq. (4-7) of Sec. 1.3.4; $S_{(\text{sing})}$ refers to those phase interfaces in the region enclosed by $S_{(s)}$ that do not coincide with $S_{(s)}$.
§ We assume here that $\mathbf{f} \equiv \sum_{A=1}^{N} \omega_{(A)}\mathbf{f}_{(A)} = -\nabla\varphi$, where $\varphi$ is not an explicit function of time.

**Table 10.4.8-3 Restricted forms of the integral overall energy balance applicable to a multiphase system†**

$$\frac{d}{dt}\int_{V_{(s)}} \rho\left(\hat{U} + \tfrac{1}{2}v^2 + \varphi + \frac{P_0}{\rho}\right) dV = \int_{S_{(\text{ent ex})}} \rho(\hat{H} + \tfrac{1}{2}v^2 + \varphi)(-\mathbf{v}\cdot\mathbf{n})\, dS + \mathcal{Q} - \mathcal{W}$$
$$+ \int_{V_{(s)}} \sum_{A=1}^{N} \mathbf{j}_{(A)} \cdot \mathbf{f}_{(A)}\, dV \quad \text{(A)‡}$$

$$\frac{d}{dt}\int_{V_{(s)}} \rho\left(\hat{U} + \tfrac{1}{2}v^2 + \frac{P_0}{\rho}\right) dV = \int_{S_{(\text{ent ex})}} \rho(\hat{H} + \tfrac{1}{2}v^2)(-\mathbf{v}\cdot\mathbf{n})\, dS + \mathcal{Q} - \mathcal{W}$$
$$+ \int_{V_{(s)}} \sum_{A=1}^{N} \mathbf{n}_{(A)} \cdot \mathbf{f}_{(A)}\, dV \quad \text{(B)}$$

$$\frac{d}{dt}\int_{V_{(s)}} \rho\left(\hat{U} + \frac{P_0}{\rho}\right) dV = \int_{S_{(\text{ent ex})}} \rho\left(\hat{U} + \frac{P_0}{\rho}\right)(-\mathbf{v}\cdot\mathbf{n})\, dS$$
$$+ \mathcal{Q} + \int_{V_{(s)}} \left[ -(P - P_0)\operatorname{div} \mathbf{v} + \operatorname{tr}(\mathbf{S}\cdot\nabla\mathbf{v}) + \sum_{A=1}^{N} \mathbf{j}_{(A)} \cdot \mathbf{f}_{(A)} \right] dV \quad \text{(C)}$$

*For an incompressible fluid:*

$$\frac{d}{dt}\int_{V_{(s)}} \rho\hat{U}\, dV = \int_{S_{(\text{ent ex})}} \rho\hat{U}(-\mathbf{v}\cdot\mathbf{n})\, dS + \mathcal{Q} + \int_{V_{(s)}} \left[ \operatorname{tr}(\mathbf{S}\cdot\nabla\mathbf{v}) + \sum_{A=1}^{N} \mathbf{j}_{(A)} \cdot \mathbf{f}_{(A)} \right] dV \quad \text{(D)}$$

*For an isobaric system:*

$$\frac{d}{dt}\int_{V_{(s)}} \rho\hat{H}\, dV = \int_{S_{(\text{ent ex})}} \rho\hat{H}(-\mathbf{v}\cdot\mathbf{n})\, dS + \mathcal{Q} + \int_{V_{(s)}} \left[ \operatorname{tr}(\mathbf{S}\cdot\nabla\mathbf{v}) + \sum_{A=1}^{N} \mathbf{j}_{(A)} \cdot \mathbf{f}_{(A)} \right] dV \quad \text{(E)}$$

---

† 1. No mass transfer across internal phase interfaces.
2. The jump mass, momentum, and energy balances of Exercises 8.3.2-1, 8.3.3-1, and 8.3.5-1 apply.
3. No mutual or external energy transmission.
4. Entrances and exits are fixed in space.
5. Work done by viscous forces (as described by the extra-stress tensor) may be neglected at entrances and exits.
6. Contributions attributable to the kinetic energies of the individual species measured with respect to the mass-averaged velocity of the material are neglected.

‡ We assume here that $\mathbf{f} \equiv \sum_{A=1}^{N} \omega_{(A)}\mathbf{f}_{(A)} = -\nabla\varphi$, where $\varphi$ is not an explicit function of time.

A large class of problems, including the one that follows, describes operations that take place at a constant pressure. This group of problems not only is of obvious practical importance but is computationally simple as well. From the form of the integral overall energy balance appropriate to an isobaric system, Eqs. (E) of Tables 10.4.8-1 to 10.4.8-3, we see that specific enthalpy is the important thermodynamic variable. It is not necessary for us to know specific internal energy as well.

Let us calculate the energy evolved when 50 lb of water and 200 lb of an aqueous solution of sulphuric acid (containing 50% $H_2SO_4$ by weight) are mixed at 70°F.

The operation is shown schematically in Fig. 10.4.9-1. If we take our system to be the fluid in the mixing tank, we see that there are two entrances and no exits. Realizing that the same external force acts on each species and neglecting the rate

## [10.4] INTEGRAL BALANCES

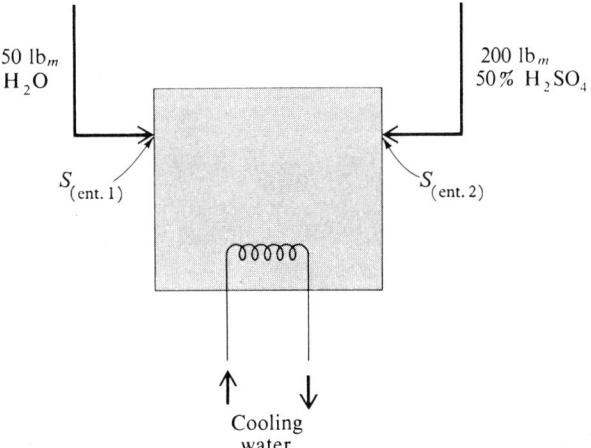

**Fig. 10.4.9-1** Unsteady-state dilution of sulfuric acid solution with water.

of dissipation of energy, the integral overall energy balance for an isobaric system in the form of Eq. (E) of Table 10.4.8-3 reduces to

$$\frac{d}{dt}\int_{V_{(s)}} \rho \hat{H}\, dV = \int_{S_{(\text{ent 1})}+S_{(\text{ent 2})}} \rho \hat{H}(-\mathbf{v}\cdot\mathbf{n})\, dS + \mathcal{Q} \tag{9-1}$$

Integrating this over time, we find that at $t = t_f$, when the diluted acid has been cooled to 70°F,

$$\text{At } t = t_f: \quad \hat{H}_{(f)}\int_{V_{(s)}} \rho\, dV = \hat{H}_{(\text{ent 1})}\int_0^{t_f}\int_{S_{(\text{ent 1})}} \rho(-\mathbf{v}\cdot\mathbf{n})\, dS\, dt$$

$$+ \hat{H}_{(\text{ent 2})}\int_0^{t_f}\int_{S_{(\text{ent 2})}} \rho(-\mathbf{v}\cdot\mathbf{n})\, dS\, dt + \int_0^{t_f} \mathcal{Q}\, dt \tag{9-2}$$

The integral overall mass balance of Sec. 10.4.6 and the integral mass balance for $H_2SO_4$ of Sec. 10.4.2 may also be integrated over time to find, respectively,

$$\text{At } t = t_f: \quad \int_{V_{(s)}} \rho\, dV = \int_0^{t_f}\int_{S_{(\text{ent 1})}} \rho(-\mathbf{v}\cdot\mathbf{n})\, dS\, dt$$

$$+ \int_0^{t_f}\int_{S_{(\text{ent 2})}} \rho(-\mathbf{v}\cdot\mathbf{n})\, dS\, dt \tag{9-3}$$

and

$$\text{At } t = t_f: \quad \omega_{(H_2SO_4,\, f)}\int_{V_{(s)}} \rho\, dV$$

$$= \omega_{(H_2SO_4,\, \text{ent 2})}\int_0^{t_f}\int_{S_{(\text{ent 2})}} \rho(-\mathbf{v}\cdot\mathbf{n})\, dS\, dt \tag{9-4}$$

From the problem statement, we know that

$$\int_0^{t_f}\int_{S_{(\text{ent 1})}} \rho(-\mathbf{v}\cdot\mathbf{n})\,dS\,dt = 50\text{ lb}_m$$

$$\int_0^{t_f}\int_{S_{(\text{ent 2})}} \rho(-\mathbf{v}\cdot\mathbf{n})\,dS\,dt = 200\text{ lb}_m \qquad (9\text{-}5)$$

This, together with Eq. (9-3), allows us to conclude from Eq. (9-4) that

$$\omega_{(\text{H}_2\text{SO}_4,\,f)} = \frac{(0.5)(200)}{250} = 0.4 \qquad (9\text{-}6)$$

Hougen, Watson, and Ragatz [1, p. 325] give an enthalpy-concentration diagram for the sulphuric acid–water system (relative to the pure components, water, and $H_2SO_4$ at 32°F and their own vapor pressures), which can be used here to estimate

$$\hat{H}_{(f)} = -85\text{ Btu/lb}_m$$
$$\hat{H}_{(\text{ent 1})} = 38\text{ Btu/lb}_m \qquad (9\text{-}7)$$
$$\hat{H}_{(\text{ent 2})} = -105\text{ Btu/lb}_m$$

Plugging this information from Eqs. (9-3), (9-5), and (9-7) into Eq. (9-2), we are able to determine the energy transferred from the solution to the cooling coil during the dilution process:

$$-\int_0^{t_f} \mathcal{Q}\,dt = (38)(50) - (105)(200) + (85)(250) = 2100\text{ Btu} \qquad (9\text{-}8)$$

**REFERENCES**

1. Hougen, O. A., K. M. Watson, and R. A. Ragatz: "Chemical Process Principles," 2d ed., pt. 1, Wiley, New York, 1958.
2. Hougen, O. A., K. M. Watson, and R. A. Ragatz: "Chemical Process Principles," 2d ed., pt. 2, Wiley, New York, 1959.
3. Smith, J. M., and H. C. Van Ness: "Introduction to Chemical Engineering Thermodynamics," 2d ed., McGraw-Hill, New York, 1959.

**10.4.10 The integral overall entropy inequality** The discussion here follows very closely that given in Sec. 7.4.7 for single-component systems. It is necessary only to substitute the differential entropy inequalities of Sec. 8.3.6 for those of Sec. 5.5.2 to obtain the following two forms of the integral overall entropy inequality for multiphase systems:

$$\frac{d}{dt}\int_{V_{(s)}} \rho\hat{S}\,dV \geq \int_{S_{(\text{ent ex})}} \rho\hat{S}(\mathbf{v}-\mathbf{v}_{(s)})\cdot(-\mathbf{n})\,dS$$

$$+ \int_{S_{(s)}} \frac{1}{T}\left(\mathbf{q} - \sum_{A=1}^{N}\mu_{(A)}\mathbf{j}_{(A)}\right)\cdot(-\mathbf{n})\,dS + \int_{V_{(s)}} \rho\frac{Q}{T}\,dV \qquad (10\text{-}1)$$

## [10.4] INTEGRAL BALANCES

and

$$\int_{V_{(s)}} \left\{ -\frac{1}{T^2} \mathbf{q} \cdot \nabla T + \frac{1}{T} \text{tr} (\mathbf{S} \cdot \nabla \mathbf{v}) \right.$$

$$\left. - \frac{1}{T} \sum_{A=1}^{N} \mathbf{j}_{(A)} \cdot \left[ T \nabla \left( \frac{\mu_{(A)}}{T} \right) - \mathbf{f}_{(A)} \right] - \sum_{A=1}^{N} \frac{1}{T} \mu_{(A)} r_{(A)} \right\} dV \geq 0 \quad (10\text{-}2)$$

In arriving at Eq. (10-1), we have assumed that the jump entropy inequality of Exercise 8.3.6-1 is valid for all phase interfaces involved. In both Eqs. (10-1) and (10-2) we have neglected terms involving the kinetic energy of diffusion ($\frac{1}{2} u_{(A)}^2$), arguing as we did in Sec. 10.4.8 that there is no experimental evidence to indicate that these terms are important experimentally in other than highly unusual circumstances.

Of the two forms, Eq. (10-1) is probably the more important. Often we will be willing to neglect $\sum_{A=1}^{N} \mu_{(A)} \mathbf{j}_{(A)}$ with respect to $\mathbf{q}$ in the second term on the right, in which case Eq. (10-1) takes on the same form as the result for single-component systems in Sec. 7.4.7.

The reader is again reminded of the caution issued in Secs. 10.4.1 and 10.4.2: $S_{\text{(ent ex)}}$ includes phase interfaces across which the individual species are carried primarily by diffusion.

### EXERCISE

**10.4.10-1** *Some additional forms of the entropy inequality* (a) Let us consider a system bounded by fixed, impermeable walls; there are no entrances and exits. The system may consist of any number of phases. Temperature is assumed to be independent of both time and position. Determine that

$$\frac{d}{dt} \int_{V_{(s)}} \rho \left( \hat{A} + \tfrac{1}{2} v^2 + \sum_{B=1}^{N} \tfrac{1}{2} \omega_{(B)} u_{(B)}^2 + \varphi \right) dV - \int_{V_{(s)}} \sum_{B=1}^{N} \mathbf{j}_{(B)} \cdot \mathbf{f}_{(B)} \, dV \leq 0$$

We assume here that

$$\mathbf{f} = \sum_{B=1}^{N} \omega_{(B)} \mathbf{f}_{(B)} = -\nabla \varphi$$

where $\varphi$ is not an explicit function of time. Discuss under what conditions Helmholtz free energy is minimized at equilibrium for a system of the type described.

(b) Consider the same system as above but require in addition that pressure be independent of time and position. Determine that

$$\frac{d}{dt} \int_{V_{(s)}} \rho \left( \hat{G} + \tfrac{1}{2} v^2 + \sum_{B=1}^{N} \tfrac{1}{2} \omega_{(B)} u_{(B)}^2 + \varphi \right) dV - \int_{V_{(s)}} \sum_{B=1}^{N} \mathbf{j}_{(B)} \cdot \mathbf{f}_{(B)} \, dV \leq 0$$

Discuss under what conditions Gibbs free energy is minimized at equilibrium for a system of the type described.

*Hint:* See Exercise 8.3.6-5.

appendix **A**
# Tensor Analysis

Tensor analysis is the language in terms of which continuum mechanics can be presented in the simplest and most physically meaningful fashion. For this reason, I suggest that those readers who are not already familiar with this subject should read at least a portion of this appendix before starting with the main text.

The degree to which tensor analysis must be mastered depends upon the aims of the student. I have written this appendix with three types of people in mind.

Many first-year graduate students in engineering are anxious to get to interesting applications as quickly as possible. I suggest that they read only those sections marked with double asterisks. They should also understand those exercises marked with double asterisks. Not all of this need be done before embarking on Chap. 1. Sometimes it is helpful to alternate between Chap. 1 and this appendix.

Those students who are somewhat more curious about the foundations of continuum mechanics will want to read the unmarked sections as well as those marked with two asterisks. The unmarked sections not only allow you to be more critical in your reading but are required for a complete understanding of the transport theorem in Sec. 1.3.2.

# [A.1] SPATIAL VECTORS

The complete appendix is recommended for anyone who wishes to do serious research in any of the subareas of continuum mechanics. The single-asterisked sections are required in order to derive the forms of various results in curvilinear coordinate systems. Without these sections, the curvilinear forms presented in Tables 2.5.1, 2.5.2, 5.6.1, etc., cannot be derived; the basis for the discussion of boundary layers on curved walls in Secs. 3.5.3 and 3.5.7 cannot be checked; and one is handicapped in working with new constitutive equations or with out-of-the-ordinary coordinate systems.

For those readers who are not sure into which category they fall, I suggest that you begin with the double-asterisked sections. As your interest in the subject grows, it is easy to turn back and read a little more.

## A.1 SPATIAL VECTORS

**\*\*A.1.1 Spatial vectors** We visualize that the real world occupies the space $E$ studied in elementary geometry. To each point of $E$ there corresponds a place in the universe.

Corresponding to each pair $(a,b)$ of the points of $E$ taken in order, there is a directed line segment denoted by $\vec{ab}$. Each directed line segment $\vec{ab}$ is characterized by a length $|ab|$ and direction (with the exception of the zero vector, which has zero length and an arbitrary direction).

Let us define the set of *spatial* vectors to be composed of the set of all directed line segments, with the understanding that two directed line segments that differ only by a parallel displacement represent the same element of the set. We define three operations for this set: addition, multiplication by real scalars, and inner product.

*Addition*  The sum $\mathbf{v} + \mathbf{w}$ of two spatial vectors $\mathbf{v}$ and $\mathbf{w}$ is defined by the familiar parallelogram rule as indicated in Fig. A.1.1-1.

*Scalar Multiplication*  Let $\alpha$ be a real number (scalar) and $\mathbf{v}$ be a spatial vector. We define the spatial vector $\alpha\mathbf{v}$ to have a length $|\alpha|\,|\mathbf{v}|$; the direction of $\alpha\mathbf{v}$ is defined to be the same as that of $\mathbf{v}$ if $\alpha > 0$ and the opposite direction if $\alpha < 0$.

*Inner product*  The inner product $\mathbf{v} \cdot \mathbf{w}$ of two spatial vectors $\mathbf{v}$ and $\mathbf{w}$ is a real number obtained by multiplying the length of $\mathbf{v}$, the length of $\mathbf{w}$, and the cosine of the angle between the directions $\mathbf{v}$ and $\mathbf{w}$.

**Fig. A.1.1-1**

These three operations satisfy the following rules.

($A_1$) $\mathbf{v} + \mathbf{w} = \mathbf{w} + \mathbf{v}$.
($A_2$) $\mathbf{u} + (\mathbf{v} + \mathbf{w}) = (\mathbf{u} + \mathbf{v}) + \mathbf{w}$.
($A_3$) $\mathbf{v} + \mathbf{0} = \mathbf{v}$. Here $\mathbf{0}$ is the zero spatial vector, which should be regarded as having zero length and arbitrary direction.
($A_4$) Given any spatial vector $\mathbf{v}$, there is another spatial vector denoted by $-\mathbf{v}$ such that

$$\mathbf{v} + (-\mathbf{v}) = \mathbf{0}$$

($M_1$) $\alpha(\beta \mathbf{v}) = (\alpha\beta)\mathbf{v}$.
($M_2$) $1\mathbf{v} = \mathbf{v}$.
($M_3$) $\alpha(\mathbf{v} + \mathbf{w}) = \alpha\mathbf{v} + \alpha\mathbf{w}$.
($M_4$) $(\alpha + \beta)\mathbf{v} = \alpha\mathbf{v} + \beta\mathbf{v}$.

($I_1$) $\mathbf{v} \cdot \mathbf{w} = \mathbf{w} \cdot \mathbf{v}$.
($I_2$) $\mathbf{u} \cdot (\mathbf{v} + \mathbf{w}) = \mathbf{u} \cdot \mathbf{v} + \mathbf{u} \cdot \mathbf{w}$.
($I_3$) $\alpha(\mathbf{v} \cdot \mathbf{w}) = (\alpha\mathbf{v}) \cdot \mathbf{w}$.
($I_4$) $\mathbf{v} \cdot \mathbf{v} \geq 0$; $\mathbf{v} \cdot \mathbf{v} = 0$, if and only if $\mathbf{v} = \mathbf{0}$.

If for any set of objects we define *addition* and *scalar multiplication* in such a way that the rules ($A_1$) to ($A_4$) and ($M_1$) to ($M_4$) hold, we define the set to be a *vector space* and we refer to the elements of the set as *vectors*.

If for any vector space an *inner product* that satisfies rules ($I_1$) to ($I_4$) is introduced, we refer to the vector space as an *inner product space*. The set of spatial vectors is an inner product space.

For spatial vectors, we adopt the following abbreviations:

$$\mathbf{v} - \mathbf{w} \equiv \mathbf{v} + (-\mathbf{w}) \tag{1-1}$$

$$v \equiv |\mathbf{v}| \equiv \sqrt{\mathbf{v} \cdot \mathbf{v}} \tag{1-2}$$

The nonnegative number $v$ (or $|\mathbf{v}|$) is the *magnitude* or length of the vector $\mathbf{v}$.

## EXERCISES

**A.I.I-1** Consider the set of all real numbers. If we understand $x + y$ and $\alpha x$ to be the ordinary numerical addition and multiplication, prove that this set constitutes a vector space.

**A.I.I-2** Consider the set of all $n$-tuples of real numbers. If $x = (\xi_1, \ldots, \xi_n)$ and $y = (\eta_1, \ldots, \eta_n)$ are elements of $R^n$, we define

$$x + y = (\xi_1 + \eta_1, \ldots, \xi_n + \eta_n)$$
$$\alpha x = (\alpha\xi_1, \ldots, \alpha\xi_n)$$
$$0 = (0, \ldots, 0)$$
$$-x = (-\xi_1, \ldots, -\xi_n)$$
$$x \cdot y = (\xi_1\eta_1 + \xi_2\eta_2 + \cdots + \xi_n\eta_n)$$

Prove that $R^n$ is an inner product space.

## A.I.2 Position vectors

Any point $z$ in $E$ may be located with respect to another point $O$ by means of the spatial vector $\mathbf{z} \equiv \overrightarrow{Oz}$. It is common practice to refer to $\mathbf{z}$ as the *position vector* of the point $z$ with respect to the *origin O*.

A particular point in $E$ having been designated as the origin $O$, the set of all position vectors, which locate points in $E$ with respect to the origin $O$, is identical to the set of all spatial vectors

When speaking in general, it rarely makes a difference whether one speaks about "the point $z$" or "the point whose position vector relative to the origin $O$ is $\mathbf{z}$." However, in computations we always find it necessary to express locations in terms of position vectors, because we are able to sense only relative locations.

## A.I.3 Spatial vector fields

Temperature, concentration, and pressure are examples of real numerically valued functions of position. We refer to any real numerically valued function of position as a *real scalar field*.

When we think of water flowing through a pipe or in a river, we recognize that the velocity of the water is a function of position. At the wall of the pipe, the velocity of the water is zero; at the center, it is a maximum. The velocity of the water in the pipe is an example of a spatial vector-valued function of position. We shall term any spatial vector-valued function a *spatial vector field*.

As another example, consider the *position vector field* $\mathbf{p}(z)$. It maps every point $z$ of $E$ into the corresponding position vector $\mathbf{z}$ measured with respect to a previously chosen origin $O$:

$$\mathbf{z} = \mathbf{p}(z) \qquad (3\text{-}1)$$

With the following definitions for addition, scalar multiplication, and inner product, the set of all spatial vector fields becomes an inner product space.

*Addition* If $\mathbf{v}$ and $\mathbf{w}$ are two spatial vector fields, we define the spatial vector field $\mathbf{v} + \mathbf{w}$ such that at every point $z$ of $E$,

$$(\mathbf{v} + \mathbf{w})(z) \equiv \mathbf{v}(z) + \mathbf{w}(z)$$

The addition on the right is that defined for spatial vectors. It is to be understood here that $\mathbf{v} + \mathbf{w}$, $\mathbf{v}$, and $\mathbf{w}$ indicate functions; $(\mathbf{v} + \mathbf{w})(z)$, $\mathbf{v}(z)$, and $\mathbf{w}(z)$ denote the values of these functions at the point $z$.

*Scalar multiplication* If $\alpha$ is a real scalar field (a real numerically valued function of position; for the general requirements of a set of scalars, see [1, p. 1]) and $\mathbf{v}$ is a spatial vector field, we define the spatial vector field $\alpha\mathbf{v}$ such that at every point $z$,

$$(\alpha\mathbf{v})(z) \equiv \alpha(z)\mathbf{v}(z)$$

The scalar multiplication on the right is that defined for spatial vectors.

*Inner product* If **v** and **w** are two spatial vector fields, we define the real scalar field **v** · **w** such that at every point $z$,

$$(\mathbf{v} \cdot \mathbf{w})(z) \equiv \mathbf{v}(z) \cdot \mathbf{w}(z)$$

The inner product indicated on the right is that defined for spatial vectors.

In the text, we have occasion to discuss many fields besides those already mentioned: stress fields, energy flux fields, mass flux fields, enthalpy fields, etc. In developing the basic concepts, we are generally concerned with real scalar fields and spatial vector fields. It is in the final results of applications that we usually become concerned with the values of real scalar fields (the average temperature in a tank or the average concentration in an exit stream) and the values of spatial vector fields (the force acting on a body or the torque exerted upon a surface).

It is common practice in the literature to refer to real scalar fields and spatial vector fields inexactly as *scalars* and *vectors*. Writers depend upon the context to clarify whether they are talking about the functions (the spatial vector fields) or their values (spatial vectors).

## REFERENCE

1. Halmos, P. R.: "Finite-dimensional Vector Spaces," 2d ed., Van Nostrand, Princeton, N.J., 1958.

**A.I.4 Bases** Let $\alpha_1$, $\alpha_2$, and $\alpha_3$ be scalars. We define the set of spatial vectors $\mathbf{e}_1$, $\mathbf{e}_2$, $\mathbf{e}_3$ to be *linearly independent* if

$$\alpha_1 \mathbf{e}_1 + \alpha_2 \mathbf{e}_2 + \alpha_3 \mathbf{e}_3 = \sum_{i=1}^{3} \alpha_i \mathbf{e}_i = 0 \qquad (4\text{-}1)$$

can hold only when the numbers $\alpha_1$, $\alpha_2$, and $\alpha_3$ are all zero. Geometrically, three vectors are linearly independent if they are not all parallel to one plane.

A *basis* for a vector space $M$ is defined to be a set $\chi$ of linearly independent vectors such that every vector in $M$ is a linear combination of elements of $\chi$.

For example, $(\mathbf{e}_1, \mathbf{e}_2, \mathbf{e}_3)$ are said to form a basis for the set of all spatial vectors if $(\mathbf{e}_1, \mathbf{e}_2, \mathbf{e}_3)$ are linearly independent as in Eq. (4-1) and if every spatial vector **v** can be written as a linear combination of them:

$$\begin{aligned}\mathbf{v} &= v_1 \mathbf{e}_1 + v_2 \mathbf{e}_2 + v_3 \mathbf{e}_3 \\ &= \sum_{i=1}^{3} v_i \mathbf{e}_i\end{aligned} \qquad (4\text{-}2)$$

The numbers $(v_1, v_2, v_3)$ are referred to as the *components* of the vector **v** with respect to the basis $(\mathbf{e}_1, \mathbf{e}_2, \mathbf{e}_3)$.

The *dimension* of a finite-dimensional vector space $M$ is defined to be the number of elements in a basis of $M$.

We accept without proof here that the number of elements in any basis of a finite-dimensional vector space is the same as that in any other basis [1, p. 13]. It

[A.I] SPATIAL VECTORS

follows as a corollary that a set of $n$ vectors in any $n$-dimensional vector space $M$ is a basis if and only if it is linearly independent, or, alternatively, if and only if every vector in $M$ is a linear combination of elements of the set [1, p. 14].

The space of spatial vectors is by definition *three dimensional*. It therefore follows that the space of spatial vector fields must be three dimensional as well.

## REFERENCE

1. Halmos, P. R.: "Finite-dimensional Vector Spaces," 2d ed., Van Nostrand, Princeton, N.J., 1958.

## EXERCISES

**A.I.4-1** Let $\mathbf{a} = \sum_{i=1}^{3} a_i \mathbf{e}_i$ and $\mathbf{b} = \sum_{j=1}^{3} b_j \mathbf{e}_j$ be two spatial vectors, where the $a_i$ ($i = 1, 2, 3$) and $b_j$ ($j = 1, 2, 3$) are real numbers (scalars). Let $\alpha$ be a real number. Express the spatial vectors $\mathbf{a} + \mathbf{b}$ and $\alpha \mathbf{a}$ as linear combinations of the $\mathbf{e}_i$ ($i = 1, 2, 3$).

**A.I.4-2** Prove that a set of $n$ vectors in an $n$-dimensional vector space $M$ is a basis if and only if it is linearly independent, or, alternatively, if and only if every vector in $M$ is a linear combination of elements of the set.

*Hint:* You may accept without proof that the number of elements in any basis of a finite-dimensional vector space is the same as in any other basis.

**A.I.4-3** Let the set $(\mathbf{m}_1, \mathbf{m}_2, \mathbf{m}_3)$ form a basis for the space of spatial vectors. Any spatial vector $\mathbf{v}$ may be expressed as

$$\mathbf{v} = \sum_{i=1}^{3} v_i \mathbf{m}_i$$

Prove that the components of $\mathbf{v}$ with respect to this basis are unique.

## A.I.5 Bases for the spatial vector fields

A basis $(\mathbf{m}_1, \mathbf{m}_2, \mathbf{m}_3)$ for the space of spatial vector fields is said to be *cartesian* [1, p. 39] if the basis fields are of unit length (at every point $z$ of $E$ the corresponding spatial vectors are of unit length):

$$\mathbf{m}_1(z) \cdot \mathbf{m}_1(z) = 1$$
$$\mathbf{m}_2(z) \cdot \mathbf{m}_2(z) = 1$$
$$\mathbf{m}_3(z) \cdot \mathbf{m}_3(z) = 1 \tag{5-1}$$

A basis is said to be *orthogonal* if the basis elements are orthogonal to one another (at every point $z$ of $E$ the corresponding spatial vectors are orthogonal to one another):

$$\mathbf{m}_i(z) \cdot \mathbf{m}_j(z) = 0 \quad \text{for } i \neq j \tag{5-2}$$

In what follows, as well as in the body of the text, we often have occasion to use an *orthogonal cartesian* basis (*orthonormal* basis).

A *rectangular cartesian* basis is the most familiar to us all. Besides being orthonormal, the basis fields have the property that for every two points $x$ and $y$ in $E$,

$$\mathbf{m}_i(x) = \mathbf{m}_i(y) \quad \text{for } i = 1, 2, 3 \tag{5-3}$$

This means that both the length and the direction of the basis fields are independent of position in $E$. We shall reserve the symbols $e_1$, $e_2$, $e_3$ for such a basis.

Every spatial vector field $\mathbf{u}$ may be written as a linear combination of rectangular cartesian basis fields $e_1$, $e_2$, $e_3$:

$$\mathbf{u} = u_1 e_1 + u_2 e_2 + u_3 e_3 = \sum_{i=1}^{3} u_i e_i \tag{5-4}$$

The quantities $u_1$, $u_2$, $u_3$ are known as the *rectangular cartesian components* of $\mathbf{u}$; in general, they are functions of position in $E$.

A special case is the position vector field defined in Sec. A.1.3:

$$\mathbf{p} = z_1 e_1 + z_2 e_2 + z_3 e_3 = \sum_{i=1}^{3} z_i e_i \tag{5-5}$$

The rectangular cartesian components $(z_1, z_2, z_3)$ of the position vector field $\mathbf{p}$ are called the *rectangular cartesian coordinates* with respect to the previously chosen origin $O$. They are naturally functions of position $z$ in $E$:

$$z_i = z_i(z) \quad \text{for } i = 1, 2, 3 \tag{5-6}$$

For this reason we will often find it convenient to think of $\mathbf{p}$ as being a function of the rectangular cartesian coordinates:

$$z = \mathbf{p}(z_1, z_2, z_3) \tag{5-7}$$

## REFERENCE

1. McConnell, A. J.: "Applications of Tensor Analysis," Dover, New York, 1957.

## EXERCISES

**A.I.5-1** If we define

$$\frac{\partial \mathbf{p}}{\partial z_1} \equiv \lim_{\Delta z_1 \to 0} \frac{1}{\Delta z_1} [\mathbf{p}(z_1 + \Delta z_1, z_2, z_3) - \mathbf{p}(z_1, z_2, z_3)]$$

determine that

$$\frac{\partial \mathbf{p}}{\partial z_1} = e_1$$

With similar definitions for $\partial \mathbf{p}/\partial z_2$ and $\partial \mathbf{p}/\partial z_3$, determine that

$$\frac{\partial \mathbf{p}}{\partial z_2} = e_2$$

and

$$\frac{\partial \mathbf{p}}{\partial z_3} = e_3$$

[A.I] SPATIAL VECTORS

**A.I.5-2** (a) Let $(z_1,z_2,z_3)$ and $(\bar{z}_1,\bar{z}_2,\bar{z}_3)$ denote two rectangular cartesian coordinate systems. If $z_i = z_i(\bar{z}_1,\bar{z}_2,\bar{z}_3)$ for $i = 1, 2, 3$, prove that

$$\frac{\partial \mathbf{p}}{\partial \bar{z}_i} = \sum_{m=1}^{3} \frac{\partial z_m}{\partial \bar{z}_i} \frac{\partial \mathbf{p}}{\partial z_m}$$

(b) Prove that

$$\bar{\mathbf{e}}_i = \sum_{m=1}^{3} \frac{\partial z_m}{\partial \bar{z}_i} \mathbf{e}_m$$

**A.I.5-3** Let u be a spatial vector field. If the $u_i$ ($i = 1, 2, 3$) are the components of **u** with respect to a set of rectangular cartesian basis fields $(\mathbf{e}_1, \mathbf{e}_2, \mathbf{e}_3)$ and the $\bar{u}_m$ ($m = 1, 2, 3$) are the components of **u** with respect to another set of rectangular cartesian basis fields $(\bar{\mathbf{e}}_1, \bar{\mathbf{e}}_2, \bar{\mathbf{e}}_3)$, prove that

$$u_m = \sum_{i=1}^{3} \frac{\partial z_m}{\partial \bar{z}_i} \bar{u}_i$$

**A.I.6 Bases for the spatial vectors** Any basis $(\mathbf{m}_1,\mathbf{m}_2,\mathbf{m}_3)$ for the spatial vector fields may be used to generate an infinite number of bases for the space of spatial vectors. The values of these functions at any point $z$ of $E$, the spatial vectors

$$\mathbf{m}_i = \mathbf{m}_i(z) \quad \text{for } i = 1, 2, 3 \tag{6-1}$$

may be used as a basis for the spatial vectors. The basis will depend upon the particular point $z$ chosen, in the sense that the magnitudes and directions of the $\mathbf{m}_i$ may vary with position. [Notice that in writing Eq. (6-1) we use the same notation both for the function and for its values.]

Of particular interest is any rectangular cartesian basis $(\mathbf{e}_1,\mathbf{e}_2,\mathbf{e}_3)$ for the spatial vector fields. The magnitude and direction of the values of these functions are independent of position in $E$. We will often find it convenient to use the values of these functions, the spatial vectors

$$\mathbf{e}_i = \mathbf{e}_i(z) \quad \text{for } i = 1, 2, 3 \tag{6-2}$$

as a basis for the spatial vectors.

For example, we will often express a particular position vector **z** and point difference $\mathbf{a} \equiv \overrightarrow{xy}$ in terms of their rectangular cartesian components:

$$\mathbf{z} = \mathbf{p}(z) = \sum_{i=1}^{3} z_i(z)\mathbf{e}_i(z) \tag{6-3}$$

$$\mathbf{a} \equiv \overrightarrow{xy} = \sum_{i=1}^{3} a_i \mathbf{e}_i(z) \tag{6-4}$$

In view of Eq. (6-2) we generally write these expressions as

$$\mathbf{z} = \mathbf{p}(z) = \sum_{i=1}^{3} z_i(z)\mathbf{e}_i \tag{6-5}$$

$$\mathbf{a} \equiv \overrightarrow{xy} = \sum_{i=1}^{3} a_i \mathbf{e}_i \tag{6-6}$$

## A.1.7 The summation convention

In writing a spatial vector field in terms of its rectangular cartesian components, notice that the summation is over a repeated index $i$:

$$\mathbf{u} = \sum_{i=1}^{3} u_i \mathbf{e}_i \tag{7-1}$$

This suggests that we adopt a simpler notation in which we understand that a summation from 1 to 3 is to be performed over every index that appears *twice* within a *single* term. This is known as the *summation convention*. It allows Eq. (7-1) to be written as

$$\mathbf{u} = u_i \mathbf{e}_i \tag{7-2}$$

With this convention, we can write the inner product of two spatial vector fields as

$$\begin{aligned}
\mathbf{v} \cdot \mathbf{w} &= (v_i \mathbf{e}_i) \cdot (w_j \mathbf{e}_j) \\
&= v_i w_j (\mathbf{e}_i \cdot \mathbf{e}_j) \\
&= v_i w_j \delta_{ij} \\
&= v_i w_i
\end{aligned} \tag{7-3}$$

In going from Eq. $(7\text{-}3)_1$ to $(7\text{-}3)_2$, rules ($I_1$) to ($I_3$) for the inner product have been employed. In Eq. $(7\text{-}3)_3$, the *Kronecker delta* $\delta_{ij}$ is introduced:

$$\delta_{ij} \equiv \begin{cases} +1 & \text{if } i = j \\ 0 & \text{if } i \neq j \end{cases} \tag{7-4}$$

Let me emphasize that the summation convention is not defined for an index that appears more than twice in a single term. If this happens, there are several possibilities.

1. In writing a relation such as

$$\begin{aligned}
(\mathbf{v} \cdot \mathbf{w})(\mathbf{q} \cdot \mathbf{n}) &= (v_i w_i)(q_j n_j) \\
&= v_i w_i q_j n_j
\end{aligned} \tag{7-5}$$

we must be careful not to confuse the summation in the expression for $(\mathbf{v} \cdot \mathbf{w})$ with that for $(\mathbf{q} \cdot \mathbf{n})$. Observe that, besides being undefined, $v_i w_i q_i n_i$ is confusing. It might mean

$$\begin{aligned}
v_i w_i q_i n_i &\stackrel{?}{=} \sum_{i=1}^{3} (v_i w_i q_i n_i) \\
&\stackrel{?}{=} \sum_{i=1}^{3} (v_i w_i) \sum_{j=1}^{3} (q_j n_j) \\
&\stackrel{?}{=} \sum_{i=1}^{3} (v_i q_i) \sum_{j=1}^{3} (w_j n_j) \\
&\stackrel{?}{=} \sum_{i=1}^{3} (v_i n_i) \sum_{j=1}^{3} (w_j q_j)
\end{aligned}$$

## [A.2] DETERMINANT

2. Sometimes a summation is intended over an index that appears more than twice in a single term. The summation sign should be used explicitly in such a case:

$$\mathbf{u} = \sum_{i=1}^{3} \sqrt{g_{ii}}\, u^i \mathbf{g}_{\langle i \rangle}$$

3. Occasionally an index may appear twice or more in a single term, although no summation is intended. For clarity, make a note to this effect next to the equation:

$$u_{\langle i \rangle} = \frac{u_i}{\sqrt{g_{ii}}} \qquad \text{no summation on } i$$

## A.2 DETERMINANT

### A.2.1 Determinant

Define $e_{ijk}$ and $e^{ijk}$ to have only three distinct values:

0, when any two of the indices are equal
$+1$, when $ijk$ is an even permutation of 123
$-1$, when $ijk$ is an odd permutation of 123

The quantities $e_{ijk}$ and $e^{ijk}$ are said to be completely skew-symmetric in the indices $ijk$; that is, interchanging any two of these indices changes the sign of the quantity. For the moment we shall have no occasion to use $e^{ijk}$, but shall return to make use of it later.

Let us introduce the notation det $(a_{ij})$ for the determinant that has as its typical entry $a_{ij}$:

$$\det(a_{ij}) = \begin{vmatrix} a_{11} & a_{12} & a_{13} \\ a_{21} & a_{22} & a_{23} \\ a_{31} & a_{32} & a_{33} \end{vmatrix} \tag{1-1}$$

When we expand the determinant det $(a_{ij})$ by rows, we find we may write

$$\det(a_{ij}) = e_{ijk} a_{1i} a_{2j} a_{3k} \tag{1-2}$$

Similarly, when we expand the determinant by columns, we have

$$\det(a_{ij}) = e_{ijk} a_{i1} a_{j2} a_{k3} \tag{1-3}$$

Equations (1-2) and (1-3) suggest that we consider the quantity $e_{ijk} a_{im} a_{jn} a_{kp}$. This quantity is completely skew-symmetric in the indices $mnp$. As a proof, we have, by relabeling indices,

$$e_{ijk} a_{im} a_{jn} a_{kp} = e_{jik} a_{jm} a_{in} a_{kp}$$
$$= -e_{ijk} a_{in} a_{jm} a_{kp} \tag{1-4}$$

In the same way, we may show that interchanging any two of the indices $mnp$ alters the sign. In view of Eq. (1-3), this suggests that we may write for an expansion by columns

$$e_{ijk} a_{im} a_{jn} a_{kp} = \det(a_{rs}) e_{mnp} \tag{1-5}$$

The same type of argument may be used to infer from Eq. (1-2) that, for an expansion by rows,

$$e_{ijk}a_{mi}a_{nj}a_{pk} = \det(a_{rs})e_{mnp} \tag{1-6}$$

As an example of the use of this notation, consider the product of two determinants:

$$\begin{aligned}
\det(a_{rs})\det(b_{xy}) &= \det(a_{rs})e_{ijk}b_{i1}b_{j2}b_{k3} \\
&= e_{mnp}a_{mi}a_{nj}a_{pk}b_{i1}b_{j2}b_{k3} \\
&= e_{mnp}(a_{mi}b_{i1})(a_{nj}b_{j2})(a_{pk}b_{k3}) \\
&= \det(a_{us}b_{sv}) \tag{1-7}
\end{aligned}$$

We wish to introduce a further concept, the cofactor. Starting with Eq. (1-5), write

$$e_{rnp}e_{ijk}a_{im}a_{jn}a_{kp} = \det(a_{st})e_{rnp}e_{mnp} \tag{1-8}$$

In Sec. A.1.7 we introduced the Kronecker delta $\delta_{rm}$, defined as

$$\delta_{rm} \equiv \begin{cases} 1 & \text{if } r = m \\ 0 & \text{if } r \neq m \end{cases} \tag{1-9}$$

An equivalent expression which is sometimes useful is

$$\delta_{rm} = \tfrac{1}{2}e_{rnp}e_{mnp} \tag{1-10}$$

We may rearrange Eq. (1-8), consequently, to read

$$(\tfrac{1}{2}e_{rnp}e_{ijk}a_{jn}a_{kp})a_{im} = \det(a_{st})\,\delta_{rm} \tag{1-11}$$

or

$$A_{ri}a_{im} = \det(a_{st})\,\delta_{rm} \tag{1-12}$$

where we define

$$A_{ri} = \tfrac{1}{2}e_{rnp}e_{ijk}a_{jn}a_{kp} \tag{1-13}$$

The quantity $A_{ri}$ is called the cofactor of the element $a_{ir}$ in the determinant $\det(a_{st})$. When the determinant is expanded in full, it is obvious that any element such as $a_{ir}$ appears once in each of a certain number of terms of the expansion; the coefficient of $a_{ir}$ in this expansion is just $A_{ri}$.

For a further discussion of determinants, see McConnell [1].

### REFERENCE

1. McConnell, A. J.: "Applications of Tensor Analysis," Dover, New York, 1957.

## EXERCISES

**A.2.1-1** Prove that

$$a_{mr}A_{ri} = \det(a_{st})\delta_{mi}$$

where $A_{ri}$ is given by Eq. (1-13).

**A.2.1-2** (a) Show that $e_{ijk}e_{mnk}$ takes the values:

+1, when $i, j$ and $m, n$ are the same permutation of the same two numbers
−1, when $i, j$ and $m, n$ are opposite permutations of the same two numbers
0, otherwise

(b) Prove that

$$e_{ijk}e_{mnk} = \delta_{im}\delta_{jn} - \delta_{in}\delta_{jm}$$

**A.2.1-3** (a) Prove Eq. (1-10).
(b) Prove that

$$e_{mnp}e_{mnp} = 2\delta_{mm} = 6$$

**A.2.1-4** If any two rows or columns are identical, prove that the determinant is zero.

**A.2.1-5** In each of the following examples, we start with an equation and proceed to derive another equation. Indicate whether each step in the derivation is valid (can be derived from the previous step) and give reasons.

Example A
Given: $b_{ijk}b_{mnk} = \delta_{im}\delta_{jn} - \delta_{in}\delta_{jm}$
Step 1: $b_{ijk}b_{mnk}\delta_{jn} = \delta_{im}\delta_{jn}\delta_{jn} - \delta_{in}\delta_{jm}\delta_{jn}$
Step 2: $b_{ink}b_{mnk} = 2\delta_{im}$
Example B
Given: $b_{ink}b_{mnk} = 2\delta_{im}$
Step 1: $b_{ijk}b_{mnk}\delta_{jn} = \delta_{im}\delta_{jn}\delta_{jn} - \delta_{in}\delta_{jm}\delta_{jn}$
Step 2: $b_{ijk}b_{mnk} = \delta_{im}\delta_{jn} - \delta_{in}\delta_{jm}$

## A.3 GRADIENT OF SCALAR

**\*\*A.3.1 The gradient of a scalar field** The gradient of a scalar field $\alpha$ is a spatial vector field denoted by $\nabla\alpha$. The gradient is specified by defining its inner product with an arbitrary spatial vector at all points $z$ in $E$:

$$\nabla\alpha(\mathbf{z}) \cdot \mathbf{a} \equiv \lim_{s \to 0} \frac{\alpha(\mathbf{z} + s\mathbf{a}) - \alpha(\mathbf{z})}{s} \tag{1-1}$$

The spatial vector **a** should be interpreted as the directed line segment or point difference $\mathbf{a} \equiv \vec{zy}$, where $y$ is an arbitrary point in $E$. In writing Eq. (1-1), we have assumed that an origin $O$ has been specified and we have interpreted $\alpha$ as a function of the position vector **z** measured with respect to this origin rather than as a function of the point $z$ itself.

Equation (1-1) may be rearranged into a more easily applied expression in the following manner.

$$\nabla \alpha(\mathbf{z}) \cdot \mathbf{a} = \lim_{s \to 0} \frac{1}{s} \{\alpha([z_1 + sa_1]\mathbf{e}_1 + [z_2 + sa_2]\mathbf{e}_2 + [z_3 + sa_3]\mathbf{e}_3)$$
$$- \alpha(z_1\mathbf{e}_1 + [z_2 + sa_2]\mathbf{e}_2 + [z_3 + sa_3]\mathbf{e}_3)\} + \lim_{s \to 0} \frac{1}{s} \{\alpha(z_1\mathbf{e}_1$$
$$+ [z_2 + sa_2]\mathbf{e}_2 + [z_3 + sa_3]\mathbf{e}_3) - \alpha(z_1\mathbf{e}_1 + z_2\mathbf{e}_2 + [z_3 + sa_3]\mathbf{e}_3)\}$$
$$+ \lim_{s \to 0} \frac{1}{s} \{\alpha(z_1\mathbf{e}_1 + z_2\mathbf{e}_2 + [z_3 + sa_3]\mathbf{e}_3) - \alpha(z_1\mathbf{e}_1 + z_2\mathbf{e}_2 + z_3\mathbf{e}_3)\}$$

$$= a_1 \lim_{a_1 s \to 0} \frac{1}{a_1 s} \{\alpha([z_1 + sa_1]\mathbf{e}_1 + [z_2 + sa_2]\mathbf{e}_2 + [z_3 + sa_3]\mathbf{e}_3)$$
$$- \alpha(z_1\mathbf{e}_1 + [z_2 + sa_2]\mathbf{e}_2 + [z_3 + sa_3]\mathbf{e}_3)\}$$
$$+ a_2 \lim_{a_2 s \to 0} \frac{1}{a_2 s} \{\alpha(z_1\mathbf{e}_1 + [z_2 + sa_2]\mathbf{e}_2 + [z_3 + sa_3]\mathbf{e}_3)$$
$$- \alpha(z_1\mathbf{e}_1 + z_2\mathbf{e}_2 + [z_3 + sa_3]\mathbf{e}_3)\}$$
$$+ a_3 \lim_{a_3 s \to 0} \frac{1}{a_3 s} \{\alpha(z_1\mathbf{e}_1 + z_2\mathbf{e}_2 + [z_3 + sa_3]\mathbf{e}_3) - \alpha(z_1\mathbf{e}_1 + z_2\mathbf{e}_2 + z_3\mathbf{e}_3)\}$$

$$= a_1 \frac{\partial \alpha}{\partial z_1}(\mathbf{z}) + a_2 \frac{\partial \alpha}{\partial z_2}(\mathbf{z}) + a_3 \frac{\partial \alpha}{\partial z_3}(\mathbf{z})$$

$$= a_i \frac{\partial \alpha}{\partial z_i}(\mathbf{z}) \qquad (1\text{-}2)$$

In arriving at this result, we have used the definition of the partial derivative in the form

$$\frac{\partial \alpha}{\partial z_2}(\mathbf{z}) \equiv \lim_{\Delta z_2 \to 0} \frac{1}{\Delta z_2} \{\alpha(z_1\mathbf{e}_1 + [z_2 + \Delta z_2]\mathbf{e}_2 + z_3\mathbf{e}_3)$$
$$- \alpha(z_1\mathbf{e}_1 + z_2\mathbf{e}_2 + z_3\mathbf{e}_3)\} \quad (1\text{-}3)$$

Since $\mathbf{a}$ is an arbitrary spatial vector, take $\mathbf{a} = \mathbf{e}_i$:

$$\nabla \alpha \cdot \mathbf{e}_i = \frac{\partial \alpha}{\partial z_i} \qquad (1\text{-}4)$$

We conclude that

$$\nabla \alpha = \frac{\partial \alpha}{\partial z_i} \mathbf{e}_i \qquad (1\text{-}5)$$

### EXERCISES

**\*\*A.3.I-I**  Prove that

$$\nabla z_i = \mathbf{e}_i$$

## [A.4] CURVILINEAR COORDINATES

**A.3.1-2**  (a) Let $(z_1,z_2,z_3)$ and $(\bar{z}_1,\bar{z}_2,\bar{z}_3)$ denote two rectangular cartesian coordinate systems such that

$$z_i = z_i(\bar{z}_1,\bar{z}_2,\bar{z}_3) \quad \text{for } i = 1, 2, 3$$

Prove that

$$\nabla z_i = \frac{\partial z_i}{\partial \bar{z}_m} \nabla \bar{z}_m$$

(b) Show that for these two coordinate systems

$$\mathbf{e}_i = \frac{\partial z_i}{\partial \bar{z}_m} \bar{\mathbf{e}}_m$$

Compare this result with that of Exercise A.1.5-2.

**A.3.1-3**  Let $\mathbf{u}$ be some spatial vector field. If the $u_i$ ($i = 1, 2, 3$) are the components of $\mathbf{u}$ with respect to a set of rectangular cartesian basis fields $(\mathbf{e}_1,\mathbf{e}_2,\mathbf{e}_3)$ and the $\bar{u}_m$ ($m = 1, 2, 3$) are the components of $\mathbf{u}$ with respect to another set of rectangular cartesian basis fields $(\bar{\mathbf{e}}_1,\bar{\mathbf{e}}_2,\bar{\mathbf{e}}_3)$, prove that

$$u_i = \frac{\partial \bar{z}_m}{\partial z_i} \bar{u}_m$$

Compare this result with that of Exercise A.1.5-3.

**A.3.1-4**  Let $(z_1,z_2,z_3)$ and $(\bar{z}_1,\bar{z}_2,\bar{z}_3)$ denote two rectangular cartesian coordinate systems such that

$$z_i = z_i(\bar{z}_1,\bar{z}_2,\bar{z}_3) \quad \text{for } i = 1, 2, 3$$

Prove that

$$\frac{\partial \bar{z}_i}{\partial z_j} = \frac{\partial z_j}{\partial \bar{z}_i}$$

## A.4 CURVILINEAR COORDINATES

**\*\*A.4.1  Curvilinear coordinates**  Consider some curve in space and let $t$ be a parameter measured along this curve. Let $\mathbf{p}$ be a position vector-valued function of $t$ along this curve. We define

$$\frac{d\mathbf{p}}{dt}(t) \equiv \lim_{\Delta t \to 0} \frac{\mathbf{p}(t + \Delta t) - \mathbf{p}(t)}{\Delta t} \tag{1-1}$$

Figure A.4.1-1 suggests that $(d\mathbf{p}/dt)(t)$ is a tangent vector to the curve at the point $t$. In Sec. A.1.3 we introduced the position vector field $\mathbf{p}$ which in Sec. A.1.5 we

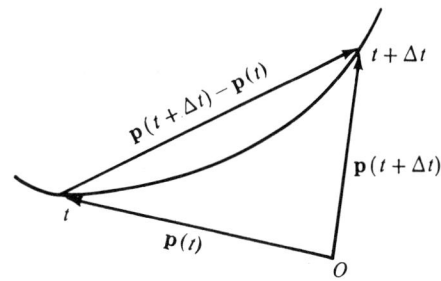

**Fig. A.4.1-1**

expressed in terms of its rectangular cartesian coordinates:

$$\mathbf{p} = z_i \mathbf{e}_i \qquad (1\text{-}2)$$

Let us assume that each $z_i$ ($i = 1, 2, 3$) may be regarded as a function of three parameters $(x^1, x^2, x^3)$, called *curvilinear coordinates*:

$$z_i = z_i(x^1, x^2, x^3) \qquad \text{for } i = 1, 2, 3 \qquad (1\text{-}3)$$

[Here we use the common notation-preserving device of employing the same symbol for both the function $z_i$ and its value $z_i(x^1, x^2, x^3)$.] For fixed values of $x^1$, $x^2$, $x^3$, respectively, these equations define surfaces called curvilinear surfaces. The curve of intersection of any two curvilinear surfaces defines a curvilinear coordinate line.

The spatial vector field $\mathbf{g}_k$ ($k = 1, 2, 3$) is defined as

$$\mathbf{g}_k \equiv \frac{\partial \mathbf{p}}{\partial x^k} = \frac{\partial(z_j \mathbf{e}_j)}{\partial x^k} = \frac{\partial z_j}{\partial x^k} \mathbf{e}_j \qquad (1\text{-}4)$$

At any point $z$, the spatial vector $\mathbf{g}_k(z)$ is tangent to the $x^k$-coordinate curve. Note that, in general, the magnitude and direction of $\mathbf{g}_k(z)$ varies with position $z$ in $E$.

May the three spatial vector fields $(\mathbf{g}_1, \mathbf{g}_2, \mathbf{g}_3)$ be regarded as a new basis for the spatial vector fields? It follows from our discussion in Sec. A.1.4 that they may, if we can demonstrate that every spatial vector field can be written as a linear combination of them. Since every spatial vector field can be written as a linear combination of the rectangular cartesian basis fields $(\mathbf{e}_1, \mathbf{e}_2, \mathbf{e}_3)$, all that it is necessary to show is that all the $\mathbf{e}_i$ may be expressed as linear combinations of the $\mathbf{g}_k$. The system of linear equations

$$\frac{\partial z_j}{\partial x^k} \mathbf{e}_j = \mathbf{g}_k \qquad (1\text{-}5)$$

can be solved for the $\mathbf{e}_j$ so long as the determinant

$$\det\left(\frac{\partial z_j}{\partial x^k}\right) = \begin{vmatrix} \frac{\partial z_1}{\partial x^1} & \frac{\partial z_1}{\partial x^2} & \frac{\partial z_1}{\partial x^3} \\ \frac{\partial z_2}{\partial x^1} & \frac{\partial z_2}{\partial x^2} & \frac{\partial z_2}{\partial x^3} \\ \frac{\partial z_3}{\partial x^1} & \frac{\partial z_3}{\partial x^2} & \frac{\partial z_3}{\partial x^3} \end{vmatrix} \neq 0 \qquad (1\text{-}6)$$

When this condition is satisfied, since

$$\frac{\partial z_m}{\partial x^k} \frac{\partial x^k}{\partial z_n} = \delta_{mn} \qquad (1\text{-}7)$$

## [A.4] CURVILINEAR COORDINATES

we may write from Eq. (1-5)

$$\frac{\partial x^k}{\partial z_m} \frac{\partial z_j}{\partial x^k} \mathbf{e}_j = \delta_{mj} \mathbf{e}_j$$

$$= \mathbf{e}_m$$

$$= \frac{\partial x^k}{\partial z_m} \mathbf{g}_k \tag{1-8}$$

We demonstrate in this way that the $\mathbf{g}_k$ ($k = 1, 2, 3$) may be regarded as a set of basis fields for the spatial vector fields.

We will refer to any set of parameters $(x^1, x^2, x^3)$ that satisfies Eq. (1-6) as a *curvilinear coordinate system*. We refer to the set $(\mathbf{g}_1, \mathbf{g}_2, \mathbf{g}_3)$ as the *natural basis* for this curvilinear coordinate system.

The natural basis fields are orthogonal if

$$\mathbf{g}_i \cdot \mathbf{g}_j = 0 \quad \text{for } i \neq j \tag{1-9}$$

When the natural basis fields are orthogonal to one another, we say that they correspond to an *orthogonal coordinate system*. Geometrically, it is clear that the cylindrical and spherical coordinate systems of Exercises A.4.1-4 and A.4.1-5 are orthogonal. One may check whether any coordinate system is orthogonal by examining

$$g_{ij} \equiv \mathbf{g}_i \cdot \mathbf{g}_j = \frac{\partial \mathbf{p}}{\partial x^i} \cdot \frac{\partial \mathbf{p}}{\partial x^j}$$

$$= \frac{\partial z_m}{\partial x^i} \mathbf{e}_m \cdot \frac{\partial z_n}{\partial x^j} \mathbf{e}_n$$

$$= \frac{\partial z_m}{\partial x^i} \frac{\partial z_n}{\partial x^j} \delta_{mn}$$

$$= \frac{\partial z_m}{\partial x^i} \frac{\partial z_m}{\partial x^j} \tag{1-10}$$

In applications, it is usually found to be more convenient to work in terms of an orthogonal cartesian (orthonormal) basis (see Sec. A.1.5). The natural basis fields defined by Eq. (1-9) may be normalized to form an orthonormal basis $(\mathbf{g}_{\langle 1 \rangle}, \mathbf{g}_{\langle 2 \rangle}, \mathbf{g}_{\langle 3 \rangle})$:

$$\mathbf{g}_{\langle i \rangle} \equiv \frac{\mathbf{g}_i}{\sqrt{\mathbf{g}_i \cdot \mathbf{g}_i}} = \frac{\mathbf{g}_i}{\sqrt{g_{ii}}} \quad \text{no summation on } i \tag{1-11}$$

This basis is referred to as the *physical basis* for the coordinate system.

In this text we will not discuss the normalized natural basis except for the case of orthogonal coordinate systems.

Any spatial vector $\mathbf{u}$ may, consequently, be expressed as a linear combination of the three physical basis vector fields associated with an orthogonal coordinate system:

$$\mathbf{u} = u_{\langle i \rangle} \mathbf{g}_{\langle i \rangle} \tag{1-12}$$

The three coefficients $(u_{\langle 1 \rangle}, u_{\langle 2 \rangle}, u_{\langle 3 \rangle})$ are referred to as the *physical components* of **u** with respect to this particular coordinate system.

In applications, we are almost always concerned with orthogonal coordinate systems and physical components of spatial vector fields. We visualize spatial vectors and spatial vector fields that have some physical interpretation (such as velocity or force) in terms of their physical components. Since we will most readily formulate boundary conditions to differential equations in terms of physical components, we will find it most natural to formulate specific problems to be solved in terms of physical components. Many engineering texts deal with physical components exclusively, never mentioning the covariant and contravariant components discussed in Sec. A.4.3.

## EXERCISES

**A.4.1-1** Discuss why, in a rectangular cartesian coordinate system, the basis vector fields $\mathbf{e}_k$ ($k = 1, 2, 3$) are what we term here the natural basis vectors.

**A.4.1-2** If the $\mathbf{g}_i$ ($i = 1, 2, 3$) are the natural basis vector fields associated with one curvilinear coordinate system $(x^1, x^2, x^3)$ and if the $\bar{\mathbf{g}}_i$ ($i = 1, 2, 3$) are the natural basis vector fields associated with another curvilinear coordinate system $(\bar{x}^1, \bar{x}^2, \bar{x}^3)$, prove that

$$\bar{\mathbf{g}}_i = \frac{\partial x^j}{\partial \bar{x}^i} \mathbf{g}_j$$

**A.4.1-3** Prove that for any rectangular cartesian coordinate system

$$g_{ij} = \delta_{ij}$$

Here $\delta_{ij}$ is a Kronecker delta.

**\*\*A.4.1-4** Given a cylindrical coordinate system

$$z_1 = x^1 \cos x^2 = r \cos \theta$$
$$z_2 = x^1 \sin x^2 = r \sin \theta$$
$$z_3 = x^3 = z$$

prove that

$$g_{11} = g_{33} = 1$$
$$g_{22} = (x^1)^2 = r^2$$
$$g_{ij} = 0 \quad \text{if } i \neq j$$

and

$$g \equiv \det(g_{ij}) = r^2$$

**\*\*A.4.1-5** Given a spherical coordinate system

$$z_1 = x^1 \sin x^2 \cos x^3 = r \sin \theta \cos \varphi$$
$$z_2 = x^1 \sin x^2 \sin x^3 = r \sin \theta \sin \varphi$$
$$z_3 = x^1 \cos x^2 = r \cos \theta$$

prove that

$$g_{11} = 1$$
$$g_{22} = (x^1)^2 = r^2$$
$$g_{33} = (x^1 \sin x^2)^2 = (r \sin \theta)^2$$
$$g_{ij} = 0 \quad \text{if } i \neq j$$

## [A.4] CURVILINEAR COORDINATES

and
$$g = \det(g_{ij}) = r^4 \sin^2 \theta$$

**A.4.1-6** In the paraboloidal coordinate system
$$z_1 = x^1 x^2 \cos x^3$$
$$z_2 = x^1 x^2 \sin x^3$$
$$z_3 = \tfrac{1}{2}[(x^1)^2 - (x^2)^2]$$

the $x^1$ surfaces and $x^2$ surfaces are paraboloids of revolution and the $x^3$ surfaces are planes through the $z^3$ axis. Prove that
$$g_{11} = g_{22} = (x^1)^2 + (x^2)^2$$
$$g_{33} = (x^1 x^2)^2$$
and
$$g_{ij} = 0 \quad \text{if } i \neq j$$

**A.4.1-7** Prove that $g_{ij}$ is symmetric in the indices $i$ and $j$.

**\*\*A.4.1-8** Let
$$\varphi(\mathbf{z}) = \text{a constant}$$
be the equation of a surface in euclidean point space. We assume here that an origin $O$ has been specified and that $\varphi$ can be interpreted as a function of the position vector $\mathbf{z}$ measured with respect to this origin rather than the point $z$ itself. Prove that $\nabla \varphi(\mathbf{z})$ is orthogonal to the surface.

*Hint:* Take an arbitrary curve on the surface and let $s$ be a parameter measured along this curve. Consider the implications of $d\varphi/ds = 0$.

**\*\*A.4.1-9** We restrict ourselves to an orthogonal curvilinear coordinate system $x^i$ ($i = 1, 2, 3$). We denote $\mathbf{v} = v_i \mathbf{e}_i$.

(a) Determine that
$$\mathbf{g}_{\langle i \rangle} = \frac{1}{\sqrt{g_{ii}}} \sum_{j=1}^{3} \frac{\partial z_j}{\partial x^i} \mathbf{e}_j \quad \text{no summation on } i$$

(b) Starting with the result of part (a), find that
$$v_j = \sum_{i=1}^{3} \frac{1}{\sqrt{g_{ii}}} \frac{\partial z_j}{\partial x^i} v_{\langle i \rangle}$$

(c) For cylindrical coordinates as defined in Exercise A.4.1-4, show that
$$v_1 = v_r \cos \theta - v_\theta \sin \theta$$
$$v_2 = v_r \sin \theta + v_\theta \cos \theta$$
$$v_3 = v_z$$

Here we introduce the common notation (see Sec. 2.5.1)
$$v_r \equiv v_{\langle 1 \rangle} \quad v_\theta \equiv v_{\langle 2 \rangle} \quad v_z \equiv v_{\langle 3 \rangle}$$

(d) For spherical coordinates as defined in Exercise A.4.1-5, prove that
$$v_1 = v_r \sin \theta \cos \varphi + v_\theta \cos \theta \cos \varphi - v_\varphi \sin \varphi$$
$$v_2 = v_r \sin \theta \sin \varphi + v_\theta \cos \theta \sin \varphi + v_\varphi \cos \varphi$$
$$v_3 = v_r \cos \theta - v_\theta \sin \theta$$

We define here (see Sec. 2.5.1)

$$v_r \equiv v_{\langle 1 \rangle} \qquad v_\theta \equiv v_{\langle 2 \rangle} \qquad v_\varphi \equiv v_{\langle 3 \rangle}$$

**A.4.1-10** We restrict ourselves to an orthogonal curvilinear coordinate system $x^i$ ($i = 1, 2, 3$).
(a) Prove that

$$\mathbf{e}_j = \sum_{i=1}^{3} \sqrt{g_{ii}} \frac{\partial x^i}{\partial z_j} \mathbf{g}_{\langle i \rangle}$$

(b) Starting with the result of part (a), find that

$$v_{\langle i \rangle} = \sqrt{g_{ii}} \sum_{j=1}^{3} \frac{\partial x^i}{\partial z_j} v_j \qquad \text{no summation on } i$$

(c) For cylindrical coordinates as defined in Exercise A.4.1-4, prove that (see Sec. 2.5.1)

$$v_r \equiv v_{\langle 1 \rangle} = v_1 \cos\theta + v_2 \sin\theta$$
$$v_\theta \equiv v_{\langle 2 \rangle} = -v_1 \sin\theta + v_2 \cos\theta$$
$$v_z \equiv v_{\langle 3 \rangle} = v_3$$

(d) For spherical coordinates as defined in Exercise A.4.1-5, prove that (see Sec. 2.5.1)

$$v_r \equiv v_{\langle 1 \rangle} = v_1 \sin\theta \cos\varphi + v_2 \sin\theta \sin\varphi + v_3 \cos\theta$$
$$v_\theta \equiv v_{\langle 2 \rangle} = v_1 \cos\theta \cos\varphi + v_2 \cos\theta \sin\varphi - v_3 \sin\theta$$
$$v_\varphi \equiv v_{\langle 3 \rangle} = -v_1 \sin\varphi + v_2 \cos\varphi$$

**A.4.1-11** (a) For cylindrical coordinates as defined in Exercise A.4.1-4, determine that the physical components of the position vector field

$$\mathbf{p} = p_{\langle i \rangle} \mathbf{g}_{\langle i \rangle}$$

are

$$p_{\langle 1 \rangle} = r \qquad p_{\langle 2 \rangle} = 0 \qquad p_{\langle 3 \rangle} = z$$

(b) For spherical coordinates as defined in Exercise A.4.1-5, determine that

$$p_{\langle 1 \rangle} = r \qquad p_{\langle 2 \rangle} = p_{\langle 3 \rangle} = 0$$

**A.4.1-12** In the ellipsoidal coordinate system

$$z_1 = \left[ \frac{(x^1 - a)(x^2 - a)(x^3 - a)}{(b - a)(c - a)} \right]^{\frac{1}{2}}$$

$$z_2 = \left[ \frac{(x^1 - b)(x^2 - b)(x^3 - b)}{(c - b)(a - b)} \right]^{\frac{1}{2}}$$

$$z_3 = \left[ \frac{(x^1 - c)(x^2 - c)(x^3 - c)}{(a - c)(b - c)} \right]^{\frac{1}{2}}$$

$a$, $b$, and $c$ are constants such that $a > b > c > 0$. The $x^1$ surfaces are ellipsoids, the $x^2$ surfaces are hyperboloids of one sheet, the $x^3$ surfaces are hyperboloids of two sheets, and all the quadrics belong to the family of confocals

$$\frac{(z_1)^2}{y - a} + \frac{(z_2)^2}{y - b} + \frac{(z_3)^2}{y - c} = 1$$

## [A.4] CURVILINEAR COORDINATES

Prove that

$$g_{11} = \frac{(x^3 - x^1)(x^2 - x^1)}{4(x^1 - a)(x^1 - b)(x^1 - c)}$$

$$g_{22} = \frac{(x^1 - x^2)(x^3 - x^2)}{4(x^2 - a)(x^2 - b)(x^2 - c)}$$

$$g_{33} = \frac{(x^1 - x^3)(x^2 - x^3)}{4(x^3 - a)(x^3 - b)(x^3 - c)}$$

and

$$g_{ij} = 0 \quad \text{if } i \neq j$$

**\*A.4.2 The dual basis** Another interesting set of spatial vector fields associated with a curvilinear coordinate system are the *dual vector fields* $\mathbf{g}^i$ ($i = 1, 2, 3$), defined as the gradients of the curvilinear coordinates:

$$\mathbf{g}^i \equiv \nabla x^i \tag{2-1}$$

It is reasonable to ask whether the dual vector fields may also be regarded as a basis for the space of spatial vector fields. Before answering this question, let us examine some of the properties of these fields.

The dual vector fields, like all other spatial vector fields, may be expressed as linear combinations of the natural basis:

$$\mathbf{g}^i = g^{ji}\mathbf{g}_j \tag{2-2}$$

The coefficients $g^{ji}$ may be regarded as being defined by this equation. Let us consider the scalar product of one of the dual vector fields with one of the natural basis fields:

$$\mathbf{g}^i \cdot \mathbf{g}_j = \nabla x^i \cdot \frac{\partial \mathbf{p}}{\partial x^j}$$

$$= \frac{\partial x^i}{\partial z_m} \mathbf{e}_m \cdot \frac{\partial z_n}{\partial x^j} \mathbf{e}_n$$

$$= \frac{\partial x^i}{\partial z_m} \frac{\partial z_n}{\partial x^j} \delta_{mn}$$

$$= \frac{\partial x^i}{\partial z_m} \frac{\partial z_m}{\partial x^j}$$

$$= \delta_j{}^i \tag{2-3}$$

Here $\delta_j{}^i$ is another form of the Kronecker delta (Secs. A.1.7 and A.2.1); the index $i$ is used in the superscript position only to preserve for the reader's eye the relative positions of $i$ and $j$ in the preceding lines. With a minimum of artificiality in the notation, indices associated with the curvilinear coordinates will maintain their relative position (superscript or subscript) in every term of an equation hereafter. We will encourage this symmetry with appropriate choices of notation where necessary, since we will find that it aids our memory and serves as a quick check for certain types of gross errors in equations. We shall elaborate further on this point shortly.

Referring to Eq. (2-2) and Sec. A.4.1, we see that another way of expressing Eq. (2-3) is to write

$$g^{ki}\mathbf{g}_k \cdot \mathbf{g}_j = g_{jk}g^{ki} = \delta_j^{\ i} \tag{2-4}$$

This should remind the reader of the discussion of cofactors in Sec. A.2.1. The discussion of determinants given there remains valid whether we use superscripts, subscripts, or any appropriate mixture. With this thought in mind, we may recognize $g^{im}$ as the cofactor of the $g_{mi}$ in det $(g_{rs})$ divided by det $(g_{rs})$:

$$g^{im} = \frac{1}{2 \det(g_{rs})} e^{ijk} e^{mnp} g_{nj} g_{pk} \tag{2-5}$$

Notice that in writing Eq. (2-5) we used $e^{ijk}$, which was defined in Sec. A.2.1 but never used. Our excuse for using it here is that we wish to foster the idea of symmetry of indices. Notice also how understood summations on indices associated with curvilinear coordinate systems occur between one superscript and one subscript. There will be more on this later.

In arriving at Eq. (2-5), we divided by det $(g_{rs})$, assuming that det $(g_{rs})$ cannot be zero. Let us prove this. Starting with the definition of $g_{mn}$, we have

$$\begin{aligned} g_{mn} = \mathbf{g}_m \cdot \mathbf{g}_n &= \frac{\partial \mathbf{p}}{\partial x^m} \cdot \frac{\partial \mathbf{p}}{\partial x^n} \\ &= \frac{\partial z_i}{\partial x^m} \mathbf{e}_i \cdot \frac{\partial z_j}{\partial x^n} \mathbf{e}_j \\ &= \frac{\partial z_i}{\partial x^m} \frac{\partial z_j}{\partial x^n} \delta_{ij} \\ &= \frac{\partial z_i}{\partial x^m} \frac{\partial z_i}{\partial x^n} \end{aligned} \tag{2-6}$$

This means that

$$\det(g_{mn}) = \det\left(\frac{\partial z_i}{\partial x^m} \frac{\partial z_i}{\partial x^n}\right) \tag{2-7}$$

Our discussion of the product of two determinants in Sec. A.2.1 allows us to write this as

$$\det(g_{mn}) = \left[\det\left(\frac{\partial z_i}{\partial x^j}\right)\right]^2 \tag{2-8}$$

But the restriction placed upon the definition of a curvilinear coordinate system in Sec. A.4.1 allows us to conclude that det $(g_{mn}) \neq 0$.

From Eqs. (2-2) and (2-3), we have that

$$\begin{aligned} \mathbf{g}^i \cdot \mathbf{g}^j &= g^{ki}\mathbf{g}_k \cdot \mathbf{g}^j \\ &= g^{ki} \delta_k^{\ j} \\ &= g^{ji} \end{aligned} \tag{2-9}$$

Since the scalar product is symmetric, $g^{ij}$ is symmetric in its indices.

## [A.4] CURVILINEAR COORDINATES

Let us now prove that the three dual vector fields $\mathbf{g}^i$ ($i = 1, 2, 3$) form another basis for the spatial vector fields. From our discussion in Sec. A.1.4, all that it is necessary to show is that every spatial vector field can be written as a linear combination of them. Since we have already shown that every spatial vector field can be written as a linear combination of the natural basis, all that we must demonstrate is that each of the natural basis fields can be expressed as a linear combination of the dual vector fields. Multiplying Eq. (2-2) by $g_{ik}$, summing on $i$, and employing Eq. (2-4)$_2$, we obtain

$$g_{ik}\mathbf{g}^i = g^{ji}g_{ik}\mathbf{g}_j$$
$$= \delta_k^j \mathbf{g}_j$$
$$= \mathbf{g}_k \qquad (2\text{-}10)$$

which completes the proof.

### EXERCISES

**A.4.2-1** Prove that in a rectangular cartesian coordinate system the natural basis vectors $\mathbf{e}_k$ and the dual basis vectors $\mathbf{e}^k$ are identical.

**A.4.2-2** If the $\mathbf{g}^i$ ($i = 1, 2, 3$) are the dual basis vector fields associated with one curvilinear coordinate system $(x^1, x^2, x^3)$ and if the $\bar{\mathbf{g}}^i$ ($i = 1, 2, 3$) are the dual basis vector fields associated with another curvilinear coordinate system $(\bar{x}^1, \bar{x}^2, \bar{x}^3)$, prove that

$$\bar{\mathbf{g}}^i = \frac{\partial \bar{x}^i}{\partial x^j} \mathbf{g}^j$$

**A.4.2-3** Prove that in orthogonal coordinate systems the dual basis fields are orthogonal to one another.

**A.4.2-4** If the curvilinear coordinates are orthogonal, prove that

$$\det(g_{mn}) = g_{11}g_{22}g_{33}$$

$$g^{11} = \frac{1}{g_{11}} \quad g^{22} = \frac{1}{g_{22}} \quad g^{33} = \frac{1}{g_{33}}$$

and

$$g_{mn} = g^{mn} = 0 \quad \text{if } m \neq n$$

### *A.4.3 Covariant and contravariant components of spatial vector fields

Given a curvilinear coordinate system, we may express every spatial vector field as a linear combination of the natural basis:

$$\mathbf{u} = u^i \mathbf{g}_i \qquad (3\text{-}1)$$

or as a linear combination of the dual basis:

$$\mathbf{u} = u_i \mathbf{g}^i \qquad (3\text{-}2)$$

The $u^i$ and $u_i$ are referred to, respectively, as the *contravariant* and *covariant* components of the spatial vector field $\mathbf{u}$.

It is because we concern ourselves with these two sets of bases in dealing with each curvilinear coordinate system that we choose to introduce superscripts as well as subscripts in our notation. We will notice hereafter that, when the summation convention is employed with covariant and contravariant components, one of the repeated indices will be a superscript and one will be a subscript. This is primarily because of our arbitrary choice of notation in Eqs. (3-1) and (3-2); it is here that the summation between superscripts and subscripts is introduced.

The notation has at least one helpful feature which should be kept in mind. We will see that any equation involving components will have a certain symmetry with respect to indices not involved in summations. For example, if the index $j$ is not repeated and if it occurs as a superscript in one term of the equation, it will occur as a superscript in all terms of the equation.

Why did we not use superscripts as well as subscripts when discussing rectangular cartesian coordinate systems? In Exercise A.4.2-1 one learns that the natural and dual basis vectors are identical in orthogonal cartesian coordinate systems. Consequently, it is pointless to distinguish between covariant and contravariant components of vectors in rectangular cartesian coordinates, and the need for superscripts as well as subscripts disappears.

Since for any spatial vector field **u**,

$$\mathbf{u} = u^i \mathbf{g}_i = u^i g_{ki} \mathbf{g}^k = u_k \mathbf{g}^k \tag{3-3}$$

we may write

$$(u^i g_{ki} - u_k) \mathbf{g}^k = 0 \tag{3-4}$$

The dual basis vector fields are linearly independent and therefore Eq. (3-4) implies that

$$u^i g_{ki} - u_k = 0 \tag{3-5}$$

or that

$$u_k = g_{ki} u^i \tag{3-6}$$

In the same way,

$$\mathbf{u} = u_i \mathbf{g}^i = u_i g^{ji} \mathbf{g}_j = u^j \mathbf{g}_j \tag{3-7}$$

so that we may identify

$$u^j = g^{ji} u_i \tag{3-8}$$

We find in this way that the $g_{ij}$ and the $g^{ij}$ may be used to "raise and lower" indices.

Let us determine the relation between the physical components $(u_{\langle 1 \rangle}, u_{\langle 2 \rangle}, u_{\langle 3 \rangle})$ of a spatial vector field **u** and its contravariant components. From Sec. A.4.1 and Eq. (3-1), we may write

$$\mathbf{u} = u^i \mathbf{g}_i = u^1 \sqrt{g_{11}} \, \mathbf{g}_{\langle 1 \rangle} + u^2 \sqrt{g_{22}} \, \mathbf{g}_{\langle 2 \rangle} + u^3 \sqrt{g_{33}} \, \mathbf{g}_{\langle 3 \rangle}$$

$$= u_{\langle i \rangle} \mathbf{g}_{\langle i \rangle} \tag{3-9}$$

## [A.4] CURVILINEAR COORDINATES

We conclude that

$$u_{\langle i \rangle} = \sqrt{g_{ii}}\, u^i \qquad \text{no summation on } i \tag{3-10}$$

A similar relation may be obtained for the physical components in terms of the covariant components. From the definition of the physical basis fields in Sec. A.4.1 and the relation between the dual basis fields and natural basis fields in Sec. A.4.2, we have

$$\mathbf{g}_{\langle i \rangle} = \frac{1}{\sqrt{g_{ii}}} g_{ii} \mathbf{g}^i = \sqrt{g_{ii}}\, \mathbf{g}^i \qquad \text{no summation on } i \tag{3-11}$$

In arriving at this result, we have taken advantage of the restriction to orthogonal coordinate systems when discussing physical basis fields. From Eq. (3-2),

$$\mathbf{u} = u_i \mathbf{g}^i = \frac{u_1}{\sqrt{g_{11}}} \mathbf{g}_{\langle 1 \rangle} + \frac{u_2}{\sqrt{g_{22}}} \mathbf{g}_{\langle 2 \rangle} + \frac{u_3}{\sqrt{g_{33}}} \mathbf{g}_{\langle 3 \rangle}$$

$$= u_{\langle i \rangle} \mathbf{g}_{\langle i \rangle} \tag{3-12}$$

We have consequently that

$$u_{\langle i \rangle} = \frac{u_i}{\sqrt{g_{ii}}} \qquad \text{no summation on } i \tag{3-13}$$

## EXERCISES

**A.4.3-1** Show the following in rectangular cartesian coordinates.
  (a) The natural basis fields and the physical basis fields are equivalent.
  (b) It is unnecessary to distinguish between covariant, contravariant, and physical components of spatial vector fields in rectangular cartesian coordinates. We may write all indices as subscripts, therefore, employing the summation convention over repeated subscripts so long as we restrict ourselves to *rectangular cartesian coordinates.*

**A.4.3-2** (a) Let **u** be some spatial vector field. If the $u^i$ ($i = 1, 2, 3$) are the contravariant components of **u** with respect to one curvilinear coordinate system ($x^1, x^2, x^3$) and if the $\bar{u}^i$ ($i = 1, 2, 3$) are the contravariant components of **u** with respect to another curvilinear coordinate system ($\bar{x}^1, \bar{x}^2, \bar{x}^3$), show that

$$u^i = \frac{\partial x^i}{\partial \bar{x}^j} \bar{u}^j$$

  (b) Similarly, show that

$$u_i = \frac{\partial \bar{x}^j}{\partial x^i} \bar{u}_j$$

**A.4.3-3** Show that the angle between two surfaces $\varphi(x^1, x^2, x^3) = $ a constant and $\psi(x^1, x^2, x^3) = $ a constant is given by

$$\cos \theta = \frac{g^{mn}(\partial \varphi / \partial x^m)(\partial \psi / \partial x^n)}{[g^{rs}(\partial \varphi / \partial x^r)(\partial \varphi / \partial x^s) g^{uv}(\partial \psi / \partial x^u)(\partial \psi / \partial x^v)]^{1/2}}$$

*Hint:* See Exercise A.4.1-8.

**A.4.3-4** Deduce that the angle $\varphi_{12}$ between the coordinate surfaces $x^1 = $ a constant and $x^2 = $ a constant is given by

$$\cos \varphi_{12} = \frac{g^{12}}{\sqrt{g^{11}g^{22}}}$$

**A.4.3-5** Establish that if two surfaces $\varphi(x^1,x^2,x^3) = $ a constant and $\psi(x^1,x^2,x^3) = $ a constant cut orthogonally, then

$$g^{mn}\frac{\partial \varphi}{\partial x^m}\frac{\partial \psi}{\partial x^n} = 0$$

**A.4.3-6** Let $\varphi$ be a scalar field. Starting with the expression for $\nabla \varphi$ with respect to a rectangular cartesian coordinate basis, show that

$$\nabla \varphi = \frac{\partial \varphi}{\partial x^i} \mathbf{g}^i$$

If we restrict ourselves to an orthogonal coordinate system, we have

$$\nabla \varphi = \sum_{i=1}^{3} \frac{1}{\sqrt{g_{ii}}} \frac{\partial \varphi}{\partial x^i} \mathbf{g}_{\langle i \rangle}$$

## A.5 SECOND-ORDER TENSORS

**\*\*A.5.1 Second-order tensor fields** A second-order tensor field $\mathbf{T}$ is a transformation (or mapping or rule) that assigns to each given spatial vector field $\mathbf{v}$ another spatial vector field $\mathbf{T} \cdot \mathbf{v}$ such that the rules

$$\mathbf{T} \cdot (\mathbf{v} + \mathbf{w}) = \mathbf{T} \cdot \mathbf{v} + \mathbf{T} \cdot \mathbf{w}$$
$$\mathbf{T} \cdot (\alpha \mathbf{v}) = \alpha(\mathbf{T} \cdot \mathbf{v}) \tag{1-1}$$

hold. By $\alpha$ we mean here a real scalar field. (Please note that we are using the dot notation in a different manner here from that used in Sec. A.1.1 when we discussed the inner product. Our choice of notation is suggestive, however, as will shortly become evident.)

We define the sum $\mathbf{T} + \mathbf{S}$ of two second-order tensor fields $\mathbf{T}$ and $\mathbf{S}$ to be a transformation such that, for every spatial vector field $\mathbf{v}$,

$$(\mathbf{T} + \mathbf{S}) \cdot \mathbf{v} = \mathbf{T} \cdot \mathbf{v} + \mathbf{S} \cdot \mathbf{v} \tag{1-2}$$

The product $\alpha \mathbf{T}$ of a second-order tensor field $\mathbf{T}$ with a real scalar field $\alpha$ is a transformation such, that for every spatial vector field $\mathbf{v}$,

$$(\alpha \mathbf{T}) \cdot \mathbf{v} = \alpha(\mathbf{T} \cdot \mathbf{v}) \tag{1-3}$$

The transformation $\mathbf{T} + \mathbf{S}$ and $\alpha \mathbf{T}$ may be easily shown to obey the rules for a second-order tensor field. If we define the zero second-order tensor field $\mathbf{0}$ by the requirement that

$$\mathbf{0} \cdot \mathbf{v} = \mathbf{0} \tag{1-4}$$

## [A.5] SECOND-ORDER TENSORS

for all spatial vector fields **v**, we see that the rules ($A_1$) to ($A_4$) and ($M_1$) to ($M_4$) of Sec. A.1.1 are satisfied and that the set of all second-order tensor fields constitutes a vector space.

If two spatial vector fields **a** and **b** are given, we can define a second-order tensor field **ab** by the requirement that it transform every vector field **v** into another vector field (**ab**) · **v** according to the rule

$$(\mathbf{ab}) \cdot \mathbf{v} = \mathbf{a}(\mathbf{b} \cdot \mathbf{v}) \tag{1-5}$$

This tensor field **ab** is called the *tensor product* or dyadic product of the spatial vector fields **a** and **b**. (Another common notation for the tensor product, preferred by some mathematicians, is **a** ⊗ **b** [1, p. 40; 2, p. 29].)

In this text, we use boldface capital letters for second-order tensor fields and boldface small letters for spatial vector fields.

### REFERENCES

1. Halmos, P. R.: "Finite-dimensional Vector Spaces," 2d ed., Van Nostrand, Princeton, N.J., 1958.
2. Lichnerowicz, A.: "Elements of Tensor Calculus," Wiley, New York, 1962.

**\*\*A.5.2 Components of second-order tensor fields** If **T** is a second-order tensor field and the $\mathbf{e}_j$ ($j = 1, 2, 3$) form a rectangular cartesian basis for the space of spatial vector fields, then we may write

$$\mathbf{T} \cdot \mathbf{e}_j = T_{ij} \mathbf{e}_i \tag{2-1}$$

The *matrix* $[T_{ij}]$ (array of components) of the second-order tensor field **T**,

$$[T_{ij}] = \begin{bmatrix} T_{11} & T_{12} & T_{13} \\ T_{21} & T_{22} & T_{23} \\ T_{31} & T_{32} & T_{33} \end{bmatrix} \tag{2-2}$$

tells how the basis fields $\mathbf{e}_j$ are transformed by **T**.

Let **v** be any spatial vector field. Equation (2-1) allows us to develop an expression for the vector field **T** · **v** in terms of the rectangular cartesian components of **v**:

$$\begin{aligned}\mathbf{T} \cdot \mathbf{v} &= \mathbf{T} \cdot (v_j \mathbf{e}_j) \\ &= v_j \mathbf{T} \cdot \mathbf{e}_j \\ &= v_j T_{ij} \mathbf{e}_i \end{aligned} \tag{2-3}$$

In this way, for each set of basis fields ($\mathbf{e}_1, \mathbf{e}_2, \mathbf{e}_3$), we may associate a matrix $[T_{ij}]$ with any second-order tensor field **T**. This association or correspondence is one-to-one (that is, for the same set of basis fields, the matrices of two different second-order tensor fields are different) [1, p. 67]. In order to prove this, observe that the matrix $[T_{ij}]$ of a second-order tensor field **T** completely determines **T** [by Eq. (2-3), **T** · **v** is determined for every **v**].

Given any second-order tensor field **T**, which transforms the rectangular cartesian basis fields according to Eq. (2-1), define a new second-order tensor **T***:

$$\mathbf{T}^* = T_{ij}\mathbf{e}_i\mathbf{e}_j \tag{2-4}$$

which is the sum of nine tensor products (see Sec. A.5.1). But

$$\mathbf{T}^* \cdot \mathbf{e}_j = (T_{ik}\mathbf{e}_i\mathbf{e}_k) \cdot \mathbf{e}_j$$
$$= T_{ik}\mathbf{e}_i \delta_{kj}$$
$$= T_{ij}\mathbf{e}_i \tag{2-5}$$

indicating that the same matrix $[T_{ij}]$ corresponds to both **T** and **T***. Since we showed above that, with the choice of a particular set of rectangular cartesian basis fields $\mathbf{e}_i$ ($i = 1, 2, 3$), there is a one-to-one correspondence between ($3 \times 3$) matrices and second-order tensor fields, we conclude that

$$\mathbf{T} = \mathbf{T}^* = T_{ij}\mathbf{e}_i\mathbf{e}_j \tag{2-6}$$

The nine coefficients $T_{ij}$ ($i = 1, 2, 3; j = 1, 2, 3$) are referred to as the *rectangular cartesian components* of **T**. We will find this representation for second-order tensor fields in terms of a sum of tensor products of basis fields to be a very useful one.

The *identity* tensor field **I** is a specific example of a second-order tensor field. It transforms every spatial vector field into itself:

$$\mathbf{I} \cdot \mathbf{e}_j = I_{ij}\mathbf{e}_i = \mathbf{e}_j = \delta_{ij}\mathbf{e}_i \tag{2-7}$$

Here $\delta_{ij}$ is the Kronecker delta defined in Sec. A.1.7. From Eq. (2-7), we have

$$(I_{ij} - \delta_{ij})\mathbf{e}_i = \mathbf{0} \tag{2-8}$$

But since the rectangular cartesian basis fields are linearly independent (Sec. A.1.4), we conclude that

$$I_{ij} = \delta_{ij} \tag{2-9}$$

Let us pause before pursuing these ideas further to say something about the notation I have chosen to use here. When we write $\mathbf{T} \cdot \mathbf{e}_i$, the dot is to remind us that **T** operates on the quantity that follows. It has a completely different significance from the dot in $\mathbf{e}_i \cdot \mathbf{e}_j$, where the dot indicates a scalar product. This is a disadvantage to the notation chosen, but I feel that, when any equation is read in context, there is little excuse for confusion. By this I mean that, having been told that **T** is a second-order tensor field and that $\mathbf{e}_i$ is a spatial vector field, the reader will have no occasion to interpret $\mathbf{T} \cdot \mathbf{e}_i$ as the scalar product of two spatial vector fields. The advantage of the notation is that it is suggestive of the operation to be carried out when **T** is written as a sum of tensor products:

$$\mathbf{T} \cdot \mathbf{e}_i = (T_{jk}\mathbf{e}_j\mathbf{e}_k) \cdot \mathbf{e}_i$$
$$= T_{jk}\mathbf{e}_j(\mathbf{e}_k \cdot \mathbf{e}_i) \tag{2-10}$$

The dot in $\mathbf{T} \cdot \mathbf{e}_i$ *reminds* the reader that, when **T** is written as the sum of tensor products, the transformation is accomplished by taking the *scalar product* between

## [A.5] SECOND-ORDER TENSORS

the second spatial vector of the tensor product and the spatial vector to be transformed, $e_i$. The notation adopted here is more common in engineering and applied science texts where considerable emphasis is placed upon working out problems in specific coordinate systems. Mathematicians adopt a slightly different notation when treating subjects where the introduction of coordinate systems is either avoided or is of secondary importance. If one understands any one system of notation, there is little difficulty in adapting to another.

Let us return to Eq. (2-6) and observe that the set of nine tensor products $e_i e_j$ ($i = 1, 2, 3; j = 1, 2, 3$) forms a basis (Sec. A.1.4) for the vector space of second-order tensor fields (Sec. A.5.1). Certainly, every element of the set of second-order tensor fields is expressible as a linear combination of the $e_i e_j$. We must show that the $e_i e_j$ are linearly independent. If

$$\mathbf{A} = A_{ij} \mathbf{e}_i \mathbf{e}_j = 0 \tag{2-11}$$

then

$$\mathbf{A} \cdot \mathbf{e}_k = (A_{ij} \mathbf{e}_i \mathbf{e}_j) \cdot \mathbf{e}_k = 0 \cdot \mathbf{e}_k = 0$$

$$A_{ij} \mathbf{e}_i \, \delta_{jk} = 0$$

as

$$A_{ik} \mathbf{e}_i = 0 \tag{2-12}$$

Since the rectangular cartesian basis fields are linearly independent, this implies that

$$A_{mk} = 0 \tag{2-13}$$

We conclude that the nine tensor products of the form $\mathbf{e}_i \mathbf{e}_j$ are linearly independent and consequently form a basis for the vector space of second-order tensors. As a by-product, we find that the vector space of second-order tensors is nine dimensional (Sec. A.1.4).

In physical applications it is often convenient to introduce an orthogonal curvilinear coordinate system. If the $\mathbf{g}_{\langle i \rangle}$ ($i = 1, 2, 3$) are the associated physical basis fields (Sec. A.4.1), then we may write by analogy with Eq. (2-1)

$$\mathbf{T} \cdot \mathbf{g}_{\langle i \rangle} = T_{\langle ji \rangle} \mathbf{g}_{\langle j \rangle} \tag{2-14}$$

By the same argument that led us to Eq. (2-6), we may consequently write

$$\mathbf{T} = T_{\langle ij \rangle} \mathbf{g}_{\langle i \rangle} \mathbf{g}_{\langle j \rangle} \tag{2-15}$$

where the nine coefficients $T_{\langle ij \rangle}$ ($i = 1, 2, 3; j = 1, 2, 3$) are referred to as the *physical components* of $\mathbf{T}$. The set of nine tensor products $\mathbf{g}_{\langle i \rangle} \mathbf{g}_{\langle j \rangle}$ ($i = 1, 2, 3; j = 1, 2, 3$) forms another basis for the vector space of second-order tensor fields.

## REFERENCE

1. Halmos, P. R.: "Finite-dimensional Vector Spaces," 2d ed., Van Nostrand, Princeton, N.J., 1958.

## EXERCISES

**A.5.2-1** (*a*) Given an orthogonal curvilinear coordinate system, prove that Eq. (2-15) holds for every second-order tensor field $\mathbf{T}$.

(b) Prove that the nine tensor products $\mathbf{g}_{\langle i \rangle}\mathbf{g}_{\langle j \rangle}$ ($i = 1, 2, 3; j = 1, 2, 3$) form a basis for the vector space of second-order tensor fields.

**\*A.5.2-2** (a) Given any curvilinear coordinate system, prove that every second-order tensor field $\mathbf{T}$ may be written as

$$\mathbf{T} = T^{ij}\mathbf{g}_i\mathbf{g}_j$$

The nine coefficients $T^{ij}$ ($i = 1, 2, 3; j = 1, 2, 3$) are referred to as the *contravariant components* of $\mathbf{T}$.

(b) Prove that the nine tensor products $\mathbf{g}_i\mathbf{g}_j$ ($i = 1, 2, 3; j = 1, 2, 3$) form a basis for the space of second-order tensor fields.

**A.5.2-3** (a) Given any curvilinear coordinate system, prove that every second-order tensor field $\mathbf{T}$ may be written as

$$\mathbf{T} = T_{ij}\mathbf{g}^i\mathbf{g}^j$$

The nine coefficients $T_{ij}$ ($i = 1, 2, 3; j = 1, 2, 3$) are referred to as the *covariant components* of $\mathbf{T}$.

(b) Prove that the nine tensor products $\mathbf{g}^i\mathbf{g}^j$ ($i = 1, 2, 3; j = 1, 2, 3$) form a basis for the space of second-order tensor fields.

**\*A.5.2-4** (a) Given any curvilinear coordinate system, prove that every second-order tensor field $\mathbf{T}$ may be written as

$$\mathbf{T} = T_{i.}^{\ j}\mathbf{g}^i\mathbf{g}_j$$

The nine coefficients $T_{i.}^{\ j}$ ($i = 1, 2, 3; j = 1, 2, 3$) are referred to as the mixed components of $\mathbf{T}$ *covariant* in $i$ and *contravariant* in $j$.

(b) Prove that the nine tensor products $\mathbf{g}^i\mathbf{g}_j$ ($i = 1, 2, 3; j = 1, 2, 3$) form a basis for the space of second-order tensor fields.

**\*A.5.2-5** (a) Given any curvilinear coordinate system, prove that every second-order tensor field $\mathbf{T}$ may be written as

$$\mathbf{T} = T^i_{\ j}\mathbf{g}_i\mathbf{g}^j$$

The nine coefficients $T^i_{\ j}$ ($i = 1, 2, 3; j = 1, 2, 3$) are referred to as the *mixed* components of $\mathbf{T}$, *contravariant* in $i$ and *covariant* in $j$.

(b) Prove that the nine tensor products $\mathbf{g}_i\mathbf{g}^j$ ($i = 1, 2, 3; j = 1, 2, 3$) form a basis for the space of second-order tensor fields.

**\*A.5.2-6** Show that

$$T_{\langle ij \rangle} = \frac{T_{ij}}{\sqrt{g_{ii}g_{jj}}} \qquad \text{no summation on } i \text{ and } j$$

$$T_{\langle ij \rangle} = T^i_{\ j}\frac{\sqrt{g_{ii}}}{\sqrt{g_{jj}}} \qquad \text{no summation on } i \text{ and } j$$

and

$$T_{\langle ij \rangle} = T_{i.}^{\ j}\frac{\sqrt{g_{jj}}}{\sqrt{g_{ii}}} \qquad \text{no summation on } i \text{ and } j$$

**A.5.2-7** Show that the identity tensor has the following equivalent forms:

$$\begin{aligned}\mathbf{I} &= \delta_i^{\ j}\mathbf{g}_j\mathbf{g}^i \\ &= \delta_j^{\ i}\mathbf{g}^j\mathbf{g}_i \\ &= g^{ij}\mathbf{g}_i\mathbf{g}_j \\ &= g_{ij}\mathbf{g}^i\mathbf{g}^j\end{aligned}$$

[A.5] SECOND-ORDER TENSORS

[The $g_{ij}$ and $g^{ij}$ are usually referred to as the covariant metric tensor (components) and contravariant metric tensor (components), respectively. These names will not be used here, since they are not consistent with the form of presentation chosen.]

*A.5.2-8  Show that it is unnecessary to distinguish between covariant, contravariant, and physical components of second-order tensor fields *in rectangular cartesian coordinate systems*. We may therefore write all indices as subscripts, employing the summation convention over repeated subscripts *so long as we restrict ourselves to rectangular cartesian coordinate systems*.

*A.5.2-9  (a) Let **T** be some second-order tensor field. If the $T^{ij}$ ($i = 1, 2, 3; j = 1, 2, 3$) are the contravariant components of **T** with respect to one curvilinear coordinate system $(x^1, x^2, x^3)$ and if the $\bar{T}^{ij}$ ($i = 1, 2, 3; j = 1, 2, 3$) are the contravariant components of **T** with respect to another curvilinear coordinate system $(\bar{x}^1, \bar{x}^2, \bar{x}^3)$, show that

$$T^{ij} = \frac{\partial x^i}{\partial \bar{x}^m} \frac{\partial x^j}{\partial \bar{x}^n} \bar{T}^{mn}$$

(b) Similarly, show that

$$T_{ij} = \frac{\partial \bar{x}^m}{\partial x^i} \frac{\partial \bar{x}^n}{\partial x^j} \bar{T}_{mn}$$

$$T_{i.}^{\;j} = \frac{\partial \bar{x}^m}{\partial x^i} \frac{\partial x^j}{\partial \bar{x}^n} \bar{T}_{m}^{\;\cdot n}$$

and

$$T^i_{\;j} = \frac{\partial x^i}{\partial \bar{x}^m} \frac{\partial \bar{x}^n}{\partial x^j} \bar{T}^{m}_{\;\cdot n}$$

*A.5.2-10  (a) If the $T_{ij}$ ($i = 1, 2, 3; j = 1, 2, 3$) are the covariant components of a second-order tensor field **T** and the $T^{ij}$ ($i = 1, 2, 3; j = 1, 2, 3$) are the contravariant components, show that

$$T_{ij} = g_{im} g_{jn} T^{mn}$$

(b) Similarly, show that

$$T_{i.}^{\;j} = g_{im} T^{mj}$$

$$T^i_{\;j} = g^{im} T_{mj}$$

and

$$T^{ij} = g^{im} g^{jn} T_{mn}$$

We find here that the $g_{ij}$ and the $g^{ij}$ may be used to raise and lower indices. (Compare with the relations between covariant and contravariant components of spatial vector fields found in Sec. A.4.3.) We use the dot in writing $T^i_{\;j}$ and $T_{i.}^{\;j}$ to remind ourselves which index has been raised or lowered; this is unnecessary when dealing with symmetric second-order tensors (Sec. A.5.3).

**A.5.2-11**  Let **T** be some second-order tensor field. If the $T_{ij}$ ($i, j = 1, 2, 3$) are the components of **T** with respect to some rectangular cartesian coordinate system $(z_1, z_2, z_3)$ and the $\bar{T}_{mn}$ ($m, n = 1, 2, 3$) are the components of **T** with respect to another rectangular cartesian coordinate system $(\bar{z}_1, \bar{z}_2, \bar{z}_3)$, show that

$$T_{ij} = \frac{\partial z_i}{\partial \bar{z}_m} \frac{\partial z_j}{\partial \bar{z}_n} \bar{T}_{mn}$$

and

$$T_{ij} = \frac{\partial \bar{z}_m}{\partial z_i} \frac{\partial \bar{z}_n}{\partial z_j} \bar{T}_{mn}$$

**A.5.2-12** (a) Let $\mathbf{T} = T_{ij}\mathbf{e}_i\mathbf{e}_j$ and $\mathbf{S} = S_{ij}\mathbf{e}_i\mathbf{e}_j$ be two second-order tensor fields, where the $T_{ij}$ and $S_{ij}$ $(i, j = 1, 2, 3)$ are real scalar fields. Let $\alpha$ be a real scalar field. Express the second-order tensor fields $\mathbf{T} + \mathbf{S}$ and $\alpha\mathbf{T}$ as linear combinations of the $\mathbf{e}_i\mathbf{e}_j$.

*(b) Express the second-order tensor fields $\mathbf{T} + \mathbf{S}$ and $\alpha\mathbf{T}$ as linear combinations of the $\mathbf{g}_i\mathbf{g}^j$.

## **A.5.3 The transpose of a second-order tensor field; symmetric, skew-symmetric, and orthogonal tensor fields

Let $\mathbf{T}$ be any second-order tensor field. We define $\mathbf{T}^T$, the *transpose* of $\mathbf{T}$, to be that second-order tensor field such that, if $\mathbf{u}$ and $\mathbf{v}$ are any two spatial vector fields,

$$(\mathbf{T} \cdot \mathbf{u}) \cdot \mathbf{v} = \mathbf{u} \cdot (\mathbf{T}^T \cdot \mathbf{v}) \tag{3-1}$$

To determine the relation of $\mathbf{T}^T$ to $\mathbf{T}$, let

$$\mathbf{u} = \mathbf{e}_i \quad \mathbf{v} = \mathbf{e}_j$$
$$\mathbf{T} = T_{mn}\mathbf{e}_m\mathbf{e}_n \quad \mathbf{T}^T = T_{rs}{}^T \mathbf{e}_r\mathbf{e}_s \tag{3-2}$$

Here the $\mathbf{e}_i$ $(i = 1, 2, 3)$ represent a set of rectangular cartesian basis fields. Then

$$(\mathbf{T} \cdot \mathbf{e}_i) \cdot \mathbf{e}_j = \mathbf{e}_i \cdot (\mathbf{T}^T \cdot \mathbf{e}_j)$$
$$(T_{mn}\mathbf{e}_m\mathbf{e}_n \cdot \mathbf{e}_i) \cdot \mathbf{e}_j = \mathbf{e}_i \cdot (T_{rs}{}^T \mathbf{e}_r\mathbf{e}_s \cdot \mathbf{e}_j) \tag{3-3}$$
$$T_{ji} = T_{ij}{}^T$$

We see that, if we represent $\mathbf{T}$ as indicated in Eq. $(3-2)_3$, we may represent its transpose by

$$\mathbf{T}^T = T_{nm}\mathbf{e}_m\mathbf{e}_n \tag{3-4}$$

Since for any spatial vector field $\mathbf{w}$,

$$\mathbf{T} \cdot \mathbf{w} = T_{ij}w_j\mathbf{e}_i$$
$$= T_{ji}{}^T w_j \mathbf{e}_i = w_j T_{ji}{}^T \mathbf{e}_i \tag{3-5}$$

we are prompted to introduce the definition

$$\mathbf{w} \cdot \mathbf{T}^T \equiv \mathbf{T} \cdot \mathbf{w} \tag{3-6}$$

One may think of this operation as being carried out in the following manner:

$$\mathbf{w} \cdot \mathbf{T}^T = (w_k \mathbf{e}_k) \cdot (T_{ji}{}^T \mathbf{e}_j \mathbf{e}_i)$$
$$= w_j T_{ji}{}^T \mathbf{e}_i \tag{3-7}$$

A second-order tensor field $\mathbf{T}$ is said to be *symmetric* if it is identical with its transpose:

$$\mathbf{T} = \mathbf{T}^T \tag{3-8}$$

In terms of their rectangular cartesian components, we have

$$T_{ij} = T_{ij}{}^T = T_{ji} \tag{3-9}$$

## [A.5] SECOND-ORDER TENSORS

A second-order tensor field **T** is said to be *skew-symmetric* if

$$\mathbf{T} = -\mathbf{T}^T \tag{3-10}$$

The relation between the rectangular cartesian components of these two tensor fields is, consequently,

$$T_{ij} = -T_{ij}{}^T = -T_{ji} \tag{3-11}$$

An *orthogonal* tensor field or transformation of the space of spatial vector fields is one that preserves lengths and angles. If **u** and **v** are any two spatial vector fields and **Q** is an orthogonal tensor field, we require

$$(\mathbf{Q} \cdot \mathbf{u}) \cdot (\mathbf{Q} \cdot \mathbf{v}) = \mathbf{u} \cdot \mathbf{v} \tag{3-12}$$

But this means that

$$[\mathbf{Q}^T \cdot (\mathbf{Q} \cdot \mathbf{u})] \cdot \mathbf{v} = [(\mathbf{Q}^T \cdot \mathbf{Q}) \cdot \mathbf{u}] \cdot \mathbf{v}$$
$$= \mathbf{u} \cdot \mathbf{v} \tag{3-13}$$

or

$$[(\mathbf{Q}^T \cdot \mathbf{Q}) \cdot \mathbf{u} - \mathbf{u}] \cdot \mathbf{v} = 0 \tag{3-14}$$

Since **u** and **v** are arbitrary spatial vector fields, we conclude that

$$(\mathbf{Q}^T \cdot \mathbf{Q}) \cdot \mathbf{u} = \mathbf{u} \tag{3-15}$$

and

$$\mathbf{Q}^T \cdot \mathbf{Q} = \mathbf{I} \tag{3-16}$$

It can be shown further that (see Exercise A.5.3-3)

$$\mathbf{Q} \cdot \mathbf{Q}^T = \mathbf{Q}^T \cdot \mathbf{Q} = \mathbf{I} \tag{3-17}$$

Here we introduce the notation **A** · **B**, where **A** and **B** are any two second-order tensor fields. If **v** is any spatial vector field, the spatial vector field **A** · (**B** · **v**) is obtained by first applying the transformation **B** to the spatial vector field **v** and then applying the transformation **A** to the result. We may think of **A** · (**B** · **v**) as being obtained from **v** as the consequence of one transformation **A** · **B**:

$$\mathbf{A} \cdot (\mathbf{B} \cdot \mathbf{v}) = (\mathbf{A} \cdot \mathbf{B}) \cdot \mathbf{v} \tag{3-18}$$

where

$$\mathbf{A} \cdot \mathbf{B} \equiv (A_{ij}\mathbf{e}_i\mathbf{e}_j) \cdot (B_{km}\mathbf{e}_k\mathbf{e}_m)$$
$$= A_{ij}B_{jm}\mathbf{e}_i\mathbf{e}_m \tag{3-19}$$

This observation gives us another view of second-order tensor fields. Any second-order tensor field **A** is a transformation that assigns to any other second-order tensor field **B** a second-order tensor field **A** · **B** defined by Eq. (3-19). However, not all transformations of the space of second-order tensor fields into itself are of this form [1, p. 69].

Let us determine the transpose of $\mathbf{A} \cdot \mathbf{B}$, where $\mathbf{A}$ and $\mathbf{B}$ are second-order tensor fields. If $\mathbf{u}$ and $\mathbf{v}$ are any two spatial vector fields,

$$[(\mathbf{A} \cdot \mathbf{B}) \cdot \mathbf{u}] \cdot \mathbf{v} = (\mathbf{B} \cdot \mathbf{u}) \cdot (\mathbf{A}^T \cdot \mathbf{v})$$
$$= \mathbf{u} \cdot [(\mathbf{B}^T \cdot \mathbf{A}^T) \cdot \mathbf{v}] \qquad (3\text{-}20)$$

We conclude that

$$(\mathbf{A} \cdot \mathbf{B})^T = \mathbf{B}^T \cdot \mathbf{A}^T \qquad (3\text{-}21)$$

## REFERENCE

1. Hoffman, Kenneth, and Ray Kunze: "Linear Algebra," Prentice-Hall, Englewood Cliffs, N.J., 1961.

## EXERCISES

**A.5.3-1** *(a) If $\mathbf{T}$ is a second-order tensor field, show that, with respect to any curvilinear coordinate system,

$$T_{ij} = T_{ji}{}^T$$

and

$$T^{ij} = T^{Tji}$$

(b) Show that, with respect to any orthogonal curvilinear coordinate system,

$$T_{\langle ij \rangle} = T^T_{\langle ji \rangle}$$

**A.5.3-2** *(a) If $\mathbf{A}$ and $\mathbf{B}$ are any two second-order tensor fields, show that, with respect to any curvilinear coordinate system,

$$\mathbf{A} \cdot \mathbf{B} = A_{ij} B^{jk} \mathbf{g}^i \mathbf{g}_k = A^{ij} B_{jk} \mathbf{g}_i \mathbf{g}^k$$
$$= A_{i\cdot}^{\;j} B_{jk} \mathbf{g}^i \mathbf{g}^k = A^{ij} B_{j\cdot}^{\;k} \mathbf{g}_i \mathbf{g}_k$$

(b) Show that, with respect to any orthogonal curvilinear coordinate system,

$$\mathbf{A} \cdot \mathbf{B} = A_{\langle ij \rangle} B_{\langle jk \rangle} \mathbf{g}_{\langle i \rangle} \mathbf{g}_{\langle k \rangle}$$

**A.5.3-3** Starting with Eq. (3-16), prove Eq. (3-17).

**A.5.3-4** *Isotropic second-order tensors* Let $\mathbf{A}$ be a second-order tensor field and $\mathbf{Q}$ an orthogonal tensor field. If

$$\mathbf{Q} \cdot \mathbf{A} \cdot \mathbf{Q}^T = \mathbf{A}$$

we refer to $\mathbf{A}$ as an *isotropic* second-order tensor. Prove that

$$\mathbf{A} = a\mathbf{I}$$

where $a$ is a scalar field.

*Hint:* Let $[\mathbf{Q}]$ denote the matrix (array) of the components of $\mathbf{Q}$ with respect to an appropriate basis.

(a) Let

$$[\mathbf{Q}] = \begin{bmatrix} 1 & 0 & 0 \\ 0 & 1 & 0 \\ 0 & 0 & -1 \end{bmatrix}$$

## [A.5] SECOND-ORDER TENSORS

to conclude that $A_{13} = A_{31} = A_{23} = A_{32} = 0$.

(b) Let

$$[Q] = \begin{bmatrix} 1 & 0 & 0 \\ 0 & -1 & 0 \\ 0 & 0 & 1 \end{bmatrix}$$

to conclude that $A_{12} = A_{21} = A_{23} = A_{32} = 0$.

(c) Let

$$[Q] = \begin{bmatrix} 0 & 1 & 0 \\ 1 & 0 & 0 \\ 0 & 0 & 1 \end{bmatrix}$$

to conclude that $A_{11} = A_{22}$.

(d) Let

$$[Q] = \begin{bmatrix} 1 & 0 & 0 \\ 0 & 0 & 1 \\ 0 & 1 & 0 \end{bmatrix}$$

to conclude that $A_{22} = A_{33}$.

**A.5.3-5** In Exercises A.4.1-9 and A.4.1-10, we studied the relations between the rectangular cartesian components of a spatial vector field **v** and the physical components of this vector field with respect to an orthogonal curvilinear coordinate system. Let us consider these relationships from a different point of view.

(a) Consider the transformations **A** and **B** such that

$$\mathbf{g}_{\langle j \rangle} = \mathbf{A} \cdot \mathbf{e}_j = A_{(ij)} \mathbf{e}_i$$

and

$$\mathbf{e}_j = \mathbf{B} \cdot \mathbf{g}_{\langle j \rangle} = B_{(ij)} \mathbf{g}_{\langle i \rangle}$$

Prove that **A** and **B** are orthogonal transformations and, therefore,

$$\mathbf{B} = \mathbf{A}^T$$

and

$$\mathbf{A} = \mathbf{B}^T$$

Note that **A** and **B** are *not* second-order tensors. They are transformations relating two sets of spatial vector fields that happen to be bases for the space of all spatial vector fields. The parentheses around the subscripts of the coefficients $A_{(ij)}$ and $B_{(ij)}$ are used to remind the reader that they are not components of second-order tensors. The definitions for an orthogonal transformation, transpose of a transformation, etc., are the same as for second-order tensors. The summation convention will continue to be used.

(b) Prove that, for any vector field

$$\mathbf{v} = v_i \mathbf{e}_i = v_{\langle j \rangle} \mathbf{g}_{\langle j \rangle}$$

we have

$$v_i = A_{(ij)} v_{\langle j \rangle} = B_{(ji)} v_{\langle j \rangle}$$

and

$$v_{\langle i \rangle} = B_{(ij)} v_j = A_{(ji)} v_j$$

(c) Starting with the results of Exercise A.4.1-9, parts (c) and (d), immediately write down the results of Exercise A.4.1-10, parts (c) and (d).

(d) Starting with the results of Exercise A.4.1-10, parts (c) and (d), immediately write down the results of Exercise A.4.1-9, parts (c) and (d).

### A.5.4 The inverse of a second-order tensor field [1, p. 62]

We say that a second-order tensor field $\mathbf{A}$ is *invertible*, when the following conditions are satisfied:

1. If $\mathbf{u}_1$ and $\mathbf{u}_2$ are spatial vector fields such that $\mathbf{A} \cdot \mathbf{u}_1 = \mathbf{A} \cdot \mathbf{u}_2$, then $\mathbf{u}_1 = \mathbf{u}_2$.
2. There corresponds to every spatial vector field $\mathbf{v}$ at least one spatial vector field $\mathbf{u}$ such that $\mathbf{A} \cdot \mathbf{u} = \mathbf{v}$.

If $\mathbf{A}$ is invertible, we define as follows a second-order tensor field $\mathbf{A}^{-1}$, called the *inverse* of $\mathbf{A}$. If $\mathbf{v}_1$ is any spatial vector field, by property 2 we may find a spatial vector field $\mathbf{u}_1$ for which $\mathbf{A} \cdot \mathbf{u}_1 = \mathbf{v}_1$. Say that $\mathbf{u}_1$ is not uniquely determined, such that $\mathbf{v}_1 = \mathbf{A} \cdot \mathbf{u}_1 = \mathbf{A} \cdot \mathbf{u}_2$. By property 1, $\mathbf{u}_1 = \mathbf{u}_2$ and we have a contradiction. The spatial vector field $\mathbf{u}_1$ is uniquely determined. We define $\mathbf{A}^{-1} \cdot \mathbf{v}_1$ to be $\mathbf{u}_1$.

In order to prove that $\mathbf{A}^{-1}$ satisfies the linearity rules for a second-order tensor field, Eqs. (1-1), we may evaluate $\mathbf{A}^{-1} \cdot (\alpha_1 \mathbf{v}_1 + \alpha_2 \mathbf{v}_2)$, where $\alpha_1$ and $\alpha_2$ are real scalar fields. If $\mathbf{A} \cdot \mathbf{u}_1 = \mathbf{v}_1$ and $\mathbf{A} \cdot \mathbf{u}_2 = \mathbf{v}_2$, we have

$$\mathbf{A} \cdot (\alpha_1 \mathbf{u}_1 + \alpha_2 \mathbf{u}_2) = \alpha_1 \mathbf{A} \cdot \mathbf{u}_1 + \alpha_2 \mathbf{A} \cdot \mathbf{u}_2$$
$$= \alpha_1 \mathbf{v}_1 + \alpha_2 \mathbf{v}_2 \tag{4-1}$$

This means that

$$\mathbf{A}^{-1} \cdot (\alpha_1 \mathbf{v}_1 + \alpha_2 \mathbf{v}_2) = \alpha_1 \mathbf{u}_1 + \alpha_2 \mathbf{u}_2$$
$$= \alpha_1 \mathbf{A}^{-1} \cdot \mathbf{v}_1 + \alpha_2 \mathbf{A}^{-1} \cdot \mathbf{v}_2 \tag{4-2}$$

It follows immediately from the definition that, for any invertible transformation $\mathbf{A}$,

$$\mathbf{A}^{-1} \cdot \mathbf{A} = \mathbf{A} \cdot \mathbf{A}^{-1} = \mathbf{I} \tag{4-3}$$

If $\mathbf{A}$, $\mathbf{B}$, and $\mathbf{C}$ are second-order tensor fields such that

$$\mathbf{A} \cdot \mathbf{B} = \mathbf{C} \cdot \mathbf{A} = \mathbf{I} \tag{4-4}$$

let us show that $\mathbf{A}$ is invertible and $\mathbf{A}^{-1} = \mathbf{B} = \mathbf{C}$. If $\mathbf{A} \cdot \mathbf{u}_1 = \mathbf{A} \cdot \mathbf{u}_2$, we have from Eq. (4-4)

$$(\mathbf{C} \cdot \mathbf{A}) \cdot \mathbf{u}_1 = (\mathbf{C} \cdot \mathbf{A}) \cdot \mathbf{u}_2$$
$$\mathbf{u}_1 = \mathbf{u}_2 \tag{4-5}$$

This fulfills property 1 of an invertible transformation. The second property is also satisfied. If $\mathbf{v}$ is any spatial vector field and if $\mathbf{u} = \mathbf{B} \cdot \mathbf{v}$, then, by Eq. (4-4),

$$\mathbf{A} \cdot \mathbf{u} = (\mathbf{A} \cdot \mathbf{B}) \cdot \mathbf{v}$$
$$= \mathbf{v} \tag{4-6}$$

Now that we have proved $\mathbf{A}$ to be invertible, from Eq. (4-4),

$$\mathbf{A}^{-1} \cdot \mathbf{A} \cdot \mathbf{B} = \mathbf{C} \cdot \mathbf{A} \cdot \mathbf{A}^{-1} = \mathbf{A}^{-1}$$
$$\mathbf{B} = \mathbf{C} = \mathbf{A}^{-1} \tag{4-7}$$

## [A.5] SECOND-ORDER TENSORS

In this way, we have shown that Eq. (4-3) is valid for some second-order tensor field $\mathbf{A}^{-1}$ if and only if $\mathbf{A}$ is invertible.

As a trivial example of an invertible second-order tensor field, we have the identity transformation, for which $\mathbf{I}^{-1} = \mathbf{I}$. Neither the zero tensor field $\mathbf{0}$ nor a tensor product $\mathbf{ab}$ is invertible.

For any orthogonal transformation $\mathbf{Q}$, we have that

$$\mathbf{Q}^{-1} = \mathbf{Q}^T \tag{4-8}$$

### REFERENCE

1. Halmos, P. R.: "Finite-dimensional Vector Spaces," Van Nostrand, Princeton, N.J., 1958.

### EXERCISES

**A.5.4-1** (a) The second-order tensor fields $\mathbf{A}$ and $\mathbf{B}$ are invertible. Show that $(\mathbf{A} \cdot \mathbf{B})^{-1} = \mathbf{B}^{-1} \cdot \mathbf{A}^{-1}$.
(b) Show that, if $\alpha \neq 0$ and $\mathbf{A}$ is invertible, $(\alpha \mathbf{A})^{-1} = (1/\alpha)\mathbf{A}^{-1}$.
(c) Show that, if $\mathbf{A}$ is invertible, $\mathbf{A}^{-1}$ is invertible and $(\mathbf{A}^{-1})^{-1} = \mathbf{A}$.

**A.5.4-2** (a) If $\mathbf{A}$ is invertible and, consequently, Eq. (4-3) holds, show that $\det(A_{ij}) \neq 0$, where the $A_{ij}$ denote the rectangular cartesian components of $\mathbf{A}$.
(b) Beginning with an equation of the same general form as Eq. (1-12) of Sec. A.2.1, show that, if $\det(A_{ij}) \neq 0$, $\mathbf{A}$ must be invertible.

**\*\*A.5.5 The trace of a second-order tensor field** Definition: Let $\mathbf{a}$ and $\mathbf{b}$ be spatial vector fields, let $\alpha$ be a scalar field, and let $\mathbf{S}$ and $\mathbf{T}$ be second-order tensor fields. An operation "tr" which assigns to each second-order tensor $\mathbf{T}$ a number tr $(\mathbf{T})$ is called a *trace* if it obeys the following rules:

tr $(\mathbf{S} + \mathbf{T}) = $ tr $(\mathbf{S}) + $ tr $(\mathbf{T})$

tr $(\alpha \mathbf{T}) = \alpha$ tr $(\mathbf{T})$

tr $(\mathbf{ab}) = \mathbf{a} \cdot \mathbf{b}$

With respect to a rectangular cartesian coordinate system, the trace of any second-order tensor $\mathbf{T} = T_{ij}\mathbf{e}_i\mathbf{e}_j$ may be written as

$$\text{tr }(\mathbf{T}) = T_{ij} \text{ tr }(\mathbf{e}_i\mathbf{e}_j) = T_{ij}(\mathbf{e}_i \cdot \mathbf{e}_j)$$
$$= T_{ij}\delta_{ij} = T_{ii} \tag{5-1}$$

The trace of a second-order tensor may be thought of as the sum of the diagonal components in the matrix $[T_{ij}]$.

### EXERCISES

**A.5.5-1** (a) Prove that the trace of a second-order tensor product does not depend upon the coordinate system being used.
(b) Use the definition to show that the trace of a second-order tensor field does not depend upon the coordinate system being used.

**A.5.5-2** The second-order tensor **T** may be expressed with respect to two different rectangular cartesian coordinate systems as

$$\mathbf{T} = T_{ij}\mathbf{e}_i\mathbf{e}_j = \bar{T}_{mn}\bar{\mathbf{e}}_m\bar{\mathbf{e}}_n$$

Without using the definition of the trace or the results of Exercise A.5.5-1, prove that

$$T_{ii} = \bar{T}_{mm}$$

**A.5.5-3** *(a) Show that, with respect to any curvilinear coordinate system,

$$\operatorname{tr}(\mathbf{T}) = T^i_{.i} = T^{.i}_{i.} = g_{ij}T^{ij} = g^{ij}T_{ij}$$

**(b) Show that, with respect to any orthogonal curvilinear coordinate system,

$$\operatorname{tr}(\mathbf{T}) = T_{\langle ii \rangle}$$

**A.5.5-4** *(a) If **A** and **B** are second-order tensor fields, show that, with respect to any curvilinear coordinate system,

$$\operatorname{tr}(\mathbf{A} \cdot \mathbf{B}) = A^{ij}B_{ji} = A_{ij}B^{ji}$$
$$= A^{.j}_{i.}B^{.i}_{j.} = A^i_{.j}B^j_{.i}$$

(b) Show that, with respect to any orthogonal curvilinear coordinate system,

$$\operatorname{tr}(\mathbf{A} \cdot \mathbf{B}) = A_{\langle ij \rangle}B_{\langle ji \rangle}$$

**A.5.5-5** (a) Let **A** and **B** be second-order tensor fields. We define

$$(\mathbf{A},\mathbf{B}) \equiv \operatorname{tr}(\mathbf{A} \cdot \mathbf{B}^T)$$

Show that $(\mathbf{A},\mathbf{B})$ satisfies the requirements for an inner product in the vector space of second-order tensor fields.

(b) We define the "length" of a second-order tensor field as

$$\|A\| \equiv \sqrt{(\mathbf{A},\mathbf{A})}$$

If $(\mathbf{A},\mathbf{B}) = 0$, we say that **A** and **B** are orthogonal to each other.

Consider an orthogonal curvilinear coordinate system. Show that the set of nine second-order tensor products $(\mathbf{g}_m\mathbf{g}_n)$ ($m = 1, 2, 3; n = 1, 2, 3$) is an *orthonormal* basis for the space of second-order tensor fields (with respect to the inner product and length as defined above). This justifies labeling the components $T_{\langle mn \rangle}$ of the second-order tensor field **T** as *physical* components.

## A.6 GRADIENT OF VECTOR

**\*\*A.6.1 The gradient of a vector field** The gradient of a spatial vector field **v** is a second-order tensor field denoted by $\nabla\mathbf{v}$. The gradient is specified by defining how it transforms an arbitrary spatial vector at all points $z$ in $E$:

$$\nabla\mathbf{v}(\mathbf{z}) \cdot \mathbf{a} = \lim_{s \to 0} \frac{1}{s}[\mathbf{v}(\mathbf{z} + s\mathbf{a}) - \mathbf{v}(\mathbf{z})] \tag{1-1}$$

The spatial vector **a** should be interpreted as the directed line segment or point difference $\mathbf{a} = \overrightarrow{zy}$, where $y$ is an arbitrary point in $E$. In writing Eq. (1-1), we have assumed that an origin $O$ has been specified and we have interpreted **v** as a function of the position vector **z** measured with respect to this origin rather than as a function of the point $z$ itself.

## [A.6] GRADIENT OF VECTOR

By analogy with our discussion of the gradient of a scalar field in Sec. A.3.1, we have that

$$(\nabla \mathbf{v}) \cdot \mathbf{a} = \frac{\partial \mathbf{v}}{\partial z_j} a_j \qquad (1\text{-}2)$$

For the particular case $\mathbf{a} = \mathbf{e}_j$,

$$(\nabla \mathbf{v}) \cdot \mathbf{e}_j = \frac{\partial \mathbf{v}}{\partial z_j} = \frac{\partial v_i}{\partial z_j} \mathbf{e}_i \qquad (1\text{-}3)$$

In reaching this result, we have noted that the magnitudes and directions of the rectangular cartesian basis fields are independent of position. Comparing Eq. (1-3) with Eqs. (2-1) and (2-6) of Sec. A.5.2, we conclude that[1]

$$\nabla \mathbf{v} = \frac{\partial v_i}{\partial z_j} \mathbf{e}_i \mathbf{e}_j \qquad (1\text{-}4)$$

The trace (Sec. A.5.5) of the gradient of a spatial vector field $\mathbf{v}$ is a familiar operation:

$$\text{tr}(\nabla \mathbf{v}) = \frac{\partial v_i}{\partial z_j} \text{tr}(\mathbf{e}_i \mathbf{e}_j)$$

$$= \frac{\partial v_i}{\partial z_i} \qquad (1\text{-}5)$$

It is more common to refer to this operation as the *divergence* of the spatial vector field $\mathbf{v}$. Several symbols for this operation are common:

$$\text{div } \mathbf{v} \equiv \nabla \cdot \mathbf{v}$$
$$\equiv \text{tr}(\nabla \mathbf{v}) = \frac{\partial v_i}{\partial z_i} \qquad (1\text{-}6)$$

## REFERENCES

1. Morse, P. M., and Herman Feshbach: "Methods of Theoretical Physics," pt. 1, McGraw-Hill, New York, 1953.
2. Bird, R. B., W. E. Stewart, and E. N. Lightfoot: "Transport Phenomena," 7th printing, Wiley, New York, 1960.

[1] Unfortunately, while I believe this to be the most common meaning for the symbol $\nabla \mathbf{v}$, some authors define [1, p. 65; 2, p. 723]

$$\nabla \mathbf{v} \equiv \frac{\partial v_i}{\partial z_j} \mathbf{e}_j \mathbf{e}_i$$

Where we would write $(\nabla \mathbf{v}) \cdot \mathbf{w}$, they say instead $\mathbf{w} \cdot (\nabla \mathbf{v})$.

**\*A.6.2 Covariant differentiation** In Sec. A.6.1 we arrived at the components of the gradient of a spatial vector field **v** with respect to a rectangular cartesian coordinate system. Here we derive an expression for the mixed components of $\nabla \mathbf{v}$ with respect to any curvilinear coordinate system (See Sec. A.5.2).

In Sec. A.6.1 we showed that

$$\nabla \mathbf{v} \cdot \mathbf{a} = \frac{\partial \mathbf{v}}{\partial z_i} a_i \qquad (2\text{-}1)$$

In terms of curvilinear coordinates, we may express this operation as

$$\nabla \mathbf{v} \cdot \mathbf{a} = \frac{\partial \mathbf{v}}{\partial x^j} \frac{\partial x^j}{\partial z_i} a_i$$

$$= \frac{\partial \mathbf{v}}{\partial x^j} \bar{a}^j \qquad (2\text{-}2)$$

where the $\bar{a}^j$ ($j = 1, 2, 3$) represent the contravariant curvilinear components of the spatial vector **a** (see Sec. A.4.3). Let us examine the quantity

$$\frac{\partial \mathbf{v}}{\partial x^j} = \frac{\partial}{\partial x^j}(v^i \mathbf{g}_i)$$

$$= \frac{\partial v^i}{\partial x^j} \mathbf{g}_i + v^i \frac{\partial \mathbf{g}_i}{\partial x^j} \qquad (2\text{-}3)$$

Unlike the basis fields $\mathbf{e}_j$ of rectangular cartesian coordinates, the natural basis fields of curvilinear coordinates are functions of position. In differentiation, they cannot be treated as constants:

$$\frac{\partial \mathbf{g}_i}{\partial x^j} = \frac{\partial}{\partial x^j}\left(\frac{\partial \mathbf{p}}{\partial x^i}\right) = \frac{\partial^2}{\partial x^j \partial x^i}(z_k \mathbf{e}_k)$$

$$= \frac{\partial^2 z_k}{\partial x^j \partial x^i} \mathbf{e}_k \qquad (2\text{-}4)$$

We saw in Sec. A.4.1 that

$$\mathbf{e}_k = \frac{\partial x^m}{\partial z_k} \mathbf{g}_m \qquad (2\text{-}5)$$

This allows us to write

$$\frac{\partial \mathbf{g}_i}{\partial x^j} = \frac{\partial^2 z_k}{\partial x^j \partial x^i} \frac{\partial x^m}{\partial z_k} \mathbf{g}_m = \begin{Bmatrix} m \\ j\ i \end{Bmatrix} \mathbf{g}_m \qquad (2\text{-}6)$$

where we define the symbol $\begin{Bmatrix} m \\ j\ i \end{Bmatrix}$ as

$$\begin{Bmatrix} m \\ j\ i \end{Bmatrix} = \frac{\partial^2 z_k}{\partial x^j \partial x^i} \frac{\partial x^m}{\partial z_k} \qquad (2\text{-}7)$$

These symbols are known as the *Christoffel symbols of the second kind.*

## [A.6] GRADIENT OF VECTOR

*Christoffel symbols of the first kind* are defined by

$$[ji,p] \equiv g_{pm} \begin{Bmatrix} m \\ j\ i \end{Bmatrix} \tag{2-8}$$

From Sec. A.4.1, we have that

$$g_{pm} = \frac{\partial z_n}{\partial x^p} \frac{\partial z_n}{\partial x^m} \tag{2-9}$$

and we express Eq. (2-8) as

$$[ji,p] = \frac{\partial z_n}{\partial x^p} \frac{\partial z_n}{\partial x^m} \frac{\partial^2 z_k}{\partial x^j \partial x^i} \frac{\partial x^m}{\partial z_k}$$

$$= \frac{\partial^2 z_k}{\partial x^j \partial x^i} \frac{\partial z_k}{\partial x^p} \tag{2-10}$$

Equation (2-8) also allows us to write

$$\begin{Bmatrix} r \\ j\ i \end{Bmatrix} = g^{rp}[ji,p] \tag{2-11}$$

While Eq. (2-7) is sufficient to define the Christoffel symbols of the second kind, it is rarely used in practice. Equation (2-9) may be differentiated to obtain

$$\frac{\partial g_{ij}}{\partial x^k} = \frac{\partial^2 z_m}{\partial x^k \partial x^i} \frac{\partial z_m}{\partial x^j} + \frac{\partial z_m}{\partial x^i} \frac{\partial^2 z_m}{\partial x^k \partial x^j} \tag{2-12}$$

Two similar expressions may be obtained by rotating the indices $i$, $j$, and $k$:

$$\frac{\partial g_{jk}}{\partial x^i} = \frac{\partial^2 z_m}{\partial x^i \partial x^j} \frac{\partial z_m}{\partial x^k} + \frac{\partial z_m}{\partial x^j} \frac{\partial^2 z_m}{\partial x^i \partial x^k} \tag{2-13}$$

and

$$\frac{\partial g_{ki}}{\partial x^j} = \frac{\partial^2 z_m}{\partial x^j \partial x^k} \frac{\partial z_m}{\partial x^i} + \frac{\partial z_m}{\partial x^k} \frac{\partial^2 z_m}{\partial x^j \partial x^i} \tag{2-14}$$

Adding Eqs. (2-13) and (2-14) and subtracting Eq. (2-12), we have

$$\frac{\partial g_{kj}}{\partial x^i} + \frac{\partial g_{ik}}{\partial x^j} - \frac{\partial g_{ij}}{\partial x^k} = 2 \frac{\partial^2 z_m}{\partial x^i \partial x^j} \frac{\partial z_m}{\partial x^k}$$

$$= 2[ij,k] \tag{2-15}$$

From Eqs. (2-11) and (2-15) we have another expression for the Christoffel symbols of the second kind which is usually found to be more convenient to use in practice than Eq. (2-7):

$$\begin{Bmatrix} m \\ i \ j \end{Bmatrix} = g^{mk}[ij,k]$$

$$= \frac{g^{mk}}{2}\left(\frac{\partial g_{kj}}{\partial x^i} + \frac{\partial g_{ik}}{\partial x^j} - \frac{\partial g_{ij}}{\partial x^k}\right) \tag{2-16}$$

Let us return to Eq. (2-3) and write, with the help of Eq. (2-6),

$$\frac{\partial \mathbf{v}}{\partial x^j} = \frac{\partial v^i}{\partial x^j}\mathbf{g}_i + v^i\begin{Bmatrix} m \\ j \ i \end{Bmatrix}\mathbf{g}_m$$

$$= \left[\frac{\partial v^i}{\partial x^j} + \begin{Bmatrix} i \\ j \ m \end{Bmatrix}v^m\right]\mathbf{g}_i$$

$$= v^i_{,j}\mathbf{g}_i \tag{2-17}$$

Here we define the symbol $v^i_{,j}$ as

$$v^i_{,j} \equiv \frac{\partial v^i}{\partial x^j} + \begin{Bmatrix} i \\ j \ m \end{Bmatrix}v^m \tag{2-18}$$

The quantity $v^i_{,j}$ is referred to as the ($j$th) *covariant derivative* of the ($i$th) contravariant component of the spatial vector field **v**. This allows us to write Eq. (2-2) as

$$(\nabla \mathbf{v}) \cdot \mathbf{a} = v^i_{,j}a^j\mathbf{g}_i \tag{2-19}$$

with the understanding that the $a^j$ represent the contravariant components of the spatial vector **a** in the curvilinear coordinate system under consideration. If we follow the practice, introduced in Sec. A.5.2, of expressing second-order tensor fields as sums of tensor products of basis fields, we may represent $\nabla \mathbf{v}$ as

$$\nabla \mathbf{v} = v^i_{,j}\mathbf{g}_i\mathbf{g}^j \tag{2-20}$$

The nine quantities of the form $v^i_{,j}$ ($i = 1, 2, 3; j = 1, 2, 3$) represent the mixed components of the second-order tensor field $\nabla \mathbf{v}$.

In Eq. (2-3), we expressed **v** as a linear combination of the natural basis vectors. How are the expressions obtained above altered when we express **v** as a linear combination of the dual basis vectors? We have in this case

$$\frac{\partial \mathbf{v}}{\partial x^j} = \frac{\partial}{\partial x^j}(v_i\mathbf{g}^i) = \frac{\partial v_i}{\partial x^j}\mathbf{g}^i + v_i\frac{\partial \mathbf{g}^i}{\partial x^j} \tag{2-21}$$

Our major problem is to obtain an expression for $\partial \mathbf{g}^i/\partial x^j$.

From Sec. A.4.2,

$$\mathbf{g}_i \cdot \mathbf{g}^j = \delta_i^j \tag{2-22}$$

Taking the derivative of this expression with respect to $x^k$, we obtain

$$\frac{\partial \mathbf{g}_i}{\partial x^k} \cdot \mathbf{g}^j + \mathbf{g}_i \cdot \frac{\partial \mathbf{g}^j}{\partial x^k} = 0 \tag{2-23}$$

or

$$\mathbf{g}_i \cdot \frac{\partial \mathbf{g}^j}{\partial x^k} = -\frac{\partial \mathbf{g}_i}{\partial x^k} \cdot \mathbf{g}^j = -\begin{Bmatrix} m \\ k \ i \end{Bmatrix} \mathbf{g}_m \cdot \mathbf{g}^j$$

$$= -\begin{Bmatrix} j \\ k \ i \end{Bmatrix} \tag{2-24}$$

The spatial vector $\partial \mathbf{g}^j/\partial x^k$, like any other element of the vector space of spatial vectors, may be written as a linear combination of the dual basis vectors:

$$\frac{\partial \mathbf{g}^j}{\partial x^k} = A_{kt}{}^j \mathbf{g}^t \tag{2-25}$$

where the coefficients $A_{kt}{}^j$ are yet to be determined. The scalar product of this equation with $\mathbf{g}_i$ yields, from Eq. (2-24),

$$\mathbf{g}_i \cdot \frac{\partial \mathbf{g}^j}{\partial x^k} = A_{ki}{}^j = -\begin{Bmatrix} j \\ k \ i \end{Bmatrix} \tag{2-26}$$

This allows us to write

$$\frac{\partial \mathbf{g}^j}{\partial x^k} = -\begin{Bmatrix} j \\ k \ i \end{Bmatrix} \mathbf{g}^i \tag{2-27}$$

Returning to Eq. (2-21), we may write

$$\frac{\partial \mathbf{v}}{\partial x^j} = \frac{\partial v_i}{\partial x^j} \mathbf{g}^i - v_i \begin{Bmatrix} i \\ j \ k \end{Bmatrix} \mathbf{g}^k$$

$$= \left[ \frac{\partial v_i}{\partial x^j} - \begin{Bmatrix} k \\ j \ i \end{Bmatrix} v_k \right] \mathbf{g}^i$$

$$= v_{i,j} \mathbf{g}^i \tag{2-28}$$

where we define the symbol $v_{i,j}$ as

$$v_{i,j} \equiv \frac{\partial v_i}{\partial x^j} - \begin{Bmatrix} k \\ j \ i \end{Bmatrix} v_k \tag{2-29}$$

The quantity $v_{i,j}$ is referred to as the (jth) *covariant derivative* of the (ith) covariant component of the spatial vector field **v**.

From Eqs. (2-2) and (2-28), we obtain

$$(\nabla \mathbf{v}) \cdot \mathbf{a} = v_{i,j} a^j \mathbf{g}^i \tag{2-30}$$

again with the understanding that the $a^j$ now represent the contravariant components of the spatial vector **a** in whatever curvilinear coordinate system is under consideration. In terms of our discussion in Sec. A.5.2, we conclude that

$$\nabla \mathbf{v} = v_{i,j} \mathbf{g}^i \mathbf{g}^j \tag{2-31}$$

The $v_{i,j}$ represent the covariant components of $\nabla \mathbf{v}$.

Equation (2-31) can be written in a form similar to that of Eq. (2-20):

$$\nabla \mathbf{v} = v_{i,j} g^{ik} \mathbf{g}_k \mathbf{g}^j$$
$$= v_{k,j} g^{ki} \mathbf{g}_i \mathbf{g}^j \qquad (2\text{-}32)$$

Equation (2-20) in turn may be written as

$$\nabla \mathbf{v} = (g^{ki} v_k)_{,j} \mathbf{g}_i \mathbf{g}^j \qquad (2\text{-}33)$$

We conclude that

$$(g^{ki} v_k)_{,j} = g^{ki} v_{k,j} \qquad (2\text{-}34)$$

(Remember here that the nine tensor products of the form $\mathbf{g}_i \mathbf{g}^j$ were shown to be linearly independent in Sec. A.5.2.) This means that the $g^{ki}$ may be treated as constants with respect to covariant differentiation.

### EXERCISES

**A.6.2-1** Show that

$$(g_{in} v^n)_{,j} = g_{in} v^n_{,j}$$

implying that the $g_{in}$ may be treated as constants with respect to covariant differentiation.

**A.6.2-2** Starting with

$$(g_{in} v^n)_{,j} = \frac{\partial(g_{in} v^n)}{\partial x^j} - \begin{Bmatrix} k \\ j\ i \end{Bmatrix} g_{kn} v^n$$

rework Exercise A.6.2-1.

**A.6.2-3** Show that in rectangular cartesian coordinates

$$\begin{Bmatrix} i \\ j\ k \end{Bmatrix} = 0$$

and covariant differentiation reduces to ordinary partial differentiation.

**A.6.2-4** Show that in cylindrical coordinates where

$$z_1 = x^1 \cos x^2 = r \cos \theta$$
$$z_2 = x^1 \sin x^2 = r \sin \theta$$
$$z_3 = x^3 = z$$

the only nonzero Christoffel symbols of the second kind are

$$\begin{Bmatrix} 1 \\ 2\ 2 \end{Bmatrix} = -r$$

and

$$\begin{Bmatrix} 2 \\ 1\ 2 \end{Bmatrix} = \begin{Bmatrix} 2 \\ 2\ 1 \end{Bmatrix} = \frac{1}{r}$$

**A.6.2-5** Show that in spherical coordinates where

$$z_1 = x^1 \sin x^2 \cos x^3 = r \sin \theta \cos \varphi$$
$$z_2 = x^1 \sin x^2 \sin x^3 = r \sin \theta \sin \varphi$$
$$z_3 = x^1 \cos x^2 = r \cos \theta$$

the only nonzero Christoffel symbols of the second kind are

$$\begin{Bmatrix} 1 \\ 2\ 2 \end{Bmatrix} = -r$$

$$\begin{Bmatrix} 1 \\ 3\ 3 \end{Bmatrix} = -r \sin^2 \theta$$

$$\begin{Bmatrix} 2 \\ 1\ 2 \end{Bmatrix} = \begin{Bmatrix} 2 \\ 2\ 1 \end{Bmatrix} = \begin{Bmatrix} 3 \\ 1\ 3 \end{Bmatrix} = \begin{Bmatrix} 3 \\ 3\ 1 \end{Bmatrix} = \frac{1}{r}$$

$$\begin{Bmatrix} 2 \\ 3\ 3 \end{Bmatrix} = -\sin \theta \cos \theta$$

and

$$\begin{Bmatrix} 3 \\ 2\ 3 \end{Bmatrix} = \begin{Bmatrix} 3 \\ 3\ 2 \end{Bmatrix} = \cot \theta$$

## A.7 THIRD-ORDER TENSORS

**A.7.1 Third-order tensor fields** Our discussion of third-order tensor fields closely parallels the treatment of second-order tensor fields in Sec. A.5.1.

A third-order tensor field $\boldsymbol{\beta}$ is a transformation (or mapping or rule) that assigns to each given spatial vector field $\mathbf{v}$ a second-order tensor field $\boldsymbol{\beta} \cdot \mathbf{v}$ such that the rules

$$\begin{aligned} \boldsymbol{\beta} \cdot (\mathbf{v} + \mathbf{w}) &= \boldsymbol{\beta} \cdot \mathbf{v} + \boldsymbol{\beta} \cdot \mathbf{w} \\ \boldsymbol{\beta} \cdot (a\mathbf{v}) &= a(\boldsymbol{\beta} \cdot \mathbf{v}) \end{aligned} \quad (1\text{-}1)$$

hold. The quantity $a$ denotes a real scalar field. (Note: We are using the dot notation in a different manner here than we used it in Sec. A.1.1 when we discussed the inner product. Our choice of notation is suggestive in the same way that it was in our treatment of second-order tensor fields, as will be clear shortly.)

We define the sum $\boldsymbol{\alpha} + \boldsymbol{\beta}$ of two third-order tensor fields $\boldsymbol{\alpha}$ and $\boldsymbol{\beta}$ to be a transformation such that, for every spatial vector field $\mathbf{v}$,

$$(\boldsymbol{\alpha} + \boldsymbol{\beta}) \cdot \mathbf{v} = \boldsymbol{\alpha} \cdot \mathbf{v} + \boldsymbol{\beta} \cdot \mathbf{v} \quad (1\text{-}2)$$

The product $a\boldsymbol{\beta}$ of a third-order tensor field $\boldsymbol{\beta}$ with a real scalar field $a$ is a transformation such that, for every spatial vector field $\mathbf{v}$,

$$(a\boldsymbol{\beta}) \cdot \mathbf{v} = a(\boldsymbol{\beta} \cdot \mathbf{v}) \quad (1\text{-}3)$$

The transformations $\boldsymbol{\alpha} + \boldsymbol{\beta}$ and $a\boldsymbol{\beta}$ may be easily shown to obey the rules for a third-order tensor field. If we define the zero third-order tensor $\mathbf{0}$ by the requirement that

$$\mathbf{0} \cdot \mathbf{v} = \mathbf{0}$$

for all spatial vector fields **v**, then rules ($A_1$) to ($A_4$) and ($M_1$) to ($M_4$) of Sec. A.1.1 are satisfied and the set of all third-order tensor fields constitutes a vector space.

If three spatial vector fields **a**, **b**, and **c** are given, we can define a third-order tensor field **abc** by the requirement that

$$(\mathbf{abc}) \cdot \mathbf{v} = \mathbf{ab}(\mathbf{c} \cdot \mathbf{v}) \tag{1-4}$$

hold for all spatial vector fields **v**. This tensor field **abc** is called the tensor product of the spatial vector fields **a**, **b**, and **c**.

### A.7.2 Components of third-order tensor fields

If **β** is a third-order tensor field and the $\mathbf{e}_k$ ($k = 1, 2, 3$) form a rectangular cartesian basis for the space of spatial vector fields, then we may write

$$\boldsymbol{\beta} \cdot \mathbf{e}_k = \beta_{ijk} \mathbf{e}_i \mathbf{e}_j \tag{2-1}$$

The set of 27 coefficients $\beta_{ijk}$ ($i, j, k = 1, 2, 3$) will hereafter be referred to as the coefficient matrix $\beta_{ijk}$, reminiscent of the nomenclature used for second-order tensor fields. The coefficient matrix $[\beta_{ijk}]$ tells us how the basis fields $\mathbf{e}_k$ are transformed by **β**. If **v** is any spatial vector field, then

$$\boldsymbol{\beta} \cdot \mathbf{v} = \boldsymbol{\beta} \cdot (v_k \mathbf{e}_k) = v_k \boldsymbol{\beta} \cdot \mathbf{e}_k$$
$$= v_k \beta_{ijk} \mathbf{e}_i \mathbf{e}_j \tag{2-2}$$

In this way, for each set of basis fields ($\mathbf{e}_1, \mathbf{e}_2, \mathbf{e}_3$), we may associate a set of 27 coefficients with any third-order tensor field **β**. This association or correspondence is one-to-one. (That is, for the same set of basis fields, two different third-order tensor fields will have two different coefficient matrices. In order to prove this, observe that the coefficient matrix $[\beta_{ijk}]$ of a third-order tensor field **β** completely determines **β**; by Eq. (2-2), **β** · **v** is determined for every **v**.)

Using the arguments applied in this discussion of second-order tensor fields (see Sec. A.5.2), we may show that every third-order tensor field **β** may be written as a sum of 27 tensor products:

$$\boldsymbol{\beta} = \beta_{ijk} \mathbf{e}_i \mathbf{e}_j \mathbf{e}_k \tag{2-3}$$

The coefficients $\beta_{ijk}$ are the same as those introduced in Eq. (2-1). They are referred to as the rectangular cartesian components of **β**.

Continuing in the same fashion, we can show that the 27 tensor products $\mathbf{e}_i \mathbf{e}_j \mathbf{e}_k$ ($i, j, k = 1, 2, 3$) are linearly independent. Since every third-order tensor field **β** can be expressed as a linear combination of the $\mathbf{e}_i \mathbf{e}_j \mathbf{e}_k$, we conclude that the 27 tensor products $\mathbf{e}_i \mathbf{e}_j \mathbf{e}_k$ form a set of basis fields for the space of third-order tensor fields. Consequently, the vector space of third-order tensor fields is 27 dimensional.

In physical applications, we often find it convenient to speak in terms of an orthogonal curvilinear coordinate system. If the $\mathbf{g}_{\langle k \rangle}$ ($k = 1, 2, 3$) are the associated physical basis fields (Sec. A.4.1), then we may write by analogy with Eq. (2-1),

$$\boldsymbol{\beta} \cdot \mathbf{g}_{\langle k \rangle} = \beta_{\langle ijk \rangle} \mathbf{g}_{\langle i \rangle} \mathbf{g}_{\langle j \rangle} \tag{2-4}$$

# [A.7] THIRD-ORDER TENSORS

By the same argument which leads us to Eq. (2-3), we may consequently write

$$\beta = \beta_{\langle ijk \rangle} \mathbf{g}_{\langle i \rangle} \mathbf{g}_{\langle j \rangle} \mathbf{g}_{\langle k \rangle} \tag{2-5}$$

where the 27 coefficients $\beta_{\langle ijk \rangle}$ ($i, j, k = 1, 2, 3$) are referred to as the physical components of $\beta$. The set of 27 tensor products $\mathbf{g}_{\langle i \rangle} \mathbf{g}_{\langle j \rangle} \mathbf{g}_{\langle k \rangle}$ ($i, j, k = 1, 2, 3$) forms another basis for the vector space of third-order tensor fields.

## EXERCISES

**A.7.2-1** Prove that every third-order tensor field $\beta$ may be written as a sum of 27 tensor products, Eq. (2-3).

**A.7.2-2** Prove that the 27 tensor products $\mathbf{e}_i \mathbf{e}_j \mathbf{e}_k$ ($i, j, k = 1, 2, 3$) are linearly independent and that these tensor products form a basis for the space of third-order tensor fields.

**A.7.2-3** (a) Given an orthogonal curvilinear coordinate system, prove that Eq. (2-5) holds for every third-order tensor field $\beta$.

(b) Prove that the 27 tensor products $\mathbf{g}_{\langle i \rangle} \mathbf{g}_{\langle j \rangle} \mathbf{g}_{\langle k \rangle}$ ($i, j, k = 1, 2, 3$) form a basis for the vector space of third-order tensor fields.

**A.7.2-4** Let $\beta$ be some third-order tensor field. If the $\beta_{ijk}$ ($i, j, k = 1, 2, 3$) are the components of $\beta$ with respect to one rectangular cartesian coordinate system ($z_1, z_2, z_3$) and the $\bar{\beta}_{ijk}$ ($i, j, k = 1, 2, 3$) are the components with respect to another rectangular cartesian coordinate system ($\bar{z}_1, \bar{z}_2, \bar{z}_3$), prove that

$$\beta_{ijk} = \frac{\partial \bar{z}_m}{\partial z_i} \frac{\partial \bar{z}_n}{\partial z_j} \frac{\partial \bar{z}_p}{\partial z_k} \bar{\beta}_{mnp}$$

and

$$\bar{\beta}_{ijk} = \frac{\partial z_i}{\partial \bar{z}_m} \frac{\partial z_j}{\partial \bar{z}_n} \frac{\partial z_k}{\partial \bar{z}_p} \beta_{mnp}$$

*__A.7.2-5__ (a) Given any curvilinear coordinate system, prove that every third-order tensor field $\beta$ may be written as

$$\beta = \beta_{ijk} \mathbf{g}^i \mathbf{g}^j \mathbf{g}^k$$

The 27 coefficients $\beta_{ijk}$ ($i, j, k = 1, 2, 3$) are referred to as the *covariant components* of $\beta$.

(b) Prove that the 27 tensor products $\mathbf{g}^i \mathbf{g}^j \mathbf{g}^k$ ($i, j, k = 1, 2, 3$) form a basis for the space of third-order tensor fields.

*__A.7.2-6__ (a) Given any curvilinear coordinate system, prove that every third-order tensor field $\beta$ may be written as

$$\beta = \beta^{ijk} \mathbf{g}_i \mathbf{g}_j \mathbf{g}_k$$

The 27 coefficients $\beta^{ijk}$ ($i, j, k = 1, 2, 3$) are referred to as the *contravariant components* of $\beta$.

(b) Prove that the 27 tensor products $\mathbf{g}_i \mathbf{g}_j \mathbf{g}_k$ ($i, j, k = 1, 2, 3$) form a basis for the space of third-order tensor fields.

*__A.7.2-7__ Prove that

$$\beta_{\langle ijk \rangle} = \frac{\beta_{ijk}}{\sqrt{g_{ii} g_{jj} g_{kk}}} \quad \text{no summation on } i, j, k$$

and

$$\beta_{\langle ijk \rangle} = \sqrt{g_{ii} g_{jj} g_{kk}} \, \beta^{ijk} \quad \text{no summation on } i, j, k$$

**\*A.7.2-8** Prove that it is unnecessary to distinguish between covariant, contravariant, and physical components of third-order tensor fields *in rectangular cartesian coordinate systems*. We may therefore write all indices as subscripts, employing the summation convention over repeated subscripts *so long as we restrict ourselves to rectangular cartesian coordinate systems*.

**\*A.7.2-9** (a) Let $\beta$ be some third-order tensor field. If the $\beta^{ijk}$ ($i, j, k = 1, 2, 3$) are the contravariant components of $\beta$ with respect to one curvilinear coordinate system ($x^1, x^2, x^3$) and if the $\bar{\beta}^{ijk}$ ($i, j, k = 1, 2, 3$) are the contravariant components of $\beta$ with respect to another curvilinear coordinate system ($\bar{x}^1, \bar{x}^2, \bar{x}^3$), prove that

$$\beta^{ijk} = \frac{\partial x^i}{\partial \bar{x}^m}\frac{\partial x^j}{\partial \bar{x}^n}\frac{\partial x^k}{\partial \bar{x}^p}\bar{\beta}^{mnp}$$

(b) Similarly, prove that

$$\beta_{ijk} = \frac{\partial \bar{x}^m}{\partial x^i}\frac{\partial \bar{x}^n}{\partial x^j}\frac{\partial \bar{x}^p}{\partial x^k}\bar{\beta}_{mnp}$$

**\*A.7.2-10** If the $\beta_{ijk}$ ($i, j, k = 1, 2, 3$) are the covariant components of a third-order tensor field $\beta$ and the $\beta^{mnp}$ ($m, n, p = 1, 2, 3$) are the contravariant components, prove that

$$\beta_{ijk} = g_{im}g_{jn}g_{kp}\beta^{mnp}$$

**A.7.2-11** We will have occasion to use a particular third-order tensor field defined by its components with respect to a rectangular cartesian coordinate system ($z_1, z_2, z_3$):

$$\boldsymbol{\epsilon} = e_{ijk}\mathbf{e}_i\mathbf{e}_j\mathbf{e}_k$$

The quantity $e_{ijk}$ is defined in Sec. A.2.1.

(a) Prove that with respect to any other rectangular cartesian coordinate system ($\bar{z}_1, \bar{z}_2, \bar{z}_3$), we have

$$\boldsymbol{\epsilon} = \left[\det\left(\frac{\partial z_m}{\partial \bar{z}_n}\right)\right]e_{ijk}\bar{\mathbf{e}}_i\bar{\mathbf{e}}_j\bar{\mathbf{e}}_k$$

(b) Starting with an expression for the components of the identity tensor, prove that

$$\left[\det\left(\frac{\partial z_m}{\partial \bar{z}_n}\right)\right]^2 = 1$$

In view of the discussion in Sec. A.9.2 and the result of Exercise 1.2.1-2, it is necessary that the rectangular cartesian coordinate system used here in defining $\boldsymbol{\epsilon}$ be right-handed.

**A.7.2-12** (a) The third-order tensor field $\boldsymbol{\epsilon}$ is defined in Exercise A.7.2-11. Prove that for an arbitrary curvilinear coordinate system we may write

$$\boldsymbol{\epsilon} = \epsilon^{ijk}\mathbf{g}_i\mathbf{g}_j\mathbf{g}_k$$

where

$$\epsilon^{ijk} \equiv \frac{1}{\sqrt{g}}e^{ijk}$$

\*(b) Prove that we may also write

$$\boldsymbol{\epsilon} = \epsilon_{ijk}\mathbf{g}^i\mathbf{g}^j\mathbf{g}^k$$

[A.7] THIRD-ORDER TENSORS 649

where

$$\epsilon_{ijk} \equiv \sqrt{g}\, e_{ijk}$$

(c) For an orthogonal curvilinear coordinate system, prove that

$$\boldsymbol{\epsilon} = e^{ijk}\mathbf{g}_{\langle i\rangle}\mathbf{g}_{\langle j\rangle}\mathbf{g}_{\langle k\rangle}$$
$$= e_{ijk}\mathbf{g}_{\langle i\rangle}\mathbf{g}_{\langle j\rangle}\mathbf{g}_{\langle k\rangle}$$

**A.7.3 Another view of third-order tensor fields** If $\boldsymbol{\beta}$ is a third-order tensor field and if $\mathbf{u}$ and $\mathbf{v}$ are spatial vector fields, then $\boldsymbol{\beta} \cdot \mathbf{u}$ is a second-order tensor field and $(\boldsymbol{\beta} \cdot \mathbf{u}) \cdot \mathbf{v}$ is a spatial vector field. For convenience, let us introduce the notation

$$\boldsymbol{\beta}:\mathbf{uv} \equiv (\boldsymbol{\beta} \cdot \mathbf{u}) \cdot \mathbf{v} \tag{3-1}$$

For example, consider the tensor product of rectangular cartesian basis fields, $\mathbf{e}_i\mathbf{e}_j$. Then

$$\boldsymbol{\beta}:\mathbf{e}_i\mathbf{e}_j = (\boldsymbol{\beta} \cdot \mathbf{e}_i) \cdot \mathbf{e}_j$$
$$= (\beta_{mni}\mathbf{e}_m\mathbf{e}_n) \cdot \mathbf{e}_j$$
$$= \beta_{mji}\mathbf{e}_m \tag{3-2}$$

This suggests that we may use $\boldsymbol{\beta}$ to define a transformation (or mapping or rule) which assigns to every tensor product of spatial vector fields, $\mathbf{uv}$, a spatial vector field $\boldsymbol{\beta}:\mathbf{uv}$ such that the rules

$$\boldsymbol{\beta}:(\mathbf{ab} + \mathbf{uv}) = \boldsymbol{\beta}:\mathbf{ab} + \boldsymbol{\beta}:\mathbf{uv}$$
$$\boldsymbol{\beta}:(\alpha\mathbf{ab}) = \alpha(\boldsymbol{\beta}:\mathbf{ab}) \tag{3-3}$$

hold. The quantity $\alpha$ denotes a real scalar field.

Since every second-order tensor field $\mathbf{T}$ may be written as a linear combination of tensor products (see Sec. A.5.2), we have that

$$\boldsymbol{\beta}:\mathbf{T} = \boldsymbol{\beta}:(T_{ij}\mathbf{e}_i\mathbf{e}_j)$$
$$= T_{ij}\boldsymbol{\beta}:\mathbf{e}_i\mathbf{e}_j$$
$$= T_{ij}\beta_{mji}\mathbf{e}_m \tag{3-4}$$

It follows immediately from the rules given in Eqs. (3-3) that, for any two second-order tensor fields $\mathbf{S}$ and $\mathbf{T}$ and for any scalar field $\alpha$, we have the rules

$$\boldsymbol{\beta}:(\mathbf{S} + \mathbf{T}) = \boldsymbol{\beta}:\mathbf{S} + \boldsymbol{\beta}:\mathbf{T}$$
$$\boldsymbol{\beta}:(\alpha\mathbf{T}) = \alpha(\boldsymbol{\beta}:\mathbf{T}) \tag{3-5}$$

If $\boldsymbol{\epsilon}$ is the third-order tensor defined in Exercise A.7.2-11 and $\mathbf{B}$ is any skew-symmetric second-order tensor, then

$$\mathbf{b} \equiv \boldsymbol{\epsilon}:\mathbf{B} \tag{3-6}$$

is known as the corresponding *axial vector*. The vorticity vector is introduced as the axial vector corresponding to the second-order vorticity tensor in Sec. 1.4.2.

## A.8 GRADIENT OF TENSOR

**A.8.1 The gradient of a second-order tensor field** The gradient of a second-order tensor field **T** is a third-order tensor field denoted by $\nabla \mathbf{T}$. The gradient is specified by defining how it transforms an arbitrary spatial vector at all points $z$ in $E$:

$$\nabla \mathbf{T}(\mathbf{z}) \cdot \mathbf{a} = \lim_{s \to 0} \frac{1}{s}[\mathbf{T}(\mathbf{z} + s\mathbf{a}) - \mathbf{T}(\mathbf{z})] \tag{1-1}$$

The spatial vector **a** should be interpreted as the directed line segment or point difference $\mathbf{a} \equiv \overrightarrow{zy}$, where $y$ is an arbitrary point in $E$. In writing Eq. (1-1), we have assumed that an origin $O$ has been specified and we have interpreted **T** as a function of the position vector **z** measured with respect to this origin rather than as a function of the point $z$ itself.

By analogy with our discussions of the gradient of a scalar field in Sec. A.3.1 and of the gradient of a spatial vector field in Sec. A.6.1, we have that

$$(\nabla \mathbf{T}) \cdot \mathbf{a} = \frac{\partial \mathbf{T}}{\partial z_j} a_j \tag{1-2}$$

For the particular case $\mathbf{a} = \mathbf{e}_k$,

$$(\nabla \mathbf{T}) \cdot \mathbf{e}_k = \frac{\partial \mathbf{T}}{\partial z_k}$$
$$= \frac{\partial T_{ij}}{\partial z_k} \mathbf{e}_i \mathbf{e}_j \tag{1-3}$$

In reaching this result, we have noted that the magnitudes and directions of the rectangular cartesian basis fields are independent of position. Comparing Eq. (1-3) with Eqs. (2-1) and (2-3) of Sec. A.7.2, we conclude that[1]

$$\nabla \mathbf{T} = \frac{\partial T_{ij}}{\partial z_k} \mathbf{e}_i \mathbf{e}_j \mathbf{e}_k \tag{1-4}$$

A common operation is the divergence of a second-order tensor field **T**:

$$\text{div } \mathbf{T} \equiv \nabla \cdot \mathbf{T}$$
$$\equiv \frac{\partial T_{ij}}{\partial z_j} \mathbf{e}_i \tag{1-5}$$

---

[1] Although I believe that this is the most common meaning for the symbol $\nabla \mathbf{T}$, some authors define [1, pp. 723 and 730]

$$\nabla \mathbf{T} \equiv \frac{\partial T_{ij}}{\partial z_k} \mathbf{e}_k \mathbf{e}_i \mathbf{e}_j$$

Where we would write $(\nabla \mathbf{T}) \cdot \mathbf{v}$, they say instead $\mathbf{v} \cdot (\nabla \mathbf{T})$.

**REFERENCE**

1. Bird, R. B., W. E. Stewart, and E. N. Lightfoot: "Transport Phenomena," 7th printing, Wiley, New York, 1960.

**\*A.8.2  More on covariant differentiation**  In Sec. A.8.1 we arrived at the components of the gradient of a second-order tensor field **T** with respect to a rectangular cartesian coordinate system. Here we derive expressions for the mixed components of $\nabla \mathbf{T}$ with respect to any curvilinear coordinate system.

In Sec. A.8.1, we showed that

$$(\nabla \mathbf{T}) \cdot \mathbf{a} = \frac{\partial \mathbf{T}}{\partial z_m} a_m \qquad (2\text{-}1)$$

In terms of curvilinear coordinates, we may express this operation as

$$(\nabla \mathbf{T}) \cdot \mathbf{a} = \frac{\partial \mathbf{T}}{\partial x^k} \frac{\partial x^k}{\partial z_m} a_m$$

$$= \frac{\partial \mathbf{T}}{\partial x^k} \bar{a}^k \qquad (2\text{-}2)$$

where the $\bar{a}^k$ ($k = 1, 2, 3$) represent the contravariant curvilinear components of the position vector **a**.

On the basis of our discussion in Sec. A.6.2, we may express

$$\frac{\partial \mathbf{T}}{\partial x^k} = \frac{\partial}{\partial x^k}(T^{ij}\mathbf{g}_i\mathbf{g}_j)$$

$$= \frac{\partial T^{ij}}{\partial x^k}\mathbf{g}_i\mathbf{g}_j + T^{ij}\frac{\partial \mathbf{g}_i}{\partial x^k}\mathbf{g}_j + T^{ij}\mathbf{g}_i\frac{\partial \mathbf{g}_j}{\partial x^k}$$

$$= \frac{\partial T^{ij}}{\partial x^k}\mathbf{g}_i\mathbf{g}_j + T^{ij}\begin{Bmatrix} r \\ k\ i \end{Bmatrix}\mathbf{g}_r\mathbf{g}_j + T^{ij}\mathbf{g}_i\begin{Bmatrix} r \\ k\ j \end{Bmatrix}\mathbf{g}_r$$

$$= \left[\frac{\partial T^{ij}}{\partial x^k} + \begin{Bmatrix} i \\ k\ r \end{Bmatrix}T^{rj} + \begin{Bmatrix} j \\ k\ r \end{Bmatrix}T^{ir}\right]\mathbf{g}_i\mathbf{g}_j$$

$$= T^{ij}{}_{,k}\mathbf{g}_i\mathbf{g}_j \qquad (2\text{-}3)$$

Here we define the symbol $T^{ij}{}_{,k}$ as

$$T^{ij}{}_{,k} \equiv \frac{\partial T^{ij}}{\partial x^k} + \begin{Bmatrix} i \\ k\ r \end{Bmatrix}T^{rj} + \begin{Bmatrix} j \\ k\ r \end{Bmatrix}T^{ir} \qquad (2\text{-}4)$$

The quantity $T^{ij}{}_{,k}$ is referred to as the ($k$th) *covariant derivative* of the ($ij$) contravariant component of the second-order tensor field **T**.

This allows us to write Eq. (2-2) as

$$(\nabla \mathbf{T}) \cdot \mathbf{a} = T^{ij}{}_{,k}\bar{a}^k\mathbf{g}_i\mathbf{g}_j \qquad (2\text{-}5)$$

If we follow the practice introduced in Sec. A.7.2 of expressing third-order tensor fields as sums of tensor products of basis fields, we may represent $\nabla \mathbf{T}$ as

$$\nabla \mathbf{T} = T^{ij}{}_{,k}\mathbf{g}_i\mathbf{g}_j\mathbf{g}^k \tag{2-6}$$

The $T^{ij}{}_{,k}$ ($i, j, k = 1, 2, 3$) are consequently the mixed components of $\nabla \mathbf{T}$.

## EXERCISES

**A.8.2-1** Let $\mathbf{T}$ be any second-order tensor field. For any curvilinear coordinate system, prove the following:

(a)
$$\nabla \mathbf{T} = T_{ij,k}\mathbf{g}^i\mathbf{g}^j\mathbf{g}^k$$

where

$$T_{ij,k} \equiv \frac{\partial T_{ij}}{\partial x^k} - \begin{Bmatrix} r \\ k\ i \end{Bmatrix} T_{rj} - \begin{Bmatrix} r \\ k\ j \end{Bmatrix} T_{ir}$$

(b)
$$\nabla \mathbf{T} = T^i{}_{.j,k}\mathbf{g}_i\mathbf{g}^j\mathbf{g}^k$$

where

$$T^i{}_{.j,k} \equiv \frac{\partial T^i{}_{.j}}{\partial x^k} + \begin{Bmatrix} i \\ k\ r \end{Bmatrix} T^r{}_{.j} - \begin{Bmatrix} r \\ k\ j \end{Bmatrix} T^i{}_{.r}$$

(c)
$$\nabla \mathbf{T} = T^{\ j}_{i..,k}\mathbf{g}^i\mathbf{g}_j\mathbf{g}^k$$

where

$$T^{\ j}_{i..,k} \equiv \frac{\partial T^{\ j}_{i.}}{\partial x^k} - \begin{Bmatrix} r \\ k\ i \end{Bmatrix} T^{\ j}_{r.} + \begin{Bmatrix} j \\ k\ r \end{Bmatrix} T^{\ r}_{i.}$$

**A.8.2-2** (a) Prove that

$$\nabla \mathbf{I} = 0$$

(b) Conclude that

$$g_{ij,k} = 0$$

and

$$g^{ij}{}_{,k} = 0$$

**A.8.2-3** We may define a fourth-order tensor field $\Theta$ to be a transformation (or mapping or rule) which assigns to each spatial vector field $\mathbf{v}$ a third-order tensor field $\beta$ such that the rules

$$\Theta \cdot (\mathbf{v} + \mathbf{w}) = \Theta \cdot \mathbf{v} + \Theta \cdot \mathbf{w}$$
$$\Theta \cdot (\alpha \mathbf{v}) = \alpha(\Theta \cdot \mathbf{v})$$

hold. Here $\alpha$ is a real scalar field.

The gradient of a third-order tensor field $\beta$ is the fourth-order tensor field denoted by $\nabla \beta$. If $\mathbf{a} = a_i\mathbf{e}_i$ indicates the directed line segment or point difference $\mathbf{a} \equiv \overrightarrow{zy}$, where y is an arbitrary point in $E$, then $\nabla \beta(z)$ is the linear transformation which assigns to $\mathbf{a}$ the third-order tensor field given by the following rule:

$$\nabla \beta(z) \cdot \mathbf{a} = \lim_{s \to 0} \frac{1}{s}[\beta(z + s\mathbf{a}) - \beta(z)]$$

## [A.9] VECTOR PRODUCT AND CURL

We conclude by an argument which is analogous to the one used in discussing the gradient of a second-order tensor field:

$$\nabla \beta = \beta^{ijk}{}_{,m} \mathbf{g}_i \mathbf{g}_j \mathbf{g}_k \mathbf{g}^m$$
$$= \beta_{ijk,m} \mathbf{g}^i \mathbf{g}^j \mathbf{g}^k \mathbf{g}^m$$

where

$$\beta^{ijk}{}_{,m} \equiv \frac{\partial \beta^{ijk}}{\partial x^m} + \begin{Bmatrix} i \\ m\ r \end{Bmatrix} \beta^{rjk} + \begin{Bmatrix} j \\ m\ r \end{Bmatrix} \beta^{irk} + \begin{Bmatrix} k \\ m\ r \end{Bmatrix} \beta^{ijr}$$

and

$$\beta_{ijk,m} \equiv \frac{\partial \beta_{ijk}}{\partial x^m} - \begin{Bmatrix} r \\ m\ i \end{Bmatrix} \beta_{rjk} - \begin{Bmatrix} r \\ m\ j \end{Bmatrix} \beta_{irk} - \begin{Bmatrix} r \\ m\ k \end{Bmatrix} \beta_{ijr}$$

(a) Prove that (see Exercises A.7.2-11 and A.7.2-12)

$$\nabla \boldsymbol{\epsilon} = 0$$

(b) Conclude that

$$\epsilon^{ijk}{}_{,m} = 0$$

and

$$\epsilon_{ijk,m} = 0$$

**A.8.2-4** (a) By writing out in full that

$$\epsilon_{rst,p} = 0$$

and putting $r = 1$, $s = 2$, $t = 3$, prove that [1, p. 155]

$$\frac{\partial \log \sqrt{g}}{\partial x^p} = \begin{Bmatrix} m \\ m\ p \end{Bmatrix}$$

(b) Writing out

$$g^{rs}{}_{,t} = 0$$

in full, deduce from part (a) that [1, p. 155]

$$\frac{1}{\sqrt{g}} \frac{\partial(\sqrt{g}\, g^{rs})}{\partial x^s} + \begin{Bmatrix} r \\ m\ n \end{Bmatrix} g^{mn} = 0$$

(c) Using the result of part (a), prove that [1, p. 155]

$$\operatorname{div} \mathbf{v} = v^r{}_{,r} = \frac{1}{\sqrt{g}} \frac{\partial}{\partial x^r} (\sqrt{g}\, v^r)$$

## REFERENCE

1. McConnell, A. J.: "Applications of Tensor Analysis," Dover, New York, 1957.

## A.9 VECTOR PRODUCT AND CURL

**\*\*A.9.1 The vector product and curl** Let **a** and **b** be two spatial vector fields. Students are often advised to remember the vector product (**a** ∧ **b**) with respect to a set of rectangular cartesian coordinate basis fields in the form of a determinant:

$$(\mathbf{a} \wedge \mathbf{b}) = \begin{vmatrix} \mathbf{e}_1 & \mathbf{e}_2 & \mathbf{e}_3 \\ a_1 & a_2 & a_3 \\ b_1 & b_2 & b_3 \end{vmatrix}$$

$$= (a_2 b_3 - a_3 b_2)\mathbf{e}_1$$
$$+ (a_3 b_1 - a_1 b_3)\mathbf{e}_2$$
$$+ (a_1 b_2 - a_2 b_1)\mathbf{e}_3 \qquad (1\text{-}1)$$

With respect to the physical basis fields of an orthogonal curvilinear coordinate system, the vector product takes a similar form:

$$(\mathbf{a} \wedge \mathbf{b}) = \begin{vmatrix} \mathbf{g}_{\langle 1 \rangle} & \mathbf{g}_{\langle 2 \rangle} & \mathbf{g}_{\langle 3 \rangle} \\ a_{\langle 1 \rangle} & a_{\langle 2 \rangle} & a_{\langle 3 \rangle} \\ b_{\langle 1 \rangle} & b_{\langle 2 \rangle} & b_{\langle 3 \rangle} \end{vmatrix}$$

$$= (a_{\langle 2 \rangle} b_{\langle 3 \rangle} - a_{\langle 3 \rangle} b_{\langle 2 \rangle})\mathbf{g}_{\langle 1 \rangle}$$
$$+ (a_{\langle 3 \rangle} b_{\langle 1 \rangle} - a_{\langle 1 \rangle} b_{\langle 3 \rangle})\mathbf{g}_{\langle 2 \rangle}$$
$$+ (a_{\langle 1 \rangle} b_{\langle 2 \rangle} - a_{\langle 2 \rangle} b_{\langle 1 \rangle})\mathbf{g}_{\langle 3 \rangle} \qquad (1\text{-}2)$$

The direction of $(\mathbf{a} \wedge \mathbf{b})$ is found by the familiar right-hand rule:

> When the index finger points in the direction $\mathbf{a}$ and the middle finger points in the direction $\mathbf{b}$, the thumb points in the direction $\mathbf{a} \wedge \mathbf{b}$.

A similar mnemonic device is often suggested for the components of the curl of a vector field $\mathbf{v}$ with respect to a set of rectangular cartesian basis fields:

$$\text{curl } \mathbf{v} = \begin{vmatrix} \mathbf{e}_1 & \mathbf{e}_2 & \mathbf{e}_3 \\ \dfrac{\partial}{\partial z_1} & \dfrac{\partial}{\partial z_2} & \dfrac{\partial}{\partial z_3} \\ v_1 & v_2 & v_3 \end{vmatrix}$$

$$= \left(\frac{\partial v_3}{\partial z_2} - \frac{\partial v_2}{\partial z_3}\right)\mathbf{e}_1$$
$$+ \left(\frac{\partial v_1}{\partial z_3} - \frac{\partial v_3}{\partial z_1}\right)\mathbf{e}_2$$
$$+ \left(\frac{\partial v_2}{\partial z_1} - \frac{\partial v_1}{\partial z_2}\right)\mathbf{e}_3 \qquad (1\text{-}3)$$

We find it convenient to adopt a more compact notation. Instead of Eq. (1-1) and (1-2), we write for the vector product

$$(\mathbf{a} \wedge \mathbf{b}) = e_{ijk} a_j b_k \mathbf{e}_i \qquad (1\text{-}4)$$

and

$$(\mathbf{a} \wedge \mathbf{b}) = e_{ijk} a_{\langle j \rangle} b_{\langle k \rangle} \mathbf{g}_{\langle i \rangle} \tag{1-5}$$

The symbol $e_{ijk}$ is defined to have only three distinct values:

0, when any two of the indices are equal
+1, when $ijk$ is an even permutation of 123
−1, when $ijk$ is an odd permutation of 123

Rather than Eq. (1-3), we prefer to write for the curl of a vector field **v**

$$\operatorname{curl} \mathbf{v} = e_{ijk} \frac{\partial v_k}{\partial z_j} \mathbf{e}_i \tag{1-6}$$

Equations (1-4) to (1-6) are suggested by our presentation of determinants in Sec. A.2.1.

## EXERCISES

**\*\*A.9.1-1** Show that the curl of the gradient of a scalar field $\alpha$ is identically zero:

curl $(\nabla \alpha) = 0$

**\*\*A.9.1-2** Show that, for any spatial vector field **v**,

div curl **v** = 0

**\*\*A.9.1-3** Show that, for any spatial vector field **v**,

curl curl **v** = $\nabla$(div **v**) − div $(\nabla \mathbf{v})$

**\*\*A.9.1-4** Three spatial vectors **a**, **b**, **c** may be viewed as forming three edges of a parallelepiped. If **a**, **b**, **c** have the same orientation as the first two fingers and thumb on the right hand, show that $(\mathbf{a} \wedge \mathbf{b}) \cdot \mathbf{c}$ determines the volume of the corresponding parallelepiped.

### A.9.2 More on the vector product and curl
Our discussion in Sec. A.9.1 suggests that we take as our formal definition for the vector product of any two spatial vector fields **a** and **b**,

$$(\mathbf{a} \wedge \mathbf{b}) \equiv \boldsymbol{\epsilon} : (\mathbf{b}\mathbf{a}) \tag{2-1}$$

where (see Exercises A.7.2-11 and A.7.2-12)

$$\boldsymbol{\epsilon} \equiv e_{ijk} \mathbf{e}_i \mathbf{e}_j \mathbf{e}_k \tag{2-2}$$

The $\mathbf{e}_m$ ($m = 1, 2, 3$) are to be interpreted as any convenient set of rectangular cartesian basis fields.

In order to find the geometrical interpretation of the spatial vector field $(\mathbf{a} \wedge \mathbf{b})$, let us fix our attention on a particular point $z$ of $E$ in Fig. A.9.2-1. For brevity, we adopt the inexact notation $\mathbf{a} = \mathbf{a}(z)$, $\mathbf{b} = \mathbf{b}(z)$ in Fig. A.9.2-2, where we have introduced a particular rectangular cartesian coordinate system, the origin of which

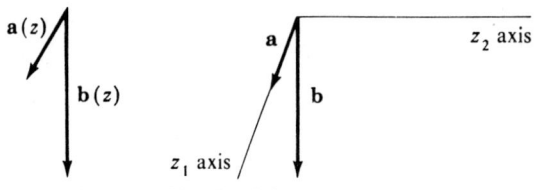

Fig. A.9.2-1　　Fig. A.9.2-2

coincides with the point $z$. We have chosen the $z_1$ axis along **a** and the $z_2$ axis perpendicular to **a**, but in the plane of **a** and **b**. In this special coordinate system, we have

$$a_i = (a, 0, 0) \qquad b_i = (b \cos \theta, b \sin \theta, 0) \tag{2-3}$$

when $a$ and $b$ denote the magnitudes of **a** and **b**, respectively. The components of $\mathbf{c} = (\mathbf{a} \wedge \mathbf{b}) = (\mathbf{a} \wedge \mathbf{b})(z)$ with respect to this coordinate system are, consequently,

$$c_i = (0, 0, ab \sin \theta) \tag{2-4}$$

We conclude that $(\mathbf{a} \wedge \mathbf{b})$ lies along the perpendicular to the plane of **a** and **b** and its magnitude is $ab \sin \theta$. We must still decide on its direction along this line.

We observe that

$$c^2 = e_{ijk} c_i a_j b_k \tag{2-5}$$

is a positive scalar. Thus suggests that we examine the sign of

$$e_{ijk} d_i a_j b_k \tag{2-6}$$

where **d** is any spatial vector. If we take the same rectangular cartesian coordinate system indicated in Fig. A.9.2-2, we see that

$$e_{ijk} d_i a_j b_k = d_3 ab \sin \theta \tag{2-7}$$

The quantity (2-6) will vanish only if $d_3 = 0$, that is, only if one vector becomes coplanar with the other two. If we continuously deform the triad (**a**,**b**,**d**) in such a way that it never becomes coplanar, we see that Eq. (2-6) varies continuously but always retains the same sign. Let us deform it continuously until **a** coincides with the positive $z_1$ axis, **b** with the positive $z_2$ axis, and **d** with the $z_3$ axis. The quantity (2-6) must be positive if **d** coincides with the positive $z_3$ axis, but negative if **d** coincides with the negative $z_3$ axis.

Returning to Eq. (2-5), we see that **a**, **b**, and $\mathbf{c} = (\mathbf{a} \wedge \mathbf{b})$ must have the same orientation with respect to one another as the coordinate system axes. In "right-handed" coordinate systems, the basis fields $\mathbf{e}_1$, $\mathbf{e}_2$, $\mathbf{e}_3$ have the same orientation as the index finger, middle finger, and thumb on the right hand. We clarify the definition of $\boldsymbol{\epsilon}$, Eq. (2-2), by requiring that the rectangular cartesian coordinate system be a right-handed one. Consequently, the direction of $(\mathbf{a} \wedge \mathbf{b})$ is found by the right-hand rule given in Sec. A.9.1.

We will generally find it convenient to limit ourselves to right-handed coordinate systems. Left-handed coordinate systems are discussed in Exercise A.9.2-4.

## [A.9] VECTOR PRODUCT AND CURL

If the tensor product **ba** in Eq. (2-1) is replaced by the gradient of a spatial vector field **v**, we have the definition of the curl of **v**:

$$\text{curl } \mathbf{v} \equiv \nabla \wedge \mathbf{v} \equiv \boldsymbol{\epsilon}:(\nabla \mathbf{v}) \tag{2-8}$$

With respect to a rectangular cartesian coordinate system, we have

$$\text{curl } \mathbf{v} = e_{ijk} \frac{\partial v_k}{\partial z_j} \mathbf{e}_i \tag{2-9}$$

## EXERCISES

**A.9.2-1**  *(a) Let **a** and **b** be any two spatial vector fields. Show that, with respect to any curvilinear coordinate system, we may write

$$(\mathbf{a} \wedge \mathbf{b}) = \epsilon^{ijk} a_j b_k \mathbf{g}_i$$

and

$$(\mathbf{a} \wedge \mathbf{b}) = \epsilon_{ijk} a^j b^k \mathbf{g}^i$$

(b) Show that, with respect to an orthogonal curvilinear coordinate system,

$$(\mathbf{a} \wedge \mathbf{b}) = e_{ijk} a_{\langle j \rangle} b_{\langle k \rangle} \mathbf{g}_{\langle i \rangle}$$

*A.9.2-2  Let **v** be any spatial vector field. Show that with respect to any curvilinear coordinate system we may write

(a)

$$\text{curl } \mathbf{v} = \epsilon^{ijk} v_{k,j} \mathbf{g}_i$$

and

(b)

$$\text{curl } \mathbf{v} = \epsilon_{ijk} v^k{}_{,m} g^{mj} \mathbf{g}^i$$

**A.9.2-3**  If **v** is a spatial vector field, show that

$$\text{curl } \mathbf{v} = \text{div } (\boldsymbol{\epsilon} \cdot \mathbf{v})$$

where $\boldsymbol{\epsilon} \cdot \mathbf{v}$ is a second-order tensor field.

**A.9.2-4**  (a) Let the $\bar{\mathbf{e}}_j$ ($j = 1, 2, 3$) be a set of left-handed rectangular cartesian basis fields. Use the right-hand rule and Exercise A.7.2-11 to show that

$$\bar{\mathbf{e}}_1 \cdot (\bar{\mathbf{e}}_2 \wedge \bar{\mathbf{e}}_3) = \det\left(\frac{\partial z_i}{\partial \bar{z}_j}\right) = -1$$

Here the coordinates $z_i$ ($i = 1, 2, 3$) refer to some right-handed rectangular cartesian basis $\mathbf{e}_i$ ($i = 1, 2, 3$); the coordinates $\bar{z}_j$ ($j = 1, 2, 3$) refer to the left-handed basis $\bar{\mathbf{e}}_j$ ($j = 1, 2, 3$).

From this result and Exercise A.7.2-11, we conclude that, when left-handed rectangular cartesian coordinate systems are employed, we must write

$$\boldsymbol{\epsilon} = -e_{ijk} \mathbf{e}_i \mathbf{e}_j \mathbf{e}_k$$

(b) Show that, when dealing with left-handed curvilinear coordinate systems, we must take the negative square root in Exercise A.7.2-12 and write

$$\epsilon^{ijk} = -\frac{1}{\sqrt{g}} e^{ijk}$$

and

$$\epsilon_{ijk} = -\sqrt{g}\, e_{ijk}$$

## A.10 DETERMINANT OF TENSOR

**A.10.1 The determinant of a second-order tensor field** [1, p. 102] Let the $\mathbf{g}_i$ ($i = 1, 2, 3$) be a set of basis fields for an arbitrary curvilinear coordinate system. At any point $z$ of $E$, the basis fields may be thought of as three edges of a parallelepiped as shown in Fig. A.10.1-1. We wish to introduce the magnitude of the determinant of a second-order tensor field $\mathbf{T}$ at the point $z$ as the ratio of the volume of the parallelepiped spanned by $\mathbf{T} \cdot \mathbf{g}_1(z)$, $\mathbf{T} \cdot \mathbf{g}_2(z)$, $\mathbf{T} \cdot \mathbf{g}_3(z)$ to the volume of the parallelepiped spanned by $\mathbf{g}_1(z)$, $\mathbf{g}_2(z)$, $\mathbf{g}_3(z)$. We choose the sign of the determinant of $\mathbf{T}$ to be positive if $\mathbf{T} \cdot \mathbf{g}_1(z)$, $\mathbf{T} \cdot \mathbf{g}_2(z)$, $\mathbf{T} \cdot \mathbf{g}_3(z)$, and $\mathbf{g}_1(z)$, $\mathbf{g}_2(z)$, $\mathbf{g}_3(z)$ have the same orientation (as the first two fingers and thumb on either the left or right hand); negative if they have the opposite orientation.

This together with Exercise A.9.1-4 suggests that we take as our formal definition of the scalar field det $\mathbf{T}$,

$$\det \mathbf{T} \equiv \frac{[(\mathbf{T} \cdot \mathbf{g}_1) \wedge (\mathbf{T} \cdot \mathbf{g}_2)] \cdot (\mathbf{T} \cdot \mathbf{g}_3)}{(\mathbf{g}_1 \wedge \mathbf{g}_2) \cdot \mathbf{g}_3} \tag{1-1}$$

In order to better appreciate the relationship of det $\mathbf{T}$ to the concept of the determinant introduced in Sec. A.2.1, let us apply this definition to a set of rectangular cartesian basis fields:

$$\begin{aligned}
\det \mathbf{T} &= \frac{\{(\mathbf{T} \cdot \mathbf{e}_1) \wedge (\mathbf{T} \cdot \mathbf{e}_2)\}(\mathbf{T} \cdot \mathbf{e}_3)}{(\mathbf{e}_1 \wedge \mathbf{e}_2) \cdot \mathbf{e}_3} \\
&= \frac{T_{j1} T_{k2} T_{i3} (\mathbf{e}_j \wedge \mathbf{e}_k) \cdot \mathbf{e}_i}{(\mathbf{e}_1 \wedge \mathbf{e}_2) \cdot \mathbf{e}_3} \\
&= T_{j1} T_{k2} T_{i3} \frac{e_{ijk}}{e_{312}} \\
&= e_{ijk} T_{i1} T_{j2} T_{k3} \\
&= \det (T_{mn})
\end{aligned} \tag{1-2}$$

When $\mathbf{T}$ is expressed in terms of its physical components with respect to an orthogonal curvilinear coordinate system, the results of Sec. A.2.1 are again directly applicable

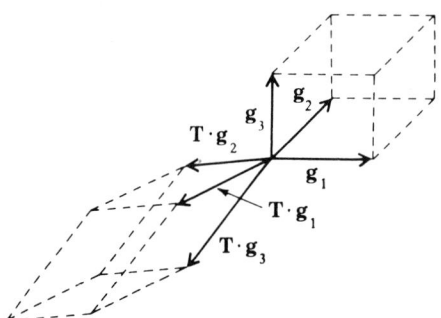

**Fig. A.10.1-1** Parallelepipeds spanned by $\{\mathbf{g}_1, \mathbf{g}_2, \mathbf{g}_3\}$ and by $\{\mathbf{T} \cdot \mathbf{g}_1, \mathbf{T} \cdot \mathbf{g}_2, \mathbf{T} \cdot \mathbf{g}_3\}$.

(Exercise A.10.1-2). But in general, minor modifications must be made (Exercise A.10.1-3).

It is easy to show (see Exercise A.10.1-4)

$$\det (\mathbf{S} \cdot \mathbf{T}) = (\det \mathbf{S})(\det \mathbf{T}) \tag{1-3}$$

$$\det (\mathbf{T}^{-1}) = \frac{1}{\det \mathbf{T}} \tag{1-4}$$

and

$$\det (\mathbf{T}^T) = \det \mathbf{T} \tag{1-5}$$

## REFERENCE

1. Coleman, B. D., H. Markovitz, and W. Noll: "Viscometric Flows of Non-Newtonian Fluids," Springer-Verlag, New York, 1966.

## EXERCISES

**A.10.1-1** In terms of the components of $\mathbf{T}$ with respect to a rectangular cartesian coordinate system, deduce that

$$e_{ijk}T_{im}T_{jn}T_{kp} = \det \mathbf{T} e_{mnp}$$

**A.10.1-2** In terms of the physical components of $\mathbf{T}$ with respect to an orthogonal curvilinear coordinate system, deduce that

$$e_{ijk}T_{\langle im \rangle}T_{\langle jn \rangle}T_{\langle kp \rangle} = \det \mathbf{T} e_{mnp}$$

**\*A.10.1-3** For an arbitrary curvilinear coordinate system,

$$\mathbf{T} = T^{ij}\mathbf{g}_i\mathbf{g}_j = T_{ij}\mathbf{g}^i\mathbf{g}^j = T_i{}^k\mathbf{g}^i\mathbf{g}_k$$

Prove that

$$\epsilon_{ijk}T^{im}T^{jn}T^{kp} = \det \mathbf{T} \epsilon^{mnp}$$

$$\epsilon^{ijk}T_{im}T_{jn}T_{kp} = \det \mathbf{T} \epsilon_{mnp}$$

and

$$\epsilon^{ijk}T_i{}^m T_j{}^n T_k{}^p = \det \mathbf{T} \epsilon^{mnp}$$

**A.10.1-4** Prove Eqs. (1-3), (1-4), and (1-5).

**A.10.1-5** Prove the following rule for differentiation of determinants:

$$\dot{\overline{\det \mathbf{T}}} = (\det \mathbf{T})[\text{tr}\, (\mathbf{T}^{-1} \cdot \dot{\mathbf{T}})]$$

*Hint:* Begin with

$$\det T_{rs} = \tfrac{1}{6} e_{ijk} e_{mnp} T_{im} T_{jn} T_{kp}$$

**A.10.1-6** (a) If $\mathbf{A}$ is invertible and consequently Eq. (4-3) of Sec. A.5.4 holds, show that $\det \mathbf{A} \neq 0$.

(b) Beginning with an equation of the same general form as Eq. (1-12) of Sec. A.2.1, show that, if $\det \mathbf{A} \neq 0$, $\mathbf{A}$ must be invertible.

**A.10.1-7** *Orthogonal tensors* Prove that, if $\mathbf{Q}$ is an orthogonal second-order tensor, then

$$\det \mathbf{Q} = \pm 1$$

If $\det \mathbf{Q} = -1$ and if $(\mathbf{e}_1, \mathbf{e}_2, \mathbf{e}_3)$ is a right-handed triad, then $\{\mathbf{Q} \cdot \mathbf{e}_1, \mathbf{Q} \cdot \mathbf{e}_2, \mathbf{Q} \cdot \mathbf{e}_3\}$ is left-handed. In the context of changes of frame in Sec. 1.2.1, $\mathbf{Q}$ may be thought of as both a rotation and a reflection when $\det \mathbf{Q} = -1$.

## A.II  INTEGRATION

**\*\*A.II.1  Integration of spatial vector fields**  A volume, surface, or line integration is an addition of quantities associated with different points in space. When we integrate a spatial vector field (a spatial vector-valued function of position), we add spatial *vectors* associated with different points in space.

Let $\mathbf{v}$ be some spatial vector field and let $z$ and $y$ denote two points in space. By the parallelogram rule for the addition of spatial vectors (Sec. A.1.2), we may write

$$\mathbf{v}(z) + \mathbf{v}(y) = v_i(z)\mathbf{e}_i + v_i(y)\mathbf{e}_i$$
$$= [v_i(z) + v_i(y)]\mathbf{e}_i \tag{1-1}$$

In applying the parallelogram rule here, we take advantage of the fact that the direction and magnitude of the rectangular cartesian basis fields are independent of position in space.

Equation (1-1) suggests how we should proceed with the integration of a spatial vector field. Let us consider an integration of $\mathbf{v}$ over some region $R$ (this might be a curve, surface, or volume):

$$\int_R \mathbf{v} \, dR = \int_R v_i \mathbf{e}_i \, dR = \left( \int_R v_i \, dR \right) \mathbf{e}_i \tag{1-2}$$

Since the magnitude and direction of the rectangular cartesian basis fields are independent of position in space, they may be treated as constants with respect to integration. By Eq. (1-2) we have transformed the problem of integration of a spatial vector field to the familiar problem of integration of three scalar fields.

### EXERCISES

**\*A.II.1-1**  Let $\mathbf{v} = v_i \mathbf{e}_i$ for some rectangular cartesian coordinate system $(z_1, z_2, z_3)$ and $\mathbf{v} = \bar{v}_j \bar{\mathbf{g}}^j$ for some curvilinear coordinate system $(\bar{x}^1, \bar{x}^2, \bar{x}^3)$.

(a) Show that

$$\int_R \mathbf{v} \, dR = \int_R v_i \, dR \mathbf{e}_i \neq \int_R \bar{v}_j \, dR \, \bar{\mathbf{g}}^j$$

(b) Show that

$$\int_R \mathbf{v} \, dR = \int_R \frac{\partial \bar{x}^j}{\partial z_i} \bar{v}_j \, dR \, \mathbf{e}_i$$

Sometimes it is most convenient to express the integration of a spatial vector field in terms of its covariant or contravariant components with respect to some curvilinear coordinate system, even though the integration is carried out with respect to some rectangular cartesian coordinate system.

**\*\*A.II.1-2**  Extend the discussion of this section to second-order tensor fields.

## A.11.2 Green's transformation [1, p. 815]

Our object here is to develop Green's transformation, a special case of which is the divergence theorem (Gauss' theorem).

Let $\varphi$ be any scalar, vector, or tensor. If **n** is understood to be the outwardly directed unit normal to the closed surface $S_m$ shown in Fig. A.11.2-1, we may approximate

$$\int_{S_m} \varphi \mathbf{n}\, dS \doteq [\varphi(z_1 + \Delta z_1, z_2, z_3) - \varphi(z_1, z_2, z_3)] \Delta z_2\, \Delta z_3\, \mathbf{e}_1$$
$$+ [\varphi(z_1, z_2 + \Delta z_2, z_3) - \varphi(z_1, z_2, z_3)] \Delta z_1\, \Delta z_3\, \mathbf{e}_2$$
$$+ [\varphi(z_1, z_2, z_3 + \Delta z_3) - \varphi(z_1, z_2, z_3)] \Delta z_1\, \Delta z_2\, \mathbf{e}_3 \quad (2\text{-}1)$$

Dividing through by $\Delta V_m \equiv \Delta z_1\, \Delta z_2\, \Delta z_3$, we see that in the limit as

$$\begin{array}{l}\Delta z_1 \to 0 \\ \Delta z_2 \to 0 : \\ \Delta z_3 \to 0 \end{array} \qquad \frac{1}{\Delta V_m} \int_{S_m} \varphi \mathbf{n}\, dS = \nabla \varphi \qquad (2\text{-}2)$$

Now consider a region of space that has a volume $V$ and a closed bounding surface $S$. Let **n** be the outwardly directed unit normal spatial vector field to the surface $S$. Referring to Fig. A.11.2-2, we may define a volume integral to be obtained as the result of a limiting process in which we visualize the region $V$ to be

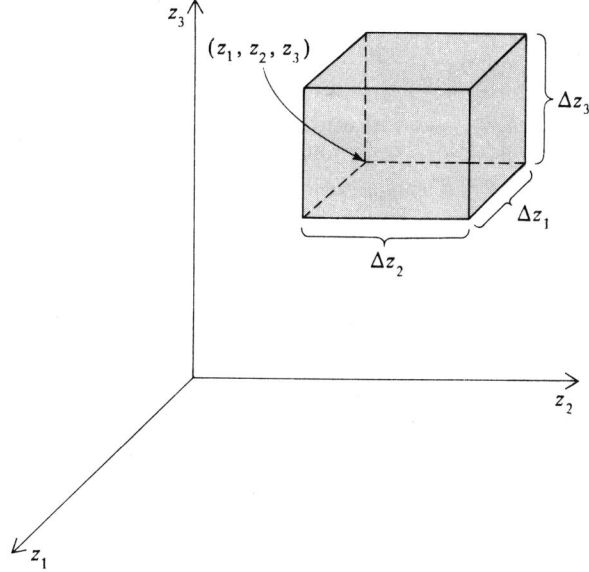

**Fig. A.11.2-1** The element of volume $\Delta V_m \equiv \Delta z_1\, \Delta z_2\, \Delta z_3$; the closed bounding surface of $\Delta V_m$ is $S_m$.

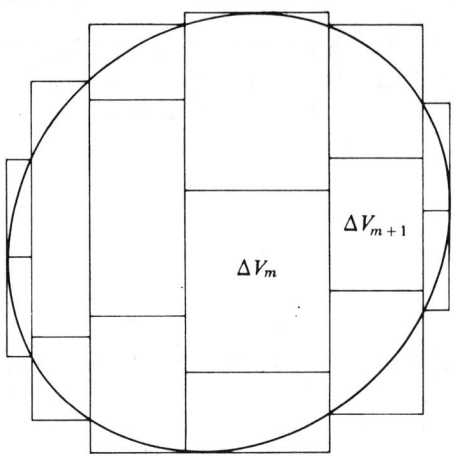

**Fig. A.11.2-2** The approximation of $V$ by $K$ parallelepipeds.

approximated by $K$ parallelepipeds:

$$\int_V \nabla \varphi \, dV = \lim_{\max \Delta V_m \to 0} \sum_{m=1}^{K} (\nabla \varphi)_m \Delta V_m \tag{2-3}$$

By $(\nabla \varphi)_m$ we mean the value of $\nabla \varphi$ evaluated at some point in the $m$th parallelepiped. From Eqs. (2-2) and (2-3), we conclude

$$\int_V \nabla \varphi \, dV = \lim_{\max \Delta V_m \to 0} \sum_{m=1}^{K} \int_{S_m} \varphi \mathbf{n} \, dS$$

$$= \int_S \varphi \mathbf{n} \, dS \tag{2-4}$$

In arriving at this result, we have noted in Fig. A.11.2-2 that the contributions to $S_m$ and $S_{m+1}$ cancel on that portion of their boundary that they share in common. We will refer to Eq. (2-4) as *Green's transformation*.

If **v** is a spatial vector field, then Green's transformation requires

$$\int_V \nabla \mathbf{v} \, dV = \int_S \mathbf{v} \mathbf{n} \, dS \tag{2-5}$$

A special case is the familiar divergence theorem (Gauss' theorem):

$$\int_V \operatorname{div} \mathbf{v} \, dV = \int_S \mathbf{v} \cdot \mathbf{n} \, dS \tag{2-6}$$

Let **T** be any second-order tensor field. Green's transformation says that

$$\int_V \nabla \mathbf{T} \, dV = \int_S \mathbf{T} \mathbf{n} \, dS \tag{2-7}$$

It is a special case of this last that we will have several occasions to use:

$$\int_V \operatorname{div} \mathbf{T} \, dV = \int_S \mathbf{T} \cdot \mathbf{n} \, dS \tag{2-8}$$

For an alternative view of Green's transformation and for a further application, see Exercises A.11.2-1 and A.11.2-2.

## REFERENCE

1. Ericksen, J. L.: In S. Flügge (ed.), "Handbuch der Physik," vol. 3/1, Springer-Verlag, Berlin, 1960.

## EXERCISES

**A.11.2-1** (a) Assuming $\varphi$ is a scalar field, write the $i$ component of Eq. (2-4) in rectangular cartesian coordinates.
(b) Assuming $\varphi$ is a spatial vector field, write the $ij$ component of Eq. (2-4) in rectangular cartesian coordinates.
(c) Assuming $\varphi$ is a second-order tensor field, write the $ijk$ component of Eq. (2-4) in rectangular cartesian coordinates.
(d) Use the result of part (b) to derive Eq. (2-6).
(e) Use the result of part (c) to arrive at Eq. (2-8).

**A.11.2-2** Let **v** be a spatial vector field. Show as another case of Green's transformation that

$$\int_V \text{curl } \mathbf{v} \, dV = \int_S (\mathbf{n} \wedge \mathbf{v}) \, dS$$

*Hint:* Use Exercise A.11.2-1, part (c).

## **A.11.3 Change of variable in volume integrations**

Let $x^1$, $x^2$, $x^3$ denote one system of coordinates and $\bar{x}^1$, $\bar{x}^2$, $\bar{x}^3$ another, such that

$$x^1 = x^1(\bar{x}^1, \bar{x}^2, \bar{x}^3)$$
$$x^2 = x^2(\bar{x}^1, \bar{x}^2, \bar{x}^3)$$
$$x^3 = x^3(\bar{x}^1, \bar{x}^2, \bar{x}^3)$$

From calculus, we know that

$$\iiint_{R_{x^1 x^2 x^3}} F(x^1, x^2, x^3) \, dx^1 \, dx^2 \, dx^3 = \iiint_{R_{\bar{x}^1 \bar{x}^2 \bar{x}^3}} F(x^1(\bar{x}^1, \bar{x}^2, \bar{x}^3), \ldots) \left| \det\left(\frac{\partial x^i}{\partial \bar{x}^j}\right) \right| d\bar{x}^1 \, d\bar{x}^2 \, d\bar{x}^3 \tag{3-1}$$

If one coordinate system above is rectangular cartesian, we have (see Exercise A.11.3-1)

$$\iiint_{R_{z_1 z_2 z_3}} F(z_1, z_2, z_3) \, dz_1 \, dz_2 \, dz_3 = \iiint_{R_{x^1 x^2 x^3}} F(z_1(x^1, x^2, x^3), \ldots) \sqrt{g} \, dx^1 \, dx^2 \, dx^3 \tag{3-2}$$

Here $g \equiv \det(g_{ij})$.

## EXERCISES

**A.11.3-1** Prove that

$$\left| \det \left( \frac{\partial z_i}{\partial x^j} \right) \right| = \sqrt{g}$$

and that, consequently, Eq. (3-2) holds.

**A.11.3-2** What form does Eq. (3-2) take when the $x^1$, $x^2$, $x^3$ denote (a) cylindrical coordinates and (b) spherical coordinates?

**A.11.3-3** (a) Consider some curve in space and let $t$ be a parameter along this curve. We know from Sec. A.4.1 that $(d\mathbf{p}/dt)(t)$ is tangent to the curve at the point $t$. Prove that

$$\left( \frac{d\mathbf{p}}{dt} \cdot \frac{d\mathbf{p}}{dt} \right)(dt)^2 = \left( \frac{dz_1}{dt} dt \right)^2 + \left( \frac{dz_2}{dt} dt \right)^2 + \left( \frac{dz_3}{dt} dt \right)^2$$

If we denote by $s$ arc length measured along the curve, we know from calculus or differential geometry that

$$(ds)^2 = \left( \frac{dz_1}{ds} ds \right)^2 + \left( \frac{dz_2}{ds} ds \right)^2 + \left( \frac{dz_3}{ds} ds \right)^2$$

This means that

$$\frac{d\mathbf{p}}{ds} \cdot \frac{d\mathbf{p}}{ds} = 1$$

(b) In any coordinate system, three coordinate curves intersect at each point in space. Choose a particular point and denote the unit tangent vectors to the three coordinate curves at this point by $d\mathbf{p}/ds_{(1)}$, $d\mathbf{p}/ds_{(2)}$, and $d\mathbf{p}/ds_{(3)}$. Here $s_{(i)}$ denotes arc length measured along the $x^i$ coordinate curve.

Let us obtain an expression for the volume $dV$ of the parallelepiped formed from $d\mathbf{p}/ds_{(1)}$, $d\mathbf{p}/ds_{(2)}$, and $d\mathbf{p}/ds_{(3)}$ with sides of length $ds_{(1)}$, $ds_{(2)}$, and $ds_{(3)}$, respectively. Starting with the definition

$$dV \equiv \left( \frac{d\mathbf{p}}{ds_{(1)}} \wedge \frac{d\mathbf{p}}{ds_{(2)}} \right) \cdot \frac{d\mathbf{p}}{ds_{(3)}} ds_{(1)} ds_{(2)} ds_{(3)}$$

show that

$$dV = \sqrt{g}\, dx^1\, dx^2\, dx^3$$

In this way, we may suggest the result of Eq. (3-2).

appendix **B**

# More on the Transport Theorem

The derivation of the transport theorem in Sec. 1.3.2 may be unsatisfactory for the reader who has chosen to read only the double-asterisked sections of Appendix $A$. He may understandably object to using Exercise A.10.1-5, which he has not proved, in order to arrive at Eq. (2-4) of Sec. 1.3.2. In the sections that follow, alternative derivations of the transport theorem are suggested.

## B.1 ALTERNATIVE DERIVATIONS

**B.1.1 First alternative proof of the transport theorem** Let us start with Eq. (2-1) of Sec. 1.3.2:

$$\frac{d}{dt}\int_{V_{(m)}} \Psi\, dV = \int_{V_{(m)}} \left(\frac{d_{(m)}\Psi}{dt} + \frac{\Psi}{J}\frac{d_{(m)}J}{dt}\right) dV \tag{1-1}$$

and let us set

$$\Psi = 1 \tag{1-2}$$

to obtain

$$\frac{dV_{(m)}}{dt} = \int_{V_{(m)}} \frac{1}{J} \frac{d_{(m)}J}{dt} dV \qquad (1\text{-}3)$$

This last gives us an expression for the time rate of change of the volume associated with the material body.

Intuitively, we can say that the rate at which the volume of the body increases can be related to the net rate at which the bounding surface of the body moves in an outward direction:

$$\frac{dV_{(m)}}{dt} = \int_{S_{(m)}} \mathbf{v} \cdot \mathbf{n}\, dS \qquad (1\text{-}4)$$

An application of Green's transformation (Sec. A.11.2) allows us to express this last as

$$\frac{dV_{(m)}}{dt} = \int_{V_{(m)}} \text{div } \mathbf{v}\, dV \qquad (1\text{-}5)$$

By eliminating the time rate of change of the volume of the body between Eqs. (1-3) and (1-5), we find

$$\int_{V_{(m)}} \left( \frac{1}{J}\frac{d_{(m)}J}{dt} - \text{div } \mathbf{v} \right) dV = 0 \qquad (1\text{-}6)$$

But this statement is true for any body or for any portion of a body (since a portion of a body is itself a body). We conclude that the integrand in Eq. (1-6) must be zero:

$$\frac{1}{J}\frac{d_{(m)}J}{dt} = \text{div } \mathbf{v} \qquad (1\text{-}7)$$

Equation (1-7) is just what we need to put Eq. (1-1) in the form of the transport theorem [Eq. (2-9) of Sec. 1.3.2]:

$$\frac{d}{dt}\int_{V_{(m)}} \Psi\, dV = \int_{V_{(m)}} \left( \frac{d_{(m)}\Psi}{dt} + \Psi\, \text{div } \mathbf{v} \right) dV \qquad (1\text{-}8)$$

From here the alternative forms of the transport theorem [Eqs. (2-10) and (2-11) of Sec. 1.3.2], as well as the generalized transport theorem [Eq. (2-12) of Sec. 1.3.2], follow as described in Sec. 1.3.2.

The advantage of this discussion over that given in Sec. 1.3.2 is that we have been able to avoid the use of Exercise A.10.1-5 in arriving at Eq. (1-7). The disadvantage is that we have adopted an intuitively motivated step, Eq. (1-4), in the course of our proof.

**B.I.2 Second alternative proof of the transport theorem** Let us consider the operation

$$\frac{d}{dt}\int_{V_{(m)}} \rho \hat{\Psi}\, dV$$

## [B.1] ALTERNATIVE DERIVATIONS

rather than

$$\frac{d}{dt}\int_{V_{(m)}} \Psi\, dV$$

which was examined in Sec. 1.3.2.

As explained in Sec. 1.3.2, if we look at this volume integration in the reference configuration $\kappa$, we may interchange differentiation and integration. Using the results of Sec. A.11.3, we may say that

$$\frac{d}{dt}\int_{V_{(m)}} \rho\hat{\Psi}\, dV = \frac{d}{dt}\int_{V_{(m)\kappa}} \rho\hat{\Psi} J\, dV$$

$$= \int_{V_{(m)\kappa}} \left[\rho\frac{d_{(m)}\hat{\Psi}}{dt} + \frac{\hat{\Psi}}{J}\frac{d_{(m)}(\rho J)}{dt}\right] J\, dV$$

$$= \int_{V_{(m)}} \left[\rho\frac{d_{(m)}\hat{\Psi}}{dt} + \frac{\hat{\Psi}}{J}\frac{d_{(m)}(\rho J)}{dt}\right] dV \qquad (2\text{-}1)$$

From Exercise 1.3.3-3,

$$\rho J = \rho_0 \qquad (2\text{-}2)$$

where $\rho_0$ denotes the density distribution in the reference configuration. Since $\rho_0$ is not a function of time (the reference configuration and consequently the density distribution in the reference configuration is fixed for all time), we have that

$$\frac{d_{(m)}(\rho J)}{dt} = \frac{d_{(m)}\rho_0}{dt} = 0 \qquad (2\text{-}3)$$

This allows us to write Eq. (2-1) in the form of the transport theorem derived in Exercise 1.3.3-1:

$$\frac{d}{dt}\int_{V_{(m)}} \rho\hat{\Psi}\, dV = \int_{V_{(m)}} \rho\frac{d_{(m)}\hat{\Psi}}{dt}\, dV \qquad (2\text{-}4)$$

Since the equation of continuity, Eq. (3-3) of Sec. 1.3.3, gives us

$$\rho\frac{d_{(m)}\hat{\Psi}}{dt} = \frac{d_{(m)}(\rho\hat{\Psi})}{dt} - \hat{\Psi}\frac{d_{(m)}\rho}{dt}$$

$$= \frac{d_{(m)}(\rho\hat{\Psi})}{dt} + \rho\hat{\Psi}\,\text{div}\,\mathbf{v}$$

$$= \frac{\partial(\rho\hat{\Psi})}{\partial t} + \text{div}\,(\rho\hat{\Psi}\mathbf{v}) \qquad (2\text{-}5)$$

we may easily express Eq. (2-4) in forms analogous to Eqs. (2-9) to (2-11) of Sec. 1.3.2:

$$\frac{d}{dt}\int_{V_{(m)}} \rho\hat{\Psi}\, dV = \int_{V_{(m)}} \left[\frac{d_{(m)}(\rho\hat{\Psi})}{dt} + \rho\hat{\Psi}\,\text{div}\,\mathbf{v}\right] dV \qquad (2\text{-}6)$$

$$\frac{d}{dt}\int_{V_{(m)}} \rho\hat{\Psi}\, dV = \int_{V_{(m)}} \left[\frac{\partial(\rho\hat{\Psi})}{\partial t} + \text{div}\,(\rho\hat{\Psi}\mathbf{v})\right] dV \qquad (2\text{-}7)$$

and

$$\frac{d}{dt}\int_{V_{(m)}} \rho\hat{\Psi}\, dV = \int_{V_{(m)}} \frac{\partial(\rho\hat{\Psi})}{\partial t}\, dV + \int_{S_{(m)}} \rho\hat{\Psi}\mathbf{v}\cdot\mathbf{n}\, dS \qquad (2\text{-}8)$$

By the argument given in Sec. 1.3.2, we can finally arrive at another form of the *generalized transport theorem*:

$$\frac{d}{dt}\int_{V_{(s)}} \rho\hat{\Psi}\, dV = \int_{V_{(s)}} \frac{\partial(\rho\hat{\Psi})}{\partial t}\, dV + \int_{S_{(s)}} \rho\hat{\Psi}\mathbf{v}_{(s)}\cdot\mathbf{n}\, dS \qquad (2\text{-}9)$$

The advantage of this discussion of the transport theorem is that everything that we have done is rigorous. No intuitive arguments are involved. The serious disadvantage, however, is that Eq. (2-4) is not the form of the transport theorem required for the derivation of the equation of continuity in Sec. 1.3.3 [Eqs. (2-6) to (2-8) were obtained from Eq. (2-4) only after we assumed the equation of continuity in Eq. (2-5)].

appendix C
# Derivation of Inequality (4-1) of Sec. 8.5.4

We wish to show here that, for *stable equilibrium* at some time $t = t_e$, inequality (4-1) of Sec. 8.5.4 is implied by inequality (2-22) of Sec. 8.5.2 and Eq. (3-1)$_1$ of Sec. 8.5.3. In order to take into account the constraints imposed by conservation of mass, Euler's first law, and the energy balance, we will evaluate the second derivative on the left of inequality (2-22) of Sec. 8.5.2 by differentiating Eq. (3-1)$_1$ of Sec. 8.5.3 with respect to time. This operation involves taking a derivative of surface integrals with respect to time. We begin by investigating this operation.

Let us consider an integral over a material surface of some quantity $B$. For a material derivative of such a surface integral, we have [1, p. 229]

$$\frac{d}{dt}\int_{S_{(m)}} B\, dS = \int_{S_{(m)}} \left(\frac{d_{(m)}B}{dt} + Bv^\alpha_{,\alpha} + \frac{B}{2a}\frac{\partial a}{\partial t}\right) dS \qquad (1\text{-}1)$$

Here $a$ is the determinant of the covariant surface metric tensor [2, p. 167]; the operation $\partial/\partial t$ denotes a derivative with respect to time holding the spatial coordinates fixed; the $v^\alpha$ ($\alpha = 1, 2$) are the contravariant surface components of the velocity vector; the comma notation indicates a surface covariant derivative [2, p. 180].

The phase interfaces $S_{(\text{sing})}$ are not material surfaces, since we allow mass transfer across them.  Equation (1-1) may be adapted to a time derivative following the nonmaterial surface $S_{(sing)}$, if in the derivation of Eq. (1-1) we replace material particles by "surface points," one of which is located at each intersection of the surface coordinate curves.  (This is in the spirit of the derivation of the generalized transport theorem given by Truesdell and Toupin [3, p. 347]).  We obtain

$$\frac{d}{dt} \int_{S_{(sing)}} B \, dS = \int_{S_{(sing)}} \left( \frac{dB}{dt} + Bu^\alpha_{,\alpha} + \frac{B}{2a} \frac{\partial a}{\partial t} \right) dS \tag{1-2}$$

where the $u^\alpha$ ($\alpha = 1, 2$) are the surface components of the velocity of a surface point. We realize that the definition of the surface coordinates has been left arbitrary and, therefore, that the definition of $u^\alpha$ is not fixed.  This is unimportant, since in our results the effect of the surface points will arise only through the speed of displacement of $S_{(sing)}$, $u_{(\xi)}$, which is independent of the choice of parametrization [3, p. 499].

Let us now take the derivative of Eq. $(3\text{-}1)_1$ of Sec. 8.5.3 and substitute it into inequality (2-22) of Sec. 8.5.2 to obtain at time $t = t_e$:

$$\int_V \left[ \frac{\partial^2 \check{S}}{\partial t^2} + \sum_{A=1}^N \lambda_{(A)} \frac{\partial^2}{\partial t^2} \left( \rho_{(A)} - \sum_{j=1}^K M_{(A)} \nu_{(A,j)} \Psi_{(j)} \right) + \boldsymbol{\lambda}_m \cdot \frac{\partial^2 (\rho \mathbf{v})}{\partial t^2} \right.$$

$$+ \lambda_e \frac{\partial^2}{\partial t^2} \left( \check{E} - \sum_{A=1}^N \sum_{j=1}^K \varphi_{(A)} M_{(A)} \nu_{(A,j)} \Psi_{(j)} \right) \Bigg] dV - \int_{S_{(sing)}} \left[ \frac{\partial \check{S}}{\partial t} + \frac{d\check{S}}{dt} \right.$$

$$+ \sum_{A=1}^N \lambda_{(A)} \frac{\partial}{\partial t} \left( \rho_{(A)} - \sum_{j=1}^K M_{(A)} \nu_{(A,j)} \Psi_{(j)} \right) + \sum_{A=1}^N \lambda_{(A)} \frac{d\rho_{(A)}}{dt} + \boldsymbol{\lambda}_m \cdot \frac{\partial (\rho \mathbf{v})}{\partial t}$$

$$+ \boldsymbol{\lambda}_m \cdot \frac{d(\rho \mathbf{v})}{dt} + \lambda_e \frac{\partial}{\partial t} \left( \check{E} - \sum_{A=1}^N \sum_{j=1}^K \varphi_{(A)} M_{(A)} \nu_{(A,j)} \Psi_{(j)} \right) + \lambda_e \frac{d\check{E}}{dt} \Bigg] u_{(\xi)} \, dS$$

$$- \int_{S_{(sing)}} \left( \frac{du_{(\xi)}}{dt} + u_{(\xi)} u^\alpha_{,\alpha} + \frac{u_{(\xi)}}{2a} \frac{\partial a}{\partial t} \right)$$

$$\times \left[ \check{S} + \sum_{A=1}^N \lambda_{(A)} \rho_{(A)} + \rho \boldsymbol{\lambda}_m \cdot \mathbf{v} + \lambda_e \check{E} \right] dS \leq 0 \tag{1-3}$$

In taking the derivative of Eq. $(3\text{-}1)_1$ of Sec. 8.5.3, we use the transport theorem for a region containing a singular surface and Eq. (1-2).

## REFERENCES

1. Aris, Rutherford: "Vectors, Tensors, and the Basic Equations of Fluid Mechanics," Prentice-Hall, Englewood Cliffs, N.J., 1962.
2. McConnell, A. J.: "Applications of Tensor Analysis," Dover, New York, 1957.
3. Truesdell, C., and R. A. Toupin: In S. Flügge (ed.), "Handbuch der Physik," vol. 3/1, Springer-Verlag, Berlin, 1960.

# Name Index

Acosta, A. J., 124, 259, 261
Acrivos, A., 116, 143
Adivarahan, P., 413
Aris, Rutherford, 9, 563, 580, 581, 669
Arnold, J. H., 505, 506
Arnold, K. R., 531
Ashare, Edward, 81, 217

Batchelor, G. K., 170
Bedingfield, C. H., Jr., 483
Beegle, B. L., 496
Bennett, G. W., 510
Bird, R. B., 4, 35, 53, 54, 57, 59, 66, 74,
 81, 85, 87, 89, 90, 92, 94, 107, 177,
 211, 217, 219, 247, 250, 262, 270,
 276, 304, 311, 329, 336, 342, 346,
 376, 384, 405, 414, 419, 420, 431,
 440, 453–455, 464, 469, 470, 474,
 475, 477, 478, 484, 510, 516, 521,
 529, 532, 536, 537, 549, 556, 566,
 574, 581, 589–590, 593, 639, 650
Birkhoff, Garrett, 101
Bizzell, G. D., 191, 193
Blasius, H., 145
Boussinesq, J., 176
Bowen, R. M., 298, 468, 474, 499
Brand, Louis, 27, 176, 203–205, 232,
 234, 246, 388, 389, 410, 411, 430,
 570, 571, 588
Brennecke, L. F., 537
Brenner, Howard, 118
Brewer, Leo, 493
Brinkman, H. C., 206
Brown, G. M., 279, 380, 432, 485, 522
Brown, Harrison, 522

Callen, H. G., 280, 498
Carley, J. F., 74
Carman, P. C., 577
Carslaw, H. S., 87, 310, 311, 316, 317,
 319–322, 509
Cash, F. M., 74
Chapman, Sydney, 298

Chen, J. D., 516
Christopher, R. H., 217
Chu, Ju Chin, 537
Churchill, R. V., 120, 136, 513
Colburn, A. P., 396, 536
Cole, J. D., 142
Coleman, B. D., 54–56, 81, 85, 104, 106,
 284, 302, 473, 498, 658
Corrsin, Stanley, 170
Cowling, T. G., 298
Crank, J., 515
Curtiss, C. F., 38, 57, 94, 469, 470, 474,
 475, 477, 521, 566, 574
Cussler, E. L., Jr., 531

Dahler, J. S., 38
Dallon, D. S., 276
Danckwerts, P. V., 513, 515
Darcy, H. P., 193
Defay, R., 498
de Groff, H. M., 336
de Groot, S. R., 462, 468, 473
Deissler, R. G., 178, 181, 390, 394, 555
Denbigh, K. G., 498
Drake, R. M., 399
Dranoff, J. S., 413
Drew, T. B., 483, 536
Dwight, H. B., 562

Eckart, Carl, 468
Eckert, E. R., 399
Ehrlich, Robert, 276
Emde, Fritz, 131
Ericksen, J. L., 27, 33, 116, 661
Evans, R. B., 575, 577

Feshbach, Herman, 639
Finn, Robert, 70
Flumerfelt, R. W., 298
Foreman, R. W., 477
Fredrickson, A. G., 53, 56, 143
Furry, W. H., 477

Gaggioli, R. A., 257
Getty, R. L., 537
Gibbs, J. W., 279, 498
Gill, W. N., 563
Ginn, R. F., 214
Goldstein, S., 66, 69, 146, 154, 177, 384, 388
Graessley, W. W., 435
Grew, K. E., 477, 522
Grossetti, E., 38
Guggenheim, E. A., 477
Gutfinger, Chaim, 143

Haase, Rolf, 498
Hallman, T. M., 382, 384
Halmos, P. R., 265, 495, 605–607, 627, 636
Happel, Job, 118
Hegde, M. G., 194
Hermes, R. A., 53, 143
Hiemenz, K., 157
Hill, R., 262, 276, 277
Hinze, J. O., 170, 176, 177
Hirschfelder, J. O., 57, 469, 470, 474, 475, 477, 521, 566, 574
Hoffman, Kenneth, 633
Hopke, S. W., 54, 270, 271, 276
Hougen, O. A., 596, 600
Howarth, L., 145, 157
Hsu, H. W., 531
Huang, Jinn-Hule, 413

Ibbs, T. L., 477, 522
Illingworth, C. R., 336
Ince, Simon, 42, 73, 129, 252
Irving, J., 94, 116–118, 131, 211, 517, 582

Jaeger, J. C., 87, 310, 311, 316, 317, 319–322, 509
Jahnke, Eugene, 131
Jakob, Max, 396, 399, 405, 407
Johnson, M. W., Jr., 262
Jones, A. L., 477
Jones, R. C., 477

Kaplan, Wilfred, 22, 117, 269
Karim, S. M., 49
Kays, W. M., 431

Kearsley, E. A., 276
Keenan, J. H., 301, 444
Kellogg, O. D., 120, 127, 310
Kittredge, C. P. 247
Krieger, I. M., 85
Kunii, Daizo, 413
Kunze, Ray, 633

Ladyzhenskaya, O. A., 70
Lamb, Horace, 116, 277
Landau, L. D., 348
Lapple, C. E., 247
Laufer, Joh, 181
Laurence, R. L., 565
Leigh, D. C., 56
Lertes, Peter, 38
Lescarboura, J. A., 81, 217
Levich, V. G., 528
Lewis, G. N., 493
Lichnerowicz, A., 627
Lifshitz, E. M., 348
Lightfoot, E. N., 4, 35, 54, 59, 66, 74, 87, 89, 90, 107, 177, 211, 247, 250, 304, 311, 336, 342, 346, 347, 376, 384, 405, 414, 419, 431, 453–455, 464, 477, 478, 484, 510, 515, 516, 529, 531, 532, 536, 537, 549, 556, 581, 589–591, 593, 639, 650,
Lin, C. C., 170
Lin, C. Y., 510, 516
Livingston, P. M., 38
Lodge, A. S., 56
London, A. L., 431

McAfee, K. B., 565
McConnell, A. J., 57, 148, 164, 362, 369, 607, 612, 653, 669
Markovitz, H., 54–56, 81, 85, 104, 106, 658
Marshall, W. R., Jr., 432
Masamune, Shinobu, 413
Mason, E. A., 575, 577
Matsuhisa, Seikichi, 85
Mazur, P., 462, 468
Meixner, J., 468
Metzner, A. B., 143, 214
Milberger, E. C., 477
Milne-Thomson, L. M., 122, 132

# NAME INDEX

Mischke, R. A. 413
Modell, M., 496
Morgan, A. J. A., 336
Morse, P. M., 639
Müller, Ingo, 298, 468, 474, 499
Mullineux, N., 94, 116-118, 131, 211, 517, 582

Newman, John, 374, 380, 526, 528
Noll, W., 12, 13, 15, 35, 45, 46, 48, 49, 54-57, 70, 81, 85, 104, 106, 169, 177, 203, 205, 231-233, 263, 297, 298, 302, 389, 390, 410, 430, 498, 569, 570, 658.

Oldroyd, J. G., 48, 56

Patel, J. G., 194
Paton, J. B., 74
Paul, Rajendra, 537
Pawlowski, J., 85, 262
Pearson, J. R., 116
Petersen, E. E., 143
Pigford, R. L., 559
Pitzer, K. S., 493
Pollard, W. G., 577
Power, G., 262, 276, 277
Prager, W., 49
Prandtl, L., 138, 139, 176
Present, R. D., 577
Prigogine, I., 498
Proudman, I., 116

Ragatz, R. A., 596, 600
Randall, Merle, 493
Ranz, W. E., 432
Reid, W. H., 170
Reik, H. G., 468
Reiner, Markus, 49, 53, 54
Rivlin, R. S., 205, 233, 299, 389, 411, 475, 476
Robinson, Ernest, 510
Rosenhead, L., 49
Rouse, Hunter, 42, 73, 129, 252
Rowley, D. S., 247

Sabersky, R. H., 125, 259, 261

Sadowski, T. J., 217
Sampson, R. A., 116
Sani, R. L., 262, 276
Sankarasubramanian, R., 563
Satterfield, C. N., 577
Savins, J. G., 218
Scheidegger, A. E., 193
Schlichting, Hermann, 87, 115, 139, 145, 151, 153, 154, 157, 161, 165, 167, 176, 177, 186, 190, 191, 232, 234, 251, 311, 358, 360, 361, 364, 367, 369, 370, 378, 379, 404
Schowalter, W. R., 143
Scott, D. S., 566, 574, 576
Scriven, L. E., 38
Serrin, J., 298, 348
Seshadri, C. V., 531
Shah, M. J., 143
Shames, I. H., 250
Shertzer, C. R., 214
Sherwood, T. K., 559, 577
Shinner, Ruel, 143
Siegel, R., 382, 384
Sisko, A. W., 53
Slattery, J. C., 24, 41, 54-57, 96, 106, 107, 121, 194, 204, 212, 217, 257, 270, 271, 276, 295, 298, 496, 498, 510, 565,
Smith, G. F., 177, 205, 233, 299, 389, 411, 475
Smith, J. M., 413, 445, 596
Sparrow, E. M., 382, 384
Spencer, A. J., 177, 205, 233, 299, 389, 411, 475, 476
Squires, P. H., 74
Stakgold, Ivar, 69
Stevenson, W. H., 509
Stewart, W. E., 4, 35, 54, 59, 66, 74, 87, 89, 90, 107, 177, 211, 247, 250, 276, 304, 311, 336, 342, 346, 376, 384, 405, 414, 419, 431, 453-455, 464, 477, 478, 484, 510, 515, 516, 529, 532, 536, 537, 549, 556, 581, 589-591, 593, 639, 650
Szynanski, Piotr, 93

Taylor, G. I., 177, 562

Taylor, T. D., 116
Tien, Chi, 143
Tietjens, O. G., 138, 139
Toor, H. L., 531
Toupin, R. A., 12, 16, 17, 19, 20, 22, 25, 27, 28, 35-38, 41, 42, 46, 50, 221, 279, 284-287, 291, 294-296, 303, 425, 440, 462, 468, 473, 499, 597, 670
Towle, W. L., 559
Townsend, A. A., 170
Truesdell, C., 12, 13, 15-17, 19, 20, 22, 23, 25, 27, 28, 35-38, 41, 42, 45, 46, 48-50, 55-57, 169, 177, 203, 205, 221, 231-233, 263, 279, 284-287, 291, 294-299, 302, 303, 389, 390, 410, 425, 430, 440, 462, 468, 473, 499, 500, 569, 570, 597, 670
Turian, R. M., 276, 336

Van Ness, H. C., 445, 596

Wasserman, M. L., 96, 276
Watson, G. M., 575, 577
Watson, K. M., 596, 600
Wehner, J. F., 564, 583
Whitaker, Stephen, 98, 207, 212, 249, 251-253, 510, 565
White, J. L., 143
Wilde, D. J., 273
Wilhelm, R. H., 564, 583
Wilke, C. R., 485
Willhite, G. P., 413
Willmore, T. J., 30
Wilson, H. A., 556, 557
Wright, W. A., 510

Yau, Joseph, 143
Yih, Chia-Shun, 28

# Subject Index

Activity, relative, 479
Angular velocity tensor, 15
Angular velocity vector, 16
Anisotropic solid (*see* Oriented solid)
Annulus:
 flow through, 74, 76, 77, 94
 helical flow in, 86
 tangential flow in, 82, 328, 329
Apparent viscosity (*see* Nonnewtonian behavior)
Approximate boundary-layer theory, 186, 400
 flow past body of revolution: for newtonian fluid, 191, 404
 for power-model fluid, 191, 405
 flow past flat plate, 190
 plane flow past curved wall, 186, 400
Archimedes, theorems of, 42
Area averaging, 183, 396, 559

Bases, 606, 607, 609
 cartesian, 607
 dual, 623
 orthogonal cartesian, 607
 orthonormal, 607
 physical, 617
 rectangular cartesian, 607
Bernoulli equation, 122, 243
 (*see also* Integral balances)
Bingham plastic model, 53
Body:
 configuration of, 3
 deformation of, 3
 individual species, 447
 motion of, 3
 multicomponent, 452
 single component, 2
Borda mouthpiece, 236
Boundary conditions:
 in energy transfer, 309
 in fluid mechanics, 69
  continuity of velocity at phase interface, 42, 69, 127

Boundary conditions (*continued*)
 in mass transfer, 502
Boundary-layer theory, 137, 352
 body of revolution, 161, 367
 convergent channel, 158
 curved wall, 145, 361
 energy transfer, 352
 entrance of falling film, 376–378
 entrance of tube, 370, 375, 378
 flat plate, 142, 143, 356
 Mangler's boundary-layer equations, 165
 Mangler's transformation, 167, 370
 mass transfer, 546
 natural convection, 355
 plane stagnation flow, 155
 solutions found by combination of variables (similar solutions), 151
 viscoelastic fluids, 143, 150, 167
 wedge, 363
 (*see also* Approximate boundary-layer theory)
Bounding principles (*see* Extremum principles)
Brinkman number, 338

Caloric equation of state, 278, 280, 284, 287
Catalytic reaction, 451, 517, 519, 532, 542, 563, 578, 582
Catherine wheel, 259
Cauchy's first law, 41
 for multicomponent bodies, 460
 in other frames of reference, 43
 in various coordinate systems, 60
Cauchy's lemma, 39
Cauchy's second law, 43, 461
Channel:
 boundary-layer flow in convergent, 158
 energy transfer, 328, 384
 flow through straight, 76
 potential flow in convergent or divergent, 133

675

Chemical potential, 280, 285
Christoffel symbols:
  of first kind, 641
  of second kind, 640
    in cylindrical coordinates, 644
    in spherical coordinates, 645
Combination of variables, solution by, 88, 151
Components:
  contravariant, 623, 630, 647
  covariant, 623, 630, 647
  physical, 617, 629, 647
  rectangular cartesian, 608, 627, 646
Compressible flow, 323
Condensation, 532, 536, 537
Conduction, 310, 323
  bar, 322
  cylinder, 313
  flat sheet, 317, 320, 322
  neglection of, 347
  semi-infinite slab, 311, 313
  sphere, 322
  wall of pipe, 317
Cone-plate viscometer, 101
Configuration, 3, 447
  reference, 3
Conservation of mass, 17
Contact energy transmission, 291
Continuity, equation of, 20, 457
Contravariant components (*see* Components)
Convection:
  diffusion induced, 503, 507, 520, 538
  forced, 323, 345, 545–551
  natural:
    at vertical plate, 378
    between concentric cylinders, 343
    between vertical plates, 338
    boundary-layer equations, 355
    general approach, 344
Coordinate systems:
  curvilinear, 615
  cylindrical, 618, 620
  paraboloidal, 619
  rectangular cartesian, 608, 625, 631, 648
  right-and left-handed, 656

Coordinate systems (*continued*)
  spherical, 618, 620
  (*see also* bases)
Corner, potential flow near, 134
Couette flow, 81
  of compressible newtonian fluid, 323, 330
Covariant components (*see* Components)
Covariant differentiation, 640, 651
Creeping flow, 99
  general solutions for, 116, 345
Curl, 653, 655
Current density, 528
Cylinder:
  potential flow past, 9, 133
  rotating in infinite field, 87
  (*see also* Conduction)

D'Alembert paradox, 126
Dam, force on, 237
Damköhler number, 538
Darcy's law, 193
  (*See also* Local volume averaging)
Deformation:
  gradient, 50
  relative, 55
  single component, 3
  species A, 448
Deissler's expression:
  for Reynolds stress tensor, 177
  for turbulent energy flux vector, 388
  for turbulent mass flux vector, 555
Determinant, 611, 658
  differentiation of, 659
Determinism, principle of, 45, 296, 472
Differential momentum balance (*see* Cauchy's first law)
Diffusion:
  coefficients:
    binary, 477, 480
    multicomponent, 476
  forced, 477, 492, 525
  ordinary, 477
  pressure, 477, 522, 525
  ternary, 528
  thermal, 477, 520, 521
Dispersion, longitudinal, 560

SUBJECT INDEX

Divergence theorem (see Green's transformation)
Drag:
 on arbitrary body, 254
 on closed torus, 277
 comparison of two bodies, 277
 (see also Sphere)
Drag coefficient, 114, 232, 233, 271
Dual basis, 621
Dyadic product, 627

Eckert number, 338
Eddy conductivity, 390
Eddy diffusivity, 555
Eddy viscosity, 176
Effectiveness factor, 578, 582
Electrochemical systems, 525
Ellis model, 53
Energy balance, 292, 461
 differential, 293, 462
  table of various forms, 305, 464
  in various coordinate systems, 306, 307
 jump, 295, 465
Energy flux principle, 291
Energy flux vector, 293
 (see also Flux of thermal energy)
Energy loss coefficient, 247
Energy transmission:
 external, 290
 mutual, 291
 contact, 291
Enthalpy, 281, 287
Entropy, 279, 283
Entropy flux vector, 300
Entropy inequality, 301, 467
 differential, 301, 467, 470
 jump, 303, 468, 469
Equation of continuity:
 in various coordinate systems, 60
 multicomponent, 456
 single component, 20
 species A, 450, 456, 457, 482
Equation of state (see Caloric equation of state)
Equilibrium, 485
Euler equation, 279, 285

Euler's first law, 37, 460
Euler's second law, 38, 461
Evaporation, 503, 529, 540
Expansion, sudden, 247
Extra stress tensor, 49
Extremum principles, 262
 application of, 271
 physical interpretation, 269

Fick's first law, 481, 574
Fick's second law, 482, 540
Film:
 energy transfer to, 376, 377
 flow in, 97, 98
 gas absorption in, with chemical reaction, 547, 551
 stagnant, 507, 510, 529, 532, 536, 537
Fin, cooling, 396, 399
Flat plate, boundary-layer flow past, 137, 143
Flux of thermal energy, 469, 474
 constitutive equations for, 296, 474
Force:
 contact, 36
 external, 35
 frame indifference of, 37
 mutual, 36
Fourier's law, 297
 in various coordinate systems, 307
Frame of reference:
 change of, 12
 moving, 237
Frame indifference:
 acceleration, 16
 definition for scalar, vector or second-order tensor, 12
 velocity, 15, 16
Friction factor (see Drag coefficient)
Friction loss coefficient, 246, 250
Froude number, 337
Fundamental equation (see Caloric equation of state)
Furnace design, 125

Gas well, 492
Gauss' theorem (see Green's transformation)

Gibbs-Duhem equation, 279, 286
Gibbs equation, 279, 286
Gibbs free energy, 281, 287
Gradient:
  of scalar field, 613
  of second-order tensor field, 650
  of third-order tensor field, 652
  of vector field, 638
Graetz problem, 380
Gravitational potential, 59
Gravity, 59
  (see also Force, external)
Green's transformation, 661

Heat capacity, 290
Helical flow, 86
Helmholtz free energy, 281, 287
Heterogeneous reaction (see Catalytic reaction)
Homogeneous thermodynamically, 279, 284
Homogeneous reaction, 451, 511, 541, 593, 594
  general solutions for first-order, 515, 516, 550

Integral balances, 218, 419, 584
  empirical correlations: for $\mathscr{E}$, 246
    for $\mathscr{F}$ and $\mathscr{G}$, 231
    for $\mathscr{J}_{(A)}$, 589
    for $\mathscr{Q}$, 430
  energy balance, 420, 595
    various forms of, 424–426, 596–598
  entropy inequality, 441, 600, 601
  mass balance, 219, 222, 224, 594
    for individual species, 584, 586
  mechanical energy balance, 238, 241, 243, 244, 595
    relation to Bernoulli equation, 241
    table of various forms, 438–441
  moment of momentum balance, 254, 258, 259, 595
  momentum balance, 225, 228, 235, 595
Integration, 660
  change of variable in, 663
Interface phase, 24

Internal energy, 279, 283
Intrinsic derivative:
  definition of, 6
  material, 6
Intrinsically stable equilibrium, 485
Intrinsically stable state of rest, 498
Irrotational flow, 33, 120
Isentropic fluid, 427
Isochoric motion, 21, 22

Jet, two-dimensional, 236
Jump mass balance, 24, 459
  species A, 451
Jump moment of momentum balance, 45
Jump momentum balance, 41, 460

Knudsen diffusion, 576
Kronecker delta, 610, 612

Laplace's equation in various coordinate systems, 121
Leibnitz rule, 22
Lift coefficient, 233
Local action, principle of, 42, 296, 472
Local volume averaging, 193, 405, 565
  averages of volume-averaged variables, 207
  Cauchy's first law, 200
  channel flow, 212
  differential energy balance, 406
  empirical correlations: for $\delta_{(A)}$ and $\Delta_{(A)}$, 568
    for g, 202
    for h, 408
  equation of continuity, 199, 577
    for species A, 565
  mass-density tortuosity vector, 567
  molar-density tortuosity vector, 567
  newtonian flow through nonoriented medium, 202
  newtonian flow through oriented medium, 205, 206
  pressure-gauge measurement, 212
  radial flow, 212
  theorem for local volume average of gradient, 196
  thermal tortuosity vector, 408

# SUBJECT INDEX

Local volume averaging (*continued*)
  tube flow, 209, 215
  viscoelastic flow through nonoriented medium, 203, 215
  (*see also* Porous media)

Macroscopic balances (*see* Integral balances)
Mangler's transformation, (*see* Boundary-layer theory)
Mass:
  balance for species A, 450
  conservation of, 17, 458
  density, 17
  fraction, 453
Mass flux vector, 454, 499
  constitutive equations for, 475, 480, 483
  table of various forms, 455
Mass transfer coefficient, 589
  (*see also* Newton's law of mass transfer)
Material coordinates:
  single component, 4
  species A, 448
Material derivative:
  intrinsic, single component, 6
  intrinsic, species A, 449
  multicomponent, 458
  single component, 5
  species A, 448
Material frame indifference, principle of, 46, 296, 472
Material objectivity, principle of (*see* Material frame indifference, principle of)
Material particle, 2
  mass-averaged, 457
Maxwell relations, 281, 287, 288
Mixing length theory, 174
Mole fraction, 453
Motion:
  single component, 3
  species A, 448
Mutual energy transmission, 291

Natural convection:
  in heat transfer, 338, 343, 355, 378, 431, 522

Natural convection (*continued*)
  in mass transfer, 522
Navier-Stokes equation, 58
  for a two-dimensional flow, 65
  in various coordinate systems, 61
Newtonian model, 49, 302, 303
Newton's law of cooling, 313
Newton's law of mass transfer, 508
Newton's law of viscosity (*see* Newtonian model)
Newton's second law (*see* Euler's first law)
Noll simple fluid (*see* Simple fluid)
Nonnewtonian behavior, 51, 54, 80, 303
Nonpolar bodies, 38
Nozzle, 238

Objectivity of material properties, principle of (*see* Material frame indifference, principle of)
Oriented solid, 298
  (*see also* Porous media)
Oscillating flow, between two flat plates, 92, 329
Ostwald-de Waele model (*see* Power model)

Partial mass variables, 282, 289
Partial molal variables, 282, 289
Path lines, 7
Peclet number:
  for energy transfer, 338
  for mass transfer, 548
Perturbation parameter, 335
Perturbation solution, 332
Phase interface, 22, 458
Physical components (*see* Components)
Plane stagnation flow (*see* Stagnation point)
Plate, moving, 87, 90, 91, 237
Point source, turbulent diffusion from, 556
Poiseuille's law, 73
Porous media:
  nonoriented (isotropic), 194, 202, 203, 409, 410, 412, 568, 570, 572
  oriented (anisotropic), 194, 205, 411, 571
  permeability, 193

Porous media (*continued*)
  permeability tensor, 194, 215
  (*see also* Local volume averaging)
Position vectors, 605
Potential energy, 59
Potential flow, 119
  convergent or divergent channels, 133
  corner, 134
  cylinder, 10, 132
  nonnewtonian fluid, 121
  sphere, 129
  stagnation point, 137
Power model, 52
Prandtl number, 337
Prandtl's mixing length theory, 174, 182, 388, 555
Pressure:
  hydrostatic, 51
  mean, 49
  thermodynamic, 49, 280, 285, 303
Pressure gauge reading, 70, 73, 212
Pump, 254, 261

Radial flow:
  between parallel plates, 107
  between porous cylindrical shells, 86
  between porous spherical shells, 86
Ram, hydraulic, 253
Rate of deformation tensor, 28
  in various coordinate systems, 63
Reversible system, 444
Reynolds analogy, 395
Reynolds number, 100
Reynolds stress tensor, 172, 174
Rigid body, motion of, 17
Rocket, 237
Rotating bucket, 94
Rotational flow, 33
Ruark number, 100

Schmidt number, 538
Second law of thermodynamics (*see* Entropy inequality *and* Integral Balances)
Similarity transformation (*see* Combination of variables, solution by)
Simple fluid, 54
Siphon, 124

Sisko model, 53
Slip at boundaries, 42, 69, 127
Soret effect (*see* Diffusion, thermal)
Sound, speed of, 348
Speed of displacement, 23
Speed of sound, 348
Sphere:
  alternate expression for force on, 253
  conduction from, 346
  diffusion, 507
  Ellis model flow past, 271
  empirical correlations for $\mathscr{F}$ and $\mathscr{G}$, 231
  newtonian flow past, 111
  potential flow past, 130
  rotating in unbounded fluid, 111
Stability of equilibrium (*see* Intrinsically stable equilibrium)
Stable state of rest (*see* Intrinsically stable state of rest)
Stagnation point:
  boundary-layer flow near, 154, 155
  potential flow near, 136
Stagnation temperature, 435
Stefan-Maxwell equations, 478, 479
Stokes' law, 114
Stokes' paradox, 100
Stoakes' relation, 49
Stratified flow, 99
Streak lines, 10
Stream function, 25–27, 65
  forms for, 66
Streamlines, 8, 123
Stress, 36
Stress deformation behavior, 47, 51, 54, 77
Stress equation of motion (*see* Cauchy's first law)
Stress principle, 36
Stress tensor, 39
  constitutive equations, 46, 473
  symmetry of, 44
Strouhal number, 100
Sublayer, laminar, 175
Substantial derivative (*see* Material derivative)
Summation convention, 610
Surface tension, 57

SUBJECT INDEX

Tangential flow
  between concentric cylinders, 82
  between concentric spheres, 110
  between parallel disks, 108
  (see also Annulus)
Taylor's vorticity transport theory, 177
Temperature, 280, 285
Tensor:
  metric, 632
  second-order field, 626
    components of, 627
  inverse of, 636
  isotropic, 634
  orthogonal, 633, 660
  skew-symmetric, 633
  symmetric, 632
  second-order field: trace of, 637
    transpose of, 632
  third-order field, 645, 649
    components of, 646
  fourth-order field, 652
Tensor product, 627
Thermal conductivity, 297, 302
  oriented (anisotropic) solid, 298
Thermodynamics, 278, 279, 283
Thermostatics, 279
Time averaging, 169, 386, 552
  Cauchy's first law, 172
  differential energy balance, 386
  empirical correlations: for Reynolds
      stress tensor, 174
    for turbulent energy flux vector, 387
    for turbulent mass flux vector, 554
  equation of continuity, 171
  for individual species, 553
  jump mass balance, 223
  jump momentum balance, 230
  (see also Turbulence; Turbulent flow)
Torque, 82, 105
Transpiration, 414
Transport theorem:
  alternative form of, 21
  derivation of, 18, 665, 666
  generalized form, 20
  for individual species, 450
  for multicomponent body, 458, 459
  for a region containing a singular
      surface, 22

Tube:
  emptying, 126, 133, 237
  energy transfer::
    downstream of entrance, 379, 381
    in entrance of, 370, 378
    in heated section of, 379, 381, 391, 395
  flow through, 70, 77, 92, 183
Tubular reactor, 563, 582
Turbulence:
  free, 169, 389
  homogeneous, 169
  isotropic, 169
  wall, 169
  (see also Time averaging; Turbulent flow)
Turbulent flow:
  buffer zone, 175, 182
  fully developed, 175, 393
  laminar sublayer, 175, 182
  in tube, 178, 251
  (see also Time averaging; Turbulence)

Ultracentrifuge, 524

Vector, 603
  addition, 603, 605
  axial, 649
  field, 605
  inner product, 603, 606
  integration, 660
  scalar multiplication, 603, 605
Velocity, 4
  mass-averaged, 452
  molar-averaged, 453
  single component, 4
  species A, 448
Viscoelastic behavior (see Nonnewtonian behavior)
Viscoelastic fluids, 54
Viscous heating:
  channel flow, 328
  Couette flow, 323, 330
  oscillatory flow, 329
  tangential annular flow, 328, 329
  when it can be neglected, 336
Viscosity:
  apparent, 51

Viscosity: (*continued*)
  apparent (*continued*)
    measurement of, 77
  bulk coefficient of, 49, 302
  eddy coefficient of, 176
  kinematic, 65
  shear coefficient of, 49, 302
Viscous dissipation, 323
  neglection of, 336
Viscous portion of the stress tensor (*see* Extra stress tensor)

Volume integration, 663
Von Karman's similarity hypothesis, 177
Vortex lines, 123
Vorticity:
  tensor, 32
  vector, 33

Wall:
  oscillating, 91
  suddenly accelerated, 87, 90
Wire-coating die, 74, 77